THE OCEAN BASINS AND MARGINS

Volume 7B:
The Pacific Ocean

THE OCEAN BASINS AND MARGINS

THE OCEAN BASINS AND MARGINS

Edited by
Alan E. M. Nairn
Earth Science and Resources Institute
University of South Carolina
Columbia, South Carolina

Francis G. Stehli
DOSECC, Inc.
Gainesville, Florida

and

Seiya Uyeda
Earthquake Research Institute
University of Tokyo
Tokyo, Japan

Volume 7B:

The Pacific Ocean

Springer Science+Business Media, LLC

Library of Congress Cataloging in Publication Data

(Revised for vol. 7B)

Nairn, A. E. M.
The ocean basins and margins.

Includes bibliographies.
Contents: v. 1. The South Atlantic. —v. 2. The North Atlantic. — [etc.] — v. 7B.
The Pacific Ocean.
1. Submarine geology. 2. Continental margins. I. Stehli, Francis Greenough, joint
author. QE39.N27   551.4′608   72-83046
ISBN 978-1-4615-8043-0   ISBN 978-1-4615-8041-6 (eBook)
DOI 10.1007/978-1-4615-8041-6

© 1988 Springer Science+Business Media New York
Originally published by Plenum Press, New York in 1988.

CONTRIBUTORS

R. N. Brothers
Department of Geology
University of Auckland
Auckland, New Zealand

Chin Chen
Graduate Program in Oceanography
  and Limnology
Western Connecticut State University
Danbury, Connecticut

J. Dupont
ORSTOM
Paris, France

James V. Eade
New Zealand Oceanographic Institute
Division of Marine and Freshwater
  Sciences, DSIR
Wellington, New Zealand
Present address:
CCOP/SOPAC, c/o Mineral Re-
  sources Department
Suva, Fija

Hiromi Fujimoto
Ocean Research Institute
University of Tokyo
Tokyo, Japan

George E. Gehrels
Division of Geological and Planetary
  Sciences
California Institute of Technology
Pasadena, California
Present address:
Department of Geosciences
University of Arizona
Tucson, Arizona

Kerry A. Hegarty
Department of Geology
University of Melbourne
Victoria, Australia

Nobuhiro Isezaki
Department of Earth Sciences
Kobe University
Nada, Kobe, Japan

R. J. Korsch

Department of Geology
Victoria University of Wellington
Wellington, New Zealand
Present address:
Bureau of Mineral Resources
Canberra, Australian Capital
   Territory
Australia

A. R. Lillie

Department of Geology
University of Auckland
Auckland, New Zealand

Robin Riddihough

Geological Survey of Canada
Ottawa, Canada

Jason B. Saleeby

Division of Geological and Planetary
   Sciences
California Institute of Technology
Pasadena, California

David B. Stone

Geophysical Institute
University of Alaska
Fairbanks, Alaska

K. Suyehiro

Department of Earth Sciences
Faculty of Science
Chiba University
Chiba, Japan

Yoshibumi Tomoda

Ocean Research Institute
University of Tokyo
Tokyo, Japan

Seiya Uyeda

Earthquake Research Institute
University of Tokyo
Tokyo, Japan

Jeffrey K. Weissel

Lamont–Doherty Geological
   Observatory of Columbia
   University
Palisades, New York

H. W. Wellman

Department of Geology
Victoria University of Wellington
Wellington, New Zealand

Makoto Yamano

Earthquake Research Institute
University of Tokyo
Tokyo, Japan

CONTENTS

Chapter 2. The Northeast Pacific Ocean and Margin

Robin Riddihough

Chapter 6. **Complexities in the Development of the Caroline Plate Region, Western Equatorial Pacific**

Kerry A. Hegarty and Jeffrey K. Weissel

Chapter 7. **The Norfolk Ridge System and Its Margins**

James V. Eade

Chapter 8. **Regional Geology of New Caledonia**

R. N. Brothers and A. R. Lillie

Chapter 9. **The Tonga and Kermadec Ridges**

J. Dupont

Chapter 10. **The Geological Evolution of New Zealand and the New Zealand Region**

R. J. Korsch and H. W. Wellman

Chapter 11. **Geophysics of the Pacific Basin**

Chapter 11A. **Gravity**

Hiromi Fujimoto and Yoshibumi Tomoda

Chapter 11B. **Heat Flow**

Makoto Yamano and Seiya Uyeda

Chapter 11C. **Controlled Source Seismology**

K. Suyehiro

Chapter 11D. **Magnetic Anomalies**

Nobuhiro Isezaki

Chapter 1

BERING SEA–ALEUTIAN ARC, ALASKA

David B. Stone

Geophysical Institute
University of Alaska
Fairbanks, Alaska 99775

I. INTRODUCTION

A. Geographic Setting

The Aleutian Island Arc forms part of the northern boundary of the Pacific Ocean, stretching from the Alaska Peninsula in the east to the Kamchatka Peninsula in the west (Fig. 1). The islands of the arc are located on the crest of a large ridge that follows the same trend as the islands. South of the Aleutian Ridge is the Aleutian Trench, which again connects from south of the Alaska Peninsula in the east through to the Kuril–Kamchatka Trench in the west. The trench is about 3400 km from end to end.

Though many of the Aleutian islands have a relatively rugged topography, they are not very high. The highest peaks are those formed by the active volcanoes, Shishaldin on Unimak Island being the tallest at 2858 m. Excluding the recent volcanoes, the topography seldom exceeds 700 m. In contrast, the Alaska Peninsula has considerably higher elevation, and the continuation of the active volcano line of the Aleutian Arc terminates with a succession of volcanoes all over 3000 m high (Iliamna, 3073 m; Redoubt, 3110 m; and Spurr, 3375 m) (Coats, 1956b).

North of the Aleutian/Alaska Peninsula Island arc system lies the Bering Sea. This sea is one of the so-called marginal seas of the Pacific Ocean, though, as will be discussed below, it appears to have been formed through a different geological process than that which formed the other marginal seas, such as the Philippine Sea.

The west side of the Bering Sea, from Cape Kamchatka, where the trend of the Aleutian Island Arc cuts the mainland, to Cape Navarin, is bordered by the Kamchatka and Koryak ranges. These mountains form narrow rugged ranges more or less parallel to the coast. Elevations are often above 1000 m, and occasionally above 2000 m. Because the trend of the ranges is parallel to the coast, there are no major drainages into the Bering Sea.

North of Cape Navarin, in the region of Anadyr Bay, several large rivers discharge into the Bering Sea, most notably the Anadyr River. These rivers drain not only part of the Koryak coastal mountains, but also the Anyui and Chukotsk ranges to the north, and thus contribute a considerable amount of sediment to the Bering Sea.

The north side of Anadyr Bay is formed by the Chukotsk Peninsula. Though generally mountainous, its relief is commonly below 1000 m. The far eastern point of the Chukotsk Peninsula forms one side of the Bering Strait, and the western end of the Seward Peninsula forms the other. The general morphology of the Seward Peninsula is very similar to that of the Chukotsk Peninsula. Together, the Chukotsk and Seward peninsulas delineate the northern edge of the Bering Sea.

From the Seward Peninsula southwards, the margins of the Bering Sea are low-lying and generally swampy as far south as the point where the Kuskokwim Mountains are truncated by the coast in the vicinity of Cape Newenham. The morphology of this low-lying area is largely due to the Yukon and Kuskokwim rivers which have drained through the region for at least the recent geological past. The sediment load carried by these two rivers is formidable, and has no doubt played a significant role in the depositional history of the Bering Sea Shelf (see Table I).

Bristol Bay, in the southeastern Bering Sea, is flanked by the Alaska Peninsula Range to the south and the Kuskokwim Mountains to the north. The Alaska Peninsula Range contains many of the active volcanoes of the Aleutian/Alaska Peninsula system; like the Kamchatka–Koryak system, it is long and narrow.

From the above it can be seen that the Bering Sea is more or less surrounded by mountainous terrain, all of which is shedding debris into it; however, none of these ranges is large enough in extent or of the appropriate morphology to provide really large amounts of sediment. The majority of the sediment entering the Bering Sea today comes from distant source areas via the major rivers. The Yukon River dominates in terms of the amount of material transported into the Bering Sea, with the Anadyr, Kuskokwim, and other assorted rivers transporting subsidiary amounts (see Table I).

Water circulation within the Bering Sea is extremely important, especially with regard to how efficiently it connects the cold Arctic water to the warmer Pacific and vice versa. This mixing of the Arctic, Bering, and Pacific waters has a dramatic effect on climate. In the recent geological past it is known that the Bering Straits were narrower, or perhaps ceased to exist (e.g., see Hopkins, 1967). Today the Bering Sea connects the Pacific through to the Arctic Ocean, though the flow is severely restricted because the Bering Strait is less than 50 m deep and only 85 km

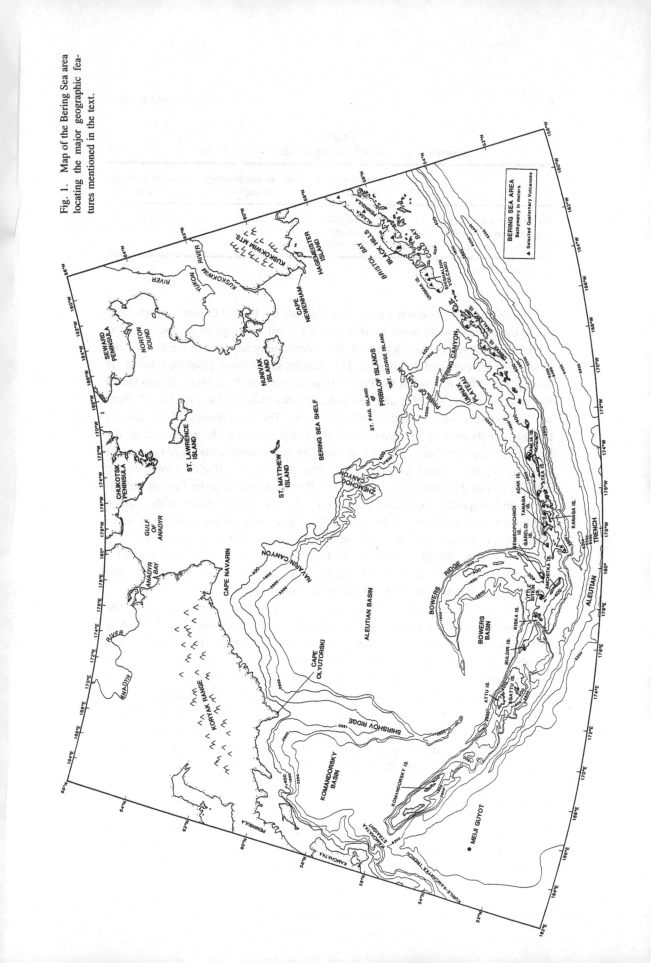

Fig. 1. Map of the Bering Sea area locating the major geographic features mentioned in the text.

TABLE I
Sources of Sediment Entering the Bering Sea Today[a]

	Drainage area (km^2)	Mean annual discharge (m^3 sec^{-1})	Sediment discharge (10^9 kg yr^{-1})
Yukon River (at Kaltag)	766,600	6220	60
Anadyr River (at mouth)	200,000	1600	10
Kuskokwim River (at Crooked Creek)	80,550	1270	4

[a]From Sharma (1979).

wide. The connection between the Bering Sea and the Pacific Ocean is much more open, the average water depth to the top of the Aleutian Ridge being about 200 m. At the far west end of the Aleutian Ridge, immediately adjacent to the Kamchatka Peninsula, is a curious deep channel. This channel slopes down from the floor of the Bering Sea (3400 m) toward the Pacific floor at about 5000 m depth. It also marks the junction of the Aleutian and Kuril–Kamchatka trenches, and thus the "channel" effectively forms a saddle between them. The morphology of the area is further complicated by the presence of the Meiji Guyot at the north end of the Emperor Seamount Chain. Notwithstanding the minor complications of the bottom topography, the channel between the Bering Sea and the Pacific Ocean at the western end of the Aleutians has no sill or other impediment to the flow of bottom water and suspended sediment loads from one side to the other; with a slope southward from the Bering to the Pacific. The significance of this in terms of water circulation is unclear, as is the origin of the channel itself. It is possible that it is an erosional feature, though its scale would argue against this. It is also possible that it holds clues as to the tectonic relationships between the Kuril–Kamchatka and Aleutian trenches and the adjacent plates (Cormier, 1975).

In terms of bathymetry, the Bering Sea subdivides into a number of parts. The most obvious are: the very large shallow shelf area, the shelf edge roughly dividing the whole Bering Sea in half along a northwest–southeast line; Bowers and Shirshov ridges; and the three basins formed by these ridges, the Komandorsky, Bowers, and Aleutian basins. In addition to these major subdivisions, there are some other notable bathymetric features such as the large canyons along the shelf edge, namely the Navarin, Zemchug, Pribilof, and Bering canyons. The Umnak Plateau, one edge of which is formed by the Bering Canyon, is also worthy of note inasmuch as it appears to be geologically allochthonous with respect to the shelf area, the Aleutian Arc, and the deep Bering Sea, as explained in a later section.

B. Tectonic Setting

Since plate tectonic concepts gained general acceptance in the 1960s and 1970s, a number of models relating to the tectonic setting of Arctic areas in general, and the Bering Sea in particular, have been put forward.

The conventional model accepts that the Aleutian/Alaska Peninsula subduction zone is the connecting link between the Kuril–Kamchatka subduction zone and the largely transform motion along the Fairweather–Queen Charlotte transform fault system in southeast Alaska (Fig. 2).

Southeast Alaska seems to be clearly attached to the North American plate, the easternmost boundary of which is the mid-Atlantic spreading ridge. The mid-Atlantic ridge can be traced northward to join the Arctic Ocean spreading center, which is itself clearly defined as far as the Siberian continental shelf edge. The Euler pole of Atlantic spreading, and by implication Arctic spreading too is inland from the point where the active spreading center joins the continental margin (Fig. 3). Conventional wisdom has consistently drawn the connection from the Euler pole to the plate

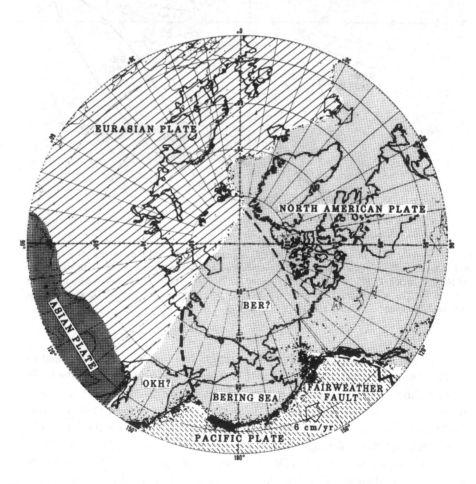

Fig. 2. A polar view of the distribution of earthquake epicenters from Tarr (1970) showing the conventional plate boundaries by shading patterns and two possible additional plates (Okhotsk and Bering) discussed by Stone (1983) shown by dashed lines. Also shown are the directions of relative motion of the plates at the Arctic spreading center and across the Aleutian Trench.

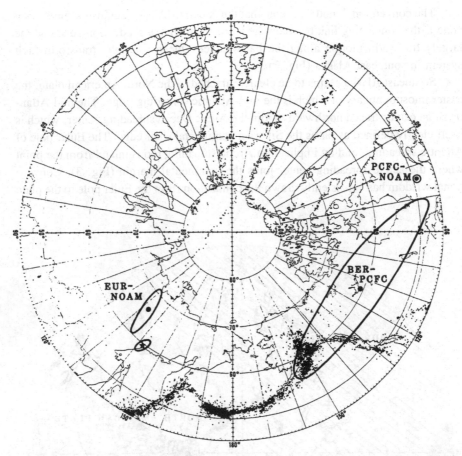

Fig. 3. A polar view of epicenters (see Fig. 2) together with the Euler poles and their ovals of confidence for relative motion determined between Europe and North America (EUR–NOAM), Pacific and North America (PCFC–NOAM), and Bering and Pacific Plates (BER–PCFC) as taken from Minster *et al.* (1974). The star represents the location of the Euler pole of opening of the Arctic Ocean alone, as taken from Savostin and Karasik (1981).

boundaries in the Pacific as a more or less straight line through Sakhalin Island to the Japan/Kuril trench system (Figs. 2 and 3). This is done in large part on the basis of the diffuse seismicity along this line. An alternative to this model has been proposed by Chapman and Solomon (1976), Savostin and Karasik (1981), Stone (1983), and Cook *et al.* (1986) in which the connection is made from the Euler pole across eastern Siberia, through Magadan to the Aleutian system, thus forming a separate Okhotsk plate. Stone (1983) also discusses the possibility of an additional plate boundary along the Canadian Arctic to join the Aleutian system in interior Alaska, thus forming a separate Bering Plate (see Fig. 2). If this plate boundary is real, the proposed motion in Canada and Alaska must be extremely slow. However, whether or not the Okhotsk and Bering plates move independently is of considerable importance to the relative motions that are taking place along the southwestern and southern margins of the Bering Sea.

A commonly used model for the Euler poles of relative motion between plates is that of Minster and Jordan (1978). This is the model they label RM-2. It, plus an earlier version labeled RM-1 (Minster *et al.*, 1974), are both based on observed directions and rates of relative plate motions on a global scale. For individual plate boundaries, such as the Pacific–North American boundary, they use the azimuths of transform faults and the directions of slip vectors determined for selected earthquake fault-plane solutions to locate the Euler pole, and magnetic anomaly data to determine rates of motion.

For the RM-1 model, Minster *et al.* (1974) noted that the data they used from the Aleutians were systematically different from those for the west coast of North America. On this basis they proposed a separate Bering Plate (which included the Okhotsk Plate). The creation of a separate Bering Plate was later criticized by Engdahl *et al.* (1977) on the basis that the differences between the Aleutian slip vectors and the western American data could be explained in terms of velocity anomalies due to the orientation of the down-going slab in the Aleutians. In the later RM-2 model, Minster and Jordan (1978) use both Kuril–Kamchatka and carefully selected Aleutian data together with the western American data. Since the only significant difference between the RM-1 and RM-2 models for this area is the addition or subtraction of the Kuril–Kamchatka (Okhotsk Plate) and the Aleutian data, and since the RM-1 model gives a significantly better fit of the predicted relative motion to the major transform system in southeast Alaska than does the RM-2 model, it would seem that the RM-1 model is perhaps a better representation of relative motion of the Pacific Plate with respect to Alaska. This carries with it the implication that there is relative motion between the Okhotsk and North American (Alaskan) plates, and that either the Aleutian data still have systematic errors associated with them, or that there is indeed relative motion taking place between the North American and Bering Sea plates. The relative motions for the Pacific/Bering Sea plate boundary are thus calculated using the RM-1 model (Minster and Jordan, 1974), with the Euler pole at 50.9°N and 293.7°E at a rate of relative motion 0.75°/m.y.

In the context of these various models, particularly with reference to the possibility of small amounts of relative motion between the Bering Sea and North America, it is of interest to note that, based on detailed bathymetric studies, the Arctic Ocean Euler pole is not coincident with that for Atlantic spreading (Savostin and Karasik, 1981). Savostin and Karasik (1981) calculate a pole for today's motion (Fig. 3) which is south and east of that calculated by Minster *et al.* (1974); they give coordinates of about 62°N and 142°E which is very similar to that calculated earlier by Zonenshayn *et al.* (1978).

In terms of understanding and interpreting the geology and geophysics of the

Bering Sea region, it is obviously desirable to know something about past plate motions. Unfortunately, much of the evidence relating to these has been subducted. In general, all that remains are the magnetic anomaly patterns, and these are incomplete inasmuch as there is only part of one side of the pattern left on the Pacific side of the arc, with the exception of a small fragment of the Kula plate at the western end (Lonsdale and Smith, 1986).

To overcome this problem, two related approaches have been taken. One is to assume that the hot spots form a fixed reference frame (Morgan, 1972; Duncan, 1981; Engebretson et al., 1985) and the other is to do successive reconstructions around a loop where more complete anomaly patterns exist. The latter approach commonly reconstructs from the Pacific Plate across the East Pacific Rise to the Antarctic Plate, to the Indian Plate to the African Plate to North America. Since the errors in each reconstruction sum, and since there is a question as to whether Antarctica is one or two pieces, the accuracy of this technique is not as high as one would like (e.g., see the error analysis of Stock and Molnar, 1983).

The hot-spot reference frame has the inherent problem that it is not known how stationary the hot spots have remained with respect to one another. It seems likely, however, that the relative motion has been small over that last few tens of millions of years which is important to Bering Sea geology (Duncan, 1981).

A number of different reconstructions of the motions of the plates making up the Pacific Basin have been made. Quantitative estimates of the motions include those of Engebretson (1982), Rea and Duncan (1986), and Engebretson et al. (1984, 1985). These reconstructions, along with most others, predict that the plates making up the Pacific underwent several thousand kilometers of northward motion relative to North America, since Cretaceous time. This motion is of great significance to the geology of the Bering Sea, since it seems likely that most of the motion was taken up by subduction along both the Aleutian and the now inactive Koryak subduction zones, with an unknown component taken up along the Bering Sea shelf edge. Actual details of the motions to be expected in the Bering Sea region are impossible to determine at present, because we lack information regarding the north side of the now subducted Kula Ridge (Atwater, 1970) and for the age of the Bering Sea floor.

A generalized outline of the motions of the major plates relative to the North American craton are shown in Fig. 4. Relative to a Bering Sea fixed with respect to the craton, all the plates of the north Pacific have a very variable, but generally northward component from about mid-Jurassic time to the beginning of the Cretaceous (Engebretson et al., 1985; Rea and Duncan, 1986). It must be emphasized that the set of reconstructions shown in Fig. 4 are highly speculative, particularly with regard to the locations of the Alaskan terranes, and are designed simply to show the dominant northward motion. The details of possible motions north of the Kula ridge are indeterminate due to lack of knowledge regarding the Kula Ridge (and other northern ridges?) and to the possibility that the Bering Sea was part of a

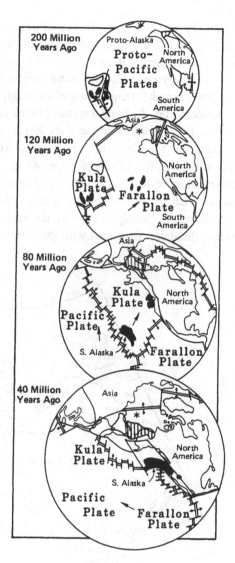

Fig. 4. A highly speculative set of reconstructions of the paleolatitudes of some of the terranes forming southern Alaska today, together with possible locations of the Pacific spreading centers. (Base data from Stone *et al.*, 1982; redrawn from "Alaska Geographic" Vol. 9, 1982.)

plate that was also moving. From about the end of the Cretaceous, the motions of the Pacific Plate itself are recorded relative to the hot spots by the Emperor–Hawaii and other Seamount chains. The published reconstructions assume that the Kula spreading rate was symmetric and the hot spots remained fixed.

Superimposed on the general northward motion are several major changes in plate motion and organization that are likely to have had significant effects on the Bering Sea. Two of these are the creation of the Kula Ridge at about 85 m.y. (Woods and Davies, 1982) and its eventual demise. A further (perhaps related) event is the change in direction of Pacific Plate motion as recorded by the bend in the Pacific hot-spot lines. This latter event is dated at about 43 m.y. (Dalrymple *et al.*, 1977). The dating of the demise of the Kula Ridge is more problematic, and depends on whether it simply ceased to operate all at once, or was progressively subducted. Byrne (1979) places the demise between anomalies 24 and 25 at 58–59 m.y. based on the straightening of the magnetic anomalies generated by the ridge at its western end. Rea and Duncan (1986) estimate an age of 42–50 m.y. for the cessation of spreading on the Kula Ridge, their model being based largely on the geometry of the plates of the ancestral North Pacific, in which they include a relatively short-lived Chinook Plate (82–50 m.y.). A 44 m.y. age is in accord with the fragment of Kula Plate containing anomaly 19 identified on the Pacific side of the Aleutian arc (Lonsdale and Smith, 1986).

Since the Bering Sea *per se* did not exist until the creation of the Aleutian Ridge, the dates associated with the demise of the Kula Plate and the bend in the

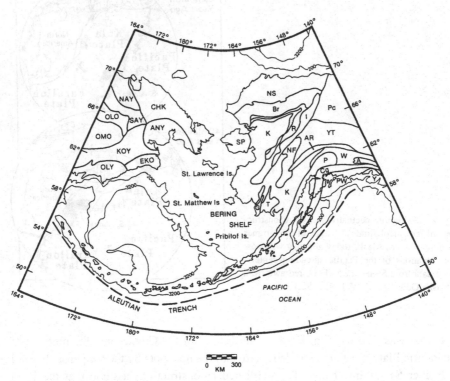

Fig. 5. A simplified tectonostratigraphic terrane map for Alaska (A, Alexander; An, Angayuchan; AR, Complex of Alaska; range terranes: Br, Composite terranes of the Brooks Range; Cg, Chugach; G, Goodnews; I, Innoko and associated terranes; K, Cretaceous overlap sequences: NF, Nixon Fork; NS, North Slope; P, Peninsular; Pc, Porcupine; Yukon related terranes: PW, Prince William; R, Ruby; SP, Seward Peninsula; T, Togiak; W, Wrangellia; Y, Yakutat; YT, Yukon-Tanana) and far eastern USSR (ANY, Anadyr; CHK, Chukotsk; EKO, Ekonay; KOY, Koryak; NAY, North Anyui; OLO, Oloi; OLY, Olyutorsk; OMO, Omolon; SAY, South Anyui). (Redrawn from Jones *et al.*, 1987, and Fujita and Newberry, 1983a,b.)

Hawaiian–Emperor Seamount chain are important to models of the development of the Bering Sea. Since there are no magnetic anomalies in the Bering Sea that can be unambiguously interpreted as having been generated by the Kula Ridge, it seems likely that the Kula Plate had to have been subducted south of the "Bering Sea Plate." The Kula Ridge was, however, about 20° south of the present latitude of the Aleutian Arc if it ceased to operate between anomalies 24 and 25 (Engebretson, 1982) which allows for, but does not require one or more plates to have existed to the north of it, and for the Aleutian Arc to have been initiated far south of its present location.

As noted above, the general motion of the plates of the Pacific Basin seems to have been northward. In terms of the geology of onshore Alaska, this northward motion is thought to have been part of the overall tectonic regime that created the collage of tectonostratigraphic terranes seen in both Alaskan and Siberian geology today (Fig. 5) (Howell *et al.*, 1985; Beck *et al.*, 1980; Stone *et al.*, 1982; Churkin and Trexler, 1981; Fujita and Newberry, 1983a,b; Jones *et al.*, 1987). It would seem probable that the same processes creating the collage would have operated to the west of mainland Alaska too, especially when it is recognized that the Bering Sea Shelf is roughly equivalent in size to mainland Alaska. That similar tectonostratigraphic terranes exist on the shelf has been proposed on the basis of magnetic anomaly patterns (McGeary and Ben-Avraham, 1981). Similarly, some of the enigmatic features of the deep Bering Sea, such as Bowers and Shirshov ridges, have been explained away as fragments from elsewhere, rafted in on the plates of the Pacific (Ben-Avraham and Cooper, 1982). Both of these concepts are discussed under sections concerning the geology of the Bering Sea Shelf and the deep Bering Sea.

II. BERING SEA SHELF

In attempting to describe the geology and geological setting of the Bering Sea Shelf, it is important to keep in mind its great size. This is demonstrated by the fact that the total shelf area is roughly twice the area of California.

The information available to develop a model of the Bering Sea Shelf geology and geological history comes from the few islands available for direct study, notably St. Lawrence Island, St. Matthew Island, Nunivak Island, and the Pribilof Islands, from extrapolation of onshore geology, from marine geophysical measurements and dredge hauls, and from a limited number of oil industry exploratory wells (Figs. 1 and 6). Except for local areas, the onshore geology in both Alaska and Siberia has not been mapped at better than the reconnaissance level. As a result, extrapolations offshore tend to be very general, and mainly concern structural trends and broad age ranges of rocks. A number of authors, including McGeary and Ben-Avraham (1981), have pointed out that if it is assumed that the Bering Sea Shelf is underlain by tectonostratigraphic terranes similar to those seen on mainland Alaska, then the extrapolation of specific geological features or formations cannot be carried very far offshore.

Although much of the geology of the Bering Sea Shelf is speculative, there is enough information available to split it into two quite different geological provinces. The rocks exposed on the Chukotsk Peninsula, St. Lawrence Island, and the Seward Peninsula include Paleozoic sequences similar enough to each other to warrant their being considered as part of the same original geological setting. Similarly, the structural trends of the Brooks Range appear to curve southward in the vicinity of Cape Lisburne, and can be extrapolated to connect through to the Seward Peninsula, swing westward through St. Lawrence Island, then northwestward into the Chukotsk Peninsula. This very large southward-bowing syntaxis effectively demarcates the older northeast corner of the Bering Sea from the younger southern part. A number of hypotheses have been put forward as to the origin of the syntaxis, perhaps the most popular being that it is the result of compression between the North American and Eurasian plates caused by the opening of the North Atlantic (Patton and Tailleur, 1977).

To the south and west of the syntaxis through St. Lawrence Island, the Bering Sea Shelf bathymetry shows an extraordinarily flat and featureless bottom topography, with an average southwestward slope of about 0.24 m/km (Sharma, 1979). The only directly visible and accessible parts of the outer shelf are St. Matthew Island and the Pribilof Islands. The existence of volcanic rocks on these islands, together with the high-amplitude short-wavelength magnetic anomalies observed more or less parallel to the shelf edge, has led several authors to identify an ancient magmatic arc connecting the Okhotsk–Chukotsk volcanic belt in Siberia through to southwestern Alaska (Zinkevich et al., 1983; Marlow et al., 1976; Parfenov and Natal'in, 1985).

A. St. Matthew Island

The geology of St. Matthew Island has been described by Patton et al. (1975, 1976) and consists of dominantly flat-lying to gently dipping flows of andesite and basalt together with volcaniclastic deposits of andesitic and basaltic tuff and conglomerate (Fig. 6). The general petrology and chemical composition of these rocks are given by Wittbrodt (1985). Potassium–argon age determinations combined with magnetostratigraphy give an age of about 79 m.y. for these extrusive rocks (Wittbrodt et al., 1987). A hornblende age from one of the granodiorite intrusions penetrating the volcanic rocks is 61 m.y. No diagnostic fossils were reported; however, leaf, wood, and pollen fossils obtained from within five ashfall tuff layers are consistent with the radiometric ages.

In addition to the dominantly volcanic rock suites, a small fault-bounded section of volcanic graywacke and argillite is mapped in the southcentral part of St.

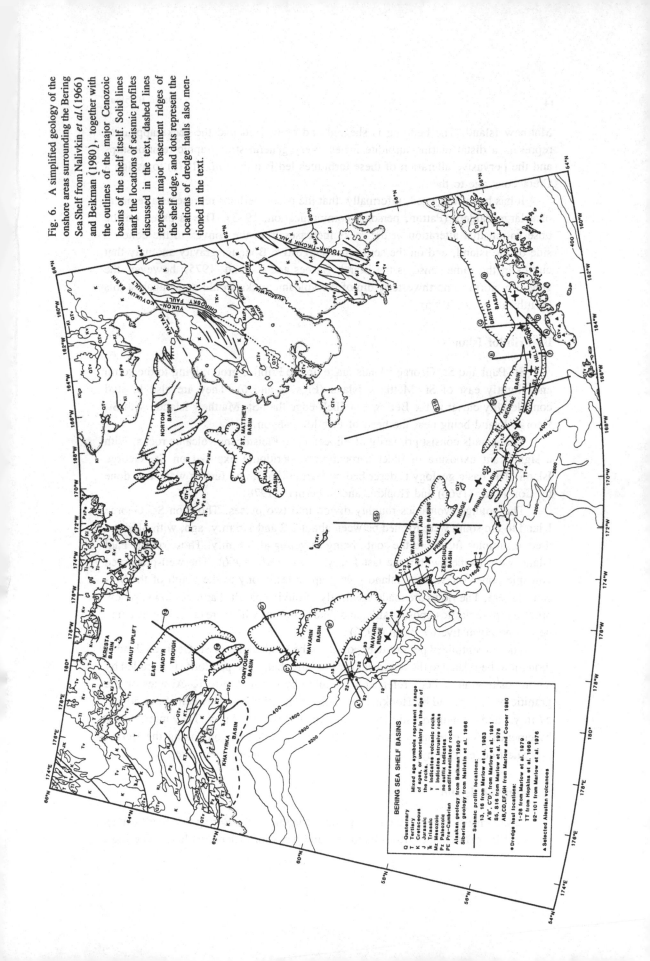

Fig. 6. A simplified geology of the onshore areas surrounding the Bering Sea Shelf from Nalivkin et al. (1966) and Beikman (1980), together with the outlines of the major Cenozoic basins of the shelf itself. Solid lines mark the locations of seismic profiles discussed in the text, dashed lines represent major basement ridges of the shelf edge, and dots represent the locations of dredge hauls also mentioned in the text.

Matthew Island. The bedding is sheared and crenulated and the rocks appear to represent a distal marine turbidite facies. Very general stratigraphic relationships and the pervasive alteration of these turbidites led Patton *et al.* (1975) to assign a Cretaceous age to them.

It has been suggested, informally, that the island perhaps represents a remnent of a large caldera (Patton, personal communication, 1983). This is based on the observation that alteration of the rocks increases from the convex to the concave side of the island, and on the somewhat suggestive Bouguer gravity anomaly that curves in the same sense as the island (Barnes and Estlund, 1975); however, the generally north to northwesterly dip of the volcanic rocks would argue against this (Wittbrodt *et al.*, 1985).

B. Pribilof Islands

St. Paul and St. George islands make up the Pribilof group, lying to the south and slightly east of St. Matthew Island (Figs. 1 and 6). They are also located considerably closer to the Bering Sea Shelf edge than St. Matthew Island, with St. George Island being near the head of Pribilof Canyon.

The islands consist primarily of Pliocene and Pleistocene volcanic rocks, with a significant exposure of older serpentinized peridotite exposed on St. George Island. The basic geology is described by Barth (1956), and further work was done by Cox *et al.* (1966) and Hopkins and Silberman (1978).

The younger volcanics roughly divide into two pulses. Those on St. George Island to the south were formed between about 2.2 and 1.6 m.y. ago, with offshore dredge samples further to the south being as young as 0.8 m.y. Those on St. Paul Island were formed within the last 1 m.y. (Cox *et al.*, 1966). The well-preserved volcanic cones on St. Paul Island give graphic testimony to the youth of the volcanics there. The flows are made up mostly of olivine basalt, basanite flows, pillow breccias, pyroclastic deposits, sills, and dikes, most of which are described as being nepheline normative by Barth (1956).

The serpentinized peridotite exposed on the south shore of St. George Island is thought to be related to the basement rocks of the shelf. The peridotite is intruded by a composite granitic body referred to as aplite by Barth (1956). K–Ar ages on the granitic body reported by Hopkins and Silberman (1978) are between about 50 and 57 m.y., and the authors estimate a minimum age of 52 ± 2 m.y. They interpret the basement rocks as being related to a Mesozoic subduction zone along the Bering Sea Shelf edge. This hypothesis will be discussed in more detail later.

C. Nunivak Island

Other clues relating to the geology and structure of the Bering Sea Shelf can be obtained by extrapolating from mainland and nearshore island geology. The most significant nearshore island in this respect, ignoring for the moment the Aleutian Islands, is Nunivak Island.

Apart from a small exposure of sedimentary rocks assigned a Cretaceous age (Hoare *et al.*, 1968), all the rocks exposed on Nunivak Island are volcanics of Recent to Miocene age (Figs. 1 and 6). By virtue of its geographic location off the Yukon–Kuskokwim deltas, and the apparent similarity of the volcanics to those seen in the delta area, Nunivak Island is considered to be part of the same volcanic terrain. This impression is enhanced by the fact that the small exposures of Cretaceous sediments could also be related to the Cretaceous sediments of the Yukon–Koyukuk province to the north and east, which extends to the southern flank of the Brooks Range (Patton, 1973; Harris *et al.*, 1987). The Cretaceous rocks seen on Nunivak Island may also tie in with the presumed Mesozoic basement under the Bering Sea Shelf.

The volcanic rocks on Nunivak Island consist of both tholeiitic and alkalic flow units, with the tholeiites being volumetrically dominant. The volcanism seems to have been episodic, with the center of volcanic activity moving eastwards with time. The oldest volcanic rocks investigated have been dated at 6.1 m.y., and are alkalic basalts. Another episode of volcanism occurred between 4.8 and 5 m.y. with both tholeiites and alkalic rocks being erupted. Episodes from 3.2 m.y. to 4.1 m.y., 1.5 m.y. to 1.7 m.y., and 0.9 m.y. to 0.3 m.y. all seem to have been dominantly tholeiitic, whereas the most recent period, which has continued into historic times, is dominantly alkalic.

D. Yukon–Kuskokwim Delta

The onshore geology in the Yukon–Kuskokwim delta area is poorly exposed, consisting mainly of Quaternary deltaic deposits. However, there are a number of exposures of Quaternary volcanic rocks, and Cretaceous felsic intrusive rocks toward the north side toward Norton Sound. Inland from the delta deposits the general structural trend is southwest–northeast, as evidenced by the trends of the major faults such as the Kaltag, the Chiroskey, and the Iditarod–Nixon Fork Faults (parallel to and south of the Kuskokwim River, Fig. 6). There is evidence that some of these faults, particularly the Kaltag Fault, continue offshore under the eastern edge of the Bering Sea (Scholl and Hopkins, 1969) but they have not been detected farther to the west. The same argument holds for the geological and structural trends in the Cape Newenham–Hagemeister Island region. Here, some of the major fault systems of Alaska such as the Iditarod–Nixon Fork system, and the Holitna–Farewell–Denali–Togiak–Tikchik and Mulchatna Fault system (labeled Togiak–Tikchik on Fig. 6) appear to terminate at or near the coastline (Pratt *et al.*, 1972; Marlow and Cooper, 1983). The general trend of all these faults and the geological grain of southwest Alaska is roughly perpendicular to, though distant from, the Bering Sea Shelf edge. This leads to one of the major problems of modeling the geology of the Bering Sea

Shelf: what is the relationship between these southwest-trending structures that are chopped off by the coastline and the northwesterly structural trends nearer the shelf edge? It should be noted in this respect that the onshore coastal rocks vary widely in type and age. The oldest rocks are Precambrian (D. L. Turner *et al.*, 1983), but the majority are Mesozoic with some Tertiary volcanic and intrusive rocks (Fig. 6). It is also of interest to note that in the Kanektok area, a little to the north of Cape Newenham (Figs. 1 and 6), a thermal event reset many of the K–Ar radiometric clocks about 100 m.y. ago (Turner, 1984; D. L. Turner *et al.*, 1983). There is circumstantial evidence that this event may be very widespread, and hence of great significance in unraveling the geological history of the Bering Sea Shelf.

E. Structural Trends

A number of authors have related rock types and trends seen on the Alaska Peninsula and associated islands, with dredge hauls and trends seen along the Bering Sea Shelf edge. Based on these relationships, they have made large-scale extrapolations to connect geological trends from the Alaska Peninsula through the Bering Sea to the Koryak Mountains of Siberia (Moore, 1973; Hopkins *et al.*, 1969; Marlow *et al.*, 1976). These correlations are by no means unequivocal (Fig. 7). The sediment transport directions in the Cretaceous flysch deposits exposed on the Shumagin Islands (in the Shumagin Formation), on Kodiak Island, and eastwards to the Gulf of Alaska (in the Kodiak Formation and the Valdez Group) are all to the southwest, more or less parallel to the trend of the Alaska Peninsula. The most westerly exposure of this flysch is on Sanak Island where the sediment transport direction has swung around to the northwest to be more or less parallel to the extrapolated trend of the Bering Sea Shelf edge (Moore, 1973). The sediments are interpreted as trench deposits with the dominant transport direction being along the length of the trench (Nilsen and Zuffa, 1982). The change in trend at Sanak Island, combined with the identification of Cretaceous sediments dredged in Pribilof Canyon, led to the conclusion that these were connected, and thus the pre-Aleutian Arc structural style continued around the corner of the Bering Sea Shelf.

Work by McLean (1979a) has shown that there is little similarity between the dredge samples from Pribilof Canyon and the Shumagin Formation, the dredge samples having more similarity to shallow water members of the Cretaceous Chignik Formation, also exposed on the Alaska Peninsula.

More compelling evidence that geological trends seen on the Alaska Peninsula can be traced to the northwest to parallel the Bering Sea Shelf edge comes from a combination of data from geophysical exploration and the study of onshore outcrops. In the Black Hills, on the Alaska Peninsula (Fig. 1), outcrops of Upper Jurassic rocks are equated with the Naknek Formation exposed to the northeast along the length of the Alaska Peninsula and along the west shore of Cook Inlet.

Offshore geophysical evidence, principally in the form of seismic reflection profiles, shows a well-defined basement high aligned with the Black Hills, but

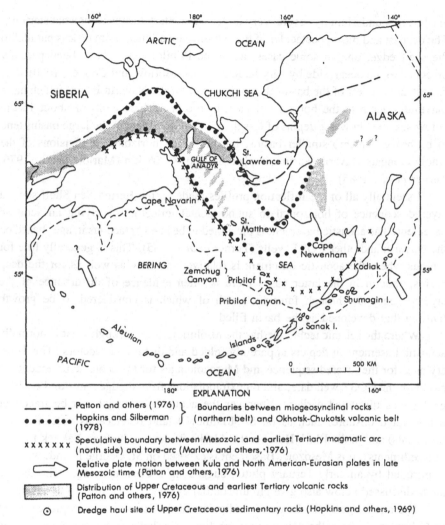

Fig. 7. Postulated boundaries between uppermost Cretaceous and lowest Tertiary volcanic rocks and pre-Upper Cretaceous of a miogeosyncline in the Bering Sea region. (From McLean, 1979a.)

trending more or less parallel to the Bering Sea Shelf edge (Fig. 6). This ridge is thus interpreted as being constructed of Upper Jurassic Naknek sediments (Marlow and Cooper, 1980).

F. Geophysical Data

Geophysical data on the geology and structure of the Bering Sea Shelf come from several thousands of kilometers of seismic reflection profiling, limited refraction profiling, and magnetic and gravimetric data. As mentioned earlier, the bathymetry of the Bering Sea Shelf is remarkably smooth, which made the discovery that it is underlain by several very large sedimentary basins somewhat startling. The deepest and most spectacular of these basins are aligned more or less parallel to the shelf edge, and in some cases, as in the Pribilof basin, have been partially breached to the ocean side by very large canyons (Marlow and Cooper, 1980). The largest and deepest of the basins discovered so far is the Navarin Basin, which has a maximum depth to the bottom reflector, or presumed basement, of about 15 km below sea level in water depths of less than 200 m. In practice, the large basins tend to be broken up into smaller basins, but the rough horizontal dimensions of the whole elongate Navarin Basin complex are 400 × 100 km (Marlow *et al.*, 1976; Turner *et al.*, 1985).

Essentially all of the reflection profiles made on the Bering Sea Shelf show a layered sequence of horizontal to subhorizontal reflectors overlying an acoustic basement. The acoustic basement is presumed to be late Cretaceous in age, based on the Navarin and other COST wells (Turner *et al.*, 1985). This is generally true for regions where the acoustic basement is relatively shallow as well as for the deep basins, though in the latter cases there is often evidence of disturbance of the layered sequence through faulting, much of which is considered to be growth faulting that developed as the basin filled.

Where the seismic technique has the resolution to see through what is normally acoustic basement, it depicts apparently folded and distorted reflectors. The Tertiary ages for the layered sequence and Mesozoic ages for the acoustic basement are based on the COST well data, and on extrapolations from dredge hauls and onshore geology as discussed earlier. Wherever the acoustic basement can be traced or extrapolated onshore, such as on St. Matthew Island (Cretaceous volcanics and sediments) and in the Black Hills on the Alaska Peninsula (Jurassic Naknek Formation sediments), it is Mesozoic. Basement is exposed on St. George Island, where it is intruded by an early Tertiary pluton suggesting a Cretaceous age. The dredge hauls discussed below also give circumstantial evidence for a Mesozoic basement.

The age range of the Cenozoic sediments of the main layered sequence is not well known. Where the lower units of the main layered sequence have been dredged, as for instance on the continental slope west of St. Matthew Island, rocks as old as Eocene have been found (Marlow, 1979), but more commonly the oldest samples are found to be Oligocene or Miocene, and range up to Recent.

The general conclusions drawn are that the acoustic basement consists of folded Mesozoic and older rocks, with the layered sequences above the basement being composed of Cenozoic and largely Neogene sediments. However, it should be reemphasized that the large dimensions of the Bering Sea Shelf are such that whole geological complexes could have been missed.

G. Dredge Hauls

A number of dredge hauls have been made in the Bering Sea (Fig. 6), and along the shelf edge; the majority of these did not contain samples that were of help in interpreting the nature of the acoustic basement (Marlow *et al.*, 1979a,b). Four localities which did provide useful information were in Pribilof Canyon (Cretaceous), on Pribilof Ridge to the east of St. George Island (Jurassic), on the shelf edge due west of the Pribilof Islands (Jurassic), and on the shelf edge southwest of the Navarin Basin (Cretaceous).

The oldest of the dated dredge hauls are Jurassic. The samples dredged from the Pribilof Ridge are from an area where the acoustic basement has no layered sequence on top of it, and is thus interpreted as a locality where the basement has been upwarped so as to be exposed (Vallier *et al.*, 1980). The dredge haul produced many samples of Upper Jurassic (Kimmeridgian) rocks, so that their presence through ice-rafting seems unlikely. The haul also contained young basalts similar to those seen on the Pribilof Islands. The Jurassic sediments are interpreted as being Naknek Formation equivalents, but unlike the Naknek Formation, whose source area consisted largely of the intrusives of the Aleutian Range batholith, these seem to have been derived from a volcanic terrane. No nearby or coastal source localities are known.

A second dredge haul that produced Jurassic rocks was from the shelf edge to the west of the Pribilof Islands. These are also interpreted as being Naknek Formation equivalent volcanogenic sediments, and contained pelecypod fossils of Kimmeridgian age (Marlow and Cooper, 1980).

The dredge hauls made in the inner gorge of Pribilof Canyon produced Cretaceous (Campanian) diatomaceous mudstones (Hopkins *et al.*, 1969; McLean, 1979a). These rocks are also interpreted as representing acoustic basement in the area. They are thought to have been dredged from bedrock at depths of more than 1000 m, but to have been deposited in neritic or upper bathyal depths (Hopkins *et al.*, 1969, p. 1473; McLean, 1979a). This is consistent with the general scenario that the main area of the Bering Sea Shelf has subsided below sea level, and that subsidence measuring from hundreds to more than a thousand meters has occurred in a number of areas (Scholl *et al.*, 1975a; Cooper *et al.*, 1979b). As mentioned earlier, correlations have been made in the literature between these dredge samples and the deep-water flysch deposits of the Shumagin Formation. In general, the early interpretation followed the model put forward by Burk (1965) that in Cretaceous time the shelf edge and/or trench extended around the corner of the Alaska Peninsula to head northwest across the Bering Sea. McLean (1979a) points out that the rock correlations are poor, and that the dredged samples are much closer to the Hoodoo and Chignik formations of the Alaska Peninsula, which are shallow-water and terrestrial deposits (Mancini *et al.*, 1978), and do not necessarily imply anything about a shelf-trench situation.

The fourth dredge locality, southwest of the Navarin Basin and west of St. Matthew Island, produced samples of micritic limestone, muddy limestone, and sandy siltstone with volcanic sandstone, all dated as being late Cretaceous (Campanian or Maestrichtian) on the basis of fossil pollen and spores. The similarity of these samples to those recovered from the Navarin Basin COST well argues strongly that they represent the acoustic basement of the adjacent area (Marlow *et al.*, 1981; Marlow and Cooper, 1980; Turner *et al.*, 1985).

The above discussion of dredge samples has been confined to the Mesozoic rocks which might represent acoustic basement dredged from the outer parts of the Bering Sea Shelf. Of the other rock samples dredged, those of Miocene and younger age are the most plentiful (Marlow *et al.*, 1976); however, Eocene ages were also obtained for samples from the continental shelf edge to the southwest of the Navarin Basin (Marlow *et al.*, 1981). These Tertiary samples are all considered to have come from the main layered series overlying the acoustic basement.

H. Siberian Geology

To complete the circuit of the geology around the Bering Sea Shelf requires a brief review of the geology of the Koryak Mountains and the Chukotsk Peninsula of eastern Siberia (Figs. 1 and 6).

The geology of eastern Siberia that has direct relevance to the Bering Sea Shelf has been reviewed by McLean (1979b) and Marlow *et al.* (1981). The geology of this region divides into two distinct belts in a similar way to that of the northeastern Alaskan side of the Bering Sea. The northernmost belt includes rock suites which appear to be similar to those found on St. Lawrence Island, in the Seward Peninsula, and in the western Brooks Range. Similar rocks of dominantly Paleozoic age (Patton and Tailleur, 1977) are mapped as occurring throughout the Chukotsk Peninsula, and form the western arm of the proposed syntaxis through St. Lawrence Island (Fig. 6).

On the southern edge of the syntaxis is an arcuate belt of calc-alkaline volcanic and plutonic rocks of Cretaceous and early Tertiary age (Marlow *et al.*, 1981), known as the Okhotsk–Chukotsk volcanic belt. These rocks have been interpreted as representing the remnants of a subduction zone, presumably the result of northward motion in the Bering Sea region prior to the formation of the Aleutian Arc.

South of this belt of volcanic rocks, and west of the Gulf of Anadyr lies the Koryak Range. Many of the rock assemblages in the Koryak fold belt, forming the Koryak Range, appear to be structurally juxtaposed, and consist of different geological sequences with varied lithologies. McLean (1979b) compares the whole assemblage on a regional scale to that of the Franciscan Formation and associated assemblages in California. He makes the analogy that the Okhotsk–Chukotsk volcanic/plutonic belt represents the Sierra plutonic belt; the Koryak Mountains the Franciscan assemblage of coastal California; and the two great basins, the Anadyr and the Khatyrka, compare with the Great Valley Sequence and the California shelf basins, respectively. The Koryak Mountains sequence, known as the Pikul'ney Series, can also be compared with the Uyak and McHugh complexes of southern Alaska.

The Okhotsk–Chukotsk volcanic/plutonic belt can be traced via St. Lawrence

Fig. 8. Basement topography of the Bering Sea Shelf area based on seismic reflection profiling.

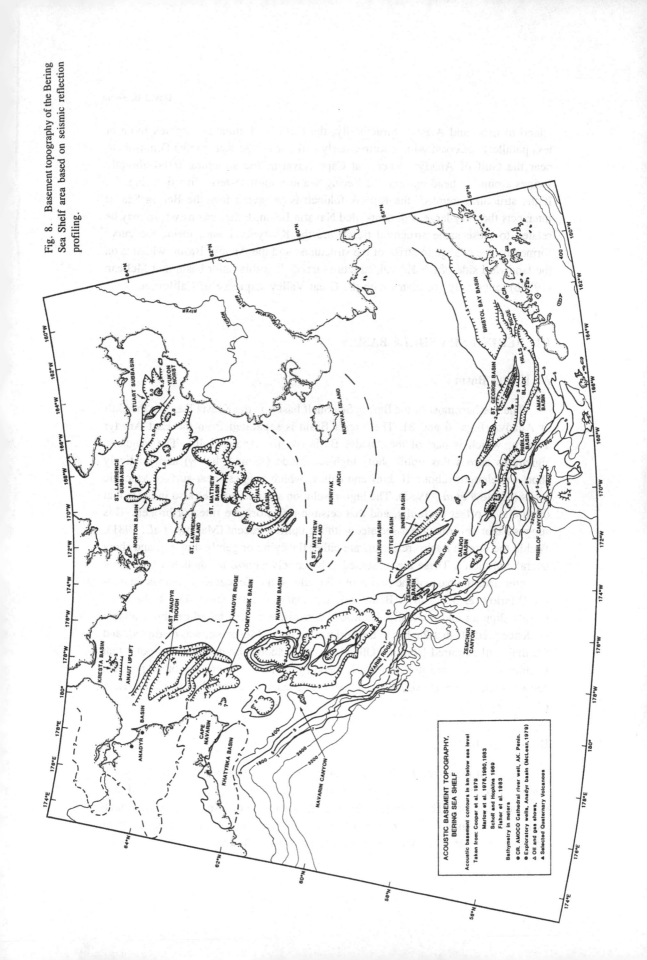

ACOUSTIC BASEMENT TOPOGRAPHY,
BERING SEA SHELF

Acoustic basement contours in km below sea level

Taken from: Cooper et al. 1979
Marlow et al. 1976, 1980, 1983
Scholl and Hopkins 1969
Fisher et al. 1983

Bathymetry in meters

● CR. AMOCO Cathedral river well, AK. Penin.
● Exploratory wells, Anadyr basin (McLean, 1979)
△ Oil and gas shows,
▲ Selected Quaternary Volcanoes

Island to mainland Alaska. Structurally, the Koryak Mountains complex more or less parallels the coast with a northeasterly strike from the Kamchatka Peninsula to near the Gulf of Anadyr, where, at Cape Navarin, the structural trend abruptly swings around to head out into the Bering Sea in a southeasterly direction (Fig. 6). If the structural trend of the Koryak foldbelt is projected into the Bering Sea, it intersects the very large sediment-filled Navarin Basin. Other basins which may be related to these same structural trends are the Khatyrka Basin, inside the curve formed by the change in strike of the structures, and the Anadyr Basin, which is on the landward side of the Koryak structural trend. It is this latter basin that McLean (1979b) tentatively correlates with the Great Valley sequence of California.

III. BERING SEA SHELF BASINS

A. Kresta Basin

The northernmost of the Bering Sea Shelf basins is the Kresta Basin in the Gulf of Anadyr (Figs. 6 and 8). The Kresta Basin is separated from the East Anadyr Trough, which is part of the Anadyr Basin, by the Anaut Uplift. The magnetic signatures across this uplift show high-amplitude (around 400 γ) and relatively short-wavelength (about 10 km) anomalies, which suggest near-surface volcanic rocks (Marlow et al., 1983). The high-resolution seismic profiles also indicate that the uplift is offset by faults, and that seismic reflectors are tilted or broken. This suggests that deformation associated with the uplift is recent (Marlow et al., 1981). Within the Kresta Basin, reflectors are either flat-lying or gently dipping, and show lateral continuity. These are presumed to be largely Cenozoic, though a prominent reflector near the base of the sediment fill could be the Cenozoic–Mesozoic boundary (Marlow et al., 1981). Below this boundary, the reflectors tend to be more steeply dipping and generally disturbed. A possible model based on the available evidence predicts that the basin itself originated in the Mesozoic, was disrupted, and the main fill acquired from the Oligocene on. It should be noted at this point that seismic reflectors are not necessarily bedding surfaces. There are other possible causes, such as diagenesis, that can give the appearance of bedding (e.g., see Hein et al., 1978).

B. Anadyr Basin

South of the Anaut Uplift lies the much larger East Anadyr Trough. The south boundary of this trough is the Anadyr Ridge separating it from the Navarin Basin even farther south. The Anadyr Ridge appears similar to the Anaut Uplift inasmuch as it has similar gravimetric and magnetic signatures indicating the presence of volcanic rocks at shallow depth (lines 13, 16, Figs. 6 and 9).

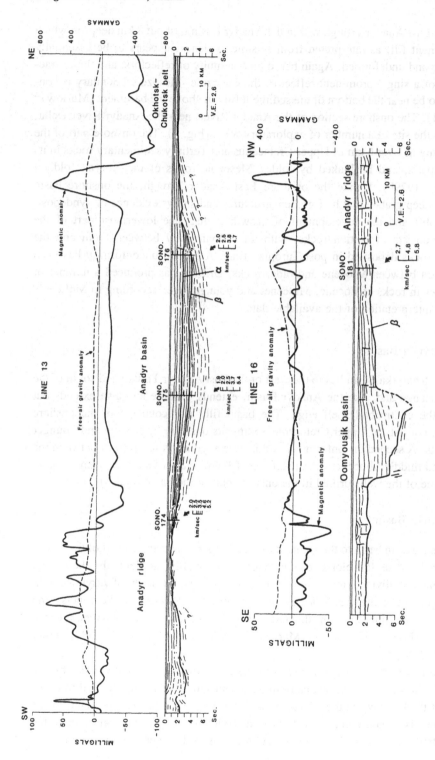

Fig. 9.   Cross sections of the Anadyr Ridge, Anadyr Basin, and the Oomyousik Basin as shown in Fig. 6. Also shown are the magnetic and gravity anomalies along the same lines (13 and 16). (From Marlow, 1983.)

The East Anadyr Trough within the Anadyr Basin is about 8 km deep, in which the sediment fill, as interpreted from seismic evidence, appears to be essentially flat-lying and undeformed. Again based on continuity of reflectors, and the characteristics of a single prominent reflector, the Cenozoic–Mesozoic boundary is considered to be near the bottom of the sediment fill but above the basement (Marlow *et al.*, 1981). The onshore section of the Anadyr River, near the Anadyr River Delta, has been the site of a number of exploratory wells (Fig. 8). The onshore part of the basin contains over 6 km of Upper Cretaceous and Tertiary sedimentary rocks in its deepest parts, and is flanked by folded Mesozoic rocks of the Koryak foldbelt (McLean, 1979b). Unlike the offshore East Anadyr Trough, the onshore basin contains deep graben and half graben structures and shows evidence of syndepositional uplift and the development of growth faults. The lowermost part of the section consists of Aptian rocks, with an unconformity between them and the younger rocks deposited in post-Turonian time. A second unconformity has been recognized between Miocene and older rocks. Drilling has produced a number of gas shows in rocks of Eocene, Miocene, and younger ages according to McLean's (1979b) interpretation of the available data.

C. Khatyrka Basin

The Khatyrka Basin lies to the west of the East Anadyr Trough and south of the Koryak Range and, like the Anadyr Basin, extends onshore and also extends out toward the continental shelf edge. The basin fill, as recognized in the onshore geology, consists of Upper Cretaceous sediments overlain by Eocene and younger sediments. A second unconformity, in this case angular, is recognized between the lower and middle Miocene rocks (McLean, 1979b). Gas shows have been reported from some of the wells drilled in the onshore part of the basin (Fig. 8).

D. Navarin Basin

The Navarin basin to the south of the Gulf of Anadyr is the largest of the basins discovered to date. It is elongate with its long axis roughly parallel to the shelf edge. The basin is subdivided into a number of smaller basins or troughs of varying depth. The major subbasins are, from north to south, the Navarinski, Pervenets, and Pinnacle Island subbasins. The deepest parts of the Navarin basin are on the order of 12–15 km (Figs. 6, 10, and 11) (Marlow, 1979; Marlow *et al.*, 1981, 1983; Turner *et al.*, 1985).

The main layered sequence seen in the basin fill can be subdivided into two parts. The lower unit is acoustically more transparent than the upper, and the beds within it thicken toward the deeper parts of the basin. The upper unit overlies the lower one discordantly in places, and extends out from the basin margins across the whole shelf. A number of high-angle faults cutting both upper and lower units have

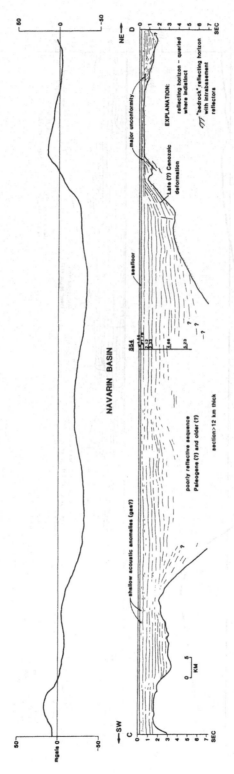

Fig. 10. Interpretive line drawing of seismic reflection profile C–D as shown in Fig. 6. (From Marlow, 1983.)

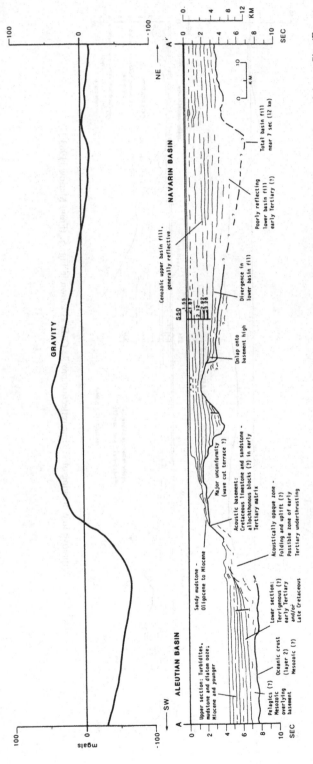

Fig. 11. Similar profile to Fig. 10 along line A–B (Fig. 6). Note that the vertical kilometer scale at right is applicable only to the shelf part of the profile. (From Marlow *et al.*, 1981.)

been recognized, and some show increasing offset with depth, perhaps indicating growth faulting (Marlow et al., 1976). The fill in the southern parts of the basin is essentially undeformed, but in the northern "lobe" the sedimentary section is folded into large (15 km wide) anticlines which, it is speculated, are due to diapiric cores (Marlow, 1979).

A COST well was drilled in 1983, located between the Pervenets and Pinnacle Island subbasins. It reached a total depth of about 5 km (16,400 ft). A composite section (Turner et al., 1985) based on biostratigraphic, seismic, and lithological data indicates a thick Tertiary section (3.9 km, 12,780 ft) including Pliocene, Miocene, Oligocene, and Eocene sediments (the main layered sequence) lying unconformably above late Cretaceous sediments (Maestrichtian and possibly Campanian or older). The 768-m (2520 ft) Maestrichtian unit is composed of nonmarine coal-bearing sediments intruded by Miocene diabase and basalt sills. Below this is 335 m (1100 ft) of Campanian (?) marine shale. The nonmarine section is interpreted as representing a deltaic environment. In general terms, the Cretaceous rocks are very similar to those dredged from the shelf edge and canyon walls discussed earlier.

The Eocene rocks which uncomformably overlie the Cretaceous section record a marine transgression, and indicate rapidly deepening water, presumably the result of basin subsidence. During the Oligocene the water became shallow enough to expose local subbasins to wave-base erosion. Shelf sands were then deposited until the late Miocene. A late Miocene uncomformity represents a depositional hiatus, but is followed by shelf sand deposition up to the present. The source areas for these sands were chiefly volcaniclastic but also included a quartz mica schist terrane (Turner et al., 1985).

E. Oomyousik Basin

In most maps of the northern basins, the Anadyr Ridge appears to form the divide between the Navarin basin and the East Anadyr Trough (e.g., see Marlow et al., 1983). This is based on extrapolating the general northwest trends of the basins and ridges. However, this is complicated somewhat by the presence of the Oomyousik Basin (Marlow et al., 1981). This basin is only seen on one profile (Figs. 6 and 9) so its configuration cannot be determined directly, but it is south of the Anadyr Ridge. To the northeast, the end of the East Anadyr Trough is also poorly constrained, and it seems possible that the two could connect. Similarly, it is possible to connect the Oomyousik Basin to the Khatyrka Basin discussed earlier. If this latter connection is real, it has implications with regard to the geological continuity between Siberia, the Bering Sea Shelf, and the rest of Alaska inasmuch as it precludes a direct continuation of the Siberian structural trends along the Bering Sea Shelf edge.

F. Zemchug, Dalnoi, Garden, Pribilof, Walrus, Otter, and Inner Basins

Continuing southwards along the Bering Sea Shelf margin, there are a number of smaller basins separated by another basement high, the Pribilof Ridge (Figs. 6 and 8). On the oceanward side are the Zemchug Basin which has been partially breached to form the T-shaped Zemchug Canyon, the Dalnoi Basin, the Garden Basin, and the Pribilof Basin. All of these basins are very elongate. The Zemchug and Pribilof basins contain about 4 km of sediment and the Dalnoi and Garden basins about 1.5 km. The Pribilof Basin is a graben bounded by high-angle faults, and the faults observed in the Zemchug Basin can be interpreted to be similar in style (Marlow and Cooper, 1980; Marlow et al., 1976). To the north of the Pribilof Ridge and northwest of the Pribilof Islands are three elongate and relatively shallow basins (about 1.5 km deep), the Walrus, Otter, and Inner basins (Figs. 6, 8, and 12).

G. St. George Basin

Still north of the Pribilof Ridge and east of the Pribilof Islands is the St. George Basin (Marlow et al., 1977). This basin is also elongate, but is considerably larger in area than the other nearby basins, being about 200 km long by about 50 km wide at its widest point. St. George Basin can be roughly subdivided into east and west segments, with the sedimentary fill reaching a thickness of about 8 km in the eastern segment and about 9 km in the western one (Figs. 6, 8, and 13). Two COST wells have been drilled on the south side of the basin and are discussed by Turner et al. (1984a,b). St. George COST well number 1 was completed in 1976 and has a total depth of 4.2 km (13,771 ft) and COST well number 2 was completed in 1982 and has a depth of 4.5 km (14,626 ft). Both wells penetrated Pleistocene through Oligocene sediments, most of which reflected a neritic to bathyal environment. COST well number 1 also encountered sediments of middle Eocene age before entering a sequence of volcanic rocks. The K–Ar ages on the volcanic rocks show considerable scatter ranging from 23 to 136 Ma. Because the Eocene sediments were deposited unconformably on the volcanics, the latter have been assigned a probable middle Eocene age.

COST well number 2 encountered Oligocene or possibly Eocene sediments lying unconformably on early Cretaceous to late Jurassic sediments. The paleoenvironment of deposition of these sediments is interpreted as inner neritic to continental, and coal is present in the uppermost parts. The unconformity at the top of the Mesozoic section corresponds to the acoustic basement in the seismic profiles. An interpretive section from the deep Bering Sea across the continental slope and the St. George Basin is shown in Fig. 14. This section was published before the COST well results were available; thus, the sediments labeled terrigenous should be neritic

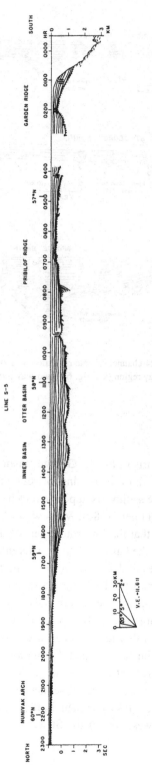

Fig. 12. Interpretive line drawing of seismic reflection profile S5 (Fig. 6) across the Pribilof Ridge and the Inner and Otter basins. Vertical scales are two-way travel time in seconds and water depth only in kilometers (at 1.5 km sec⁻¹). (From Marlow et al., 1976.)

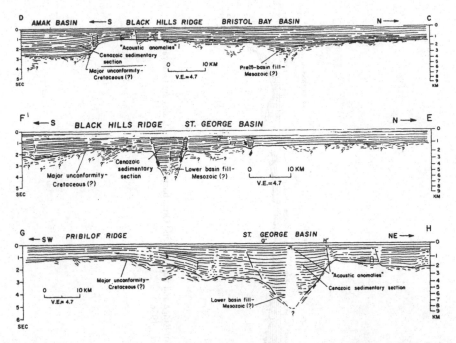

Fig. 13. Interpretive line drawing of 24-channel seismic reflection data along profiles C–D, E–F, and G–H in the St. George Basin–Bristol Bay region (see Fig. 6). Two-way travel time is in seconds. (From Marlow and Cooper, 1980.)

to bathyal. In the easternmost parts of the St. George Basin, faulting was observed along the flanks. These faults show increasing offset with depth, and are thus interpreted as growth faults. The sediments appear not to have been disturbed other than by faulting (Marlow and Cooper, 1980). Marlow *et al.* (1976, 1977) interpret the seismic records as showing that the lowermost part of the basin fill consists of coarse debris deposited early in the basin's history. This interpretation is borne out by the presence of conglomerates in the Mesozoic section. The south margin of the St. George Basin is formed in part by the Pribilof Ridge (at the westernmost end) and in part by the Black Hills Ridge (at the eastern end) (Fig. 13). The trend of these two ridges is similar, but they are apparently separated by a shallow saddle-shaped depression (Fig. 8). The Black Hills Ridge can be traced onshore, as was discussed earlier, and correlates with the Jurassic Naknek Formation seen in the Black Hills. The surface of the acoustic basement of the Black Hills Ridge is unconformably overlain by the sediments of the main layered sequence. The erosional unconformity appears to represent a wave-cut surface which, if real, indicates subsidence on the order of 2 km on the southwest side of the St. George Basin.

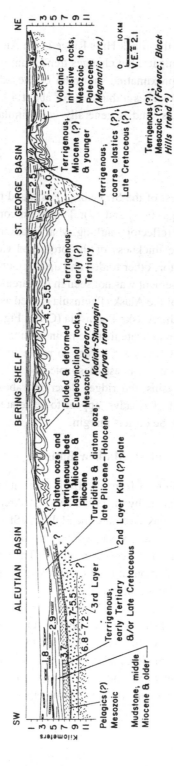

Fig. 14. Generalized cross section across the Bering Sea margin. (From Marlow et al., 1977.)

H. Amak Basin

On the southwest side of the Black Hills Ridge is the Amak Basin (Figs. 6, 8, and 13). This basin is also elongate, but changes trend from roughly parallel to the Bering Sea Shelf edge to approximating the trend of the Alaska Peninsula. The Amak Basin sedimentary fill reaches a thickness of about 4 km, displays apparent growth faults along its margins, and is generally very similar in character to the other basins.

I. Bristol Bay Basin

To the north and northwest of the St. George Basin and the Black Hills Ridge lies the Bristol Bay Basin (Figs. 6, 8, and 13). This basin consists of a very broad depression with a prominent reflector marking what is assumed to be the Mesozoic/Cenozoic boundary. The thickness of the presumed Cenozoic sediments is about 3 km. Below this reflector, other folded and discontinuous reflectors are seen as deep as 7 km. Acoustic basement was not seen in this area (Marlow and Cooper, 1980). If the known geology of the Alaska Peninsula is used as a guide, then the fact that a well drilled nearby on the Alaska Peninsula (CR in Fig. 8) bottomed at 4 km in Oligocene sediments, would indicate that the main layered sequence is also about this age. The older basement rocks are harder to relate to the Alaska Peninsula geology, but are presumed to be of latest Mesozoic age.

The major and minor basins and ridges described above, with the possible exception of those in the Gulf of Anadyr, appear to be related in some way to the tectonic history of the Bering Sea Shelf margin.

J. Nunivak Arch and Hall and St. Mathew Basins

Toward the Alaskan mainland from the Bering Sea margin basins described above, the subsurface is dominated by the broad Nunivak Arch (Figs. 8 and 15). On the northwest side of the Nunivak Arch are the Hall and St. Matthew basins (Fig. 8). These are both relatively shallow and contain sections of the main layered sequence about 1 km thick in their deepest portions. Their principal significance seems to lie in their possible relationships with the seaward extension of the Kaltag Fault (Fig. 6), one of the major structural features of western Alaska. If this fault extends far out onto the Bering Sea Shelf, it puts important constraints on the possible geological histories of the area. At the present time, the only reason for relating the St. Matthew Basin to the Kaltag Fault is their similarity of trend and alignment. The Hall Basin is smaller, and its long axis is oriented perpendicular to the Kaltag–St. Matthew basin trend and parallel to the Bering Sea Shelf margin features, perhaps indicating where the transition from onshore to shelf-margin structural trends takes place.

K. Norton Basin

Essentially all of the above discussion has been confined to those parts of the Bering Sea Shelf south of the hypothesized syntaxis from the Chukotsk Peninsula south through St. Lawrence Island and then north through the Seward Peninsula to the Brooks Range.

The Norton Basin lies within the syntaxis, north of St. Lawrence Island, and includes Norton Sound lying between the Seward Peninsula and the Yukon Delta. The Norton Basin is very large in areal extent, but generally shallower than the basins associated with the shelf edge. Its western margin is close to the Siberian coast, and the overall basin is elongated more or less parallel to the Seward Peninsula coast and the length of St. Lawrence Island. It terminates near the Yukon Delta close to the extension of the Kaltag Fault.

Recent geophysical work on the Norton Basin associated with oil exploration (Fisher, 1982) and the drilling of two stratigraphic test wells have produced a relatively detailed picture of the area. The basin can be subdivided structurally in two parts, the St. Lawrence subbasin north and east of St. Lawrence Island and the Stuart subbasin north and east of the Yukon Delta. These two areas are divided by the Yukon Horst. Figure 8 shows the generalized structure as redrawn from Fisher *et al.* (1982). Both parts of the basin show considerably deeper basement depths than were initially estimated by Scholl and Hopkins (1969). The deepest parts of the St. Lawrence subbasin are over 5 km, and the Stuart subbasin as deep as 6.5 km. Farther west than the St. Lawrence subbasin, the basement depths are subdued, seldom exceeding 2 km.

The two COST wells drilled in the St. Lawrence subbasin (COST No. 1) and in the Stuart subbasin (COST No. 2) shared generally similar stratigraphy, though the Stuart subbasin showed more nearshore and onshore facies (Turner *et al.*, 1983a,b). The Tertiary rocks show a generally neritic marine environment back through Miocene time, with a possible disconformity in the mid-Miocene. In the Oligocene section there are both marine and continental deposits, both wells showing coal-bearing units. The lowermost sedimentary unit is thought to be of Eocene age and is bounded above and below by unconformities. The lower of the two is the major unconformity which is also the acoustic basement. Below the acoustic basement to the total depth of the wells are metamorphic rocks very similar to those seen in the York Mountains on the Seward Peninsula, consisting of phyllites, quartzites, and marbles.

The fact that both wells bottom out in metamorphic rocks fits with the hypothesis that most of the basin is underlain by the rocks that form the syntaxis, which in large part consist of Paleozoic carbonates (Patton and Tailleur, 1977).

The age of the formation of the basins is not known, but may be related to an early Paleogene extensional regime recognized on the Seward Peninsula. The basins could also represent earlier pull-apart features associated with late Cretaceous motion on the Kaltag Fault.

IV. THE ALEUTIAN ISLAND ARC

A. Physiography

The Holocene geological history of the Aleutian Island Arc has been dominated by sea-level changes, both eustatic and tectonic, and by volcanism. Evidence of ash layers, lahars, nuée ardente deposits, volcanic breccias, and calderas is abundant. The Holocene history of the Aleutian Island Arc is important, inasmuch as it bears on the Bering Land Bridge (Hopkins, 1967) and the time of occupation of the islands. Black (1974, 1976) makes a strong case for the various migrations and movements of the Aleuts to be tied to geological events, particularly changes in sea level and large volcanic eruptions. There are more than 40 volcanoes in the Arc that have been active in Holocene time, and possibly many more, as several subsea cones are recognized, but their activity is difficult to determine. It is also difficult to determine the activity of the onshore volcanoes due to the low population density and the pervasive overcast weather conditions which often prohibit satellite and aircraft observations. A compilation of known historic eruptions is given by Kienle and Swanson (1982) and selected volcanoes are shown in Fig. 1.

The detailed dimensions of the roughly 2200-km-long Aleutian Island Arc depend on how it is defined. The west end can be considered as either the end of the bathymetrically well-defined Aleutian Ridge, or the intersection of the arc trend with the Kamchatka Peninsula, across the deep channel connecting the Bering Sea with the Pacific Ocean. In practical terms the choice of one or the other is of little consequence, though the gap itself may be significant (Cormier, 1975). The east end of the arc is more problematic. If defined as strictly an island arc, then the east end is in the vicinity of Unimak Island. This break more or less marks the change in geology from the "ocean–ocean" arc system of the western end to the "ocean–continent" system farther east. No terrestrial rocks older than Tertiary have been identified in the western part, whereas the Alaska Peninsula has large sections of Mesozoic rocks exposed with occasional outcrops as old as Paleozoic (Burk, 1965; Beikman, 1980). For the purposes of this review, it is largely the section from Unimak Island west that will be discussed, though it should be kept in mind that in terms of today's subduction zone, there is no obvious break at this location. The volcanoes, seismicity, and bathymetric trench all continue eastwards from Unimak Island, though the tectonic setting gets progressively more complex toward the Gulf of Alaska and Interior Alaska, which is where the bathymetric trench and Benioff Zone trend, respectively (Kienle and Swanson, 1982; Stone, 1983).

The western Aleutian Arc consists of a thick welt of material rising more than 3000 m from the Bering Sea floor on the north and more than 6000 m from the Aleutian Trench on the south. The morphology of the ridge can be subdivided into the main ridge, about 100 km wide at its base and topped by the Aleutian Islands, and the forearc basin between the main ridge and the trench, also about 100 km wide, making the whole ridge system about 200 km wide at its base (Fig. 16). The main ridge is like a guyot in cross section with relatively smooth topography across the summit, which appears to represent recent deposition on a previously wave-cut platform.

The islands represent only a small percentage of the area of the ridge crest, and it is only the shorelines and recent volcanoes that offer good rock exposures, the remainder being largely tundra covered. In addition, much of the present island area is covered by the recent and presently active volcanoes and their products, which severely limits the possibilities for direct observation of the older rocks. Population centers are few and far between, the most western nongovernment population center in the U.S. Aleutian Islands being Atka Island. From here to the west there are a number of military and Coast Guard bases, but these too are very spread out. There is a significant population on the USSR-owned Komandorsky Islands, but at the present time these islands are inaccessible to Western scientists.

Offshore, the bathymetry is reasonably well known, particularly from the work of Nichols and Perry (1966) and a later compilation by Scholl et al. (1974). As mentioned above, the bathymetry shows the Aleutian Arc as consisting of the main ridge with its wave-cut flat top and a forearc basin between the ridge and the trench. It also shows that the main ridge is scarred by a number of large canyons. These are considered to be mainly erosional features caused by sediment scouring and slumping. In some cases the canyons appear to be structurally controlled in terms of location, the most notable cases being Amlia and Adak canyons, which appear to be related to the fracture zones of the same names.

Particularly notable features of the bathymetry on the north side of the western part of the ridge are the sections of very straight shelf edge, dropping from the ridge to the deep part of the Bering Sea (Figs. 1 and 16). One of these sections is located between the Komandorsky Islands and about 172°E longitude. Another trends east-southeast from a point adjacent to Attu Island to near Buldir Island. This latter trend has been suggested to be fault controlled on the basis of its linearity (Panuska, 1980; Gibson and Nichols, 1953). A possible eastward continuation of this trend can be recognized on the ridge crest. The two together make a linear feature from Attu Island along the shelf edge to Buldir Island, thence via a ridge crest depression, the Buldir Depression, to a graben-like feature on Kiska Island and thence along a linear bathymetric feature adjacent to Amchitka Island and Rat Island to the Pacific side of the ridge. If the features forming this lineation are tectonically related, they may be associated with a shortening of the arc system, and/or with the fact that Bowers Ridge joins the Aleutian Arc in this general location (Fig. 16).

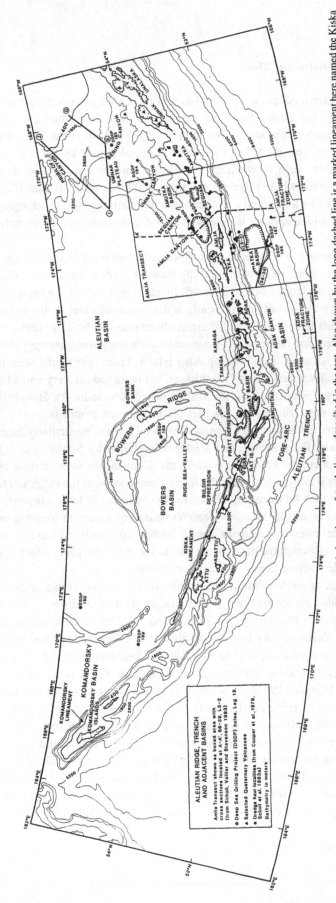

Fig. 16. Aleutian ridge bathymetry and ridge crest basins together with locations of track lines described in the text. Also shown by the long dashed line is a marked lineament here named the Kiska lineament.

Another physiographic observation of interest is that west of Seguam Volcano (Fig. 16), the volcanoes of the present active arc are all perched on the northernmost edge of the ridge. The exception to this occurs where the situation is confused by the junction of Bowers Ridge with the Aleutian Arc.

In addition to physiographic studies, sources of information used to deduce the geological history of the Aleutian Arc and its relationship to the Bering Sea, include seismicity studies aided by investigations related to nuclear tests on Amchitka Island (Engdahl, 1971, 1972; Jacob, 1972), marine geophysical investigations, dredge hauls, and a series of Deep Sea Drilling Project (DSDP) holes.

B. Deep Structure

At present there is no single accepted model for the structure beneath the Aleutian Arc, or Ridge, in large part because the available data come from different locations along the arc. However, the range of models that fit all the available data along the arc is limited by the common assumption that the structure remains reasonably similar along its length.

The most obvious constraints that have to be satisfied are the observed natural seismicity, the observed velocities obtained from refraction profiles, and the observed gravimetric and magnetic variations across the arc.

The Aleutian Arc is one of the most seismically active areas of the world and has been the site of a number of great earthquakes (M ⩾ 7). If the Aleutian Arc earthquakes are plotted against both location and time, it is apparent that seismic energy is not released uniformly along the arc, but is released episodically from discrete blocks some of which can be distinguished in Fig. 17 (Kelleher, 1970; Sykes, 1971). This is most clearly demonstrated by the Rat Island earthquake of 1965 (M = 7.5), and its aftershocks (Jordan et al., 1965; Stauder, 1968b). The aftershock zone is easily distinguished on epicenter plots as a block-shaped area of high earthquake density largely to the west of the initial shock at 179°E. Other similar, apparently fault-bounded areas of strain release are observed throughout the Aleutian Arc and adjacent areas (Spence, 1977) and in general, the main event is located on the eastern edge of the eventual aftershock zone (Sykes, 1971). To a large extent, the boundaries of these individual blocks undergoing strain accumulation and release coincide with mapped fracture zones on the Pacific Plate, and also with changes in the chemistry of the present volcanic arc (Kay and Kay, 1982). That the seismicity gets deeper from the south toward the Bering Sea side of the arc has been known for some time (Gutenberg and Richter, 1954), but the first systematic study of the distribution of earthquakes with depth was that of Engdahl (e.g., see Engdahl, 1977). This study arose from the observations needed in relation to the nuclear testing sites on Amchitka Island. The nuclear tests Longshot in 1965, Milrow in 1969, and Cannikin in 1971 allowed the seismometer networks to be calibrated and better velocity models for the north Pacific, Bering Sea and Aleutian

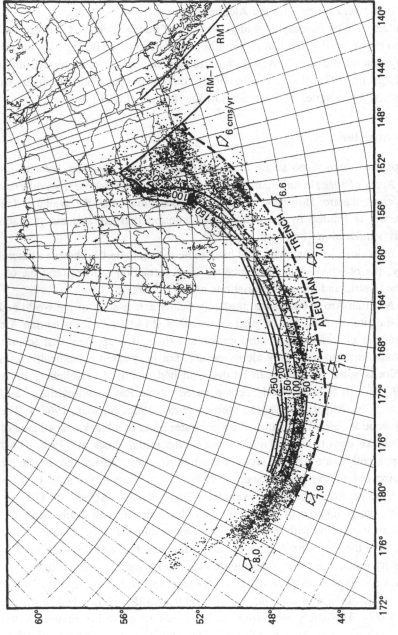

Fig. 17.   Seismicity of the Bering Sea–Alaska region for the period 1962–1969 together with hypocenter contours (in km). The relative motion vector for the Pacific Plate with respect to North America is shown. Segmentation of Aleutian seismicity can be seen in the epicenter distribution.

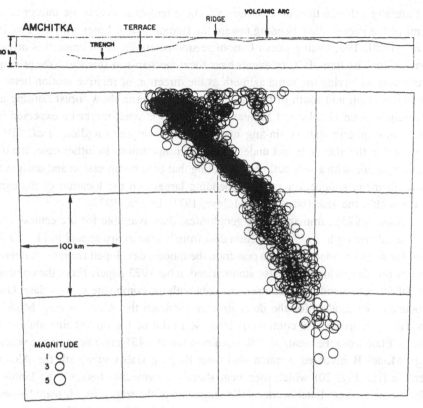

Fig. 18. Projection of hypocenters of earthquakes beneath the Aleutian Island Arc in the vicinity of Amchitka Island. (From Engdahl, 1977.)

region to be constructed (Jacob, 1972; Engdahl, 1971, 1972). The velocity models still need refining to allow for the location of the downgoing subducting slab, since a high-velocity slab can give rise to significant errors of location of seismic events, especially with regard to their depth. The hypocenter distribution seen beneath the Aleutian Island Arc appears fairly typical for an island arc, inasmuch as there is a well-defined zone of earthquakes increasing with depth toward the concave side of the arc. Cross sections, as shown in Fig. 18, reveal a pronounced bend in the Benioff Zone as defined by the dipping hypocenters. The reality of this bend is now being reconsidered, and may be in large part due to the distribution of seismometer stations and to the velocity models used, but the overall dipping plane is not in question.

Among the earliest studies relating to the tectonic regime of the Aleutian Island Arc are those of Stauder (1968a,b). Stauder points out the systematic difference in the earthquake mechanisms observed near the Aleutian trench and those associated with the Benioff Zone. The former are mostly tensional events (normal faulting?)

and are aligned with the curve of the arc. These tensional events are interpreted in terms of the Pacific Plate bending toward the Benioff Zone. Later work by Frohlich *et al.* (1980, 1982) using ocean bottom seismometers has confirmed this original conclusion. The Benioff Zone events have focal mechanisms that show the principal stress axes as having the same azimuth as the direction of relative motion between the plates north and south of the Arc. The stress patterns show thrust faulting and downdip tension (Isacks and Molnar, 1971). This is what might be expected if a dense lithospheric slab is sinking under its own weight. Engdahl *et al.* (1977) suggest that the slab is in fact under downdip compression. In either case, the data are compatible with a subducting slab, noting that both compression and tension are found in many subduction zones, depending largely on the location of the earthquake within the slab (Isacks and Molnar, 1971; Uyeda, 1977).

Grow (1973) summarized the geophysical data available for the central Aleutian Island Arc up to 1973. Refraction data from his summary appear in Fig. 19. He used the seismic velocity data to constrain the models developed from the numerous gravity profiles, which were also summarized in his 1973 paper. From the combination of reflection and refraction seismic data with the composite gravity data, Grow produced two models for the deep structure beneath the Aleutian Arc. Model A involved a more or less continuous increase in dip of the downgoing slab of the Pacific Plate from the location of the trench to about a 45° angle beneath the volcano line. Model B involved a much shallower dipping slab underneath the Aleutian Terrace (see Fig. 20) which then bent sharply downwards beneath the landward edge of the terrace. Later work on the sediments of the trench and forearc basins in the Amlia transect (Fig. 20) tends to support model A of Grow (1973) (Scholl *et al.*, 1983a,b; McCarthy and Scholl, 1985) whereas the seismicity profiles of Engdahl (e.g., see Engdahl, 1977) support model B. One point to note in these various data compilations and models is the low level of seismic activity associated with and beneath the outer terrace (Frohlich *et al.*, 1980). This possibly implies that the Pacific Plate is decoupled from the overriding arc in this region. Similarly, the lack of pelagic sediments near the trench, and the generally undisturbed trench fill imply that much of the sedimentary cover on the subducting plate is also being subducted. It has been postulated that high pore pressure in water-saturated sediments can in fact act as a ''lubricant'' for the slip plane in situations of this sort (Grow, 1973; Moore *et al.*, 1982).

C. Geology

The only geological information available for many of the U.S. Aleutian Islands, and the source of almost all the geological base maps is contained in reconnaissance maps reported in the USGS 1028 series of bulletins. In addition to the bulletins, some local mapping has been reported for a few individual islands. The exception to this is for Atka and Amlia islands (Hein *et al.*, 1981) which have

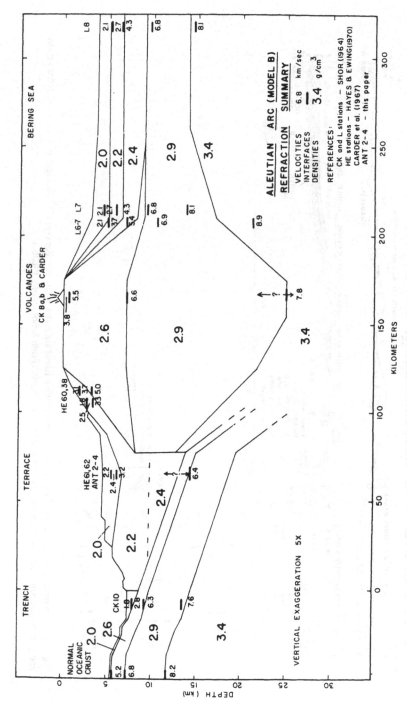

Fig. 19. A summary of seismic velocities determined by refraction studies across the central Aleutian Island arc, together with the densities used in gravity modeling (model B of Grow, 1973). (From Grow, 1973.)

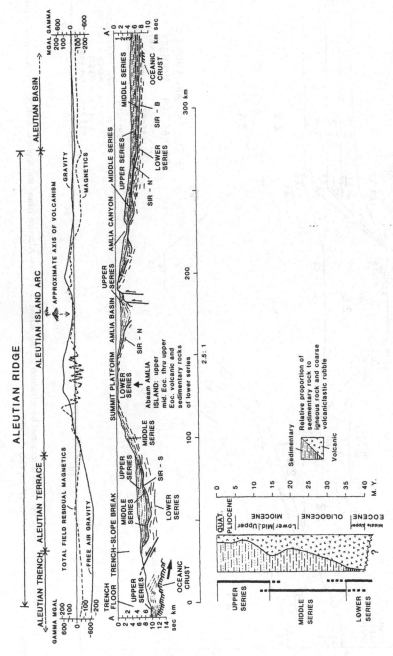

Fig. 20. A composite cross section of the geology of the Aleutian Island arc in the Amlia corridor (see Fig. 16) together with associated magnetic and gravity anomalies and idealized chronostratigraphic framework. (From Scholl *et al.*, 1983b.)

been mapped as part of an Aleutian Arc transect project (Fig. 16). For the USSR Komandorsky Islands, the geology is available in the Russian literature as reported by Shmidt (1974) and Shmidt *et al.* (1973).

One of the difficulties that arose in all the early mapping, and is a continual problem with interpreting the geology reported in the USGS 1028 series, is the lack of age control for the older units mapped. As discussed below, no rocks older than Eocene have been documented. However, some of the rocks now known to be of Eocene age were originally assigned ages as old as Paleozoic (?) (Coats, 1956a). The original ages were assigned on the basis of degree of alteration, very long-range lithological correlation, and some leaf impressions. The leaf impressions in the Andrew Lake Formation of Adak Island have since been reinterpreted in the light of new data as Eocene (Scholl *et al.*, 1970b). Radiometric age dating techniques have also been undergoing development since the USGS 1028 series was first published (1955), and confirm that the oldest rocks found so far are of mid to late Eocene age in the U.S. Aleutian Islands.

The basement geology of the western Aleutian Ridge is interpreted as consisting of an initial welt of extrusive, intrusive, and sedimentary rocks that presumably formed at the time the original Aleutian subduction zone started. The timing of this event is not known, but on the basis that the oldest well-dated rocks are mid to late Eocene, a K–Ar age of 42.3 ± 4.6 m.y. from a basalt from Ulak Island (DeLong and McDowell, 1975; DeLong *et al.*, 1978), and that a major change in the relative motions of the Pacific and North American plates took place at about 43 m.y. (Dalrymple *et al.*, 1977), it is often assumed that these events are related (Wallace and Engebretson, 1982). Pickthorn *et al.* (1984) consider that the middle and late Eocene rocks with minimum K–Ar ages of approximately 40–42 m.y. indicate an earlier time of formation; however, this could still be directly related to the changes in direction of the Pacific plates.

The mechanics and processes involved in the initiation of subduction zones are not well known; similarly, the rock types involved in the basic ridge-building phase of development of the Aleutian Arc are poorly known. However, the few samples available from dredge hauls and drilling, combined with the observed seismic velocities and densities, indicate that the bulk of the geanticlinal mass of the ridge is composed of igneous rocks (Scholl *et al.*, 1983a). This conclusion is supported by studies made on the xenoliths included in several plutons along the arc. These plutons include the oldest known, the Finger Bay and Hidden Bay plutons on Adak Island and the Kagalaska Pluton. In none of the xenoliths studied were sedimentary protoliths found. The range of xenolith compositions indicates that the arc crust is igneous in origin, with the lower crust composed of gabbros crystallized from mantle-derived melts (Conrad and Kay, 1983; Kay *et al.*, 1983; Debari *et al.*, 1987; Conrad *et al.*, 1983). These rocks comprise the informally named lower series of Scholl *et al.* (1983a,c). In general, the degree of metamorphism noted in Aleutian rocks is low; however, Coats (1956b) reported finding rounded boulders of hornblende gneiss, slate, and schist together with intrusive rock boulders on the wave-cut platforms of Tanaga and Ogliuga islands. The source of the metamorphic rocks is not known, but Coats (1956b) considers them older than the intrusives. Since they are so localized, they may represent an exotic fragment caught up in the arc. Ice rafting is also possible, but the large numbers found make it unlikely.

The initial ridge built up above sea level ". . . where a tree-covered crown of eruptive centers issued mafic to felsic products similar to those erupted today" (Scholl et al., 1983b). This scenario is based at least in part on the Andrew Lake Formation exposed on Adak Island, which consists of terrestrially derived sediments with organic remains which have been interpreted as fossil leaf impressions (Coats, 1956a; Scholl et al., 1970b). Some of the organic remains have since been reinterpreted as being algal in origin (Scholl, personal communication, 1984), similar remains having since been discovered on Amlia Island.

Scholl et al. (1983a) recognize the sediments covering the lower series on the flanks of the initial ridge as being of Oligocene and younger in age, and informally name them the middle series. The middle series was presumably derived from the original ridge structure. Thermal or tectonic alteration of the early and middle series appears to be limited to the immediate vicinity of later intrusive bodies, and to the forearc region.

The upper series of Scholl et al. (1983a) consists of basin-filling and slope-mantling deposits which are in general little deformed. This division of the observed rocks into the three informally named series (lower, middle, upper) was based on the inferred formational and depositional history of the arc, and although ages are assigned to each of these series, they are not considered to be exactly time-stratigraphic (Fig. 20; Scholl et al., 1983a,b,c).

Earlier workers divided the rocks of the Aleutian Arc into the early and late marine series (Wilcox, 1959) which was later modified to simply the early and late series (Marlow et al., 1973), but the recognition that there have been three major episodes in the formation of the present arc has led to the current designation.

Because of the large dimensions of the arc, the limited exposure afforded by the islands, and the limited number of marine geophysical profiles available, a transect study was initiated. The location of the transect corridor studied in detail by the U.S. Geological Survey and others is shown in Fig. 16. The combined studies include geological mapping of Atka and Amlia islands (McLean et al., 1983; Hein et al., 1981), marine geophysical studies, including reflection and refraction seismic lines in the vicinity of the Amlia Fracture Zone (Scholl et al., 1983c), the arc, forearc, and trench (Scholl et al., 1983a; McCarthy and Scholl), and the Amlia Summit Basin (Scholl et al., 1983c).

In addition to the studies mentioned above, two DSDP holes (186, 187) were drilled on the Aleutian Terrace within the boundaries of the corridor, and one (184) was drilled on the Bering Sea side on the flanks of Umnak Plateau (Fig. 16).

Compiling all of the information available for the transect has allowed a composite cross section through the arc from the Pacific Ocean to the Bering Sea to be assembled (Fig. 20).

The available geology for the arc east of about Kiska Island indicates that the Amlia transect cross section is a good approximation for the arc in general, with the known differences arising more from detail rather than basic concept. From Kiska Island to the west, the ridge may have had a somewhat different geological history. This is based on the observation that the chemistry and petrology of the far western islands are somewhat different from those to the east (Kay et al., 1985, 1986). Whether or not these differences are important in terms of the overall geology and tectonics of the arc is not known, but it may be of significance in the context that the Komandorsky and Bowers basins, which adjoin the west end of the arc, are different both structurally and thermally from the Aleutian Basin which adjoins the east end of the arc. On the Pacific side, there are marked differences in the seismicity between the eastern and western ends. The lack of seismicity, or perhaps more important the lack of a well-defined Benioff Zone at the western end of the arc, is usually interpreted as the result of Pacific–Bering motion being almost parallel to the arc trend. It has also been suggested that it is the result of a different thermal regime and/or a different structural setting (e.g., Cormier, 1975). However, a recent reevaluation of the seismicity of the area shows a small but significant component of convergence in the region (Newberry et al., 1986), and seismic reflection studies indicate the presence of subduction related structures including a small fore-arc basin (Vallier et al., 1984).

V. ALEUTIAN RIDGE SUMMIT BASINS

There are numerous summit basins delineated by the bathymetry on the otherwise relatively smooth and apparently wave-planed ridge crest (Nichols and Perry, 1966). One of these, Amlia Basin, is located within the transect mentioned earlier (Figs. 16 and 20). Some of these summit basins are considered erosional, some have not been explored sufficiently to speculate on their origins, and others are considered structural in origin. Of this latter group, five are much larger and better defined than the others. Of these, the Buldir Depression, the Pratt Depression, and the Amlia and Amukta basins are on the northern edge of the Aleutian Ridge, with the Buldir Depression being elongate and generally aligned with the Holocene volcanic centers. The Sunday Basin, on the northeast side of Amchitka Island, is located in an apparently structurally complex area at the junction of Bowers Ridge and the Aleutian Arc perhaps indicating a different structural origin (Fig. 16).

A. Sunday and Buldir Basins

Reflection profiles across the roughly 2-km-deep Sunday Basin (Marlow et al., 1970) show that it contains about 1.5 km of sediment, which is considerably less

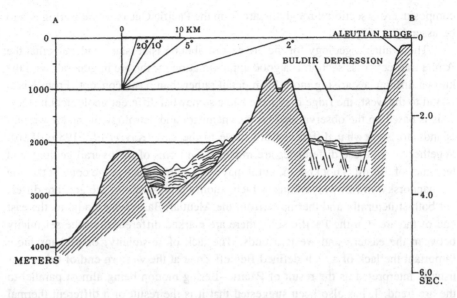

Fig. 21.   A schematic cross section of Buldir depression based on seismic reflection profiles. Two-way travel times are in seconds and water depths in meters. (From Marlow *et al.*, 1970.)

than that seen in all but the Buldir Basin. The upper parts of the section are highly reflective and largely undeformed. The lower parts are mildly deformed, with the deformation being interpreted as due to differential compaction. The whole section is assigned to the late series, and is thought to be made up of sandy and silty turbidites on the basis of its reflectivity. The structural style of the basin is not clear from the profiles of Marlow *et al.* (1970) or from the earlier profile of Ewing *et al.* (1965); thus, more exploration will have to be done before conclusions can be drawn regarding its origin.

Of the other four major ridge-top basins, only the Buldir Basin has less sediment fill than the Sunday Basin, with Marlow *et al.* (1970) reporting about 300 m (Fig. 21). The sediment is again assigned to the late series. Marlow *et al.* (1970) and Scholl *et al.* (1975b) report that the formation of the Buldir Depression, a roughly 1-km-deep, 50-km-long by 20-km-wide bathtub-shaped graben, was accompanied by minor volcanism, including a small bathymetrically defined cone at the eastern end (Hydro Cone) (Fig. 22). This, together with the observation that the long axis of the graben aligns with the volcano line from Buldir to Little Sitkin, implies a volcano-tectonic origin. They also note that Buldir Depression is still in the process of formation.

B. Pratt, Amlia, and Amukta Basins

The other three summit basins are also described as grabens, though for Pratt Depression, between Little Sitkin and Semisopochnoi islands, there are no pub-

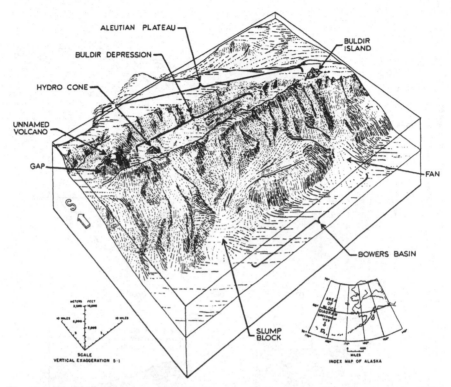

Fig. 22. Physiographic diagram of Buldir depression by Tau Alpha. (From Marlow *et al.*, 1970.)

lished geophysical data. Morphologically, the Pratt Depression is elongate along the trend of the volcano line, and the center of the depression is disrupted by the Kay Sea Cone, a submerged volcanic center, again pointing to a volcano-tectonic origin.

The Amlia and Amukta basins are adjacent to one another on the northern shelf edge of the Aleutian Ridge between Amlia and Amukta islands (Fig. 16). Both basins are more properly described as half grabens, with the steep scarplike walls forming the northern edge of the basins in alignment with each other and with the volcano line to the west (Atka to Gareloi volcanoes, Figs. 1 and 16). To the east, the trend of the volcano line from Seguam to Umnak Island is offset from the adjacent line to the west, and is aligned with a scarp which can be interpreted as the south margin of the Amukta Basin. The location of these basins may thus be related to this change of trend of the volcano line. It will also be noted that the Pratt Basin is located at a similar change in trend between the volcanoes west of Little Sitkin and east of Semisopochnoi islands.

The sediment fill in both the Amlia and Amukta basins is interpreted as being largely turbidites of the late series (Fig. 23). In the case of the Amlia Basin, the fill is level with breaches in the north wall at a water depth of about 1 km. The excess

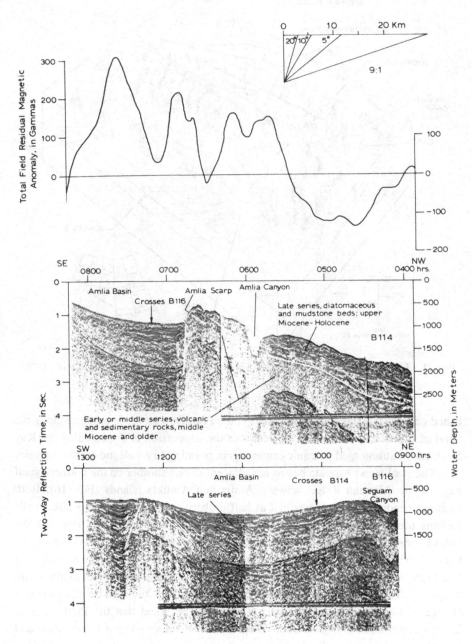

Fig. 23. An interpreted seismic reflection profile and magnetic anomaly along a line paralleling the NE–SW axis of the Amlia basin. (From Scholl *et al.*, 1975b.)

sediments flow into the Bering Sea through these breaches and have carved significant canyons, notably the Amlia Canyon and the Seguam Canyon.

The Amukta Basin has approximately the same water depth as the Amlia Basin, the deepest portion being at the base of the north wall. In this case the sediments have not yet breached to the Bering Sea; however, the Amukta Canyon has almost cut through the north wall.

The depths of sediment ponded in the Amlia and Amukta basins are not well determined, but are on the order of 2–3 km (Scholl *et al.*, 1975a,b). The age of basin formation and filling can be inferred from a combination of DSDP data from holes 184 and 185 on the flanks of Umnak Plateau, from the onlapping of the basin fill over the early or middle series rocks of the faulted blocks, from the fact that the canyon cutting involves late series sediments, and from the available dredge haul and shallow core data. The preferred interpretation from Scholl *et al.* (1975a,b) is that the basin formation was related to the onset of late Cenozoic volcanism, which fits with the constraints given above. K–Ar dating of igneous rocks at various localities along the Aleutian Arc points to a late Miocene onset of volcanism, with the present volcano line being built up in Pliocene time (K–Ar ages are summarized in Marlow *et al.*, 1973). An alternative hypothesis is that the basins were the result of the plutonism that occurred in mid-Miocene times. The argument against this is the observation that the mid-Miocene volcanics are dominantly found along the centerline of the Aleutian Ridge, whereas the late Miocene–Pliocene volcanic centers and basins are mainly located on the northern edge and align with the basins.

As mentioned earlier, the major summit basins are located at changes in trend of the volcano line (Marlow *et al.*, 1973) and are possibly related to the fracture zones that are commonly associated with these changes in trend (Spence, 1977). Kay and Kay (1982) also note that the chemical compositions and petrology of the Holocene volcanics also relate to their position with respect to the major segmentation of the arc. They note that the large tholeiitic volcanoes are at the junctions where changes of trend or segmentation occur, and the smaller calc-alkaline centers tend to be in the centers of the segments. They relate this distribution of magma types to the style of magma chamber present, which is in turn related to access to the mantle.

VI. ALEUTIAN TERRACE AND TRENCH

It is beyond the scope of this review to discuss the details of subduction beneath the Aleutian Island Arc, but since the geological history and tectonics of the Bering Sea are directly related to this subduction zone, a brief review is in order.

The basic setting of the central part of the Aleutian Island Arc is that of a well-defined bathymetric trench, a marked terrace between the ridge crest and the trench, and the ridge itself which is crowned by a well-defined chain of volcanoes. The geophysical signature is equally well defined, with a seismically active Benioff Zone, dipping such that the volcano line lies above the 110-km hypocenter contour, and typical arc-trench gravimetric and magnetic anomalies (Ewing *et al.*, 1965; Watts, 1975; Scholl *et al.*, 1983a; Stone, 1968; Engdahl, 1977). The volcanic arc is segmented, with short chains of volcanoes being aligned with one another, and adjacent chains being offset and/or having slightly different trends (Marsh, 1979; Spence, 1977). The overall effect is to produce a smooth curve centered on a pole located in the Gulf of Anadyr; however, the volcanic segmentation combined with the discrete blocks defined by the release of seismic energy through time belie this apparent smoothness.

The relatively shallow effects of subduction, and the details of the morphology, age, and structure of the Aleutian Trench and Terrace, are best known in the vicinity of the Amlia transect (Scholl *et al.*, 1983a), while the details of the seismicity are best known between Kiska and Adak islands, partly as the result of the seismometer networks initially set up for the nuclear tests on Amchitka Island.

Looking arcwards from the Pacific side of the trench, the ocean floor with its covering of 200–300 m of pelagic sediments dips irregularly toward the trench via a sequence of east–west-trending asymmetric ridges (Grim and Erickson, 1969). The floor of the trench contains a landward-thickening wedge of upper series deposits. Against the landward wall, the upper series attain thicknesses on the order of 4 km, and consist mainly of silty, and possibly sandy turbidite beds thought to be younger than about 0.5 m.y. (Scholl *et al.*, 1983a,b).

The inner or landward slope to the trench again consists of a series of asymmetric ridges and basins, often with steep trench-facing slopes of up to 30–35° (Grow, 1973). Beneath this, and traceable in seismic records for several tens of kilometers, is the north-dipping oceanic basement. This basement also appears to be faulted to produce a series of ridges parallel to the strike of the trench (McCarthy and Scholl, 1985).

Inland from the top of the arc side of the trench slope is the Aleutian Terrace which is interpreted as a forearc basin. Multichannel profiling reported by Scholl *et al.* (1983a) (Fig. 24) shows the basin to have been formed by repeated uplift at the outer (trench) edge, and then filled with turbidites flowing down from the ridge crest. The basin fill contains both middle and upper series sediments which attain thicknesses of about 4–5 km (Scholl *et al.*, 1983a; Grow, 1973). The upper series constitutes about two-thirds of the basin fill. The two DSDP holes, 186 and 187, were drilled on the outer edge of the Aleutian Terrace, and penetrated chiefly diatomaceous clay and sand rich in volcaniclastic detritus (Stewart, 1978). The oldest rocks encountered in DSDP hole 186 were Pliocene, with a displaced mass of Miocene within the late Pliocene section. The fossils in this latter section imply deposition in much shallower water than their present location which led to the interpretation of their being a displaced block. DSDP hole 187 encountered late Miocene beds as part of the terrace basin "basement."

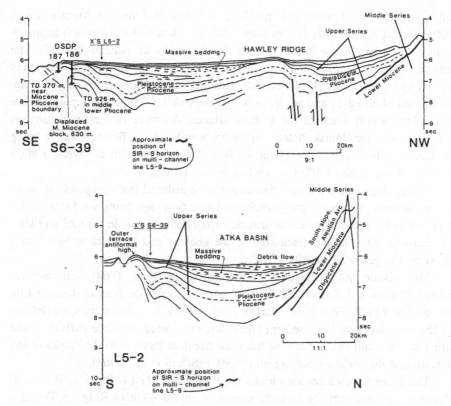

Fig. 24. Two interpreted seismic reflection profiles (S6-39 and L5-2) across the Atka basin on the Aleutian terrace (see Fig. 16 for location of profiles). The normal faults under Hawley Ridge are probably high-angle reverse faults (Scholl, personal communication, 1984). Two-way travel time is in seconds. (From Scholl et al., 1983b.)

The flank of the Aleutian Ridge above the Aleutian Terrace is underlain by about 1 km of stratified slope deposits. These have been interpreted, from dredge haul data, as being middle series rocks and are early Oligocene through mid–Miocene (Scholl et al., 1983a). The data described above are shown in Figs. 20 and 24.

VII. UMNAK PLATEAU

Umnak Plateau is the name given to an area to the north and west of Umnak Island. The plateau itself is flat-topped, and at about a depth of 2000 m, thus making it a distinct physiographic province. It is shallower than the Aleutian Basin (3000 m), but considerably deeper than either the Aleutian Ridge, which forms its southern boundary, or the Bering Sea Shelf to the east of the plateau (Fig. 16). The plateau itself is bounded by the Umnak and Bering canyons, but its areal extent (as defined by Cooper *et al.*, 1980) extends eastwards across the Bering Canyon to the Bering Sea Shelf edge, and northwards past the Pribilof Canyon to a point on the shelf edge roughly south of the Pribilof Islands. All three of the main canyons associated with the Umnak Plateau region are very large, the Bering Canyon being the largest with a volume of about 4300 km^3 compared with an average large canyon volume of about 500 km^3 (Scholl *et al.*, 1970a).

The geology of the Umnak Plateau has been deduced from geophysical information (seismic, magnetic, gravimetric, and heat flow) and from two DSDP holes (184 and 185), with peripheral information from dredge hauls in Pribilof and Umnak canyons. Most of the discussion of the geology and structure is based on a report by Cooper *et al.* (1980).

The seismic profiles include two-ship refraction (Shor, 1964), sonobuoy refraction (Childs and Cooper, 1979), and many reflection profiles as described by Scholl *et al.* (1968, 1970a) and Marlow *et al.* (1977). Four main units are defined for the area: the acoustic basement (AB), a main layered sequence (MLS), a unit called the rise unit (RU) traceable from the Aleutian Basin onto the flanks of the plateau, and the surface-mantling unit (SMU) (Scholl *et al.*, 1968).

The acoustic basement seen under the Bering Sea Shelf is considered to be of Mesozoic age, whereas the acoustic basement on the Aleutian Ridge is Tertiary. Both of these acoustic basements can be traced to the Umnak Plateau. To the west, the acoustic basement is the basement of the Aleutian Basin of presumed Mesozoic age, and it can be traced beneath the western edge of the plateau. Refraction studies show that the eastern parts of the plateau have a different velocity structure at depth, and may represent an ocean–continent transition (Fig. 25).

Above the acoustic basement, the main layered sequence forms the bulk of the plateau and is commonly several kilometers thick. It has a number of units distinguishable within it, and is considered to be Cenozoic in age (Fig. 26). The rise unit consists of Pleistocene turbidites that cover the main layered sequence at abyssal depths, and die out on the flanks of the plateau. The uppermost unit, the surface-mantling unit, is made up of diatomaceous and siliceous sediments, much of which appears to have been derived from the Aleutian Ridge.

The DSDP holes penetrated the top of the main layered sequence and encountered late Miocene rocks (DSDP reports and Bujak, 1984), but penetration was less than 1 km in both cases. Seismically, the main layered sequence has been subdivided, on the basis of velocity, into three units which can be traced over most of the area. Basement depth and sediment thickness vary across the plateau, as can be seen from Figs. 25 and 26. The velocity structure of the igneous crust beneath the

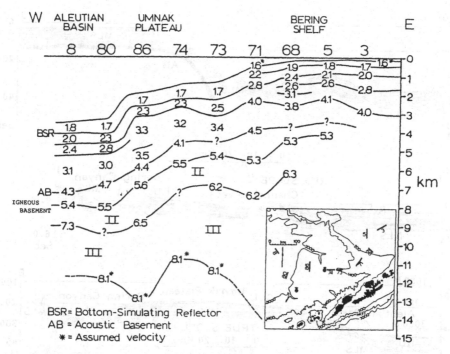

Fig. 25.   A composite seismic velocity profile across the Umnak plateau region derived from sonobuoy data. (From Cooper *et al.*, 1980.)

plateau can also be seen to change. A two-layer oceanic crust is seen on the Aleutian Basin side, with a more continental or transitional crust to the east.

Deformation of the sediments of the plateau is not extensive, and is mainly confined to areas of canyon cutting. There are, however, a number of basement faults that have been recognized with cumulative offsets of several kilometers. None of these faults reach the seafloor, but they are considered to be late Cenozoic in age.

Diapirs are seen in a number of localities, and are considered to be shale diapirs. This conclusion is based on arguments involving a sedimentary layer of variable thickness that could be the source of the shale (Cooper *et al.*, 1980).

Perhaps the most important point that can be drawn from this discussion of the Umnak Plateau is that it is dissimilar from each of its three neighboring areas, the Bering Sea Shelf, the Aleutian Ridge, and the Aleutian Basin. It is most similar to the ocean floor of the Aleutian Basin and a model involving compression and deformation of an ocean floor and adjacent continental margin has been proposed (Fig. 27 from Cooper *et al.*, 1980). The differences in the geology of the Umnak Plateau with respect to the surrounding area have also led to the hypothesis that it is an allochthonous oceanic plateau (Ben-Avraham and Cooper, 1981) as discussed later.

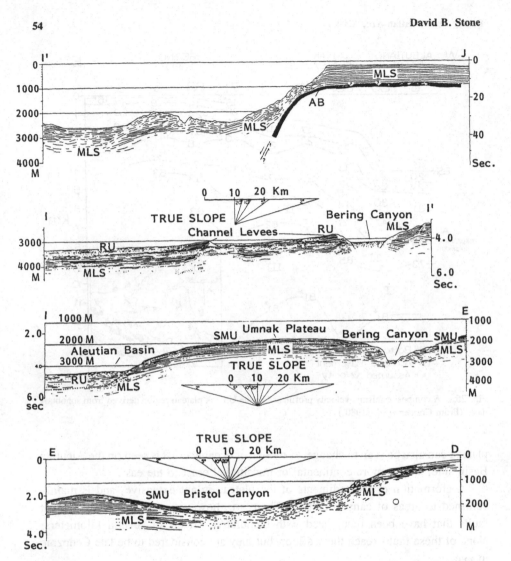

Fig. 26.   Interpreted seismic reflection profiles along lines J–I–E–D as shown on Fig. 16. AB, acoustic basement; MLS, main layered sequence; SMU, surface-mantling unit; RU, rise unit. Two-way travel time is in seconds. (From Scholl *et al.*, 1968.)

VIII. DEEP BERING SEA

The deep Bering Sea, defined as that part below a depth of about 3000 m, can be subdivided into three areas as shown in Figs. 1 and 28. These are the Aleutian Basin (sometimes known as the Bering Basin), Bowers Basin, and the Komandorsky Basin (also known as the Kamchatka Basin). The names used here are those used by Scholl *et al.* (1974) and Beikman (1980). The Aleutian Basin is the largest of the basins encompassing the area between the Alaskan Bering Sea Shelf edge,

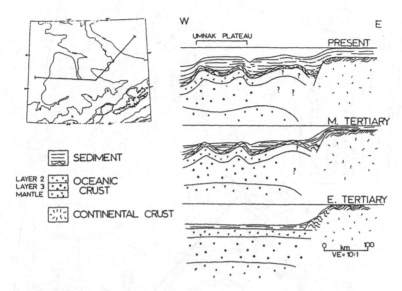

Fig. 27. Cartoon sections showing the possible evolution of the Umnak plateau along the line indicated. (From Cooper *et al.*, 1980.)

Bowers Ridge, which loops north and west from the central Aleutian Ridge, and Shirshov Ridge, which trends south from Cape Olyotorsky in eastern Siberia. The area enclosed by the arcuate Bowers Ridge and the Aleutian Ridge forms Bowers Basin, and the area between Shirshov Ridge, the Aleutian Ridge and Kamchatka Peninsula forms the Komandorsky Basin.

The bathymetry of all three basins shows extremely flat floors all nominally at the same depth. When looked at in detail, there is a gentle slope giving the whole area a "half saucer" shape, with the deepest part located in Bowers Basin. This is consistent with the floors of all three basins being the result of turbidity current deposits transported from the north and northwest (Ludwig *et al.*, 1971a).

As is the case for most studies in marine geology, geological models have to be based on scattered observations, since large-scale coverage is not commonly available. In the case of the deep Bering Sea, the available observations include six DSDP holes, less than 50 heat flow measurements, a few dredge hauls, and a few thousand kilometers of seismic reflection and/or refraction records. A few of the seismic surveys collected gravity data at the same time and the majority collected magnetic data. The distribution of data available can be seen from the track lines shown in Fig. 29. The seismic profiles are dominantly derived from reflection techniques but two-ship and sonobuoy refraction data have also been collected. The types of data available are listed in Cooper *et al.* (1979a). The quality of the seismic data is variable, and the improvement with time is very marked, presumably due to improved instrumentation and processing. The contrast between the early profiles and the two published by Cooper *et al.* (1982), and reproduced as Fig. 30, is quite

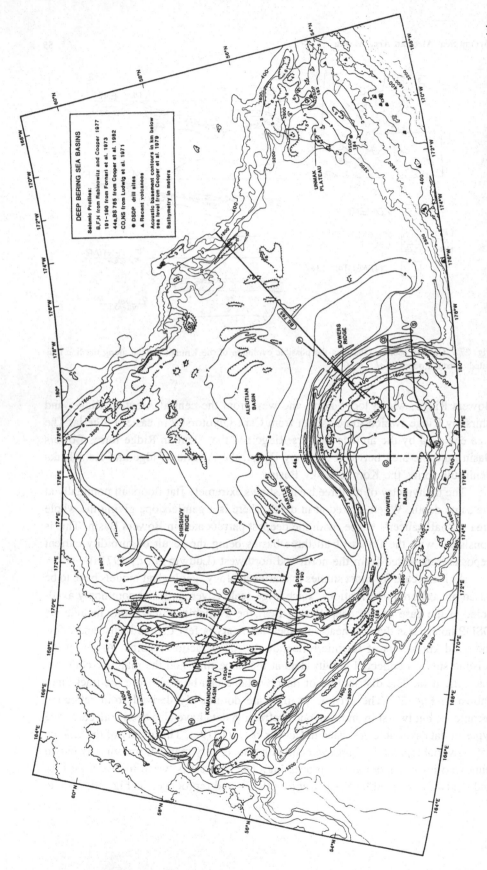

Fig. 28. Basement topography of the oceanic parts of the Bering Sea. Solid lines locate seismic reflection profiles, the dashed line the location of the gravity profile described in the text, and the dashed extension of line 44a a composite section across the whole area.

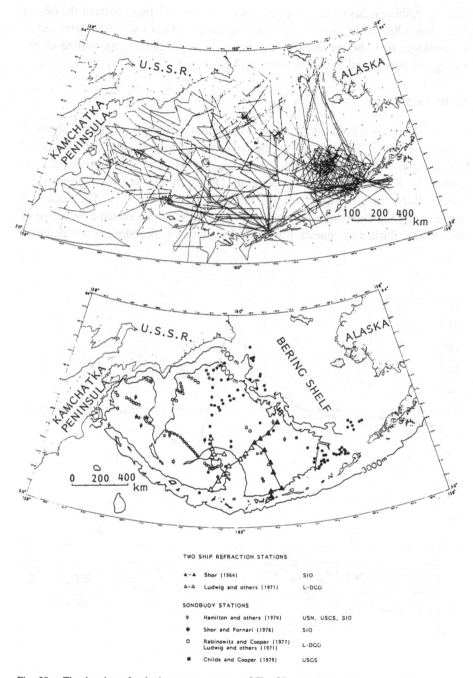

Fig. 29.   The data base for the basement contours of Fig. 28 come from reflection profiling along the tracks shown in the upper figure and for refraction data as shown in the lower figure. (From Cooper *et al.*, 1979a.)

striking. Although this section is nominally concerned with those parts of the Bering Sea below 3000 m, the two ridges separating the three basins are an integral part of the geology; thus, before considering the deep basins further, an outline of the geology of Bower and Shirshov ridges will be given.

A. Bowers Ridge

As can be seen from Fig. 1, Bowers Ridge has a pronounced arcuate shape. It is also asymmetric in cross section, the flanks being steeper to the north and east. Gravity data, as reported by Kienle (1971) and Watts (1975), and magnetic data as reported by Kienle (1971) and Cooper *et al.* (1976a,b) have been used to interpret Bowers Ridge as a fossil or inactive subduction zone (Fig. 31). This model is supported by the observation of Cooper *et al.* (1982) that the acoustic basement of the Aleutian Basin dips beneath Bowers Ridge, with a thick sediment wedge above

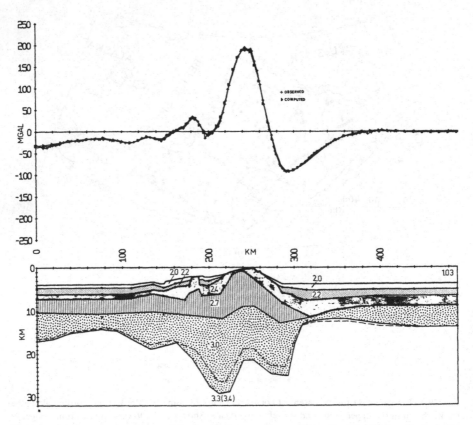

Fig. 31.   A crustal section computed for line N–P across Bowers Ridge (see Fig. 28) based on gravity models using seismic refraction data (dots) as control. (From Kienle, 1971.)

the dipping zone. Both the dipping basement and the sediment wedge are clearly seen in Fig. 30, and are described in more detail below.

The summit area of Bowers Ridge has a magnetic anomaly pattern very reminiscent of the magnetic anomalies seen over ridges associated with known active subduction zones, such as the Aleutian Ridge itself. On this basis, the main body of the ridge is interpreted as being made up of a typical volcanic arc assemblage of rocks. This interpretation is supported by a few samples of albitized volcanic rocks dredged from the north face of the ridge (Scholl *et al.*, 1975a). These rocks are also reminiscent of those found on the flanks of the Aleutian Ridge.

Cooper *et al.* (1982) note that on the Aleutian Basin side of Bowers Ridge there may well be a thick wedge of deformed sediment, upslope from the now well-defined trench-fill type deposits. This suggestion is made on the basis that the acoustic basement may represent altered sediments as opposed to volcanic oceanic crust and is thus masking the expected fore-arc sediments. This interpretation fits well with the observation that the magnetic signal is much more subdued over the flank facing the Aleutian Basin, and is thus more easily interpreted as being over a sedimentary rather than an igneous basement.

The uppermost sedimentary layers mantle the flanks of Bowers Ridge and appear to rise conformably with the topography from the basins on either side (Figs. 30 and 32). In some profiles, such as those quoted by Ludwig *et al.* (1971b), from west of an area on Bowers Ridge known as Bowers Bank (Fig. 16), the sediments appear to have been deposited conformably on both basement peaks and valleys. An exception to this is seen in the vicinity of Rude Sea Valley (Figs. 16 and 32) where the sediments are disturbed, presumably due to turbidity currents and slumping (Fig. 32).

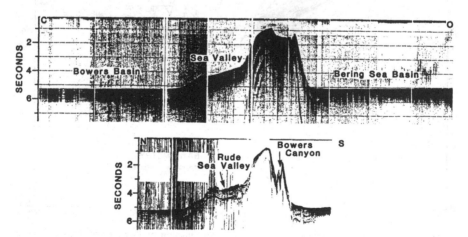

Fig. 32. Two profiles (C–O, S–N; see Fig. 28) across Bowers Ridge showing the distribution of sediment fill with respect to the acoustic basement. Two-way travel time is in seconds. (From Ludwig *et al.*, 1971a.)

The summit region of Bowers Ridge consists of a series of relatively flat-topped platform areas separated by shallow depressions. The topmost parts of the ridge are almost sediment-free or have only a thin mantle of sediment. The commonly accepted interpretation is that the summit areas represent wave-cut platforms. At the southeastern end of the ridge, where it connects to the Aleutian Ridge, these summit areas are on the order of 200 m below present sea level. To the north and west, around the curve of the ridge, the water depth above the summit becomes consistently deeper, reaching over 1000 m on the northernmost crest of the ridge. If these sediment-poor platforms do represent wave action, subsidence of parts of the ridge of over 1000 m is called for.

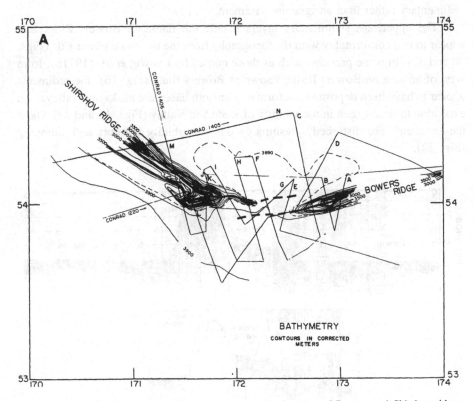

BATHYMETRY
CONTOURS IN CORRECTED METERS

Fig. 33.   Part A shows the bathymetry in meters near the junction of Bowers and Shirshov ridges together with a series of track lines. Part B shows the seismic reflection profiles obtained along these track lines, adjusted so that the axes of Bowers (solid line) and Shirshov (dotted line) ridges align from profile to profile. This clearly shows a separation of the two ridges, which is also shown by the heavy dashed line in part A. (From Rabinowitz, 1974.)

B. Bowers–Shirshov Ridge Connection

Traced to the west, the topographic expression of Bowers Ridge becomes very subdued; however, it is still clearly visible in the subsurface. In a similar fashion, Shirshov Ridge when traced to the south and east becomes subdued, and again can be traced in the subsurface. The two subsurface traces, those of Bowers and Shirshov ridges, trend toward one another, but whether they are in fact connected is the subject of some debate. Rabinowitz (1974) shows that the signatures of each of these ridges as defined by the acoustic basement, parallel one another for several tens of kilometers, separated by roughly 10 km (Fig. 33). Unfortunately, the profiles do not have the resolution to show whether or not the two features merge at depth; however, Rabinowitz (1974) concluded that the two ridges have different structural origins.

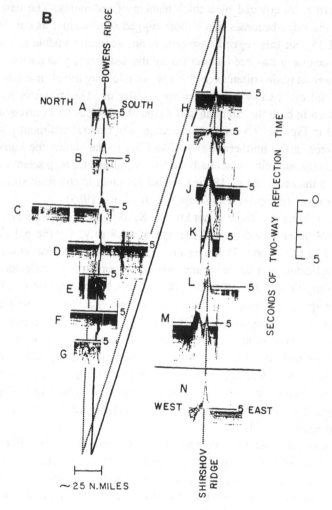

Fig. 33. *(continued)*

C. Shirshov Ridge

The principal sources of data on the structure and geology of Shirshov Ridge are two reflection profiles reported by Ludwig *et al.* (1971a), a series of reflection and refraction (sonobuoy) lines reported by Rabinowitz and Cooper (1977) and seismic sounding and dredging reported by Neprochnov *et al.* (1985). The gravity and magnetic data for two profiles were interpreted by Kienle (1971). Dredge hauls of palagonitized tuff testify to the volcanic nature of the ridge (Scholl *et al.*, 1975a) as do the high-amplitude, short-wavelength magnetic anomalies.

Shirshov Ridge, as defined by its acoustic basement, is generally assumed to be made up of volcanic rocks, is asymmetric in cross section, and also changes in character along its length. The northern part of Shirshov Ridge is characterized by steep, sediment-poor basement scarps on its western flank, with a more gently dipping eastern flank covered by a thick blanket of sediments. The basement topography of the ridge becomes much more rugged southwards, as can be seen in Figs. 34 and 35, but this rugged basement is not generally visible in the surface topography because it has been smoothed by the sedimentary deposits. The sediment-filled summit basins often exceed 1.5 sec of two-way travel time (on the order of 1.5 km), and often show a strong internal reflector. This reflector may be the reflector P, seen in both the Aleutian and Komandorsky Basins as discussed below and indicated in Fig. 34. The crest of the ridge is in general sediment-poor. Some sediment is seen in the northernmost profiles, but to the south, the narrower and more rugged ridge summit is essentially devoid of sediment. Deep seismic sounding data indicate a thickened crust (18 km) and an asymmetric structure similar to that seen in the seismic reflection data (Neprochnov *et al.*, 1985).

The age of Shirshov Ridge is not known. Scholl and co-workers consider it to be pre-early Miocene based on a K–Ar age of 16.8 m.y. for the palagonite tuff dredged from its flank about 75 km east of DSDP hole 191 (Scholl *et al.*, 1975a). Also, if the reflector seen in the summit basins is equivalent to reflector P seen in the deep basins, then this implies that the summit basins are pre-middle to late Miocene, the age of reflector P in the abyssal basins. This age is based on tracing reflector P to DSDP holes 190 and 191 (Fig. 35). Dredging, as reported by Neprochnov *et al.* (1985), produced considerable ice-rafted debris plus some samples interpreted to be bedrock. These bedrock samples include three rock associations: (1) amphibolites, dolerites, and quartz dolerites; (2) Triassic (two samples), late Cretaceous to Paleocene, and Oligocene deep-water siliceous, siliceous–argillaceous, and tuffaceous sedimentary deposits; (3) Neogene poorly consolidated siliceous and terrigenous deposits.

The origin and time of development of Shirshov Ridge is enigmatic. The acoustic basement of the Aleutian Basin dips toward the northern end of the ridge and the adjacent Siberian Shelf and is covered by a thick prism of later sediment. The acoustic basement in this case may not be volcanic but rather lithified sediment,

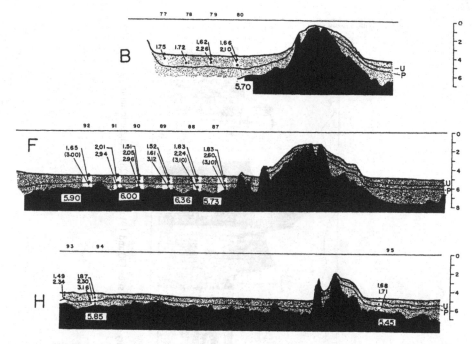

Fig. 34.   Composite profiles B, F, and H (see Fig. 28) across Shirshov Ridge. These profiles clearly show the morphology of the acoustic basement and the distribution of the overlying sediments. Two-way travel time for the reflection profiles is in seconds. Dots represent locations of seismic refraction measurements. (From Rabinowitz and Cooper, 1977.)

as it has a slightly lower seismic velocity than is seen elsewhere for volcanic basement. If this is so, then an interpretation in terms of crustal subsidence due to sediment loading from onshore sources may be a more plausible explanation of the basement dip than an explanation in terms of a fossil subduction zone. No such dipping basement is seen along the Komandorsky Basin margin of the Shirshov Ridge. The eastward-dipping basement structures of the ridge itself are interpreted by Neprochnov *et al.* (1985) to represent sections of the ocean floor that have been thrust over one another in response to spreading in the Komandorsky Basin.

Two heat flow measurements, reported by Langseth *et al.* (1980) from the western flanks of Shirshov Ridge, give values of about 50 mW cm^{-2}. These relatively low values contrast sharply with the high values determined within the Komandorsky Basin itself (138 mW m^{-2}), and are discussed below in the context of the age of the deep basin.

D. The Aleutian (Bering Sea) Basin

The Aleutian Basin is by far the largest of the three basins of the Bering Sea. In most respects it appears to be underlain by normal oceanic crust, which is covered

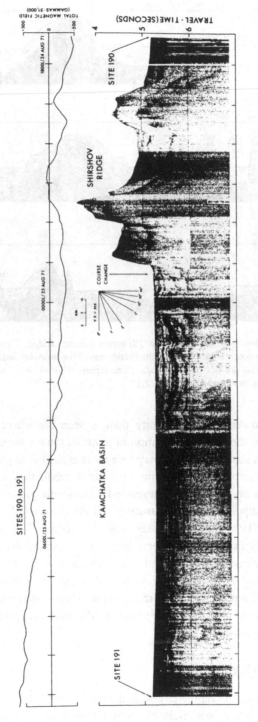

Fig. 35. A seismic reflection profile obtained by the *Glomar Challenger* in the vicinity of DSDP sites 190 and 191 (see Fig. 28). Magnetic anomaly values along the track line are also given. (From Scholl *et al.*, 1973.)

by a thicker-than-average sedimentary sequence. The acoustic basement in the central part of the basin, which is inferred to represent the top of the volcanic rocks forming the ancient ocean floor, is about 4 km below the seafloor (Cooper et al., 1979a, 1982), though it deepens to about 9 km in places along the edges of the basin (Fig. 28). The acoustic basement is rough and hummocky. This roughness is smoothed out by a lowermost thin sedimentary unit (a few hundred meters thick) characterized by strong internal reflectors. This unit has been interpreted as being a highly lithified pelagic limestone or chert. Throughout most of the basin (noting that there are a limited number of track lines), the sedimentary section also contains an uppermost set of prominent reflectors. These can be traced laterally to DSDP hole 190 where they appear to correlate with a diatomaceous silty clay and mudstone of mid-Miocene to Holocene age. Below this, the reflectors are less clear, especially toward the edges of the basin in the vicinity of Bowers Ridge and the Bering Sea Shelf. This loss of clarity in the reflectors may simply represent a change in the style of sedimentation, and/or the effects of a regionally extensive diagenetic front (Hein et al., 1978).

The single 600-km-plus line (BS765) located in Fig. 28 and shown in Fig. 30 clearly demonstrates the stratified nature of the sediments and the rough acoustic basement in the center of the Aleutian Basin. It also clearly shows the acoustic basement dipping down toward Bowers Ridge to the southwest, and beneath the Bering Sea Shelf edge to the northeast. The sediment thickness in the wedges formed by the dipping basement are in excess of 9 km. Prior to the two reflection surveys reported by Cooper et al. (1982), it had been suspected that Bowers Ridge was an extinct island arc. This inference was drawn on the basis of seismic reflection profiles with limited penetration (Ludwig et al., 1971b), seismic refraction data which showed an "arclike" root under Bowers Ridge (Ludwig et al., 1971a; Ewing et al., 1965; Shor, 1964), and the magnetic and gravity signatures (Kienle, 1971). These data also indicate that the arc is inactive, an observation supported by the lack of seismicity. Detailed studies of the reflection profiles across the sediment wedge above the now inactive trench show that the lower one-third to one-half of the section is deformed and faulted. These sediments, when traced away from the deformed zone, lie with apparent conformity on the pelagic sediments covering the acoustic basement. The acoustic basement is, in turn, interpreted as representing the top of the oceanic crust. The whole sedimentary package is thus interpreted as being roughly the age of the ocean floor itself. Reflection profiles across the wedge also show a "fanning" of the reflectors with increasing thickness of many of the mid to upper wedge units. This fanning is presumed to be due to subsidence during sedimentation, with maximum subsidence taking place under the thickest part of the wedge. The uppermost layers, interpreted to be Miocene or younger on the basis of correlations with DSDP 190, are of more uniform thickness and cross the fossil trench to abut against Bowers Ridge, indicating cessation of tectonic activity in the trench prior to this time.

The interpretation of the acoustic basement on the northeast end of the profile BS765 shown in Fig. 30, which dips under the Bering Sea Shelf edge, is not so clear since the other geophysical signatures are more ambiguous. There are no clear-cut magnetic or gravimetric signatures, and some of the rocks dredged along this margin, together with physiographic interpretations, suggest that the margin has subsided at least 1000 m since Eocene time. Current interpretations of the tectonic history of the Alaskan Bering Sea Shelf edge involve limited subduction along this margin, with a strong component of strike-slip motion. This aspect of the Bering Sea Shelf margin is discussed in more detail below.

The age of the uppermost sedimentary section in the Aleutian Basin, extrapolated from DSDP 190 located between Bowers and Shirshov ridges, is from mid-Miocene to Holocene. Below the dated sediments of mid-Miocene age, the center of the Aleutian Basin has a minimum thickness of 2.5 km of sediments above the acoustic basement. Since there is little evidence regarding sedimentation rates, a wide range of ages is possible for the rocks forming the acoustic basement. To obtain an age for the formation of the oceanic crust is even more problematic. Cooper *et al.* (1976b) interpret the magnetic anomalies found in the Aleutian Basin as representing the magnetic reversal anomalies M1 through M13 of Mesozoic age (e.g., see Larson and Hilde, 1975), but this age assignment is not clear-cut particularly in the light of new data (Marlow *et al.*, 1986; Cooper, personal communication) (Fig. 36).

The age analysis done by Cooper *et al.* (1976b) on the magnetic data available for the deep Bering Sea is very thorough, but in the final analysis relies very heavily on a particular profile (A6A-A6). The correlation from the profile to the model profiles based on the work of Larson and Hilde (1975) and others is suggestive, but not compelling. Similarly, the correlations from the key profile to the adjusted profiles elsewhere in the Aleutian Basin are only suggestive. Cooper *et al.* (1976b) support their choice of correlation by comparing the depth of the acoustic basement, adjusted for the anomalous sediment load, with the age versus depth curves of Sclater (1972) and find that they are consistent with the age selected. They also show that plausible choices of magnetization can be made, that plausible spreading rates are predicted, and that the ages determined do not conflict with the surrounding geology. Available evidence favors a Cretaceous age for the oceanic crust of the Aleutian Basin, but this conclusion probably should not yet be used as a fundamental foundation for tectonic or other models.

Though high-resolution data on the basement topography of the Aleutian Basin in particular, and the deep Bering Sea in general, are limited, basement topography maps have been compiled (Cooper *et al.*, 1979a,c). For the Aleutian Basin, the large-scale topography of the acoustic basement is relatively featureless over much of the area; however, there is a curious trend of highs and lows that more or less cuts the basin in half. Toward the southwest end of the trend a basement high known as Bartlett Ridge is located between Bowers and Shirshov ridges (Cooper *et al.*,

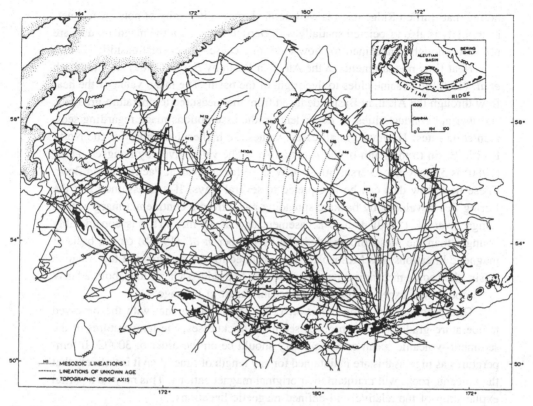

Fig. 36.   Magnetic data for the Bering Sea Basin plotted with the ship track lines as zero and positive anomalies facing north. The heavy line follows a central magnetic anomaly associated with Bowers and Shirshov ridges. Dotted lines show the correlation of anomalies proposed by Cooper *et al.* (1976).

1979a). In addition to Bartlett Ridge, a series of other linear basement ridges have recently been discovered (Marlow *et al.*, 1986). Bartlett Ridge has a relief of about 3 km with respect to the surrounding basement floor. From Bartlett Ridge southwest to the point where Bowers and Shirshov ridges meet, a somewhat subdued and ill-defined basement ridge can be seen. This latter ridge could equally well be interpreted as being related to the buried trench system associated with Bowers Ridge if it were not for its appearance as an elongated extension of Bartlett Ridge. To the immediate northeast of Bartlett Ridge the basement topography shows no noticeable preferred orientation; however, farther to the northeast are a series of rises and depressions generally elongated parallel to the projected trend of Bartlett Ridge. These basement features have elevation differences on the order of 1 km. The overall trend intersects the Bering Sea Shelf edge on a line coincident with the basement highs that separate the Navarin, Khatyrka, and Anadyr basins.

The significance, if any, of the apparent linear feature formed by Bartlett Ridge and the basement topography to the northeast is not clear, but its existence, noting that Bartlett Ridge itself is well-defined, makes it worthy of note. A magnetic anomaly is also associated spatially with Bartlett Ridge, but the magnetic data are not well enough constrained to show a definite cause–effect relationship.

Heat flow measurements in the Aleutian Basin by Foster (1962) and Langseth et al. (1980) give some clues to the origin of the basin. The mean value of the heat flow through the Aleutian Basin, deduced from the measured temperature gradients and thermal conductivities, is dependent on the assumptions made regarding sedimentation rates. However, in all cases, values are higher than those for the North Pacific Basin to the south of the Aleutian Arc. The value given by Langseth et al. (1980) is 55 mW m^{-2} versus 40–45 mW m^{-2} for equivalent Pacific Ocean floor. These heat flow data can be interpreted in several ways. If it is assumed that the Bering Sea developed by backarc spreading away from a continental margin, then Langseth et al. (1980) deduce a basement age of 44 m.y. This is considerably younger than the Mesozoic age deduced by Cooper et al. (1976b) on the basis of magnetic lineations. If a Mesozoic age is assumed for the basement of the Aleutian Basin, then some mechanism has to be invoked to produce a heat flow about 20 mW m^{-2} above that expected for trapped oceanic crust of that age.

One problem that Langseth et al. (1980) point out is that with the observed temperature gradients and observed sediment thicknesses, the temperatures at the sediment–volcanic ocean crust boundary should be on the order of 300°C. If temperatures as high as this are maintained for any length of time, then it is unlikely that the volcanic rocks will maintain their original magnetizations. This may be a partial explanation of the relatively ill-defined magnetic lineations.

E. Bowers Basin

Though the water depth in Bowers Basin is almost identical to that in the Aleutian Basin, depth to the acoustic basement is markedly shallower (Fig. 28). Acoustic basement is interpreted to be the top of the volcanic oceanic crust, and is about 6 km below sea level or 2–3 km below the ocean floor for most of the basin, dropping to over 7 km near the Aleutian Ridge.

Several two-ship and sonobuoy refraction lines have been made in Bowers Basin, together with reflection profiles of variable quality. The basic structure as given by Ewing et al. (1965) and Ludwig et al. (1971a,b) is summarized in Fig. 37. The most recently published high-quality reflection data are those of Cooper et al. (1982) (Fig. 30) which show that the general morphology of the subbottom is similar in the central parts of both the Aleutian and Bowers basins, except for the contrasting depths to acoustic basement. The morphology of the structures seen on either side of Bowers Ridge itself are, however, very different. An interpretive section across the two basins is shown in Fig. 37. The differences between the basins are in accord with the interpretations that the flank of Bowers Ridge facing the Aleutian Basin represents a fossil subduction zone, whereas the Bowers Basin

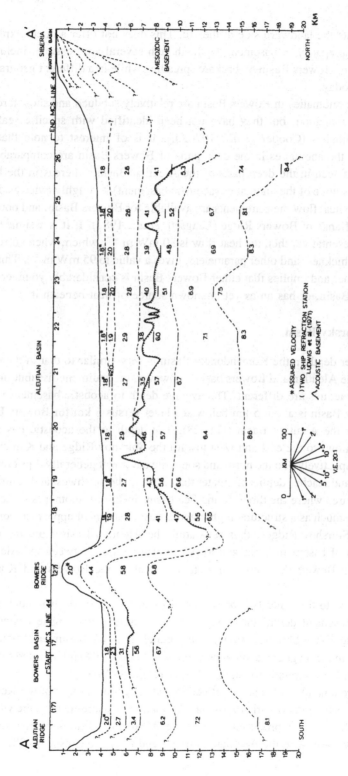

Fig. 37. A composite structure for the Bering Sea along the north–south line including line 44a as shown in Fig. 28. (From Cooper et al., 1979c.)

flank represents the landward side of the arc. This does not offer a direct explanation of the discrepancy in basement depth, though several are possible, including the formation of Bowers Basin by backarc spreading while in a different geographic setting from today.

Magnetic anomalies in Bowers Basin are relatively subdued and show a rough north–south elongation, but they have not been identified with specific seafloor spreading anomalies (Cooper *et al.*, 1976a,b). It is of interest to note that the amplitudes of the anomalies in the deep parts of Bowers Basin are comparable to the amplitudes seen in the deep parts of the Aleutian Basin, whereas in the latter the presumed source of the anomalies (acoustic basement) is roughly twice as deep.

Only two heat flow measurements are available for Bowers Basin, and both are on the lower flanks of Bowers Ridge (Langseth *et al.*, 1980). If it is assumed that these are representative, then the heat flow is 80 mW m^{-2}, which, when corrected for sediment thickness and other parameters, gives a value to 95 mW m^{-2}. This is a very high value, and implies that either Bowers Basin is considerably younger than the Aleutian Basin, or has an as yet unknown source of heat beneath it.

F. Komandorsky Basin

The water depth in the Komandorsky Basin is very similar to that in the other two basins, the Aleutian and Bowers basins; however, the sediment distribution and subbottom structures are different. The average depth to acoustic basement in the Komandorsky Basin is about 5 km below sea level versus 6 km for Bowers Basin and 8 km for the Aleutian Basin (Fig. 28). The depth to the acoustic basement generally shallows to the east and west toward the Shirshov Ridge and Kamchatka Peninsula, respectively. To the north and south the basement generally dips down to below 6 km, and reaches depths of greater than 7 km in the northwest and southwest corners. The area where the three Bering Sea basins meet one another is somewhat confused, inasmuch as a structural high in the acoustic basement appears to connect Bowers and Shirshov Ridges, thus separating the Aleutian Basin from the other two. A series of basement highs and lows can also be seen between the Aleutian Ridge and the Bowers–Shirshov connection separating the Bowers and Komandorsky basins.

In contrast to the other two basins, the Komandorsky Basin has much more variation in basement depth, including a crude alignment of elongate basins and troughs parallel to the Shirshov Ridge. This parallelism of the basement structure is also reflected in the magnetic anomalies observed, though it should be noted that the magnetic data base is relatively small (Cooper *et al.*, 1976a).

There are a number of seismic reflection and refraction studies of the Komandorsky Basin. Shor and Fornari (1976) made refraction measurements in the vicinity of DSDP hole 191 and produced velocity versus depth sections comparing the Komandorsky Basin with the Aleutian, Bowers, and Pacific basins (Fig. 38). They

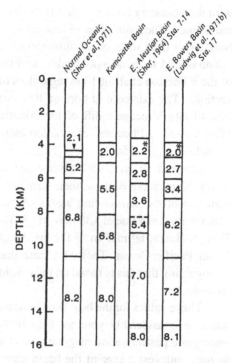

Fig. 38. Composite crustal structures for the Bering Sea. (Redrawn from Shor and Fornari, 1976.)

concluded that the Komandorsky Basin has a velocity structure much closer to normal oceanic crust than the other two basins, and postulated a model for its origin involving a period of active seafloor spreading about a roughly north–south axis. The timing of the spreading episodes is constrained, as it has to fit with the known ages of the rocks cored in DSDP 191. These cores showed basaltic basement of Oligocene age (29 ± 6 m.y.) overlain by undisturbed sediments of the same age at a depth of about 900 m subbottom.

Various other reflection and refraction profiles have been published for the Komandorsky Basin and clearly show a much rougher subbottom topography than was observed in the other basins (Shor and Fornari, 1976; Fornari et al., 1973; Rabinowitz and Cooper, 1977). In some cases hills in the acoustic basement reach the present seafloor surface. These data are summarized by Rabinowitz and Cooper (1977), who have compiled a series of cross sections across the Komandorsky Basin, the Shirshov Ridge, and into the Aleutian Basin. Rabinowitz and Cooper (1977) also note two velocity transitions in the sedimentary section which are seen as prominent reflections. They label these P and U after the notation of Ewing et al. (1965) who found similar reflections in the Aleutian Basin. From the DSDP holes, it appears that both of these reflections coincide with lithological changes, the upper one, U, being an internal reflecting horizon within a well-bedded unit of diatomaceous silty clay and interbedded turbidites. The lower reflector, P, coincides with the boundary between late Miocene diatomaceous sediments and mid-Miocene mudstones. The main point of interest regarding these reflectors, particularly the P horizon, is their wide extent throughout the western Bering Sea, including both the Aleutian and Komandorsky basins, and the more or less constant subbottom depth of the P horizon at about 1 km regardless of the overall differences in the subbottom sections. The existence of horizon P is ascribed to changes in sedimentation due to mid to late Miocene uplift of the Aleutian Arc, and it also coincides with mid-Miocene–early Pliocene deformation along the Siberian margin of the Bering Sea (Scholl *et al.*, 1975a).

Langseth *et al.* (1980) discuss the measured heat flow in the Komandorsky Basin (Kamchatka Basin in their terminology), and conclude that it is much higher than any of the surrounding basins (Aleutian, Bowers, and Pacific basins) with a mean corrected value of 138 mW m^{-2}. This is in comparison with 95 mW m^{-2} for Bowers Basin, 69 mW m^{-2} for the Aleutian Basin, and 40–45 mW m^{-2} for the North Pacific Ocean. They also note that the heat flow increases systematically (recognizing that this is based on only eight measurements) from the northeast to the southwest.

These values for the heat flow indicate that the crust in the Komandorsky Basin must be considerably younger than that in the other two basins, or that it has undergone postformation magmatic or volcanic activity. The extremely high values in the southwest corner of the basin may indicate that there has been recent magmatic activity. If true, this has significant implications in terms of the plate interactions and seismicity seen in the area today (Langseth *et al.*, 1980; Cormier, 1975; Newberry *et al.*, 1986).

IX. ORIGIN OF THE BERING SEA AND ITS MARGINS

At the present time there is no wholly satisfactory explanation for the observed geology and morphology of the Bering Sea. As has been mentioned in earlier sections, it is superficially similar to many of the other backarc basins of the Pacific Basin; however, the basement structure seen in the long reflection profiles and the magnetic anomaly data do not seem to fit a backarc spreading model. Similarly, models involving the isolation of a section of a plate from an ancient Pacific Basin, perhaps the Kula Plate, leave many questions unanswered. One of the key questions involves the origin of Bowers and Shirshov ridges, and another, the nature of the Bering Sea Shelf edge basins.

The clearest way to approach the origin of the Bering Sea is to list some of the major constraints and questions, then outline some possible scenarios.

1. The age of the seafloor in the abyssal parts of the Bering Sea is obviously critical to any model. Cooper *et al.* (1976b) assess the age of the Aleutian Basin to include the magnetic reversal anomalies M1 through M13 (117–132 m.y.). This

age call is based on correlations of magnetic anomaly models with the observed anomalies. The published magnetic data (Cooper *et al.*, 1976a,b) are suggestive, but not compelling, and care should be taken not to lean too heavily on these data. More recent magnetic data (Marlow *et al.*, 1986, and Cooper, personal communication) verify the orientation and continuity of the magnetic lineations, but do not confirm or deny the M-series age call. It should be pointed out that there is other circumstantial evidence for a Mesozoic age, such as the similarity of the pelagic sediments above the rough igneous crust in the Bering Sea with the known Mesozoic units in the Pacific (Marlow *et al.*, 1982). Similarly, it is possible to interpret the strata that thin and dip gently away from the Alaskan Bering Sea Shelf edge as being associated with the Cretaceous rocks known to underlie the shelf. In this context it should be noted that the heat flow data (Langseth *et al.*, 1980), if taken at face value, indicate a much younger age (44 m.y.).

2. The magnetic lineations in the Aleutian Basin described by Cooper *et al.* (1976a,b) were not well defined; however, new data have confirmed the original alignments (Marlow *et al.*, 1986). The amplitudes of the anomalies are low, as would be expected from the high thermal gradient observed (Langseth *et al.*, 1980). It is possible that they do not, in fact, represent magnetic lineations derived from steady seafloor spreading at an ocean ridge system, but have a more complex origin such as is observed in some of the marginal basins along the western Pacific.

3. The age of the Aleutian Island Arc is important to any model for the development of the Bering Sea. The oldest documented age is 42.3 ± 4.6 m.y. (DeLong *et al.*, 1978) based on K–Ar dating of a basalt from Ulak Island. The fossil record gives ages as old as mid–late Eocene (< 50 m.y.) (Scholl *et al.*, 1970b, 1975a). The fossils are from shallow-water deposits, so that the origin of the arc predates them by at least the time needed to build the arc from oceanic depths to sea level. This could be anything from a few million years to several tens of millions of years.

4. The existence of Bowers and Shirshov ridges with the associated Bowers and Komandorsky basins has to be accounted for. The two basins are substantially different in structure from the Aleutian Basin. The Komandorsky Basin appears considerably younger than either of the other basins based on sediment thickness and heat flow, and has a basaltic acoustic basement age of about 30 m.y. (Creager *et al.*, 1973). It is possible that the Komandorsky Basin was formed independently of the rest of the Bering Sea.

The existence of a fossil subduction zone along the northern and eastern edges of Bowers Ridge also places severe constraints on possible tectonic models. Unfortunately, there is essentially no age control on Bowers Ridge or the subduction complex other than the age of the seafloor dipping beneath it.

5. Along the northern and eastern margins of the deep Bering Sea are the Koryak Mountains and the Bering Sea Shelf. The Koryak Mountains consist in large part of mélange formed during the Cretaceous. This is consistent with their

origin in a Cretaceous subduction zone environment similar to that envisioned for
the Franciscan Formation of California. More problematic in terms of constraints on
models, is the observation that the Siberian margin of the Bering Sea has undergone
repeated uplift and broad compressional deformation in Cenozoic time, particularly
in middle Miocene through early Pliocene time (Gladenkov, 1964; Tilman, 1969;
and Drabkin, 1970; as quoted by Marlow *et al.*, 1982). In most models for the
development of the Bering Sea region, all relative motion should have transferred to
the Aleutian Subduction Zone as soon as it developed. The Alaskan Bering Sea
Shelf edge poses a similar problem inasmuch as it too appears to have undergone
deformation throughout Cenozoic time. In this case the principal deformation is
seen as subsidence, much of the continental margin having subsided at least 1500–
1700 m in post-Eocene time (post-Aleutian Arc formation). The fact that the
Koryak Mountains underwent uplift and compression while the Alaskan Bering Sea
Shelf subsided and formed a series of large basins also needs explanation.

Most of the constraints mentioned above apply primarily to the oceanic parts of
the Bering Sea and their margins. There are also a number of questions and constraints that apply to the shelf areas, such as the significance, if any, of the NE–SW-
trending belt of recent volcanism extending from the Seward Peninsula to the south
shore of Norton Sound, through Nunivak Island and possibly as far out across the
shelf as the Pribilof Islands. Similarly, what is the significance of the apparent
change of trend of the continental shelf-edge structures from parallel to the margin
of the deep Bering Sea to parallel to the trend of the Alaska Peninsula? A corollary
to this is how do the NE–SW geological trends of onshore Alaska, particularly
those represented by major faults (Fig. 6), interact with the NW–SE trends along
the continental margin?

In addition to what might be termed the "local" constraints listed above, any
model has to take into account the tectonic setting of the North Pacific–Arctic
region as a whole. The history of the Bering Sea region we see today is confined to
Mesozoic and Cenozoic times, with considerable activity taking place during the
Tertiary period.

During this same period of time the North Atlantic Ocean opened, starting in
mid-Cretaceous time, followed in early to mid-Tertiary time by seafloor spreading
in the Arctic Ocean about the Nansen–Gakkel Ridge [initiated about 65 m.y.b.p. at
the Greenland end, and perhaps as late as 40 m.y.b.p. in more northern parts,
according to Sweeney (1978) and Vogt *et al.* (1981)]. Both of these opening oceans
require relative rotation between North America and Eurasia, and their effects on
the Bering Sea depend critically on where the boundary between the North American and Eurasian plates is placed. Pitman and Talwani (1972) place the boundary
between Alaska and Siberia, and use the relative motion to create a subduction zone
along Bowers Ridge. The apparent continuity of the Paleozoic geology from the
Brooks Range in Alaska to at least Wrangel Island in Siberia argues against this
possibility. A major plate boundary existed along the edge of the Siberian craton,

parallel to the Verkhoyansk Mountains. Many other possible boundaries are associated with the margins of the accreted terranes, depending on their time of accretion (Watson and Fujita, 1985; Parfenov and Natal'in, 1985).

To the south of the Bering Sea, the northern margins of the several plates of the Pacific Basin are difficult to track backwards in time in detail, because much of the evidence has already been lost by subduction. However, essentially all the models show a dominant northward motion throughout latest Mesozoic and Tertiary time. Perhaps the most comprehensive compilation of Pacific Plate motion data is that of Engebretson *et al.* (1985), based on the assumption of a fixed hot-spot reference frame. The paleogeographies for the Pacific Ocean floor based on Engebretson's reconstructions (Engebretson *et al.*, 1984, 1985) require several thousand kilometers of ocean crust to be subducted north of the known positions of the Kula, Farallon, and Pacific spreading centers. Depending on whether or not parts of Alaska and Siberia were also moving north, all or some of this subduction had to take place along the Alaskan–Siberian margins.

Paleomagnetic data from the major southern Alaskan terranes (principally the Peninsular and Wrangellia terranes) show that in earliest Mesozoic time they were at equatorial latitudes (Stone, 1981; Hillhouse, 1977; Stone *et al.*, 1982; Panuska and Stone, 1985). Between then and the present, the various terranes presumably rode north carried by the plates of the Pacific region. Engebretson *et al.* (1985) show a number of plausible plate motion models which could achieve the required movement; however, the timing of the arrival of the displaced terranes is more difficult to determine. The paleomagnetic data available for the Alaska Peninsula Terrane and the adjacent terranes to the south, particularly the Chugach terrane, indicate a relatively late arrival (Panuska and Stone, 1985; Coe *et al.*, 1985; Plumley *et al.*, 1982). The existence of a large Eocene turbidite fan in the Gulf of Alaska, the Zodiac Fan, which had an apparent Alaskan source, may also require that parts of southern Alaska were farther south at that time (Stevenson *et al.*, 1983). If the Chugach terrane was farther south, the question arises whether at least parts of the Aleutian Island Arc might not have been farther south, too. Certainly today the whole trench–volcano complex appears to have continuity from the Aleutian Islands through the Alaska Peninsula, but there is a question as to whether this was also true in Eocene and later Tertiary times. The paleomagnetic data available for the Aleutian Islands *per se* are sparse (Stone *et al.*, 1983; Harbert *et al.*, 1987) and can be interpreted as indicating no displacement or a slight southward displacement of the Aleutian Arc with respect to North America in Oligocene time. None of the following models claim to explain or deal with all the constraints listed above; in fact, they are all conceptual models which have to call on various unexplained processes to account for many of the observations. They are put forward largely to show where current models are leading us, and also perhaps to highlight how much we do not know about the Bering Sea.

Although models involving some form of backarc spreading cannot be ruled out, particularly for the case of the Komandorsky Basin, most of the current models involve some form of trapping of a piece of older ocean floor behind the growing Aleutian Arc. The differences between the models arise principally in terms of when and where the arc developed.

The basic scenario, starting in latest Cretaceous–earliest Tertiary time, involves the generally northward-moving plates of the Pacific region being subducted beneath what are now the Koryak Mountains in Siberia and along the southern margin of ancestral Alaska. The plate being subducted is commonly assumed to be the Kula Plate, but unless the Kula Plate was extraordinarily large, it seems plausible to assume that at least one other plate existed north of it at the end of Cretaceous time.

If the scenario does not allow for significant changes in the paleogeography of the various terranes making up southern and interior Alaska today, then it must also be assumed that the Alaskan Bering Sea Shelf edge was undergoing dominantly strike-slip motion but with a significant component of subduction (e.g., see Cooper et al., 1976b; Scholl et al., 1975a; Marlow et al., 1982; Marlow and Cooper, 1984). This combination of strike-slip motion and subduction could account for the large pull-apart features such as the Navarin basin. In early Tertiary time the present Aleutian Subduction Zone developed, possibly in response to the changing direction of plate motion in the Pacific Basin as recorded by the changes in direction of the hot-spot traces. Development of the Aleutian Arc isolated part of a Pacific Basin plate which now forms the oceanic part of the Bering Sea. On this model, the post-Eocene deformation within the Bering Sea region is the result of unknown intraplate stresses.

This very basic scenario does not explain either Shirshov or Bowers ridges. A modification proposed by Ben-Avraham and Cooper (1981) considers both of these ridges as having been rafted in on the oceanic plate that now forms the Bering Sea and then isolated by the Aleutian Arc. These authors also raft in Umnak Plateau and use its interaction with the previous active subduction zone to trigger the formation of the Aleutian Arc.

A further modification of this scenario could involve the formation of the Aleutian Subduction Zone in early Eocene time somewhat to the south of its present latitude. Since that time, the relative motion of the plates of the Pacific Basin, with respect to the northern land masses, could have been taken up at both the Aleutian subduction zone and farther north beneath the Koryak Mountains. This does not answer the question of where to take up the required relative motion between the various parts of southern Alaska and the Bering Sea Shelf, but does allow an explanation of the Cenozoic deformation along the north and east margins of the deep Bering Sea. Models of this type are discussed by Marlow and Cooper (1983) and raise the question of how much of Alaska was at southerly latitudes in Eocene time. The Zodiac Fan, if derived from Alaska, requires a large land mass to be able to produce the required volume of debris (Stevenson et al., 1983), but the larger the land mass, the farther inland the boundary of the southern terranes has to be. It is also possible that much of the relative motion needed by the various models can be accommodated by the deformation.

Resolution of the complexities of the geology of the Bering Sea may be difficult to achieve, if for no other reason than there appear to be several interacting parts. However, if key questions such as the age of the oceanic crust beneath the Aleutian Basin, the age of the Komandorsky and Bowers basins, or the age of Bowers and Shirshov ridges could be answered, this would put considerable constraints on the existing models. Similarily, as more information on the timing and mode of development of the major basins on the Bering Sea Shelf becomes available, it will become easier to restrict the possible relative motion models. The answers to these and other questions are within range of current technology, so hopefully will not be too long in materializing.

ACKNOWLEDGMENTS

I would like to acknowledge J. N. Davies, C. Helfferich, J. Kienle, L. Shapiro, and G. D. Sharma from the University of Alaska, and M. Marlow, D. Scholl, and T. Vallier from the U.S. Geological Survey for their scientific and editorial comments. I would also like to thank the Geophysical Institute of the University of Alaska for supporting my efforts to compile and write this review.

REFERENCES

Atwater, T., 1970, Implication of plate tectonics for the Cenozoic tectonic evolution of western North America, *Geol. Soc. Am. Bull.* **81**:3513–3536.

Barnes, D. F., and Estlund, M. B., 1975, Gravity map of St. Matthew Island, in: Patton *et al.*, Recon. Geol. Map St. Matthew Is., Map MF-642.

Barth, T. F. W., 1956, Geology and petrology of the Pribilof Islands, Alaska, *U.S. Geol. Surv. Bull.* **1028F**.

Beck, M., Cox, A., and Jones, D. L., 1980, Mesozoic and Cenozoic microplate tectonics of western North America, *Geology* **8**:454–456.

Beikman, H. M., 1980, Geologic Map of Alaska, U.S. Geol. Surv. Map 1:250,000.

Ben-Avraham, Z., and Cooper, A. K., 1982, The early evolution of the Bering Sea by collision of ocean rises and north Pacific subduction zones. *Geol. Soc. Am. Bull.* **92**:485–495.

Black, R. F., 1974, Geology and ancient Aleuts, Amchitka and Umnak Islands, Aleutian, *Arct. Anthropol.* **11**(2):126–140.

Black, R. F., 1976, Geology of Umnak Island, eastern Aleutian Islands as related to the Aleuts, *Arct. Alp. Res.* **8**:7–35.

Bujak, J. P., 1983, Cenozoic dinoflagellate cysts and acritarchs from the Bering Sea and northern North Pacific, *Micropaleontology* **30**:180–214.

Burk, C. A., 1965, Geology of the Alaska Peninsula Island arc and continental margin, *Geol. Soc. Am. Mem.* **99**.

Byrne, T., 1979, Late Paleocene demise of the Kula–Pacific spreading center, *Geology* **7**:341–344.

Carder, D. S., Tocher, D., Bufe, C., Stewart, S. W., Eisler, J., and Berg, E., 1967, Seismic wave arrivals from LONGSHOT 0°–27°: *Seismol. Soc. Am. Bull.* **57**: 573–590.

Chapman, M. E., and Solomon, S. C., 1976, North American–Eurasian Plate Boundary in Northeast Asia, *J. Geophys. Res.* **81**:921–930.

Childs, J. R., and Cooper, A. K., 1979, Marine seismic sonobuoy data from the Bering Sea region, U.S. Geol. Surv. Open-File Rep. 79-371.

Churkin, M., and Trexler, J. H., 1981, Continental plates and accreted oceanic terranes in the Arctic, in: *The Ocean Basins and Margins*, Vol. 5 (A. E. M. Nairn, M. Churkin, and F. G. Stehli, eds.), Plenum Press, New York, pp. 1–20.

Coats, R. R., 1956a, Geology of northern Adak Island, *U.S. Geol. Surv. Bull.* **1028-C.**

Coats, R. R., 1956b, Reconnaissance geology of some western Aleutian Islands, Alaska, *U.S. Geol. Surv. Bull.* **1028-E.**

Coe, R. S., Globerman, B. R., Plumley, P. W., Thrupp, G. A., 1985, Paleomagnetic Results from Alaska and their Tectonic Implications, in: *Tectonostratigraphic Terranes of the Circum-Pacific Region*, Circum-Pacific Council for Energy and Mineral Resources, *Earth Sci. Ser.* **1**:85–108.

Conrad, W. K., and Kay, R. W., 1983, Utramafic and mafic inclusions from Adak Island: Crystallization history, and implications for the nature of primary magmas and crustal evolution in the Aleutian arc, *J. Petrol.* **25**:88–125.

Conrad, W. K., Kay, S. M. and Kay, R. W., 1983, Magma mixing in the Aleutian Arc: Evidence from cognate inclusions and composite xenoliths, *J. Volcanol. Geotherm. Res.* **18**:279–295.

Cook, D. B., Fujita, K., McMullen, C. A., 1986, Present-day plate interactions in northeast Asia: North American, Eurasian and Okhotsk plates, Jour. Geodynam. **6**:33–52.

Cooper, A. K., Bailey, K. A., Howell, J., Marlow, M. S., and Scholl, D. W., 1976a, Preliminary residual magnetic map of the Bering Sea Basin and Kamchatka Peninsula, U.S. Geol. Surv. Map MF-713.

Cooper, A. K., Marlow, M. S., and Scholl, D. W., 1976b, Mesozoic magnetic lineations in the Bering Sea marginal basin, *J. Geophys. Res.* **81**:1916–1934.

Cooper, A. K., Marlow, M. S., Parker, A. W., and Childs, J. R., 1979a, Structure-contour map on acoustic basement in the Bering Sea, U.S. Geol. Surv. Map MF-1165.

Cooper, A. K., Marlow, M. S., and Scholl, D. W., 1979b, Cenozoic collapse of the outer Bering Sea continental margin, *Geol. Soc. Am. Pac. Sec. Abstr. Progr.* **11**:73.

Cooper, A. K., Scholl, D. W., Marlow, M. S., Childs, J. R., Redden, G. R., Kvenvaolden, K. A., and Stevenson, A. J., 1979c, Hydrocarbon potential in the Aleutian Basin, Bering Sea, *Bull. Am. Assoc. Pet. Geol.* **63**:2070–2087.

Cooper, A. K., Scholl, D. W., Vallier, T. L., and Scott, E. W., 1980, Resource report for the deep-water areas of proposed OCS Lease Sale No. 70, St. George Basin, Alaska, U.S. Geol. Surv. Open-File Rep. 80-246.

Cooper, A. K., Marlow, M. S., and Ben-Avraham, Z., 1982, Multichannel seismic evidence bearing on the origin of Bowers Ridge, Bering Sea, *Geol. Soc. Am. Bull.* **92**:474–484.

Cormier, V. F., 1975, Tectonics near the junction of the Aleutian and Kuril–Kamchatkan arcs and a mechanism for Middle Tertiary magmatism in the Kamchatka basin, *Geol. Soc. Am. Bull.* **86**:443–453.

Cox, A., Hopkins, D. M., and Dalrymple, G. B., 1966, Geomagnetic polarity epochs Pribilof Islands, Alaska, *Geol. Soc. Am. Bull.* **77**:883–910.

Creager, J. S., Scholl, D. W., Boyce, R. E., Echols, R. J., Fublam, T. J., Grow, J. A., Koizumi, I., Lee, H. J., Ling, H. Y., Stewart, R. J., Supko, P. R., Worsley, T. R., (P. R. Supko, ed.) 1973, *Initial Reports of the Deep-Sea Drilling Project* **19**:913.

Dalrymple, G. B., Claugue, D. A., and Lanphere, M. A., 1977, Revised age for Midway Volcano, Hawaiian volcanic chain, *Earth Planet. Sci. Lett.* **37**:107–116.

Debrai, S., Kay, S. M., Kay, R. W., 1987, Ultramafic xenoliths from Adagdak Volcano, Adak, Aleutian Islands, Alaska: Deformed igneous cumulates from the make of an island arc, *Jour. Geol.* **95**:329–342.

DeLong, S. E., and McDowell, F. W., 1975, K–Ar ages from the Near Islands, Western Aleutian Islands, Alaska: Indication of a mid-Oligocene thermal event, *Geology* **3**:691–694.

DeLong, S. E., Fox, P. J., and McDowell, F. W., 1978, Subduction of the Kula Ridge at the Aleutian Trench, *Geol. Soc. Am. Bull.* **89**:83–95.

Duncan, R. A., 1981, Hotspots in the southern oceans—An absolute frame of reference for motion of the Gondwana continents, *Tectonophysics* **74**:29–42.

Engdahl, E. R., 1971, Explosion effects and earthquakes in the Amchitka Island region, *Science* **173**:1232–1235.

Engdahl, E. R., 1972, Seismic effects of the MILROW and CANNIKIN nuclear explosions, *Bull. Seismol. Soc. Am.* **62**:1411–1423.

Engdahl, E. R., 1977, Seismicity and plate subduction in the central Aleutians, in: *Island Arcs, Deep Sea Trenches and Back-Arc Basins* (M. Talwani and W. C. Pitman, eds.), American Geophysical Union Maurice Ewing Series 1, pp. 259–271.

Engdahl, E. R., Sleep, N. H., and Lin, M. T., 1977, Plate effects in North Pacific subduction zones, *Tectonophysics* **37**:95–116.

Engebretson, D. C., 1982, Relative motions between oceanic and continental plates in the Pacific basin, Ph.D. thesis, Stanford University.

Engebretson, D. C., Cox, A., and Gordon, R. G., 1984, Relative motions between oceanic plates of the Pacific Basin, *J. Geophys. Res.* **89**:2625–2637.

Engebretson, D. C., Cox, A., and Gordon, R. G., 1985, Relative motions between oceanic and continental plates in the Pacific Basin, *Geol. Soc. Am. Spec. Pap.* **206**.

Ewing, M., Ludwig, W. J., and Ewing, J., 1965, Oceanic structural history of the Bering Sea, *J. Geophys. Res.* **70**:4593–4600.

Fisher, M. A., 1982, Petroleum geology of Norton Basin, Alaska, *Am. Assoc. Pet. Geol. Bull.* **66**:286–301.

Fisher, M. A., Patton, W. W., and Holmes, M. L., 1982, Geology of Norton Basin and Continental Shelf beneath northwestern Bering Sea, Alaska, *Am. Assoc. Pet. Geol. Bull.* **66**:255–285.

Fornari, D. J., Iuliucci, R. J., and Shor, G. G., 1973, Preliminary site surveys in the Bering Sea for the DSDP Leg 19, *Initial Reports of the Deep Sea Drilling Project* **19**:569–613.

Foster, T. D., 1962, Heat flow measurements in the N.E. Pacific and Bering Sea, *J. Geophys. Res.* **67**:2991–2993.

Frohlich, C., Caldwell, J. G., Malahoff, A., Latham, G. V., and Lawton, J., 1980, Ocean bottom seismography measurements in the central Aleutians, *Nature* **286**:144–145.

Frohlich, C., Billington, S., Engdahl, E. R., and Malahoff, A., 1982, Detection and location of earthquakes in the central Aleutian subduction zone using island and ocean bottom seismograph stations, *J. Geophys. Res.* **87**:3679–3690.

Fujita, K., and Newberry, J. T., 1983a, Tectonic evolution of northeastern Siberia and adjacent regions, *Tectonophysics* **89**:337–357.

Fujita, K., and Newberry, J. T., 1983b, Accretionary terranes and tectonic evolution of northeast Siberia, in: *Accretion Tectonics in the Circum-Pacific Regions* (M. Hashimoto and S. Uyeda, eds.), Terra. Sci. Publ. Co., Tokyo, pp. 43–57.

Gibson, W. M., and Nichols, H., 1953, Configuration of the Aleutian Ridge, Rat Islands–Semisopochnoi Island to west of Buldir Island, Alaska, *Geol. Soc. Am. Bull.* **64**:1173–1187.

Grim, P. J., and Erickson, B. H., 1969, Fracture zones and magnetic anomalies south of the Aleutian Trench, *J. Geophys. Res.* **76**:1488–1494.

Grow, J. A., 1973, Crustal and upper mantle structure of the central Aleutian arc, *Geol. Soc. Am. Bull.* **84**:2169–2192.

Gutenberg, B., and Richter, C. F., 1954, *Seismicity of the Earth and Associated Phenomena*, Princeton University Press, Princeton, N.J.

Hamilton, E. L., Moore, D. G., Buffington, E. C., Sherrer, P. L., and Curray, J. R., 1974, Sediment velocities from sonobuoys—Bay of Bengal, Bering Sea, Japan Sea and North Pacific, *J. Geophys. Res.* **79**:2653–2668.

Harris, R. A., Stone, D. B., Turner, D. L., 1987, Tectonic implications of Paleomagnetic and Geochronologic data from the Yukon–Koyukuk Province, Alaska, *Geol. Soc. Amer. Bull.*, **99**:362–375.

Hayes, D. E., and Ewing, M., 1970, Pacific boundary structure, in: *The Sea*, Vol. 4 (A. E. Maxwell, ed.), Wiley–Interscience, New York, pp. 29–72.

Hein, J. R., Scholl, D. W., Barron, J. A., Jones, M. G., and Miller, J., 1978, Diagenesis of Late

Cenozoic diatomaceous deposits and formation of the bottom simulating reflector in the southern Bering Sea, *Sedimentology* **25**:155–181.

Hein, J. R., McLean, H., and Vallier, T. L., 1981, Reconnaissance geologic map of Atka and Amlia Islands, Alaska, U.S. Geological Survey Open-File Report 81-159.

Hillhouse, J. W., 1977, Paleomagnetism of the Triassic Nikolai Greenstone, McCarthy Quadrangle, Alaska, *Can. J. Earth Sci.* **14**:2578–2592.

Harbert, W., 1987, New Paleomagnetic data from the Aleutian Islands: Implications for Terrane Migration and Deposition of the Zodiac Fan, *Tectonics,* **6**:585–602.

Hoare, J. M., Conden, W. H., Cox, A., and Dalrymple, G. B., 1968, Geology, paleomagnetism and K–Ar ages of basalts from Nunivak Island, Alaska, *Geol. Soc. Am. Mem.* **116**:377–413.

Hopkins, D. M. (ed.), 1967, *The Bering Land Bridge,* Stanford University Press, Stanford.

Hopkins, D. M., and Silberman, M. L., 1978, Potassium–argon ages of basement rocks from Saint George Island, Alaska, *J. Res. U.S. Geol. Surv.* **6**:435–438.

Hopkins, D. M., Scholl, D. W., Addicott, W. O., Pierce, R. L., Smith, P. B., Wolfe, J., Gershanovich, D., Kotenev, B., Lohman, K. E., Lips, J. H., and Obradovich, J., 1969, Cretaceous, Tertiary and Early Pleistocene rocks from the continental margin in the Bering sea, *Geol. Soc. Am. Bull.* **8**:1471–1480.

Howell, D. G., Jones, D. L., Schermer, E. R., Tectonostratigraphic Terranes of the Circum–Pacific Region, in: Tectonostratigraphic Terranes of the Circum–Pacific Region (D. G. Howell, ed.), Circum–Pacific Council for Energy and Mineral Resources, *Earth Sci. Ser.* **1**:3–30.

Isacks, B., and Molnar, P., 1971, Distribution of stresses in the descending lithosphere from a global survey of focal mechanism solutions of mantle earthquakes, *Rev. Geophys. Space Phys.* **9**:103–174.

Jacob, K. H., 1972, Global tectonic implications of anomalous seismic P travel times from the nuclear explosion Longshot, *J. Geophys. Res.* **77**:2556–2573.

Jones, D. L., and Silberling, N. J., 1979, Mesozoic stratigraphy—The key to tectonic analysis of southern and central Alaska, U.S. Geological Survey Open File Report 79-1200.

Jones, D. L., Silberling, N. J., Coney, P. J., and Plafker, G., 1987, Lithotectonic Terrane Map of Alaska (west of the 141st Meridian), U.S. Geol. Surv. Map MF–1874–A.

Jordan, J. N., Lander, J. F., and Black, R. A., 1965, Aftershocks of 4 February, 1965 Rat Island earthquakes, *Science* **148**:1323–1325.

Kay, R. W., Rubenstone, J. L., Wasserburg, G. J., and Mahlburg, K. S., 1986, Aleutian terranes from Nd isotopes, *Nature* **322**:605–609.

Kay, S. M., and Kay, R. W., 1982, Tectonic controls on tholeiitic and calc-alkaline magmatism in the Aleutian Arc, *J. Geophys. Res.* **87**:4051–4072.

Kay, S. M., Kay, R. W., Bruekner, H. K., and Rubenstone, J. L., 1985, Tholeiitic Aleutian Arc Plutonism: The Finger Bay pluton, Adak, Alaska, *Contrib. Mineral. Petrol.* **82**:99–116.

Kelleher, J. A., 1970, Space–time seismicity of the Alaska–Aleutian seismic zone, *J. Geophys. Res.* **75**:5745–5756.

Kienle, J., 1971, Gravity and magnetic measurements over Bowers Ridge and Shirshov Ridge, Bering Sea, *J. Geophys. Res.* **76**:7138–7153.

Kienle, J., and Swanson, S. W., 1982, Volcanism in the eastern Aleutian arc: Later Quaternary and Holocene centers, tectonic setting and petrology, *J. Volcanol. Geotherm. Res.* **17**:393–432.

Langseth, M. G., Hobart, M. A., and Horai, K., 1980, Heat flow in the Bering Sea, *J. Geophys. Res.* **85**:3740–3750.

Larson, R. L., and Hilde, T. W. C., 1975, A revised time-scale of magnetic reversals for the Early Cretaceous–Late Jurassic, *J. Geophys. Res.* **80**:2586–2596.

Lonsdale, P., Smith, D., 1986, Kula plate not Kula, *EOS* **67**:1199.

Ludwig, W., Houtz, R. E., and Ewing, M., 1971a, Sediment distribution in the Bering Sea: Bowers Ridge, Shirshov Ridge, and enclosed basins, *J. Geophys. Res.* **76**:6367–6375.

Ludwig, W. J., Murauchi, W., Den, N., Ewing, M., Hotta, H., Houtz, R. E., Yoshii, T., Ansanuma, T., Hagiwasa, K., Sata, T., and Ando, S., 1971b, Structure of Bowers Ridge, Bering Sea, *J. Geophys. Res.* **76**:6350–6366.

McCarthy, J., and Scholl, D. W., 1985, Mechanism of subduction accretion along the central Aleutian Trench, *Geol. Soc. Am. Bull.* **96**:691–701.

McGeary, S. E., and Ben-Avraham, Z., 1981, Allochthonous terranes in Alaska: Implication for the structure and evolution of the Bering Sea Shelf, *Geology* **9**:608–615.

McLean, H., 1979a, Pribilof segment of the Bering Sea continental margin, a reinterpretation of Upper Cretaceous dredge samples, *Geology* **7**:307–310.

McLean, H., 1979b, Review of petroleum geology of Anadyr and Khatyrka basins, USSR, *Am. Assoc. Pet. Geol.* **63**:1467–1477.

McLean, H., Hein, J. R., and Vallier, R. L., 1983, Reconnaissance geology of Amlia Island, Aleutian Islands, Alaska, *Geol. Soc. Am. Bull.* **94**:1020–1027.

Mancini, E. A., Dexter, T. M., and Wingate, F. H., 1978, Upper Cretaceous arc–trench gap sedimentation on the Alaska Peninsula, *Geology* **6**:437–439.

Marlow, M. S., 1979, Hydrocarbon prospects in the Navarin Basin Province, northwest Bering Sea shelf, *Oil Gas J.* **Oct. 29**:190–196.

Marlow, M. S., and Cooper, A. K., 1980, Mesozoic and Cenozoic structural trends beneath the southern Bering Sea shelf, *Am. Assoc. Pet. Geol. Bull.* **64**:2139–2155.

Marlow, M. S., and Cooper, A. K., 1983, Wandering terranes in southern Alaska: The Aleutian microplate and implications for the Bering Sea, *J. Geophys. Res.* **88**:3439–3446.

Marlow, M. S., and Cooper, A. K., 1985, Regional geology of the Bergingian Continental Margin, Proc. in: *Formation of Active Ocean Margins*, (N. Narv, eds.), Terra Sci. Pub. Co., Tokyo, 497–515.

Marlow, M. S., Scholl, D. W., Buffington, E. C., Alpha, R. R., Smith, D. B., and Shipek, C. J., 1970, Buldir depression—Late Tertiary graben on the Aleutian Ridge, Alaska, *Mar. Geol.* **8**:85–108.

Marlow, M. S., Scholl, D. W., Buffington, E. C., and Alpha, T. R., 1973, Tectonic history of the central Aleutian arc, *Bull. Geol. Soc. Amer.* **84**:1555–1574.

Marlow, M. S., Scholl, D. W., Cooper, A. K., and Buffington, E. C., 1976, Structure and evolution of the Bering Sea shelf south of St. Lawrence Island, *Am. Assoc. Pet. Geol. Bull.* **60**:161–183.

Marlow, M. S., Scholl, D. W., and Cooper, A. K., 1977, St. George Basin, Bering Sea Shelf: A collapsed Mesozoic margin, in: *Island Arcs, Deep-Sea Trenches and Back-Arc Basins* (M. Talwani and S. C. Pitman, eds.), Maurice Ewing Series 1, American Geophysical Union, pp. 211–220.

Marlow, M. S., Cooper, A. K., Scholl, D. W., Vallier, T. L., and McLean, H., 1979a, Description of dredge samples from the Bering Sea continental margin, U.S. Geological Survey Open-File Report 79-1139.

Marlow, M. S., Scholl, D. W., Cooper, A. K., and Jones, D. L., 1979b, Shallow water Upper Jurassic rocks dredged from Bering Sea continental margin, *Am. Assoc. Pet. Geol. Bull.* **63**:490–491.

Marlow, M. S., Carlson, P., Cooper, A. K., Karl, H., McLean, H., McMullin, R., and Lynot, M. B., 1981, Hydrocarbon resource report for proposed OCS Sale No. 83, Navarin Basin, Alaska, U.S. Geological Survey Open-File Report 81-252.

Marlow, M. S., Cooper, A. K., Scholl, D. W., and McLean, H., 1982, Ancient plate boundaries in the Bering Sea region, *Geol. Soc. London Spec. Pub.* **10**:210–211.

Marlow, M. S., Cooper, A. K., and Childs, J. R., 1983, Tectonic evolution of the Gulf of Anadyr and formation of Anadyr and Navarin Basins, *Am. Assoc. Pet. Geol. Bull.* **67**:646–665.

Marlow, M. S., Parson, L. M., Carlson, P. R., and Cooper, A. K., 1986, Buried basement ridges and linear magnetic anomalies in the western Aleutian Basin, Bering Sea, *Eos* **67**:1227.

Marsh, B. D., 1979, Island arc volcanism, *Am. Sci.* **67**:161–172.

Minster, J. B., and Jordan, T. H., 1978, Present-day plate motions, *J. Geophys. Res.* **83**:5331–5354.

Minster, J. B., Jordan, T. H., Molnar, P., and Haines, E., 1974, Numerical modelling on instantaneous plate tectonics, *Geophys. J. R. Astron. Soc.* **36**:541–576.

Moore, J. C., 1973, Cretaceous continental margin sedimentation, southwestern Alaska, *Geol. Soc. Am. Bull.* **84**:595–614.

Moore, J. C., et al., 1982, Offscraping and underthrusting of sediment at the deformation front of the Barbados Ridge, DSDP Leg 78A, *Geol. Soc. Am. Bull.* **93**:1065–1077.

Morgan, W. J., 1972, Deep mantle convection plumes and plate motions, *Am. Assoc. Pet. Geol.* **56**:203–213.

Nalivkin, D. V., Markóvskiy, A. P., Muzylev, S. A., and Shatalov, E. T., 1966, Geological map of the Union of Soviet Socialist Republics, Scale 1 : 5,000,000, Ministry of Geology of the USSR, All Union Geological Research Institute.

Neprochnov, Y. P., Sedov, V. V., Merklin, L. R., Zinkevich, V. P., Levchenko, O. V., Baranov, B. V., and Rudnik, G. B., 1985, Tectonics of the Shirshov Ridge, Bering Sea, *Geotectonics* **19**:194–207.

Newberry, J. T., Laclair, D. L., Fujita, K., 1986, Seismicity and tectonics of the far western Aleutian Islands, *Journ. Geodynam.* **6**:13–32.

Nichols, H., and Perry, R. B., 1966, Bathymetry of the Aleutian Arc, Alaska, scale 1 : 400,000, U.S. Coast and Geodetic Survey, Monograph 3.

Nilsen, T. H., and Zuffa, G. G., 1982, The Chugach terrane, a Cretaceous trench-fill deposit, southern Alaska, *Geol. Soc. London Spec. Publ.* **10**:213–227.

Panuska, B. C., 1980, Stratigraphy and sedimentary petrology of the Kiska Harbor Formation, unpublished M.S. thesis, University of Alaska.

Panuska, B. C., and Stone, D. B., 1985, Latitudinal motion of the Wrangellian and Alexander Terranes and the southern Alaska superterrane, in: *Tectonostratigraphic Terranes of the Circum-Pacific Region* (D. G. Howell, ed.), Circum-Pacific Council for Energy and Mineral Resources, pp. 109–180.

Parfenov, L. M., and Natal'in, B. A., 1985, Mesozoic accretion and collision tectonics of Northeastern Asia, in: *Tectonostratigraphic Terranes of the Circum-Pacific Region* (D. G. Howell, ed.), Circum-Pacific Council for Energy and Mineral Resources, pp. 363–373.

Patton, W. W., 1973, Reconnaissance geology of the northern Yukon–Koyukuk province, Alaska, *U.S. Geol. Surv. Prof. Pap.* **774A**.

Patton, W. W., and Tailleur, I. L., 1977, Evidence in the Bering Straight region for differential movement between North America and Eurasia, *Geol. Soc. Am. Bull.* **88**:1298–1304.

Patton, W. W., Miller, T. P., Berg, H. C., Gryc, G., Hoare, J. M., and Ovenshine, A. T., 1975, Reconnaissance geologic map of St. Matthew Island, Bering Sea, Alaska, U.S. Geological Survey Miscellaneous Field Studies Map MF-642.

Patton, W. W., Lanphere, M. A., Miller, T. P., and Scott, R. A., 1976, Age and tectonic significance of volcanic rocks of St. Matthew Island, Bering Sea, Alaska, *J. Res. U.S. Geol. Surv.* **4**:67–73.

Pickthorn, L. B. G., Vallier, T. L., and Scholl, D. W., 1984, Geochronology of igneous rocks from the Aleutian Island Arc, *Geol. Soc. Am. Abstr. Progr.* **16**:328.

Pitman, W. C., and Talwani, M., 1972, Sea floor spreading in the North Atlantic, *Geol. Soc. Am. Bull.* **83**:619–646.

Plumley, P. W., Coe, R. S., Byrne, T., Reid, M. R., and Moore, J. C., 1982, Palaeomagnetism of Paleocene volcanic rocks from Kodiak Island, *Nature* **300**:50–52.

Pratt, R. M., Rutstein, M. S., Walton, F. W., and Buschur, J. A., 1972, Structural trends beneath Bristol Bay, Bering Shelf, Alaska, *J. Geophys. Res.* **77**:4996–4999.

Rabinowitz, P. D., 1974, Seismic profiling between Bowers Ridge and Shirshov Ridge in the Bering Sea, *J. Geophys. Res.* **79**:4977–4979.

Rabinowitz, P. D., and Cooper, A. K., 1977, Structure and sediment distribution in the western Bering Sea basin, *Mar. Geol.* **24**:308–320.

Rea, D. K., and Duncan, R. A., 1986, North Pacific plate convergence, a quantitative record of the past 140 million years, *Geology* **14**:373–376.

Savostin, L. A., and Karasik, A. M., 1981, Recent plate tectonics of the Arctic basin and of northeastern Asia, *Tectonophysics* **74**:111–145.

Scholl, D. W., and Hopkins, D. M., 1969, Newly discovered Cenozoic basin, Bering Sea shelf, Alaska, *Am. Assoc. Pet. Geol.* **53**:2067–2078.

Scholl, D. W., Buffington, E. D., and Hopkins, D. M., 1968, Geologic history of the continental margin of North America in the Bering Sea, *Mar. Geol.* **6**:297–330.

Scholl, D. W., Buffington, E. D., Hopkins, D. M., and Alpha, T. R., 1970a, The structure and origin of the large submarine canyons of the Bering Sea, *Mar. Geol.* **8**:187–210.

Scholl, D. W., Greene, H. G., and Marlow, M. S., 1970b, Eocene age of the Adak Paleozoic(?) rocks, Aleutian Islands, Alaska, *Geol. Soc. Am. Bull.* **81**:3583–3592.

Scholl, D. W., and others (Shipboard Scientific Party), 1973, DSDP Site 191 report, *Initial Reports of the Deep Sea Drilling Project* **19**:413–424.

Scholl, D. W., Alpha, T. R., Marlow, M. S., and Buffington, E. D., 1974, Base map of the Aleutian–Bering Sea region, U.S. Geological Survey Miscellaneous Investigation Series Map I87.

Scholl, D. W., Buffington, E. D., and Marlow, M. S., 1975a, Plate tectonics and the structural evolution of the Aleutian–Bering Sea region, *Geol. Soc. Am. Spec. Pap.* **131**:1–31.

Scholl, D. W., Marlow, M. S., and Buffington, E. D., 1975b, Summit basins of the Aleutian Ridge, North Pacific, *Am. Assoc. Pet. Geol. Bull.* **59**:799–816.

Scholl, D. W., Vallier, T. L., and Stevenson, A. J., 1983a, Arc, forearc, and trench sedimentation and tectonics: Amlia corridor of the Aleutian Ridge, *Am. Assoc. Pet. Geol. Bull.* Mem 34:105-134. in: Studies in Continental Margin Geology, (J. S. Watkins and C. L. Drake, eds.)

Scholl, D. W., Vallier, T. L., and Stevenson, A. J., 1983b, Geologic evolution of the Aleutian Ridge—Implications for petroleum resources, *J. Alaska Geol. Soc.* **3**:33–46.

Scholl, D. W., Vallier, T. L., and Stevenson, A. J., 1983c, Sedimentation and deformation in the Amlia fracture zone sector of the Aleutian Trench, *Mar. Geol.* **48**:105–134.

Sclater, J. G., 1972, Heat flow and elevation of the marginal basins of the western Pacific, *J. Geophys. Res.* **77**:5705–5720.

Sharma, G. D., 1979, *The Alaskan Shelf: Hydrographic, Sedimentary and Geochemical Environment,* Springer-Verlag, Berlin.

Shmidt, O. A., 1974, Tectonic development of the Komandorski Islands: Geotectonics, *Akad. Nauk SSR* **6**:377–383.

Shmidt, O. A., Serova, M. Y., and Dolmatova, L. M., 1973, Stratigraphy and paleontological features of the volcanic rock series on the Komandorski Islands, *Akad. Nauk SSR Izv. Ser. Geol.* **11**:77–87.

Shor, G. G., Jr., 1964, Structure of the Bering Sea and the Aleutian Ridge, *Mar. Geol.* **1**:213–219.

Shor, G. G., Jr., and Fornari, D. J., 1976, Seismic refraction measurements in the Kamchatka Basin, W. Bering Sea, *J. Geophys. Res.* **81**:5260–6266.

Spence, W., 1977, The Aleutian arc: Tectonic blocks, episodic subduction, strain diffusion, and magma generation, *J. Geophys. Res.* **82**:213–230.

Stauder, W., 1968a, Tensional character of earthquake foci beneath the Aleutian trench with relation to seafloor spreading, *J. Geophys. Res.* **73**:7693–7701.

Stauder, W., 1968b, Mechanism of the Rat Island earthquake sequence of February 4, 1965, with relation to island arcs and sea-floor spreading, *J. Geophys. Res.* **73**:3847–3858.

Shor, G. G., Jr., 1971, Site Surveys in the north Pacific and Bering Sea from Antipode expedition, Mar. Phys. Lab. Tech., Mem., Scripps Inst. Oceanog. **218**:149.

Stevenson, A. J., Scholl, D. W., and Vallier, T. L., 1983, Tectonic and geologic implications of the Zodiac Fan, Aleutian Abyssal Plain, northeast Pacific, *Geol. Soc. Am. Bull.* **96**:259–273.

Stewart, R. J., 1978, Neogene volcaniclastic sediments from Atka Basin, Aleutian Ridge, *Am. Assoc. Pet. Geol. Bull.* **62**:87–97.

Stock, J. M., and Molnar, P., 1983, Some geometrical aspects of uncertainties in combined plate reconstructions, *Geology* **11**:697–701.

Stone, D. B., 1968, Geophysics in the Bering Sea and surrounding areas: A review, *Tectonophysics* **6**:433–460.

Stone, D. B., 1981, Triassic paleomagnetic data and paleolatitudes for Wrangellia, Alaska, Alaska Div. Geol. Geophys. Surv., Geol. Rep. 73, pp. 55–62.

Stone, D. B., 1983, Present day plate boundaries in Alaska and the Arctic, *J. Alaska Geol. Soc.* **3**:1–14.

Stone, D. B., Panuska, B. C., and Packer, D. R., 1982, Paleolatitudes versus time for southern Alaska, *J. Geophys. Res.* **87**:3697–3708.

Stone, D. B., Harbert, W., Vallier, T., and McLean, H., 1983, Eocene paleolatitudes for the Aleutian Islands. *Trans. Am. Geophys. Union (Eos)* **64**:87.

Sweeney, J. F. (ed.), 1978, *Arctic Geophysical Review,* Publication of the Earth Physics Branch, Ottawa, Canada, **45**(4):108.

Sykes, L. R., 1971, Aftershock zones of great earthquakes, seismicity gaps and earthquake prediction for Alaska and the Aleutians, *J. Geophys. Res.* **76**:8021–8041.

Tarr, A. C., 1970, New maps of polar seismicity, *Bull. Seismol. Soc. Am.* **60**:1745–1747.

Turner, D. L., 1984, Tectonic implications of widespread Cretaceous overprinting of K–Ar ages in Alaskan metamorphic terranes, *Geol. Soc. Am. Abstr. Progr.* **16**:338.

Turner, D. L., Forbes, R. B., Aleinikoff, J. N., Hedge, C. E., and McDougall, I., 1983, Geochronology of the Kilbuck Terrane of southwestern Alaska, *Geol. Soc. Am. Abstr. Progr.* **15**:407.

Turner, R. F., Bolm, J. G., McCarthy, C. M., Steffy, D. A., Lowry, P., and Flett, T. O., 1983a,

Geological and operational summary: Norton Sound COST no. 1 well, Norton Sound, Alaska, U.S. Geological Survey Open-File Report 83-124.

Turner, R. F., Bolm, J. G., McCarthy, C. M., Steffy, D. A., Lowry, P., Flett, T. O., and Blunt, D., 1983b, Geological and operational summary: Norton Sound COST no. 2 well, Norton Sound, Alaska, U.S. Geological Survey Open-File Report 83-557.

Turner, R. F., McCarthy, C. M., Comer, C. D., Larson, J. A., Bolm, J. G., Banet, A. C., and Adams, A. J., 1984a, Geological and operational summary: St. George Basin COST no. 1 well, Bering Sea, Alaska, Outer Continental Shelf Report MMS 84-0016.

Turner, R. F., McCarthy, C. M., Comer, C. D., Larson, J. A., Bolm, J. G., Flett, T. O., and Adams, A. J., 1984b, Geological and operational summary: St. George Basin COST no. 2 well, Bering Sea, Alaska, Outer Continental Shelf Report MMS 84-0018.

Turner, R. F., McCarthy, C. M., Steffy, D. A., Lynch, M. B., Martin, G. C., Sherwood, K. W., Flett, T. O., and Adams, A. J., 1984c, Geological and operational summary, Navarin Basin COST no. 1 well, Outer Continental Shelf Report MMS 84-0031.

Turner, R. F., Martin, G. C., Flett, T. O., and Steffy, D. A., 1985, Geologic report for the Navarin Basin planning area, Alaska, Outer Continental Shelf Report MMS 85-0045.

Uyeda, S., 1977, Some basic problems in the trench-arc back arc system, in: *Island Arcs, Deep Sea Trenches and Back-Arc Basins* (M. Talwani and W. C. Pitman, eds.), American Geophysical Union Maurice Ewing Series 1, pp. 1–14.

Vallier, T. L., Underwood, M. B., Jones, D. L., and Gardner, J. V., 1980, Petrography and geological significance of Upper Jurassic rocks dredged near Pribilof Islands, southern Bering Sea continental shelf, *Am. Assoc. Pet. Geol. Bull.* **64**:845–950.

Vallier, T. L., McCarthy, J., Scholl, D. W., Stevenson, A., and O'Connor, R., 1984, Offshore structures in the Near Islands region, western Aleutian Island Arc, Alaska, *Geol. Soc. Am. Abstr. Progr.* **16**:328.

Vogt, P., Bernero, C., Kovacs, L., and Taylor, P., 1981, Structure and plate tectonic evolution of the marine Arctic as revealed by aeromagnetics, *Oceanologica Acta*, Suppl. to Vol. 4, 26th Geol. Congr., Paris, pp. 25–40.

Wallace, W. K., and Engebretson, D., 1982, Correlation between plate motions and Late Cretaceous to Paleogene magmatism in southwestern Alaska, *Eos* **63**:915.

Watson, B. F., Fujita, K., 1985, Tectonic Evolution of Kamchatka and the Sea of Okhotsk and implications for the Pacific Basin, Tectonostratigraphic Terranes of the Circum–Pacific Region, (D. G. Howell, ed.), Circum–Pacific Council for Energy and Mineral Resources, *Earth Sci. Ser.* **1**:333–348.

Watts, A. B., 1975, Gravity field of the northwest Pacific Ocean Basin and its margin: Aleutian Island Arc–Trench system, Geol. Soc. Am. Map Chart Ser. MC-10.

Wilcox, R. E., 1959, Igneous rocks of the Near Islands, Aleutian Islands, Alaska, Int. Geol. Congr. 20th, Mexico, 1956, Proc. Sect. 11A, pp. 365–378.

Wittbrodt, P. R., 1985, Paleomagnetism and petrology of St. Matthew Island, Bering Sea, Alaska, M.S. thesis, University of Alaska.

Wittbrodt, P. R., Stone, D. B., and Turner, D. L., 1987, Paleomagnetism and geochronology of St. Matthew Island, Bering Sea (in preparation).

Woods, M. T., and Davies, G. F., 1982, Late Cretaceous genesis of the Kula plate, *Earth Planet Sci. Lett.* **58**:161–166.

Zinkevich, V. D., Kazimirov, A. D., and Peyve, A. A., 1983, Tectonics of the continental margins of the Bering Sea, *Geotectonics* **17**:513–525.

Zonenshayn, L. P., Natapov, L. M., Savostin, L. A., and Stavskii, A. P., 1978, Recent plate tectonics of Northeastern Asia in connection with the opening of the North Atlantic and the Arctic Ocean Basins, *Oceanology* **18**:550–555.

Chapter 2

THE NORTHEAST PACIFIC OCEAN AND MARGIN

Robin Riddihough

Geological Survey of Canada
Ottawa, Canada

I. INTRODUCTION

The northeast Pacific Ocean floor and the adjacent continental margin has occupied a critical position in the development of plate tectonic theory. The magnetic surveys conducted from the *USS Pioneer* in the late 1950s provided (Fig. 1), with their publication in 1961 (Raff and Mason, 1961), the first glimpse of the remarkable magnetic anomaly "stripes" which were to lead to a coherent time scale of magnetic field reversals, confirmation of seafloor spreading processes, and a solution to the enigma of continental drift–plate tectonics. J. Tuzo Wilson and F. Vine in 1965 first identified the Juan de Fuca Ridge as a spreading ridge and recognized the symmetry in the magnetic anomaly pattern which was the key to its interpretation (Vine and Wilson, 1965). The enormous apparent offsets required by these anomalies led Wilson (1965a,b) to propose the mechanism of transform faults. Throughout the 1960s the region was the subject of numerous studies and investigations, culminating in the thesis and publication of Atwater (1970), which effectively provided the framework and foundation of almost all subsequent work on the tectonics of the region.

The region contains active examples of almost all aspects of the plate tectonic process. Because, at least for the last 100 m.y., the complex tectonic style evident today has been the dominant process along the margin, it provides an almost unique

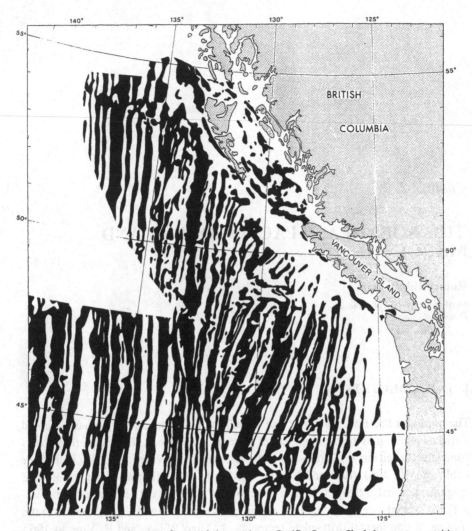

Fig. 1. Magnetic anomaly map of part of the northeast Pacific Ocean. Shaded areas are positive anomalies. Data from Raff and Mason (1961), Potter *et al.* (1974), and Currie *et al.* (1983).

laboratory for the application of the geological principle that the present is the key to the past.

II. THE JUAN DE FUCA RIDGE SYSTEM

Figure 2 shows the principal features of the Juan de Fuca Ridge system. The central section of the ridge (Juan de Fuca Ridge) is connected to the north and south through major transform faults [the Sovanco Fracture Zone (FZ) and the Blanco FZ]

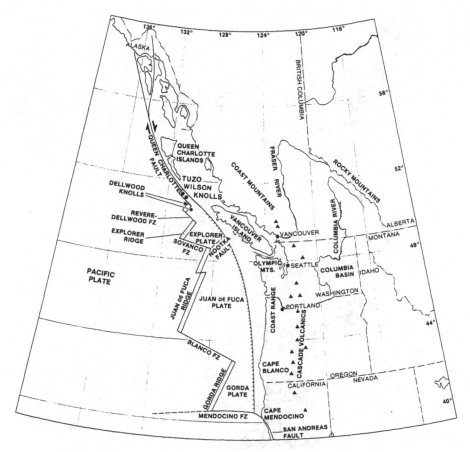

Fig. 2. Location map of the northeast Pacific and features referred to in the text.

to smaller sections of ridge, which become progressively more complex toward the northern and southern triple junctions. This progression is reflected in the seismicity (Fig. 3), which shows little or no activity on the central section of the ridge between 45° and 48°N, but concentrations of seismic activity along the major fracture zones and near the northern and southern triple junctions (Chandra, 1974; Rogers, 1983a; Riddihough et al., 1984).

Spreading rates on the ridge system have been established by magnetic anomaly analysis (e.g., Vine and Wilson, 1965; Atwater, 1970; Atwater and Mudie, 1973; Carlson, 1976; Riddihough, 1977; Riddihough, 1984). On the central ridge section they vary from 60 mm/yr near the northern intersection with the Sovanco FZ to 56 mm/yr just north of the Blanco FZ. Spreading rates on the Explorer Ridge are probably lower (Riddihough, 1977, 1984; Wilson et al., 1984) at around 40 mm/yr. Rates at the Dellwood Knolls are suspected to be low, but are largely indeterminate (Riddihough et al., 1980). The southern part of the Gorda Ridge has extremely low

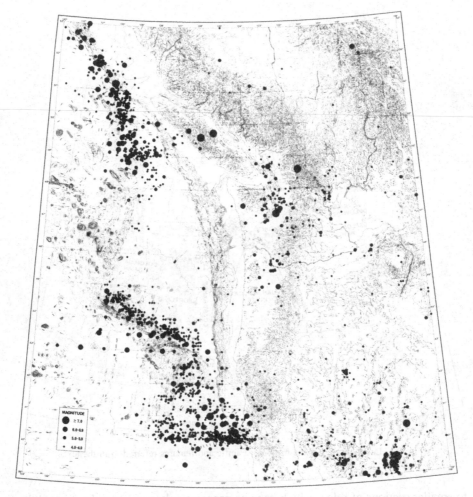

Fig. 3.   Seismicity of the Juan de Fuca Plate region, all events of magnitude > 4 to the end of 1977.
(From Riddihough *et al.*, 1984.)

spreading rates (20 mm/yr) (Atwater and Mudie, 1973; Riddihough, 1980), while
the northern section is probably spreading at rates comparable with the southern
Juan de Fuca Ridge.

The maximum rate of spreading is regarded as median between fast and slow
spreading rates on a global scale, and the resulting morphology of the ridge system
exhibits features which have been regarded as characteristic of both. Standard
bathymetric surveys have shown that the Juan de Fuca system can vary between
sections with a deep axial rift valley (the Gorda Ridge) to sections consisting of a
broad ridge with little or no axial rift. However, a new generation of side-scan and
swath mapping techniques demonstrates that in detail, variations in ridge mor-

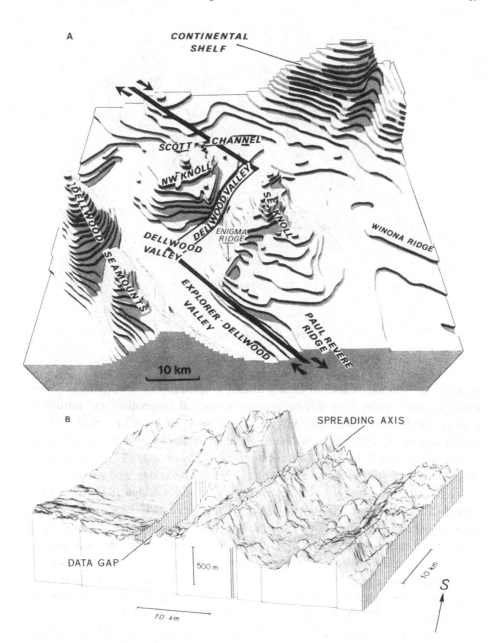

Fig. 4.   Morphology of the Juan de Fuca Ridge system. (A) Dellwood Knolls: contour levels are 100 m.
Double lines outline probable spreading axis; solid lines are transform faults. (Modified from Riddihough
et al., 1980.) (B) A section of Explorer Ridge; computer-generated 3D image from Seabeam data
(Malahoff *et al.*, 1984.) (C) Portion of northern Juan de Fuca ridge showing conjugate spreading system
of Endeavour Ridge and West Valley enclosing Endeavour Seamount; computer-generated 3D image
from Seabeam data (Malahoff *et al.*, 1984).

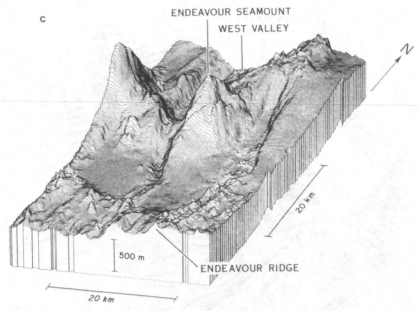

Fig. 4. (*continued*)

phology are widespread and not clearly related to spreading rate. In particular, the spreading style varies from diffuse areas of equivocal magmatism and intrusion such as may be seen at the Tuzo Wilson and Dellwood Knolls (Chase *et al.*, 1975; Riddihough *et al.*, 1980) through deep, flat-floored valleys with steep, rifted walls (northern Explorer Ridge and Gorda Ridge) (Srivastava *et al.*, 1971), to strongly lineated systems with widespread faulting, a central rifted ridge, and striking symmetry [Endeavour segment of the northern Juan de Fuca Ridge (Karsten *et al.*, 1986)] (Fig. 4). The genesis of these styles is not yet clearly understood, but is generally felt to be a symptom of a combination of adjustments to variable plate motions and variations in magma supply, both longitudinally along the ridge and in time. Conjugate rift systems have been identified at the north end of the Juan de Fuca Ridge, and a major propagator—the Cobb offset (Johnson *et al.*, 1983)— seems to be continuing a complex history of ridge propagation which is evident in the region (Hey, 1977; Hey and Wilson, 1982; Wilson *et al.*, 1984).

The two major transform faults, the Blanco FZ and the Sovanco FZ, are known to be far from simple. The Blanco FZ can be divided into a number of segments and contains at least two overdeepened pull-apart basins which may contain spreading. The eastern part of the Sovanco FZ consists of a series of uplifted blocks (the Sovanco Ridge) and the western part, an arcuate series of faults converging on the southern Explorer valley (Malahoff *et al.*, 1984).

Mineralization on the Juan de Fuca Ridge system has been of particular interest in the last few years, and a number of sulfide occurrences have now been identified. Bottom photography, dredging, and submersible operations have located a number of active vents, which are the subject of both geological and biological investigations (Normark *et al.*, 1983). Detailed heat flow investigations on the northern end of the ridge (Davis and Lister, 1977; Davis *et al.*, 1980; Hyndman *et al.*, 1978) have provided the basis for a fuller understanding of the role of seawater in circulation through fractured ocean crust near a ridge system (Lister, 1977).

III. CONTEMPORARY PLATE MOTIONS

The detailed magnetic anomaly maps of the Juan de Fuca area have enabled the spreading history and details of the region to be examined very closely. McKenzie and Parker (1967) and Atwater (1970) first suggested that if spreading on the Juan de Fuca Ridge was parallel to the Blanco FZ and on the order of 50–60 mm/yr, then vector addition of Pacific–America motion would imply convergence of the Juan de Fuca Plate with North America along the margins of British Columbia, Washington, Oregon, and northern California. Atwater (1970) estimated a convergence rate of 20 mm/yr in a northeasterly direction. Since then, other estimates (e.g., Carlson, 1976; Riddihough, 1977, 1984; Nishimura *et al.*, 1984) have calculated a present convergence rate of between 45 mm/yr in the north and 32 mm/yr near Cape Blanco. A present pole of rotation between the Juan de Fuca and Pacific Plates has been calculated (Nishimura *et al.*, 1984; Riddihough, 1984) as being near 146°W, 10°S at 0.6°/m.y. A summary of the motions of the system is given in Fig. 5.

The slower spreading rate of the Explorer Ridge estimated from magnetic

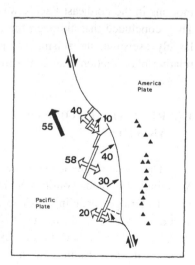

Fig. 5. Contemporary plate motions in the northeast Pacific relative to America Plate. Solid arrows are relative motions; open arrows are spreading. Numbers are mm/yr or km/m.y.b.p.

anomalies (Riddihough, 1977, 1984) implies that the portion of the Juan de Fuca system presently being created at this part of the ridge, must be moving independently of the remainder of the Juan de Fuca Plate. This subplate was first named the Explorer Plate by Barr and Chase (1974), and the transform fault boundary with the Juan de Fuca Plate established and named the Nootka Fault by Hyndman *et al.* (1979). A zone of seismicity is now regarded as associated with the 30 mm/yr left lateral motion of this fault.

Within the Gorda Plate, the slow spreading of the southern part of the ridge similarly implies that at least the southern part of the plate may be partially independent. Riddihough (1980) proposed the existence of a transform isolating the southern Gorda plate. A complex deformation mechanism in response to the geometry of the Gorda Ridge and Mendocino FZ has been suggested by Carlson and Stoddard (1981), following Silver (1971a). Intense seismicity within the plate with appropriate left lateral northeast-trending shear (Smith and Knapp, 1980) seems to support this interpretation.

The vector calculations which predict the motion of all sections of the Juan de Fuca Plate system with respect to North America are critically dependent upon accurate calculation of contemporary Pacific–America motion. The majority of calculations have used the solution of Minster and Jordan (1978) for this vector and produce a minimum resultant error in the Juan de Fuca–America convergence velocity of $\pm$ 7 mm/yr and $\pm$ 7° (Nishimura *et al.*, 1984; Riddihough, 1984). Motions of the Explorer and southern Gorda Plates with respect to the Pacific Plate are even more uncertain, so that little or no convergence may occur here. Riddihough (1984) calculated a local pole of motion for the Explorer Plate which implies considerably reduced convergence along the northern Vancouver Island margin with extreme lateral variability.

An assessment of contemporary regional seismicity in terms of relative plate motions in the northeast Pacific was attempted by Hyndman and Weichert (1983). They concluded that although the main Juan de Fuca–America convergence was largely aseismic, the seismicity observed was consistent with the magnitude of relative plate motions calculated from magnetic anomaly analyses.

IV. RECENT HISTORY OF THE RIDGE SYSTEM AND PLATE MOTIONS

The oldest magnetic anomalies on the Juan de Fuca Plate (now just beneath the Washington–Oregon continental slope) have been dated at near 8–9 m.y.b.p. The record of ridge spreading back to this time can thus be reconstructed with some accuracy. Atwater and Menard (1970) first examined the magnetic anomaly pattern in detail and noted that the ridge system had rotated clockwise 10–15° during this period and that spreading rates had slowed. They suggested that this rotation, which resulted in the ridge being closer to perpendicular to the Queen Charlotte and San Andreas Faults, was part of an adjustment to the diminishing size of the Juan de Fuca Ridge and interaction with the continental America Plate. Riddihough (1977) and Carlson (1976) attempted to show how interaction rates at the margin had changed during the last 10 m.y., the general pattern being one of progressive reduction in convergence rate.

The calculation of rotation poles (Riddihough, 1984; Wilson *et al.*, 1984) now suggests that the changes in spreading pattern may have been a result of varying resistance at the subduction zone to the downgoing Juan de Fuca Plate. The idea [first suggested for this area by Atwater (1970)] that the youngest, most bouyant part of the plate would show the most resistance to subduction was developed by Menard (1978) in the concept of "pivoting subduction." In the northeast Pacific prior to 5 m.y.b.p., the youngest material of the Juan de Fuca Plate was subducting beneath Vancouver Island; as a consequence, the Juan de Fuca–America pole of rotation lay to the north, and convergence rates increased southwards along the margin. Between 5 and 3 m.y.b.p. the northern, youngest portion of the plate broke off as the independent Explorer Plate. The youngest part of the Juan de Fuca Plate was now to the south where the Gorda Ridge was closest to the margin, and motions adjusted so that convergence now increased northwards. The Explorer Plate began to move about a local pole so that its convergence was severely reduced. A cartoon of this plate history is shown in Fig. 6.

Throughout this complex adjustment of plate breakup, ridge reorientations, changes in spreading rates, and poles of motion, the Juan de Fuca Ridge remained approximately the same distance offshore from the continental margin (Engebretson

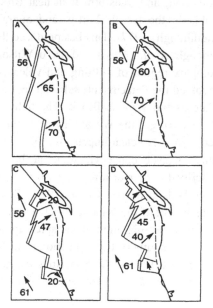

Fig. 6. Chronology of plate breakup and movements of the Juan de Fuca Plate system relative to North America, (A) 6 m.y.b.p., (B) 4 m.y.b.p., (C) 2 m.y.b.p., (D) 0 m.y.b.p.

et al., 1985). All plate and ridge rotations were also clockwise in conformity with the general regional right-lateral shear imposed by the much larger Pacific and America Plates (Fig. 6). A general picture of adjustment and interaction between the plates of the region is emerging, confirming the view that the ridge (and spreading) is essentially tensional in response to plate motions.

The adjustment of the ridge to changes in spreading has been suggested by Hey (1977) and Wilson *et al.* (1984) as being achieved through the mechanism of ridge propagation. Hey (1977) has shown how such a process results in V-shaped pseudofaults intersecting at the ridge and marked as discontinuities in the magnetic anomaly pattern. Such discontinuities in the Juan de Fuca area were first noted by Vine (1966) and examined by Silver (1971b). A detailed chronological reconstruction for the northeast Pacific by Hey and Wilson (1982) has shown with remarkable accuracy how the observed complex pattern of magnetic anomalies can be reproduced using this mechanism. Although many aspects of this process are not understood, the circumstantial evidence for its application to this area seems overwhelming.

V. TRIPLE JUNCTIONS

The northern and southern extremities of the Juan de Fuca Plate system are both triple plate junctions and have provided unique opportunities for the examination of the tectonic complexities of such features.

Off northern Vancouver Island (Fig. 7), an intersection between ridge, subduction zone, and transform fault near Brooks Peninsula, seems to have become unstable in the last 1–2 m.y. and migrated northwestwards (Chase *et al.*, 1975; Riddihough, 1977). This has produced the spreading centers of the Dellwood (Riddihough *et al.*, 1980) and Tuzo Wilson Knolls (Chase, 1977), together with a complex region of faulting and volcanism. It seems probable that this migration involved the detachment, subsidence, and tilting of the youngest part of the Pacific Plate as the Winona Block. The uplifted edge of this block now forms the Paul Revere Ridge, the Winona Basin to the northeast containing up to 5 km of recent sediments. No clear impetus for the recent change in geometry at the northern triple junction has become evident; it may be part of the regional plate "breakup" process described earlier or a response to the influence of a possible "hot spot" near the Tuzo Wilson Knolls (Chase, 1977). Davis and Riddihough (1982) suggest that the detachment of the Winona Block was the result of resistance to oblique subduction, and that the migration of the spreading centers was a response to this process.

The southern triple junction (Fig. 8), now at Cape Mendocino, is marked by the intersection of the San Andreas transform fault, the Mendocino FZ, and the convergence zone along the California–Oregon continental slope. The geometry of

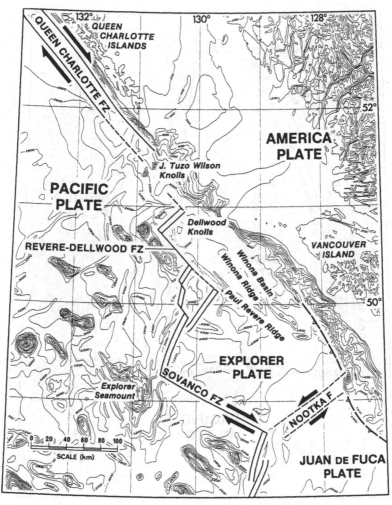

Fig. 7. Features and topography of the Explorer Plate and northern triple junction region. Motion across transform faults is shown by split arrows. (Modified from Riddihough *et al.*, 1984.)

the motion between the Pacific–America and Juan de Fuca (locally Gorda) Plates requires that this junction migrate northwards along the margin at approximately 50 km/m.y. The local tectonics of the region are again very complex (e.g., Fox, 1976), involving northwesterly oriented, right lateral strike slip faulting both inland and north of Cape Mendocino (Herd, 1978), and northeasterly, left lateral strike slip faulting within the southern Gorda Plate (Smith and Knapp, 1980). The orientation of magnetic anomalies along the Gorda Ridge shows that spreading of the Gorda Plate away from the Pacific Plate has been nonparallel to the Mendocino FZ since at least 5 m.y.b.p. This seems to have led to some subduction of the Gorda Plate

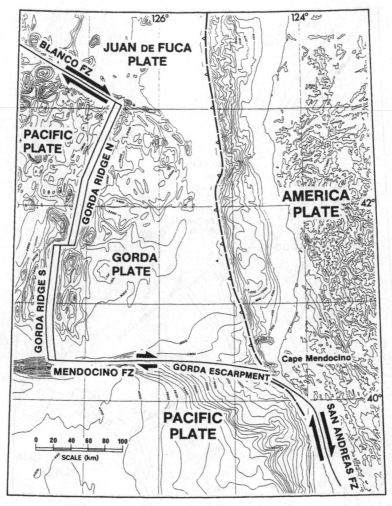

Fig. 8.   Features and topography of the Gorda Plate and southern triple junction. Motion across transforms is shown by split arrows. (Modified from Riddihough *et al.*. 1984.)

beneath the Mendocino FZ and is regarded as the principal factor behind the complex pattern of magnetic anomalies and contemporary deformation and seismicity within the southern Gorda Plate (Silver, 1971a; Couch, 1980; Carlson and Stoddard, 1981; Jachens and Griscom, 1983).

The southernmost part of the Gorda Ridge is now perpendicular to the Mendocino FZ and has slowed its spreading rate to less than 10 mm/yr (Atwater and Mudie, 1973; Riddihough, 1980) in the last 2 m.y. This may be part of a process whereby the triple junction migrates northwards.

VI. SEAMOUNTS AND HOT SPOTS

The influence of hot spots in the plate tectonics of the northeast Pacific is not clear. Nevertheless, the ocean floor west of the Juan de Fuca Ridge is characterized by a large number of seamount chains with a consistent northwesterly trend (Fig. 9). Many of the chains are short, may have been generated at the ridge, and are of a similar age to the seafloor on which they are now located (Barr, 1974b). Vogt (1981) and Vogt and Hey (1976) noted that apart from an orientation in conformity with the motion of the Pacific Plate over the mantle, these small seamount chains become progressively smaller along their length toward their youngest members. The authors suggested that they may be the product of a single pulse of magma as the overlying plate moved across the mantle. An association with complex conjugate spreading at the ridge has been suggested for the Endeavour Seamount (47°N) and the Heck Seamount Chain (Karsten *et al.*, 1986).

There are at least two major chains in the region which appear to be related to a fixed, long-lived, mantle plume, i.e., Cobb–Eickelberg Seamount Chain and the Kodiak–Bowie Seamount Chain. These stretch up to 1500 km northwestwards

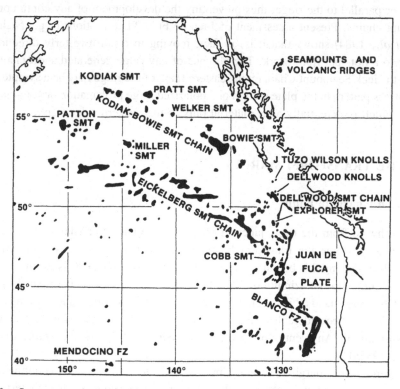

Fig. 9. Seamounts and volcanic ridges in the northeast Pacific Ocean. (From Riddihough *et al.*, 1984.)

across the Pacific Plate, young toward the Juan de Fuca Ridge system, and have formed the basis of Pacific Plate absolute motion determinations (e.g., Silver *et al.*, 1974; McDougall and Duncan, 1980). The present position of the hot spot in the Kodiak–Bowie Chain is close to the Queen Charlotte Islands and may relate to the Tuzo Wilson Knolls. Chase (1977) speculated that it could have influenced spreading at this center, the petrochemistry of the Tuzo Wilson lavas being characteristic of neither normal ocean ridge basalts nor hot spots. A reinterpretation by Turner *et al.* (1980) suggests, however, that Bowie Seamount is more likely to be the youngest member of the chain.

The influence of the Cobb "hot spot" on the geometry, geochemistry, and history of the Juan de Fuca Ridge was first suggested by Vogt and Byerly (1976). Detailed investigations now show that the ridge may be composed of a number of different segments and may have jumped westwards by 20–30 km toward Brown Bear Seamount, the easternmost member of the Eickelberg–Cobb Seamount Chain (Johnson *et al.*, 1982).

The absence of any significant seamount chains on the Juan de Fuca Plate remains a striking feature which lacks any satisfactory explanation. Silver *et al.* (1974) suggested that the absolute motion of this plate could be such that it moved toward or parallel to the ridge, thus preventing the development of any continuous seamount chains. Present assessment of Juan de Fuca Plate motions (e.g., Nishimura *et al.*, 1984) shows that it is, in fact, moving in a northwesterly direction relative to a hot-spot framework. The absence of any ridge-generated short chains similar to Heck Seamount Chain on the eastern flank of the ridge may be a symptom of the stress pattern in the plate system or a polarized ridge propagation or conjugate system which preferentially adds material only to the western ridge flank.

VII. THE TRANSFORM MARGIN

A. Queen Charlotte Fault

Northwest from the triple junction off northern Vancouver Island, the continental margin forms the transform boundary between the Pacific and America Plates. Motion along this boundary from Minster and Jordan (1978) is estimated to be 50–60 mm/yr at an azimuth of N 20°E. At least along the southern part of the Queen Charlotte Islands this is oblique to the principal topographic and bathymetric features. There is considerable historic seismicity along the margin with right lateral strike slip motion. An event of magnitude 8.0 occurred in 1949 (Milne *et al.*, 1978; Rogers, 1983a).

In general, the margin is strongly lineated with steep scarps and no continental rise (Fig. 10). Immediately offshore, the seafloor drops steeply to a narrow terrace at a depth of approximately 1500 m. The outer edge of the terrace is similarly linear

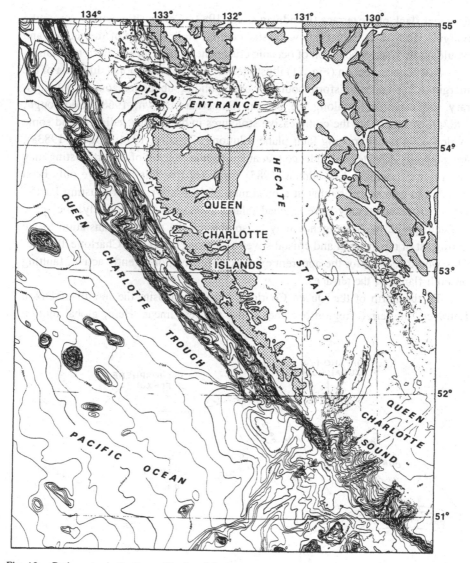

Fig. 10.   Bathymetry in the Queen Charlotte Islands region showing the Queen Charlotte terrace. (From Seemann, 1982.)

and drops steeply to depths of 2500–3000 m. A combination of seismic profiling, seismic refraction, heat flow, and gravity interpretation (Hyndman *et al.*, 1982; Horn *et al.*, 1984) suggests that the 20- to 30-km-wide terrace is composed of folded and indurated sediments of up to 5 km in thickness which overlie a strip of depressed oceanic crust. The magnetic anomalies of the Pacific Ocean floor (Currie *et al.*, 1983) terminate at the outer scarp of the terrace, although microseismicity indicates that, at least during the period of observation, faulting occurred only along the inner flank of the terrace (Hyndman *et al.*, 1982). Composite sections across the margin (Horn *et al.*, 1984) show crustal faulting on both sides of the terrace and would allow some subduction of oceanic crust beneath the Queen Charlotte Islands.

A schematic section (Fig. 11) illustrates the key interpretational dilemma of the margin—is it a pure transform boundary or an oblique subduction zone? Contemporary and recent uplift along the margin (Sutherland-Brown, 1968; Riddihough, 1982a), together with the compressive structures in the terrace sediment and some evidence for underthrusting fault-plane solutions (Chandra, 1974; Rogers, 1983a), are generally accepted as evidence that at least some element of underthrusting may be present. Yorath and Hyndman (1983) suggested that this may have only been initiated at 6 m.y.b.p. so that at a maximum of 10 mm/yr, underthrusting Pacific Ocean crust would only have reached inland for 60 km. Alternatively, the space problem posed by convergence may be soluble by the periodic translation to the northwest of the material and crustal sliver underlying the Queen Charlotte Terrace. This would account for the apparent existence of active, or recently active, faulting on both flanks of the terrace.

To the north of the Queen Charlotte Islands, opposite the mouth of Dixon Entrance, the terrace region becomes wider and less linear. Here, global motion

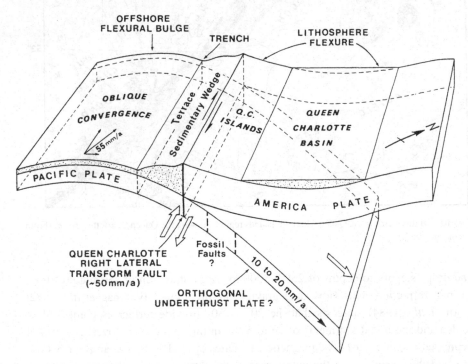

Fig. 11. Schematic diagram of convergence along the Queen Charlotte Islands margin. (Modified from Yorath and Hyndman, 1983.)

solutions suggest that there may be only a minor amount of compression, and perhaps even extension. Von Heune *et al.* (1978) interpret seismic profiles across the margin of Dixon Entrance as showing a depositional basin beneath the slope of up to 5 km of sediments, with ocean crust dipping toward the continent.

B. Geology of Queen Charlotte Islands

On land, the oldest rocks of the Queen Charlotte Islands are upper Triassic and include over 4 km of pillow lavas, breccias, and basalt flows (Karmutsen Formation), overlain by 1 km of Upper Triassic to Lower Jurassic limestones and argillites. This sequence has been assigned to the Wrangellia terrane (Jones *et al.*, 1982; Yorath and Chase, 1981), an allochthonous terrane which was formed many thousands of kilometers to the south (e.g., Yole and Irving, 1980) and which collided with the Alexander terrane, probably in late Jurassic times. A suture assemblage has been identified (Yorath and Chase, 1981) which includes conglomerates and intermediate to acidic plutons of Jurassic age, probably derived through anatexis of parts of the underlying Karmutsen Formation (Sutherland-Brown, 1968). Subsequent to the suturing of the two terranes, a mid to upper Cretaceous sequence of terrigenous sediments was laid down in a shelf to slope environment (Haida, Honna, and Skidegate Formations; Sutherland-Brown, 1968). The two terranes are estimated to have docked with the continental margin prior to the early Tertiary, a convergence which may be related to the plutonic uplift of the British Columbia Coast Mountains (Yorath and Chase, 1981).

The Tertiary period in the Queen Charlotte Islands is characterized as one of rifting with a major accumulation of subaerial volcanic flows and pyroclastic rocks comprising the Masset Formation. Nonmarine terrigenous clastic rocks were laid down over this formation during the Oligocene and Miocene. Most recently, Yorath and Hyndman (1983) postulate a period of uplift for the Queen Charlotte Islands which may be associated with flexural uplift due to the oblique convergence of the Pacific Plate. In many ways, the present environment of the Islands can be regarded as a continuation of 200 m.y. of complex accretion, convergence, suturing, and continental margin tectonics.

VIII. CONVERGENT MARGIN: VANCOUVER ISLAND, WASHINGTON, AND OREGON

The absence of a bathymetric trench along the foot of the continental slope beneath which the Juan de Fuca Plate is converging is a major difference between the Juan de Fuca Plate subduction zone and other subduction zones around the globe (Fig. 12). However, seismic profiling (e.g., McManus *et al.*, 1972; Barr, 1974a; Chase *et al.*, 1975), seismic refraction data (Ellis *et al.*, 1983), and gravity in-

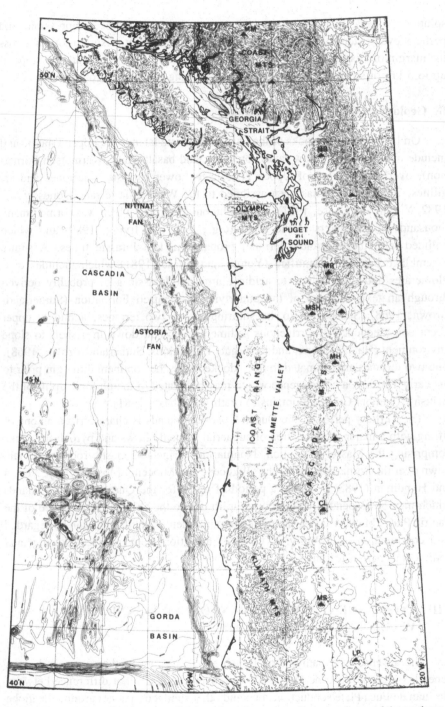

Fig. 12.   Topography and bathymetry of the continental margin and arc–trench gap of the northeast
Pacific Ocean. Notable Cascade volcanic centers from north to south are: MM, Meager Mountain; MB,
Mount Baker; MR, Mount Rainier; MSH, Mount St. Helens; MH, Mount Hood; CL, Crater Lake; MS,
Mount Shasta; LP, Lassen Peak. Contour interval: on land, 500 m; beneath sea, 200 m.

terpretations (e.g., Srivastava, 1973; Couch and Braman, 1979) show that the ocean crust does dip toward the margin at 5° or greater. High sedimentation rates (e.g., up to 5 km in 1 m.y. in the Winona Basin) in the Pliocene–Pleistocene may have filled any topographic depression that existed (Von Heune and Kulm, 1973; Kulm and Fowler, 1974). This, combined with the low rigidity of the young Juan de Fuca Plate, is thought to explain the absence of a marked trench.

Extending from the foot of the continental slope to the Juan de Fuca Ridge is a blanket of recent terrigenous sediments forming the Cascadia Basin and Nitinat and Astoria Fans (Kulm and Fowler, 1974). The ridge has acted as a sediment barrier, broken only by sea channels through the Sovanco Ridge (the Revere Channel), the Blanco FZ (Cascadia Channel), and across the southern part of the Gorda Ridge (Escanaba Trough). Apparently, sediment supply has been high enough to ensure that the Cascadia Basin has been filled to these overflow levels for much of its recent history.

The nature of accretion and deformation at the foot of the continental slope is generally in conformity with the concept of convergence. A number of detailed studies along the continental margin off Vancouver Island (Tiffin et al., 1972; Yorath, 1980), Washington (Silver, 1972; Carson et al., 1974; Barnard, 1978), Oregon (Kulm and Fowler, 1974; Seely, 1977; Kulm and Scheidegger, 1979), and California (Silver, 1971c) have shown compression, imbricate thrusting, and uplift in late Tertiary and younger sediments. The geometry of an anticlinal ridge at the foot of the slope which forms a deformation front was used by Carson et al. (1974) and Von Heune and Kulm (1973) to estimate convergence rates similar to those predicted from magnetic anomalies. However, calculations by Barnard (1978) suggest that either not all the horizontal convergence is shown in the overlying sediments or there has been a recent reduction in convergence rate.

Seismicity is minimal beneath the continental shelf and slope, and despite experiments to determine its nature, no thrusting events have been identified. It is possible that the underthrusting of the Juan de Fuca Plate may be predominantly aseismic (e.g., Hyndman and Weichert, 1983) and that the presence and supply of large amounts of sediments may be a contributory factor.

IX. THE ARC–TRENCH GAP

The region of the continent that lies between the coast and the Cascade volcanic belt (Fig. 12) is characterized by general seismicity, with events up to magnitude 7.3 (Fig. 3) (Crosson, 1972, 1983; Chandra, 1974; Milne et al., 1978). Detailed hypocenter determinations have shown that this seismicity falls into two suites, that associated with the downgoing Juan de Fuca Plate at depths up to 90 km, and that occurring within the overlying crust.

The absence of an easily identifiable Benioff zone of earthquakes extending to hundreds of kilometers beneath the Pacific Northwest was at one time advanced as a

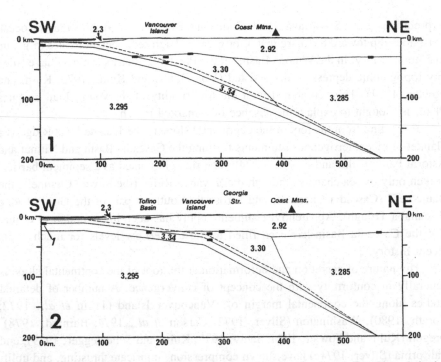

Fig. 13.   Vertical sections (no vertical exaggeration) across the descending Juan de Fuca Plate, derived
from gravity data. Sections are: (1) central Vancouver Island; (2) southern Vancouver Island; (3)
Olympic Peninsula and Puget Sound; (4) Washington coast north of the Columbia River. Numbers are
densities in g/cm^3; bars represent seismic control. (Modified from Riddihough, 1979.)

significant argument against the existence of a subducted slab. However, as first
suggested by Atwater (1970), and later quantified by Riddihough and Hyndman
(1977), the absence of earthquakes can be ascribed to the rapid reheating of the
young Juan de Fuca Plate as it descends relatively slowly into the mantle. Some
recent hypocenters have been located as deep as 90 km (Taber and Smith, 1984),
but in general, the prediction from the observation of Deffeyes (1972) from global
studies that 70 km should be the maximum depth seems to hold true. The geometry
of the dipping slab has been investigated both by hypocenter locations (Smith and
Knapp, 1980; Crosson, 1976, 1983; Michaelson and Weaver, 1983) and by travel
time delays and teleseisms (Langston, 1981; Michaelson, 1983). In general terms, it
seems to be characterized by an upper, shallower-dipping section which reaches a
depth of 40–60 km, a "knee bend," and then a deeper, steeper-dipping section
which continues down to an approximate depth of 100–120 km beneath the volcanic
arc. Dips increase along the arc–trench gap from north to south (Riddihough,
1979), with possible segmentation in the descending plate (Hughes et al., 1980;
Michaelson, 1983, 1984) and variations in arc–trench distance (Fig. 13).
      Seismicity within the downgoing slab has generally been identified as ten-

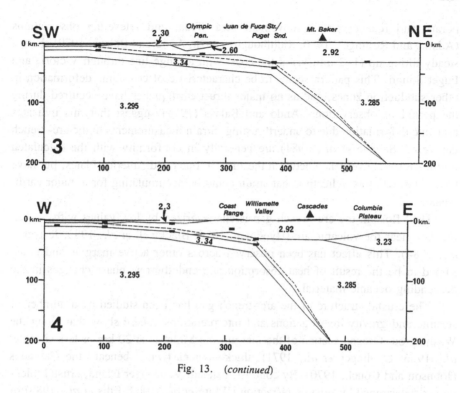

Fig. 13. (continued)

sional, with some events showing eastward-dipping normal faulting (McKenzie and Julian, 1971; Chandra, 1974). Rogers (1983a) notes that slab seismicity is concentrated near the "knee bend" between the shallower- and deeper-dipping plate segments and may be associated with phase changes occurring in the descending oceanic crust and mantle. The observed rate of seismic moment is comparable with the magnitude of such phase changes, the thickness of the Juan de Fuca Plate, and its rate of descent into the mantle.

    Seismicity in the overlying crust of the arc–trench gap is principally concentrated in the Puget Sound area (Milne et al., 1978), with the northern limit near Texada Island, B.C. (48°30′N), and a southern limit near Portland, Oregon (45°N). Fault plane solutions (Crosson, 1972; Chandra, 1974; Milne et al., 1978; Rogers, 1979a,b) are dominantly northwest–southeast, right-lateral strike slip, interpreted as the result of north–south compression. Recent detailed epicenter locations have identified at least one north-northwest–south-southwest feature which is active in the Puget Sound region (Weaver and Smith, 1983). Association with the oblique northeasterly convergence of the Juan de Fuca Plate beneath the north–south Washington margin has been suggested by both Rogers (1983b) and Weaver and Smith (1983), and seems to be in accord with the concepts of Fitch (1972) and Dewey (1980).

    The aseismic nature of Juan de Fuca Plate subduction is supported by contemporary and recent vertical movement data. Tidal and releveling observations (Adams and Reilinger, 1980; Riddihough, 1982a; Ando and Balazs, 1979) show a steady tilting up (1–2 mm/yr) at the coast with a hinge line through Victoria and Puget Sound. This pattern seems to be characteristic of coseismic deformation in other subduction zones, and as no major thrust earthquakes have occurred during the period of observations, Ando and Balazs (1979) suggest that this indicates aseismic deformation due to underthrusting. Strain measurements in the arc–trench gap (e.g., Savage *et al.*, 1981) are generally in conformity with the calculated northeasterly convergence between the Juan de Fuca and America Plates, but have been interpreted as indicating that strain could be accumulating for a major earthquake.

Heat flow in the arc–trench gap is characterized by low values with a steep increase near the volcanic arc (Hyndman, 1976; Blackwell *et al.*, 1982; Jessop *et al.*, 1984). This effect has been observed across other active margins and is assumed to be the result of heat absorption and endothermic phase changes in the descending oceanic material.

The crustal structure of the arc–trench gap has been studied by a number of seismic and gravity investigations and interpretations. These show that along the Washington–Oregon section of the margin, the Moho is at 20 km or less (Berg *et al.*, 1966; Dehlinger *et al.*, 1971), thickening eastwards beneath the Cascades (Johnson and Couch, 1970). By contrast, beneath Vancouver Island, crustal thicknesses determined by seismic refraction (White *et al.*, 1968; Ellis *et al.*, 1983) of 30–40 km are in conflict with gravity interpretations (Dehlinger *et al.*, 1971; Stacey, 1973; Riddihough, 1979) which indicate mantle densities occurring at depth as shallow as 25 km. Riddihough (1979) suggested that low-velocity high-density material may occur over the downgoing subducting slab.

Between the Coast Ranges and the Cascade volcanic arc, there is a characteristic forearc depression. In British Columbia it is occupied by Georgia Strait, but is continuous to the south through Puget Sound and the Willamette Valley. In the latter areas, it is occupied by up to 10 km of sediments, ranging in age from Miocene to Recent. Rogers (1983a) speculates that the origin of this depression is associated with the phase changes in the downgoing slab.

Geology of Arc–Trench Region

The oldest rocks exposed on Vancouver Island are seen only in the south and are metamorphic schists and gneisses which may be as old as the Precambrian. They are succeeded by 3 km of late Paleozoic submarine lava flows with interbedded deep-water marine sediments (the Sicker Group; Muller, 1977). In the middle Triassic, a similar thickness of submarine basalts was accumulated (the Karmutsen Formation, also seen on the Queen Charlotte Islands). The late Triassic to Jurassic periods saw the accumulation of marine limestones, followed by a more violent volcanic episode which produced the andesitic breccias and tuffs, interbedded with graywacke, of the 4-km-thick Bonanza Formation. Up to this time, the structure and history of Vancouver Island was thus largely oceanic or island arc in tectonic style, and it has been assigned to the Wrangellia terrane. As discussed for the Queen Charlotte Islands, this terrane was formed to the south and collided with North America in Cretaceous time.

The collision has been related to the uplift and intrusion of granitic batholiths (middle Jurassic) as part of the emplacement of the Coast Mountains crystalline complex. On Vancouver Island, the erosion products of this orogeny accumulated in the marine sandstones and shales of the Cretaceous Nanimo Group. The Tertiary period on Vancouver Island is represented by submarine basalts and marine sediments now exposed along the south and west coasts.

To the south, in the Coast Ranges of Washington and Oregon, there are virtually no rocks older than Tertiary, a situation thought to be explicable by the fact that they occupy a major embayment in the pre-Tertiary continental margin (the Columbia embayment; e.g., Dickinson, 1976). The rocks have been characterized as a seamount province and consist of exceptionally thick sequences of Eocene submarine basalts. Pillow structures are widespread, and marine fossils occur in the interbedded sediments. Higher flows may be subaerial, and a number of features suggest comparisons with the Hawaiian Islands and seamount chain (Duncan, 1982). Overlying these lavas are marine sediments probably derived from the surrounding land areas. The basin in which they accumulated had, nevertheless, largely disappeared by the Miocene epoch and the time of the eruption of the Columbia River basalt. By this time, the coastline was probably close to its present position. In the Olympic Peninsula, younger marine sedimentary rocks and basalts (the Soleduck Formation) are thrust beneath the Eocene Crescent Volcanics in an older example of the accretion and convergence tectonics seen along the margin today (McKee, 1972).

In southern Oregon and northern California, the arc–trench gap is occupied by the Klamath Mountains. Their geology is complex and in considerable contrast to the Coast Ranges to the north. In California, up to 7 km of Paleozoic and 5 km of Lower Mesozoic eugeosynclinal strata occur. In the core of the mountains, the sequences have been highly deformed and metamorphosed, and there is a series of late Jurassic dioritic and granodioritic plutons. Cenozoic rocks are sparse. Irwin (1981) regards the region as consisting of a series of tectonic slices or fragments of oceanic crust and island arc that accreted in sequence from east to west.

X. THE VOLCANIC ARC

The Quaternary and late Tertiary volcanic arc of the High Cascades which is associated with the subduction of the Juan de Fuca Plate stretches from Meager

Mountain in British Columbia to Lassen Peak in California. It has passed through a number of surges and declines, but continues to be active into the present. Seven of the volcanoes have been active since 1800, the most recent eruptions being at Lassen Peak (1914–1917) and Mt. St. Helens (1980). After its initial explosive phase (USGS, 1982), the Mt. St. Helens eruption is continuing with lava dome building in its central crater (e.g., Cashman and Taggart, 1983). The petrochemistry of the arc is characterized as calc-alkaline and typical of Benioff zone magmatism (Souther, 1977); however, the proportion of andesitic to basaltic materials seems to have steadily decreased over the last 40 m.y. McBirney (1978) asserts that analyses of composition and trace elements in the rocks point to a primary mantle origin rather than origin from the melting of subducted material.

The Cascades south of the Columbia River in Oregon have been shown (Couch *et al.*, 1982; Blakely *et al.*, 1985) to lie in a downfaulted graben structure. In Washington, the present High Cascades are constructed on a basement of an older Cascade arc, and no graben structure has been detected. This contrast in the arc tectonics may be due to differences in style of convergence between the two areas (Rogers, 1983a): segmentation in the downgoing plate (Hughes *et al.*, 1980), differing basement geology, or the influence of Basin and Range extension.

XI. REGIONAL PLATE TECTONIC HISTORY

As discussed in the Introduction, the regional history of the northeast Pacific and the adjacent continent is in many ways a continuation into the past of the tectonic diversity that is evident today—a margin that ranges from transform to convergent, with complex offshore plate systems and a continual history of rapid changes in both time and place. The elucidation of this history has been the subject of many investigations. Its most important aspect, the reconciliation of offshore plate tectonics with onshore geology, is still at the stage where there are considerably more questions than answers.

The major problem in reconstructing plate tectonic geometry for the western margin of North America is that for much of the past 200 m.y. it has been the site of plate consumption and subduction. There is thus little or no direct evidence for the nature and geometry of the plates which interacted with the margin during this period. Estimation of their positions, size, and motions has to be based on reconstructions using the magnetic reversal record of the remaining Pacific Plate. Such reconstructions have to be based on the critical assumption of symmetrical seafloor spreading, an assumption which presumes that spreading rates and directions between two plates can be calculated from the distances between dated magnetic anomalies on one plate and the orientation of appropriate fossil fracture zones.

The Pacific Ocean basin has generally been the site of three plates—the Pacific Plate (now occupying most of the basin), the Farallon Plate (of which the Juan de

Fuca Plate is a remnant), and the Kula Plate (now almost completely disappeared). The history of the basin for the last 120 m.y. is summarized in Fig. 14, which shows how North America, moving westwards, has progressively overridden most of the Farallon Plate and, eventually, parts of the Pacific–Farallon spreading ridge. This simplified history (from Riddihough, 1982b) is based on the reconstructions of Atwater (1970), Atwater and Molnar (1973), Coney (1977), and Stone (1977).

The more detailed reconstruction that is needed for a close examination of the northeast Pacific during this period depends on a number of factors and has been attempted by a number of methods. While the magnetic anomaly pattern allows a comprehensive history of Pacific–Farallon and Pacific–Kula movements to be reconstructed, the critical information needed to determine interactions along the

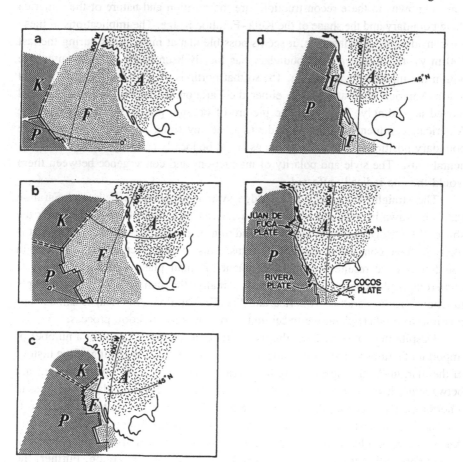

Fig. 14. Plate tectonic reconstructions in the eastern Pacific Ocean from 120 m.y.b.p. to the present ((a) 120 m.y.b.p.; (b) 80 m.y.b.p.: (c) 40 m.y.b.p.; (d) 20 m.y.b.p.; (e) 0 m.y.b.p.). A, North American Plate; F, Farallon Plate; K, Kula Plate; P, Pacific Plate. Principal sources: Coney (1977), Atwater (1970), Atwater and Molnar (1973), and Stone (1977). (Reproduced by permission of Geoscience Canada.)

margin is the motion of the Pacific Plate relative to the America Plate. Atwater and Molnar (1973) traced spreading between the Pacific–Antarctic–Indian–African–America Plates to determine this. Coney (1971, 1977) used the motion of the Pacific and African Plates over the ''hot-spot'' framework connected through mid-Atlantic, Africa–America spreading. Paleomagnetic data, though sparse for the predominantly oceanic Pacific Plate, can also be used (e.g., Irving, 1981).

The combined history of plate interactions for the northeast Pacific margin for the last 100 m.y. is summarized in Fig. 15, which attempts to show a generalized version of the consensus of recent reconstructions, with particular attention to the detailed work of Engebretson (1982). Directions of motion shown are no more accurate than ± 20°, and rates of motion vary as shown. Some of the most important unknowns in these reconstructions are the position and nature of the America Plate boundary and the shape of the Kula–Farallon Ridge. The implications of these are considerable. For instance, it seems possible that at many times during the last 100 m.y., the America Plate boundary has locally been located almost anywhere within the unshaded zone in Fig. 15, so that portions of what are now continental North America were attached to either the Farallon, Kula, or Pacific Plates and moved accordingly. The simplistic picture of oceanic plates converging with the American continent is also misleading, as at any time, depending on the plate boundary positions, each of these plates could be locally carrying oceanic or continental crust. The style and polarity of interactions and convergence between them would thus be radically affected.

The straight and simple lineal ridge system between the Kula and Farallon Plates is drawn in most constructions as a convenience. In fact, all evidence for the shape of the ridge has been destroyed. Judging by the Pacific–Farallon Ridge, it is likely to have contained a number of transforms. Its intersection with the margin could thus have resulted in the detachment of small plates and the presence of numerous triple junctions, as in Fig. 16. Finally, the stability of the major triple junctions (Kula–Farallon–America and Pacific–Farallon–America) is almost impossible to predict, given the uncertainties of the reconstruction process.

Despite these reservations, the reconstructions of Fig. 15 contain a number of important features which are essential to any geological understanding of the history of the margin. With respect to North America, all the ''oceanic'' plate motions are between northwest and northeast. More specifically, the Farallon Plate converges in a northeasterly direction, the Kula Plate either almost northerly or to the northwest, and the Pacific Plate in a northwesterly direction. If the orientation of the North American margin lies between north and northwest, interactions thus vary between almost perpendicular convergence and right-lateral strike slip motion. Further, the speeds of motion are high enough that material on the oceanic plates could move up to 150 km/m.y. relative to North America. One other important feature is that the motion of the Farallon Plate relative to North America slowed dramatically about 40 m.y. ago. This was the time when the Pacific Plate changed its direction of motion

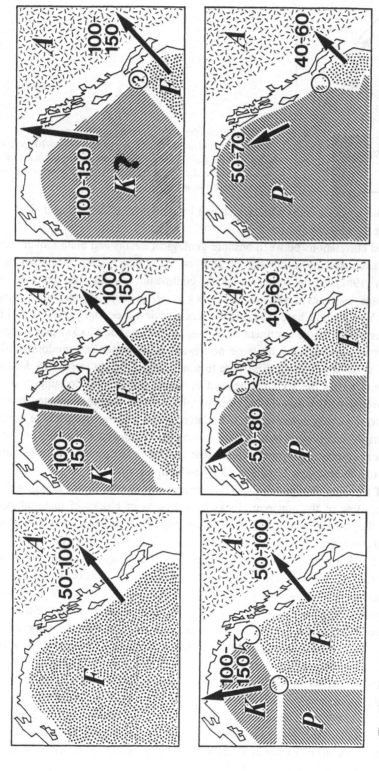

Fig. 15. Plate interactions off western Canada and southeast Alaska during the last 100 m.y. from a consensus of sources (as in Fig. 14). Arrows and numbers represent movement relative to America Plate in mm/m.y.b.p. or km/m.y.b.p. Open circles are triple junction positions with movement where suggested. The independent existence of the Kula Plate beyond 50 m.y.b.p. has been questioned. (Reproduced by permission of Geoscience Canada.)

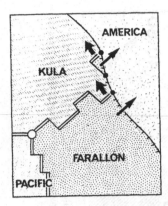

Fig. 16. Illustration of the complexity of interactions of the American margin that could be produced between 100 and 40 m.y.b.p. if the Kula–Farallon ridge were segmented in a similar manner to the Pacific–Farallon ridge. (Reproduced by permission of Geoscience Canada.)

relative to the hot-spot framework (as shown by the bend in the Emperor–Hawaii Seamount at 42 m.y. ago).

Detailed correlations between onshore geological events and plate movements have been attempted by a number of authors, particularly Carlson (1976) and Engebretson (1982) for the northeast Pacific continental margin. The general picture of collage tectonics developed by Coney et al. (1980), Monger and Irving (1980), and Yorath and Chase (1981) (for a recent review, see Jones et al., 1982) is of the western edge of the continent as an assembly of exotic blocks, pieces, and slivers of crust characterized by large northward movements and predominantly right lateral strike slip faulting. The plate tectonic picture shown in Fig. 15 does in fact represent exactly the style of regional movements necessary to support this. Material moving northwards on the oceanic plates could become accreted to the margin by northeasterly convergence to be later faulted and translated to the northwest as a triple junction moves along the margin. Throughout the process, the detailed, local complexity of interactions that is seen in the present plate tectonic framework of the northeast Pacific is likely to be applicable.

One of the most fascinating aspects of recent investigations of the region has been the detailing of the apparent clockwise rotations of blocks of material now exposed in the Washington and Oregon Coast Range and the Cascades. By comparison of paleomagnetic directions in these rocks with similar aged material to the east and on the North American craton, a number of researchers (e.g., Simpson and Cox, 1977; Beck, 1980; Bates et al., 1981; Magill et al., 1982) have shown that for material accreted within the last 100 m.y., there has been a consistent clockwise rotation. Rotations of up to 2°/m.y. back to at least Miocene time seem to be implied, and individual blocks of material (as small as a hundred kilometers in extent) may have rotated independently. Attempts at reconstructing the accretion history (e.g., Simpson and Cox, 1977; Duncan, 1982) or proposing a more general strain system which would result in the observed rotations (Beck, 1976) have all relied on the northeasterly, partially oblique convergence that is implied in the history shown in Fig. 15. The details of the process are certainly not as yet understood; in particular, the detachment faults bounding the rotated blocks have not generally been identified or are apparently covered by undisturbed sediments.

XII. SUMMARY COMMENTS

The northeast Pacific is an area which was germinal in developing some of the key concepts of plate tectonics. Continuing investigations of the plates involved and their complex boundaries have revealed examples of almost all aspects of the plate tectonic process, from the volcanism of the Cascades, the conjugate ridges and propagating rifts of the Juan de Fuca Ridge, to the lineated offshore terrace of the Queen Charlotte Fault. The complexity of the tectonic environments that can be seen within this small area is probably typical of plate tectonics in action and shows clearly that the diversity of environments observed over short distances within the geological record is a normal consequence of the processes involved. Of importance in the understanding of plate tectonics is the demonstration in the northeast Pacific that "feedback" between the subduction zone and the ridge, resistance to convergence, and the influence of adjacent plates all affect the plate motions observed.

The longer history of the region has been predominantly one of plate convergence with an almost continuous record of a spreading ridge system intersecting the margin and of migrating triple junctions. While the northerly plate movements relative to North America are exactly in conformity with the growing body of evidence from paleomagnetic studies, the local complexity shown by the present regime in the northeast Pacific must be considered as a probable analogue of past conditions which applied throughout this history.

REFERENCES

Adams, J., and Reilinger, R., 1980, Time behavior of vertical crustal movements measured by releveling in North America; a geologic perspective, Proceedings, Second International Symposium on Problems related to the redefinition of the North American Vertical Geodetic Networks (NAD), Can. Inst. Surveying, Ottawa, pp. 327–339.

Ando, M., and Balazs, E. I., 1979, Geodetic evidence for aseismic subduction of the Juan de Fuca Plate, J. Geophys. Res. 84:3023–3028.

Atwater, T., 1970, Implications of plate tectonics for the Cenozoic tectonic evolution of western North America, Geol. Soc. Am. Bull, 81:3513–3536.

Atwater, T., and Menard, H. W., 1970, Magnetic lineations in the northeast Pacific, Earth Planet Sci. Lett. 7:445–450.

Atwater, T., and Molnar, P., 1973, Relative motion of the Pacific and North American Plates deduced from sea-floor spreading in the Atlantic, Indian, and South Pacific Oceans, Proceedings, Conference on Tectonic Problems of the San Andreas Fault System, Stanford University, California, pp. 136–148.

Atwater, T., and Mudie, J. D., 1973, Detailed near-bottom geophysical study of the Gorda Rise, J. Geophys. Res. 78:8665–8686.

Barnard, W. D., 1978, The Washington continental slope: Quaternary tectonics and sedimentation, Mar. Geol. 27:79–114.

Barr, S. M., 1974a, Structure and tectonics of the continental slope west of Vancouver Island, Can. J. Earth Sci. 11:1187–1199.

Barr, S. M., 1974b, Seamount chains formed near the crust of the Juan de Fuca Ridge, northeast Pacific Ocean, Mar. Geol. 17:1–19.

Barr, S. M., and Chase, R. L., 1974, Geology of the northern end of Juan de Fuca Ridge and sea-floor spreading, Can. J. Earth Sci. 11:1384–1406.

Bates, R. G., Beck, M. E., and Burmester, R. F., 1981, Tectonic rotations in the Cascade Range of southern Washington, *Geology* **9**:184–189.

Beck, M. E., Jr., 1976, Discordant paleomagnetic pole positions as evidence of regional shear in the western Cordillera of North America, *Am. J. Sci.* **276**:694–712.

Beck, M. E., Jr., 1980, Paleomagnetic record of plate-margin tectonic processes along the western edge of North America, *J. Geophys. Res.* **85**:7115–7131.

Berg, J. W., Trembly, L., Emilia, D. A., Hutt, J. R., King, J. M., Long, L. T., McKnight, W. R., Sarmah, S. K., Souders, R., Thiruvathukal, J. V., and Vossler, D. A., 1966, Crustal refraction profile, Oregon Coast Range, *Bull. Seismol. Soc. Am.* **56**:1357–1362.

Blackwell, D. D., Bowen, R. G., Hull, D. A., Riccio, J., and Steele, J. L., 1982, Heat flow, arc volcanism and subduction in northern Oregon, *J. Geophys. Res.* **87**:8735–8754.

Blakely, R. J., Jachens, R. C., Simpson, R. W., and Couch, R. W., 1985, Tectonic setting of the southern Cascade Range as interpreted from its magnetic and gravity fields, *Geol. Soc. Am. Bull.* **96**:43–48.

Carlson, R. L., 1976, Cenozoic plate convergence in the vicinity of the Pacific Northwest: A synthesis and assessment of plate tectonics in the northeastern Pacific, Ph.D. thesis, University of Washington, Seattle.

Carlson, R. L., and Stoddard, P. R., 1981, Deformation of the Gorda Plate: A kinematic view, *EOS Trans. Am. Geophys. Union* **62**:1035.

Carson, B. J., Juan, P. B., Myers, P. B., and Barnard, W. D., 1974, Initial deep-sea sediment deformation at the base of the Washington continental slope: A response to subduction, *Geology* **3**:561–564.

Cashman, K. V., and Taggart, J. E., 1983, Petrologic monitoring of 1981 and 1982 eruptive products from Mount St. Helens, *Science* **221**:1385–1387.

Chandra, U., 1974, Seismicity, earthquake mechanisms and tectonics along the western coast of North America from 42°N to 61°N, *Bull. Seismol. Soc. Am.* **64**:5129–1549.

Chase, R. L., 1977, J. Tuzo Wilson Knolls: Canadian hot spot, *Nature* **266**:344–346.

Chase, R. L., Tiffin, D. L., and Murray, J. W., 1975, The western Canadian Continental Margin, *Can. Soc. Petrol. Geol. Mem.* **4**:701–722.

Coney, P. J., 1971, Cordilleran tectonic transitions and motion of the North American Plate, *Nature* **233**:462–465.

Coney, P. J., 1977, Mesozoic–Cenozoic Cordilleran plate tectonics, in: *Cenozoic Tectonics and Regional Geophysics of the Western Cordillera*, (R. B. Smith and G. P. Eaton, eds.), *Geol. Soc. Am. Mem.* **152**:33–50.

Coney, P. J., Jones, D. L., and Monger, J. W. H., 1980, Cordilleran suspect terranes, *Nature* **288**:329–333.

Couch, R. W., 1980, Seismicity and crustal structure near the north end of the San Andreas Fault system, in: *Studies of the San Andreas Fault Zone in northern California*, California Division of Mines and Geology Special Report 140, pp. 139–151.

Couch, R. W., and Braman, D., 1979, Geology of the continental margin near Florence, Oregon, *Oreg. Geol.* **41**:171–179.

Couch, R. W., Pitts, G. S., Gemperle, M., Braman, D. E., and Veen, C. A., 1982, Gravity anomalies in the Cascade Range in Oregon: Structural and thermal implications, Open File Report 0-82-9, Department of Geology and Mineral Resources, Portland, Oreg.

Crosson, R. S., 1972, Small earthquakes, structure and tectonics of the Puget Sound Region, *Bull. Seismol. Soc. Am.* **62**:1133–1171.

Crosson, R. S., 1976, Crustal structure modeling of earthquake data. 2. Velocity structure of the Puget Sound Region, Washington, *J. Geophys. Res.* **81**:3047–3054.

Crosson, R. S., 1983, Review of seismicity in the Puget Sound Region from 1970 through 1978, in: *Earthquake Hazards of the Puget Sound region, Washington*, Proceedings of Workshop IV, U.S. Geological Survey Open File Report 83-19, pp. 6–18.

Currie, R. G., Cooper, R. V., Riddihough, R. P., and Seemann, D. A., 1983, Multiparameter geophysical surveys off the west coast of Canada: 1973–1982, in: Current Research, Part A, Paper 83-1A, Geological Survey of Canada, pp. 207–212.

Davis, E. E., and Riddihough, R. P., 1982, The Winona Basin: Structure and tectonics, *Can. J. Earth Sci.* **19**:767–788.

Davis, E. E., and Lister, C. R. B., 1977, Tectonic structures on the Juan de Fuca Ridge, *Geol. Soc. Am. Bull.*, **88**:346–363.

Davis, E. E., Lister, C. R. B., Wade, U. S., and Hyndman, R. D., 1980, Detailed heat flow measurements over the Juan de Fuca ridge system and implications for the evolution of hydrothermal circulation in young ocean crust, *J. Geophys. Res.* **85**:299–310.

Deffeyes, K. S., 1972, Plume convection with an upper mantle temperature inversion, *Nature* **240**:539–544.

Dehlinger, P., Couch, R. W., McManus, D. A., and Gemperle, M., 1971, Northeast Pacific structure, in: *The Sea,* Vol. 4, Part II (A. E. Maxwell, ed.), Wiley, New York, pp. 133–189.

Dewey, J. F., 1980, Episodicity, sequence and style at convergent plate boundaries, *Geol. Assoc. Can. Spec. Pap.* **20**:553–573.

Dickinson, W. R., 1976, Sedimentary basins developed during the evolution of the Mesozoic–Cenozoic arc-trench system in North America, *Can. J. Earth Sci.* **13**:1268–1287.

Duncan, R. A., 1982, A captured island chain in the Coast Range of Oregon and Washington, *J. Geophys. Res.* **87**:10,827–10,837.

Ellis, R. M., Spence, G. D., Clowes, R. M., Waldron, D. A., Jones, I. F., Green, A. G., Forsyth, D. A., Mair, J. A., Berry, M. J., Mereau, R. F., Kanasewich, E. R., Cumming, G. L., Hajnal, Z., Hyndman, R. D., McMechan, G. A., and Loncarevic, B. D., 1983, The Vancouver Island seismic project: A co-crust onshore-offshore study of a convergent margin, *Can. J. Earth Sci.* **20**:719–741.

Engebretson, D. C., Cox, A., and Gordon, R. G., 1982, Relative motions between oceanic and continental plates in the Pacific Basin, Ph.D. thesis, Stanford University, California.

Fitch, J. J., 1972, Plate convergence, transcurrent faults and internal deformation adjacent to southeast Asia and the western Pacific, *J. Geophys. Res.* **77**:4432–4460.

Fox, K. F., 1976, Melanges in the Franciscan Complex, a product of triple-junction tectonics, *Geology* **4**:737–740.

Herd, D. G., 1978, Intracontinental plate boundary east of Cape Mendocino, California, *Geology* **6**:721–725.

Hey, R., 1977, A new class of "pseudofaults" and their bearing on plate tectonics: A propagating rift model, *Earth Planet Sci. Lett.* **37**:321–325.

Hey, R. N., and Wilson, D. S., 1982, Propagating rift explanation for the tectonic evolution of the northeast Pacific—The pseudomovie, *Earth Planet Sci. Lett.* **58**:167–188.

Horn, J. R., Clowes, R. M., Ellis, R. M., and Bird, D. N., 1984, The seismic structure across an active oceanic/continental transform fault zone, *J. Geophys. Res.* **89**:3107–3120.

Hughes, J. M., Stoiber, R. E., and Carr, M. J., 1980, Segmentation of the Cascade volcanic chain, *Geology* **8**:15–17.

Hyndman, R. D., 1976, Heat flow measurements in the inlets of southwestern British Columbia, *J. Geophys. Res.* **81**:337–349.

Hyndman, R. D., and Weichert, D. H., 1983, Seismicity and rates of relative motion on the plate boundaries of western North America, *Geophys. J. R. Astron. Soc.* **72**:59–82.

Hyndman, R. D., Rogers, G. C., Bone, M. N., Lister, C. R. B., Wade, U. S., Barrett, D. L., Davis, E. E., Lewis, T., Lynch, S., and Seemann, D., 1978, Geophysical measurements in the region of the Explorer Ridge off western Canada, *Can. J. Earth Sci.* **15**:1508–1525.

Hyndman, R. D., Riddihough, R. P., and Herzer, R., 1979, The Nootka Fault Zone—A new plate boundary off western Canada, *Geophys. J. R. Astron. Soc.* **58**:667–683.

Hyndman, R. D., Lewis, J. J., Wright, J. A., Burgess, M., Chapman, D. S., and Yamano, M., 1982, Queen Charlotte fault zone: Heat flow measurements, *Can. J. Earth Sci.* **19**:1657–1669.

Irving, E., 1981, Phanerozoic continental drift, *Phys. Earth Planet Inter.* **24**:197–2041.

Irwin, W. P., 1981, Tectonic accretion of the Klamath Mountains, in: *The Geotectonic Development of California,* Rubey Vol. I (W. G. Ernst, ed.), Prentice-Hall, Englewood Cliffs, N.J., pp. 29–49.

Jachens, R. C., and Griscom, A., 1983, Three-dimensional geometry of the Gorda Plate beneath northern California, *J. Geophys. Res.* **88**:9375–9392.

Jessop, A. M., Lewis, T. J., Judge, A. S., Taylor, A. E., and Drury, M. J., 1984, Terrestrial heat flow in Canada, *Tectonophysics* **103**:239–261.

Johnson, H. P., Vance, T. C., Delaney, J. R., and Karsten, J. L., 1982, Characterization of the symmetrical and offset portions of the Juan de Fuca Ridge between 44°N and 47°N, *Trans. Am. Geophys. Union (EOS)* **63**:1146.

Johnson, H. P., Karsten, J. L., Delaney, J. R., Davis, E. E., Currie, R. G., and Chase, R. L., 1983, A detailed study of the Cobb offset of the Juan de Fuca Ridge: Evolution of a propagating rift, *J. Geophys. Res.* **88**:2297–2315.

Johnson, S. H., and Couch, R. W., 1970, Crustal structure in the North Cascade Mountains of Washington and British Columbia from seismic refraction measurements, *Bull. Seismol. Soc. Am.* **60**:1259–1269.

Jones, D. L., Cox, A., Coney, P., and Beck, M., 1982, The growth of western North America, *Sci. Am.* **247**(5):70–84.

Karsten, J. L., Hammond, S., Davis, E. E., and Currie, R. G., 1986, Detailed geomorphology and neotectonics of the Endeavour segment of the Juan de Fuca Ridge: evolution of a propagating rift, *Bull. Geol. Soc. Am.* **97**:213–221.

Kulm, L. D., and Fowler, G. A., 1974, Cenozoic sedimentary framework of the Gorda–Juan de Fuca Plate and adjacent continental margin—A review, *Soc. Econ. Paleontol. Mineral. Spec. Publ.* **19**:212–229.

Kulm, L. D., and Scheidegger, K. F., 1979, Quaternary sedimentation on the tectonically active Oregon continental slope, *Soc. Econ. Paleontol. Mineral. Special Publ.* **27**:247–263.

Langston, C. A., 1981, Evidence for the subducting lithosphere under Vancouver Island and western Oregon from teleseismic P wave conversions, *J. Geophys. Res.* **86**:3857–3866.

Lister, C. R. B., 1977, Qualitative models of spreading center processes including hydrothermal penetration, *Tectonophysics* **37**:203–218.

McBirney, A. R., 1978, Volcanic evolution of the Cascade Range, *Annu. Rev. Earth Planet. Sci.* **6**:437–456.

McDougall, I., and Duncan, R. A., 1980, Linear volcanic chains–recording plate motions? *Tectonophysics,* **63**:275–295.

McKee, B., 1972, *Cascadia: The Geologic Evolution of the Pacific Northwest,* McGraw–Hill, New York.

McKenzie, D. P., and Julian, B., 1971, The Puget Sound, Washington, earthquake and the mantle structure beneath the northwestern U.S., *Geol. Soc. Am. Bull.* **82**:3519–3524.

McKenzie, D. P., and Parker, R. L., 1967, The North Pacific: An example of tectonics on a sphere, *Nature* **216**:1276–1280.

McManus, D. A., Holmes, M. L., Carson, B., and Barr, S. M., 1972, Late Quaternary tectonics, northern end of Juan de Fuca Ridge, *Mar. Geol.* **12**:141–164.

Magill, J. R., Wells, R. E., Simpson, R. W., and Cox, A. V., 1982, Post 12 my rotation of southwest Washington, *J. Geophys. Res.* **87**:3761–3776.

Malahoff, A., Hammond, S. R., Embley, R. W., Curry, R. G., Davis, E. E., Riddihough, R. P., and Sawyer, B. S., 1984, Juan de Fuca Ridge Atlas: Preliminary Seabeam Bathymetry, Earth Physics Branch, Open File 84-6, Department of Energy, Mines & Resources, Ottawa, Canada.

Menard, H. W., 1978, Fragmentation of the Farallon Plate by pivoting subduction, *J. Geol.* **86**:99–110.

Michaelson, C. A., 1983, Three-dimensional velocity structure of the crust and upper mantle in Washington and northern Oregon, M.Sc. thesis, University of Washington, Seattle.

Michaelson, C. A., and Weaver, C. S., 1984, Upper mantle structure from teleseismic P-wave arrivals in Washington and northern Oregon, *J. Geophys. Res.* **91**:2077–2094.

Milne, W. G., Rogers, G. C., Riddihough, R. P., McMechan, G. A., and Hyndman, R. D., 1978, Seismicity of western Canada, *Can. J. Earth Sci.* **15**:1170–1193.

Minster, J., and Jordan, T. H., 1978, Present day plate motions, *J. Geophys. Res.* **83**:5331–5354.

Monger, J. W. H., and Irving, E., 1980, Northward displacement of north-central British Columbia, *Nature* **285**:289–294.

Muller, J. E., 1977, Evolution of the Pacific margin, Vancouver Island and adjacent regions, *Can. J. Earth Sci.* **14**:2062–2085.

Nishimura, C., Wilson, D. S., and Hey, R. N., 1984, Pole of rotation analysis of present day Juan de Fuca Plate motion, *J. Geophys. Res.* **89**:10,283–10,290.

Normark, W. R., Lupton, J. E., Murray, J. W., Delaney, J. R., Johnson, H. P., Koski, R. A., Clague, D. A., and Morton, J. L., 1982, Polymetallic sulfide deposits and water column of active hydrothermal vents on the southern Juan de Fuca Ridge, *Mar. Technol. Soc. J.* **16**:46–53.

Normark, W. R., Morton, J. L., Koski, R. A., Clague, D. A., and Delaney, J. R., 1983, Active hydrothermal vents and sulfide deposits on the southern Juan de Fuca Ridge, *Geology* **11**:158–163.

Potter, K., Morley, J., and Elvers, D., 1974, NOS Sea Map Series: Magnetic maps 12042-12, 13242-12, National Ocean and Atmospheric Administration, U.S. Department of Commerce, Washington, D.C.

Raff, A. D., and Mason, R. G., 1961, Magnetic survey off the west coast of North America, 40°N to 52°N latitude, *Geol. Soc. Am. Bull.* **72**:1267–1270.

Riddihough, R. P., 1977, A model for recent plate interactions off Canada's west coast, *Can. J. Earth Sci.* **14**:384–396.

Riddihough, R. P., 1979, Structure and gravity of an active margin—British Columbia and Washington, *Can. J. Earth Sci.* **16**:350–363.

Riddihough, R. P., 1980, Gorda Plate motions from magnetic anomaly analysis, *Earth Planet. Sci. Lett.* **51**:163–170.

Riddihough, R. P., 1982a, Contemporary movements and tectonics on Canada's west coast: A discussion, *Tectonophysics* **86**:319–341.

Riddihough, R. P., 1982b, One hundred million years of plate tectonics in Western Canada, *Geosci. Can.* **9**:28–34.

Riddihough, R. P., 1984, Recent movements of the Juan de Fuca Plate system, *J. Geophys. Res.* **89**:319–341.

Riddihough, R. P., and Hyndman, R. D., 1977, Canada's active western margin: The case for subduction, *GeoSci. Can.* **3**:269–278.

Riddihough, R. P., Currie, R. G., and Hyndman, R. D., 1980, The Dellwood Knolls and their role in triple junction tectonics off northern Vancouver Island, *Can. J. Earth Sci.* **17**:577–593.

Riddihough, R. P., Beck, M. E., Chase, R. L., Davis, E. E., Hyndman, R. D., Johnson, S. H., and Rogers, G. C., 1984, Geodynamics of the Juan de Fuca Plate, in: *Geodynamics of the Eastern Pacific Region, Caribbean and Scotia Arcs* (R. Cabre, ed.), American Geophysical Union, Geodynamics Series 9, pp. 5–21.

Rogers, G. C., 1979a, Earthquake fault plane solutions near Vancouver Island, *Can. J. Earth Sci.* **16**:523–531.

Rogers, G. C., 1979b, Juan de Fuca Plate Map: Fault Plane Solutions, JFP-6, Earth Physics Branch, Energy, Mines and Resources, Open File.

Rogers, G. C., 1983a, Seismotectonics of British Columbia, Ph.D. thesis, University of British Columbia, Vancouver.

Rogers, G. C., 1983b, Some comments on the seismicity of the northern Puget Sound–southern Vancouver Island region, in: *Earthquake Hazards of the Puget Sound Region, Washington*, Proceedings of Workshop IV, U.S. Geological Survey, Open File Report 83-19, pp. 19–39.

Savage, J. C., Lisowski, M., and Prescott, W. H., 1981, Geodetic strain measurements in Washington, *J. Geophys. Res.* **86**:4929–4940.

Seely, D. R., 1977, The significance of landward vergence and oblique structural trends of trench inner slopes, in: *Island Arcs, Deep Sea Trenches and Back Arc Basins* (M. Talwani and W. C. Pitman, eds.), Maurice Ewing Ser. 1, American Geophysical Union, pp. 187–198.

Seemann, D. A., 1982, Bathymetry off the coast of British Columbia, Open File 82-25, Earth Physics Branch, Department of Energy, Mines & Resources, Ottawa, Canada.

Silver, E. A., 1971a, Tectonics of the Mendocino triple junction, *Geol. Soc. Am. Bull.* **82**:2965–2978.

Silver, E. A., 1971b, Small plate tectonics in the northeastern Pacific, *Geol. Soc. Am. Bull.* **82**:3491–3495.

Silver, E. A., 1971c, Transitional tectonics and late Cenozoic structure of the continental margin of northernmost California, *Geol. Soc. Am. Bull.* **82**:1–22.

Silver, E. A., 1972, Pleistocene tectonic accretion of the continental slope off Washington, *Mar. Geol.* **13**:239–249.

Silver, E. A., von Heune, R., and Crouch, J. K., 1974, Tectonic significance of the Kodiak–Bowie Seamount chain, northeastern Pacific, *Geology* **2**:147–150.

Simpson, R. W., and Cox, A., 1977, Paleomagnetic evidence for tectonic rotation of the Oregon Coast Range, *Geology* **5**:585–589.

Smith, S. W., and Knapp, J. S., 1980, The northern termination of the San Andreas Fault, in: *Studies of the San Andreas Fault Zone in northern California* (R. Streitz and R. Sherborne, eds.), California Division of Mines and Geology, Special Report 140, pp. 153–164.

Souther, J. G., 1977, Volcanism and tectonic environments in the Canadian Cordillera—A second look, *Geol. Assoc. Can. Spec. Pap.* **16**:3–24.

Srivastava, S. P., 1973, Interpretation of gravity and magnetic measurements across the continental margin of British Columbia, Canada, *Can. J. Earth Sci.* **10**:1664–1677.

Srivastava, S. P., Barrett, D. L., Keen, C. E., Manchester, K. S., Shih, K. G., Tiffin, D. L., Chase, R. L., Thomlinson, A. G., Davis, E. E., and Lister, C. R. B., 1971, Preliminary analysis of geophysical measurements north of Juan de Fuca Ridge, *Can. J. Earth Sci.* **8**:1265–1281.

Stacey, R. A., 1973, Gravity anomalies, crustal structure and plate tectonics in the Canadian Cordillera, *Can. J. Earth Sci.* **10**:615–628.

Stone, D. B., 1977, Plate tectonics, paleomagnetism, and the tectonic history of the northeastern Pacific, *Geophys. Surv.* **3**:3–37.

Sutherland-Brown, A., 1968, Geology of the Queen Charlotte Islands, British Columbia, Department of Mines and Petroleum Resources Bulletin 54.

Taber, J. J., and Smith, S. W., 1984, Seismicity and focal mechanisms associated with the subduction of the Juan de Fuca Plate beneath the Olympic Peninsula, Washington, *Bull. Seismol. Soc. Am.* **75**:237–249.

Tiffin, D. L., Cameron, B. E. B., and Murray, J. W., 1972, Tectonic and depositional history of the continental margin off Vancouver Island, B.C., *Can. J. Earth Sci.* **9**:280–296.

Turner, D. L., Jarrard, R. D., and Forbes, R. B., 1980, Geochronology and origin of the Pratt–Welker seamount chain, Gulf of Alaska: A new pole of rotation for the Pacific Plate, *J. Geophys. Res.* **85**:6547–6556.

USGS, 1982, *The 1980 Eruptions of Mount St. Helens* (P. W. Lipman and D. R. Mullineaux, eds.), *U.S. Geol. Surv. Prof. Pap.* **1250**.

Vine, F. J., 1966, Spreading of the ocean floor: New evidence, *Science* **154**:1405–1415.

Vine, F. J., and Wilson, J. T., 1965, Magnetic anomalies over a young oceanic ridge off Vancouver Island, British Columbia, *Science* **150**:485–489.

Vogt, P. R., 1981, On the applicability of thermal conduction models to midplate volcanism: Comments on a paper by Gass *et al.*, *J. Geophys. Res.* **86**:950–960.

Vogt, P. R., and Byerly, G. R., 1976, Magnetic anomalies and basalt composition in the Juan de Fuca–Gorda Ridge area, *Earth Planet. Sci. Lett.* **33**:185–207.

Vogt, P. R., and Hey, R. N., 1976, Evidence for very small hot-spots in the eastern Pacific and the problem of basalt discharge episodicity, *Trans. Am. Geophys. Union (EOS)* **57**:1015.

Von Heune, R., and Kulm, L. D., 1973, Tectonic summary of Leg 18, *Initial Reports of the Deep Sea Drilling Project* **18**:961–976.

Von Heune, R., Shor, G. O., and Wageman, J., 1978, Continental margins of the eastern Gulf of Alaska and boundaries of tectonic plates, *Am. Assoc. Pet. Geol. Mem.* **29**:273–290.

Weaver, C. S., and Smith, S. W., 1983, Regional tectonic and earthquake hazard implications of a crustal fault zone in southwestern Washington, *J. Geophys. Res.* **88**:10,371–10,383.

White, W. R. H., Bone, M. N., and Milne, W. G., 1968, Seismic refraction surveys in British Columbia, 1964–1966: A preliminary interpretation, *Am. Geophys. Union Monogr.* **12**:81–93.

Wilson, D. S., Hey, R. N., and Nishimura, C., 1984, Propagation as a mechanism of ridge reorientation: A model for the tectonic evolution of the Juan de Fuca Ridge, *J. Geophys. Res.* **89**:9215–9225.

Wilson, J. T., 1965a, A new class of faults and their bearing on continental drift, *Nature* **207**:343–347.

Wilson, J. T., 1965b, Transform faults, oceanic ridges, and magnetic anomalies southwest of Vancouver Island, *Science* **150**:482–485.

Yole, R. W., and Irving, E., 1980, Displacement of Vancouver Island: Paleomagnetic evidence from the Karmutzen Formation, *Can. J. Earth Sci.* **17**:1210–1288.

Yorath, C. J., 1980, The Apollo structure in Tofino Basin, Canadian Pacific continental shelf, *Can. J. Earth Sci.* **17**:758–775.

Yorath, C. J., and Chase, R. L., 1981, Tectonic history of the Queen Charlotte Islands and adjacent areas—A model, *Can. J. Earth Sci.* **18**:1717–1739.

Yorath, C. J., and Hyndman, R. D., 1983, Subsidence and thermal history of Queen Charlotte Basin, *Can. J. Earth Sci.* **20**:135–159.

Chapter 3

THE INTERPLAY OF ACCRETIONARY AND ATTRITIONARY TECTONICS ALONG THE CALIFORNIA MARGIN

Jason B. Saleeby and George E. Gehrels*

Division of Geological and Planetary Sciences
California Institute of Technology
Pasadena, California 91125

I. INTRODUCTION

The North American Cordilleran orogen consists of a structurally complex mountainous region which runs along the western edge of the continent for over 5000 km between Alaska and Mexico. Recent syntheses of structural, stratigraphic, paleobiogeographic, and paleomagnetic data have demonstrated that much of the orogen was constructed in Jurassic through early Tertiary time by the tectonic accretion of an assortment of crustal fragments with length scales ranging up to 1000 km (Monger and Ross, 1971; Davis *et al.*, 1978; Beck, 1980; Coney *et al.*, 1980; Saleeby, 1983). The California segment of the orogen is unique in that the volume of accreted materials is significantly less than what is typically preserved to the north, and the western limits of Proterozoic North American crust extend nearly to the present continent edge. The limits of such ancient crust coincide with major structural breaks which represent a lineage of plate junctures. Such junctures include the modern transform system as well as ancient suture and truncation zones whose activity extends back into the Paleozoic. These unique aspects of the California segment are displayed in Fig. 1 which shows a division of southwestern North

*Present address: Department of Geosciences, University of Arizona, Tuscon, Arizona 85721.

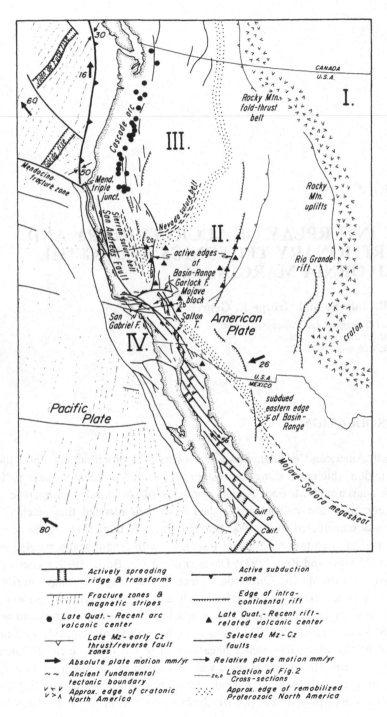

Fig. 1.   Map showing selected tectonic features of Cordilleran orogen and offshore region in the vicinity of California. (Modified after Drummond *et al.*, 1981, and Speed, 1979.)

America into four major tectonic domains. Domain I includes sialic North America which has escaped Phanerozoic deformation. This domain constitutes the present craton and lies east of the Rio Grande rift and the Rocky Mountain uplifts in the United States and the Rocky Mountain fold-thrust belt in Canada. Domain II includes sialic North America which has undergone moderate to substantial Phanerozoic deformation and regions containing Mesozoic igneous rocks which bear the isotopic imprint of Proterozoic North America. The western limits of this domain are defined by suture belts composed of ancient island arc and oceanic sedimentary and basement rock complexes present in the central Nevada and western Sierra Nevada regions. In southeastern California and northern Mexico, such limits are defined by the Mojave–Sonora megashear which constitutes a major continental truncation zone (Silver and Anderson, 1974; Anderson and Schmidt, 1983). Domain III includes accretionary crustal and upper mantle fragments that are presently attached to the North American plate as a result of mid-Paleozoic and later collisional and active margin tectonism. Boundary structures between domains II and III can be located along the Mojave–Sonora megashear and the Sierran and Nevada suture belts, but are for the most part lost in the Oregon, Idaho, and Washington regions beneath Cenozoic volcanic fields. Such boundary structures reappear and continue northward near the Canadian border. Domain IV represents an actively displacing continental fragment that is presently fixed to the Pacific plate. The eastern boundary of this domain is defined by the San Andreas fault and the Gulf of California rift system. To the west of domain IV lies Pacific plate oceanic crust. Domain IV is terminated to the south where the East Pacific Rise enters the Gulf of California, and to the north at the Mendocino triple junction. Continuing northward, domain III is seen actively growing with the subduction of the Gorda and Juan de Fuca plates and the formation of a thick accretionary prism.

The region of Fig. 1 can be used to demonstrate some of the modern concepts of accretionary tectonics and continental margin evolution. Tectonic accretion denotes the structural emplacement of mainly crustal fragments against cratonic crust or cratonic cored accretionary masses. Such a structural arrangement is seen within and between domains III and II with major accretionary interfaces lying along the Sierran and Nevada suture belts. As noted above, tectonic accretion is presently active along the western margin of domain III north of the Mendocino Triple Junction, where a thick sedimentary trench fill is being structurally imbricated by subduction. Tectonic attrition denotes the removal of continental margin fragments by rifting and transform faulting. Such attrition is seen in the removal of domain IV along the modern San Andreas–Gulf of California plate juncture system. The early phases of continental fragmentation processes, which may ultimately lead to tectonic attrition, are exhibited in the Basin–Range and Rio Grande rifts. Tectonic attrition results in the dispersal of the dislodged fragments to new, perhaps distant sites of accretion. Such attrition may also result in the igneous generation of new oceanic crust by seafloor spreading along the attritionary zone, as demonstrated by the Gulf of California rift system. Tectonic dismemberment of lithospheric and crustal fragments by attritionary, dispersive, and subsequent accretionary processes results in the production and structural juxtaposition of multiple fragments, many of which show unique geological histories relative to adjacent fragments. Such fragments are termed tectonostratigraphic or suspect terranes (Coney *et al.*, 1980). The regional pattern of interdigitated terranes which characterizes domains III and IV is termed an "orogenic collage" (Helwig, 1974; Davis *et al.*, 1978). In general, the crustal fragments which compose the collage at the present crustal level only represent the vestiges of once significantly larger lithospheric slabs (plates and microplates) which interacted dynamically with typical transport rates of 10 to 100 mm/yr.

The tectonic evolution of the California segment of the Cordillera can be broken into four periods based on the types of information that are available for its interpretation. These are: (1) the present San Andreas–Gulf of California transform–rift system for which detailed reconstructions can be made with some confidence back into Miocene time, and which can be related in some detail to the marine record of ocean floor spreading. (2) A period of time for which the migration and accretion of terranes along the margin can be loosely related to the marine record of ocean floor spreading, although kinematic patterns along terrane boundaries and the extent to which terrane boundaries actually represent lithospheric plate junctures are obscure. This time interval runs from early Miocene back through Cretaceous time. (3) A period of time for which the migration and accretion of displaced terranes along the margin cannot be directly related to a marine record. The original kinematic patterns which affected terrane boundaries formed during this time period are further obscured by deformation and remobilization arising from subsequent accretionary and attritionary episodes. This time period extends from Jurassic back to Mississippian time and encompasses the early phases of Cordilleran active margin tectonics which saw the most voluminous phases of accretion. (4) The preactive margin history of the Cordillera which is recorded in the late Proterozoic rifting formation of a passive margin and the early Paleozoic history of a stable shelf. Again, California latitudes of the Cordilleran orogen are somewhat unique with regard to this early time interval. Within domain II lies a relatively complete record of the formation of the Cordilleran stable shelf (Stewart, 1972; Speed, 1983). To the north, domain II thins markedly, particularly as it crosses the Canadian border. In these northern segments the early history of the Cordilleran stable shelf is highly obscured by later accretionary and perhaps attritionary events. The presence of a relatively complete early margin history in the Nevada–California latitudes provides a useful tectonic frame of reference for later Phanerozoic active margin tectonics. In earlier literature the stable shelf was often referred to as the miogeocline.

This chapter focuses on the evolution of the California segment of the Cordillera with emphasis on events which dramatically altered the configuration of the margin. Major questions in the minds of the authors are also emphasized. The prime question concerns the relative importance of tectonic accretion versus attrition.

Tectonic attrition by transform faulting and seafloor spreading characterize the modern California margin. The present plate juncture system is dominated by transform links. In the Gulf of California, such links constitute a family of boundary transforms which extend into juvenile oceanic crust and connect with short spreading ridges of the actively growing Gulf of California (Moore, 1973). Boundary transforms constitute the transform edges and resulting marginal offsets in rifted crustal blocks which bound the respective spreading ocean basin (LePichon and Hayes, 1971). The Gulf of California boundary transforms for the most part define the rifted edges of the northwestern Mexican mainland and eastern Baja California. In Alta California the stepping pattern of boundary transforms passes northwestward into a nonideal intracontinental transform and rift system marked by the San Andreas fault system and the Basin–Range extensional province. In southern Alta California the transition between the Gulf system and the San Andreas–Basin–Range system consists of a confused family of ephemeral shallow crustal strike-slip faults and midcrustal to possible upper mantle decollements, which together dissipate transform motion between the Pacific and North America plates. In central and northern Alta California the San Andreas system consists of a myriad of active and inactive strike-slip faults which, in an obscure fashion, tie into the Mendocino triple junction.

The complex family of Gulf boundary transforms and spreading ridges and Alta California strike-slip faults and related decollements is here termed a hybrid boundary transform. This is derived from the view that this complicated system will ultimately lead to a truncated continental margin with predominately boundary transform edges, yet a number of complex kinematic patterns are operating together toward this end. Taking this perspective and looking back into the geological record, one must ask how far back into geological time hybrid forms of transform tectonics can be seen to have operated along the California segment of the orogen. The view expressed in this chapter is that the unique characteristics of the California segment of the orogen are in large part due to numerous transform episodes which resulted in tectonic attrition working both in series and in parallel with accretion. Major transform episodes are suggested for several time periods back to Jurassic time by ocean floor plate kinematic models and by paleomagnetic data on displaced terranes. Additional episodes are further suggested in early Mesozoic and late Paleozoic times by the occurrence of transform affinity ophiolites in the western Sierra suture belt, and by the coincidence of the suture belt with a continental truncation zone. This perspective for the development of the California margin must also be considered in the light of accretionary events which are widely viewed as collisional in origin. Central to this question is the origin of major structural breaks and the uniqueness in their implications for plate kinematic patterns, the characteristic length scales for terranes which directly reflect collisions, and the modification of collisional masses by transform tectonics. Consideration of the modern plate juncture system will help pose these questions.

II. IDEAL AND NONIDEAL ASPECTS OF THE MODERN PLATE JUNCTURE AND A GLIMPSE INTO THE FUTURE

The theory of plate tectonics and seafloor spreading was advanced considerably by the kinematic analysis of large lithospheric plates from an idealized point of view. In this view, lithospheric plates remain rigid during transport with little or no internal deformation, while deformation and translation are concentrated along narrow plate boundaries (Wilson, 1965; Morgan, 1968; LePichon, 1968). Such a view has fostered a number of fairly successful models for the movements of large plates both at the present time, and for the geological past extending into latest Mesozoic time (Atwater and Molnar, 1973; Minster and Jordan, 1978; Engebretson et al., 1984; Page and Engebretson, 1984). Length scales for the application of such kinematic analyses are in most cases substantially greater than 1000 km. Ideal behavior along the eastern margin of the Pacific Plate can be traced northward along the East Pacific Rise and into the Gulf of California rift system (Moore, 1973). Such behavior degenerates markedly along the juncture as it passes into the San Andreas transform link.

Nonideal plate behavior is manifest along the San Andreas system primarily by the wide array of strike-slip faults that are dissipating the dextral sense of motion between the North American and Pacific plates. The main members of this fault system are shown by dark lines in Fig. 1 which are braided over a zone which is typically on the order of 100 km (Crowell, 1979; Graham, 1979; Lubetkin, 1980; Fuis et al., 1982; Yeats, 1983; Sieh and Jahns, 1984). In addition to the system of right-slip faults, a possible dextral component of about 7 mm/yr in extension across the Basin–Range may also account for a significant fraction of the relative plate motion (Thompson and Burke, 1973; Sieh and Jahns, 1984). Minster and Jordan (1978) calculate an overall rate of 55 ± 5 mm/yr between the North American and Pacific plates, of which between 10 and 35 mm/yr can be documented along different segments of the San Andreas fault proper (Sieh and Jahns, 1984). Relative motion vectors for the Pacific versus North American plate are shown in the Gulf of California (Fig. 1). In Alta California latitudes, such vectors cannot be applied across a well-defined boundary structure due to the effective width of the plate juncture system which lies between the California coast and perhaps as far east as the eastern margin of the Basin–Range province.

The dissipation of transform motion between the Pacific and North American plates along the San Andreas system is complex in time as well as space. For example, the San Gabriel fault acted as a major link in the San Andreas system from late Miocene to mid-Pliocene time with about 60 km of right-lateral offset (Crowell,

1979). In mid-Pliocene time, right-slip along the San Gabriel fault was for some unknown reason terminated, and transferred to the modern San Andreas trace which forms the northeast boundary between the Mojave block and the Transverse Ranges (Fig. 2B). Along the northern segment of the system there is some suggestion that different branches have changed their slip rates through late Quaternary time up to the present.

Nonideal plate behavior is further manifest at the northern termination of the San Andreas system. The transition between the San Andreas and the Mendocino triple junction is characterized by a wide shear zone and right-slip fault system which extends from the offshore trace of the San Andreas fault over 100 km inland across the northern California Coast Ranges (Blake *et al.*, 1985, 1987). This region is also characterized by east-directed thrusts which root into the triple junction structural complex. The Mendocino triple junction is a transform–transform–trench junction (Fig. 1) which is migrating northward at about 55 mm/yr (Atwater, 1970). The Gorda plate is a small oceanic plate being generated along the southeast flank of the Gorda Rise and subducted along the northern California–southern Oregon

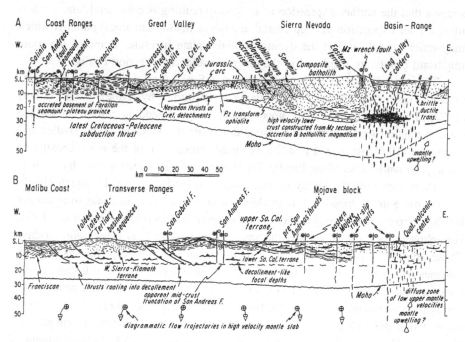

Fig. 2. Simplified cross sections showing contrasting crustal structure and neotectonic patterns of central California and southern California areas. Topographic profile accentuated 2 : 1 for identification purposes. (A) Monterey Bay area to California–Nevada border after Saleeby *et al.* (1986). (B) Malibu offshore to eastern Mojave. Geology simplified after Yeats (1981) and Silver (1983). Focal depths on decollement-like seismic events projected on cross-section plane from vicinity of line after Webb and Kanamori (1985). Crustal structure and depth to Moho after Hearn (1984). High-velocity mantle slab after Hadley and Kanamori (1977), Humphreys *et al.* (1984), and Bird and Rosenstock (1984).

Coast. This "microplate" is behaving in a nonrigid manner as manifest by the deformation of its magnetic lineaments, and internal left-slip faults which roughly parallel the Gorda Rise (Silver, 1971). Focal mechanisms for earthquakes along the Mendocino fracture zone which show thrusting as well as strike-slip components (Bolt et al., 1968; Seeber et al., 1970) further attest to the nonideal plate behavior of this region.

The complex pattern of multiple strike-slip faults differentially operating in time and space along the Alta California transform juncture produces crustal fragments with length scales of 100 km and less. The rapidly evolving configuration of strike-slip movement along the modern juncture leads to the question of future first-order configurations. This question may be divided into shallow crustal versus lithospheric phenomena. In terms of shallow crustal phenomena, one might ask what in the early Pliocene geological record suggests the subsequent shift of major strike-slip movement from the San Gabriel to the present Mojave–San Andreas branch of the system. In a search of the literature, little can be found in terms of substantial dynamic or even kinematic models for this shift. Such questions lead to a number of important recent contributions to southern California tectonics which suggest that the surface expression of strike-slip faulting is ephemeral, and furthermore a poor expression of deep crustal and upper mantle kinematic patterns which may more directly reflect the dynamics of the plate juncture. Some of the more important points are displayed in the diagrammatic cross section of Fig. 2.

The main shallow crustal features shown in Fig. 2 were constructed after Yeats (1981). This author emphasized P-wave residual studies by Hadley and Kanamori (1977) showing a high-velocity upper mantle ridge extending across the surface trace of the San Andreas fault and eastward beneath the Mojave region. Hadley and Kanamori (1977) suggested that the northeast termination of the ridge constitutes the upper mantle level of the Pacific–North America plate juncture. Such a view is consistent with crustal velocity studies conducted by Hearn (1984), which clearly resolve the San Andreas fault at shallow crustal levels, yet fail to do so at deeper levels. Deep crust and mantle seismic tomography studies (Humphreys et al., 1984; Humphreys, 1985) have refined the shape of the anomalous ridge considerably. These studies map the ridge as a vertical to steeply north-dipping slab with a thickness of up to 70 km. The axis of the slab is nearly east–west and is close to the cross-section line of Fig. 2B. The high-velocity slab may include deep crustal rocks and is seen to extend as deep as 250 km into the mantle. Humphreys et al. (1984) suggest that the slab is descending, or subducting steeply northward beneath the Transverse Ranges and western Mojave block. A similar model suggesting mantle downwelling beneath the Transverse Ranges was constructed mainly from crustal kinematic patterns along surface faults (Bird and Rosenstock, 1984). These kinematic analyses suggest that the shallow crustal rocks of the western Mojave are riding as a semicoherent flap across the top of the high-velocity slab with the San Andreas fault truncating the western edge of the flap. Yeats (1981) also emphasized the midcrustal flattening of north-dipping thrusts in the Transverse Ranges and Malibu coast based on focal mechanism studies of Hadley and Kanamori (1978). He suggested that such thrust faults root into a regional decollement which underlies the Transverse Ranges and western Mojave regions. Focal mechanism studies by Webb and Kanamori (1985) generally support the hypothesis of one or more regional decollements under the Transverse Ranges and western Mojave block. Such a decollement is shown diagrammatically in Fig. 2B; it presumably decouples the brittle upper crust from ductilely deforming lower crust and the descending upper mantle slab. The midcrust decollement is also thought to terminate at or near the cryptic Pacific–North America lithospheric juncture beneath the central Mojave. Above the decollement the brittle upper crust is partitioned into strike-slip and thrust-bounded nappes. Zones of deep crustal seismicity and high velocities also suggest that domains of brittle, cool lower crustal materials are being imbricated at depth beneath the midcrustal decollement (Sibson, 1983; Humphreys, 1985; Webb and Kanamori, 1985). This suggests the existence of additional lower crustal decollements, and perhaps even the Moho is behaving as a decollement. The Moho decollement could conceivably be a delamination zone between lower crustal materials and the descending upper mantle slab.

Crustal structure and kinematic patterns shown in Fig. 2B differ substantially from those shown in Fig. 2A. The Fig. 2A cross section traverses the central California Coast Ranges, Great Valley, Sierra Nevada, and westernmost Basin–Range (after Saleeby *et al.*, 1986). Much of the crustal structure, including the Moho, beneath the Great Valley and Sierra Nevada is considered to be static and inherited from Mesozoic tectonics. Such a view is based on extremely low heat flow (Lachenbruch and Sass, 1978) in conjunction with seismic focal depth studies (Wong and Savage, 1983) and rheological models of the crust and upper mantle (Sibson, 1982; Smith and Bruhn, 1984). These considerations along with crustal structure and compositional patterns suggest the existence of a relatively cool, rigid, and brittle crustal column between the axial Great Valley and axial Sierra Nevada. Sierra Nevada crustal structure reflects mainly Cretaceous batholithic layering, whereas Mesozoic thrust or perhaps detachment structures constitute much of the western Sierra–Great Valley structure. To the west of the Great Valley area, San Andreas transform motion is distributed across a series of right-slip faults which cut through the offshore region as well as a major section of the Coast Ranges. To the east the Sierran block is abruptly truncated by the western edge of the Basin–Range. In general, the truncation zone corresponds with large fault scarps and fault-line scarps, abrupt shallowing of the Moho, increased heat flow and seismicity, and a string of late Quaternary volcanic centers. In the section line of Fig. 2B the truncation zone is defined by the active Long Valley caldera. The Basin–Range province in general is characterized by attenuated lithosphere and relatively slow upper mantle seismic velocities. Smith and Bruhn (1984) have modeled a regional brittle–ductile transition within Basin–Range crust at a depth of about 10 km. Brittle rocks above the transition fracture primarily by normal faulting, whereas deeper ductile layers are undergoing aseismic deformation. In Fig. 2A the brittle–ductile transition is shown to shallow westward to the Long Valley caldera, which in turn forms a steep crustal-scale boundary with the thick brittle crust of the axial Sierra Nevada. The distinct north-to-south variation in crustal structure between sections 2A and 2B occurs over a transition zone in the region of the Garlock fault (Fig. 1), which constitutes the shallow-crustal tectonic boundary between the Sierra Nevada–western Basin–Range domain and the Mojave block (Davis and Burchfiel, 1973).

The cross sections of Fig. 2 are stacked for the purpose of direct comparison between the two. Except for their eastern edges, tremendous contrasts in crustal structure are most evident. Humphreys *et al.* (1984) have resolved an apparent low-velocity upper mantle zone centered beneath the intersection of the eastern Sierra–western Basin–Range boundary and the Garlock fault. A diffuse mantle low-velocity zone extends southeastward from this zone beneath the eastern Mojave apparently truncating the eastern end of the high-velocity mantle slab (Humphreys *et al.*, 1984). Crustal-level geology contrasts in this portion of the Mojave from that to the west by an array of late Quaternary volcanic centers (Fig. 1), higher seismicity (Hileman, 1978), and a more pronounced tectonomorphology (Bull, 1978). The mantle in this zone is suggested in Fig. 2B to be in a state of upwelling. The next known low-velocity zone to the south lies beneath the Salton Trough (Humphreys *et al.*, 1984), the northernmost recognized segment of the Gulf of California rift system.

Low-velocity upper mantle of the eastern Mojave and western Basin–Range lies in a position that is consistent with a succession of dextral steps in the Gulf of California rift systen (Fig. 1). Hadley and Kanamori (1977) suggested that northwest striking right-slip faults of the central Mojave (e.g., Helendale) are a subtle surface expression of the upper mantle level plate juncture. The aggregate rate of slip on these faults is about 6.5 mm/yr (Anderson, 1979; Yeats, 1981), similar to the component of the Pacific–North America relative motion suggested across the Basin–Range (Thompson and Burke, 1973; Sieh and Jahns, 1984). Detailed studies of the surface break resulting from the great 1892 Independence earthquake along the eastern Sierra Nevada indicate a dominance of right- over normal-slip (Lubetkin, 1980). These relations suggest an imminent change in the future locus of the Alta California plate juncture from the present trace of the San Andreas fault to an eastern Mojave–eastern Sierra Nevada trace. Such a change will be of much greater geological consequence than the late Pliocene shift from the San Gabriel to the San Andreas fault.

Late Quaternary volcanic centers shown in Fig. 1 as being rift-related are characterized by basaltic associations, including bimodal and alkalic suites and sporadic voluminous silicic-ignimbrites derived from large caldera complexes. Long Valley represents such a caldera. These volcanic centers are concentrated along the eastern Mojave–eastern Sierra Nevada (western Basin–Range edge) locus and along the north segment of the eastern Basin–Range edge. The south segment of the eastern Basin–Range edge (Arizona–northern Mexico; Fig. 1) lacks young volcanism and possesses a more subdued tectonomorphology. It faces the Gulf of California rift system, and is suggested to have waned in its volcanotectonic activity with the successful propagation of the Gulf rift system. The eastern Mojave–eastern Sierra zone of rift volcanism, seismicity, and low-velocity upper mantle is suggested here to be the actively propagating Gulf rift system. This model predicts the northward propagation of Salton Trough-like spreading centers through the eastern Mojave–eastern Sierra trace followed by the northward-growing Gulf of California. Concomitant waning in volcanotectonic activity along the northern segment of the eastern Basin–Range edge might be expected as the northern Basin–Range would ultimately face the northern reaches of the enlarged "Gulf of California." One might consider the question of the rate of rift propagation relative to northward migration of the Mendocino triple junction, and whether the rift will continue up the east flank of the Cascade arc or choose a San Andreas-like transform tie to a coastal triple junction.

The discussions and speculations presented above are directed toward several ends. Of primary interest are the active tectonics of the modern plate juncture, and the marked deviations from ideal plate behavior. The modern system also demonstrates some of the structural complexities of tectonic attrition. The Alta California attritionary terrane is being cleaved by emphemeral strike-slip faults and horizontally layered with a regional decollement(s) and shallow-level thrusts by the attritionary processes alone. Based on plate motion patterns and rates, one might predict that within 100 m.y. the California attritionary terrane will collide and become an accretionary terrane in the Pacific northwest. In such accretionary terranes it is common to interpret major flat-lying structures and intraterrane fragments or nappes as a direct result of collisional accretionary processes. However, the crustal and upper mantle kinematics of the modern plate juncture system force us to consider major horizontal tectonics as a crustal behavior pattern linked to plate juncture kinematics, but with existence not uniquely dependent on a specific plate kinematic pattern. As we look back into what is thought to be a relatively straightforward plate tectonic history for the pre-San Andreas Tertiary, we will be struck again by the tendency of the crust to partition itself into both flat-lying and strike-slip bounded structural units.

III. THE NOT SO STRAIGHTFORWARD PRE-SAN ANDREAS TERTIARY

Discussion of the pre-San Andreas Tertiary centers first on Fig. 3 which diagrammatically shows the plate configuration off the California coast at about 20 Ma, as well as several outstanding tectonic features present along the continental

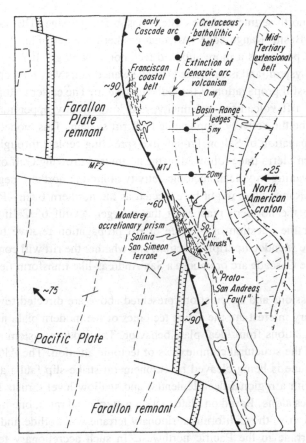

Fig. 3. Map showing selected major tectonic features for about 20 Ma along the California margin. Symbols similar to those in Fig. 1. Farallon–Pacific Plate geometry and subduction of East Pacific Rise after Atwater and Molnar (1973). Extinction pattern in Cenozoic arc volcanism after Dickinson (1971). Mid-Tertiary extensional belt after Coney (1979). Curved arrows show mid-Tertiary rigid rotations after Greenhaus and Cox (1979). Luyendyk et al. (1980), Magill and Cox (1981), and Kanter (1983).

margin. The most important feature shown is the subduction of a segment of the East Pacific Rise which resulted in the initiation of the San Andreas transform system. The new transform system is shown in a phase of lengthening accomplished by the southward migration of a ridge–trench–transform triple junction and the northward migration of the Mendocino triple junction (transform–trench–transform). Isopleths showing the extinction dates of arc volcanism are shown in the central California region (Dickinson, 1981). Such extinction dates correspond to the termination of subduction by spreading ridge consumption and triple junction migration. Note that 310 km of right-slip has been removed from the San Andreas fault (Huffman, 1972; Nilsen, 1984), and the Gulf of California has been closed by a comparable amount. Such palinspastic restorations appear grossly incomplete when viewed at a scale which includes the entire California margin. Plate kinematic modeling by Atwater (1970) suggests that about 1800 km of post-30 Ma right-lateral motion is required between North America and the Pacific plate in order to satisfy major plate motion patterns. Furthermore, the 310-km offset noted above does not satisfy a minimum of 500 to 600 km of offset needed to transport the Salinia terrane to its present position in the central California Coast Ranges. Salinia represents a large fragment of primarily Mesozoic granitoid material which is notably out of place as can be seen by its structural juxtaposition between the Franciscan complex and the San Simeon terrane both representing trench and oceanic assemblages.

The discrepancy between the observed 310-km offset on the San Andreas fault and the offsets suggested by Salinia and plate kinematic models has led a number of workers to suggest the existence of a proto-San Andreas fault. An actual fault surface has not been identified, although it conceivably ran along the eastern border of Salinia. Paleomagnetic data on marine strata lying depositionally on Salinia crystalline rocks suggest that any substantial pre-San Andreas offset occurred prior to late Eocene time (Kanter, 1983; Nilsen, 1984). Upper Cretaceous marine strata thought to lie depositionally on Salinia have paleomagnetic signatures suggesting northward displacement of up to 2500 km (Champion et al., 1984). Acceptance of these paleomagnetic and stratigraphic arguments implies latest Cretaceous to early Eocene migration of Salinia at a rate of about 100 mm/yr. Such movements cannot be related to the movement discrepancy implied by the kinematic models of Atwater (1970), and as discussed below late Cretaceous–Paleocene apparent northward displacements of almost unbelievable magnitude are widespread throughout the North American Cordillera. Nevertheless, an alternative model, which would ignore the paleomagnetic data and derive Salinia from the southern California region, can be developed by integrating right-slip components throughout all branches of the San Andreas system (Graham, 1979). One could readily argue for about 500 km of northward displacement for Salinia with such a restoration. The appeal in deriving Salinia from 500 to 600 km south of its present position comes from the desire to match the northern end of the terrane with the southern termination of the Sierra Nevada batholith (Ross, 1978). However, the Salinia granitoids seem to resemble the eastern margins of the Cretaceous batholithic belt which shows distinct transverse petrochemical zonation over a vast region (Moore, 1959; Saleeby et al., 1986). Thus, a direct fit for Salinia at the southern end of the Sierras is not readily found. The southern end of the Sierra Nevada batholith is cut by major west-directed thrusts of early to mid-Tertiary age, and such structures continue and are widespread throughout the Mojave region (Silver, 1983). On the cross section of Fig. 2B the upper crustal rocks of the Mojave block and eastern Transverse Ranges are depicted as a series of nappes which were previously tectonically transported and are now riding en masse above the modern midcrustal decollement. Silver

(1983) suggests that eastern batholithic rocks of Salinia were transported westward as one or more such nappes prior to northward offset along the San Andreas system.

Two contrasting models for Salinia transport can be envisaged from Fig. 3. First, the "proto-San Andreas fault" can be thought of as a major boundary along which outboard terranes migrated northward at 1000-km scales. In this case Salinia is exotic to California and was transported into the region between latest Cretaceous and early Eocene time. In contrast, as suggested by Silver (1983), Salinia was thrust westward in the early to mid-Tertiary, and what is shown as the "proto-San Andreas fault" simply approximates the future locus of the San Andreas system, about where it happened to slice off the Salinia nappe(s). These two contrasting models give rise to two distinctly different interpretations of the crustal structure beneath Salinia shown at the extreme western end of the Fig. 2A cross section. In the first case, Salinia could be a lithospheric fragment retaining its subbatholithic roots. In the second case, Salinia would presumably be a regionally flat-lying nappe, possibly above Franciscan trench and oceanic assemblages. Geophysical data presently available for Salinia do not help distinguish between the two models (Saleeby et al., 1986).

The discussion above does little toward resolving the relative plate motion problem posed by Atwater (1970). If the entire San Andreas system is considered, along with liberal components of northward motion distributed in the Basin–Range, a discrepancy of about 1000 km remains. This suggests that large components of right-slip are hidden in the offshore region of California. Such slip could have occurred along distinct transcurrent faults, or could have been distributed within accretionary prism complexes that formed along the pre- to early San Andreas subducting trench. One such prism complex has been studied off the Monterey Bay area (D. S. McCulloch, in Saleeby et al., 1986). Growth of the prism by imbricate thrusting of active trench deposits is constrained to between Paleocene and mid-Miocene time. Within the prism, there is a major north-directed thrust oriented across the grain of the prism indicating shortening of the prism from south to north. Such structural relations suggest that the Monterey prism was accreted in an oblique subduction setting with a strong northward component in underthrusting. Tertiary accretionary prism-like strata of the Franciscan coastal belt, north of San Francisco, also show structural complications suggestive of right-lateral shear (Blake et al., 1985, 1987). An additional component of right-lateral shear was distributed along the continental margin in Oligocene–Miocene time by the clockwise rigid rotation of 10- to 100-km-scale crustal blocks. The general locations of such rotated blocks are shown in Fig. 3 by circular arrow symbols (references given in the legend). The scale of the rotated blocks and the magnitude of the rotations, typically about 90°, draw attention once again to the likely role of midcrustal detachment tectonics. Finally, another possible source for mid-Tertiary right-slip could reside in strike-slip faults along the western edge of the Basin–Range that have been obscured by extensional structures, or components of dextral-sense strain during the development of the major mid-Tertiary extensional belt (discussed below). In summary, one is tentatively forced to accept the likelihood that about 1000 km of post-30 Ma right-lateral displacement between the Pacific and North American plates is hidden in offshore transcurrent faults, accretionary prisms, and rotated crustal blocks along the coastal ranges. It is possible, but difficult to test, that some component of the displacement was distributed across the Basin–Range.

From the above discussion it is evident that a fundamental problem in resolving major transform-related offsets along the continental margin is that tectonic attrition leaves little geological evidence of having operated. An interesting aspect of the pre-San Andreas Tertiary is the possibility of yet another type of tectonic attrition in addition to transform or rift tectonics. West-directed low-angle shears of the southern California region appear to have attenuated the continental crust, and perhaps as Silver (1983) suggests, shed crystalline nappes westward to be caught and carried northward by oceanic plate movement. The driving force for such low-angle tectonics is unclear. From spatial arguments one may argue that the early Miocene phase may have been related to subduction of the East Pacific Rise (Fig. 3). In contrast, earlier phases may in some way be related to low-angle tectonics widespread along a major extensional belt also shown in Fig. 3.

The mid-Tertiary extensional belt runs from the southern Canadian Cordillera to northwest Mexico (Armstrong, 1982). The belt generally corresponds in time with the eruption of voluminous silicic ignimbrites. The extensional belt represents one of the major enigmas in Cordilleran tectonics. Individual exposures of the belt are often called metamorphic core complexes (Coney, 1979). Such complexes are typically characterized by domical exposures of pre- to mid-Tertiary plutonic and metamorphic basement with a strong mylonitic or cataclastic overprint. The overprinted fabrics are often low-dipping and plunging, and are regionally subparallel to a major low-angle detachment fault or faults. Above the detachment fault rides a brittle veneer of supracrustal rocks typically attenuated by imbricate sets of normal faults. Dominant slip directions in the normal faults are often parallel or subparallel to linear shape fabrics in the lower plate metamorphic core. Such kinematic indicators generally suggest overall east–west extension. Contrasting models for core complex development range from pure shear stretching of the crust with the detachment representing an exhumed brittle–ductile transition having little regionally integrated translation (Miller et al., 1983) versus the detachment surface representing a crustal-scale normal shear zone with significant translation (Wernicke, 1981). In either case the belt of core complexes clearly represents the locus of significant crustal extension. The core complex problem gives us another glimpse into the structural complexity of continental crust, and the tendency for the crust to partition itself into generally flat-lying structural units. One might ask if the core complex structural style might be typical of extensional environments in continental crust. In the case of the mid-Tertiary belt we are privileged to examine the midcrustal levels due to subsequent Basin–Range extension, and due to the fact that the subsequent extension has not (yet?) successfully rifted the crust completely apart.

A major problem posed by the mid-Tertiary extensional belt concerns its relations with plate tectonic processes. Numerous core complexes formed along the belt in Oligocene time, and the range of ages spans Eocene to Miocene time (Coney, 1979; Gans and Miller, 1983). Thus, the belt cannot be related directly to the subduction of the East Pacific Rise and the onset of Basin–Range tectonics, even though an overlap with Basin–Range tectonics is indicated by the age span. In the search for a link with plate tectonic processes, the preextended state of the crust and its plate tectonic framework must first be considered. Prior to the onset of mid-Tertiary extension, the North American Cordillera had undergone a major late Cretaceous to Paleocene compressional deformation event termed the Laramide orogeny (Coney, 1978). As discussed below, the Laramide orogeny represents the culmination of a major Jurassic–Cretaceous accretionary phase in the orogen which resulted in the construction of much of the Cordilleran basement between Alaska and southern California. The initiation of mid-Tertiary extension immediately following the Laramide culmination suggests a dilational relaxation of the imbricated and thickened crust and perhaps a direct link between the two very different tectonic regimes. Possible controls on Tertiary extensional detachment surfaces by earlier thrusts are discussed in Smith and Bruhn (1984) and Saleeby et al. (1986).

Figure 4 is a map of part of the North American Cordillera covering the British Columbia to northern Mexico segment. The major features of the continent are plotted on a palinspastic base approximated for the later part of the Paleocene (construction discussed below). Pertinent features here include the axis of the mid-Tertiary extensional belt, the Sevier and Canadian Rockies fold-thrust belt, and the Rocky Mountain uplifts. Laramide deformation throughout much of the United States consisted of the thick-skinned Rocky Mountain uplifts which in time followed mainly thin-skinned thrusting of the Sevier belt. Sevier belt thrusting extended back into Jurassic time. The Rocky Mountain uplifts and the Sevier belt appear to merge northward into the Canadian Rockies fold-thrust belt. Jurassic and Cretaceous phases of thrusting along the Cordilleran wide belt are generally related to major accretionary episodes and/or the emplacement of large batholiths which in most cases immediately followed such crustal building episodes (Allmendinger, 1983; Hamilton, 1983). Laramide deformation is generally related to the geometry and rate of plate convergence (Coney and Reynolds, 1977; Engebretson et al., 1984). Specifically, partial coupling between continental lithosphere and the shallow-dipping, rapidly subducting Farallon plate is envisaged as driving the Laramide compressional deformations. Major plate motion changes which occurred at about 40 Ma resulted in a rapid decrease in convergence rates between the Farallon plate and North America (Engebretson et al., 1984). The resulting steepening and retreating of the downgoing slab presumably pulled subcontinental asthenosphere into the

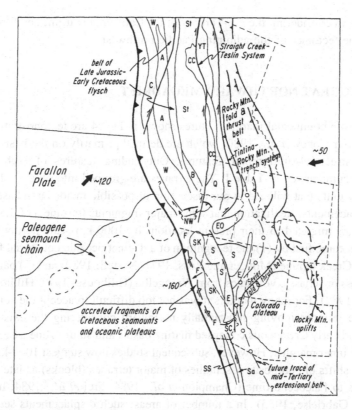

Fig. 4. Map showing proposed configuration of the California and adjacent Pacific northwest margin near the time of the Paleocene to early Eocene. Major symbols similar to Fig. 1. Geographic boundaries used as base for palinspastic restoration of Cenozoic extension, major right-lateral faulting, and final phases of Laramide orogeny, particularly in Pacific northwest. Restoration discussed in text. Tectonostratigraphic terranes modified after Coney *et al.* (1980), with symbols and constituents summarized in Table I.

zone of earlier partial coupling thereby promoting crustal extension along the mid-Tertiary belt (Coney, 1979). This phase of regional extension was followed closely by Basin–Range extension and the San Andreas–Gulf of California plate juncture system.

The philosophy and procedures used in the construction of Fig. 4 focus mainly on the outstanding features and problems of the Mesozoic California margin. The Mesozoic Era was a time of both major accretionary and attritionary tectonic episodes along the California margin. Tremendous apparent northward latitudinal shifts in much of the Cordilleran basement framework as late as Cretaceous–Paleocene time stand out as a major obstacle in understanding the Mesozoic tectonics. Such apparent northward shifts also pose substantial problems in constructing the palinspastic base used in Fig. 4. The apparent northward shifts are of first-order importance in considering the pre-Tertiary of the California margin, and force us to consider the tectonics of the adjoining Pacific northwest.

IV. THE GREAT NORTHWARD MEGADRIFT

The main continental margin features shown in Fig. 4 are tectonostratigraphic terranes (after Coney *et al.*, 1980) which are defined primarily on the basis of pre-Cenozoic structural-stratigraphic features. Outstanding features of each terrane shown are summarized in Table I. The terranes are grouped into blocks which are operational units that will aid in our discussion of possible major late-phase north-ward displacements. Realization that some major fragments (terranes) of the orogen have perhaps migrated to their present locations by 1000-km-scale northward dis-placements came first by (1) the recognition of a distinct oceanic equatorial fauna in the Cache Creek terrane (Monger and Ross, 1971; Nestell, 1980) and (2) paleomag-netic studies on Triassic volcanics from Wrangellia (Hillhouse, 1977; Hillhouse and Gromme, 1982). For many workers it was not too difficult to accept the concept of these large terranes moving substantially northward considering the existence of Jurassic and early Cretaceous collapsed marine basins and suture zone assemblages along their inner margins. However, subsequent studies now suggest 1000-km-scale northward shifts of composite assemblies of major terranes (blocks) as late as mid-Cretaceous to Paleocene time (Champion *et al.*, 1984; Frei *et al.*, 1984; Irving *et al.*, 1985; Gabrielse, 1985). In a number of areas, such displacements seem geo-logically unreasonable.

The first-order palinspastic restorations used in Fig. 4 involve the relocation of the Canadian and California blocks. The position of the California block is modified from the reconstruction of Frei *et al.* (1984) for approximately 50 Ma. Modifica-tions in this reconstruction include (1) straightening of the northwest Klamath Mountains by differential counterclockwise rotation diminishing southwards based on the rotation of a posttectonic Cretaceous pluton and regional field relations (Schultz and Levi, 1983; Saleeby, 1984); (2) deletion of the Oregon Coast Ranges, which, as discussed below, are shown in the early stages of being accreted along the subducting trench; and (3) southward restoration of all rocks west of the eastern Sierra Nevada edge. It is possible that the Sierra Nevada (probably entire California block) migrated northward up to perhaps 500 km since mid-Cretaceous time (Frei *et al.*, 1984). An unknown and perhaps significant component of such displacement could have been distributed across the Basin–Range and mid-Tertiary extensional provinces. Alternatively, all or most of such displacement was related to Cordillera-wide megadrift, and accordingly it is shown in its late stages in Fig. 4. An indirect argument for the Sierra Nevada having moved such a distance during or since mid-Cretaceous time can be presented from spatial relations with the Canadian block that are discussed below.

Major right-slip faults of the Canadian Cordillera are shown in the midst of displacing and further fragmenting the Canadian block. For the purpose of Fig. 4 the Tintina and northern Rocky Mountain Trench zone has had about 450 km of displacement restored and the Straight Creek fault 190 km. Offsets as high as 900 km for the Tintina–Rocky Mountain system, and 300 km for the Straight Creek and related faults suggested for mid-Cretaceous to Eocene time (Gabrielse, 1985), are considered to be in part related to earlier phases of Canadian block megadrift. The amounts of slip restored in Fig. 4 are based arbitrarily on displacement values summarized in Davis et al. (1978). Such restorations required widening or narrowing of basement terranes along the sinuous paths of the strike-slip zones. We have chosen to widen the terranes implying subsequent contractile strains imposed during right-transpressive transport. Such strains are also assumed across the inner border of the Canadian block and continuing into the Rocky Mountain thrust belt, and are considered to represent the final phases of the Laramide orogeny. The width of the northern Canadian block has been additionally increased by restoration of about 150 km of right-slip on the Catham Straight fault in southeastern Alaska (Hudson et al., 1980). In the southwest, the Salinia–San Simeon block is shown in a state of active displacement from parentlands perhaps far to the south (Champion et al., 1984). The block is shown as moving into the southern California region, but in a pre-San Andreas position (Kanter, 1983; Nilsen, 1984). As discussed earlier, a viable alternative to the megadisplacement of Salinia is its westward overthrusting across Franciscan–San Simeon assemblages within a time interval which may roughly correspond to that depicted in Fig. 4.

The Farallon–Pacific ridge was located west of the area shown in Fig. 4 from about 50 Ma and back in time. The Kula–Farallon ridge is considered to have passed northward of the figure area in latest Cretaceous or Paleocene time. The fracture zone trends shown in the offshore region are extrapolated eastward from ridge reconstructions presented by Atwater and Molnar (1973) and Engebretson (1982). Approximate relative and absolute plate motions are taken from Engebretson (1982). They show North America moving WSW at about 50 mm/yr and the Farallon plate moving about 120 mm/yr NE relative to a hot-spot reference frame. In the central California region a resultant of about 160 mm/yr convergence is shown between the Farallon and North America plates oriented along a NE–SW trend. This high rate of convergence has been related to the Laramide orogeny (Engebretson et al., 1984). Also shown in the offshore region is a Paleogene-age seamount chain which is being underthrust along the northern California–Oregon coast region. Much of the Oregon Coast Ranges are presently underlain by accreted remnants of the chain (Duncan, 1981) which underwent major clockwise rotation subsequent to accretion (Simpson and Cox, 1977; Magill et al., 1981).

The subduction and accretion of the Oregon coast seamount complex brings to mind a series of important questions. The first is the possibility that the accretion and subsequent rotation were kinematically linked to continued northward move-

ment of the Canadian block. Furthermore, in the Fig. 4 reconstruction it was
necessary to restore an equivocal amount of Canadian Rockies thrust belt deformation in order to avoid major overlaps with California, Nevada, and Idaho basement
terranes in both space and time. We must then ask if seamount accretion could also
help drive thrust belt deformation. This brings us to a yet larger-scale question.
Fragments of pelagic limestone and submarine basalt which constitute one of the
main late components accreted to the Franciscan have faunal affinities, paleomagnetic signatures, and geological history remnants which suggest that they were
derived from the "mirror image" of the Shatsky Rise, Hess Rise, and Mid-Pacific
Mountains (Alvarez *et al.*, 1980; Vallier *et al.*, 1983; Sliter, 1984). This series of
bathymetric ridges and plateaus form anomalous highlands in the modern west-
central Pacific Ocean. The implication is that fragments of the highlands, perhaps
major ones, were split off eastward along the Pacific–Farallon ridge, traveled with
the Farallon plate, and converged on the North American margin. Underthrust high-
velocity lower crust shown under the California Coast Ranges in the Franciscan
(Fig. 2A) could represent partially subducted remnants of the oceanic highlands
(Saleeby *et al.*, 1986). The estimated arrival of the highland fragments to the
California margin corresponds roughly to the onset of the Laramide orogeny. Thus,
an additional mechanism for the Laramide orogeny may involve the rapid subduction of excessively thick oceanic crust along a low-dipping subduction zone. It
should also be noted that the last major oceanic highland fragments accreted along
the margin (Oregon coast chain) roughly correspond in time with the cessation of
the Laramide orogeny and the onset of regional mid-Tertiary extension.

Reconstruction of the plate tectonic geometry along the California margin for
pre-Cenozoic time brings us to even greater uncertainty. One insurmountable problem is our inability to reconstruct the geometry of oceanic plate junctures and
movement patterns adjacent to the continental margin. In a recent synthesis, Page
and Engebretson (1984) attempted to correlate events along the California margin
with computational models of major plate motions. We feel that both the synthesis
of the geology and the plate motion models are worthy in this work, but for much of
the pre-Cenozoic the continental margin geology and major oceanic plate motions
(Pacific, Farallon, Kula) are at best indirectly related. This view is based on the
likelihood that the Canadian block and/or major fragments (terranes) of it migrated
by and perhaps at times "stalled" outboard of the California margin throughout
much of Mesozoic time. Thus, the California margin was probably buffered from
the motions of major oceanic plates by the outboard terranes and marginal ocean
basin(s). In order to consider plate tectonic phenomena along the Mesozoic California margin, we must therefore rely primarily on paleomagnetic data from displaced
terranes, and the motion of the North American continent, in addition to the geological record. One is forced here to make a fundamental decision: whether to
accept the late Cretaceous–Paleocene megadisplacements suggested by recent paleomagnetic studies (Beck *et al.*, 1981; Champion *et al.*, 1984; Irving *et al.*, 1985),

or to dismiss them as a strange coincidence. In the second case it would be difficult to improve significantly upon the synthesis given by Davis *et al.* (1978) for the Mesozoic evolution of the margin and the possible relation between California basement terranes and major displaced fragments such as Wrangellia and the Cache Creek terrane. For this reason the model presented below will attempt to accommodate the late Cretaceous–Paleocene megadisplacements. In doing so, analytical uncertainties in paleomagnetic data and geological uncertainties in field relations will be stretched to barely overlapping limits. The model is intended to present an endmember possibility in late-phase mobility along the continental margin. Such mobility is certainly within the realm of plate transport rates. Future tests of such an endmember view will lie in better resolution of geological relations and paleomagnetic signatures.

The implications and alternatives to the northward megadisplacement of the Salinia–San Simeon block were discussed above. The possibility of such a large fragment of the California Coast Ranges being so profoundly exotic is trivial, however, if compared to the implications for major displacements cast by the data of Irving *et al.* (1985) on the Canadian block. These data suggest that most terranes west of the Canadian Rockies thrust belt were displaced northward from California paleolatitudes as late as late Cretaceous–Paleocene time, and that the block underwent a significant clockwise rotation during its displacement. However, numerous geological relations discussed in Davis *et al.* (1978) suggest accretion of the Cache Creek, Stikine, and Wrangellia terranes sequentially in late Triassic, Jurassic, and mid-Cretaceous times, respectively. Furthermore, geological relations suggest that the Eastern terrane (Fig. 4, Table I) was a western marginal basin facies of the Paleozoic North American shelf (Monger, 1977; Davis *et al.*, 1978). Finally, there is no record of a sizable late Cretaceous–Paleocene marginal basin between the Canadian block and the Canadian Rockies thrust belt. Thus, if late-stage megadisplacements of the Canadian block did in fact occur, either a totally cryptic marginal basin has closed or a tremendous tract of previously accreted crust was destroyed. Possible destruction mechanisms include the northward transcurrent driving of crustal fragments to perhaps central Alaska and the Bering shelf, or compressional shortening and uplift of a major mountain chain that would likely exceed in size the modern Canadian Rockies.

The proposed time of Canadian block megadisplacement could conceivably correspond to rapid northward migration of the Kula plate along California paleolatitudes. Engebretson (1982) computed rates of between 100 and 150 mm/yr with trajectories nearly due north for the Kula plate during late Cretaceous–Paleocene time. This would require a strong coupling between the Kula plate and the block in order to drive it on the order of 2000 km northward within a 30 m.y. time interval.

Turning attention toward California, one should ask if there is any evidence for the migration of the Canadian block through California paleolatitudes in latest

TABLE I
Outline of Tectonostratigraphic Terranes[a]

Symbol	Terrane	Major basement constituents	Terrane groupings
SS	San Simeon terrane	Late Cretaceous accretionary prism	
Sa	Salinia	Cretaceous batholithic complex	
F	Franciscan	Cretaceous to mid-Tertiary accretionary prism complex with Jurassic tectonitic schist blocks	
SK	Western Sierra-Klamath	Jurassic arc terrane with late Paleozoic and Jurassic transform and rifted arc ophiolites and Cache Creek affinity accretionary prism rocks	California Block
SC	Southern California	Upper plate Precambrian to Mesozoic sialic rocks, lower plate Mesozoic basinal tectonitic schists	
S	Sonomia	Mid- to late-Paleozoic arc built on composite basement	
E	Eastern terranes	Composite of early Paleozoic to early Mesozoic marginal basin, continental rise and fragmented shelf rocks	Fragments contributed to
EO	Eastern Oregon	Composite of SK, CC, and W-type rocks	
NW	Northern Washington	Composite of SK, CC, Q and B-type rocks	
CC	Cache Creek	Late Paleozoic seamounts with early Mesozoic pelagic strata	Stikinian superterrane
St	Stikine	Late Paleozoic arc complex, basement unknown	
Q	Quesnellia	Late Paleozoic arc complex, basement unknown	
YT	Yukon-Tanana	Mid- to late-Paleozoic arc rocks, ensialic metamorphic rocks, and nappes of late Paleozoic ophiolite	Canadian block
B	Bridge River	Upper Triassic to mid-Jurassic accretionary prism complex	
W	Wrangellia	Late Paleozoic arc terrane with Triassic rift and oceanic plateau sequence	Central and southern wrangellian superterrane
A	Alexander	Early Paleozoic arc complex, late Paleozoic marine sequence, Triassic rift complex	
C	Chugach	Mid-Cretaceous to Tertiary accretionary prism complex	

[a]Table modified after Coney et al., 1980. More detailed references in text.

Cretaceous–Paleocene time. The first point to consider is the possibility of about 500 km of northward displacement of the California block during or following latest Cretaceous time (Frei *et al.*, 1984). Greater displacements of about 2000 km have been suggested for the Peninsular Ranges for a similar time interval by Beck *et al.* (1981). As with the Canadian block, such late-stage megadisplacements raise major geological complications. Such displacements of California fragments could have occurred in conjunction with Canadian block displacement. This would imply that nearly the entire accretionary mass of the North American Cordillera was translated or distorted northward during roughly Laramide time and perhaps in conjunction with strong northward movement of the Kula plate. The critical question in attempting to relate motions of major oceanic plates to the proposed megadisplacement or the Laramide orogeny is the position in time of the Kula–Farallon ridge along the California margin. Direct interaction with the Kula plate would yield only the strong northward driving force for the megadisplacements, whereas Farallon plate interaction could only yield the strong convergent component to drive Laramide deformation.

Some possible geological marks of late Cretaceous–Paleocene northward megadrift in the California region are as follows: (1) a major right-lateral shear system within the Franciscan central mélange belt which is constrained to latest Cretaceous–early Eocene time (Blake *et al.*, 1985, 1987). Much of the mélange fragmentation and mixing traditionally attributed to subduction deformation may be attributed to major right-lateral shear. (2) The final major phases of magmatism within the composite Sierra Nevada batholith are represented by large 85–90 Ma batholith-scale plutons that are arranged spatially in a dextral-stepping fashion. At least two of these batholiths are known to have vented ignimbritic eruptions, and it seems likely that all did. The batholithic plutons are suggested to have been the roots of large pull-apart calderas (Saleeby, 1981), much like the great Toba depression in northern Sumatra which has formed along a major dextral shear zone (Williams and McBirney, 1968). The Sierran pull-apart caldera system could have conceivably formed during the northward migration of the Canadian block, or alternatively as a pure response to strong oblique subduction of the Kula plate. By using a simple pull-apart geometry, one may derive about 100 km of northward migration of the western Sierra and forearc region along the batholithic axis. (3) Added components of northward migration derived from mapped dextral faults and cross-fold sets of late Cretaceous age within the batholith and its wall rocks could be on the order of 100 km as well (Saleeby, 1981; Saleeby and Busby-Spera, 1986). (4) An additional and conceivably significant component of dextral transport may have occurred along the axis of late Cretaceous magmatism at about 100 Ma. Throughout both the Sierra Nevada and Peninsular Ranges batholith, a magmatic gap and reorganization of emplacement patterns occurred between 105 and 100 Ma (Silver *et al.*, 1979; Saleeby *et al.*, 1986). Confident ties across the strike of the composite Sierra Nevada batholith cannot be made for times prior to this event. This hypothetical boundary, as well as the batholithic "pull-apart structures," also coincides with cryptic or poorly preserved Mesozoic dextral faults that are locally observed in metamorphic pendants, and are also derived from offsets in batholithic geochemical patterns. These may be in excess of 300 km (R. W. Kistler, in Saleeby *et al.*, 1986). (5) Finally, as noted above, moderate components of latest Cretaceous–Paleocene dextral offset of the entire Sierra Nevada could be hidden in cryptic faults along the western Basin–Range. It is possible that an undetectable component of dextral offset could have occurred on precursors to the Basin–Range structures. Such displacements could add only minor components to the western Nevada orocline which had already formed by mid-Cretaceous time (Geissman *et al.*, 1984; Saleeby *et al.*, 1986), and thus was related to the early phases of megadrift as discussed below.

As mentioned above, the more geologically acceptable views for the northward megadrift envisage the accretion of major Cordilleran terranes between Triassic and mid-Cretaceous times. Great batholithic masses (Sierra Nevada of California, Coast Plutonic and Omenica of British Columbia) are thought to have been emplaced across the major suture zones following the accretionary events (Saleeby, 1981, 1983; Monger *et al.*, 1982). The only other significant modifications in the Mesozoic accretionary masses are generally believed to be Cenozoic strike-slip faulting and extension. In studying Fig. 4, one is struck by the greater expanse of accretionary terranes present in the Pacific northwest relative to the California region. Terrane assemblies of the Pacific northwest, south of central Alaska, can be grouped into three major families or superterranes: (1) the Eastern terrane, whose paleogeographic affinities seem to reflect broad facies relations with the outer reaches of the North American stabile shelf; (2) the Stikine–Cache Creek assembly, which is thought to have undergone a complex accretionary history between late Triassic and mid-Jurassic time (Davis *et al.*, 1978); and (3) the Wrangellian–Alexander superterrane (including the Peninsular terrane of southern Alaska), which is thought to have been accreted in mid-Cretaceous time (Davis *et al.*, 1978; Coney *et al.*, 1980; Saleeby, 1983; Gehrels and Saleeby, 1985). Additional major terranes of the Pacific northwest include Quesnellia (which has some affinity to Stikine) and the Yukon–Tanana terrane (whose affinities could be to the Eastern terrane, Quesnellia, or Stikine). The tectonic boundary between the Eastern terrane and the Stikinian superterrane is obscured by Upper Triassic to Lower Jurassic volcanic arc strata of the Takla–Nicola belt and the late Jurassic–early Cretaceous Omenica crystalline belt. The boundary between the Wrangellian and Stikinian superterranes is obscured by a belt of Upper Jurassic to Lower Cretaceous flysch and the mid-Cretaceous to Eocene Coast plutonic complex.

Widening our attention to include the California region as well, we find that there are southern analogues for the Eastern terrane and the Stikinian superterrane, but none for the Wrangellian superterrane. Stikinia- and Quesnellia-affinity rocks in

California form a major constituent of the western Sierra–Klamath terrane and rocks to the east included in Sonomia (Speed, 1979). Eastern terrane rocks in California and Nevada are not necessarily direct analogues to Eastern terrane rocks of British Columbia, but they are in a similar tectonic position and partly consist of similar rock assemblages. These Eastern terranes are thought to represent the remnants of Paleozoic to early Mesozoic outer shelf, marginal basin, and continental rise sequences against which the Stikinia superterrane and its relatives to the south were accreted.

The late Cretaceous–early Tertiary Franciscan complex forms an outer "armor" along the western Sierra–Klamath terrane in a similar fashion as the Chugach terrane along southern and southeastern Alaska. Both Franciscan and Chugach represent inner trench wall accretionary prism complexes of Cretaceous to Tertiary age which formed following the more voluminous accretion of inboard terranes. A major contrast between the Franciscan and Chugach geometry is the presence of Wrangellian superterrane sandwiched between Chugach and Stikinia in the Pacific northwest. As mentioned above, there is strong evidence indicating that much of the Franciscan complex was sheared by dextral motion during and following its trench wall accretion, and paleomagnetic data on the Chugach terrane suggest substantial northward migration along its inner trench wall environment (Stone *et al.*, 1982). Franciscan and Chugach accretionary prism complexes represent the only Mesozoic terranes that can be linked to kinematic patterns of the Farallon, Pacific, or Kula plates in terms of an accretionary interface with the North American margin. Dextral shearing within Franciscan rocks can be related primarily to Kula and Pacific interactions, with major convergence related to Farallon interactions (Page and Engebretson, 1984). Chugach northward displacement is likely to be related to Kula interactions as well as overall movement patterns of the Canadian block.

The Stikinian superterrane and its relatives constitute the first assemblages thought to have moved northward to Cordilleran accretionary sites. The Cache Creek terrane is an important constituent of the Stikinian superterrane. It is notable for a Carboniferous to Permian fusulinid fauna (verbeekinid) which is clearly exotic to North America, and which is found only in carbonate reef buildups (or blocks) which formed on subsiding seamounts in the equatorial proto-Pacific and Tethys (Monger and Ross, 1971; Monger, 1977; Yancy, 1979; Ben-Avraham *et al.*, 1981). The seamount-reef fragments are depositionally overlain or structurally encased within Carboniferous to Triassic radiolarian chert, undoubtedly representing pelagic deposits which buried the seamounts subsequent to their volcanic and shallow-water histories. The Stikine terrane contains its own exotic late Paleozoic fauna preserved within carbonate buildups. Most notable are schwagerinid fusulinids which are distinct from North American shelf schwagerinids and akin to those found in Quesnellia and Sonomia (Monger and Ross, 1971; Monger, 1977; Ross and Ross, 1983). As discussed below, these distinct faunas appear to be an east Pacific form which diverged during partial isolation from the North American shelf.

Returning focus to the California region, we consider the setting of the Cache Creek-affinity rocks. The western Sierra–Klamath terrane is a composite of Paleozoic to Jurassic ophiolitic and Jurassic–Triassic basinal marine and volcanic arc sequences. The tectonic interface between such assemblages and Eastern terrane Mesozoic shelf rocks, possibly built on North American Proterozoic sial, is termed the Sierra foothills suture (Saleeby *et al.*, 1978). In the northern Sierra and Klamath Mountains, rocks of Sonomia lie along the inner edge of the suture. Cache Creek-like seamount fragments and blocks are encased within chert-rich mélange units of both the Sierras and Klamaths oceanward of the suture (Davis *et al.*, 1978; Saleeby, 1981; Ando *et al.*, 1983). Such blocks are also recycled as olistoliths within chaotic rocks beneath Jurassic island arc deposits of the Sierra–Klamath terrane (Douglass, 1967; Behrman, 1978; Saleeby, 1979). Within the same and very similar mélange units and basal arc deposits, clasts and blocks of Stikine–Sonomia schwagerinid-bearing carbonates occur (Behrman, 1978; Saleeby, 1983; Standlee, 1985). Thus, the western Sierra–Klamath terrane records the joining of the two very different superterrane assemblages in early Mesozoic time. The joining of these superterranes and their accretion to North America presumably represents a major collisional event. One must then ask what the collisional geometry was and where it occurred.

V. CRYPTIC COLLISIONAL OROGENIES

Major collisional orogenies have been hypothesized in the California region for Mississippian (Antler), early Triassic (Sonoma), and late Jurassic (Nevadan) times (Schweickert and Cowan, 1975; Dickinson, 1977; Speed, 1979). In each of these orogenies, east-facing exotic or reversed fringing island arcs are hypothesized to have collided with the passive or active North American margin. However, only for the Sonoma orogeny can an island arc terrane of reasonable size be identified outboard of a collisional suture which exhibits clear geological contrast (Sonomia of Speed, 1979). As discussed below, a distinct collisional mass for the Antler orogeny has not been identified. Jurassic island arc rocks deformed during the Nevadan orogeny are widespread along the western Sierra–Klamath terrane, but neither a collisional interface of the right age nor the collided arc can be resolved with confidence (Saleeby, 1981; Snoke *et al.*, 1982; Harper and Wright, 1984; Sharp, 1985). Nevertheless, Nevadan structures seem to suggest a collisional event (Day *et al.*, 1985; Saleeby *et al.*, 1986; Blake *et al.*, 1987). The theme developed below is that Mesozoic collisional processes were intimately related to the early stage of northward megadrift, and by virtue of strong syn- and post-collisional strike-slip movements the collisional masses are grossly incomplete. In this context the concept of collision seems useful for the most part in a passing sense inasmuch as the collision events did not terminate or strongly influence the overall kinematic pattern of the orogen. The dominant pattern from early Mesozoic time on has been northward translation of terranes during their eastward impingement and shortening against North America. The terranes are believed to have in general decreased in size throughout the multistage history by erosion of uplifted high-level nappes, by tectonic burial of underthrust or subducted fragments, and by strike-slip attrition along terrane margins.

In the discussion below, a scenario is developed whereby major terranes, which are believed to have interacted intimately with the California margin, are moved in the simplest pattern possible given the major constraints. Such constraints include mainly geological history and tectonic position, paleobiogeographic data, and paleomagnetic data. Inconsistencies between these different data sets are reconciled by deleting paleomagnetic data on possible outlier data points, or on plutons which have clearly been transported along intraterrane thrust faults. In some cases it is also necessary to stretch geological relations nearly to the limit of reason, and paleomagnetic data to the limits of analytical uncertainty, in order to force an overlap in such constraints. However, it is felt that a generally acceptable model can be developed which accounts for first-order tectonic relations and kinematic patterns. In-depth discussions of the data used to assemble the model are beyond the scope of this chapter. The model is shown as a series of time slice frames in Fig. 5. Political boundaries of western North America along with paleolatitude estimates are used for geographic reference.

Figure 5A shows the central Cordilleran margin for Permo-Carboniferous time. This time interval serves to depict the pre-megadrift state of the margin and the first events believed to be closely related to megadrift accretionary tectonics. The spatial relations of terranes and plate junctures for this time slice are highly speculative, but plausible, and are used primarily for identification purposes. The Cordilleran shelf is shown lapping westward across the eroded and subsided Antler orogenic belt in Nevada and southern California. Although highly obscure, the nature of the Antler orogeny poses important initial constraints for the resolution of subsequent late Paleozoic and early Mesozoic tectonic events. The Antler event is defined by the early Mississippian eastward overthrusting of continental slope-rise strata across the Cordilleran stable shelf with the concomitant development of a foredeep depositional wedge within the shelf region. The Antler event has been explained as a result of backarc thrusting along the rear of a west-facing fringing system (Burchfiel and Davis, 1972, 1975; Silberling, 1973), or collision of an east-facing exotic arc with a passive margin (Dickinson, 1977; Speed, 1979). The only possible island arc remnants observed yet for the Antler event include Devonian arc-type igneous suites of the northern Sierra and eastern Klamaths, but these are included in Sonomia (after Speed, 1979) and appear to record continuous deposition throughout Antler time. Furthermore, the occurrence of Lower Paleozoic Eastern assemblage-like rocks within the basement complex of the Sierran–Klamath arc sequences suggest that the Sonomia arc was built at least in part on the Cordilleran continental slope-rise. Thus, the straightforward interpretation for the Antler event

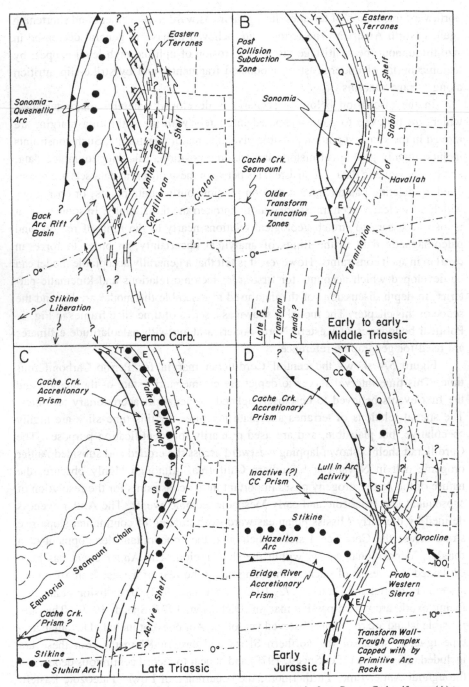

Fig. 5. Tectonic model showing evolution of the California margin from Permo-Carboniferous (A) to mid-Cretaceous (H) time highlighting relationships of major Cordillera-wide accretionary events and California attritionary events. Major tectonic symbols similar to those in Fig. 1. Tectonostratigraphic terrane symbols given in Table I. Explanation of sequence given in text.

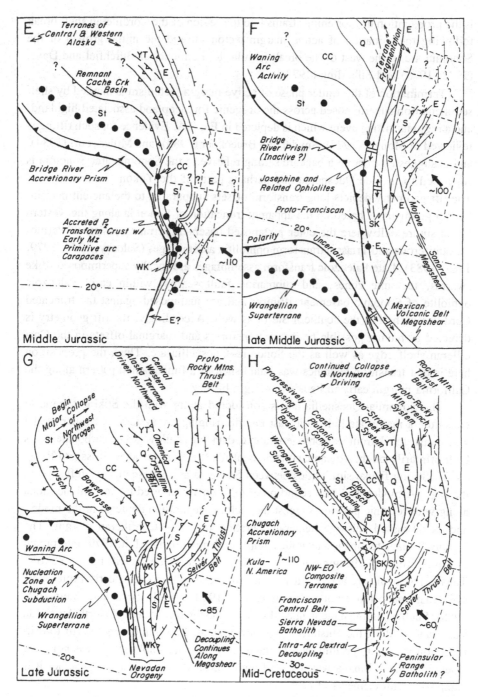

Fig. 5. *(continued)*

is the propagation of shortening strains into the region of the Cordilleran shelf break following the initiation of active margin tectonism and the initial growth of the Sonomia arc to the west of the break [similar to the model of Burchfiel and Davis (1972, 1975) and Silberling (1973)].

Termination of the Antler phase of active margin tectonism is marked by shelf sedimentation superimposed across the orogenic foredeep and associated highland, and a regional rifting event which is depicted in Fig. 5A. Evidence for such rifting is reviewed in Monger (1977), Snyder and Brueckner (1983), and Miller *et al.* (1984), and is suggested to be in a backarc setting relative to the Sonomia arc. Sonomia is suggested to include Quesnellia and perhaps parts of the Yukon–Tanana terrane. The rift geometry depicts long transform offsets subparallel to the ancient margin. Such a geometry is based on the apparent truncation of Sonomia along the western Sierra suture belt where the outer juxtaposed basement terrane consists of Permo-Carboniferous and Triassic transform ophiolitic assemblages (Saleeby, 1978, 1979, 1981, 1983). Late Paleozoic transform assemblages are locally superimposed (dike swarms, tectonite zones, and submarine fault scarps) within early Paleozoic ophiolite of Sonomic basement affinity, and are juxtaposed against the truncated Cordilleran shelf in the southern Sierra as well. Accordingly, the rift geometry is depicted as having possible boundary transforms and marginal offsets in the Cordilleran shelf edge as well as the Sonomia–Quesnellia arc. Thus, the proposition here is that transform tectonics was established as a fundamental pattern along the California margin early in its active margin history.

The Sonomia–Quesnellia arc is considered along with the Stikine terrane to constitute a major late Paleozoic east Pacific fringing arc system. A similar model was constructed from paleobiogeographic data on shallow-water carbonate reefs built primarily on the volcanic constructs of the three arc terranes (Ross and Ross, 1983). Strong faunal similarities between Sonomia and Quesnellia are reflected in the Fig. 5A model, and the second-order differences reported for the Stikine fauna along with tectonic position suggest the tectonic separation depicted in the model, and as discussed below ephemeral liberation from the Cordilleran margin. The Yukon–Tanana terrane is a tectonic complex (Tempelmann-Kluit, 1979) interpreted as a mixture of the fringing arc system and the late Paleozoic rift basin. Yukon–Tanana Paleozoic arc activity at least locally occurred within an early Proterozoic sialic basement framework (Dillon *et al.*, 1980; Dusel-Bacon and Aleinikoff, 1985). Such a basement element could mark a vestigial fragment of North American sial stranded along the outskirts of the newly created margin during the rift-to-drift transition stages in the initiation of the early Paleozoic stable shelf. It is not clear whether Yukon–Tanana was accreted with Quesnellia or Stikine, or whether it is composite with elements of each.

During the early Triassic, following the late Paleozoic rifting event, Sonomia was accreted to the Nevada–eastern California region of the North American shelf (Fig. 5B). The Sonomia arc is believed to have undergone a reversal resulting in the rapid growth of an accretionary prism along its inner edge, and the thrust emplacement of the prism eastward across the shelf (similar to Speed, 1979). The absence of a substantial foredeep and related orogenic welt related to the Sonoma orogeny may reflect a relatively "loose" closure of the backarc basin. Mesozoic and early Cenozoic magmatism, thrusting, and strike-slip faulting obscure the early Mesozoic history of Quesnellia, but at least partial impingement on the North American margin is shown in conjunction with Sonomia accretion. As the Sonomia–Quesnellia marginal basin closed, east-dipping subduction was reestablished along the outer edge of the collided arc system as well as along the transform truncation zones to the south. A major seamount chain, represented by the Cache Creek terrane, is shown moving closer to the Cordilleran margin by the eastward subduction of equatorial oceanic crust.

Figure 5C shows the late Triassic accretion of Cache Creek and affinity oceanic assemblages along western North America. Accretion is envisaged primarily along a west-facing subducting trench resulting in a borderland arc system which in the southern California region was built on disrupted North American shelf (Saleeby et al., 1978) and to the north across the eroded Sonomia orogenic zone and its successor basin (Speed, 1978, 1979). North of the California region, the arc (Takla–Nicola) was built on Quesnellia and Eastern terrane rocks (Davis et al., 1978; Templemann-Kluit, 1979). The Stikine terrane is shown moving northward as an active arc (Stuhini), but of uncertain polarity. We have chosen an initial northeastward polarity with the possible accretion of additional Cache Creek-affinity ocean floor assemblages along the northern Stikine edge. Thus, the "Cache Creek" domain of the protoeast Pacific is shown rapidly closing by opposite polarity subduction zones. Migration of Stikine along the margin is shown to be partly controlled by transform faults which may be inherited from the late Paleozoic marginal rifting event. The marginal rifting event is thought to have liberated Stikine from the Cordilleran margin, relative to Sonomia or Quesnellia, because of the amalgamation of Cache Creek assemblages along the inner Stikine border.

Collision of the Stikine terrane along the California margin is shown progressing along a trench–trench–transform triple junction in early to middle Jurassic time (Fig. 5D). Cache Creek-affinity accretionary prism rocks (Calaveras) emplaced along the Sierran Foothills suture in the Triassic show strong northward (dextral-sense) shortening superimposed over suture-related structures (Sharp, 1985; Saleeby et al., 1986), which are envisaged as the mark of Stikine "collision" and continued northward migration. The cryptic suture resulting from this "collision" consists of late Paleozoic–early Mesozoic oceanic transform assemblages of the western Sierra–Klamath terrane (Saleeby, 1978, 1979, 1982) accreted outboard of a Cache Creek-like prism (Calaveras). The original transform assemblages and structures may have originated in conjunction with the breakup and dispersal of the Stikine–Sonomia–Quesnellia arc complex. This is suggested by the ubiquity of late Paleozoic ophiolitic assemblages at virtually every major structural position outboard (as well as diked into the apparent edge) of the Sierra Nevada Sonomia fragment. Late Triassic–early Jurassic arc sequences built within such serpentinite mélange belts of the western Sierra–Klamath terrane (Saleeby, 1978, 1982; Wright, 1982) are interpreted as arc carapaces formed along the transform wall and trough complexes. The northward "collision" and continued migration of Stikine is also shown to impose a dextral-sense oroclinal bend across structures and tectonic boundaries of the Nevada suture belt. Initiation of a broad zone of east-directed foreland thrusting inland from the California margin is also shown (Speed, 1983; Saleeby et al., 1986). The impact of Stikinia along the California margin is suggested to trigger a reversal in its polarity with the initiation of Bridge River accretion and Hazelton arc magmatism above a northeast-dipping subduction zone.

Closure of the Cache Creek ocean basin and the northward accretion of the Stikine terrane (Fig. 5E) resulted in a westward shift in Pacific northwest continental margin arc activity from the Takla–Nicola belt to the Hazelton belt (Davis et al., 1978). In contrast, middle Jurassic arc magmatism in California followed generally the same trace as that of the late Triassic–early Jurassic. An apparent late Early to early Middle Jurassic lull in California arc activity is thought to correspond in time to Stikine migration along the California margin. The contrast in time–space patterns of continental margin arc activity between California and regions to the north resulted from the lack of a major accretionary mass (Stikine) being added to the California margin. Note also that the orogen in the Pacific northwest widens significantly northward, and is shown actively undergoing shortening during Stikine accretion. Subsequent major shortening in the Pacific northwest region is envisaged for Laramide time with the northward driving of crustal fragments (terranes) into central and western Alaska and the Bering shelf. The northwest-driven terranes consist primarily of fragmented Cordilleran (?) shelf assemblages and late Paleozoic–early Mesozoic marginal basin-arc associations (Coney et al., 1980).

The next time slices (Fig. 5F,G) are tightly compressed relative to other slices since major changes occurred along the California margin at a rapid pace. First, between 170 and 155 Ma a major oblique-spreading system rifted the mid-Jurassic Sierra–Klamath arc (Sharp, 1980; Saleeby, 1981, 1982, 1983; Saleeby et al., 1982; Harper et al., 1985), which disrupted and over large regions terminated magmatic arc activity. The system is believed to have been dominated by transform links oriented along the continental margin. The transform system is shown to parallel and perhaps work sympathetically with the Mojave–Sonora megashear (Silver and Anderson, 1974; Anderson and Schmidt, 1983; Harper et al., 1985). The eastern branch of the megashear system cut into the North American Paleozoic shelf and across its boundary with the Nevada suture belt. Shelf, Eastern terrane, and Sonomia fragments were dispersed southeastward along the eastern megashear branch for distances in excess of 500 km. The main rift system is shown running through the western Sierra–Klamath terrane as a chain of ophiolite-floored rift basins (Saleeby, 1981, 1982; Harper et al., 1985). Igneous accretion of these ophiolites produced the final major basement constituent of the western Sierra–Klamath terrane. The Coast Range ophiolite segment of the rift basin complex may have formed in direct proximity to the Sierra–Klamath segments or in part farther south in conjunction with the Mexican volcanic belt megashear. The California

Great Valley is suggested to have originated as part of the rift–transform system. The geometry of the rift basin complex is thought to more closely resemble Cayman Trough-like structures (Perfit and Heezen, 1978) than the modern Gulf of California, with rift segments forming within large leaky-transform zones.

The Stikinian superterrane is shown decoupled from the North American margin along the leaky-transform system. The North American margin is shown sliding northward relative to the superterrane as did the craton relative to western California and southwestern Mexico. The position of the decoupling zone relative to the superterrane is based on the preference of extending the main trace of the megashear system through the western Sierra–Klamath terrane as the ophiolitic assemblages, and also on an apparent need to move the Stikinian superterrane to a more southerly position (relative to California) to be later assembled into the Canadian block. The Stikinia superterrane is thus shown separated from the western Sierra–Klamath terrane by middle to late Jurassic rift basin crust with a major boundary transform. Such transforms (possibly including additional rifts) are suggested to have extended northward causing fragmentation of earlier accreted terranes (e.g., Quesnellia).

Figure 5F also shows the arrival of a major Jurassic island arc system to the region of the California continental margin. The newly arrived arc represents the Wrangellian superterrane. At this time slice, the superterrane consists of a Jurassic arc built on the composite basement of Wrangellia, the Alexander terrane, and pre-Jurassic ensimatic fragments of the Peninsular terrane. The Wrangellian superterrane is known to be an exotic element of the Cordillera based on its outer position relative to the Stikinian superterrane, and on paleomagnetic–stratigraphic data from the Triassic of Wrangellia indicating substantial northward displacement during its residence in the proto-Pacific as an oceanic plateau (Hillhouse, 1977; Jones et al., 1977; Ben-Avraham et al., 1981). Furthermore, geological history and paleobiogeographic data from the Alexander terrane (Ross and Ross, 1983; Gehrels and Saleeby, 1985; Gehrels, 1985) suggest migration from an early Paleozoic Gondwana orogenic belt through equatorial paleolatitudes, and finally to the North American margin as part of the Wrangellian superterrane. The Jurassic arrival of the superterrane to the California margin is based on geological history and the interpretation of the Stone et al. (1982) paleomagnetic data in the light of Tipper and Smith's (1985) paleobiogeographic data. Reed et al. (1983) suggest a northeasterly-facing direction for at least part of the Jurassic superterrane arc. Gehrels and Saleeby (1985) suggest primarily a transform migration pattern for the arc from the South American to the California margin. In Figure 5F we take an intermediate position and migrate the north-facing arc along a boundary transform adjacent to the actively rifting Mexican volcanic belt megashear system (Anderson and Schmidt, 1983). Figure 5F also shows magmatism along the Hazelton arc of Stikine in a waning phase. This could be related to the regional rifting and transform displacement event as is more clearly seen in time–space patterns in arc and ophiolitic magmatism of the California region.

The nature of the ocean basin which closed between the Stikinian and Wrangellian superterranes is one of the outstanding problems in Cordilleran tectonics.

The vestigial trace of the basin is locally marked by remnants of Jura-Cretaceous flysch in addition to the Bridge River terrane, but in general the trace is obliterated by the Coast plutonic complex. Two major possibilities are suggested for the basin: (1) It was a major ocean basin which separated the two superterranes until their Jura-Cretaceous impingement upon one another and together upon North America, or (2) it was an interarc basin formed in early Mesozoic time along the southeast Pacific margin implying some degree of early phase (late Paleozoic?) coupling between the two superterranes. Regardless of the origin, it appears to have undergone substantial closure during the formation of the Bridge River accretionary prism and the subsequent deposition and deformation of Jura-Cretaceous flysch. Furthermore, the presence of Bridge River-affinity assemblages in the western Klamath Mountains (Davis et al., 1978) indicates some form of physical continuity with the Sierra–Klamath belt. In Fig. 5E and F, such continuity is depicted along the transtensional/transpressional suture belt of the Sierra–Klamath terrane bounding the ocean basin between the Stikinian and Wrangellian superterranes.

Consumption of young marginal basin crust of the Sierra–Klamath terrane and the remnant Bridge River ocean basin led to the late Jurassic collision between the Wrangellian superterrane and the Stikinia–western Sierra–Klamath continental margin (Fig. 5G). Collision in the California region resulted in the imbrication of arc and remnant arc segments and young rift basin floor ophiolites along the western Sierra–Klamath terrane in what local workers term the Nevadan orogeny (Schweickert and Cowan, 1975; Saleeby, 1981; Harper and Wright, 1984; Day et al., 1985). Nevadan thrusting in the Klamath Mountains was mainly westward, whereas in the Sierra Nevada–Great Valley both eastward and westward transport of nappes are suggested. West-dipping midcrustal reflectors shown under the eastern Great Valley in Fig. 2A may be Nevadan thrusts or alternatively Cretaceous detachments related to the Sierra Nevada batholith. The apparent change in Nevadan transport direction between the Klamaths and Sierras calls for a transverse tear of Nevadan age between the ranges. Such a transverse structure may have been inherited from the mid-Jurassic rift–transform geometry.

The Fig. 5G geometry suggests the Nevadan orogeny is a time-transgressive Cordillera-wide event. Collision of the Wrangellian superterrane with Stikine is thought to have been progressive extending from late Jurassic (Nevadan of the California region) through early Cretaceous time ("Nevadan" of the Pacific northwest). The first phases of the Wrangellian–Stikinian collision are marked by the final closure of the Cache Creek remnant ocean basin with the west-directed thrusting of Cache Creek prism rocks over Stikine and with the intensification of east-directed thrusts distributed toward the foreland. Cache Creek uplift shed molasse westward into the Bowser Basin. Cache Creek-type and reworked Sonomia detritus is also present as a synorogenic flysch within and adjacent to ophiolite-floored basins of the western Sierra–Klamath terrane. Crustal thickening related to the final closure of the Cache Creek basin led to widespread anatexis and the development of the Omenica crystalline belt (Monger et al., 1982). After the initial impact of the Wrangellian superterrane it slid northward with a series of transtensional basins opening along the suture zone. These basins filled with flysch as well, and were rapidly deformed along transpressional welts (Gehrels and Saleeby, 1985). The presence of arc-type pyroclastic apron sequences within the flysch basins indicates continuing arc volcanism during basin closure. The Wrangellian and most of the Stikinian (Bridge River) accretionary prisms were presumably destroyed by uplift and erosion or perhaps underthrusting during final closure and suturing.

As the Wrangellian–Stikinian basin system closed, northeastward subduction nucleated along the southwestern margin of the outer superterrane, and growth of the Chugach accretionary prism commenced (Pavlis, 1982). At about the same time the main phases of eastward Franciscan subduction initiated beneath the western Sierra–Klamath terrane, during and/or following the northward evacuation of the trailing Wrangellian fragments. This phase of subduction is believed to mark the first direct interaction between the California margin and the Farallon–Kula system of oceanic plates. The Chugach prism is shown migrating northward along the inner-trench wall environment as discussed above.

Figure 5H shows the Canadian block congealed in terms of the suturing of the Wrangellian and Stikinian superterranes. Batholithic activity of the Coast plutonic complex has initiated along the trace of the "Nevadan" suture zone just as the early Cretaceous batholiths of the western Sierra Nevada formed along the Sierran segment of the suture belt. The Canadian block is shown moving rapidly northward and the late phases of the Sierra Nevada batholith are shown to be emplaced along a zone of dextral decoupling with North America. As discussed above, the decoupling zone is thought to coincide primarily with the axis of late Cretaceous magmatism, with the possible last decoupling zone following the eastern margin of the batholith along the future boundary with the Basin–Range province. Northward movements along the Sierra Nevada and Canadian block represent the late phases of the Cordilleran megadrift and the onset of the Laramide orogeny. The initiation of the major Canadian dextral fault systems discussed by Gabrielse (1985) is shown as an integral part of late-phase megadrift. The Canadian block is thought to have been at least partially coupled to the Kula plate. Additionally, or alternatively, a major seamount–oceanic plateau system may have begun its collision with the Cordilleran margin as discussed above. If this mechanism was involved in the northward displacement of the Canadian block, one might find Franciscan seamount fragment analogues in the Chugach terrane. The critical unknown is the time–space relations of the Kula–Farallon Ridge relative to the California margin. Additionally, possible major northward shifts in Salinia, the San Simeon terrane, and the Peninsular Ranges batholith conceivably commenced during the Fig. 5H time interval.

The next distinct state of the margin is that depicted in Fig. 4. During the time interval between Fig. 5G (late Jurassic) and Fig. 4 (Paleocene–early Eocene), composite aggregates of small terranes, with affinities covering nearly all the major terranes discussed above, were assembled and emplaced into the eastern Oregon and northern Washington areas. Figure 5H shows the fragmentation of these composite terranes along the southern end of the Canadian block. These composite terranes are thought to in essence represent "footprints" of the Canadian block left during the late phases of northward megadrift. A critical link to California terranes present in northern Washington is an array of middle to late Jurassic ophiolitic nappes (Whetten *et al.*, 1978, 1980) which are very similar to the rift-related ophiolites of the western Sierra–Klamath terrane. The Sierra–Klamath ophiolites clearly mark a major attritionary zone that developed within the accretionary mass. The position of the western Washington ophiolites relative to the Canadian block and analogous, yet distinctly less voluminous array of accretionary terranes present in the California region also gives the California terranes the appearance of being "footprints" left from both early- and late-phase megadrift movements. The "footprint" phenomenon is suggested to be primarily the result of accretionary and attritionary processes working simultaneously or closely in series. The final question that we consider is how many different forms and episodes of attritionary tectonics have affected the California margin in pre-Cenozoic time, and how atypical is the modern regime?

VI. THE CALIFORNIA MARGIN—A HYBRID BOUNDARY TRANSFORM SINCE WHEN?

As defined above, boundary transforms constitute the transform edges and resulting marginal offsets in rifted crustal blocks which bound the respective spreading ocean basin. In the discussion of the modern California transform system, much attention was placed on hybrid forms of transform tectonics. In such tectonics the transform decoupling zones may take on forms of other plate tectonic boundaries or entities (e.g., rift segments, subduction zones, decollements, caldera–batholith chains). As a modern example, major decollement tectonics of the southern California region were discussed as a hybrid form of the San Andreas transform system. The view presented here is that hybrid forms of transform tectonics were important and perhaps dominant along the California margin throughout Mesozoic time.

Two major "collisional" masses appear to have impacted the California margin and then continued their migration northward. Both "collisional" suture traces have left ophiolitic basement belts which formed along transform faults or oblique rifts. The Mojave–Sonora megashear represents a major zone of intracontinental and magmatic arc transform decoupling which also greatly affected the California margin. Accretionary prism complexes of the western Sierra and Franciscan complex show signs of strong dextral shearing during and after their inner trench wall accretion, and much of the Chugach prism of the Pacific northwest may have formed close to the California region. The Sierra Nevada batholith appears to represent a zone of transform-like decoupling along the magmatic arc axis.

As we look back into pre-Mesozoic time, we must ask if the sharp bend in the distribution of stable shelf strata from the Nevada region into eastern California is solely a result of Mesozoic transform truncations, or was there perhaps a boundary transform inherited from the late Precambrian rift generation of the Cordilleran passive margin? We are also intrigued by the Permo-Carboniferous rifting episode recorded in Eastern terrane rocks that presumably occurred in a marginal basin system. We have depicted an oblique rift geometry in Fig. 5A and have related late Paleozoic transform ophiolites of the Sierran suture belt to the rift system. Such considerations lead us to conclude that the modern plate juncture system may not be atypical for the Phanerozoic California margin, and that transform-dominated attritionary events have nearly kept pace with tectonic accretion. The California margin at this point in geological time is somewhat unique to the North American Cordillera in this regard.

ACKNOWLEDGMENTS

The ideas set forth in this chapter are a synthesis of our own research, the literature, and a number of personal communications with Cordilleran geoscientists. The synthesis was compiled in early 1985, and thus numerous additional constraints are expected by publication time. We would particularly like to thank J. W. H. Monger and J. O. Wheeler for field trips and stimulating discussions on British Columbia geology. Funds for field and geochronological studies in California, Oregon, and southeastern Alaska which formed the basis of our synthesis and ideas, were provided by the National Science Foundation, U.S. Geological Survey, Alfred P. Sloan Foundation, Geological Society of America, and Sigma Xi. Division of Geological and Planetary Sciences Contribution No. 4436.

REFERENCES

Allmendinger, R. W., 1983, Relations of Mesozoic foreland deformation to continental margin tectonics, central U.S. Cordillera, *Geol. Soc. Am. Abstr. Progr.* **15**:271.

Alvarez, W., Kent, D. V., Premoli Silva, I., Schweickert, R. A., and Larson, R. A., 1980, Franciscan complex limestone deposited at 17° south paleolatitude, *Geol. Soc. Am. Bull.* **91**:476–484.

Anderson, J. G., 1979, Estimating the seismicity from geological structure for seismic risk studies, *Seismol. Soc. Am. Bull.* **69**:135–158.

Anderson, T. H., and Schmidt, V. A., 1983, The evolution of Middle America and the Gulf of Mexico–Caribbean Sea region during Mesozoic time, *Geol. Soc. Am. Bull.* **94**:236–252.

Ando, C. J., Irwin, W. P., Jones, D. L., and Saleeby, J. B., 1983, The ophiolitic North Fork terrane in the Salmon River region, central Klamath Mountains, California, *Geol. Soc. Am. Bull.* **94**:236–252.

Armstrong, R. L., 1982, Cordilleran metamorphic core complexes—from Arizona to southern Canada, *Annu. Rev. Earth Planet. Sci.* **10**:129–154.

Atwater, T., 1970, Implications of plate tectonics for the Cenozoic tectonic evolution of western North America, *Geol. Soc. Am. Bull.* **81**:3513–3536.

Atwater, T., and Molnar, P., 1973, Relative motion of the Pacific and North American plates deduced from sea-floor spreading in the Atlantic, Indian, and south Pacific Oceans, *Stanford Univ. Publ. Geol. Sci.* **13**:136–148.

Beck, M. E., Jr., 1980, Paleomagnetic record of plate-margin tectonic processes along the western edge of North America, *J. Geophys. Res.* **85**:7115–7131.

Beck, M. E., Burmester, R. F., Engebretson, D. C., and Schoonover, R., 1981, Northward translation of Mesozoic batholiths of western North America: Paleomagnetic evidence and tectonic significance, *Geofis. Int.* **20**:144–162.

Behrman, P. G., 1978, Pre-Callovian rocks west of the Melones Fault Zone central Sierra Nevada foothills, in: Mesozoic Paleogeography of the western United States, Pac. Sect. Soc. Econ. Paleontol. Mineral. Pacific Coast Paleogeography Symposium 2, pp. 337–348.

Ben-Avraham, Z., Nur, A., Jones, D., and Cox, A., 1981, Continental accretion: From oceanic plateaus to allochthonous terranes, *Science* **213**:7–54.

Bird, P., and Rosenstock, R. W., 1984, Kinematics of present crust and mantle flow in southern California, *Geol. Soc. Am. Bull.* **95**:946–957.

Blake, M. C., Jr., Jayko, A. S., and McLaughlin, R. J., 1985, The central belt Franciscan—An oblique transform melange? *Geol. Soc. Am. Abstr. Progr.* **17**:342.

Blake, M. C., and 17 others, 1987, Continent–Ocean Transect C1: Mendocino triple junction to North American craton: 3 sheets 1 : 500,000 scale and explanatory text, *Geol. Soc. Am.* (in press).

Bolt, B. A., Lomnitz, C., and McEvilly, T. V., 1968, Seismological evidence on the tectonics of central and northern California and the Mendocino Escarpment, *Seismol. Soc. Am. Bull.* **58**:1725–1767.

Bull, W. B., 1978, Tectonic geomorphology of the Mojave Desert, California, U.S. Geological Survey National Earthquake Hazards Reduction Program Summaries of Technical Reports **5**:6–9.

Burchfiel, B. C., and Davis, G. A., 1972, Structural framework and evolution of the southern part of the Cordilleran orogen, western United States, *Am. J. Sci.* **272**:97–118.

Burchfiel, B. C., and Davis, G. A., 1975, Nature and controls of Cordilleran orogenesis, western United States: Extensions of an earlier synthesis, *Am. J. Sci.* **275-A**:363–396.

Champion, D. E., Howell, D. G., and Gromme, C. S., 1984, Paleomagnetic and geologic data indicating 2500 km of northward displacement for the Salinian and related terranes, *J. Geophys. Res.* **89**:7736–7752.

Coney, P. J., 1978, Mesozoic–Cenozoic Cordilleran plate tectonics, *Geol. Soc. Am. Mem.* **152**.

Coney, P. J., 1979, Tertiary evolution of Cordilleran metamorphic core complexes, in: *Cenozoic Paleogeography of the Western United States*, Pacific Coast Paleogeography Symposium 3 (J. M. Armentrout, M. R. Cole, and H. TerBest, Jr., eds.), pp. 15–28.

Coney, P. J., and Reynolds, S. J., 1977, Cordilleran Benioff zones, *Nature* **270**: 403–406.

Coney, P. J., Jones, D. L., and Monger, J. W. H., 1980, Cordilleran suspect terranes, *Nature* **288**: 329–333.

Crowell, J. C., 1979, The San Andreas fault system through time, *J. Geol. Soc. London* **136**: 293–302.

Davis, G. A., and Burchfiel, B. C., 1973, Garlock fault: An intracontinental transform structure, southern California, *Geol. Soc. Am. Bull.* **84**:1407–1422.

Davis, G. A., Monger, J. W. H., and Burchfiel, B. C., 1978, Mesozoic construction of the Cordilleran 'collage', central British Columbia to central California, in: Mesozoic Paleogeography of the Western United States, Pacific Coast Paleogeography Symposium 2, Pac. Sect. Soc. Econ. Paleontol. Mineral., pp. 1–32.

Day, J. W., Moores, E. M., and Tuminas, A. C., 1985, Structure and tectonics of the northern Sierra Nevada, *Geol. Soc. Am. Bull.* **96**: 436–450.

Dickinson, W. R., 1971, Clastic sedimentary sequences deposited in shelf, slope and trough settings between magmatic arcs and associated trenches, *Pac. Geol.* **3**:15–30.

Dickinson, W. R., 1977, Paleozoic plate tectonics and the evolution of the Cordilleran continental margin, in: Soc. Econ. Paleontol. Mineral. Pacific Coast Paleogeography Symposium 1, Paleozoic Paleogeography of the Western U.S., pp. 137–157.

Dickinson, W. R., 1981, Plate tectonics and the continental margin of California, in: *The Geotectonic Development of California* **1**:2–28.

Dillon, L. T., Pessel, G. H., Chen, J. H., and Veach, N. C., 1980, Middle Paleozoic magmatism and orogenesis in the Brooks Range, Alaska, *Geology* **8**:338–343.

Douglass, R. C., 1967, Permian Tethyan fusulinids from California, *U.S. Geol Survey Prof. Paper* 593A.

Drummond, K. J., Moore, G. W., Golovchenko, X., Larson, R. L., Pitman III, W. C., Rinehart, W. A., Simkin, T., and Siebert, L., 1981, Plate tectonic map of the circum Pacific region-northeast quadrant, *Am. Assoc. Pet. Geol., Circum-Pacific Council for Energy and Mineral Resources*.

Duncan, R. A., 1981, A captured island chain in the Coast Range, Oregon and Washington, *EOS Trans. Am. Geophys. Union* **61**:947.

Dusel-Bacon, C., and Aleinikoff, J. N., 1985, Petrology and tectonic significance of augen gneiss from a belt of Mississippian granitoids in the Yukon–Tanana terrane, east-central Alaska, *Geol. Soc. Am. Bull.* **96**:411–425.

Engebretson, D. C., 1982, Relative motions between oceanic and continental plates in the Pacific basin, Unpublished Ph.D. thesis, Stanford University.

Engebretson, D. C., Cox, A., and Thompson, G. A., 1984, Correlation of plate motions with continental tectonics: Laramide to Basin–Range, *Tectonics* **3**:115–119.

Frei, L. S., Magill, J. R., and Cox, A., 1984, Paleomagnetic results from the central Sierra Nevada: Constraints on reconstructions of the western United States, *Tectonics* **3**:157–177.

Fuis, G. S., Mooney, W. D., Healy, J. H., Mechan, G. A., and Lutter, W. J., 1982, Crustal structure of the Imperial Valley region, *U.S. Geol. Surv. Prof. Pap.* **1254**:25–50.

Gabrielse, H., 1985, Major dextral transcurrent displacements along the Northern Rocky Mountain Trench and related lineaments in north-central British Columbia, *Geol. Soc. Am. Bull.* **96**:1–14.

Gans, P. B., and Miller, E. L., 1983, Style of mid-Tertiary extension in east-central Nevada, Geologic Excursions in the Overthrust Belt and Metamorphic Core Complexes of the Intermountain Region, The Geological Society of America, Guidebook, Part I, pp. 107–139.

Gehrels, G. E., 1985, Geologic and tectonic evolution of Annette, Gravina, Duke, and Southern Prince of Wales Islands, southeastern Alaska, Ph.D. thesis, California Institute of Technology, Pasadena.

Gehrels, G. E., and Saleeby, J. B., 1985, Constraints and speculations on the displacement and accretionary history of the Alexander–Wrangellia–Peninsular superterrane, *Geol. Soc. Am. Abstr. Progr.* **17**:356.

Geissman, J. W., Callian, J. T., Oldow, J. S., and Humphreys, S. E., 1984, Paleomagnetic assessment of oroflexural deformation in west-central Nevada and significance for emplacement of allochthonous assemblages, *Tectonics* **3**:179–200.

Graham, S. A., 1979, Tertiary paleotectonics and paleogeography of the Salinian block, in: Cenozoic Paleogeography of the western United States, Pac. Sect. Soc. Econ. Paleontol. Mineral., Pacific Coast Paleogeography Symposium 3, pp. 45–51.

Greenhaus, M. R., and Cox, A., 1979, Paleomagnetism of the Morro Rock–Islay Hill complex as evidence for crustal block rotations in central coastal California, *J. Geophys. Res.* **84**:2392–2400.

Hadley, D. M., and Kanamori, H., 1977, Seismic structure of the Transverse Ranges, California, *Geol. Soc. Am. Bull.* **88**:1461–1478.

Hadley, D. M., and Kanamori, H., 1978, Recent seismicity in the San Fernando region and tectonics in the west-central Transverse Ranges, California, *Seismol. Soc. Am. Bull.* **68**:1449–1457.

Hamilton, W., 1983, The relation to plate motions of United States Cordilleran foreland deformation, *Geol. Soc. Am. Abstr. Progr.* **15**:271.

Harper, G. D., and Wright, J. E., 1984, Middle to Late Jurassic tectonic evolution of the Klamath Mountains, California–Oregon, *Tectonics* **3**:759–772.

Harper, G. D., Saleeby, J. B., and Norman, E., 1985, Geometry and tectonic setting of sea-floor spreading for the Josephine ophiolite, and implications for Jurassic accretionary events along the California margin, in: *Tectonostratigraphic Terranes of the Circum-Pacific Region*, Circum-Pacific Council for Energy and Mineral Resources, Earth Science Series No. 1, pp. 239–257.

Hearn, T. H., 1984, Crustal structure in southern California from array data, Ph.D. thesis, California Institute of Technology, Pasadena.

Helwig, J., 1974, Eugeosynclinal basement and a collage concept of orogenic belts, *Soc. Econ. Paleontol. Mineral. Spec. Publ.* **19**:359–376.

Hileman, J. A., 1978, A contribution to the study of the seismicity of southern California, Part I, Ph.D. thesis, California Institute of Technology, Pasadena.

Hillhouse, J., 1977, Paleomagnetism of the Triassic Nikolai greenstone, south-central Alaska, *Can. J. Earth Sci.* **14**:2578–2592.

Hillhouse, J. W., and Gromme, C. S., 1982, Paleomagnetism and Mesozoic tectonics of the Seven Devils volcanic arc in northeastern Oregon, *J. Geophys. Res.* **87**:3777–3794.

Hudson, T., Plafker, G., and Dixon, K., 1980, Horizontal offset history of the Chatham Strait fault, in: *The United States Geological Survey in Alaska: Accomplishments during 1980*, pp. 128–132.

Huffman, O. F., 1972, Niocene and post-Miocene offset on the San Andreas fault in central California, *Geol. Soc. Am. Abstr. Progr.* **2**:104–105.

Humphreys, E. D., 1985, Studies of the crust-mantle system beneath southern California, Ph.D. thesis, California Institute of Technology, Pasadena.

Humphreys, E., Clayton, R. W., and Hager, B. H., 1984. A tomographic image of mantle structure beneath southern California, *Geophys. Res. Lett.* **11**:625–627.

Irving, E., Woodsworth, G. J., Wynne, P. J., and Morrison, A., 1985, Paleomagnetic evidence for displacement from the south of the Coast Plutonic Complex, British Columbia, *Can. J. Earth Sci.* **22**:584–598.

Jones, D. L., Silberling, N. J., and Hillhouse, J., 1977, Wrangellia–A displaced terrane in northwestern North America, *Can. J. Earth Sci.* **14:**2565–2677.

Kanter, L. R., 1983, Paleomagnetic constraints on the movement history of Salinia, Ph.D. thesis, Stanford University.

Lachenbruch, A. H., and Sass, J. H., 1978, Models of an extending lithosphere and heat flow in the Basin and Range province, *Geol. Soc. Am. Mem.* **152:**209–250.

LePichon, X. L., 1968, Sea-floor spreading and continent drift, *J. Earth Sci.* **73:**3661–3697.

LePichon, X., and Hayes, D. E., 1971, Marginal offsets, fracture zones, and the early opening of the South Atlantic, *J. Geophys. Res.* **76:**6283–6293.

Lubetkin, L., 1980, Late Quaternary activity along the Lone Pine Fault, Owens Valley Fault Zone, CA, Master's thesis, Stanford University.

Luyendyk, B. P., Kamerling, M. J., and Terres, R., 1980, Geometric model for Neogene crustal rotations in southern California, *Geol. Soc. Am. Bull.* **91:**211–217.

Magill, J. R., and Cox, A., 1981, Post-Oligocene tectonic rotation of the Oregon Western Cascade Range and the Klamath Mountains, *Geology* **9:**127–131.

Magill, J. R., Cox, A., and Duncan, R., 1981, Tillamook volcanic series: Further evidence for tectonic rotation of the Oregon Coast Range, *J. Geophys. Res.* **86:**2953–2970.

Miller, E. L., Gans, P. B., and Garing, J. D., 1983, The Snake Range decollement: An exhumed mid-Tertiary ductile-brittle transition, *Tectonics* **2:**239–263.

Miller, E. L., Holdsworth, B. K., Whiteford, W. B., and Rodgers, D., 1984, Stratigraphy and structure of the Schoonover Sequence, northeastern Nevada: Implications for Paleozoic plate-margin tectonics, *Geol. Soc. Am. Bull* **95:**1063–1076.

Minster, J. B., and Jordan, T. H., 1978, Present-day plate motions, *J. Geophys. Res.* **83:**5331–5354.

Monger, J. W. H., 1977, Upper Paleozoic rocks of the Western Canadian Cordillera and their bearing on Cordilleran evolution, *Can. J. Earth Sci.* **14:**1832–1859.

Monger, J. W. H., and Ross, C. A., 1971, Distribution of fusulinaceans in the western Canadian Cordillera, *Can. J. Earth Sci.* **8:**259–278.

Monger, J. W. H., Price, R. A., and Tempelman-Kluit, D. J., 1982, Tectonic accretion and the origin of the two major metamorphic and plutonic belts in the Canadian Cordillera, *Geology* **10:**70–75.

Moore, J. G., 1959, The quartz diorite boundary line in the western United States, *J. Geol.* **67:**197–210.

Moore, D. G., 1973, Plate-edge deformation and crustal growth, Gulf of California structural province, *Geol. Soc. Am. Bull* **84:**1883–1906.

Morgan, J. W., 1968, Rises, Trenches, Great faults, and crustal blocks, *J. Earth Sci.* **73:**1959–1982.

Nestel, M. K., 1980, Permian fusulinacean provinces in the Pacific northwest are tectonic juxtapositions of ecologically distinct faunas, *Geol. Soc. Am. Abstr. Progr.* **12:**144.

Nilsen, T. H., 1984, Offset along the San Andreas fault of Eocene strata from the San Juan Baustista area and western San Emidgio Mountains, California, *Geol. Soc. Am. Bull.* **95:**599–609.

Packer, D. R., and Stone, D. L., 1974, Paleomagnetism of Jurassic rocks from southern Alaska and the tectonic implications, *Can. J. Earth Sci.* **11:**976–997.

Page, B. M., and Engebretson, D. C., 1984, Correlation between the geologic record and computed plate motions for California, *Tectonics* **3:**133–155.

Pavlis, T. L., 1982, Origin and age of the Border Ranges fault of southern Alaska and its bearing on the Late Mesozoic tectonic evolution of Alaska, *Tectonics* **1:**343–368.

Perfit, R., and Heezen, B. C., 1978, The geology and evolution of the Cayman Trench, *Geol. Soc. Am. Bull.* **89:**1155–1174.

Reed, B. L., Miesch, A. T., and Lanphere, M. A., 1983, Plutonic rocks of Jurassic age in the Alaska–Aleutian range batholith: Chemical variations and polarity, *Geol. Soc. Am. Bull.* **94:**1232–1240.

Ross, C. A., and Ross, J. R. P., 1983, Late Paleozoic accreted terranes of western North America: Pre-Jurassic rocks in western North American Suspect Terranes, pp. 7–22.

Ross, D. C., 1978, The Salinian block—a Mesozoic granitic orphan in the California Coast Ranges, in: Mesozoic Paleogeography of the Western United States, Pac. Sect. Soc. Econ. Paleontol. Mineral. Pacific Coast Paleography Symposium 2, pp. 509–522.

Saleeby, J. B., Goodwin, S. E., Sharp, W. D., and Busby, C. J., 1978, Early Mesozoic paleotectonic-paleogeographic reconstruction of the southern Sierra Nevada region, in: Mesozoic Paleogeography of the western United States, Pacific Coast Paleogeography Symposium 2, Pac. Sect. Soc. Econ. Paleontol. Mineral., pp. 311–336.

Saleeby, J. B., 1978, Kings River Ophiolite, Southwest Sierra Nevada Foothills, California, *Geol. Soc. Am. Bull.* **89**:617–636.

Saleeby, J. B., 1979, Kaweah Serpentinite Melange, Southwest Sierra Nevada Foothills, California, *Geol. Soc. Am. Bull.* **90**:29–46.

Saleeby, J. B., 1981, Ocean floor accretion and volcano-plutonic arc evolution of the Mesozoic Sierra Nevada, in: *The Geotectonic Development of California, Rubey Volume 1* (W. E. Ernst, ed.), Prentice–Hall, Englewood Cliffs, N.J., pp. 132–181.

Saleeby, J. B., 1982, Polygenetic ophiolite belt of the California Sierra Nevada—Geochronological and tectonostratigraphic development, *J. Geophys. Res.* **87**:1803–1824.

Saleeby, J. B., Harper, G. D., Sharp, W. D., and Snoke, A. W., 1982, Time relations and structural stratigraphic patterns in ophiolite accretion, west-central Klamath Mountains, California, *J. Geophys. Res.* **87**:3831–3848.

Saleeby, J. B., 1983, Accretionary tectonics of the North American Cordillera, *Annu. Rev. Earth Planet. Sci.* **15**:45–73.

Saleeby, J. B., 1984, Pb/U Zircon ages from the Rogue River area, Western Jurassic Belt, Klamath Mountains, Oregon, *Geol. Soc. America Abst. with Prgms.*, no. 5, **16**:331.

Saleeby, J. B., Speed, R. C., Blake, M. C., Allemendinger, R. W., Gans, P. B., Kistler, R. W., Ross, D. C., Stauber, D. A., Zoback, M. L., Griscom, A., McCulloch, D. S., Lachenbruch, A. H., Smith, R. B., and Hill, D. P., 1986, Centennial Continent/Ocean Transect #10, C2 Central California Offshore to Colorado Plateau, 2 sheets 1 : 500,000, Geological Society of America.

Saleeby, J. B., and Busby-Spera, C. J., 1986, Fieldtrip Guide to the Metamorphic Framework Rocks of the Lake Isabella Area, Southern Sierra Nevada, California: Mesozoic and Cenozoic Structural Evolution of Selected Areas, East-Central California, *Geol. Soc. Am. Cordilleran Section Guidebook Volume*, 81–94.

Schweickert, R. A., and Cowan, D. S., 1975, Early Mesozoic tectonic evolution of the western Sierra Nevada, California, *Geol. Soc. Am. Bull.* **86**:1329–1336.

Schultz, K. L., and Levi, S., 1983, Paleomagnetism of Middle Jurassic plutons of the north-central Klamath Mountains, *Geol. Soc. Am. Abstr. Progr.* **15**:427.

Seeber, L., Barazangi, M., and Nowroozi, A. A., 1970, Microearthquake seismicity and tectonics of coastal northern California, *Seismol. Soc. Am. Bull.* **60**:1669–1699.

Sharp, W. D., 1980, Ophiolite accretion in the northern Sierra, *EOS Trans. Am. Geophys. Union* **61**:1122.

Sharp, W. D., 1985, The Nevadan orogeny of the Foothills metamorphic belt, California: A collision without a suture? *Geol. Soc. Am. Abstr. Progr.* **17**:407.

Sibson, R. H., 1982, Fault zone models, heat flow, and the depth distribution of earthquakes in the continental crust of the United States, *Seismol. Soc. Am. Bull* **72**:151–163.

Sibson, R. H., 1983, Continental fault structure and the shallow earthquake source, *J. Geol. Soc. London* **140**:741–767.

Sieh, K. E., and Jahns, R. H., 1984, Holocene activity of the San Andreas fault at Wallace Creek, California, *Geol. Soc. Am. Bull.* **95**:883–896.

Silberling, N. J., 1973, Geologic events during Permian–Triassic time along the Pacific margin of the United States, *Alberta Soc. Pet. Geol. Mem.* **2**:345–362.

Silver, E. A., 1971, Tectonics of the Mendocine Triple Junction, *Geol. Soc. Am. Bull.* **82**:2965–2978.

Silver, L. T., 1983, Paleogene overthrusting in the tectonic evolution of the Transverse Ranges, Mojave and Salinian Regions, California, *Geol. Soc. Am. Abstr. Progr.* **15**:438.

Silver, L. T., and Anderson, T. H., 1974, Possible left-lateral early to middle Mesozoic disruption of the southwestern North American craton margin, *Geol. Soc. Am. Abstr. Progr.* **6**:955.

Silver, L. T., Taylor, H. P., Jr., and Chappell, B., 1979, Some petrological, geochemical and geochronological observations of the Peninsular Ranges batholith near the international border of the U.S.A. and Mexico: Mesozoic crystalline rocks: Peninsular Ranges Batholith and Pegmatites Point Sal Ophiolite, pp. 83–110.

Simpson, R. W., and Cox, A., 1977, Paleomagnetic evidence for tectonic rotation of the Oregon Coast Range, *Geology* **5**:585–589.

Sliter, W. V., 1984, Foraminifers from Cretaceous limestone of the Franciscan Complex, northern California, in: Franciscan Geology of Northern California, Pac. Sect. Soc. Econ. Paleontol. Mineral., pp. 149–162.

Smith, R. B., and Bruhn, R. L., 1984, Intraplate extensional tectonics of the eastern Basin–Range: Inferences on structural style from seismic reflection data, regional tectonics, and thermal-mechanical models of brittle-ductile deformation, *J. Geophys. Res.* **89**:5733–5762.

Snoke, A. W., Sharp, W. D., Wright, J. E., and Saleeby, J. B., 1982, Significance of mid-Mesozoic periodotic to dioritic intrusive complexes, Klamath Mountains–Western Sierra Nevada, California, *Geology* **10**:160–166.

Snyder, W. S., and Brueckner, H. K., 1983, Tectonic evolution of the Golconda Allochthon, Nevada: Problems and perspectives, in: Pre-Jurassic Rocks in Western North American Suspect Terranes, Pac. Sect. SEPM.

Speed, R. C., 1978, Paleogeographic and plate tectonic evolution of the early Mesozoic marine province of the western Great Basin, in: *Mesozoic Paleography of the Western United States*, Pacific Coast Paleogeography Symposium 2, pp. 253–270.

Speed, R. C., 1979, Collided Paleozoic microplate in the western United States, *J. Geol.* **87**:279–292.

Speed, R. C., 1983, Evolution of the sialic margin in the central-western United States, in Watkins, J. S. and Drake, C. L., eds., Studies in Continental Margin Geology, *Am. Assoc. Petrol. Geol. Mem.* 34, Hedberg Volume, 457–468.

Standlee, L. A., 1985, Age and tectonic significance of terranes adjacent to the Melones fault zone, N. Sierra Nevada, California, *Geol. Soc. Am. Abstr. Progr.* **17**:410.

Stewart, J. H., 1972, Initial deposits in the Cordilleran geosyncline: Evidence of a late Precambrian (<850 m.y.) continental separation, *Geol. Soc. Am. Bull.* **83**:1345–1360.

Stone, D. B., Panuska, B. C., and Packer, D. R., 1982, Paleolatitudes versus time for southern Alaska: *J. Geophys. Res.*, **87**:3697–3707.

Tempelmann-Kluit, D. J., 1979, Transported cataclastite, ophiolite and granodiorite in Yukon: Evidence of arc–continent collision, *Geol. Surv. Can. Pap.* 79–14.

Thompson, G. A., and Burke, D. B., 1973, Rate and direction of spreading in Dixie Valley, Basin and Range province, Nevada, *Geol. Soc. Am. Bull.* **84**:627–632.

Tipper, H. W., and Smith, P. L., 1985, Paleobiogeography and displaced Jurassic terranes, *Geol. Soc. Am. Abstr. Progr.* **17**:413.

Vallier, T. L., Dean, W. E., Rea, D. K., and Thiede, J., 1983, Geologic evolution of Hess Rise, central North Pacific Ocean, *Geol. Soc. Am. Bull.* **94**:1289–1307.

Webb, T. H., and Kanamori, H., 1985, Earthquake focal mechanisms in the eastern Transverse Ranges and San Emigdio Mountains, southern California and evidence for a regional decollement, *Seismol. Soc. Am. Bull.* **75**:737–757.

Wernicke, B., 1981, Low-angle normal faults in the Basin and Range Province: Nappe tectonics in an extending orogen, *Nature* **291**:645–648.

Whetten, J. T., Jones, D. L., Cowan, D. S., and Zartman, R. E., 1978, Ages of Mesozoic terranes in the San Juan Islands, Washington, in: Mesozoic Paleogeography of the western United States, Paleogeography Symposium 2, Pac. Sect. Soc. Econ. Paleontol. Mineral., pp. 117–132.

Whetton, J. T., Zartman, R. E., Blakely, R. J., and Jones, D. L., 1980, Allochthonous Jurassic ophiolite in northwest Washington, *Geol. Soc. Am. Bull.* **91**:359–368.

Williams, H., and McBirney, A., 1968, Geologic and Geophysical Features of Calderas, Center for Volcanology, University of Oregon, Eugene.

Wilson, J. T., 1965, A new class of faults and their bearing on continental drift, *Nature* **207**:343.

Wong, I. G., and Savage, W. U., 1983, Deep intraplate seismicity in the western Sierra Nevada, Central California, *Seismol. Soc. Am. Bull.* **73**:797–812.

Wright, J. E., 1982, Permo-Triassic accretionary subduction complex, southwestern Klamath Mountains, northern California, *J. Geophys. Res.* **87**:3805–3818.

Yancy, T. E., 1979, Permian positions of the northern hemisphere continents as determined from marine biotic provinces, in: *Historical Geography, Plate Tectonics and the Changing Environment* (A. J. Boucot, ed.), Oregon State University Press, Corvallis, pp. 239–247.

Yeats, R. S., 1981, Quaternary flake tectonics of the California Transverse Ranges, *Geology* **9**:16–20.

Yeats, R. S., 1983, Large-scale quaternary detachments in Ventura Basin, Southern California, *J. Geophys. Res.* **88**:569–583.

Chapter 4

THE STRATIGRAPHY AND TECTONICS OF CHINA

Chin Chen

Graduate Program in Oceanography and Limnology
Western Connecticut State University
Danbury, Connecticut 06810

I. INTRODUCTION

Until the appearance in 1962 of a 12-volume series based on the first symposium on the stratigraphy of China (1959), the principal source of information was the 2-volume classic of Grabau (1924, 1928). That the latter is still an invaluable reference is a tribute to Grabau's broad understanding. The information contained in the newer volumes reflects great advances since Grabau's time, but the volumes are poorly disseminated outside China. The series covers all aspects of stratigraphy: distribution of beds, standard stratigraphic columns, biostratigraphic zonation and correlation, facies, change and depositional environment, paleogeography and correlations with other parts of the world, economic mineral deposits, and discussion of significant problems. In the post-Cultural Revolution period, further activity has been marked by a second symposium on stratigraphy and the meeting of the third general assembly of the Paleontological Society of China in 1979. Following the first of these meetings, an important series of correlation charts with explanatory notes was published by the Nanjing Institute of Geology and Paleontology of Academia Sinica (Lu *et al.*, 1974, 1976). English versions of studies on stratigraphy, paleontology, and biostratigraphy have been published by the International Union of Geological Societies (1980, 1981), the Geological Society of America (1981, 1984), *Scientia Sinica* (1973), and *Geological Magazine* (1981).

In this chapter, an attempt will be made to summarize the present status of stratigraphic and tectonic knowledge of China drawn principally from little-known Chinese literature. Taiwan will not be considered, and readers are referred to the chapter by Biq *et al.* (1985) in Volume 7A of this series.

The transliteration of Chinese words used herein follows the phonetic Pinyin system, although some traditional Wade–Giles spellings of well-known localities will appear in parentheses (e.g., Beijing = Peking). The geographic names are taken from *The Atlas of the Provinces of the People's Republic of China* (Zhonghua Renmin Gongheguo Fen Sheng Dituji) (1977).

In the tectonic field, Lee (1939) was the classical work until recently. The traditional "geosynclinal" view was brought up to date with the publication of *The Geotectonic Evolution of China* by Ren *et al.* (1980) and Huang (1978). The role of fault block movements was exemplified in a volume edited by Zhang Wen-You (1980), while Chen Guoda (1977) concentrated on subsidence history, and the development of fracture systems was presented by Zhang and Zhong (1977). Plate tectonics first appeared in Chinese literature in 1975 (Li Chun-Yu, 1975), and this was followed by Li *et al.* (1980), Li and Tang (1983), and Guo *et al.* (1983). There are several references in English that bear on plate tectonics and apply to China, especially in the Himalayan region (Allen *et al.*, 1984; Gansser, 1964, 1980; Meyerhoff, 1978; Molnar, 1977; Molnar and Tapponnier, 1975, 1981; Tapponnier and Molnar, 1977, 1979; Tapponier *et al.*, 1982).

Concerning the terminology and symbols for ages of strata and formations,

TABLE I
Letter Symbols for Geological
Systems in China

System (or series)	Symbol
Archeozoic (Archean)	Ar
Proterozoic	Pt
Sinian (Upper Proterozoic)	Z
Cambrian	ε
Ordovician	O
Silurian	S
Devonian	D
Carboniferous	C
Permian	P
Triassic	T
Jurassic	J
Cretaceous	K
Tertiary	Te
Paleogene	E
Neogene	N
Quaternary	Q

letters are used for the major geological systems (Table I). The numbers are used for subdivisions of each system throughout this chapter.

II. GEOLOGICAL AND GEOPHYSICAL SETTING

A. Geomorphological Divisions

China occupies an area of about 9,600,000 km^2 on the eastern margin of the Eurasian continent. The topography is complex and diverse, with elevations ranging from the peak of Qomolangma Feng (Mt. Everest) at 8848 m to −154 m in the Turpan Depression (43°N, 89.5°E). Between these extremes, there are mountain ranges, plateaus, deserts, canyons, alluvial plains, deltas, irregular coasts, glacial, karst, and loess topography. In general, most elevations drop from west to east, and the morphology can be visualized as three steps. The Qinghai–Xizang (Tibet) Plateau of southwest China forms the highest step, with altitudes ranging from 3000 to 4000 m (Fig. 1). This plateau, "the roof of the world," is bounded by the Kunlun Shan and Qilian Shan on the north, the Daxue Shan and Hengduan Shan to the east, and the Himalaya Shan to the south and west, with the Pamir Plateau occupying the northwest corner.

The intermediate step, with an average elevation of about 2000 m, occupies most of central and northwest China. It consists mainly of loess-covered plateaus in northcentral and northwest China, and of inland basins, such as the Tarim, Junggar, and Sichuan, which are separated by mountain ranges. The lowest step, with elevations generally below 1000 m, lies in east China and is separated from the middle step by a series of north–south-trending mountain ranges, the Da Hinggan Ling (Greater Khingan Range), Taihang Shan, Wuling Shan, and Xuefeng Shan (Fig. 1). The plains of the lower step are generally below 500 m, interspersed with hilly regions. The major plains are the Northeast Plain (Songliao Basin), the North China Plain, and the Yangzi Plain. Farther to the east are several elevated regions such as the Changbai Shan (east of the Northeast Plain), the mountainous regions of Shandong and Liaodong peninsulas, and the hilly coastal regions of southeast China. Still farther east, beyond the 20,000-km-long Chinese coast, are the marginal seas—Bo Hai, Yellow Sea, and East and South China Seas.

The boundaries between these morphological units are formed by well-developed fracture systems. The high Qinghai–Xizang Plateau is separated from the middle step of mountains, basins, and loess plateaus by several well-defined, deep fracture zones—the Himalayan fracture zone on the south and southwest, the West Kunlun Shan–Altun fracture to the north and northwest, the Qilian Shan fracture zone to the northeast, and the Longmen Shan in the area to the east and southeast. There are, in addition, several deep fracture zones within the plateau.

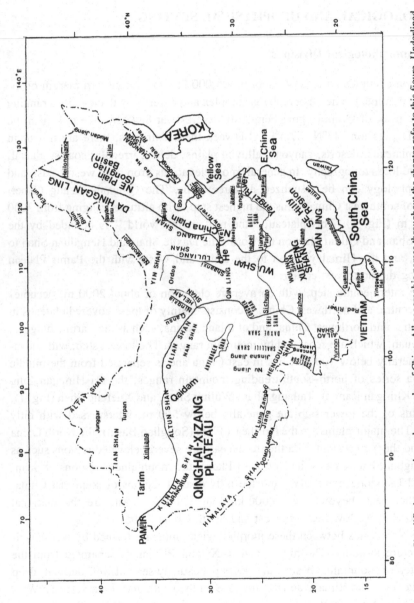

Fig. 1. Simplified topography and geography of China. All geographic names throughout the chapter refer to this figure. Underlined letters refer to province names.

The boundary between the middle step and the lower elevations of the coastal belt coincides with two northeast–southwest fracture zones: the northern along the Da Hinggan Ling and Taihang Shan, the southern following the Wuling Shan and Xuefeng Shan and extending to the Sino-Vietnam border. These zones are interrupted by an east–west-trending fracture zone.

The significance of these morphological divisions is further underlined by gravity and seismological data. The depth of the Mohorovicic discontinuity can be correlated with geological features of the topographic units and the intervening fracture zones. The lowest Bouguer gravity anomalies, from −400 to −500 mgal, coincide with the Qinghai–Xizang Plateau (Fig. 2). The steep gravity gradient at the edge of the plateau coincides, in turn, with the deep fracture zones bounding the plateau. Another zone of steep gravity gradient (from −100 to −50 mgal) lies along the Da Hinggan Ling–Taihang Shan–Wuling Shan trend dividing the middle and lower steps. The anomaly over the middle step ranges from −100 to −200 mgal, the zero anomaly line occurs along the coast and over the coastal plains.

The topography of the Mohorovicic discontinuity (Fig. 3) shows that the subsurface is deepest (about 50–70 km) below the Qinghai–Xizang Plateau, occurs at 40–50 km beneath the middle step, and at 40–30 km below the coastal region. The upper mantle is thus relatively high under east China and is depressed in southwest China. Compressive forces associated with subduction may be responsible for this depressed Moho (Ren *et al.*, 1980). The low shear-wave velocities (4.1–

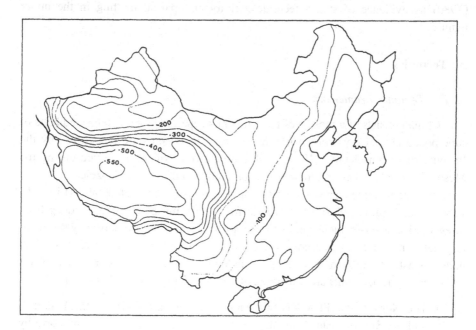

Fig. 2.   Map of average Bouguer anomalies in China (in milligals).

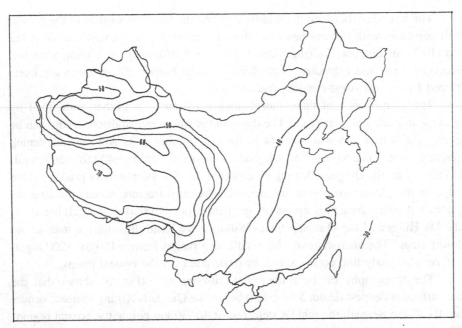

Fig. 3. The depth of the Mohorovicic discontinuity in China (in kilometers).

4.2 km/sec) below the Qinghai–Xizang Plateau have been interpreted by Ren *et al.* (1980) as evidence of strong tectonic activity and partial melting in the upper mantle.

B. Tectonics

1. *The Tectonic Framework*

China consists of a mosaic of tectonic units separated by the deformed belts of superposed tectonic systems. Three major tectonic systems are recognized by the Institute of Geomechanics (1978), and although most of these formed during the Mesozoic or later, their beginnings can be recognized in the Paleozoic.

In plate tectonic terms, the units can be organized into paleoplates (Li *et al.*, 1980; Li and Tang, 1983) using criteria for identifying plate margins, deep fractures, ophiolite zones, mélange, and metamorphic belts, and supported by sedimentary data on facies, paleontological differences, and paleomagnetic results. Each plate consists of one or more nuclei surrounded by mio- and eugeoclines. The plates and their constituent units are illustrated in Fig. 4 and are described briefly below.

- The North China Plate has two nuclei, the Tarim Platform (Fig. 4, unit 3) and the Sino-Korean Paraplatform (Fig. 4, unit 1), bounded to the north by the Tian Shan (Fig. 4, unit 5a) and Nei Monggol–Da Hinggan fold systems

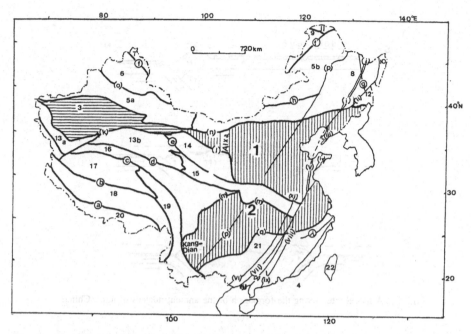

Fig. 4.   Simplified tectonic units and major fracture systems of China. The numbers refer to tectonic units and correspond to text references; lowercase letters refer to fracture systems also corresponding to references in the text and Table III.

(Fig. 4, unit 5b), and on the south by the Kunlun Shan (Fig. 4, unit 13), Qilian Shan (Fig. 4, unit 14), and Qinling (Fig. 4, unit 15) fold belts. The line of suture to the Siberian Plate on the north is the Xilamolum line (Fig. 4, h) and Borohoro–mid Tian Shan line (Fig. 4, o), while the southern suture follows the line of deep lithospheric fractures of East Kunlun Shan–South Qinling (Fig. 4, d) and Daba Shan (Fig. 4, m). The Jiamusi block, which is the eastern part of the Jilin–Heilongjiang fold system (Fig. 4, unit 8), is regarded as a median massif by Li *et al.* (1980), separated from the rest of unit 8 by the Mudanjiang fracture which probably is a subduction zone (Fig. 4, g).

The Sino-Korean massif (Zhang *et al.*, 1980) is an Archean protolith (Zhang and Zhang, 1980) bounded north and south by ancient subduction zones; its mode of origin is illustrated in Fig. 5.

The South China Plate consists of the nuclear Yangzi Paraplatform (Fig. 4, unit 2) bounded to the East by the South China geosynclinal fold system (Fig. 4, unit 21) and to the west and southwest by the Xizang–Yunnan geosynclinal fold system (Fig. 4, units 16–19). The Yarlung–Zangbo fracture system (Fig. 4, a) forms the suture between the South China Plate and the Himalayan fold system of the Indian Plate. The subduction process has

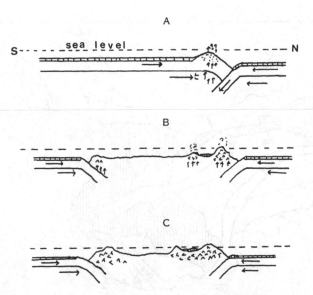

Fig. 5.   A model interpreting the formation of the ancient nucleus of north China.

been modeled by Huan *et al.* (1980), as shown in Fig. 6. The suture with the
Philippine subplate lies along the Taiwan longitudinal valley.

An overall four-stage evolutionary pattern of the Mesozoic–Cenozoic active
continental margin and island arc tectonics of the western Pacific has been
presented by Guo *et al.* (1983). Taiwan, the Korean Peninsula, Sea of Japan,
and South China Sea are included in this survey. Of the four stages, the oldest,
the Hercynian–Indosinian, represents the consolidation of the eugeosynclinal
fold belts from the margin of the ancient Asian mass. The second, Yanshanian, stage marks the formation of an Andean coastal mountain–trench system
along the Tethyan margin in the western Pacific. During the third, early
Himalayan, stage a system of offshore island arc and backarc basins formed
due to orogeny and subduction of the Pacific Plate. The final, late Himalayan,
phase was followed by the modern plate tectonic events with the composite
trench–island arc–backarc basin systems of the western Pacific.

The Himalayan fold system forms the northern part of the Indian Plate, while
the Nei Monggol–Da Hinggan (Fig. 4, unit 5b), Junggar (Fig. 4, unit 6), and
Altay (Fig. 4, unit 7) are systems of arcuate folds at the margin of the
Siberian Plate.

The process of evolution leading to the present organization of plates is most
easily represented by considering the evolution through time of three tectonic domains (Fig. 7) and the migration of the loci of tectonic activity illustrated in Fig. 8.
For convenience of reference, the subdivisions of orogenic cycles used in China are

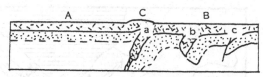

Fig. 6. An interpretive model of the subduction zone between the Indian and South China Plates. A, Indian Plate; B, South China Plate; C. Himalaya Mountains; a, Yarlung–Zangbo Deep Fracture System; b, Bangongco Deep Fracture System; c, Jinshajiang Deep Fracture System.

shown in Table II with their European equivalents. These tectonic domains—the Pal-Asiatic, Marginal Pacific, and Tethys–Himalayan—are identified by their age of formation and by the fracture systems which bound them.

The Pal-Asiatic domain developed during the Paleozoic. The folding of the Ergun–North Mongolian geosyncline at the end of early Cambrian (Xingkaiian) time was followed by the disruption of the Chinese protoplatform and the formation of a series of geosynclines, including the Tian Shan, Nan Shan, Qinling, and others. The folding and consolidation of these geosynclines during the Variscan orogeny resulted in the "cementing together" of the Siberian, Tarim, Sino-Korean, and Yangzi Platforms. This giant craton has been called Pal-Asia.

The record of Mesozoic and Cenozoic activity centers around the Marginal Pacific and Tethys–Himalayan domains. Three zones can be identified in the Marginal Pacific domain: the inner (Mesozoic), central (late Mesozoic), and outer (Cenozoic) belts. The Taiwan fold system is the only representative of the latter in China (see Biq *et al.*, Volume 7A of this series). The inner and central belts are superposed on older tectonic units of different ages.

Tectonic activity of the Marginal Pacific domain also involved some of the older Variscan tectonic units, the Nei Monggol–Da Hinggan and the Jilin–Heilongjiang Variscides, and the Caledonian South China fold system, as well as the Sino-Korean and Yangzi Paraplatforms. The domain is characterized by large, northeast- to north-northeast-trending swells and depressions. Igneous activity was

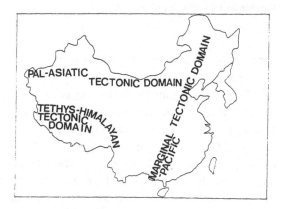

Fig. 7. Tectonic domains of China.

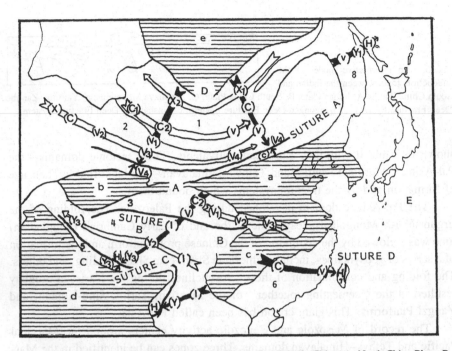

Fig. 8. The tectonic systems and the migration of tectonic activity in China. A, North China Plate; B, South China Plate; C, Indian Plate; D, Siberian Plate; E, Pacific Plate; a, Sino-Korean Paraplatform; b, Tarim Platform; c, Yangzi Paraplatform; d, Indian Platform; e, Siberian Platform; 1, Ergun–Altay–Junggar Geosynclinal Fold System; 2, Tianshan–Nei Monggol–Da Hinggan Geosynclinal Fold System; 3, Kunlunshan–Qinling–Qilianshan Geosynclinal Fold System; 4, Xizang–Yunnan Geosynclinal Fold System; 5, Himalayan Geosynclinal Fold System; 6, South China Geosynclinal Fold System; 7, Taiwan Geosynclinal Fold System; 8, Marginal Pacific Geosynclinal Fold System; Suture Line A, Borohoro-Mid-Tianshan and Xilamolun Deep Fracture Systems; Suture Line B, East Kunlunshan–South Qinling and Dabashan Deep Fracture System; Suture Line C, Yarlung Zangbo Deep Fracture System; Suture Line D, Taiwan Longitudinal Valley Fault; (X), Xingkaiian fold; (X$_1$), Early Xingkaiian (Late ϵ_1) fold; (X$_2$), Late Xingkaiian (Late ϵ_2) fold; (C), Caledonian fold; (C$_1$), Early Caledonian (Late O$_2$) fold; (C$_2$), Late Caledonian (Late S) fold; (V), Variscan; (V$_1$), Early Variscan (Late D) fold; (V$_2$), Middle Variscan (Late C$_1$) fold; (V$_3$), Late Variscan (Late C$_3$) fold; (V$_4$), End Variscan (Late P$_1$) fold; (I), Indosinian fold; (Y), Yanshanian fold; (Y$_1$), Early Yanshanian fold; (Y$_2$), Middle Yanshanian fold; (Y$_3$), Late Yanshanian fold; (H), Himalayan fold.

appreciable, with large-scale intrusion of granite and extensive outpourings of Pacific-type volcanics.

The Tethys–Himalayan domain is separated into two parts by the Yarlung Zangbo fracture zone. The orogenic activity in the southern or Himalayan part is Cenozoic, whereas in the northern part, which includes the Xizang–Yunnan, the Songpan–Garze and Qinling fold systems belong to the Indosinian and Yanshanian orogenic episodes of Mesozoic age. The tectonic activity in this domain, particularly during the Himalayan cycle, had a great effect on the old Pal-Asiatic domain, especially in the southwestern part. The Tian Shan, Kunlun Shan, and Nan Shan

TABLE II.
Subdivision of Orogenic Cycles and Important Events of Tectonic Development of China

	Geological chronology	Isotopic age (m.y.)	Subdivision of orogenic cycles and important events of tectonic development in china		Orogenic cycles of Europe
Cenozoic	Quaternary	15	Himalayan		Alpine
	Tertiary	67	Yanshanian	Formation and development of M-P & T-H	Cimmerian
Mesozoic	Cretaceous	137			
	Jurassic	190	Indosinian	Destruction and disintegration of parts of P-A; intensive activity of M-P and T-H	
	Triassic	230	Variscan	Consolidation of Central Asiatic–Mongolan . . . geosynclines; cementation of Siberian Platform with Sino-Korean and Tarim Platforms; formation of P-A	Variscan
Paleozoic	Permian	280			
	Carboniferous	350		Formation and development of P-A	Caledonian
	Devonian	405	Caledonian	Formation of South China Platform	
	Silurian	440		Disintegration of Chinese Protoplatform; formation of Kunlun, Qinling, Beishan, Tianshan (central and southern parts), and other geosynclines	
	Ordovician	550	Xingkaiian		
	Cambrian	570		Formation of Yangzi and Tarim Platforms; combination of these platforms with Sino-Korean Paraplatform to form the Chinese Protoplatform	
Late Proterozoic (Sinian Subera)[a]	Eocambrian	700	Yangziian		Assyntian
	Sinian s.s.				
	Qingbaikou	1000	?		Dalslandian
	Jixian	1400			
	Changcheng	1700	Zhongtiaoian	Formation of Sino-Korean Paraplatform	Svecofennian
Early Proterozoic	Hutuo	2000	Wutaiian		Karelian
	Wutai	2500	Fupingian		Belomorian
Archean	Fuping				

[a]According to a report by the Institute of Geochemistry, Academia Sinica, the lower limit of the Sinian Subera is fixed at 1950 ± 50 m.y. (June 17, 1976).

were transformed into elongated and uplifted fault blocks fronted by molassic foredeeps.

Tectonic activity during the Indosinian orogeny ranges from middle Triassic into the Jurassic, with folding and uplift of the Qinling, Songpan–Garze, and Sanjiang geosynclines which, together with contemporaneous fold belts in Indochina, form the Indosinian system. Indosinian orogenic activity is of paramount importance for it marks the breakdown of the Pal-Asiatic domain and the inception of the Marginal Pacific domain. As mentioned earlier, the domain is characterized by Indosinian folds and faults and associated magmatism.

The parallel to Indosinian orogenic activity is the coeval Cordilleran system in North America, which also marks an important change in the tectonic framework. The Tethyan geosyncline west of the Pamirs developed upon a Variscan or older basement during late Triassic or early Jurassic time.

Two features of the history of the various fold belts are emphasized in Fig. 8. The first and most obvious is that the age of orogenic activity in any belt tends to migrate systematically along it. Good examples of this can be seen, for instance, affecting the Ergun–Altay–Junggar (Fig. 8, 1) and Tian Shan–Nei Monggol–Da Hinggan (Fig. 8, 2) belts which lie between the Siberian Platform to the north and the Tarim–Sino-Korean Platforms in the south, although migration of tectonic activity is in opposite senses (as indicated by open arrows). There is also a migration of tectonic activity from one belt to another (shown by solid arrows in Fig. 8). Where migrations converge, the convergence is marked by a suture (Fig. 8, lines A to D). In the zone described above, the convergence, suture A, lies close the Tarim–Sino-Korean Platforms and occurred in late Variscan time. Similar patterns of orogenic migration can also be seen affecting both the Marginal Pacific and the Tethys–Himalayan domains.

2. Tectonic Units

As is clear from the preceding section, the tectonic pattern of China can be simplified to three plates, the North China Plate, the South China Plate, and the marginal zones of the Siberian and Indian Plates. The principal plates considered here are the North and South China Plates, each consisting of a nuclear zone bordered by one or more orogenic belts. These divisions are closely reflected by the morphological divisions.

Within the North China Plate are two nuclei (see above), the Sino-Korean Paraplatform and the Tarim Platform. The nucleus of the South China Plate is the Yangzi Paraplatform. A possible fourth nucleus, the South China Sea Platform, may exist as the easternmost part of the Indosinian massif. These will be described briefly, followed by short accounts of the fold belts grouped according to their locations around these nuclei.

(a) Sino-Korean Paraplatform. The Sino-Korean Paraplatform occupies north China, the southern part of northeastern China, northern Korea, Bo Hai Bay, and the northern Yellow Sea. This triangular platform is separated from neighboring units by deep fracture zones. The oldest platform in China, its shape was acquired about 1700 Ma, during the Zhongtiao orogeny (Table II). The marginal parts, such as the Alxa region in the northwest, were consolidated by the end of the Protero-zoic.

Within the basement sequence, three major unconformities are recognized: the first between the Archean (Fuping Group) and the Lower Proterozoic (Wutai Group), reflecting orogenic activity referred to as the Fuping or Anshan orogeny, at 2350–2550 Ma; a second between the lower (Wutai Group) and the upper part of the Lower Proterozoic (Hutuo Group), the time of the Wutai orogeny at about 2000 Ma; and finally the unconformity associated with the Luliang or Zhongtiao orogeny, separating the Hutuo Group from the Upper Proterozoic Sinian suberathem. The basement sediments consist of Sinian and Cambro-Ordovician neritic sedimentary, largely carbonate, rocks. These are overlain by Permo-Carboniferous continental rocks with some marine horizons followed by continental Meso-Cenozoic rocks. Silurian, Devonian, and Lower Carboniferous rocks are missing. In some areas, as the Yanshan–Liaoning (40.5°N, 118°E), or Yanliao, depression and the Shandong Peninsula, there are considerable volumes of continental rocks, volcanics, and granitoid intrusives. Cenozoic volcanics are widespread. Indosinian disturbances are restricted to the Nei Monggol and Yanliao depressions.

(b) *Tarim Platform.* The Tarim Platform is essentially a median massif bounded to the north by the Tian Shan and by the Kunlun Shan to the south. The surface cover is principally of Cenozoic age. The basement and Paleozoic rocks crop out mostly along the northern border, in such areas as Kalpin (40°30'N, 79°E) and Kuruktag (41°N, 88–92°E) in Xinjiang. In the latter area, two of the three major Precambrian unconformities mentioned above (between Archean and Lower Pro-terozoic and between Lower and Upper Proterozoic) are found, and an additional one between Lower and Upper Sinian. A diamictite (tillite) is found in the Upper Sinian; in the Lower Sinian, stromatolites occur. On the basis of the latter, it is assumed that the Tarim Platform acquired its present form near the end of the Proterozoic, contemporaneously with the Yangzi Paraplatform.

(c) *Yangzi Paraplatform.* The Yangzi Paraplatform, the nuclear region of the South China Plate, is the second most important platform in China. It covers the greater part of the Yangzi Basin from eastern Yunnan to Jiangsu and includes the southern part of the Yellow Sea. Although formerly believed to have consoli-dated contemporaneously with the Sino-Korean Paraplatform, recent isotopic stud-ies, together with investigation of microflora and stromatolites in southwestern and central China, now place the event at the end of the Proterozoic (about 700 Ma) during the Yangzi orogeny. The sedimentary cover consists mainly of Sinian (s.s.) to Triassic rocks. In Sichuan and northern Guizhou, Devonian and Carboniferous rocks are generally absent. Continental beds of Jurassic, Cretaceous, and Cenozoic age are found in Sichuan, central Yunnan, Hubei, Jiangsu, and other regions. The principal tectonism is associated with the Yanshanian cycle when well-developed folds formed within the sedimentary cover. In the western part of the paraplatform, the Kang–Dian axis shows evidence of polycyclic tectonism, for here there is evidence of earlier Variscan activity. In the lower Yangzi, there are also indications of Indosinian as well as Yanshanian activity.

(d) *Central Asiatic Fold Region.* This term is used to cover the systems of folds which separate the Siberian and Tarim–Sino-Korean Platforms. The component members are the Tian Shan (Fig. 4, unit 5a), Nei Monggol–Da Hinggan (Fig. 4, unit 5b), Junggar, Altay, Jilin–Heilongjiang, and Ergun geosynclinal fold systems. The region is divided into two parts by the Derbugan fracture zone, which is an extension of the Middle Mongolian fracture system, with the Ergun folds to the north and all the remaining zones to the south. Although the Ergun belongs to Xingkaiides (Table II), the remainder are Caledonian or Variscan, and all except the southern Tian Shan are formed of eugeoclinal deposits. Some recognizable parts of ophiolitic suites, such as the Caledonian in western Junggar (Fig. 1), Variscan, and later orogenic activity, mark stages in cratonization.

(e) *Marginal Pacific Fold Region.* Three fold belts, the Nadanhada, Upper Heilongjiang, and Yanbian fold belts, comprise the Marginal Pacific fold region which lies east of the Siberian Platform. The Nadanhada and Yanbian are eugeoclinal belts related to the Sikhote Alin system; the miogeoclinal Upper Heilongjiang belt belongs to the Mongol–Okhotsk belt. The Yanbian is the older belt, of Variscan age; the other two belong to the Yanshanian cycle.

(f) *Tethyan Fold Region.* This general term is used to cover those fold belts that form the wide zone between the Indian Plate to the south and the North China and South China Plates to the north and east, respectively. As seen in Fig. 1, they wrap around and help define the topographic Qinghai–Xizang Plateau. As in the case of the Central Asiatic fold region, a division can be made between those component elements north and east of the Yarlung Zangbo fracture zone and those to the south. Those fold belts to the north and east are the Kunlun Shan, Qilian Shan (Nan Shan), Qinling, Songpan–Garze, Tanggula, Lhasa, and Sanjiang; those to the south and west include, in addition to the Himalayas, the Sulaiman (Pakistan) and Arakan Yoma (Burma). Immediately south of the Yarlung Zangbo fracture zone, a discontinuous belt of ophiolites is exposed; these represent eugeoclinal rocks absent in the Himalayas (s.s.) at the northern margin of the Indian craton (Chang and Cheng, 1973).

The tectonic history of the ranges to the north and east is considerably more complex; in essence, those chains closest to the Tarim–Sino-Korean Paraplatform have a polycyclic history. This history began during the Caledonian, continued through the Himalayan cycle, and even extended into recent geological time—for in Qinghai, Cambrian volcanics are thrust over late Tertiary red beds. The gradual incorporation of these ranges into the cratonic structure of China is illustrated in Fig. 8.

(g) *South China Fold Region.* This region covers the tectonic units lying south of the Yangzi Paraplatform. According to the Geotectonics Group (G. Chen, 1977), it consists of the larger South China Caledonian fold belt and a smaller Variscan element, the southeast coastal belt, present in the coastal regions of Zhejiang, Fujian, Guangdong, the neighboring continental shelves, and Hainan Island. The latter was strongly deformed during the Mesozoic and is largely covered by Jurassic and Cretaceous continental volcanic rocks.

(h) *Taiwan Fold System.* The Taiwan fold system forms the major link with the island arc system of the western Pacific Ocean (see Biq *et al.*, Volume 7A of this series).

3. *Fracture Systems*

Fracture systems are not necessarily persistent or limited in time; different parts may come into existence at different times and one type of fracture may pass into another. Some important deep fractures have been reactivated on more than one occasion. From satellite pictures of China and geological and geophysical data, more than 100 deep fractures have been recognized and have been divided into three categories according to their depth (Ren *et al.*, 1980). The translithospheric fractures are the deepest and are commonly still active. They are the sites of Benioff zones and are the foci of deep earthquakes; at the surface, they are associated with ophiolitic suites and mélange. This first category includes the Yarlung–Zangbo, Bangongco–Nujiang, Jinshajiang–Honghe (Gold Sand River–Red River), East Kunlun Shan–South Qinling, Qinghaihu–North Huaiyang, Mid-Mongolian, Mudanjiang, Xilamolun, and Jiangshan–Shaoxing fracture systems. The second class consists of faults which cut the lithosphere but do not extend deeper; they are referred to as lithospheric fractures and in China are usually found in geosynclinal regions. They are characterized by mafic and ultramafic complexes, but the ophiolitic suite is poorly developed. Deep crustal fractures, the third category, are numerous, particularly in southeastern China. They are usually associated with extensive volcanics and zones of strong compression, often followed by tensional growth faults. Ultramafic rocks are absent. The most important of the second class of faults are the Tancheng–Lujiang, West Kunlun Shan–Altay–Beishan, Qilian Shan, Longmen Shan–Daba Shan, North Sino-Korean, Borohoro–Mid-Tian Shan, and the Da Hinggan–Taichang–Wuling fracture systems. Brief notes on these important fracture systems are given in Table III.

In addition to the above-mentioned fracture systems, there are many other crustal fractures, particularly in southeastern China. Often associated with extensive volcanics and zones of strong shear followed by tension, ultramafics are absent.

TABLE III
Principlal Translithospheric and Deep Fracture Zones

Name[a]	Location	Principal features and events
Translithospheric fractures		
(a) Yarlung-Zangbo [part of Indus fracture zone (Gansser, 1964)]	Xizang (Tibet); Yarlung-Zangbo R., E–W, 200 km; west along Indus R. Valley; east, bends south into Zayu region of Burma	Zone of suture between the Indian and Eurasian plates. Characterized by ophiolitic suite, glaucophane schist, and mélange. Late Cretaceous closure of Tethys (about 80 Ma). Late Eocene (about 45 Ma) initial collision of Indian and Eurasian subcontinents. Main Himalayan movement about 2 Ma
(b) Bangongco–Nujiang	Subparallel to Yarlung-Zangbo fracture system; approx. 32°N, extending westward into India and Pakistan; eastward near Dengqen (approx. 32°N, 96°E) follows Nujiang Valley south into Burma; length 2800 km	Boundary between Lhasa fold system (Fig. 4, unit 18) and Tanggula (Fig. 4, 17) and Sanjiang (Fig. 4, 19). Ultramafic rocks, greenschist, spilitic keratophyres, and mélange(?) are found. Region of strong tectonic movement and magmatic and metamorphic activity since Indosinian (Jurassic) orogenic cycle
(c) Jinshajiang–Honghe (Gold Sand R.–Red R.)	Subparallel to Yarlung-Zangbo and Bangongco fracture zones; approx. 35–35.5°N in western segment; bends south and is displaced in area 34.5°N, 81.5°E, following the Jinshajiang Valley to the Red R. Valley to Beibu Wan (northern Gulf of Tonkin) and Yinggehai, southwestern Hainan Is.; length 4000 km	Boundary between the Tanggula (Fig. 4, 17) and Songpan-Garze (Fig. 4, 16) fold systems in west. Boundary between the Sanjiang (Fig. 4, 19) and the south China (Fig. 4, 21) fold system and the Yangzi Paraplatform (Fig. 4, 2). Ophiolitic suite, mélange, glaucophane schist, greenschist, ultramafic, and mafic rocks are found. Oldest mélange in the Ailaoshan are a (24°N, 101.3°E) is of Proterozoic to early Cambrian age. Strong activity since the Indosinian (Jurassic) orogenic cycle. Probably an ancient suture

[a]Letters in parentheses correspond to letters in Fig. 4.

TABLE III   (Continued)

Name	Location	Principal features and events
(d) E. Kunlunshan–S. Qinling	Strikes approx. E–W; western part in 35–37°N; eastern end about 35–35.5°N; western end cut off by a lithospheric fracture; length in excess of 2500 km	Boundary between Songpan-Garze (Fig. 4, 16) and Qinling (Fig. 4, 15) fold systems Ophiolitic suite and mélange in the Burhan Budai and Anyemaqen Shan areas are of Permo-Triassic age and associated with the Variscan orogenic cycle
(e) Qinghaihu–N. Huaiyang	Extends 2200 km from 39.5°N, 95°E southeast to 31.3°N, 117.3°E; truncated to NW by Altun–Beishan and to SE by the Tancheng–Lujiang system	NW part boundary between Qilian (Fig. 4, 14) and the Qinling (Fig. 4, 15) fold systems The SE part separates the Sino-Korean Paraplatform (Fig. 4, 1) from the Yangzi (Fig. 4, 2) SE segment has been active since Proterozoic–Paleozoic time; in the NW part near Qinghai Hu (37°N, 100°E), mélange is found in the Triassic geosyncline High-pressure, low-temperature metamorphics, glaucophane, polycyclic granites of the Caledonian, Variscan, Indosinian, and Yanshanian orogenic cycles, and ultramafic rocks found in SE segment
(f) Mid-Mongolian	Although mostly lying in Mongolia, it is found also in NW and NE China	Boundary between the Central Asian and Mongolian geosynclines An important boundary for Paleozoic paleotectonics and paleogeography A translithospheric fault in the NW [between the Altay and Junggar fold systems (Fig. 4, units 6 and 7)], it is confined to the lithosphere separating the Ergun (Fig. 4, 9) and Upper Heilongjiang (Fig. 4, 11) fold belts from the Da Hinggan (Fig. 4, 5b) fold belt Ultramafic and mafic rocks as-

(continued)

TABLE III   (*Continued*)

Name	Location	Principal features and events
		sociated with translithospheric segment
(g) Mudanjiang	Follows 130°E between 49°N and 43°N	Lies within the Jilin–Heilongjiang fold system (Fig. 4, 8), probably the boundary of the Jiamusi Plate (Ren *et al.*, 1980)
		Paleozoic in age, with ultramafic and granitoid plutons and glaucophane schist
(h) Xilamolun	Located in the Nei Monggol–Da Hinggan fold system; stretches about 1100 km E–NE	Important Upper Paleozoic paleogeographic boundary separating warm Pacific fauna and Cathay flora from cold Arctic fauna and Angara flora
		Ophiolite, glaucophane schist, mélange, and ultramafic rocks are found
(i) Jiangshan–Shaoxing	Located in region of south China fold system	A Proterozoic fracture system. Associated with ultramafic and basic rocks
Lithospheric fractures		
(j) Tancheng–Lujiang	Extends from northern Vietnam NE into Siberia; length in China about 3600 km	Possibly the most important deep fracture system in eastern Asia belonging to the Marginal Pacific domain
		Cuts through Jilin–Heilongjiang and south China fold systems and through the Sino-Korean and Yangzi Paraplatforms, and is important site of Mesozoic and Cenozoic activity
		An important volcanic, seismic, and metallogenic zone. Made up of 11 zones (i–xi of Fig. 4) (Ren *et al.*, 1980); considerable differences occur in depth and age within each zone, or even in faults within a zone
		Faults decrease in depth of penetration toward the south
		Faults in north range from pre-Sinian or Paleozoic to Mesozoic and Cenozoic, but are generally Mesozoic in the south; in NE, faults show less

TABLE III   (*Continued*)

Name	Location	Principal features and events
		evidence of compressional behavior; further south, sheet and tensional shear predominate, with an est. 520-km displacement in the Dabie Shan (Wu *et al.*, 1981)
(k) W. Kunlunshan–Altun– Beishan		Divides the Kunlunshan fold system and forms the boundary between the Tarim Platform (Fig. 4, 3) and the Kunlunshan (Fig. 4, 13) and Qilian Shan fold systems— the boundary coinciding with a steep gravity gradient (Fig. 4, k)
		A left lateral shear fracture active since the Paleozoic
		Ultramafic and mafic rocks associated with fractures
(l) Qilianshan		Boundary between Qilianshan (Fig. 4, 14) and the Sino-Korean Paraplatforms (Fig. 4, 1)
		Consists of several subparallel fracture zones in Qilianshan fold system
		Mafic and ultramafic rocks found along fracture system
(m) Longmenshan– Dabashan	Western zone trends NE; eastern zone strikes NW–SE	Composite of western Longmenshan and eastern Dabashan fracture zones
		Both show shear and compression and are associated with mafic and ultramafic rocks
		Western zone has undergone repeated movement from Archean to Cenozoic times
		Western zone boundary between Yangzi Paraplatform (Fig. 4, 2) and the Songpan-Garze fold system (Fig. 4, 16); also associated with a steep gravity gradient and morphologic break (see pp. 163, 165).
		Reversal of shear motion along Longmenshan fracture (sinistral during Indosinian

TABLE III   (*Continued*)

Name	Location	Principal features and events
		orogeny to dextral in modern times) results from movement from SW of Qinghai–Xizang Plateau
(n) N. Sino-Korean		Boundary between the Tarim Platform (Fig. 4, 3) and the Sino-Korean Paraplatform (Fig. 4, 1) on the south and the Tianshan–Nei Monggol–Da Hinggan fold system to the north
		Associated with ultramafic, mafic, and granitoid rocks
(o) Borohoro–Mid-Tianshan	Location Fig. 8, line A	Boundary between the Tianshan (Fig. 4, 5a) and Junggar (Fig. 4, 6) fold systems
		Characterized by mafic, ultramafic, and granitoid rocks, with a few ophiolites
		Considered to be a paleoplate suture (Li and Tang, 1983)
(p) Da Hinggan–Taihang–Wuling		Two zones: Da Hinggan–Taihang in north China, and Wuling in south China
		Coincides with a steep gravity gradient
		Boundary between the Nei Monggol–Da Hinggan (Fig. 4, 5b) and the Jilin–Heilingjiang (Fig. 4, 8) fold systems
(q) Nanling		Boundary between the south China fold systems (Fig. 4, 21) and Yangzi Paraplatform (Fig. 4, 2)

Tensional motion has led to the formation of faults, producing basins now filled with Cretaceous–Tertiary red beds such as, for example, the Shaowu (northwest Fujiian, 27.3°N, 117.5°E)–Heyuan (east Guandong, 23.6°N, 114.6°E) Basin.

Purely tensional basins in eastern China developed during the Mesozoic and Cenozoic and normally have a clastic continental fill, as for example the Cangzhou (38.3°N, 116.8°E in eastern Hebei), the Liaocheng (northwest Shandong, 36.5°N, 116°E), the Lankao (northeast Henan, 34.7°N, 114.7°E), and the Weihe–Fenhe (Shaanxi–Henan, 34.5°N, 110.5–111.5°E) graben. They may be associated with basalt flows. Geophysical data show that the bounding faults do not penetrate far into the upper mantle.

III. STRATIGRAPHY OF CHINA

A. Precambrian

1. Pre-Sinian Sequence

The Precambrian rocks of China have been split up into the pre-2.2-b.y. Archeozoic (Archean), with ages ranging up to 3 b.y., and the Proterozoic, the top of which lies at 570–600 Ma (Compilation Group, 1976). In China, it is usual to consider the Lower Proterozoic along with the Archean as a pre-Sinian complex of predominantly igneous and metamorphic rocks (Fig. 9). These rocks crop out mainly in northeastern, western, and southeastern China (Research Section, 1962). As is seen in Fig. 10, the most extensive outcrops are in northeastern China where the best developed section (Fig. 9) is seen in the Wutai–Taihang Shan.

The thick Wutai–Taihang Shan section is divided into three major groups separated by unconformities reflecting major tectonic episodes at 2.4–2.5 b.y. (Anshan movements) and 2.2 b.y. (Wutai movements), and terminated at about

Age (b.y.)	Geologic time		Column	Major lithology	Thickness (m)	Tectonic movement and magmatic activities	Degree of metamorphism	Mineral deposits
1.7–1.9	Sinian			Limestones		~ Luliang ~ Movement	None	Building materials
	Early Protero- zoic	(Pt₁) Hutuo Group	Upper / Middle / Lower	Low metamor-phosed dolomitic conglomerates, feldspar, and quartz ss, slate. Dolomites, limestones with phyllites, mafic volcanic rocks with stromatolites. Dolomitic marbles with stromatolites. Bottom: meta-morphosed gold-bearing cgl. & ss.	5,800 to 8,000	Intermediate to mafic volcanic activities and pegmatitic intrusions	Light metamorphism, local granitization	Copper, uranium, gold, building materials
2–2.1		(Ar₂)				~ Wutai ~ Movement		
	Late Archeo- zoic	Wutai Group		All types granulites. amphibolites. hornblende schists, plagioclase gneisses, ferruginous quartzites	8,200 to 16,500	Intermediate mafic to mafic volcanic rocks, granitic and pegmatitic intrusions	Medium metamorphism, weak migmatization	"Anshan-type" iron ore, copper, pyrite, beryl, micas
2.4–2.5		(Ar₁)				~ Anshan ~ Movement		
	Late Archeo- zoic	Fuping Group		All types of gneisses. amphibolites. granulites. marbles, quartzites, and few magnetic iron quartzite	7,000 to 9,000	Large basic volcanic rocks, ultramafic to mafic intrusions and granites	High metamorphism, widely distri-buted, and strong migma-tization and granitization	Aluminum, zinc, "Anshan-type" iron ore, graphite, phosphorus, corundum, chromite

Fig. 9. Composite stratigraphy of the pre-Sinian in the Wutai–Taihang Shan area of north China (approx. 39°N, 113°E).

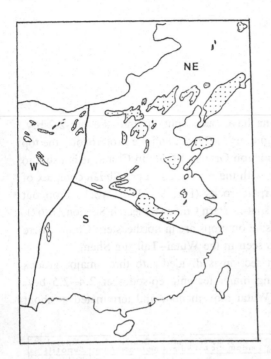

Fig. 10. Distribution of pre-Sinian rocks in China. Most occur in the northeast on the Sino-Korean Paraplatform or on the Tarim Platform lying to the west, both elements of the North China Plate.

1.7–1.9 b.y. by the Luliang movements. The differences in lithology and degree of metamorphism/migmatization are summarized in Fig. 9.

The oldest rocks are assigned to the Fuping Group (Ar_1) which consists of gneiss, granulite, marble, and quartzite showing strong migmatization and granitization. The rocks were originally marine sandstones and limestones and mafic (oceanic?) volcanics. They were assigned to the Lower Archeozoic by Fang *et al.* (1979), but according to the Compilation Group (1976), the rocks have undergone two periods of metamorphism—the earlier at 3.1 b.y. as well as the late Anshan event at 2.5 b.y. The Fuping Group is considered as equivalent to the Pilbara and Kalgoorlie of Australia or the North American Coutchiching, Keewatin, and Timiskaming Groups of the Canadian Shield.

The earliest Proterozoic, the Hutuo Group (Pt_1) consists mainly of low-grade, metamorphosed sedimentary and sedimentary-volcanic rocks. Within the Hutuo Group, Fang *et al.* (1979) recognized three subgroups separated by angular unconformities. The lowest group is an auriferous, uraniferous conglomerate with dolomitic marble, the middle group is stromatolitic magnesium limestones and phyllites showing evidence of formation in a stable neritic environment, and the upper group is formed by slightly metamorphosed conglomeratic sandstone and shale (now slate). Granitic and pegmatitic intrusions are common within the Hutuo Group. The top is cut by an unconformity reflecting erosion which followed the Luliang movements at approximately 1.7–1.9 b.y. This unconformable surface is widely developed and is used to separate the Lower Proterozoic from the remainder of the Proterozoic. The Hutuo Group is correlated with the Upper Nullagin of Australia and the Animikie Group of the Canadian Shield.

It is difficult to make meaningful paleoenvironmental assessments of these Archeozoic–Lower Proterozoic rocks, some of which have been affected by three or even four orogenic events. The presence of bacteria and of algal (stromatolitic) limestones indicates that a stable neritic environment was established certainly no later than late Archeozoic.

2. Sinian Succession

It is standard Chinese practice to refer to the later Proterozoic rocks as the Sinian Suberathem (Compilation Group, 1976); these rocks cover the age range 1.7–1.9 b.y. to 576–600 Ma and lie above the Luliang unconformity. At the top, Cambrian rocks lie disconformably upon the Sinian. The term was introduced by von Richthofen to cover all unmetamorphosed rocks below the dated Cambrian. Grabau (1924) assigned the system to the Paleozoic, but recognized that it was older than the Cambrian; it has always been implicitly regarded as correlated with the Riphean of the USSR and the Belt Supergroup of the USA. A type section was established by Li (1924) in the Chang Jiang (Yangtze = Yangzi) gorges of south-central China. In Jixian County near Beijing, Gao (= Kao) et al. (1931) assigned the Sinian to latest Proterozoic–earliest Paleozoic.

Some confusion has resulted from attempted correlation of Sinian strata in southern China with those in the north (see Research Section, 1962); for example, the Kunyang Group was correlated with the Hutuo Group of northeast China. The features that are used to separate the Sinian from the Cambrian are:

- A disconformity exists between the two.
- Sinian sequences show greater volcanicity and a greater degree of alteration by crustal movements.
- The Sinian flora is algal, but with a worm and medusa fauna in the upper part; the rich fauna of the Cambrian is absent.
- The Sinian carbonates tend to be more silicic and magnesium-rich than their Cambrian equivalents.

Characteristically, the Sinian is a thick sequence of unmetamorphosed to slightly metamorphosed marine carbonates and terrigenous clastic rocks. Recently, two type sections have been designated, for northern China (Fig. 11) and for southern China. Different formational names are used, but correlation with northern China is indicated on the southern section (Fig. 12). However, neither type section is complete, and for the complete section through the youngest Sinian, the section in the Chang Jiang Gorge, western Hubei, must be included (Fig. 13).

Rocks of the Sinian suberathem are widely distributed in northern China and

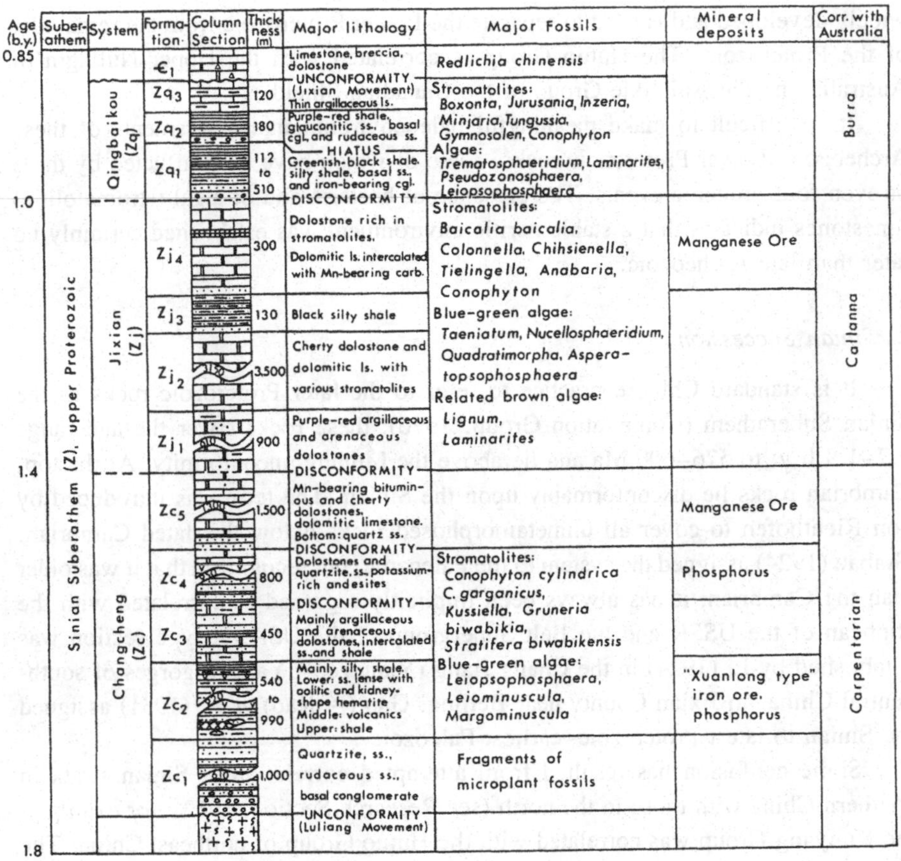

Age (b.y.)	Suberathem	System	Formation	Column Section	Thickness (m)	Major lithology	Major Fossils	Mineral deposits	Corr. with Australia
0.85	Sinian Suberathem (Z), upper Proterozoic	Qingbaikou (Zq)	ϵ_1			Limestone, breccia, dolostone	*Redlichia chinensis*		Burra
						UNCONFORMITY (Jixian Movement)			
			Zq_3		120	Thin argillaceous ls.	Stromatolites: *Boxonia, Jurusania, Inzeria, Minjaria, Tungussia, Gymnosolen, Conophyton*		
			Zq_2		180	Purple-red shale, glauconitic ss., basal cgl. and rudaceous ss.			
						HIATUS	Algae: *Tremgtosphaeridium, Laminarites, Pseudozonosphaera, Leiopsophosphaera*		
1.0			Zq_1		112 to 510	Greenish-black shale, silty shale, basal ss. and iron-bearing cgl.			
		Jixian (Zj)				DISCONFORMITY	Stromatolites: *Baicalia baicalia, Colonnella, Chihsienella, Tielingella, Anabaria, Conophyton*	Manganese Ore	Callanna
			Zj_4		300	Dolostone rich in stromatolites. Dolomitic ls. intercalated with Mn-bearing carb.			
			Zj_3		130	Black silty shale	Blue-green algae: *Taeniatum, Nucellosphaeridium, Quadratimorpha, Aspera-topsophosphaera*		
			Zj_2		3,500	Cherty dolostone and dolomitic ls. with various stromatolites	Related brown algae:		
1.4			Zj_1		900	Purple-red argillaceous and arenaceous dolostones	*Lignum, Laminarites*		
		Changcheng (Zc)				DISCONFORMITY			
			Zc_5		1,500	Mn-bearing bituminous and cherty dolostones. dolomitic limestone. Bottom: quartz ss.		Manganese Ore	Carpentarian
						DISCONFORMITY	Stromatolites: *Conophyton cylindrica, C. garganicus, Kussiella, Gruneria biwabikia, Stratifera biwabikensis*	Phosphorus	
			Zc_4		800	Dolostones and quartzite, ss., potassium-rich andesites			
						DISCONFORMITY			
			Zc_3		450	Mainly argillaceous and arenaceous dolostones, intercalated ss. and shale	Blue-green algae: *Leiopsophosphaera, Leiominuscula, Margominuscula*		
			Zc_2		540 to 990	Mainly silty shale. Lower: ss. lense with oolitic and kidney-shape hematite. Middle: volcanics. Upper: shale		"Xuanlong type" iron ore, phosphorus	
1.8			Zc_1		1,000	Quartzite, ss., rudaceous ss., basal conglomerate	Fragments of microplant fossils		
						UNCONFORMITY (Luliang Movement)			

Fig. 11.   Stratigraphy of the Lower Sinian Suberathem in Jixian, Hubei, northern China (approx. 40°N, 117.3°E).

have been separated into four systems, the lower three of which (the Changcheng, Jixian, and Qingbaikou) are well developed in Jixian County of Hubei Province. The thickest section, about 10,000 m thick, is exposed in the center of the basin; a summary of the major lithological features is shown with the stratigraphic column in Fig. 11. The fourth Sinian system is present as a thin section at the margin of the Sinian basin on the Liaodong Peninsula (approximately 40°N, 123°E). In most areas, the Lower Cambrian rests on rocks of the Qingbaikou system, with the Sinian system largely absent.

In the Changcheng system, a fining-up sequence of continental clastic rocks finally grades up from coarse clastics into fine-grained rocks and carbonates, marking a slow transgression to a neritic to littoral environment. There are iron-bearing beds in the lower part of the sequence, and locally, mafic volcanic rocks are

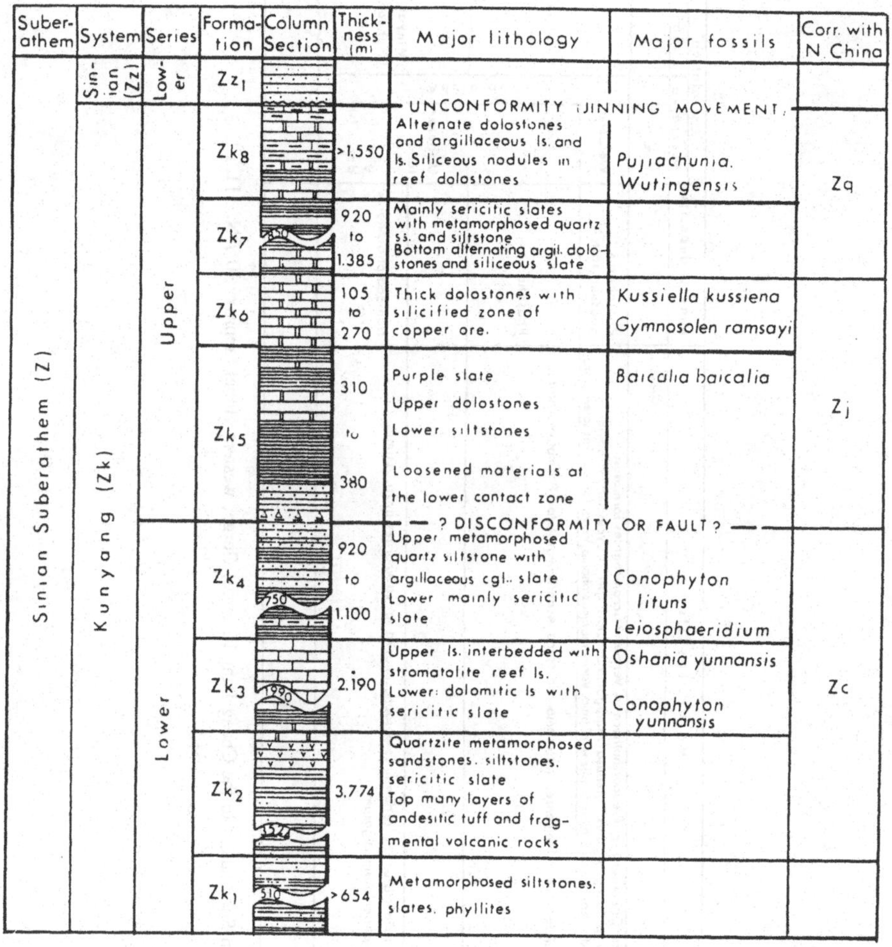

Suber-athem	System	Series	Forma-tion	Column Section	Thick-ness (m)	Major lithology	Major fossils	Corr. with N.China
Sinian Suberathem (Z)	Sinian (Zz)	Lower	Zz₁					
	Kunyang (Zk)	Upper	Zk₈		>1,550	— UNCONFORMITY (JINNING MOVEMENT) — Alternate dolostones and argillaceous ls. and ls. Siliceous nodules in reef dolostones	Pujiachunia. Wutingensis	Zq
			Zk₇		920 to 1,385	Mainly sericitic slates with metamorphosed quartz ss. and siltstone Bottom alternating argil. dolo-stones and siliceous slate		
			Zk₆		105 to 270	Thick dolostones with silicified zone of copper ore.	Kussiella kussiena Gymnosolen ramsayi	Zj
			Zk₅		310 to 380	Purple slate Upper dolostones Lower siltstones Loosened materials at the lower contact zone	Baicalia baicalia	
		Lower	Zk₄		920 to 1,100	— ? DISCONFORMITY OR FAULT ? — Upper metamorphosed quartz siltstone with argillaceous cgl. slate Lower mainly sericitic slate	Conophyton lituns Leiosphaeridium	
			Zk₃		2,190	Upper ls. interbedded with stromatolite reef ls. Lower dolomitic ls with sericitic slate	Oshania yunnansis Conophyton yunnansis	Zc
			Zk₂		3,774	Quartzite metamorphosed sandstones. siltstones. sericitic slate Top many layers of andesitic tuff and frag-mental volcanic rocks		
			Zk₁		>654	Metamorphosed siltstones. slates. phyllites		

Fig. 12. Stratigraphy of the Lower Sinian Suberathem, Kunyang System, in eastern Yunnan, southwest China (approx. 24.5°N, 102.5°E).

intercalated in the upper part. Stromatolites and blue-green algae have been identified. The carbonates tend to be siliceous and magnesium-rich.

In contrast, the Jixian system is made up of rocks deposited in a neritic environment during a regressive cycle. The sequence passes from siliceous magnesium limestones to an upper part consisting of argillaceous beds and iron-, manganese-, and phosphate-bearing carbonates. In addition to the flora found in the Changcheng system, brown algae appear for the first time. The brown algae become abundant and characterize the succeeding Qingbaikou system. The latter is a transgressive system with coarse terrigenous clastics at the base passing to carbonates in the upper part.

Suberathem	System	Series	Formation	Column section	Thickness (m)	Major lithology	Major fossils	Mineral deposits	Correlation Australia	Correlation USSR
	Paleozoic	Cambrian Lower	€1			Shales with limestone; phosphorus at bottom — DISCONFORMITY —	Hyolithes: Circotheca sp., Pseudorthotheca sp. Vermes: Micronemalites formosus. Medusa: Madigania annulata sprigg, Cyclomedusa davidi. Saarina sp. Spong spicules. Algae: Manicosiphonia, Nanamanicosiphonia, Varicamanicosiphonia, Vendotaenia, Praesolenopora.	Phosphorus, gypsum, salt	Welpena	€
Sinian Suberathem (Z)	Sinian System (Zz)	Upper	Zz4		>60	Chert-bearing and nodular dolostones and dolomitic ls.				Vendian
			Zz3		145 to 320	Mainly limestone, dolomitic limestone, with pyrite, black shale, phosphorus, glauconite.	Lophosphaeridium ichangense, Nostocomorpha prisca, Acanthomorphitae, Prismatomorphitae.	Phosphorus	Adeloide	
		Lower	Zz2		0 to 100	Purplish-red and greyish-green glacial cgl., various size, no bedding, striations — DISCONFORMITY —			Umberatana (with glacial deposits)	
			Zz1		50 to 260	Purplish-red feldspar and quartz sandstones with tuffaceous sandstone and shale. Basal cgl. and rudaceous sandstone. — UNCONFORMITY —	Trematosphaeridium holtedahlii, Leiopsophosphaera sp., Laminarites antiguissimus. Trachysphaeridium rugosum.	Iron and phosphorus ores		Minyar (Riphean)
Pre-Sinian						Gneiss and schist				

Fig. 13. Stratigraphy of the Upper Sinian Suberathem in the Chang Jiang (Yangzi) Gorge, western Hubei (approx. 30.7°N, 111.3°E).

From scattered outcrops along the margin of the basin, the Sinian system appears to be a glauconitic sandstone near the base, passing to neritic carbonates and terrigenous clastics at higher levels. There are three dominant stromatolitic horizons, and the system marks the first appearance of metazoans and red algae.

In south China, where Sinian rocks have an extensive outcrop around the Tarim and Qaidam basins and south of the Qinling, only two systems are recognized, separated by an unconformity. The sequence and its correlation with north China are shown in Fig. 12. The lower, Kunyang, system in the type section in eastern Yunnan consists of low-grade, metamorphosed, terrigenous clastics and siliceous magnesium limestones interbedded with intermediate to mafic, or soda-rich, volcanic rocks and ferruginous bands. The Sinian above the unconformity is divided into two series and four formations in the type section in the Chang Jiang (Yangzi) Gorge. The lowest formation, Zz1 (no name), consists of red fluvio-lacustrine beds with some volcaniclastic horizons. It is followed by Zz2, a glacial horizon which serves as a widespread key horizon in southern China, in Henan Province, the northwestern Tian Shan, and the Tarim basin. It is regarded as a probable continental glaciation following the tectonic/magmatic Chengjiang stage. Possible correlation with glacial beds of approximately the same age in Korea, the USSR, Greenland, northern Europe, South Africa, and Australia has been suggested. It is dated in the 600–700 Ma range.

The basal beds of the upper series, Zz3, are limestone-dominated neritic deposits. Glauconitic sandstone, phosphatic rocks, and black shale also make an appearance. A microflora with Acanthomorphitae and Prismatomorphitae has been identified. The higher beds in the Sinian Zz4 are neritic to littoral, siliceous carbonates with evaporites and phosphate beds. In addition to red algae, a soft-bodied fauna has been found, with *Hyolithes*, worms, and medusae.

It is clear from the preceding descriptions that tectonic and magmatic activity occurred sporadically throughout the Sinian suberathem, differing in time and intensity according to location. On the whole, magmatic activity was less prominent in northern China than in the south, and is wholly nontectonic in northeastern China (Compilation Group, 1976). Three phases of magmatic activity are recognized: the oldest (Sibao phase) at 1400 ± 50 Ma was followed by the Jinning stage, with which ages group around 1000 to around 900 Ma. The final phase of activity is dated at about 700 Ma.

The magmatic activity of the Sibao phase is the most heterogeneous; rapakivi granites (wiborgites) were intruded near Beijing, and pegmatites in the Wutai and Luliang Shan of northern China, which can be contrasted with an ultramafic and mafic assemblage with diorites found in the Guangxi region of southern China. The alkaline rocks intrude the lower series of the Kunyang system, but are overlain unconformably by the upper series.

The Jinning intrusives are associated with an orogenic (Jinning) movement, which is reflected in the unconformity separating the Kunyang from the Sinian system. In southern China, the acid, granitic intrusives are characteristic, but are less abundant in northern China and are rare in northeastern China. The tectonic movements, which in the south produced a rugged topography and fault basins, had little effect in northern China.

The Chengjiang phase of intrusion was associated with mild tectonism in southern China and is represented mostly by granites. The glaciation of the Sinian Zz2 postdates this intrusive and tectonic event.

The Sinian ended with further mild crustal movements in northern China, which produced an unconformity and the superposition of Cambrian on beds as low as the Hutuo Group. In the south, a marine regression is marked by a disconformity.

B. Paleozoic Erathem

1. Sedimentary Regions

Paleozoic rocks are widely spread and well developed in China. As a broad generalization, marine beds with rich faunas predominate during the Lower Paleozoic, whereas in the Upper Paleozoic continental deposits play a very conspicuous role. The Compilation Group (1976), recognizing the significance of the tectonic framework of China (Figs. 4 and 7), divided the country into three tectonic–sedimentary belts and eight sedimentary regions, as illustrated in Fig. 14. The tectonic development provides an explanation for the coincidence of the tectonic sedimentary belts with the margins of the two gigantic east–west mountain ranges: the northern tectonic belt, or Junggar–Hinggan belt, north of the Tian Shan–Yin Shan; and the southern tectonic belt, the Qinghai–Xizang (Tibet)–South China belt, south of the Kunlun Shan–Qinling. The central tectonic belt, the Tarim–North China belt, occupies a position between the east–west ranges.

Each of these tectonic–sedimentary belts (see Fig. 7) can be broken down into sedimentary regions, as shown in Fig. 14. These are virtually identical with the tectonic units described previously (see Fig. 4). Differences arise when the complex region of the young Tethyan folds and their relation to the North and South China Plates are considered. Here, it becomes necessary to consider the Tethyan fold region in terms of three units: the Kunlun–Qilian–Qinling, the Xizang–West Yunnan, and the Himalayan regions.

A very brief outline of the Paleozoic history of each region is presented prior to a systematic, period-by-period approach to the stratigraphic column. The ordering of the regions follows that of the tectonic divisions.

(a) The North China Region (= Sino-Korean Paraplatform). The marine sequence of Cambro-Ordovician neritic limestones and dolomites, with some intercalated sandstone and shale, is more complete and richly fossiliferous in western Shandong and in the Yan Shan region. For this reason, type sections were chosen there. Nineteen zones, characterized by benthonic trilobites, are found in the

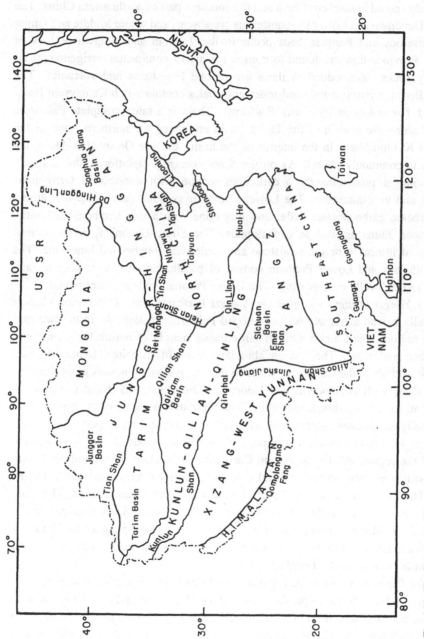

Fig. 14. The Paleozoic sedimentary regions of China showing the subdivisions of the three tectonic–sedimentary belts: northern, central, and southern (see also Figs. 4 and 7).

Cambrian. Actinocerids characterize the Ordovician. Recently, middle Lower Cambrian with *Palaeolenus* (beds of the Canglanpu Formation) has been found to be widely spread in northern China and the southern part of northeastern China. The Siluro-Devonian and Lower Carboniferous are absent, and where Middle and Upper Carboniferous and Permian beds occur in the Taiyuan and Ningwu regions of Shanxi Province they are found to consist mainly of continental terrigenous sedimentary rocks. Interbedded in these are marine limestones and volcanics. The Upper Permian purplish red sandstones and shales contain a rich Cathaysian flora.

  (b) Tarim Region (= Tarim Platform). There is a fairly complete Paleozoic section along the margin of the Tarim Basin and along the northern slope of the western Kunlun Shan. In the interior of the basin, the late Ordovician–Devonian section is commonly absent. As on the Sino-Korean Paraplatform, the Cambro-Ordovician is predominantly a carbonate section with intercalated terrigenous clastics and volcanic rocks. The Lower Cambrian contains phosphatic and arenaceous rocks, carbonaceous shale, and limestone with a rich benthonic trilobite–brachiopod fauna as well as cephalopods. The Siluro-Devonian, where found, consists of littoral to neritic sandstone and shale with interbedded limestone. The Carboniferous and Lower Permian consist of paralic carbonates, sandstone, and shale, and is locally coal-bearing. The Upper Permian beds are continental.

  (c) Yangzi (Yangtze) Region (= Yangzi Paraplatform). This region characteristically has an extremely well-developed Paleozoic section. As in the two preceding regions, the Cambro-Ordovician consists mainly of neritic limestone with sandstone and shale. There is an abundant fauna of trilobites, hyolithids, and archaeocyathids in the Cambrian, while in the Ordovician numerous graptolites are found along with *Vaginoceras* and *Sinoceras*. Once again, the basal Cambrian is phosphatic and has carbonaceous shale. The Silurian is made up of graptolitic shale and shelly sandstones intercalated with red beds in the upper part. The latter, surprisingly, are rich in corals, graptolites, and brachiopods. In the southwestern part of the region, the Devonian and Carboniferous section is predominantly marine, with limestone, sandstone, and shale and locally intercalated volcanic horizons. The Devonian has a varied shelly fauna with fish and ammonoids. These are usually absent in the southeastern part of the Sichuan basin. Some redefinition has occurred; the Maoshan Formation of the lower Yangzi is now assigned to the Lower and Middle Devonian on the basis of newly discovered antiarchs, and is not Upper Silurian as was formerly thought.

  The Permian is widespread, passing up from limestones with fusulinids, corals, and brachiopods, into a paralic sequence with limestone and coal, to a continental sequence with a well-known Cathaysian flora with *Gigantopteris*. There is, in the upper Yangzi region, an extensive basalt horizon (Emeishan basalt), and in central Guangxi intermediate to silicic volcanic rocks are present.

  (d) Southeast Region (= South China Sea Platform). The Lower Paleozoic sequence consists of thick, slightly metamorphosed rocks. The Cambro-Ordovician

is mainly flysch, flyschoid, and graptolitic beds with carbonaceous shale and coal beds. The fauna, although predominantly Chinese, also contains a mixture of Atlantic and Australian forms. The Devonian and Lower Carboniferous section consists of paralic beds, followed by a Permo-Carboniferous succession with interbedded coal and marine horizons with fusulinids and corals. Basal Devonian in Guangxi is indicated by the occurrence of *Monograptus uniformis* and *M. yukonensis*.

(e) Junggar–Hinggan Region (= Central Asiatic Fold Region). The region lies within the Northern Tectonic Belt and, in addition to the folded chains of the Tian Shan, Yin Shan, also contains the Junggar and Songhua Jiang basins (Fig. 14). The region was invaded from the north by seas in which were deposited a thick, complex marine carbonate and clastic rock sequence with intercalated volcanics which have been slightly metamorphosed. The fauna of the "Beishan complex" indicates it represents mainly the Ordovician and Silurian. The Upper Silurian is generally a red terrigenous clastic sequence. The Upper Devonian, which is widely developed, consists of an admixture of continental and littoral deposits. The Upper Paleozoic in general consists of marine to paralic terrigenous clastics and volcaniclastic rocks, with a shelly (brachiopod, coral) fauna showing Boreal–Pacific affinities, interbedded with strata carrying an Angaran flora. The Upper Permian, largely continental but with occasional marine horizons, also carries an Angaran flora.

(f) Kunlun–Qilian–Qinling Region (= Tethyan Fold Region s.l. in part). This is a region which actually forms the southern part of the Pal-Asiatic tectonic domain (Fig. 7). It contains thick, weakly metamorphosed Paleozoic beds. The Lower Paleozoic section of marine carbonates and terrigenous clastics has interbedded volcanic horizons. The Upper Paleozoic consists of marine and paralic beds, with the rocks of the Upper Devonian and Upper Permian of continental origin. The total Paleozoic sequence attains a thickness of between 10,000 and 20,000 m.

(g) Xizang (Tibet)–West Yunnan Region (= Tethyan Fold Region s.s. in part). Unlike the Kunlun–Qilian–Qinling region, in this region west of Jinsha Jiang and Ailao Shan, the Lower Paleozoic, found mainly in western Yunnan, has a Cambrian of flysch and flyschoid deposits. The Ordovician and Silurian consist of graptolitic shales with interbedded shelly limestone with a coral, cephalopod, and graptolite fauna. Fossiliferous Cambrian and Ordovian crop out in eastern Xizang. The Upper Paleozoic is widely distributed, with shallow-water marine sandstones, shale, and limestones in which are interbedded local extrusives and occasional coal seams.

(h) Himalayan Region (= Tethyan Fold Region in part). In this region, the well-developed Paleozoic has been studied recently on the north slope of Qomolangma Feng (Everest) where neritic sandstones, shale, and limestones range in age from Ordovician to Permian. This includes an important Lower Devonian monograptid fauna found in beds formerly referred to as the "transitional beds." The fauna, which contains *Neomonograptus himalayensis* and *Monograptus thomasi*,

has also been found in western Sichuan, in the western part of the Junggar Basin in Xingjiang, in Yunnan, and in Guangxi, showing its wide distribution in western China.

Correlation of the main fossil zones of the Chinese Paleozoic with the rest of the world is summarized in Figs. 15 and 16.

2. Cambrian System

Cambrian rocks are found in China in all but three areas: Taiwan, the northern part of the Nei Monggol (Inner Mongolia), and Junggar (northwest China) basins. The principal deposits are fossil-rich marine sediments. From the faunal assemblage and lithology, Lu (1962, 1981) recognized five Cambrian regions (Fig. 17) belonging to the north and the southeast China provinces. The north China province consists of north and northeast China, separated into two by the Nei Monggol Massif, and the Yangzi, and the southwest China region. The former has deposits reflecting a shallow, turbulent depositional environment, the latter a fauna indicative of a benthic environment. The fauna of both regions is of Indo-Pacific type. The southeast China province consists of two areas, one in northwest China and Jiangnan where volcanic rocks alternate with marine deposits, and the other in southeast China and Zhujiang where the succession grades up from fine clastics and carbonates into a predominantly marine carbonate sequence. The fauna shows a relationship to Atlantic type. The two areas are separated by the north China province, but in the intervening area several regions of transitional (mixed fauna) facies crop out.

(a) North China Type Section. The north China type section (Fig. 17) is located in Shandong Province where the section, which measures 810 m (Fig. 18), rests with angular unconformity on the Archean. Only ϵ_3 of the Lower Cambrian is present, however, and this consists of purplish-red shales with halite pseudomorphs interbedded with argillaceous limestone. It is followed by a Middle Cambrian transitional sequence passing from predominantly purple shale below to limestones above. The top of the Middle Cambrian is marked by a highly fossiliferous oolitic limestone.

The Upper Cambrian begins with a basal conglomerate and contains intraformational limestone conglomerates interbedded with thin, nonconglomeratic limestones and purplish-red shale. The intraformational (edgewise) conglomerates consist of flat pebbles now cemented by calcareous material.

(b) Southwest China Type Section. The type section is located in eastern Yunnan Province (Fig. 17, 2) where the Lower Cambrian is particularly well developed (Fig. 19). The position of the lower boundary of the Cambrian has been debated (see Lu, 1962). Below ϵ_1 (Fig. 19) lies a horizon of thin, dark gray, fine-grained terrigenous clastics and chert in which *Hyolithes* is found, with a phosphatic bed at the base. In some areas, the *Hyolithes* horizon may be slightly disconforma-

System	Series	Form-ation	Fossil Zonation or Representative Fossils	Correlation with foreign countries
Silurian	Upper	S_3	*Pristiograptus tumescens* zone *Bohemigraptus bohemicus* *Pristiograptus nilssoni* zone	Pridolian Ludlovian
Silurian	Middle	S_2	*Monograptus flexilis* zone *Cyrtograptus rigidus* zone *Monograptus riccartonensis* zone *Cyrtograptus murchisoni* zone	Wenlockian
Silurian	Lower	S_1	*Oktavites spiralis-* *Glyptograptus* *Persculptus* zones	Llandoverian
Ordovician	Upper	O_6	*Dicellograptus szechuanensis* zone *Dicellograptus complanatus* zone	Ashgillian
Ordovician	Upper	O_5	*Pleurograptus lui* zone *Sinoceras chinense* (Foord) zone	Caradocian
Ordovician	Middle	O_4	*Nemagraptus gracilis* zone *Glyptograptus teretiusculus* zone	Llandeilian
Ordovician	Middle	O_3	*Pterograptus elegans, Amplexograptus* *confertus, Glyptograptus* *austrodentatus* three zones	Llanvirnian
Ordovician	Lower	O_2	*Didymograptus hirundo* zone *Didymograptus deflexus* zone *Dichograptus separatus* zone	Arenigian
Ordovician	Lower	O_1	*Clonograptus tenellus* zone *Dictyonema flabelliforme* zone	Tremadocian
Cambrian	Upper	ϵ_9	*Tellerina-Calvinella* zone *Quadraticephalus-Dictyella* zone *Ptychaspis-Tsinania* zone	Trempealeauan
Cambrian	Upper	ϵ_8	*Kaolishania* zone *Changshania* zone *Chuangia* zone	Franconian
Cambrian	Upper	ϵ_7	*Drepanura* zone *Blackwelderia* zone	Dresbachian
Cambrian	Middle	ϵ_6	*Damesella, Amphoton, Crepicephalina* *Liaoyangaspis* four zones	Albertian
Cambrian	Middle	ϵ_5	*Bailiella, Poriagraulos abrota* *Sunaspis, Kochaspis hsuchuangensis* four zones	Albertian
Cambrian	Middle	ϵ_4	*Shantungaspis* zone	Albertian
Cambrian	Lower	ϵ_3	*Redlichia murakamii-* *Hoffetella* zone	Wacoubian
Cambrian	Lower	ϵ_2	*Megapalaeolenus, Palaeolenus,* *Drepanuroides, Malungia* four zones	Wacoubian
Cambrian	Lower	ϵ_1	*Eoredlichia, Hyolithus* two zones	Wacoubian

Fig. 15.   Zonation of the Lower Paleozoic of China and the equivalents in the western hemisphere.

System	Series	Form-ation	Fossil Zonation or Representative Fossils		Correlation with foreign country
Permian	Upper	P₄	*Palaeofusulina* zone		Tatarian
		P₃	*Codonofusiella* zone		Kazanian
	Lower	P₂	*Yabeina* zone *Neoschwagerina* zone		Kungurian
		P₁	*Parafusulina* zone	*Cancellina* subzone	Artinskian Sakmarian
				Misellina subzone	
Carboniferous	Upper	C₄	*Pseudoschwagerina* zone		Asselian Gzhelian Kasimovian
			Triticites simplex zone		
	Middle	C₃	*Fusulina – Fusulinella* zone		Moscovian Bashkirian
			Gastrioceras – Eostaffella subsolana zone		Namurian
	Lower	C₂	*Yuanophyllum* zone *Kueichouphyllum sinensis* zone		Viséan
		C₁	*Pseudouralinia* zone *Cystophrentis* zone		Tournaisian Etroeungtian
Devonian	Upper	D₈	*Yunnanella– yunnanellina* zone		Famenian
		D₇	*Cyrtospirifer sinensis* zone		Frasnian
	Middle	D₆	*Stringocephalus* zone *Parabornharatina* zone		Givetian
		D₅	*Acrospirifer houershanonensis– Utaratuia sinensis* zones		Eifelian
	Lower	D₄	*Euryspirifer paradoxus– Trepezophyllum cystosum* zone *Otospirifer daleensis* zone *Subcuspidalla trigonata* zone		Emsian
		D₃	*Euryspirifer tonkinensis– Xystriphyllum nobilis* zones		Siegenian
		D₂	*Orientospirifer nakaolingensis* zone		
		D₁	*Polybranchiaspis– Yunnanolepis* zones		Gedinian

Fig. 16.   Zonation of the Upper Paleozoic of China and the equivalents in the western hemisphere.

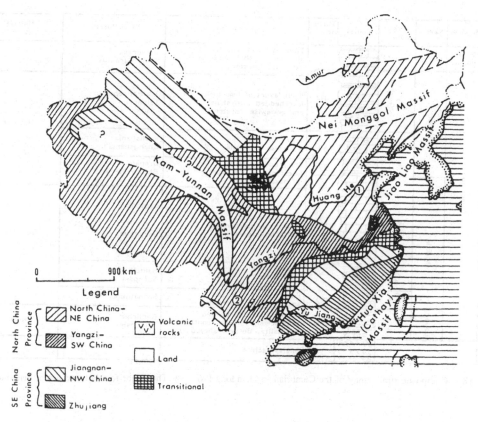

Fig. 17. Cambrian paleozoogeography of China (Lu, 1981). The location of the type sections is indicated (1 and 2). (Reprinted by permission of the Geological Society of America and Dr. Lu.)

ble up topmost Sinian Zz4. The beds have been variously regarded as early Cambrian, transitional, or Sinian.

The lowest divisions of the Lower Cambrian gradually wedge out toward the northeast (Fig. 20) and, in central and eastern China ϵ_2 and ϵ_3 overstep onto older Proterozoic and Archean, respectively, indicating transgression from the southwest.

(c) Paleoecology and Paleogeography. Lu *et al.* (1974) and Lu (1981) consider that the Cambrian trilobite fauna can be divided into two paleogeographic realms, Oriental and Occidental, with a transitional zone. The Oriental realm is characterized by redlichiids, dorypygids, and damesellids which have wide distribution in Asia, Oceania, and Antarctica. The representatives of the Occidental realm, with Olenellids and Paradoxidids, occupy much of western Europe, eastern North America, and South America. The transitional zone runs through Siberia, Tethys, North Africa, and western North America.

The depositional environment of the north China province, including the regions of Yangzi, northeast, southwest, and north China of the Oriental realm, was

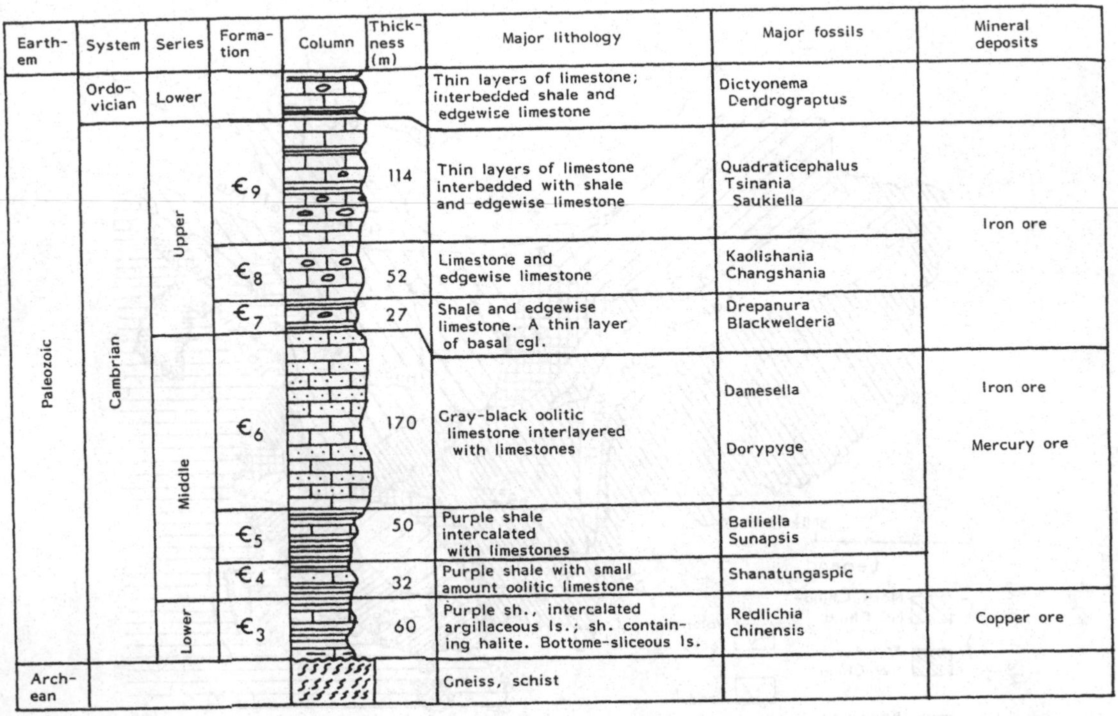

Earth-em	System	Series	Forma-tion	Column	Thick-ness (m)	Major lithology	Major fossils	Mineral deposits
Paleozoic	Cambrian	Ordo-vician	Lower			Thin layers of limestone; interbedded shale and edgewise limestone	Dictyonema Dendrograptus	
		Upper	ϵ_9		114	Thin layers of limestone interbedded with shale and edgewise limestone	Quadraticephalus Tsinania Saukiella	Iron ore
			ϵ_8		52	Limestone and edgewise limestone	Kaolishania Changshania	
			ϵ_7		27	Shale and edgewise limestone. A thin layer of basal cgl.	Drepanura Blackwelderia	
		Middle	ϵ_6		170	Gray-black oolitic limestone interlayered with limestones	Damesella Dorypyge	Iron ore Mercury ore
			ϵ_5		50	Purple shale intercalated with limestones	Bailiella Sunapsis	
		Lower	ϵ_4		32	Purple shale with small amount oolitic limestone	Shanatungaspic	
			ϵ_3		60	Purple sh., intercalated argillaceous ls.; sh. containing halite. Bottome-sliceous ls.	Redlichia chinensis	Copper ore
	Arch-ean					Gneiss, schist		

Fig. 18.   Composite stratigraphy of the Cambrian in Shandong Province. The major fossils listed are all trilobites.

that of shallow shelf and epeiric seas with their benthic fauna. The southeast China province, including Zhujiang, Jiangnan, and the northwest China region, consisted of geosynclinal deep troughs or basins with a fauna of floating and swimming forms (Figs. 21 and 22).

3. *Ordovician System*

Chang (1962) claimed that the best and most complete Ordovician sequences lie in China, and certainly there are well-developed, highly fossiliferous carbonate sequences. Transgression reached a maximum during the Ordovician, and the pattern of distribution shown in Fig. 23 also indicated the eight distinctive regions which can be defined on the basis of fauna and lithology (Chang, 1962; Sheng, 1980). It is a more complex pattern than that of the Cambrian, although the area of deposition remained much the same (see Fig. 17).

The three principal platform areas, north China (Fig. 23-2), Yangzi (Fig. 23-5), and southeast China (Fig. 23-7), have characteristic shallow-water marine facies, with some interbedded volcanics in southeast China. The North China Platform has a Pacific fauna, whereas the remaining two have a European-type fauna with additionally, a few Australian elements in southeast China.

System	Series	Forma-tion	Column. Section	Thick-ness (m)	Major lithology	Major fossils	Mineral deposits
Silurian	Mid-dle				Argillaceous ls. and sh. Basal conglomerate		
Cambrian	Middle	ϵ_5		28 – 323	limestones, argillaceous ls., dolomitic ls., interbedded sandy shale	Manchuriella Protahedinia yunnanensis	Mercury ore
Cambrian	Middle	ϵ_4		14 – 67	sandy shale interbedded with dolomitic ls.	Kutsingocephalus Douposiella	
Cambrian	Lower	ϵ_3		48	gray argillaceous and dolomitic limestones	Redlichia sp Hoffetella transversa	
Cambrian	Lower	ϵ_2		155	Upper: gray-green sandy sh. interbedded with thin sandstone layers. Lower: gray thick quartz ss. interbedded with sandy shale	Palaeolenus Redlichia mai	
Cambrian	Lower	ϵ_1 Upper member		127	Dark gray-black shale and sandy sh. interbedded with thin fine sandstone	Yunanocephalus Eoredlichia	
Cambrian	Lower	ϵ_1 Lower member		60	Gray-black thin layer of fine sandstone and siltstone, phosphatic layer at the bottom	Hyolithes	Phosphate ore
Sinian		Z_z			Fault(?) Limestone		

Fig. 19. Composite stratigraphy of the Cambrian in eastern Yunnan Province, southwest China. The major fossils listed, with the exception of *Hyolithes*, are all trilobites.

The remaining areas are geosynclinal. The Da Hinggan Ling to the north (north of the Junggar–Hinggan land area which persisted from the Cambrian as the Nei Monggol massif, Fig. 17) has a faunal assemblage similar to those of Siberia and North America. The flysch regime in northwest China has a southern fauna, but with an admixture of northern elements similar to those found in the Qin Ling. In western Yunnan, part of the Yunnan–Burma geosyncline, the fauna is of southern type resembling the European, but no northern elements are indicated.

The Ordovician of the Xizang (Tibet) region (Fig. 23-8) appears to provide an important link, for the cephalopod fauna in the Lower Ordovician is of a northern (Siberia–North America) type, but the Middle and Upper Ordovician show a change toward southern (European related) biofacies. This Ordovician appears to be part of a continuous change from the Cambrian below and Silurian above.

(a) Central China Type Section. A complete Ordovician sequence, without any faunal breaks, if found in Yichang in the Yangzi Gorge of central China (Fig.

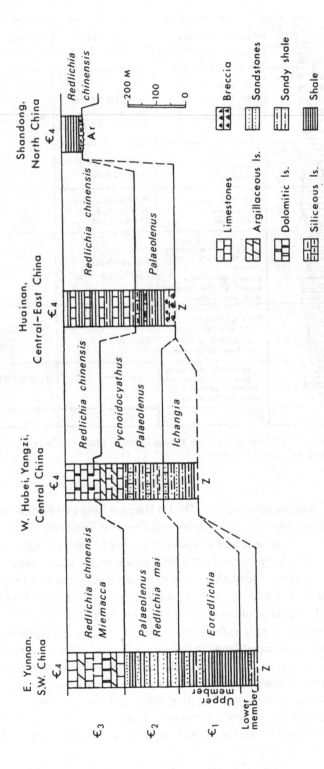

Fig. 20.   Correlation of the early Cambrian from southwest to northeast China based upon trilobite fauna.

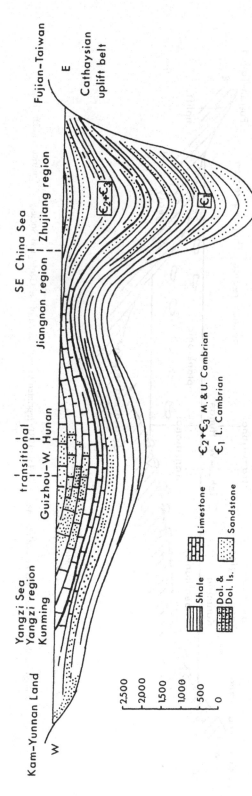

Fig. 21. Cambrian lithofacies changes in south China from Yunnan to Fujian. (After Lu, 1981; reprinted by permission of the Geological Society of America and Dr. Lu.)

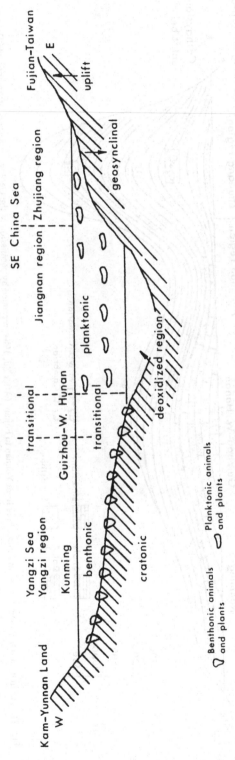

Fig. 22. Ecology of the Cambrian seas in south China. (After Lu, 1981; reprinted by permission of the Geological Society of America and Dr. Lu.)

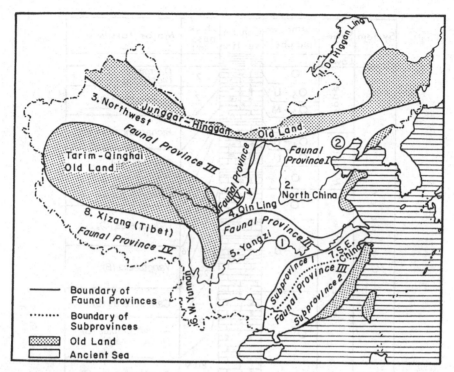

Fig. 23.   Distribution of the Ordovician system and early Ordovician faunal provinces. Two type sections are located by number in the circles.

23-1). It is conformable with Cambrian below and Silurian above. The succession in Lower and lower and middle Middle Ordovician consists mainly of carbonates (Fig. 24), with a few thin shale bands, containing a rich trilobite, brachiopod, cephalopod fauna with graptolites. The upper part of the Middle Ordovician, which is a black shale sequence, has abundant graptolites and is lithologically very similar to the Upper Ordovician. A *Dalmanitina* horizon occurs in the uppermost Upper Ordovician (Sheng, 1974).

To the west the sequence passes to terrigenous shale, while to the east the sections are predominantly limestones, indicating a clastic supply from a highland to the west into a seaway deepening toward the east.

(b) North China Type Section. Slow submergence of the region during the Ordovician is reflected by a limestone sequence. Lithologically, the lowest Ordovician is similar to the Upper Cambrian; however, the faunal change is distinct— *Saukia* and other trilobites of the Cambrian being replaced by a graptolite and brachiopod fauna. The intercalation of intraformational limestone breccias is an indication of periodic interruptions in the progress of transgression. The Lower Ordovician passes up through richly fossiliferous limestones to poorly fossiliferous dolostones.

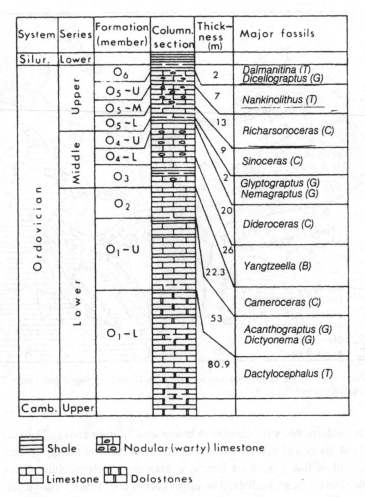

System	Series	Formation (member)	Column. section	Thick-ness (m)	Major fossils
Silur.	Lower				
Ordovician	Upper	O₆		2	Dalmanitina (T) Dicellograptus (G)
		O₅-U		7	Nankinolithus (T)
		O₅-M			
		O₅-L		13	Richarsonoceras (C)
	Middle	O₄-U		9	
		O₄-L			Sinoceras (C)
		O₃		2	Glyptograptus (G) Nemagraptus (G)
		O₂		20	Dideroceras (C)
	Lower	O₁-U		26	
				22.3	Yangtzeella (B)
					Cameroceras (C)
		O₁-L		53	Acanthograptus (G) Dictyonema (G)
				80.9	Dactylocephalus (T)
Camb.	Upper				

Shale Nodular (warty) limestone

Limestone Dolostones

Fig. 24. Composite stratigraphy of the Ordovician in Yichang, Hubei Province (34.7°N, 111.3°E). B, brachiopods; C, cephalopods; G, graptolites; T, trilobites.

The Middle Ordovician consists of thick limestones with gastropods and cephalopods interbedded with dolostones, dolomitic limestone, and cherty limestone. There is a disconformity but no dip change within the Lower Ordovician, but uplift at the end of the middle Ordovician occurred accompanied by erosion. No further deposition is recorded until middle Carboniferous time (Fig. 25).

(c) Paleoecology and Paleogeography. In the absence of major crustal movement between the Cambrian and Ordovician, the distribution and character of the marine basins of the Ordovician is similar to that which existed during the late Cambrian. Transgression, which persisted throughout the Lower Ordovician, reached a maximum during middle Ordovician time, ending with uplift which brought about a marine regression.

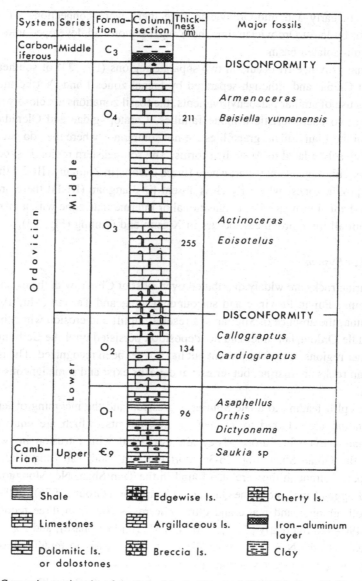

System	Series	Forma-tion	Column section	Thick-ness (m)	Major fossils
Carbon-iferous	Middle	C₃			
Ordovician	Middle	O₄		211	DISCONFORMITY Armenoceras Baisiella yunnanensis
Ordovician	Middle	O₃		255	Actinoceras Eoisotelus
Ordovician	Lower	O₂		134	DISCONFORMITY Callograptus Cardiograptus
Ordovician	Lower	O₁		96	Asaphellus Orthis Dictyonema
Camb-rian	Upper	€₉			Saukia sp

Shale   Edgewise ls.   Cherty ls.

Limestones   Argillaceous ls.   Iron-aluminum layer

Dolomitic ls. or dolostones   Breccia ls.   Clay

Fig. 25. Composite stratigraphy of the Ordovician in Tangshan, Hebei Province (39.5°N, 118.2°E).

Lu *et al.* (1976) recognized four faunal provinces based upon the environmental habitats of trilobites, graptolites, and nautiloid cephalopods. These are shown in Fig. 23. Their faunal province I is primarily a shallow-water cephalopod–trilobite fauna, with grapolites and ostracods, which shows its closest affinities with North America. Faunal province II in the Yangzi River valley is a graptolite–black shale fauna with a few trilobites, brachiopods, and cephalopods in carbonate horizons. Most of the graptolites are also found in the Middle East, Europe, and North

America. In early and mid-Ordovician time there were connections with other basins, but by latest Ordovician time the Yangzi region had developed into a quiet water, semi-isolated basin.

Faunal province III occurs in two separate regions (Fig. 23) of southeast and northwest China, and although separated by faunal zones I and IV (the transition zone) because of similar paleoenvironments, the fossil zonations are closely related. There are two characteristic trilobite families, Ceratopygidae and Olenidae, and apart from the Llanvirnian, graptolites are not common—where they do occur they are most closely related to Australian forms. The southeastern region is subdivided based upon lithological character; where flysch occurs [in Zhujiang (III-2)] there are more graptolite zones, where flysch is absent [in Jiangnan (III-1)] there are more trilobites. Faunal province IV, a transitional or intermediate zone with a mixture of fauna from all three provinces, occurs in Xizang and Qinling (Fig. 23).

4. *Silurian System*

Silurian rocks are widely distributed over most of China, with the exception of north China, Fujian Province and surrounding area, and Taiwan (Mu, 1962). In north China, the absence of Silurian is a result of uplift and erosion which began in post-Middle Ordovician time, and nondeposition persisted until the Carboniferous; in the other regions, it may be present but has not yet been recognized. The majority of Silurian rocks are marine, but terrestrial deposits exist and some igneous activity occurred.

The uplift leading to a late Silurian regression and the incoming of terrestrial sedimentation are referred to as Guangxi movements, which are equivalent to Caledonian activity in Europe. They are associated with intrusion in two main regions, the Qilian Shan (Fig. 26-3a, northern part) and Nan Ling (Fig. 26-5b, eastern part), although they are also found in the Tian Shan, Nei Monggol, along the Da Hinggan Ling, and in the Qin Ling (Compilation Group, 1976). In the Qilian Shan, both granites and mafic and ultramafic rocks occur, in three recognizable phases: 490–520 Ma (early phase), 430–460 Ma (middle phase), and 380–410 Ma (late phase). In the Nan Ling, migmatized granite of early to middle phases, and probably late-phase granites and granodiorites are known.

The six domains are basically the same as those recognized during the Ordovician, with two stable domains—north and south (Fig. 26-2,4)—bounded by northern, central, and southern mobile domains (Fig. 26-1,3,5). The sixth Gondwana (?) domain is represented by carbonates deposited on a stable area which may represent the northern margin of the southern continent, thus forming a southern shallow-water tract of Paleotethys.

Wang and Ho (1981) identified six sedimentary domains, subdivided into 15 regions in various tectonic settings (Fig. 26). They also identified three biogeographic realms: Boreal, Tethyan, and Australo-Pacific. The northern stable

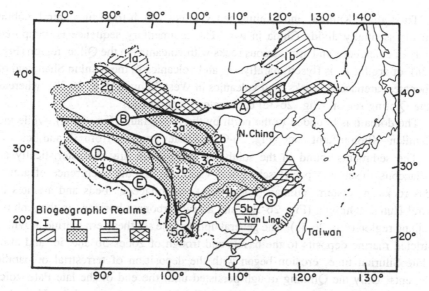

Fig. 26. Silurian marine sedimentation domains and regions and biogeographic provinces of China. The three biogeographic realms are: I, Boreal; II, Tethyan; III, Australo-Pacific—with a fourth transitional domain. The domains are: 1, North Mobile Domain—1a, Altay; 1b, Hinggan; 1c, Tianshan–Beishan; 1d, Nei Monggol–Jiliao regions. 2, North Stable Domain—2a, Tarim; 2b, Shaan-Gan-Ning (Ordos) regions. 3, Median Mobile Domain—3a, Qilian; 3b, Kunlun–West Sichuan; 3c, Qinling regions. 4, South Stable Domain—4a, South Xizang; 4b, Yangzi regions. 5, South Mobile Domain—5a, West Yunnan; 5b, Central Hunan–Qinfang; 5c, Anhui–Zhejiang regions. 6, Gondwana (?) Domain—Himalaya region. Major fault zones (circled): A, Nei Monggol fault; B, Altyn fault; C, Kolamilan–Xiugou fault; D, Bangonghu–Nujiang fault; E, Yarlung Zangbo fault; F, Jinsha–Honghe fault; G, Yichuan–Shaoxing fault. (From Wang and Ho, 1981; reprinted by permission of the Geological Society of America.)

domain is split into two by the Altyn fault (Fig. 26-B) which separates the Tarim (Fig. 26-2a) to the north from the Shaanganning (Fig. 26-2b) on the south. The sedimentary sequence is made up of clastic and argillaceous rocks with intermediate to mafic volcanics. In contrast, the south stable domain consists primarily of carbonates in south Xizang (Fig. 26-4a) and a mixed carbonate-clastic sequence in the Yangzi area (Fig. 26-4b). According to Wang (1978), northern Xizang during this time may have been a large median massif. South Xizang is bounded to the north by the Bangonghu–Nujiang (Fig. 26-D) and by the Yarlung Zangbo fault zone (Fig. 26-E) on the south.

The northern mobile domain, separating from the northern stable domain by the Nei Monggol fault zone (Fig. 26-A), has a predominantly argillaceous flysch and paraflysch sequence during the Silurian. The Hinggan region (Fig. 26-1b) and Nei Monggol region (Fig. 26-1d) have graywacke associated with volcanic materials. Andesite porphyries and tuff are abundant in central and northern Tian Shan (Fig. 26-1c) and western Altay (Fig. 26-1a).

The Qilian, Kunlun, and Qinling geosynclines which form the central mobile domain essentially divide China in two. The sedimentary sequence is composed mainly of calcareous and argillaceous rocks with volcanics in the Qilian region (Fig. 26-3a); of argillaceous flysch, paraflysch, and volcanics in the Kunlun Shan; and of carbonates, arenaceous rocks, and volcanics in West Sichuan (Fig. 26-3b), whereas in the Qinling region (Fig. 26-3c) carbonates predominate.

The domain is bounded to the north by the Altyn fault (Fig. 26-B), while the Kolamilan–Xiugou fault zone (Fig. 26-C) separates zone 3a from 3b and 3c.

The sediments found in the south mobile domain are characteristically an argillaceous flysch and paraflysch. Local variations are the occurrence of arenaceous rocks in western Yunnan (Fig. 26-5a), carbonaceous beds and arenites in central Hunan–Qinfang (Fig. 26-5b), and a few carbonate horizons in the Anhui–Zhejiang region (Fig. 26-5c). The general uplift of the southeastern marine province restricted marine deposits to the basins and trough of zones 5b and 5c, and after middle Silurian time, erosion began with the deposition of terrestrial or paralic sediments. Only the Qinfang trough persisted until the end of the late Paleozoic. The Jinsha–Honghe fault (Fig. 26-F) bounded 5a to the west and the Yichuan–Shaoxing fault (Fig. 26-G) lies south of zone 5c.

(a) *Silurian Type Section.* The type section is in Yichang in the upper Yangzi (shown in Fig. 27) where a transition up from the Ordovician is apparent. The lithological succession represents deposits of a complete depositional cycle from transgression to regression. The change from dominantly graptolitic shale at the base to shelly and coralline limestones in the Middle Silurian reflects the change from quiet and stagnant basin conditions to a wide, well-oxidized neritic shelf. During the Upper Silurian, shale, mudstone, siltstone, and sandstone point to nearshore deposition during a marine regression. The overlying Devonian is disconformable, reflecting movements of the Guangxi episode.

Fang *et al.* (1979) proposed a different subdivision of the type section, referring units S_1–S_{3m} (see Fig. 27) to the lower series, and S_{3u}, which is unfossiliferous, to the Middle Silurian.

(b) *Paleoecology and Paleogeography.* Transgression from the south occurred during the early Silurian, expanding from existing depressions leading to the maximum spread of neritic seas characterized by abundant shelly limestones in middle Silurian time. There is no break at the Ordovician–Silurian boundary, and the first indications of the Guangxi movement are apparent in late Silurian time. Uplift led to regression and the spread of littoral deposits, and continental deposits mark the basin edges.

Using rugose corals and brachiopods, Wang and Ho (1981) defined three biogeographic realms with transitional zones (Fig. 26).

The characteristic fossils of the Boreal realm are the coral *Tungussophyllum* and the brachiopods *Tuvaella* and *Tqunuspirifer*. This fauna is found in the northern mobile belt (Fig. 26-1a,1b), and is well developed from Hinggan and Mongolia to

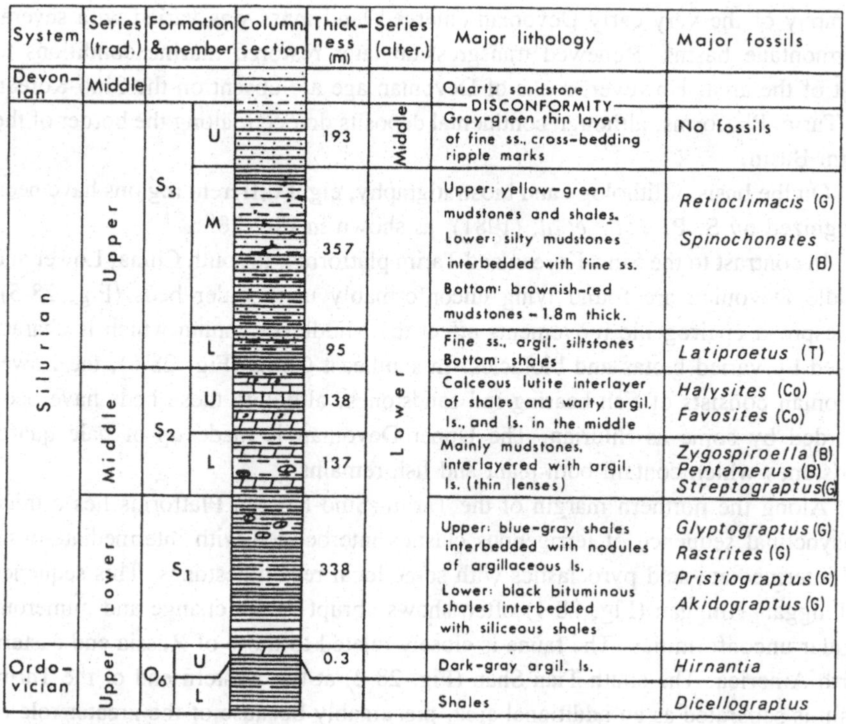

System	Series (trad.)	Formation & member	Column section	Thick-ness (m)	Series (alter.)	Major lithology	Major fossils
Devon-ian	Middle				Middle	Quartz sandstone	
Silurian	Upper	S₃	U	193		─DISCONFORMITY─ Gray-green thin layers of fine ss., cross-bedding ripple marks	No fossils
Silurian	Upper	S₃	M	357		Upper: yellow-green mudstones and shales. Lower: silty mudstones interbedded with fine ss. Bottom: brownish-red mudstones - 1.8m thick.	Retioclimacis (G) Spinochonates (B)
Silurian	Upper	S₃	L	95	Lower	Fine ss., argil. siltstones Bottom: shales	Latiproetus (T)
Silurian	Middle	S₂	U	138	Lower	Calceous lutite interlayer of shales and warty argil. ls., and ls. in the middle	Halysites (Co) Favosites (Co)
Silurian	Middle	S₂	L	137	Lower	Mainly mudstones, interlayered with argil. ls. (thin lenses)	Zygospiroella (B) Pentamerus (B) Streptograptus(G)
Silurian	Lower	S₁		338	Lower	Upper: blue-gray shales interbedded with nodules of argillaceous ls. Lower: black bituminous shales interbedded with siliceous shales	Glyptograptus (G) Rastrites (G) Pristiograptus (G) Akidograptus (G)
Ordo-vician	Upper	O₆	U	0.3		Dark-gray argil. ls.	Hirnantia
Ordo-vician	Upper	O₆	L			Shales	Dicellograptus

Fig. 27.   Silurian stratigraphy in Yichang, Hubei Province (34.7°N, 111.3°E). Co, corals.

Altay, the Sayan Mountains, and Siberia. The Tethyan fauna is much more widely developed and occupies zones 2a,b, 3a,b,c, 4a,b, and 5a (see Fig. 26). The principal rugose coral fauna occurs in the Yangzi region (Fig. 26-4b) with Lower Silurian *Donophyllum* and *Kodonophyllum* and Middle Silurian *Kyphophyllum* and *Ketophyllum*. In the Yangzi region, within the Tethyan realm in an epicontinental sea connected with the Qinling (3c) and Sanjiang troughs (5a), a pronounced endemic fauna developed.

The third zone (Fig. 26-5b,5c) contains an Australo-Pacific fauna, the few late Silurian corals being related to Australian forms.

In the Tianshan–Beishan (Fig. 26-1c) there is a mixed Tethyan and Boreal fauna, with rugose corals related to Uralian and western European faunas. The mixed fauna of the Nei Monggol and Jiliao zone (Fig. 26-1d), however, represents Asiatic and Australian forms.

5. *Devonian System*

Further uplift at the end of the Silurian associated with the Caledonian orogeny resulted in a marked marine regression so that marine conditions were restricted to Xinjiang, Xizang, and Yunnan in western China (Wang and Yu, 1962). The topography of the very early Devonian enlarged land mass was varied with several intermontane basins. Renewed transgression later restored marine conditions to most of the area. However, rocks of Devonian age are absent on the Sino-Korean and Tarim Platforms, although continental deposits do occur along the border of the Tarim Basin.

On the basis of lithology and biostratigraphy, eight different regions have been recognized by S. P. Yang *et al.* (1981), as shown in Fig. 28.

In contrast to the Sino-Korean and Tarim platforms, in south China, Lower and Middle Devonian are found lying unconformably upon older beds (Fig. 28-5). Widespread epeirogenic movements affect the Middle Devonian which is characterized by varied biotas and biofacies. In southeast China (Fig. 28-6), the Lower Devonian consists of fish-bearing red sandstones, although these beds have been regarded by some as Silurian. The Upper Devonian is made up of pale quartz sandstones which contain both plant and fish remains.

Along the northern margin of the Tarim–Sino-Korean Platforms lies a thick geosynclinal sequence of terrigenous clastics interbedded with intermediate-basic and intermediate-acid pyroclastics with some local reef limestones. This sequence of Junggar–Hinggan (Fig. 28-1) often shows abrupt facies change and numerous angular unconformities. The fauna is closely related to those of Russia and eastern North America. The south Tian Shan (Fig. 28-2) at the western end of the Tarim Basin is separated as an additional area, presumably because of the greater role of limestone and shale within a thick sequence of volcanic rocks conformably overly-

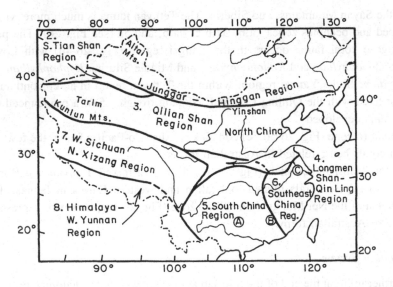

Fig. 28.   The Devonian tectonostratigraphic regions. The locations of the three type sections are indicated: A, B (south China); C (southeast China). (From S. P. Yang *et al.*, 1981; reprinted by permission of *Geology Magazine*.)

ing the Silurian. The Lower Devonian fauna of tabulate corals and brachiopods is comparable with that of Ural–Tianshan region. The Qilian Shan region (Fig. 28-3) lying between the Tarim and Sino-Korean Platforms has a sequence of continental red molassic sandstones and conglomerates thickening from east to west. Middle Devonian rocks are widely spread and may rest upon beds as old as the Cambrian. Early movements (early Variscan) resulted in an unconformity between Lower and Middle Devonian.

The Longmen Shan and Qin Ling (Fig. 28-4) lie south of the Tarim–Sino-Korean Platforms and have a sequence variable in thickness and lithology and locally metamorphosed. It is a mixture of terrestrial deposits with agnathan fish and plant remains and shallow marine beds with corals and brachiopods. The plant and fish remains are of a type common in south and southeast China (Fig. 28-5,6).

The west Sichuan and north Xizang region (Fig. 28-7) has an extensive but metamorphosed Lower Devonian succession of limestones, marbles, schists, and slates. There are numerous corals, graptolites, and tentaculites. The Middle and Upper Devonian are commonly absent, but locally thick reef limestones of these ages are found. The Himalaya–West Yunnan region (Fig. 28-8) to the south has a dominantly limestone sequence and a fauna of corals, brachiopods, graptolites, and tentaculites which has affinities with the Hercynian type of the European Devonian.

(a) Devonian Type Section. There are three types of depositional environment—continental, marine, and mixed (Fang et al., 1979)—and all three are found in three type sections set up in the south China region (see Fig. 28-A,B,C). The one illustrated in Fig. 29 is from Guilin, Guangxi Province. The lowest Devonian (D_1) is a deltaic deposit resting unconformably upon older rocks, with *Yunnanolepsis* as the characteristic fossil. This quickly gives way to a more neritic environment as a result of transgression, and shelly carbonate facies with brachiopods and corals becomes progressively more important upwards. A mild regression at the end of the Lower Devonian resulted in a repetition of the shale and subordinate limestone to limestone sequence in the Middle Devonian. Middle Devonian rocks as a result of continuing transgression may overstep older beds and rest in places on Caledonian granites. The limestones of the Lower and Middle Devonian are commonly argillaceous. Dolostones appear in the upper Middle Devonian.

There is a hiatus between the Middle and Upper Devonian, and another, marked by a disconformity, separates the Upper Devonian from the Carboniferous. Two facies are reported: a quiet neritic environment of moderate water depth where limestones with nektonic cephalopods occur, and a more energetic environment characterized by oolitic limestones and interbedded dolostones.

The second type section in the southern part of Jiangxi Province (Fig. 28-B) consists of an alternation of marine and nonmarine beds where the Lower and lower Middle Devonian (D_{1-5}) are missing and upper Middle Devonian (D_6) rests unconformably upon metamorphosed Lower Paleozoic rocks. The lithological succession consists of quartz sandstones and conglomerates interbedded with plant-bearing

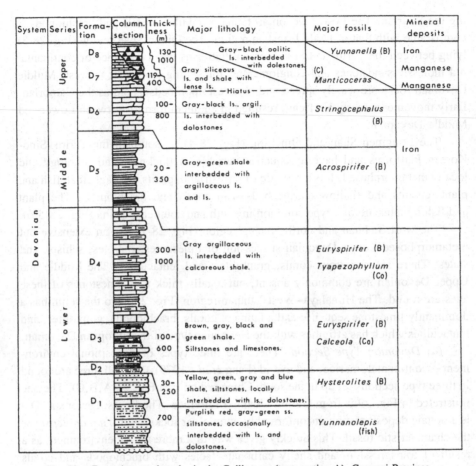

System	Series	Forma-tion	Column. section	Thick-ness (m)	Major lithology	Major fossils	Mineral deposits
Devonian	Upper	D₈		130-1010	Gray-black oolitic ls. interbedded with dolostones.	Yunnanella (B)	Iron
		D₇		119-400	Gray siliceous ls. and shale with lense ls. —Hiatus—	(C) Manticoceras	Manganese Manganese
	Middle	D₆		100-800	Gray-black ls., argil. ls. interbedded with dolostones	Stringocephalus (B)	
		D₅		20-350	Gray-green shale interbedded with argillaceous ls. and ls.	Acrospirifer (B)	Iron
	Lower	D₄		300-1000	Gray argillaceous ls. interbedded with calcareous shale.	Euryspirifer (B) Tyapezophyllum (Co)	
		D₃		100-600	Brown, gray, black and green shale. Siltstones and limestones.	Euryspirifer (B) Calceola (Co)	
		D₂					
		D₁		30-250	Yellow, green, gray and blue shale, siltstones, locally interbedded with ls., dolostones.	Hysterolites (B)	
				700	Purplish red, gray-green ss. siltstones, occasionally interbedded with ls. and dolostones.	Yunnanolepis (fish)	

Fig. 29.   Devonian stratigraphy in the Guilin area (type section A), Guangxi Province.

silty mudstones (*Lepidodendropsis, Protolepidodendron*). The overlying lowest Upper Devonian (D₇) rocks are fine-grained, with fine to medium clastics containing *Leptophloeum rhombicum* and *Bothriolepis*. *Bothriolepis*, a placoderm fish, is found in both marine and nonmarine beds. Interbedded with these nonmarine beds are horizons with *Lingula* and *Tenticospirifer*. The rest of the Upper Devonian consists of varicolored, fine- to medium-grained terrigenous clastic strata with some interbedded limestone horizons containing brachiopods. In the lower Yangzi River valley in the third type section (Fig. 28-C), the Devonian is entirely nonmarine and apparently was deposited in an intermontane basin. The Lower Devonian is primarily arenaceous with purplish-red sandstones and sandy shales which contain Antiarch placoderms. It is followed by a sequence of sandstones, black shales, and gray claystones with both plant and placoderm fish remains, all assigned to the Middle Devonian.

   (b) Paleoecology and Paleogeography. The land area of the early Devonian

reflects very strongly the influence of Caledonia movements. Some part of the continental realm was occupied by mountains and plateaus, the source of sediments now found around the margins of the old land mass, or in intermontane basins. Only the Tianshan geosynclinal remained marine (Wang and Yu, 1962). Marine transgression onto the land mass dates from the early Devonian, with seas advancing from the Indo-Pacific through the Indochina peninsula to south China or from the Sea of Okhotsk into northern China. The fauna of the seas trangressing from the north is closely related to the Boreal-Pacific realm, whereas in south China the fauna shows its affinities with that of western Europe. Lacunae occur in the Lower Devonian as a result of minor earth movements.

The scale of transgression increased in middle Devonian time, and seas advancing from south China reach into northwest China. Submarine volcanism affected the northern part of northeast China, northwest China, and is found in the Nei Monggol and in southwest China.

Further earth movements result in a hiatus between the Middle and Upper Devonian in the south China region, terminated by a renewed but weak Upper Devonian transgressive phase. Marine sandstones barely reach into southeast China which was then apparently an elevated region.

6. Carboniferous System

The Carboniferous system is very well developed in China. Traditionally, it was divided into three series, in contrast to the two divisions in North America; but following later revision, two series were adopted (Yang et al., 1962, 1979) and the country was divided into the five regions illustrated in Fig. 30. These divisions essentially repeat the pattern established in the Devonian (Fig. 28) with the north and south China regions having marine shelf carbonates with intercalated nonmarine horizons carrying a typical fusulinid, coral, and brachiopod fauna. The flora of north China is Cathaysian in type. In the north, too, the Lower Carboniferous is absent. The Junggar–Hinggan region persists with a geosynclinal character found throughout the Paleozoic to this point with a thick series of limestones and interbedded volcanics particularly well developed in the Upper Carboniferous. The Lower Carboniferous is sometimes absent. The fauna is related to that in Siberia.

Northwest China is a region of mixed lithofacies and biofacies, the latter closely related to those found in south China. A geosynclinal regime is said to predominate, with the development of a thick carbonate and clastic sequence. In Xizang–West Yunnan, the Lower Carboniferous consists mainly of limestones with a fusulinid, coral, and brachiopod fauna interbedded with basalts and succeeded, in the Upper Carboniferous, by clastics and basalts.

(a) South China Type Section. The type section for southern China lies in southern Guizhou Province (Fig. 30-1). The section, illustrated in Fig. 31, was deposited in the central part of a marine basin. It generally rests conformably upon

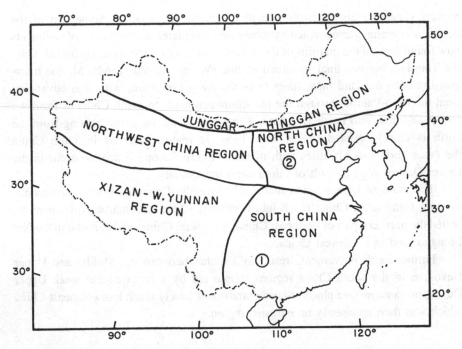

Fig. 30. Distribution of the Carboniferous system in China. The locations of the type sections are indicated: 1, south China region; 2, north China region.

the Devonian, and the occurrence of a local hiatus suggests that regression at the end of the Devonian was limited. The earlier Carboniferous (C_1), consisting mainly of limestone, was laid down in the neritic environment of a transgressing sea. This was followed by the deposition of increasing quantities of terrigenous clastics indicative of a temporary regression which persisted into the Lower Visean (C_{2L}); however, by Upper Visean time (C_{2U}), fully marine conditions with limestones predominate (Fig. 31).

Elsewhere in southeast China the sequence is predominantly continental with a few marine horizons. The section consists of grayish-white and purple-red terrigenous clastics with plants such as *Sublepidodendron, Rhodea, Asterocalamites, Mesocalamites, Archaeopteris,* and *Neuropteris.* In the marine horizons, *Eochiristites, Echinoconchus,* and *Chonetes* are found in deposits from seas transgressing from southwest China.

(b) North China Type Section. From late Ordovician time until the middle Carboniferous, north China was an emergent region undergoing erosion, and only in the middle Carboniferous (C_3) did sedimentation resume, associated with subsidence. The type section near Taiyuan in Shanxi Province (Fig. 30-2) has a typical cyclothem sequence shown in Fig. 32. The basal part of the sequence is an unstratified, massive, reddish-brown iron ore ("Shanxi" type) consisting of reddish-

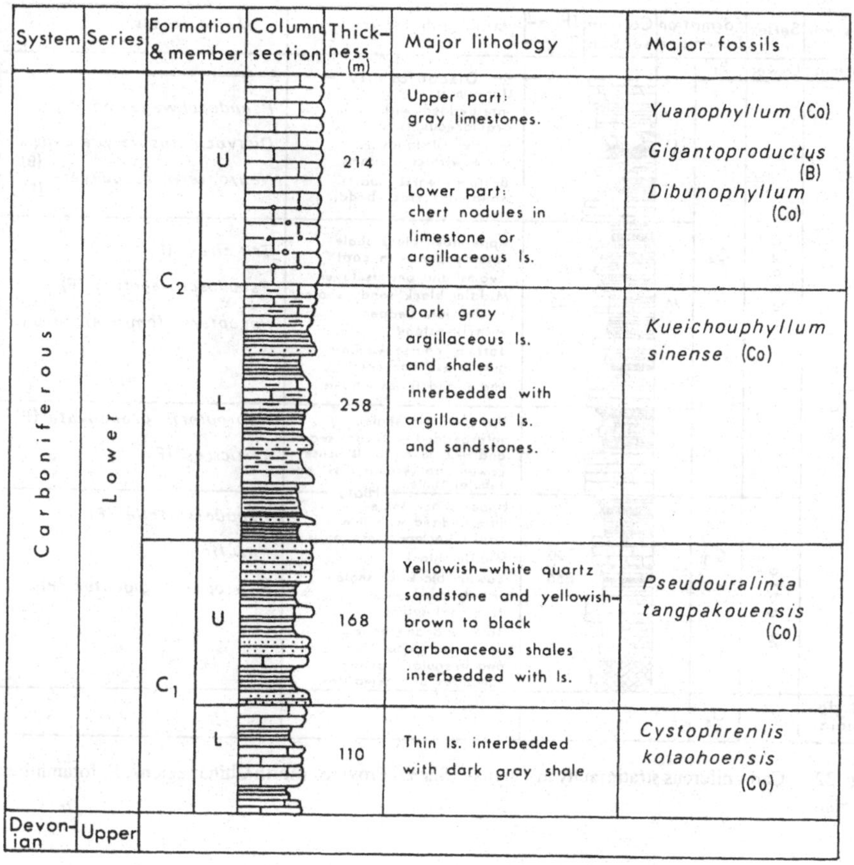

System	Series	Formation & member	Column section	Thickness (m)	Major lithology	Major fossils
Carboniferous	Lower	C₂ — U		214	Upper part: gray limestones. Lower part: chert nodules in limestone or argillaceous ls.	*Yuanophyllum* (Co) *Gigantoproductus* (B) *Dibunophyllum* (Co)
		C₂ — L		258	Dark gray argillaceous ls. and shales interbedded with argillaceous ls. and sandstones.	*Kueichouphyllum sinense* (Co)
		C₁ — U		168	Yellowish-white quartz sandstone and yellowish-brown to black carbonaceous shales interbedded with ls.	*Pseudouralinta tangpakouensis* (Co)
		C₁ — L		110	Thin ls. interbedded with dark gray shale	*Cystophrenlis kolaohoensis* (Co)
Devonian	Upper					

Fig. 31. Lower Carboniferous stratigraphy in southern Guizhou Province (south China region).

brown hematite and limonite disseminated in clay overlain by an oolitic aluminous clay. Within the latter horizon, some brachiopods and plant fragments are found, suggesting transgression over a low-lying, deeply weathered, continental surface. The remainder of C_3 reflects the effects of small oscillations of sea level in a coastal environment, with the alternation of sandstone, shale, and coal and limestone horizons. Limestones, more important in the middle of C_3, contain fusulinids. The Upper Carboniferous (C_4) lies disconformably upon C_3, the result of emergence and a brief period of erosion. It is subdivided into three, each subdivision reflecting a depositional cycle. The middle cycle has the best development of limestone, which forms an important marker horizon. This Upper Carboniferous is of economic importance for the three thick coal seams which occur within it.

The flora of the Upper Carboniferous is of Cathaysian type with *Neuropteris pseudovata, Lepidodendron posthumi, Annularia pseudostellata,* and *Pecopteris.* This floral assemblage continues into the Permian. Yang *et al.* (1962) and Sheng

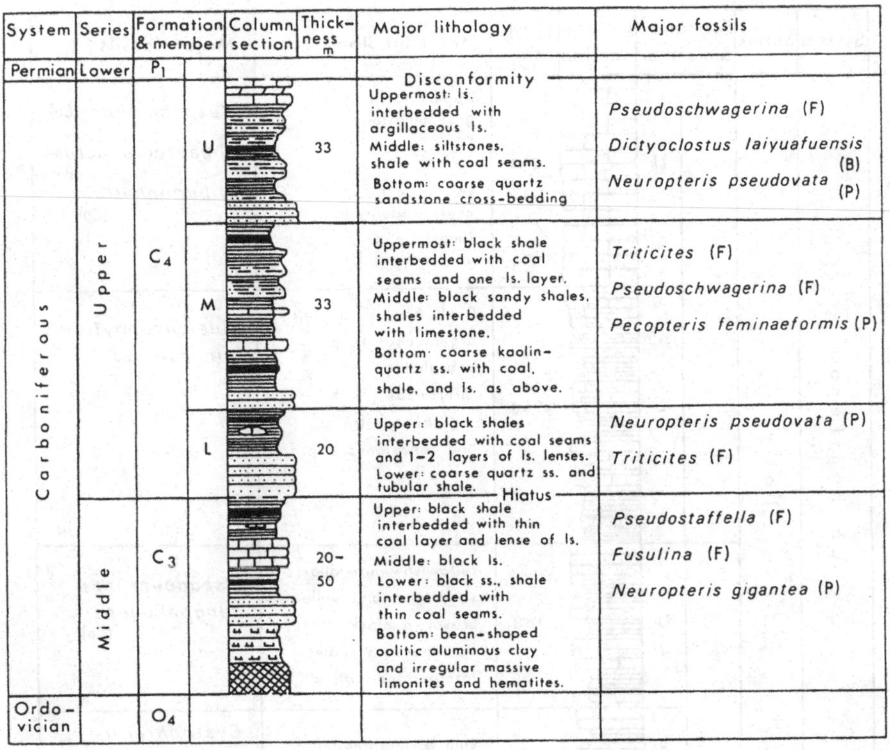

System	Series	Formation & member	Column section	Thick-ness m	Major lithology	Major fossils
Permian	Lower	P₁			— Disconformity —	
Carboniferous	Upper	C₄	U	33	Uppermost: ls. interbedded with argillaceous ls. Middle: siltstones, shale with coal seams. Bottom: coarse quartz sandstone cross-bedding	Pseudoschwagerina (F) Dictyoclostus laiyuafuensis (B) Neuropteris pseudovata (P)
			M	33	Uppermost: black shale interbedded with coal seams and one ls. layer. Middle: black sandy shales, shales interbedded with limestone. Bottom: coarse kaolin-quartz ss. with coal, shale, and ls. as above.	Triticites (F) Pseudoschwagerina (F) Pecopteris feminaeformis (P)
			L	20	Upper: black shales interbedded with coal seams and 1–2 layers of ls. lenses. Lower: coarse quartz ss. and tubular shale. —— Hiatus ——	Neuropteris pseudovata (P) Triticites (F)
	Middle	C₃		20–50	Upper: black shale interbedded with thin coal layer and lense of ls. Middle: black ls. Lower: black ss., shale interbedded with thin coal seams. Bottom: bean-shaped oolitic aluminous clay and irregular massive limonites and hematites.	Pseudostaffella (F) Fusulina (F) Neuropteris gigantea (P)
Ordo-vician		O₄				

Fig. 32. Carboniferous stratigraphy in Taiyuan, Shanxi Province (north China region). F, foraminifera; P, plants.

(1962) discussed the Permo-Carboniferous boundary problem. The *Pseudo-schwagerina* zone in the USA, USSR, and Japan is placed as the lowest Permian zone; however, in China it is regarded as Carboniferous and the boundary placed above it at the disconformity which resulted from Yunnan earth movements (Fig. 32). In consequence, Fang *et al.* (1979) equate C₄ to Permian Wolfcampian!

(c) Paleoecology and Paleogeography. The paleogeography of the early Carboniferous was similar to that of the Upper Devonian except that the Xizang (Tibetan) region had become part of the marine realm following transgression from the Indo-Pacific region through Burma (Yang *et al.*, 1962). The same sea also invaded the South China Basin. Transgression in northwest China was from the west (European) via the Tianshan geosyncline to the Qilian Shan or from the Himalayan region to the present Kunlun Shan. Southeast China, north China, and the southern part of northeast China remained emergent. Following crustal movements at the end of the early Carboniferous, local regression occurred; however, transgression was regionally renewed during mid and late Carboniferous time, and seas covered a more extensive area of China.

The Sino-Korean Paraplatform of north China, a land area since the middle

Ordovician, became a shallow basin by the middle Carboniferous, invaded by seas from the northeast transgressing toward the southwest (Fang *et al.*, 1979). This conclusion is based upon the thinning of the C_3 beds from the southern part of northeast China toward the central part of north China, beds which pinch out completely in the southern and western parts of north China. The stratigraphic position of the iron and aluminum ores at the base of the C_3 sequence also rises in the direction of thinning, to basal C_4 in southern and western parts of north China.

The slow transgression in this region, and the migration backwards and forwards of the shoreline, favored the development of coastal swamps now reflected by the presence of economically important coal deposits in Shanxi.

7. Permian System

The Permian system in China is well developed and generally fossiliferous. Sections in the south tend to be mainly marine, while those in the north become basically nonmarine. Sheng (1962) and Sheng *et al.* (1979) recognize four regions in the Permian of China (Fig. 33). In the Himalayan region, a thick, strongly folded limestone with Tethyan marine faunas closely related to those found in south China, gives way upward to beds containing a Gondwana flora with *Glossopteris communis* near Qomolangma Feng. The south China region contains a uniform marine

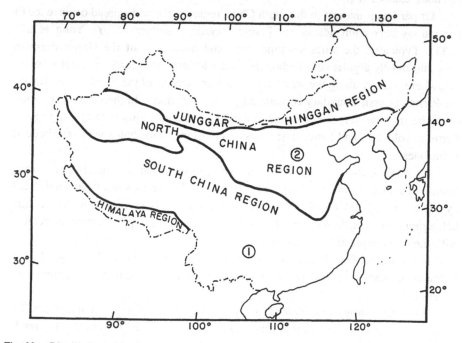

Fig. 33. Distribution of the Permian system in China. The locations of the type sections in the south China (1) and north China (2) regions are shown.

section with abundant Tethyan representatives of the fusulinids, corals, and brachiopods, as well as rare ammonites and conodonts. The north China region exhibits dominantly continental deposits containing a Cathaysian flora. The Junggar–Hinggan region is subdivided into an eastern facies of sedimentary, metamorphic, and volcanic rocks with marine fossils in the lowermost part and plants toward the top, and a western facies in Xinjiang Province consisting predominantly of continental strata with a few marine interbeds and volcanics and a flora closely related to that of the Soviet Union.

(a) *South China Type Section.* The Permian of China is divided into two series: the lower, correlated with the Leonardian and Guadalupian of North America, and the upper, correlated with the Ochoan (Fang *et al.*, 1979). As indicated earlier, the Wolfcampian equivalents are regarded as Carboniferous.

The type section for the marine rocks typical of south China is located in Guizhou Province (Figs. 33 and 34) where it lies disconformably on the Carboniferous. The lower series consists of alternating marine sandstones with coral and nonmarine shales with coal seams and plant fossils, though the top includes a massive, dark limestone with fusulinids and chert (P_1) and a massive, fossiliferous, light-colored limestone (P_2). The Upper Permian which is disconformable on P_2 is composed of alternating fossiliferous, marine limestones and terrigenous clastics with coal seams (P_3) and cherty limestones followed by colloform chert with terrigenous clastics (P_4).

Of particular interest in the south China region is the widespread occurrence of continuous deposition across the Permo-Triassic boundary (Z. Y. Yang *et al.*, 1981). Typically, the siliceous limestones and mudstones of the Upper Permian pass into mostly argillaceous carbonates in the lowermost Triassic without a break. Elsewhere, mostly along the western and eastern margins of the South China Basin, the Permo-Triassic boundary is marked by a disconformity. In the northwest corner of the basin, however, the contact is typically a nonconformity between uppermost Permian volcanics and lowermost Triassic sandstones and shales with interbedded limestones.

(b) *North China Type Section.* The type section for north China is in Taiyuan, Shanxi Province, and is entirely marine (Fig. 35). The section has been subdivided by H. H. Li (1962) on the basis of its Cathaysian flora into a lower section belonging to the *Emplectopteris–Taeniopteris* zone, and an upper section correlated with the *Gigantopteris–Lobatannularia*.

In the Lower Permian, coal is associated with the clastic rocks; in the Upper Permian, clastics coarsen and become arkosic upwards, with the occurrence of gypsum.

(c) *Paleoecology and Paleogeography.* The Yunnan movements of earliest Permian time produced some topographic relief in south China resulting in lowlands with lacustrine deposits and marginal swamps, though in other areas, marine deposition continued (P_1). Late–Early Permian rocks suggest marine transgression in

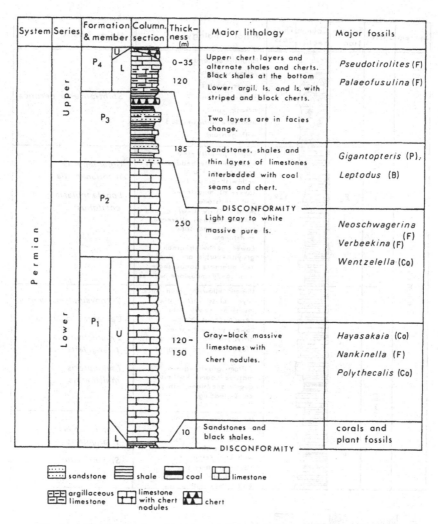

System	Series	Formation & member	Column. section	Thickness (m)	Major lithology	Major fossils
Permian	Upper	P₄ U / L		0-35 / 120	Upper: chert layers and alternate shales and cherts. Black shales at the bottom. Lower: argil. ls. and ls. with striped and black cherts. Two layers are in facies change.	Pseudotirolites (F) / Palaeofusulina (F)
		P₃				
		P₂		185	Sandstones, shales and thin layers of limestones interbedded with coal seams and chert.	Gigantopteris (P), Leptodus (B)
					DISCONFORMITY	
				250	Light gray to white massive pure ls.	Neoschwagerina (F) / Verbeekina (F) / Wentzelella (Co)
	Lower	P₁ U		120-150	Gray-black massive limestones with chert nodules.	Hayasakaia (Co) / Nankinella (F) / Polythecalis (Co)
		L		10	Sandstones and black shales.	corals and plant fossils
					DISCONFORMITY	

Fig. 34. Permian stratigraphy in central Guizhou Province, south China region.

southwest China, with limestones of the *Cancellina* zone being widely distributed (P₂). The Upper Permian deposits tend toward siliceous shales, and overall depositional environments in south China seem to become nonmarine eastward. A mild uplift accompanied by limited basaltic volcanism caused regression near the early Late Permian transition. After an interval of oscillation, evidenced by alternating marine and nonmarine beds (especially P₃), the sea transgressed toward the end of Late Permian time and neritic conditions were widely distributed.

The north China region was uplifted by Yunnan movements and was nonmarine during Permian time. Swampy conditions are indicated by abundant coal seams. Associated plant fossils and sediments suggest a warm, humid environment

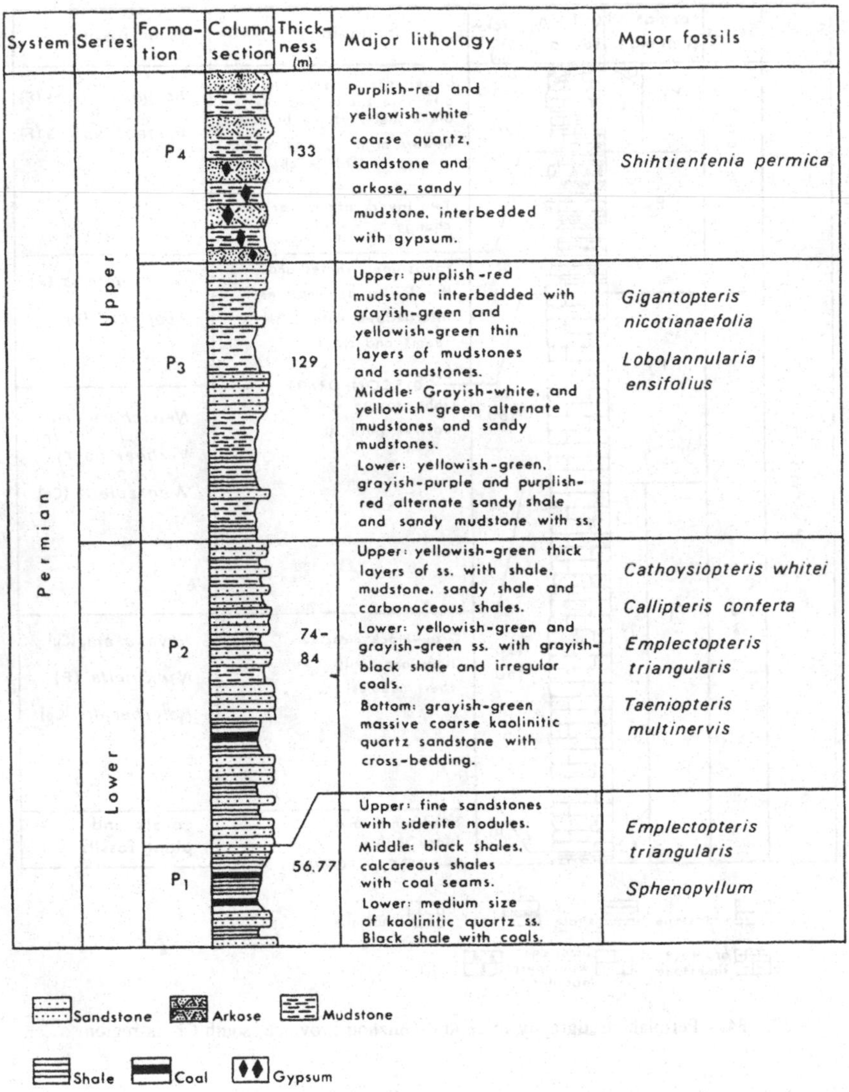

System	Series	Forma-tion	Column section	Thick-ness (m)	Major lithology	Major fossils
Permian	Upper	P₄		133	Purplish-red and yellowish-white coarse quartz, sandstone and arkose, sandy mudstone, interbedded with gypsum.	*Shihtienfenia permica*
		P₃		129	Upper: purplish-red mudstone interbedded with grayish-green and yellowish-green thin layers of mudstones and sandstones. Middle: Grayish-white, and yellowish-green alternate mudstones and sandy mudstones. Lower: yellowish-green, grayish-purple and purplish-red alternate sandy shale and sandy mudstone with ss.	*Gigantopteris nicotianaefolia* *Lobolannularia ensifolius*
	Lower	P₂		74–84	Upper: yellowish-green thick layers of ss. with shale, mudstone, sandy shale and carbonaceous shales. Lower: yellowish-green and grayish-green ss. with grayish-black shale and irregular coals. Bottom: grayish-green massive coarse kaolinitic quartz sandstone with cross-bedding.	*Cathoysiopteris whitei* *Callipteris conferta* *Emplectopteris triangularis* *Taeniopteris multinervis*
		P₁		56.77	Upper: fine sandstones with siderite nodules. Middle: black shales, calcareous shales with coal seams. Lower: medium size of kaolinitic quartz ss. Black shale with coals.	*Emplectopteris triangularis* *Sphenopyllum*

:::::: Sandstone   Arkose   Mudstone

Shale   Coal   Gypsum

Fig. 35. Permian stratigraphy in Taiyuan, Shanxi Province, north China region. All fossils are plants.

with lakes and swamps. Labyrinthodonts are found in cross-bedded fine- to medium-grained, variegated clastics. In early Late Permian time, the climate became less humid, and toward the end of the Permian arkose, shale and gypsum as well as rare plant fossils suggest the existence of a hot, dry climate.

These movements, which can be broadly assigned to late-phase Variscan, occurred over much of north China in the late early Permian. Evidence is widespread in the main tectonic zones, such as Qin Ling–Kunlun Shan, the Tian Shan–Yin Shan, and Monggol Arc regions (Compilation Group, 1976).

Early phase igneous activity is seen in mafic and ultramafic rocks in the Tian Shan, in Sichuan, and in western Yunnan. Associated granitic rocks have been dated at about 350 Ma. A middle igenous phase, with granites emplaced in the Tian Shan, Altay, Hinggan, and Bei Shan and more basic and ultrabasic rocks in northwestern and northern China, yields isotopic ages around 300 Ma. Late-phase granitic intrusions in northern and northeastern China, Nei Monggol, Qilian, and western Yunnan and mafic and ultramafic rocks in northwestern, northeastern, and western China have yielded ages of 230–260 Ma.

C. Mesozoic Erathem

1. General Tectonic and Sedimentary Framework

In western China, the principal Mesozoic–Cenozoic tectonic features, as in the Paleozoic, strike generally east–west. These features consist mainly of folded, geosynclinal sediments, associated with complex faulting and magmatic and metamorphic phenomena. The Sichuan–Yunnan region of southcentral China exhibits a north–south-trending system of close folds and faults and also shows tectonomagmatic and metamorphic effects. These structures, associated with the Tethys Belt, extend into Thailand and Malaysia to the south, while westward they swing toward the Qinghai–Xizang Plateau. The entire Tethys area is characterized by arcuate folds and faults which are closely compressed at either end and relatively widespread in the middle region (see Figs. 37 and 40). South of the Qinghai–Xizang Plateau in the Himalayas and Pamirs, the tectonic pattern is one of large-scale, plunging folds giving in plan view a ''Z''-shaped structural grain (Institute of Geomechanics, 1977).

In eastern China, Mesozoic–Cenozoic Neocathaysian trends are oriented north-northeast. The structures are primarily horsts and graben, bounded by normal or strike-slip faults. There were both magmatic and metamorphic events associated with this deformation. Intersection of the generally north–south Neocathaysian trends with the dominantly east–west Paleozoic trends produced medium- to large-sized tectonic basins with en echelon arc-shaped belts of uplift with their apices directed to the southeast.

Mesozoic rocks in China are primarily nonmarine sediments with some interbedded volcanogenic rocks and are widely distributed. Some marine Triassic rocks occur in southern China and there are marine Jurassic and Cretaceous rocks in parts of south and west China. The nonmarine deposits are mainly red, terrigenous clastics and coals with some gypsum and halite. Lithofacies and thicknesses vary markedly from one area to another. Stratigraphic correlations are mainly done with plant fossils, and Sze and Chou (1962) recognize five useful assemblages ranging in age from late Triassic to late Cretaceous–Paleocene (Fig. 36). The five floral assemblages are:

- *Danaeopsis–Bernoullia:* late Triassic (Keuper) in age. A flora found mainly in north and northwest China and closely related to the flora of Kazakhstan and Euro-American forms. Mixed with elements of the *Dictyophyllum–Clathropteris* flora, it is also found in some parts of south and southwest China and Vietnam.
- *Dictyophyllum–Clathropteris:* latest Triassic (Rhaetian) to Liassic in age. A flora showing a close resemblance to eastern Greenland, Swedish, and German flora.
- *Coniopteris–Phoenicopsis:* flora of Liassic–Dogger age, with a wide dis-

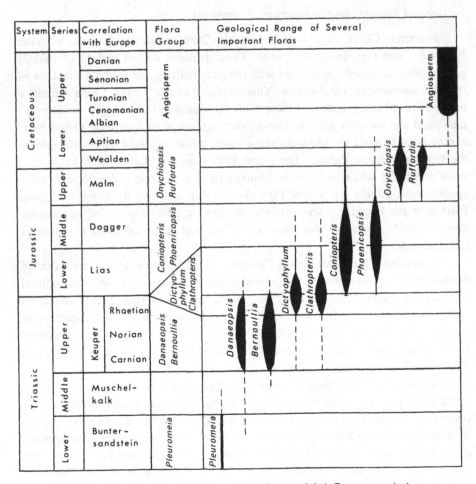

Fig. 36. Floral zonation of the Mesozoic of China and their European equivalents.

tribution in the southern part of northeast China, Nei Monggol, north and northwest China. The flora shows marked similarities to those of Siberia and Kazakhstan.

- *Ruffordia–Onychiopsis:* Late Jurassic to early Cretaceous in age, found mainly in northeast China and showing the closest resemblance to floras in the Ussuri River region near the Sino-Russian border and to the Tetori flora of Japan. There are also similarities to Korean flora.
- *Angiosperms:* from early to late Cretaceous or Paleocene, are found in northeast China.

2. Triassic System

Triassic rocks in China are mainly nonmarine but marine rocks do occur in southern, western, and extreme northeastern China. The boundary between the two follows the line of the Kunlun Shan–Qin Ling–Dabie Shan. Wang *et al.* (1981) have divided the marine Triassic into six regions and twenty-three districts (Fig. 37) based on facies, fauna, and development. These divisions roughly coincide with the fracture zones and will be used in the ensuing discussion.

Tethyan rocks occur in the Himalayan region in two belts. In the southern belt (I_1), the total thickness is less than 2000 m, with the Lower and Middle Triassic rocks being mainly richly fossiliferous, miogeosynclinal carbonates, while the Upper Triassic consists of interbedded clastics and carbonates. The northern belt (I_2) contains thick, eugeosynclinal sections of deep-water flysch and radiolarian-bearing siliceous rocks interbedded with basic volcanics. The southernmost belt (II_4) of the Gandise–Hengduan region adjoins the Himalayan region of the north and contains a thick sequence of fossiliferous marine sediment with large amounts of intercalated acid and intermediate volcanic rock. The northern belts of the Gandise–Hengduan $(II_3, II_5, II_6, II_7, II_8)$ all contain Lower and Middle Triassic marine clastics and carbonates, while the Upper Triassic rocks are intercalated marine beds and coals with basic and intermediate volcanics. Triassic rocks reach a thickness of more than 6000–7000 m where they are best developed (II_5, II_6, II_7).

Farther to the north and east in the Qilian–Bayanhar region, rocks of Ladinian age are absent in the northern sections (III_9, III_{10}), which consist of abundantly fossiliferous platform carbonates. Southward in this region, the geosynclinal, clastic flysch and volcanic sequence reaches great thicknesses and contains a few pelagic fossils in the extreme south (III_{12}). Southeastward, one encounters the Yangzi region where platform rocks of complex facies occur. Lower and Middle Triassic rocks are marine, while Carnian rocks are of alternating marine and non-marine facies. In the eastcentral part of the region (IV_{17}), the facies are transitional, and in the southernmost belt (IV_{16}) the rocks contain open-sea, shallow-water faunas, while to the northeast $(IV_{15}, IV_{18}, IV_{19})$ the rocks become sparsely fossiliferous and of littoral or lagunal facies. The Sichuan Basin (IV_{18}) in the northwest

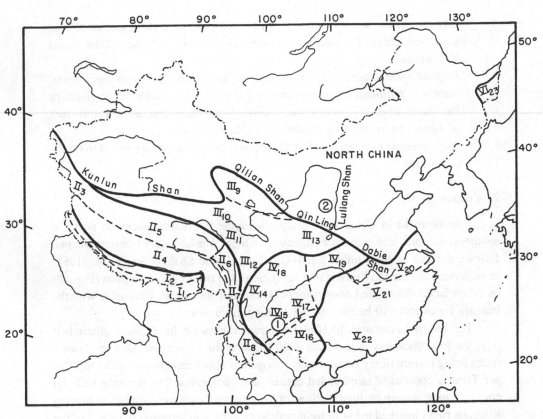

Fig. 37.   Distribution of marine Triassic in China. I, Himalayan region; II, Gandise–Hengduan region; III, Qilian–Bayanhar region; IV, Yangzi region; V, southeast China; VI, Nadanhada region. Type section locations are numbered: 1, Yangzi region; 2, north China region. (From Wang *et al.*, 1981; reprinted by permission of IUGS.)

exhibits carbonates and evaporites. In southeast China, platform rocks of partly marine and partly nonmarine facies continue. Far to the northeast in the Nadanhada Ling region in Heilongjiang Province, there are uppermost Triassic rocks consisting of volcanic ash, tuff, and diabase, interbedded with marine shales and sandstones of geosynclinal facies which contain the western Pacific pelecypod *Entomonotis*.

Nonmarine Triassic sandstones, shales, and coal seams are widely distributed in the inland basins of north China and contain extensive volcanics and occasional marine interbeds. Vertebrate fossils, including *Lystrosaurus*, *Chasmatosaurus*, and *Kanneria*, are found in these beds.

(a) South China Type Section. The type section (Fig. 38) is located in the southwestern part of Guizhou Province (Fig. 37). It is a cyclothemic development of fine-grained, variegated clastic deposits interbedded with limestones in the lowest Triassic (T_1) which rests disconformably upon the Permian. It gives way to

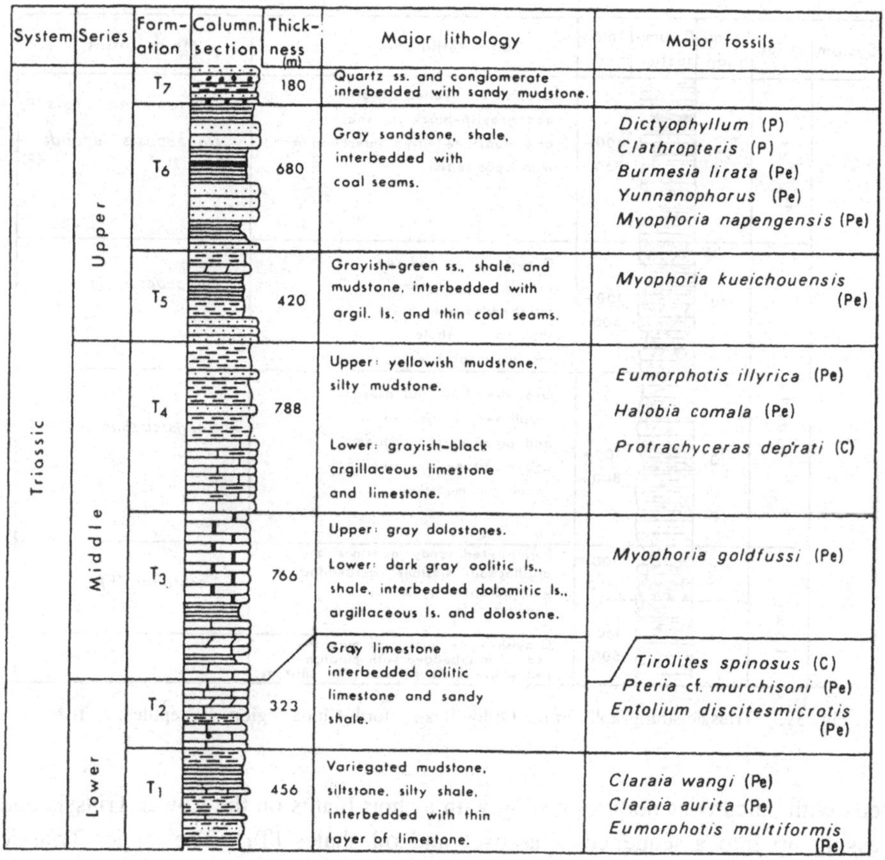

System	Series	Formation	Column section	Thickness (m)	Major lithology	Major fossils
Triassic	Upper	T_7		180	Quartz ss. and conglomerate interbedded with sandy mudstone.	
		T_6		680	Gray sandstone, shale, interbedded with coal seams.	Dictyophyllum (P) Clathropteris (P) Burmesia lirata (Pe) Yunnanophorus (Pe) Myophoria napengensis (Pe)
		T_5		420	Grayish-green ss., shale, and mudstone, interbedded with argil. ls. and thin coal seams.	Myophoria kueichouensis (Pe)
	Middle	T_4		788	Upper: yellowish mudstone, silty mudstone. Lower: grayish-black argillaceous limestone and limestone.	Eumorphotis illyrica (Pe) Halobia comala (Pe) Protrachyceras deprati (C)
		T_3		766	Upper: gray dolostones. Lower: dark gray oolitic ls., shale, interbedded dolomitic ls., argillaceous ls. and dolostone.	Myophoria goldfussi (Pe)
	Lower	T_2		323	Gray limestone interbedded oolitic limestone and sandy shale.	Tirolites spinosus (C) Pteria cf. murchisoni (Pe) Entolium discitesmicrotis (Pe)
		T_1		456	Variegated mudstone, siltstone, silty shale, interbedded with thin layer of limestone.	Claraia wangi (Pe) Claraia aurita (Pe) Eumorphotis multiformis (Pe)

Fig. 38. Triassic stratigraphy in the southwestern part of Guizhou Province, Yangzi region. Pe, pelecypods.

oolitic limestones with some terrigenous clastics (T_2). These limestones contain neritic cephalopods and some pelecypods indicating more marine conditions. The basal Middle Triassic (T_3), however, provides evidence in the form of poorly fossiliferous dolomites of a highly saline environment passing up into a more terrigenous sequence of argillaceous limestone below and fine-grained clastics above (T_4). The Upper Triassic consists of alternating marine and nonmarine beds including several coal seams. These Upper Triassic coal-bearing beds are unconformably overlain by the Lower Jurassic.

(b) North China Type Section. The best developed nonmarine sequence (Fig. 39) lies in the Shaan-Gan-Ning (Ordos) basin of North China (Fig. 37). The Lower Triassic beds conformably overlying the Permian consist of red, terrigenous clastics, separated into two units by an intervening conglomerate horizon. The Middle Triassic has a lower (T_3) division of medium- to fine-grained terrigenous

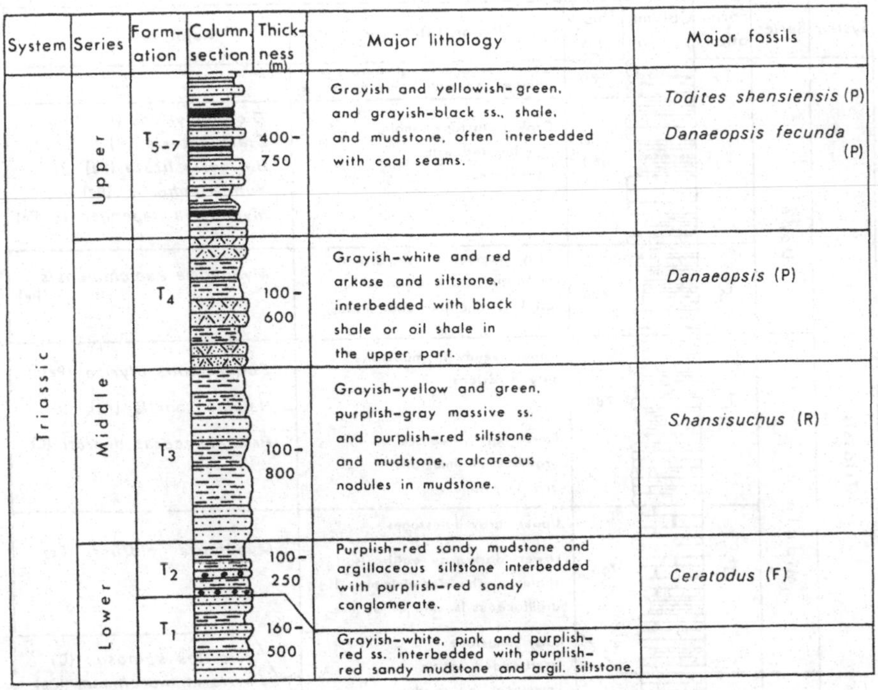

System	Series	Form-ation	Column section	Thick-ness (m)	Major lithology	Major fossils
Triassic	Upper	T_{5-7}		400-750	Grayish and yellowish-green, and grayish-black ss., shale, and mudstone, often interbedded with coal seams.	Todites shensiensis (P) Danaeopsis fecunda (P)
	Middle	T_4		100-600	Grayish-white and red arkose and siltstone, interbedded with black shale or oil shale in the upper part.	Danaeopsis (P)
		T_3		100-800	Grayish-yellow and green, purplish-gray massive ss. and purplish-red siltstone and mudstone, calcareous nodules in mudstone.	Shansisuchus (R)
	Lower	T_2		100-250	Purplish-red sandy mudstone and argillaceous siltstone interbedded with purplish-red sandy conglomerate.	Ceratodus (F)
		T_1		160-500	Grayish-white, pink and purplish-red ss. interbedded with purplish-red sandy mudstone and argil. siltstone.	

Fig. 39.   Triassic stratigraphy in the Ordos Basin, north China region. R, reptiles; F, fish.

beds with calcareous nodules resting with a short hiatus on the Lower Triassic and passing up into a sequence of arkoses and oil shales (T_4). The Upper Triassic contains several coal seams interbedded in terrigenous clastics.

(c) *Paleoecology and Paleogeography.* Earth movements loosely assigned to the Variscan toward the end of the Paleozoic resulted in marine regression from the region north of the Kunlun Shan–Qilian Shan–Qin Ling–Dabie Shan with the exception of small regions in the southernmost Qilian Shan and the Nadanhada Ling connected with the West Pacific. South of the mountains, seas transgressed from the Indian Ocean with the development of a geosyncline in west China.

There are thus two marine realms, with the Boreal restricted to the Nadanhada Ling region while the rest in south and west China can be assigned to the Tethys. In the Yangzi region (IV_{16}) there is a typical normal marine fauna. Lower in the region $IV_{15,18,19}$ (see Fig. 37), abnormal salinity is assumed to account for the rare and monotonous fauna.

Many inland basins developed in north China, and a morphological distinction can be made between northwest China and northeast China separated by the Luliang Shan in western Shanxi Province. The difference is marked by greater basin depth and continuity in the northwest in contrast to the greater erosion experienced in the northeastern basins. The depositional environment in these basins is interpreted as one of fluvial lakes in a hot and dry climate perpetuating conditions established during the Permian. During middle Triassic time, the lithological change to black shale and oil shale is commonly interpreted as climatic, but may represent reduced relief and drainage for the organic matter accumulated in a reducing environment in quiet lacustrine and swamp conditions. The flora contains *Glossopteris*.

There was some intrusive activity in the eastern Qinghai–Xizang Plateau, Qin Ling, Nan Ling, and Dabie Shan, and isotopic ages in the 190–230 Ma range have been recorded. More recently, intrusive rocks of this phase have been discovered in north China and in the Da Hinggan Ling of northeast China. Although consisting mainly of granite and diorite, there are local mafic, ultramafic, and alkalic rocks. The intrusives are generally covered by Upper Triassic or Lower Jurassic sedimentary rocks. This intrusive activity is generally related to the Indosinian orogeny (Compilation Group, 1976).

3. Jurassic System

Three broad belts based upon depositional environments, with as many as 61 subregions, were recognized by Ku (1962) in the Jurassic and Cretaceous systems (Fig. 40): the Pacific volcanic belt, which is a broad band parallel to the coast; the continuing marine Tethyan geosyncline along the border region with India; and the zone of the Inland Basin. Type sections can be defined in northern Hebei and eastern Liaoning Provinces for the Pacific coastal belt (Fig. 40 J-1), while the Inland Basin can be typified by a section in the Sichuan Basin (Fig. 40 J-2).

Two depositional regions can be identified in the Pacific volcanic belt, in the inner, uplifted part of the Neocathaysian system, and in the areas east of it (Compilation Group, 1976). The Lower and Middle Jurassic are characterized by fluvio-lacustrine deposits with some coal and horizons of intermediate to mafic volcanic rocks. There is considerably more volcanicity in the Upper Jurassic where intermediate to mafic and intermediate to acidic lavas are found interbedded with some lacustrine deposits. A few marine horizons are found in the most northeasterly areas in the Nadanhada Ling region, where bathyal silicic rocks and volcanic rocks occur, and near Guangzhou in south China. Rare late Jurassic to Cretaceous ammonoids are known from the subsurface in Taiwan but no Jurassic is known in Hainan.

There are several internal basins in the Inland Basin region, of which the best known are the Shaan-Gan-Ning (Ordos), the Tarim, Junggar, and the Qaidam north of the Kunlun Shan–Qin Ling, while to the south lie the Yunnan and Sichuan Basins. In the northern basins, the sedimentary sequence, with fluvio-lacustrine, variegated terrigenous clastic beds, is more varied than the predominantly red bed sequence of the southern basin. Marine and volcanic beds are of limited extent, a few possible early Jurassic marine fossils are reported from Sichuan (Ku, 1962), and some early Jurassic volcanics are known in Qinghai Province.

The major zone for marine deposits lies in the Tethyan belt of southwest

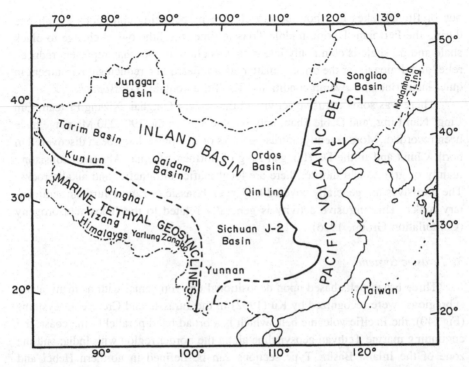

Fig. 40. Distribution of the Jurassic and Cretaceous in China. The location of the Jurassic type sections (J-1, Pacific volcanic belt; J-2, Sichuan Basin) and the Cretaceous type section C in the Songliao Basin are shown.

China. The Tethyan domain, which includes Qinghai–Xizang and the Himalayas of southwest China, is characterized by neritic, shelly limestones, shales, and littoral sandstones which carry pelecypod, brachiopod, ammonite, belemnite, and echinoid fauna.

(a) Northeast China Type Section. In northern Hebei and western Liaoning Provinces there is a thick section of alternating volcanic and nonmarine deposits (Fig. 41). There are few principal volcanic horizons, with one in the Lower, one in the Middle, and two in the Upper Jurassic. The two lower horizons consist of andesitic and basaltic lavas and volcaniclastic beds. In the Lower Jurassic, the volcanic rocks are underlain by fluvial deposits, and overlain by swampy coal deposits. The sequence is reversed in the Middle Jurassic, with the volcanic horizon resting on swampy deposits and being covered by fluvioclastic deposits.

The volcanic horizon at the base of the Upper Jurassic is split by a thick horizon of lacustrine beds, and these are overlain by organic-rich shales with coal, some sandstones, and even conglomerates. Once more, the volcanics are andesitic, but trachytic varieties also occur, along with a variety of volcaniclastic deposits.

(b) Central China Type Section. As representative of the succession of the

System	Series	Form- ation		Column section	Thick- ness (m)	Major lithology	Major fossils
		K₁					
Jurassic	Upper	J₇			84– 1,400	Yellowish-green sandy shale, congl. and coal.	*Coniopteris* (P) *Nippononaia* (Pe)
		J₆	U		300– 1,700	Grayish-black shale, oil shale, coal, sandstone.	*Coniopteris* *Lycoptera* (F) *Ferganoconcha* (Pe)
			L		200– 3,500	Red-gray-purple andesitic trachyte, andesitic breccia and tuff with sedimentary rocks.	
		J₅	U		10– 2,000	Grayish-white tuffaceous shale, sandstone, conglomerate.	*Lycoptera* *Ephemeropsis* *Nakamuranaia* (Pe)
			L		200– 2,500	Pyroxene andesite, volcanic breccia, tuff, and agglomerate with sandstone and shale.	
	Middle	J₄			200– 2,000	Red, yellow conglomerate, sandy conglomerate, sandstone, siltstone.	
		J₃			230– 970	Black andesitic basalt, andesitic breccia, yellow argil. congl.	*Coniopteris* *Podozamites* (P)
		J₂			284–543	Yellow congl., ss., siltstone.	*Cladophlebis* (P) *Ferganocencha*
	Lower	J₁	U		627– 1,313	Yellowish-brown sandstone, grayish-black shale and sandstone with coals.	*Cladophlebis* (P) *Dictyophyllum* (P) *Podozamites* (P)
			M		130– 900	Upper: dark andesitic, basalt, volcanic clastics. Lower: tuffaceous sandy shale.	*Neocalamites* (P) *Equisetites* (P) *Cladophlebis* (P)
			L		20–280	Yellow ss., congl. with coal.	*Coniopteris*

Fig. 41. Jurassic stratigraphy in north Hebei and west Liaoning, north China Region. F, fish.

Inland Basin region, one type section (Fig. 42) is chosen from eastern Sichuan (Fig. 40 J-2) and a second from the Shaan-Gan-Ning (Ordos) Basin (see Fig. 40).

In the Sichuan Basin, the Lower Jurassic, consisting of dark, plant-bearing, terrigenous, sideritic clastic beds with coal, rests disconformably upon the Upper Triassic. The lithology and fossils both are indicative of deposition in a hot, humid, swampy environment. In the Middle Jurassic the occurrence of gray clastics and variegated mudstones with conchostracans and freshwater pelecypods suggests a

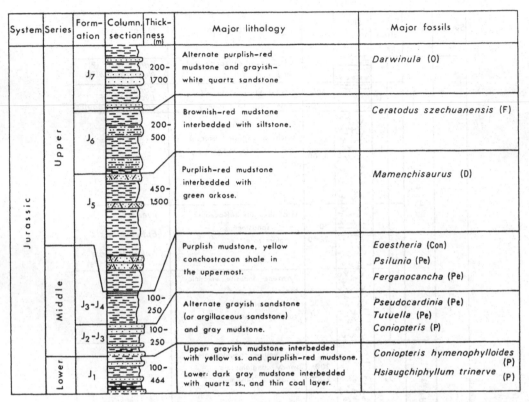

System	Series	Form- ation	Column. section	Thick- ness (m)	Major lithology	Major fossils
Jurassic	Upper	J₇		200-1,700	Alternate purplish-red mudstone and grayish- white quartz sandstone	Darwinula (O)
		J₆		200-500	Brownish-red mudstone interbedded with siltstone.	Ceratodus szechuanensis (F)
		J₅		450-1,500	Purplish-red mudstone interbedded with green arkose.	Mamenchisaurus (D)
	Middle	J₃-J₄		100-250	Purplish mudstone, yellow conchostracan shale in the uppermost.	Eoestheria (Con) Psilunio (Pe) Ferganocancha (Pe)
					Alternate grayish sandstone (or argillaceous sandstone) and gray mudstone.	Pseudocardinia (Pe) Tutuella (Pe) Coniopteris (P)
		J₂-J₃		100-250		
	Lower	J₁		100-464	Upper: grayish mudstone interbedded with yellow ss. and purplish-red mudstone. Lower: dark gray mudstone interbedded with quartz ss., and thin coal layer.	Coniopteris hymenophylloides (P) Hsiaugchiphyllum trinerve (P)

Fig. 42. Jurassic stratigraphy in east Sichuan. O, ostracods; D, dinosaur; Con, conchostracans; F, fish.

deeper lacustrine environment which apparently dried up, for the Upper Jurassic is made up of red mudstones interbedded with arkosic and quartzose sandstones. The fauna, in addition to ostracods and conchostracans, also includes vertebrates.

(c) *North China Type Section.* In the Ordos Basin, the Lower and Middle Jurassic consist of plant-bearing, lacustrine mudstones which grade upwards into coarser-grained sandstone with mudstone and oil shales and some argillaceous lacustrine limestones near the top. Subsequent uplift resulted in the nondeposition of the Upper Jurassic.

(d) *Southwest China Type Section.* The Tethyan type sequence is on Qomo- langma Feng (Everest) and consists of ammonite-rich, thick-bedded, Lower and Middle Jurassic limestones interbedded with sandstone and shale, followed by black and gray, sandy shale which also yields ammonites and belemnites.

(e) *Paleoecology and Paleogeography.* The Indosinian movements which oc- curred during the middle late Triassic changed the basic paleogeographic pattern of China which had persisted little changed through the Paleozoic. The change was most apparent with the recognition of the generally meridional trend of the Pacific volcanic belt. As a result, three major types of environment are recognizable: inland basins associated with volcanic activity parallel to the present coast; the inland, intermontane basins of central and northwest China with their lacustrine and fluvial sequences; and the shallow sea and bay environments in southwest and south China. Most of the basins in south China began to develop either in the late Triassic or in the early Jurassic, extending to as late as the middle Jurassic following the Indosinian movements.

The large basins of central China were characterized by quiet lacustrine environments (Fang *et al.*, 1979) with depocenters in the western part of the basins. The basins trend north-northeast. Climatic differences are suggested between the basin south of the Kunlun Shan–Qin Ling ranges where climate changed from humid to arid, and those to the north where climate remained humid.

Marine transgression from the Tethyan region affected southwest China during the Early and Middle Jurassic, but during the Late Jurassic regression took over. In the northeastern part of China, the Nadanhada Ling deep marine basin was connected to the west Pacific realm.

Orogenic movements (Yanshanian movements) began during the course of the Bajocian–Bathonian, and these continued until late Cretaceous time (Ku, 1962), affecting the entire country although most intense in eastern China (Compilation Group, 1976). The movements were associated with folding and rifting, with two phases of intensive magmatic activity as recorded in the Pacific volcanic belt. The Jurassic phase of magmatic activity, with granites, granodiorites, diorites, and mafic and ultramafic rocks, yields ages in the 150–190 Ma range. This activity is found mainly in southeast China, in the middle and lower Yangzi Valley, but also occurs in parts of north and northeast China, in Xizang, and in western Yunnan. Many of the endogenic metallic ore deposits are related to this igneous activity.

The present structural framework was imposed at this time.

4. *Cretaceous System*

The distribution pattern of Cretaceous rocks follows that of the Jurassic (see Fig. 40). In the Pacific volcanic belt, Cretaceous rocks, mainly volcanic, are found in a number of north-northeast to northeast-trending medium- to small-sized basins in east China (Compilation Group, 1976). Interbedded with intermediate to silicic volcanics are nonmarine, terrigenous clastics with some coal horizons and occasional marine beds, for a few marine fossils have been recorded in the Songliao Basin as well as in south China.

The Cretaceous in the Inland Basin region consists largely of red, nonmarine, terrigenous clastics with some intercalated mudstone and gypsum horizons, found in a few large- to medium-sized basins in central and west China. In the western part of northwest China, minor amounts of intermediate volcanics and a few marine horizons have been recorded. Nonmarine fauna and flora are abundant.

Marine deposits are mostly restricted to the Himalayas, Xizang, and Kunlun

Shan, as well as to Taiwan. They consist of neritic to paralic terrigenous clastic s interbedded with ammonitic limestones which also contain pelecypods and foraminifera. In the Yarlung Zangbo River of Xizang there is a thick flysch sequence of sandy shale, and black shale containing pelagic ammonites and belemnites with submarine volcanics in the Upper Cretaceous is reported.

(a) *Northeast China Type Section.* The section in the Songliao Basin (Fig. 43) may be taken as typical (Chen, 1980). There is a change from about 6000–7000 m of fine-grained clastics in the central part of the basin, to coarser-grained clastic rock with a thickness of only 200–500 m around the basin margins. The coarse molasse-type rocks of the Lower Cretaceous rest disconformably upon Jurassic beds.

Within the basin, thick lacustrine beds intercalate with deltaic facies, and locally littoral to marine glauconitic and phosphatic rocks develop. There are occasional thin but highly fossiliferous, coquinoid limestone horizons. The sedimentation pattern is dominated by fining-up sequences, but reverse cycles also occur. Except in the deepest part of the basin, there is an unconformity between the Lower and Upper Cretaceous. Here, the Cretaceous is succeeded by unconformable Cenozoic deposits.

(b) *North China Type Section.* A type section can be located in eastern Shandong where it consists of a volcano-sedimentary sequence bordering the present Pacific coast. Here, the Lower Cretaceous consists of lacustrine clastics and intermediate extrusions overlain by red fluvial and lacustrine beds of the Upper Cretaceous with no volcanics.

(c) *Southwest China Type Section.* Like the Jurassic, the Cretaceous type section is on Qomolangma Feng (Everest) in Xizang Province. It is in the Lower Cretaceous, a continuation of the shaley, ammonite-bearing lithologies of the Jurassic, with some interbedded argillaceous limestones and sandstone. In the Late Cretaceous, there is an upward transition from limestone and shale to terrigenous clastic deposits.

(d) *Paleoecology and Paleogeography.* The Cretaceous was a period of increased land area with the decreased marine realm restricted to Xizang in the southwest (Tethys), the southern part of the Tarim basin in the northwest, the northeastern part of northeast China, and Taiwan (Pacific). A few regions of brackish and saltwater lakes existed in south China, but the rest of China was entirely nonmarine.

The Songliao Basin, formed during the initial stages of Yanshanian movements, is a north-northeast horst and graben complex. The initial small, shallow basin with swamp, fluvial, and lacustrine environments, with some volcanic activity in the middle and late Jurassic, gradually expanded and broadened to its maximum extent in the early late Cretaceous. Deltas formed around the margin of the lake. There was one marine transgressive episode which is thought to have coincided with the global Albian–Cenomanian transgression. Toward the end of the Cretaceous the basin decreased in size.

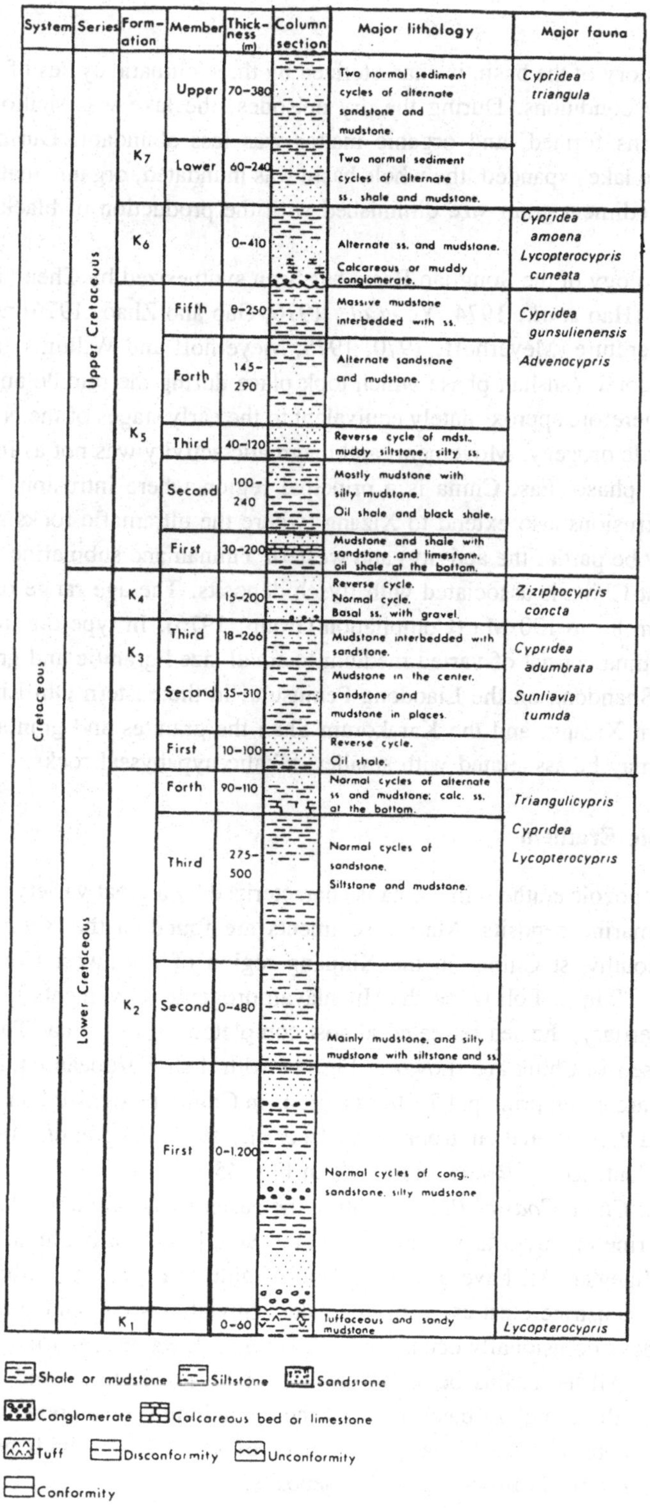

System	Series	Form-ation	Member	Thick-ness (m)	Column section	Major lithology	Major fauna
Cretaceous	Upper Cretaceous	K7	Upper	70–380		Two normal sediment cycles of alternate sandstone and mudstone.	Cypridea triangula
			Lower	60–240		Two normal sediment cycles of alternate ss. shale and mudstone.	
		K6		0–410		Alternate ss. and mudstone. Calcareous or muddy conglomerate.	Cypridea amoena Lycopterocypris cuneata
		K5	Fifth	8–250		Massive mudstone interbedded with ss.	Cypridea gunsulienensis
			Forth	145–290		Alternate sandstone and mudstone.	Advenocypris
			Third	40–120		Reverse cycle of mdst., muddy siltstone, silty ss.	
			Second	100–240		Mostly mudstone with silty mudstone. Oil shale and black shale.	
			First	30–200		Mudstone and shale with sandstone and limestone. oil shale at the bottom.	
	Lower Cretaceous	K4		15–200		Reverse cycle. Normal cycle. Basal ss. with gravel.	Ziziphocypris concta
		K3	Third	18–266		Mudstone interbedded with sandstone.	Cypridea adumbrata
			Second	35–310		Mudstone in the center, sandstone and mudstone in places.	Sunliavia tumida
			First	10–100		Reverse cycle. Oil shale.	
		K2	Forth	90–110		Normal cycles of alternate ss and mudstone; calc. ss. at the bottom.	Triangulicypris
			Third	275–500		Normal cycles of sandstone. Siltstone and mudstone.	Cypridea Lycopterocypris
			Second	0–480		Mainly mudstone, and silty mudstone with siltstone and ss.	
			First	0–1.200		Normal cycles of cong., sandstone, silty mudstone	
		K1		0–60		Tuffaceous and sandy mudstone	Lycopterocypris

Shale or mudstone   Siltstone   Sandstone

Conglomerate   Calcareous bed or limestone

Tuff   Disconformity   Unconformity

Conformity

Fig. 43. Cretaceous stratigraphy in the Songliao Basin, northeast China. All fossils are freshwater ostracods. (From Chen, 1980; reprinted by permission of *Petroleum Geology*.)

The history of the basin is characterized by three climatic cycles of alternating wet and dry conditions. During the dry episodes, the lake was shallow, several isolated basins formed, and organic matter was less abundant. During the wet episodes, the lake expanded, the whole basin was inundated, organic matter proliferated, and sediment grain size diminished with the production of black mudstone and shale.

The geology of the Songliao Basin has been synthesized by Chen (1980) from the Chinese (Hao *et al.*, 1974; Ye *et al.*, 1976; Gao and Zhao, 1976) and English language literature (Meyerhoff, 1970, 1975; Meyerhoff and Willums, 1976).

The second Yanshan phase which took place during the middle and late Cretaceous is therefore approximately equivalent to the early stages of the North American Laramide orogeny. Movement, and magmatic activity was not as intense as in the Jurassic phase. East China is a principal region where intrusions are found, although intrusions also extend to Xizang (where the ultramafic rocks of northern Xizang may be part of the activity) and western Yunnan and submarine volcanicity in southwest China is associated with flysch deposits. The age range of the intrusives is from 80 to 130 Ma (Compilation Group, 1976). In type the intrusions in southeast China consist of varied medium to small-sized granitic and granodioritic bodies. In Shandong on the Liaodong Peninsula, in the eastern Qin Ling, eastern and southern Xizang, and the Karakorum Shan the granites and granodiorites are larger and may be associated with smaller, alkalic hypabyssal rocks.

D. Cenozoic Erathem

The Cenozoic erathem in China is characterized by a great variety of predominantly nonmarine deposits. Marine sediments are found in the Xizang–Qinghai region of southwest China, in the Xinjiang region of northwest China, and on Taiwan and Hainan. Following the Himalayan orogenic movements in the middle and Late Tertiary, the sea retreated almost completely from China. The Cenozoic divisions used in China are shown in Fig. 44 with their European equivalents.

There are seven principal Tertiary regions in China, recognized on the basis of tectonic and depositional environments (Pei *et al.*, 1963; Fang *et al.*, 1979). These zones are identified by Roman numerals in Fig. 45.

I. *East China Coastal Plain*—Within the eastern coastal plains there are four large lacustrine or fluvio-lacustrine basins, the Songliao, North China, Jiang Han, and north Jiangsu. All have a complex block-fault structure and sedimentary sequences of lacustrine clastics, with oil shales, fluvial clastics, and some gypsum. Volcanic rocks occasionally occur, and a few brackish water or marine faunas have been found. All the basins began to develop in Mesozoic time, and all with the exception of the Songliao basin reached their maximum development during the Cenozoic. There are, additionally, a number of small basins with generally incomplete and variable fluvio-lacustrine deposits.

Erathem	System	Series		Major fossils	European stages
Cenozoic	Quaternary		Holo-cene	Homo sapiens	
			Pleistocene	Homo erectus pekinensis (V)	Wurmian Riss-Wurmian Rissian Mindel-Rissian Mindelian Gunz-Mindelian Gunzian
	Tertiary		Pliocene	Ilyocypris (O) Candoniella (O) Cyprinotus (O) Candona (O)	Astian Plaisancian Pontian
			Miocene	Alnus (P) Salix (P) Secuoia (P) Stephanocemas (V)	Sarmatian Vindobonian Burdigalian Aquitanian
			Oligocene	Dongyingia (O) Huabeinia (O) Cyprinotus (O)	Chattian Stampian Sannoisian
			Eocene	Liminocythere (O) Cypris (Q) Eucypris (O)	Ludian Bartonian Auversian Lutetian Cuisian Sparnacian
			Paleocene	Prionessus (V) Sphenopsalis (V) Palaeostylops (V) Bemalambda (V)	Thanetian Montian

Fig. 44. Cenozoic divisions in China showing correlation with the European stages. V, vertebrates; O, ostracods; P, plants.

II. *Nei Monggol Peneplain*—Several small- to medium-sized fluvio-lacustrine basins are found in the central and western parts of Nei Monggol. The deposits within them generally consist of thin-bedded, grayish-green sandstone and mud-stone, commonly with an abundant vertebrate fauna. They formed at various stages of the Tertiary, as a result either of structural activity or of erosion.

A widely distributed Hannobar basalt is found in the eastern part of Nei Monggol and western Liaoning Province. It occurs in a sandy oil shale and marl sequence which has yielded some Miocene vertebrate fossils such as *Trilaphodon* sp. and *Stephenocemas thomsoni;* however, according to the Compilation Group (1976) the age of the basalt may range from Paleogene to early Neogene.

III. *Eastern Liaoning Fault Basin*—This consists of several small fault basins,

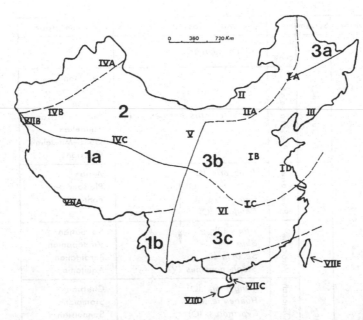

Fig. 45.   Distribution of the Tertiary in China and Miocene floristic regions. The Tertiary provinces, identified by Roman numerals. are: I, the coastal plains of east China—A, Songliao; B, north China; C, Jiang Han; D. Jiangsu. II, Nei Monggol peneplain—A, Hannobar. III, eastern Liaoning. IV, northwest China—A, Junggar; B, Tarim; C, Qaidam. V, north central China. VI, south China. VII, southwest China and Taiwan—A, Himalaya; B, west Kunlun Shan; C, Leizhou Peninsula; D, Hainan; E, Taiwan. The Miocene floristic regions are: 1, Qinghai–Xizang region—a, Plateau province; b, Hengduan mountain province. 2, inland region. 3, eastern monsoon region—a, northern temperate province; b, central warm-temperate-subtropic province; c, south subtropic and tropic province. Province boundaries are shown by light lines; heavy lines mark boundaries of floristic regions; dashed lines are uncertain Miocene floristic data. Te, general location of the Tertiary type section in the Bohai coastal area; Q, general location of the Quaternary stratigraphy on the loess plateau. (From Song *et al.*, 1981; reprinted by permission of the Geological Society of America.)

extremely variable in both thickness and lateral extent. The organic-rich sediments consist of oil shale, mudstone, and coal, with basalt horizons.

   IV. *Northwest China Piedmont*—In some of the larger basins (Tarim, Junggar, Qaidam) a Tertiary sequence of piedmont and lacustrine red sediments is found. These represent the first deposits resulting from the uplift of the mountain ranges as a result of the Yanshan orogeny (Pei *et al.*, 1963). Later, inland lakes developed in which finer-grained sediments accumulated. The effects of local movements are reflected by the development of fanglomerates and coarser stream deposits.

   V. *North Central China Intermontane*—Medium- to small-sized intermontane basins are common in northcentral China and in western and southern China. The basins with a general northeast–southwest trend formed during a late stage of the

Yanshan orogeny. They contain Eocene red lacustrine terrigenous clastics, with some local gypsum deposits which grade laterally into fluvial or piedmont facies around the margins. This pattern is repeated in the younger, overlying sediments.

VI. *South China Complex*—There is sedimentary continuity from the Cretaceous through the Paleogene in a series formerly called the ''New Red Beds'' (Compilation Group, 1976), with the Mesozoic–Cenozoic boundary determined paleontologically. The sequence is a thick, fining-up sequence passing from a red basal breccia to felspathic and micaceous sandstone to oil shale, or marl, or occasionally evaporites near the top. The environments represented range from fluvial, lacustrine, intermontane, to piedmont.

VII. *Southwest China and Taiwan Marine Basin*—The Tethyan marine domain covers what are now the Himalayas, the western Tarim Basin, and the western Kunlun Shan. Deposits in the Pacific marine region are found in Taiwan, Hainan, the Leizhou Peninsula, and limited areas of the eastern coastal plains and the southeastern continental shelf (South China Sea).

The Himalayan sequence consists of nummulitic limestone, sandstone, shale, and volcanic rocks. Only the Lower Tertiary of the western Tarim Basin is marine, with *Ostrea*, the Upper Tertiary being nonmarine.

The marine Tertiary of Taiwan consists of mainly littoral and neritic terrigenous clastics, with limestone and mafic extrusives and a mollusk and foraminifera fauna. On both Hainan Island and the Leizhou Peninsula, the sequences consist of marine, terrigenous, and bioclastic rocks with basalts. Brief marine incursions in the North China, Jiang Han, and Jiangsu Basins seem indicated by the discovery of marine foraminifera.

1. *North China Type Section*

A composite Tertiary stratigraphic section for north China is shown in Fig. 46. The sequence is composed mainly of red, gray, and yellow, medium- to fine-grained, lacustrine, terrigenous clastic rocks with gypsum, rock salt, oil shale, and limestone. The Paleocene is absent, for the basal conglomerate lies within the Lower Eocene. Oligocene beds are well developed and widely distributed and contain a few thin marine horizons. There is sedimentary continuity across the Eocene–Oligocene boundary which cannot be defined with the sparse fauna found. For the same reason, the Miocene and Pliocene cannot be differentiated.

Paleogene fossils have been recorded in the coastal region of Bohai, with gastropods (Yu *et al.*, 1978), ostracods (Hou *et al.*, 1978), foraminifera (He and Hu, 1978), charophytes (Wang *et al.*, 1978), pollen and spores (Song *et al.*, 1978a), and dinoflagellates and acritarchs (Song *et al.*, 1978b). Most species are nonmarine, a few are brackish water, and some are shallow marine.

A Paleocene mammalian fauna has been found in southern, southwestern, northwestern, and central China.

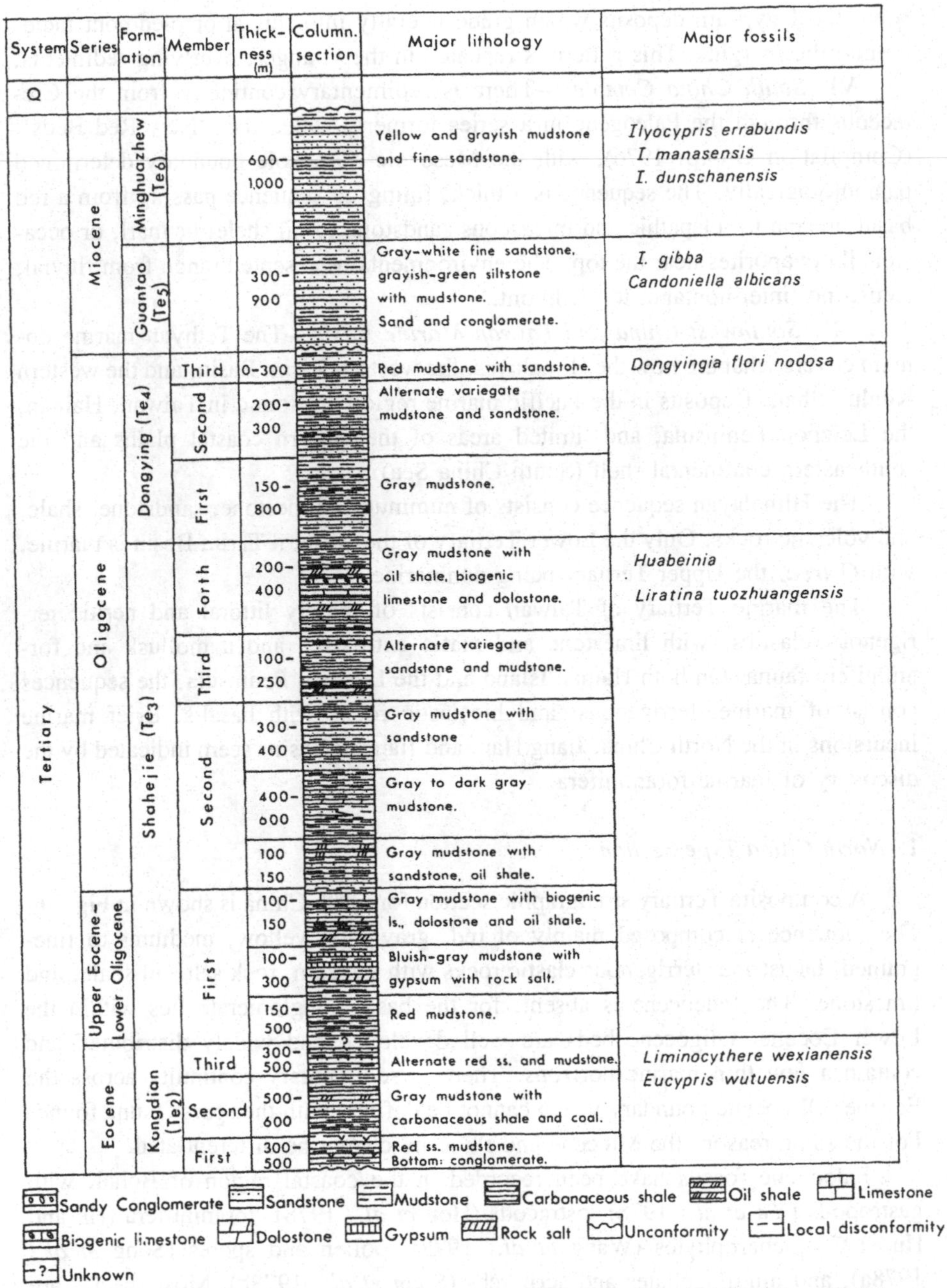

System	Series	Formation	Member	Thickness (m)	Column. section	Major lithology	Major fossils
Q							
Tertiary	Miocene	Minghuazhew (Te6)		600–1,000		Yellow and grayish mudstone and fine sandstone.	*Ilyocypris errabundis* *I. manasensis* *I. dunschanensis*
		Guantao (Te5)		300–900		Gray-white fine sandstone, grayish-green siltstone with mudstone. Sand and conglomerate.	*I. gibba* *Candoniella albicans*
	Oligocene	Dongying (Te4)	Third	0–300		Red mudstone with sandstone.	*Dongyingia flori nodosa*
			Second	200–300		Alternate variegate mudstone and sandstone.	
			First	150–800		Gray mudstone.	
		Shahejie (Te3)	Forth	200–400		Gray mudstone with oil shale, biogenic limestone and dolostone.	*Huabeinia* *Liratina tuozhuangensis*
			Third	100–250		Alternate variegate sandstone and mudstone.	
			Second	300–400		Gray mudstone with sandstone	
				400–600		Gray to dark gray mudstone.	
	Upper Eocene–Lower Oligocene		First	100–150		Gray mudstone with sandstone, oil shale.	
				100–150		Gray mudstone with biogenic ls., dolostone and oil shale.	
				100–300		Bluish-gray mudstone with gypsum with rock salt.	
				150–500		Red mudstone.	
	Eocene	Kongdian (Te2)	Third	300–500		Alternate red ss. and mudstone.	*Liminocythere wexianensis* *Eucypris wutuensis*
			Second	500–600		Gray mudstone with carbonaceous shale and coal.	
			First	300–500		Red ss. mudstone. Bottom: conglomerate.	

Sandy Conglomerate   Sandstone   Mudstone   Carbonaceous shale   Oil shale   Limestone   Biogenic limestone   Dolostone   Gypsum   Rock salt   Unconformity   Local disconformity   Unknown

Fig. 46. Composite stratigraphy of the Tertiary system in the coastal region of Bohai, north China. Ostracods are the major fossils.

2. *Paleoecology and Paleogeography*

During the Tertiary, seas withdrew from most of China with the exception of west and southwest China, for after the last stage of Yanshan movements in the Pacific realm the sea persisted only in Taiwan, Hainan, and in small, scattered areas along the coast, as well as in the Tethyan Himalayan geosyncline. The western part of the Tarim Basin was connected to the seaway of Soviet Central Asia.

The palynological and paleobotanical evidence makes it possible to define three Miocene floral assemblages (Song *et al.*, 1981) (Fig. 45). In the Qinghai–Xizang floral province, two upland florules can be recognized: the plateau province, with *Quercus, Pinus,* and monolete spores of *Polypodiaceae;* and the Hengduan Mountain province, with tropical to subtropical, evergreen, woody plants in the warm, moist valleys and pines and alpine plants on the mountains. The inland floral province is characterized by abundant herbaceous plants with shrubby pollen and with some woody angiosperm pollen.

The third floral assemblage is the eastern monsoon flora in which there are northern temperate, a central warm temperate–subtropical, and a southern subtropical and tropical florules. The temperate florule consists of broad-leaved deciduous oaks and conifers, with a few subtropical genera. The central warm temperate–subtropical florule contains forms such as Betulaceae, Juglandaceae, and Ulmaceae, with deciduous oaks, chestnuts, and evergreen oaks. The Betulaceae disappear and are replaced by Fagaceae in the southern subtropical and tropic florule.

The magmatic rocks associated with the Himalayan movements occur mainly in the Himalayan and Sichuan–Yunnan regions and in Taiwan. Isotopic ages of 10–20 Ma (Chang and Cheng, 1973) are found in associated pegmatites. Mafic and ultramafic rocks are present in the longitudinal valley of Taiwan, and Pliocene basalts occur along the East China Sea coast and on the Bohai Islands. The uplift associated with the Himalayan movements resulted in laterite weathering of the Shanxi and Shaanxi Plateau.

E. Quaternary System

Quaternary deposits are widespread in China, especially in the larger basins north of the Qin Ling–Kunlun Shan tectonic zone (Compilation Group, 1976). These are six distinctive environments:

- The north China loess deposits distributed along the Huang He (Yellow River), Shaanxi, Gansu, Shanxi, Ningxia, Nei Monggol, and some other areas.
- The eolian sand and Gobi pavement of northwest China in the Tarim, Qaidam, and Junggar Basins and the Gansu corridor.
- The glacial and fluvioglacial deposits of west, southwest, and central China,

found in middle and lower reaches of the Yangzi and generally in all high mountain locations. Glaciation has been best studied in the Lu Shan (Jiangxi Province). Holocene glacial deposits also are found at high altitudes on the Qinghai–Xizang Plateau.

• Lacustrine fluvial and glacial deposits of east and north China across the plains of China.

• Cave deposits of various stages of south and north China, although early and middle Pleistocene cave deposits are limited to north China.

• Coral reefs of the South China Sea, Taiwan, and Hainan, and guano deposits on the offshore islands.

1. *Beijing Region Type Section*

The Beijing section consists of an alternation of glacial and interglacial sediments such as lacustrine, fluvial, cave, or loess deposits. The deposits rest unconformably upon Tertiary, with disconformities between Lower and Middle and between Middle and Upper Pleistocene. Vertebrate fossils are common in interglacial horizons. In addition to *Homo erectus pekinensis* found in a cave at Zhoukoudian near Beijing, since 1949 hominid fossils have been reported from Guangdong, Shaanxi, Yunnan, Guangxi, and Hubei Provinces.

2. *North China Loess Plateau Type Section*

The second type section of the Quaternary is the loess plateau sequence in Shaanxi Province, an alternating succession of nearly horizontal loess and paleosol horizons (Fig. 47). Sedimentologically, three loess types and five paleosol types are recognized, (Lu and An, 1979).

3. *Paleoecology and Paleogeography*

The principal geological event of the Quaternary was glaciation, with glacial stages which can be correlated with those in Europe and North America. Studies of the loess succession show two environmental end members, a warm, humid forest and a cold, arid, desert–steppe climatic type. These are confirmed by mammalian fauna and pollen analyses.

During the Quaternary, only local but massive uplift or subsidence have occurred, a 3000- to 4000-m uplift of the Himalayas and Xizang Plateau, and about 1000 m of subsidence of the Tarim and Qaidam Basins. Uplift has also occurred in the Yunnan–Guizhou Plateau, the Shaan-Gan-Ning Plateau, and the Loess Plateau. Subsiding areas include the North China Plain, the East China Coastal Plain, the Jiang Han Plain, and Dongtinghu Lake (about 29°N, 112–113°E). These structures are usually fault bounded.

There are widely distributed basalts, evidence of igneous activity, found main-

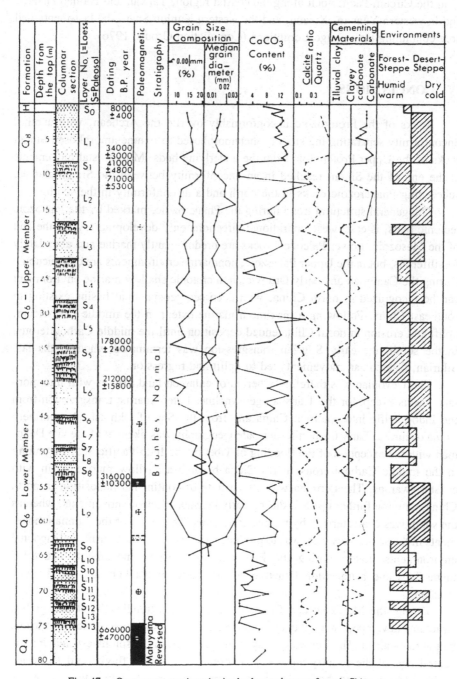

Fig. 47.   Quaternary stratigraphy in the loess plateau of north China.

ly in the Circum-Pacific belt along the coastal region, Taiwan and Hainan Islands, and in western Yunnan, Xizang, and the western Kunlun Shan, the last-mentioned area containing the youngest eruptions (Compilation Group, 1976).

IV. CONCLUSIONS

Rocks of the Proterozoic unconformably overlie the Archean, with another unconformity separating the slightly metamorphosed to nonmetamorphosed Sinian rocks of the Upper Proterozoic from the underlying beds. Mild crustal movements at the end of the Sinian resulted in an unconformity between the Sinian and the succeeding Phanerozoic rocks in the north and a disconformity in the south.

The general structural trend during the Phanerozoic, marked by the Paleozoic tectonic belts, is east–west, meridional directions only developing toward the end of the Mesozoic. Lower Paleozoic rocks are predominantly marine and abundantly fossiliferous, but in the Upper Paleozoic, nonmarine environments are conspicuous. During the Cambrian and early Ordovician, a cratonic shallow marine environment can be recognized in north China, with deep sea geosynclinal basins existing in southeast China. Following a maximum marine extent in the middle Ordovician, uplift and erosion in north China ended deposition until the middle Carboniferous. In the south, the early Silurian transgression was terminated when in the late Silurian, Caledonian movements led to uplift and regression.

The Devonian was therefore a period of enlarged land masses with intermontane basins except for the Tianshan geosyncline. Later, marine transgression from the Indo-Pacific into southern China and from the Sea of Okhotsk into northern China occurred. Early Carboniferous paleogeography was like that of the late Devonian with the exception of the flooding of Tibet by the Indo-Pacific Sea. Beginning in the middle Carboniferous, north China became a shallow marine basin with coastal swamps. The Permian was also a time of continental deposition in north China, a region further uplifted during early Permian Yunnan movements, whereas the south was characterized by marine conditions. The effect of the Yunnan movements in the south, however, was to produce variety of shallow marine and lagoonal environments. Later, during the Permian, transgressive movements reasserted themselves and a transitional marine contact between the Permian and Triassic resulted.

During the Mesozoic and Cenozoic, north–south-trending horst and graben structures developed, with abundant magmatic activity such that a Pacific volcanic belt is recognized. The Tethyan geosyncline persisted in the southwest, leaving the central part of the country an area of intermontane, inland fluvial and lacustrine basins. These inland basins received a great variety of sediments during the Cenozoic. However, the Himalayan events beginning in the Middle and extending to the late Tertiary mark the gradual closing of the Tethyan geosyncline, with major uplift occurring during the Quaternary.

In terms of plate tectonics, the geological evolution of China can be presented as the interaction of two plates, the North China Plate and the South China Plate, and the marginal region of the Siberian, Indian, and Pacific Plates.

ACKNOWLEDGMENTS

I am grateful to Dr. A. Nairn, Dr. A. Meyerhoff, E. Samodai, and M. Chen for editing the manuscript; to K. Meyerhoof for drafting all type sections; and to the Geological Society of America, the Cambridge University Press, the International Union of Geological Sciences, the Scientific Press, and Winston, and Drs. Lu Yanhao, Song Zhi-chen, and A. Meyerhoff for permission to use their figures.

REFERENCES

Allen, C. R., Gillespie, A. R., Han, Y., Sieh, K. E., Zhang, B. C., and Zhu, C. N., 1984, Red River and associated faults, Yunnan Province, China: Quaternary geology, slip rates, and seismic hazard, *Geol. Soc. Am. Bull.* **95**:686–700.

Atlas Press, 1977, Atlas of Provinces of the People's Republic of China [English and Chinese editions], Xinhua Book Company and Atlas Publication, Beijing.

Biq, C., Shyu, C. T., Chen, J. C., and Boggs, S., Jr., 1985, Taiwan: Geology, geophysics, and marine sediments, in: *The Ocean Basins and Margins*, Vol. 7A (A. E. M. Nairn, F. G. Stehli, and S. Uyeda, eds.), Plenum Press, New York, pp. 503–550.

Chang, C. F., and Cheng, H. L., 1973, Some tectonic features of the Mt. Jolmo Lungma area, southern Tibet, China, *Sci. Sin.* **16**:257–265.

Chang, W. T., 1962, *Ordovician of China* [in Chinese], Science Press, Beijing.

Chen, C., 1980, Non-marine setting of petroleum in the Sungliao basin of northeastern China, *J. Pet. Geol.* **2**:233–264.

Chen, G., 1977, *Outline of Geotectonics of China: Explanation of the Tectonic Map* [in Chinese], Seismology Press, Beijing.

Compilation Group of Geological Map of China, Chinese Academy of Geological Sciences, 1976, *An Outline of the Geology of China*, Geology Press, Beijing.

Fang, T. C., Xia, T. L., and Liu, H. L., 1979, *Paleontological Stratigraphy* [in Chinese], Geology Press, Beijing.

Gansser, A., 1964, *Geology of the Himalayas*, Interscience, New York.

Gansser, A., 1980, The significance of the Himalayan suture zone, *Tectonophysics* **62**:37–57.

Gao, J. S., Xiong, Y. S., and Gao, P., 1931, Preliminary notes on the Sinian stratigraphy of North China, *Geol. Soc. China Bull.* **13**:257–275.

Gao, R. Q., and Zhao, C. P., 1976, *Pollen and Spore Assemblage of Late Cretaceous from the Songliao Basin* [in Chinese], Science Press, Beijing.

Grabau, A. W., 1924, *Stratigraphy of China*, Part 1, Geological Survey of China, Nanking.

Grabau, A. W., 1928, *Stratigraphy of China*, Part 2, Geological Survey of China, Nanking.

Guo, L. Z., Shi, Y. S., and Ma, R. S., 1983, On the formation and evolution of the Mesozoic–Cenozoic active continental margin and island arc tectonics of the western Pacific Ocean [in Chinese], *Acta Geol. Sin.* **57**:11–21.

Hao, Y. C., Su, T. Y., Li, Y. G., Ruan, P. H., and Yuan, F. T., 1974, *Fossil Ostracods of Cretaceous and Tertiary from the Songliao Plain* [in Chinese], Geology Press, Beijing.

He, Y. T., and Hu, L. Y., 1978, *Cenozoic Foraminifera from the Coastal Region of Bohai* [in Chinese], Science Press, Beijing.

Hou, Y. T., Huang, B. L., Li, Y. P., Geng, L. Y., and Dan, G. B., 1978, *Paleogene Ostracods from the Coastal Region of Bohai* [in Chinese], Science Press, Beijing.

Huan, W. L., Wang, S. Y., Shi, Z. L., and Yan, J. Q., 1980, The distribution of earthquake foci and plate motion in the Qinghai–Xizang Plateau [in Chinese], *Acta Geophys. Sin.* **23:**267–280.

Huang, T. K., 1978, An outline of the tectonic characteristics of China, *Ecol. Geol. Helve.* **71:**611–635.

Institute of Geomechanics, Chinese Academy of Geological Sciences, 1977, *An Outline of the Tectonic Characteristics of China*, Geological Press, Beijing.

Institute of Geomechanics, Chinese Academy of Geological Sciences, 1978, Geotectonic system, in: *Professional Papers of International Scholarly Communication*, Vol. 1 [in Chinese], Geological Press, Beijing.

Ku, C. W., 1962, *Jurassic and Cretaceous of China* [in Chinese], Science Press, Beijing.

Lee, G. S., 1939, *Geology of China*, Murby, London.

Li, C. Y. [Lee, C. Y.], 1975, Tectonic evolutions of some mountain ranges in China, as tentatively interpreted on the concept of plate tectonics [in Chinese], *Acta Geophys. Sin.* **18:**52–76.

Li, C. Y., and Tang, Y. Q., 1983, Some problems on subdivision of paleoplates in Asia [in Chinese], *Acta Geol. Sin.* **57:**1–10.

Li, C. Y., Wang, Q., Zhang, Z., and Liu, X., 1980, A preliminary study of plate tectonics of China [in Chinese], *Chin. Acad. Sci. Bull.* **2:**11–22.

Li, H. H., 1962, *Continental Strata of Late Paleozoic of China* [in Chinese], Science Press, Beijing.

Li, S. G., 1924, Geology and gorge history of eastern Yangzi, *Geol. Soc. China Bull.* **3:**351–391.

Lu, Y. H., 1962, Cambrian of China [in Chinese], Science Press, Beijing.

Lu, Y. H., 1981, Provincialism, dispersal, development and phylogeny of trilobites, *Geol. Soc. Am. Spec. Pap.* **187:**143–152.

Lu, Y. H., Zhu, Z. L., Chien, Y. Y., Lin, H. L., Chow, T. Y., and Yuan, K. K., 1974, Bio-environmental control hypothesis and application to the Cambrian biostratigraphy and paleozoogeography [in Chinese], Nanjing Institute of Geology and Paleontology, Academic Sinica, Vol. 5, pp. 27–112.

Lu, Y. H., Zhu, Z. L., Chien, Y. Y., Zhou, Z. Y., Chen, J. Y., Liu, G. W., Yu, W., Chen, X., and Xu, H. K., 1976, Ordovician biostratigraphy and paleozoogeography of China [in Chinese], Nanjing Institute of Geology and Paleontology, Academia Sinica, Vol. 7, pp. 1–83.

Lu, Y. S., and An, Z. S., 1979, Natural environmental changes in the loess plateau during the Brunhes epoch [in Chinese], *Sci. Rep.* **24:**221–224.

Meyerhoff, A. A., 1970, Development in mainland China, 1949–1968, *Am. Assoc. Pet. Geol. Bull.* **54:**1567–1580.

Meyerhoff, A. A., 1975, China's petroleum potential, *World Pet. Rep. 1975* **21:**18–21.

Meyerhoff, A. A., 1978, Petroleum in Tibet and the Indo-Asia suture (?) zone, *J. Pet. Geol.* **1:**107–112.

Meyerhoff, A. A., and Willums, J. O., 1976, Petroleum geology and industry of the People's Republic of China, *U.N. ESCAP, CCOP Tech. Bull.* **10:**103–212.

Molnar, P., 1977, The collision between India and Eurasia, *Sci. Am.* **236:**31–41.

Molnar, P., and Tapponnier, P., 1975, Cenozoic tectonics of Asia: Effects of a continental collision, *Science* **189:**419–426.

Molnar, P., and Tapponnier, P., 1981, A possible dependence of tectonic strength on the age of the crust in Asia, *Earth Planet. Sci. Lett.* **52:**107–114.

Mu, E. C., 1962, *Silurian of China* [in Chinese], Science Press, Beijing.

Pei, W. C., Chou, M. C., and Zheng, J. J., 1963, *Cenozoic of China* [in Chinese], Science Press, Beijing.

Ren, J. S. [Jen, C. S.], Jiang, C. F., Zhang, Z. K., and Qin, D. Y., 1980, *The Geotectonic Evolution of China* [in Chinese], Science Press, Beijing.

Research Section of Precambrian and Metamorphic Geology, Institute of Geology, Academy of Geological Sciences, Ministry of Geology, 1962, *Precambrian of China* [in Chinese], Science Press, Beijing.

Sheng, J. Z., 1962, *Permian of China* [in Chinese], Science Press, Beijing.

Sheng, J. Z., Jin, Y. X., Rui, L., Zhang, L. X., Zheng, Z. G., Wang, Y. J., Liao, Z. T., and Zhao, J. M., 1979, Correlation charts and explanatory notes of the Permian system in China [in Chinese], Nanjing Institute of Geology and Paleontology, Academia Sinica, pp. 111–127.

Sheng, S. F., 1974, *Classification and Correlation of Ordovician in China* [in Chinese], Geology Press, Beijing.

Sheng, S. F., 1980, The Ordovician system in China, International Union of Geological Sciences Publication No. 1, pp. 1–7.

Song, Z. C., Cao, L., Zhou, H. Y., Guan, X. T., and Wang, K. D., 1978a, *Paleogene Pollens and Spores from the Coastal Region of Bohai* [in Chinese], Science Press, Beijing.

Song, Z. C., He, C. C., Qian, Z. S., Pan, Z. L., Zheng, G. G., and Zheng, Y. H., 1978b, *Paleogene Dinoflagellates and Acritarchs from the Coastal Region of Bohai* [in Chinese], Science Press, Beijing.

Song, Z. C., Li, H. M., Zheng, Y. H., and Liu, G. W., 1981, Miocene floristic regions of China, in: *Geol. Soc. Am. Spec. Pap.* **187:**249–254.

Sze, H. J., and Chou, C. Y., 1962, *Continental Strata of Mesozoic of China* [in Chinese], Science Press, Beijing.

Tapponnier, P., and Molnar, P., 1977, Active faulting and Cenozoic tectonics of China, *J. Geophys. Res.* **82:**2905–2930.

Tapponnier, P., and Molnar, P., 1979, Active faulting and Cenozoic tectonics of the Tien Shan, Mongolia, and Baikal regions, *J. Geophys. Res.* **84:**3425–3459.

Tapponnier, P., Peltzer, P., Le Dain, A. Y., Armijo, R., and Cobbold, A., 1982, Propagating extrusion tectonics in Asia: New insights from simple experiments with plasticine, *Geology* **10:**611–616.

Wang, H. Z., 1978, On the subdivision of the stratigraphic provinces of China [in Chinese], *Acta Stratigr. Sin.* **2:**81–104.

Wang, H. Z., and Ho, X. Y., 1981, Silurian rugose coral assemblages and paleobiogeography of China, *Geol. Soc. Am. Spec. Pap.* **187:**55–63.

Wang, S., Huang, L. J., Yang, C. Q., and Li, H. N., 1978, *Paleogene Charophytes from the Coastal Region of Bohai* [in Chinese], Science Press, Beijing.

Wang, Y., and Yu, C. M., 1962, *Devonian of China* [in Chinese], Science Press, Beijing.

Wang, Y. G., Chen, C. C., He, G. X., and Chen, J. H., 1981, An outline of the marine Triassic in China, International Union of Geological Sciences Publication No. 7, pp. 1–21.

Wu, F. T., Zhang, Y. M., Fang, Z. J., and Zhang, S. L., 1981, On the activity of the Tancheng–Lujiang fault zone in China [in Chinese], *Seimol. Geol.* **3:**15–26.

Yang, J. Z., Sheng, C. C., Wu, W. S., and Liu, L. H., 1962, *Carboniferous of China* [in Chinese], Science Press, Beijing.

Yang, J. Z., Wu, W. S., Zhang, L. X., and Wang, K. L., 1979, Classification and correlation of Carboniferous system in China [in Chinese], Nanjing Institute of Geology and Paleontology, Academia Sinica, pp. 98–110.

Yang, S. P., Pan, K., and Hou, H. F., 1981, The Devonian system in China, *Geol. Mag.* **118:**113–224.

Yang, Z. Y., Wu, S. B., and Yang, F. Q., 1981, Permo-Triassic boundary in the marine regime of South China [in Chinese], *Earth Sci.* **1:**4–15.

Ye, T. D., Ting, L. S., and Chang, Y., 1976, *Fossil Ostracods of Cretaceous from Songliao Basin* [in Chinese], Science Press, Beijing.

Yu, W., Mao, X. L., Chen, Z. Q., and Huang, L. S., 1978, *Paleogene Gastropods from Coastal Region of Bohai* [in Chinese], Science Press, Beijing.

Zhang, W. Y. [Chang, Wen-You] (ed.), 1980, *Formation and Development of the North China Fault Block Region* [in Chinese], Science Press, Beijing.

Zhang, W. Y. [Chang, W. Y.], and Zhong, J. Y., 1977, *On the Development of Fracture Systems in China* [in Chinese], Science Press, Beijing.

Zhang, W. Y., Wang, Y., and Li, Z., 1980, On the formation and development of the North China fault block region, in: *Formation and Development of the North China Fault Block Region* (W. Y. Zhang, ed.) [in Chinese], Science Press, Beijing, pp. 1–8.

Zhang, Z. H., and Zhang, W. H., 1980, Geological significance of the formation of the Archean protolithocrust in the North China fault block region, in: *Formation and Development of the North China Fault Block Region* (W. Y. Zhang, ed.) [in Chinese], Science Press, Beijing, pp. 36–48.

Chapter 5

THE GEOLOGY OF THE CHINA SEA

Chin Chen

Graduate Program in Oceanography and Limnology
Western Connecticut State University
Danbury, Connecticut 06810

I. INTRODUCTION

China has a long shoreline and extensive marginal seas. The coastline extends 20,000 km from the Yalu River on the Chinese–Korean border southward to the estuary of the Beilun River on the Chinese–Vietnam border (Fig. 1). The marginal seas, with a total area of about 3,336,000 km², can be divided into four geographic regions: the South China Sea (Nan Hai), the East China Sea (Dong Hai), the Yellow Sea (Huang Hai), and Bo Hai (Fig. 1). These seas are shallow, with the exception of the South China Sea and the eastern part of the East China Sea.

The Chinese coastline can be classified according to whether it is erosional or depositional or shows a complex of emergence and submergence features (Zhao, 1984). Depositional features predominate in the region north of Hangzhou Bay where straight barrier sands, tidal flats, and alluvial fans are interrupted by the deltas of the Yangzi and Yellow Rivers and the rocky Shandong and Liaodong Peninsulas (Fig. 1). Erosion predominates south of Hangzhou Bay where irregular rocky cliffs and strings of islands can be recognized. Interspersed with these two coastal types is evidence of submergence in the form of drowned valleys, irregular headlands, and bays, and of emergent features such as wave-cut terraces, coquina, and shallow-water offshore deposits. In the southeast in particular, there is evidence of coastal instability with Tertiary submergence and Quaternary emergence in the same regions (Hsieh, 1973).

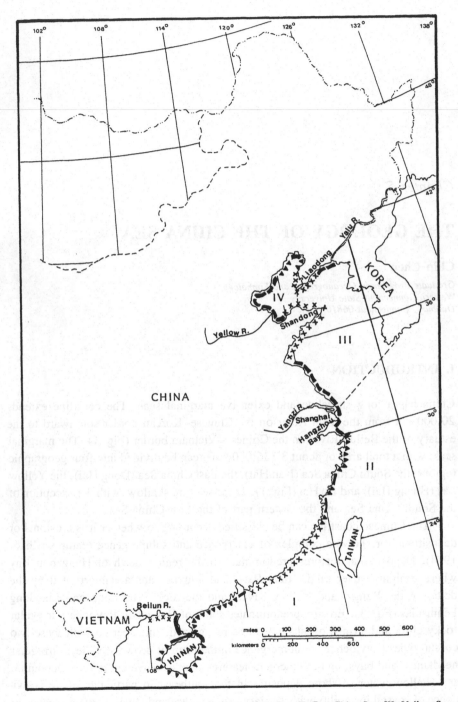

Fig. 1.   China coasts and four marginal seas: I, South China Sea; II, East China Sea; III, Yellow Sea; IV, Bohai Gulf. Triangles, erosional features; bars, depositional features; crosses, compound features of emergence and submergence.

The principal sources for the geological review of these regions, mostly Chinese, are *Marine Geological Research, Acta Oceanologica Sinica, Oceanologia et Limnologia Sinica*, and *Acta Geophysica Sinica*. There are a number of important symposium volumes, particularly the two-volume *Proceedings of the International Symposium on Sedimentation on the Continental Shelf, with Special Reference to the East China Sea* (*Acta Oceanologica Sinica*, 1983), which is based on a joint PRC/USA study, and *The Geology of the Yellow Sea and East China Sea* (Qin, 1982). Data on the structures and isopachs of the marginal sea sediments are drawn from *The Tectonic Map of China and Adjacent Areas* (Zhang, 1983) and from the results of a cooperative geophysical program carried on between the Ministry of Geology and the Lamont–Doherty Geological Observatory between 1979 and 1985.

II. SOUTH CHINA SEA

The South China Sea is the largest marginal sea in the southwest Pacific. It is bounded by the Chinese mainland to the north, the Taiwan Strait to the northeast, the Indochina–Malay Peninsula to the west, the Philippines to the east, and Borneo to the south. With an area of about 2,100,000 km² (Fig. 2), this sea represents nearly two-thirds of the area of the Chinese marginal seas.

A. Tectonics

The continental margins of the South China Sea Basin to the north and west are passive margins. To the east lies the active Manila Trench system, while to the south the Reed Bank and other shoal areas separate the basin from an extinct subduction zone along the Sunda Shelf, Borneo, and Palawan (Taylor and Hayes, 1980). The basin has complicated structural characteristics (Ren *et al.*, 1980) for it is on the trend of the Tethyan–Himalayan extension and its intersection with the marginal Pacific domain. The Neocathaysian trend of the marginal Pacific domain is characterized by northeast-striking grabens, in contrast to the northwest-trending fractures of the Tethyan–Himalayan region.

The greater part of the South China Sea can be considered as a platform (Ren *et al.*, 1980) floored by Cambro-Ordovician rocks (Hainan Island) or older rocks (Precambrian basement on Xisha Island) and the eastward extension of the Indosinian Massif from the Indochina Peninsula onto the floor of the South China Sea. The evolution of the region has recently been investigated by Taylor and Hayes (1980, 1982) and Holloway (1982).

1. Tectonic Provinces

Using the distribution of fault blocks (Zhang, 1983), the nature of the crust from seismic studies (Z. W. Li, 1984), gravity and magnetic anomalies (S. Q. Chen

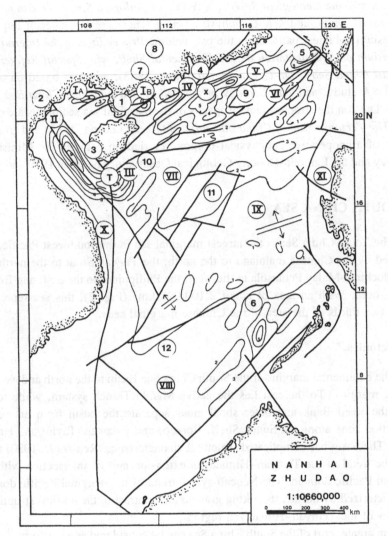

Fig. 2. Tectonic provinces with Cenozoic isopach (km) of the South China Sea. IA, Beibu Wan
(northern part of the Gulf of Tonkin); IB, Zhanjiang Gulf Depression; II, Yinggehai (southern part of the
Gulf of Tonkin); III, Southeast Hainan Basin; IV, Wanshan–Hainan Uplift; V, Zhujiang (Pearl River)
Estuary–Qiong Depression; VI, Dongsha (Pratas) Fault Swell; VII, Xisha (Paracel)–Zhongsha (Mac-
clesfield) Banks Uplift; VIII, Nansha (Reed Bank) Uplift; IX, Central Oceanic Basin; X, East Vietnam
Depression; XI, West Philippines Fault Block; T, triple junction. 1, Leizhou; 2, Hong He (Red River); 3,
Yingge; 4, Wanshan Islands; 5, Taiwan Shoal; 6, Liyue Bank; 7, Maoming; 8, Sanshui; 9, Dongsha
(Pratas) Island; 10, Xisha (Paracel) Islands; 11, Zhongsha (Macclesfield) Islands; 12, Nansha (Reef
Bank) Islands.   x , drilling site; ⚡ , spreading center; ⌒ , isopach in kilometers of Cenozoic
sediments; \ , fracture.

et al., 1981; Second Marine Geological Team, 1975), as well as sedimentary characteristics, a total of 11 tectonic provinces can be recognized (Fig. 2), each of which is discussed in turn. Most Chinese studies have concentrated on the northern and central parts of the South China Sea.

(a) Beibu Wan (Northern Part of the Gulf of Tonkin)–Zhanjiang Gulf Depression. The continuity of this province is broken by the Leizhou Peninsula which separates the Beibu Wan to the west from the Zhanjiang Gulf east of the peninsula. The sedimentary content of the depression parallels that of the coastal basins such as Maoming and Sanshui which contain Mesozoic red beds and oil shales.

The principal structure in Beibu Wan trends east to east-northeast and is marked by a central depression bordered by two uplifts. The basement to the north consists of Carboniferous carbonates and low-grade metamorphosed rocks of Lower Paleozoic age, whereas underlying the central and southern parts of the depression are higher-grade metamorphic rocks regarded as folded Caledonian basement (S. Q. Chen *et al.*, 1981). The two contrasting zones, clearly distinguishable on magnetic charts, are separated by a fault. North of the fault, the magnetic field is marked by a low-anomaly (10–60 gamma) field with local high values, whereas anomalies are in the 60–100 gamma range south of the fault. The·distinction has also been seen in the gravity data, with anomalies of −5 to −30 mgal to the north and a minimum value of −47.6 mgal to the south. Seismic data indicate more than 4 km of Cenozoic sediments filling the depression.

Within the sedimentary section in Beibu Wan, two unconformities can be recognized, one between the Carboniferous carbonates and the Cretaceous red clastics, and the other between the Lower Tertiary nonmarine clastics, which contain rare limestone layers, and the Upper Tertiary marine clastics. A thick (500 m) black shale within the Lower Tertiary is a good petroleum source rock.

In the Zhanjiang Gulf east of the Leizhou Peninsula, the depression has the form of two basins, with about 3.5 km of sediment in the western part, but only 2 km in the eastern part, according to seismic data. The magnetic field is characterized by alternate positive and negative anomalies with large gradients. The gravity anomalies, which range from −25.4 to −32.8 mgal, decrease from northeast to southwest.

(b) Yinggehai (Southern Part of the Gulf of Tonkin) Depression. This depression is a continuation of the graben which extends from Hong He (Red River) in Vietnam to Yinggehai on the southwest coast of Hainan Island. The trend of the graben is that of the Tethyan–Himalayan domain. The northern boundary fault, the Jinshajiang–Hong He Fault, is a translithospheric fault which intersects a northeast-trending fault of the Cathaysian system and a north–south-trending fault south of Hainan Island, resulting in a triple junction (D. Li, 1984).

A lithospheric fracture asymmetrically divides the depression into a northeastern part with up to 9 km of Cenozoic sediments, and a southeastern part with 6 km of sediment. Drilling and seismic data from Vietnam show that the Tertiary sequence changes from nonmarine, lacustrine, and coarse, fluvial clastics and coal in the northwest, to marine, cyclic, clastic deposits fining upward in the southeast. These rest on a basement of metamorphosed Indosinian metamorphic and intrusive rocks located 3.5–4.5 km below the surface.

(c) The Southeast Hainan Basin. This basin has grabens with a typical Neo-cathaysian northeasterly strike (Meyerhoff *et al.*, 1984) and is bounded to the north by a translithospheric fault. Basement is formed by Mesozoic granite overlain by a 9-km stratigraphic sequence. The basin is separated from the Yinggehai Depression by a north–south-trending translithospheric fracture and differs from the Yinggehai in the nature of the basement and in the structures developed.

(d) Wanshan–Hainan Uplift. The uplifted area forms a narrow strip between the small Wanshan Island (about 22°N, 114°E) to the northeast and the northeast corner of the larger Hainan Island to the southwest. The basement is of Mesozoic sedimentary rocks resting on Paleozoic igneous rocks. These crop out in Wanshan. The uplift is characterized geophysically by a sinusoidal pattern of alternating positive and negative magnetic anomalies as well as gravity anomalies of +8.6 to −18.6 mgal, which distinguish it from the adjoining depressions.

(e) Zhujiang (Pearl River) Estuary–Qiong Depression. This depression on the continental shelf at the northern margin of the South China Sea is bounded by the Wanshan–Hainan Uplift to the north and by the Dongsha (Pratas) Faults on the south. It strikes northeast and has been divided into the following four morphological areas or zones (from northwest to southeast): the Taiwan Shoal, the Zhujiang Estuary, the Shengu Submerged Bank (19.3–20.3°N, 112–113.3°E), and Southwest Qiong (off Hainan Island at 18–19°N, 110.5–112°E). In all of these zones except the last, the sediments reach 5–6 km in thickness.

Magnetic anomalies over the depression range from −147 to +50 gammas. The negative values are attributed to a Hercynian nonmagnetic metamorphic basement, the positive values to intrusive Mesozoic granites (S. Q. Chen *et al.*, 1981). The Bouguer anomaly increases from zero to +50 mgal as the shelf break is approached.

(f) Dongsha (Pratas) Fault Swell. This swell lies on the continental slope in the northeastern part of the South China Sea (Fig. 2). As indicated above, it is bounded on one side by the Zhujiang Estuary Depression, and is separated from the oceanic basin to the south by a fault striking approximately east–west. The zone is characterized by a series of step faults, becoming less steep toward the east passing into a broad geocline.

The crust is interpreted as transitional. The magnetic anomalies recorded vary from +200 to −150 gamma, with the high positive values thought to be due to mafic or ultramafic crust intrusive into nonmagnetic metamorphic rocks along the fault zone (S. Q. Chen *et al.*, 1981). The Yanshanian granites which are the common basement rocks give low positive anomalies. The Bouguer gravity anomalies range from +50 mgal on the upper continental slope to +200 mgal on the lower slope. Z. W. Li (1984) concluded from sonobuoy measurements that continental crust extends to the base of the continental slope east of Dongsha Island, whereas oceanic crust exists in the lower continental slope in the area west of the island. The basement is covered by an average thickness of 1.5 km of Cenozoic sediments, but in some regions this thickness may be twice that amount.

(g) *Xisha (Paracel)–Zhongsha (Macclesfield) Bank Uplift.* This unit actually consists of two uplifts, Xisha and Zhongsha to the west and east, respectively, and an intervening small depression. It lies in the central part of the South China Sea. The unit has a basement of nonmagnetic or weakly magnetic metamorphic rocks at relatively shallow depth. Large magnetic anomalies (greater than 200 gamma) over a few small islands on the Xisha Uplift are attributed to mafic volcanics. Granite and gneiss also occur on the islands. Gravity values over the uplift are on the order of 100–150 mgal. In excess of 1000 m of mostly coral debris has accumulated over the uplift since Miocene times.

(h) *Nansha (Reed Bank) Uplift.* The Nansha is formed of three fault blocks with intervening depressions, with a basement believed to be continental and formed of Precambrian granite. The oldest sediments penetrated in wells are paralic to neritic clastics and carbonates. They are separated by a major hiatus from the Cenozoic section. The Paleogene sequence consists of a thin limestone, neritic shale, and quartz sand. The Neogene consists of a thick limestone sequence and, at the present day, the area is characterized by the extensive development of coral reefs.

(i) *Central Oceanic Basin.* This basin is bounded by marginal faults against the uplifts (Dongsha, Zhongsha, and Nansha) and by the Manila Trench to the east (Fig. 2). The basin stratigraphic sequence consists of a sedimentary layer 2 km thick resting on oceanic Layer 2 (seismic velocities 4.5–6.3 km/sec) and Layer 3 (velocities 6.4–7.4 km/sec) (Taylor and Hayes, 1980; Z. W. Li, 1984). The sediments thin to 1 km in the central part of the basin. Gravity anomalies of 200 and 300 mgal have been recorded.

Two spreading centers are mapped: one, with an east–west trend, is located in the Scarborough Shoals (15°N, 117–119°E); the other, trending northeastward, lies in the southwestern part of the basin (11.5–12°N, 112.5–113°E) (Zhang, 1983). Magnetic anomalies 50 to 11 have been identified at the Scarborough spreading center, placing the time of activity between 17 and 32 Ma (mid-Oligocene to early Miocene) (Taylor and Hayes, 1980, 1982). The other center was proposed by Z. W. Li (1984) in the South China Sea (Fig. 3).

(j) *East Vietnam Depression.* This depression, off the east coast of Vietnam, is floored by Precambrian overlain by a 1.5-km thickness of sediments. It is bounded by a north–south-trending translithospheric fault approximately along 109.5°E.

(k) *West Philippine Fault Block.* This zone includes both the Manila Trench and the Luzon Trough. The bathymetry, sediment distribution, and seismicity have been studied by Ludwig (1970) and Ludwig *et al.* (1967) and are not discussed here.

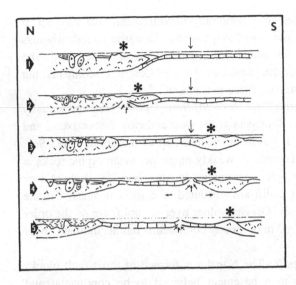

Fig. 3. Historical development of South China Sea. 1, middle Eocene; 2, middle Oligocene; 3, late Oligocene; 4, early Miocene; 5, early late Miocene. Asterisk: Liyue Bank (Northern Reed Bank); present position: 11.5°N, 116.8°E. (After Z. W. Li, 1984.)

2. Tectonic History

Several periods of tectonic movement and accompanying magmatic activity have occurred since the Proterozoic. Each is briefly outlined below.

(a) Jinning Movements. The oldest tectonic activity recorded in the South China Sea region occurred during the late Proterozoic, between 1000 and 800 Ma. Rocks belonging to these Jinning movements, granite and gneiss, are found in the basement of Xisha and possibly Nansha Islands.

(b) Caledonian Movements. During the Caledonian Early Paleozoic Movements (about 370–450 Ma), the South China Foldbelt was formed. Caledonian granites crop out along the South China coast and on Hainan Island. The alignment of the intrusion parallels the regional, northeasterly tectonic trend.

(c) Hercynian Movements. In southeast China, Hainan, the Philippine Islands, and the Indochina Peninsula, magmatic activity assigned to the Hercynian is recorded.

(d) Indosinian Movements. These were strong movements along the western margin of the South China Sea which resulted in the folding and faulting of geosynclinal sediments during early Mesozoic time. The effects are best seen in Indochina where there are also granitic and granodioritic rocks as well as acid volcanics.

(e) Yanshanian Movements. D. Li (1984) reported five episodes of Yanshanian movement during the Jurassic and Cretaceous and Early Paleogene in East and South China and in the offshore. The first began with basalt flows in an early Jurassic rift and continued until the second episode which began with the formation of depressions in the middle and late Jurassic. The third episode began andesitic and basaltic flows into an early Cretaceous rift. These continued during the fourth episode, marked by the formation of a late Cretaceous depression. The fifth and last episode of rifting continued through the Paleocene and Eocene until the formation of a depression in the early Oligocene marked the onset of the Himalayan orogeny.

(f) *Himalayan Movements.* These, the latest movements, peaking in the middle Miocene (Ren *et al.*, 1980), range through the middle and late Tertiary. Their effects are seen in the uplift and subsidence of fault blocks in the Himalayas, Taiwan, and the South China Sea. Toward the end of the tectonic activity, a marine transgression over the northern margin of the zone made itself felt. Li (1984) proposed two further episodes of rifting in the South China Sea (Fig. 3). The second episode occurred between the middle and late Oligocene; the present Liyue Bank separated from the edge of the Chinese continental margin and migrated southward. A third rifting episode occurred during the early Miocene and subduction of oceanic crust under Borneo developed during the Pliocene and Pleistocene.

B. Sedimentation and Stratigraphy

1. *Distribution of Surface Sediments*

Three types of surface sediments were recognized by Li *et al.* (1980) in the area off the Zhujiang (Pearl River) Estuary between 19.5–22.5°N and 109.5–115.5°E. Type 1 consists of recent inner neritic sediments, composed mostly of silt and clay, with a heavy mineral content indicating a short distance of transport from the continent. Type 2 consists of gravel and coarse sands with about 5% of mica, hornblende, glauconite, zircon, and titaniferous iron which form the Lower Holocene foreset beds of the Zhujiang Estuary. Type 3 sediments are Pleistocene outer neritic sediments. They comprise coarse gravel and sand with some clay. Heavy minerals such as zircon, limonite, and titaniferous ores may constitute up to 40% of the sediments.

Descriptions of the deep sea sediments and of the biostratigraphy of the South China Sea are available in English. Damuth (1980) studied Quaternary sedimentation, mainly turbidites and mass wasting as seen in echo-soundings and coring. The nature of the sediments has been confirmed by study of a long piston core (Samodai *et al.*, 1986) which revealed four turbidite layers laid down in the last 120,000 yr. Clay mineral distributions have been studied by Chen (1973, 1978). Shelf sediments from the northern part of the South China Sea have been discussed by Chen and Chen (1971), Chou (1971), and Niino and Emery (1961). Polski (1959), Waller (1960), Cheng and Cheng (1964), and Qin (1980) have reported on the foraminifera of the surficial sediments, and Rottman (1977, 1978) described the planktonic foraminifera and pteropods from the South China Sea and sediments.

2. *Stratigraphy*

Table I is a composite stratigraphy of the northern margin of the South China Sea based on drill samples (Fan, 1982). The drill site is shown in Fig. 2.

3. Paleoenvironment

The area of the drilling sites near the margin of the South China Sea reflects the effects of several phases of tectonic and magmatic activity from late Proterozoic to Mesozoic time. This region and the adjoining South China continent were profoundly affected by Yanshanian Movements. In Oligocene and early Miocene time, a warm, fluviolacustrine environment existed, but transgression gradually supervened following Himalayan tectonic activity. As a result, marine and nonmarine environments alternated during the middle Miocene. In the late Miocene, a progressive deepening resulted in a deeper marine environment during the Pliocene and Quaternary.

TABLE I
Composite Stratigraphy of the Northern Margin, South China Sea

Age	Thickness (m)	Lithology	Fauna
Holocene–Pleistocene	162–200	Greenish-gray, silty clay and sandy gravels	Planktonic foraminifera and marine ostracods
Pliocene	124–356	Gray, argillaceous sediments with sandstone, conglomerate and clay	Planktonic foraminifera and ostracods, pelecypods and gastropods
Miocene			
Upper	300–700	Pebbly conglomerate, sandstone, and mudstone, with calcareous sandstone, bituminous shale, and lignite	Marine ostracods, benthic foraminifera with some pollen and spores
Middle	400–550	Conglomerate, dolomitic sandstone, limestone, bituminous shale, and kaolinite	Pollen, spores, with a few foraminifera
Lower	650–925	Dark mudstone, kaolinitic sandstone and conglomerate, dolomitic sandstone, bituminous shale, siltstone, and lignite	Pollen, spores
		Disconformity	
Oligocene			
Upper	210–730	Black shale, coal, dolomitic sandstone, and sandy gravels	Pollen, spores
Lower	340	Mainly gray sandstone, siltstone, with conglomerate and calcareous sandstone and dark shale	Pollen, spores
		Unconformity	
Mesozoic		Granite porphyry, granodiorite or metamorphic rocks	

III. EAST CHINA SEA

The continental shelf of the East China Sea averages 740 km in width and is bordered to the east by the narrow Okinawa Trough and Ryukyu Islands. The Taiwan Strait lies to the southwest. The arbitrary boundary with the Yellow Sea (Fig. 4) is taken as the line from the Korean (Tsushima) Strait through Cheju Island to the mouth of the Yangzi River. The area thus defined is about 752,000 km².

A. Tectonics

The East China Sea is relatively stable; the few recorded shallow earthquakes have foci located in the outer shelf, Taiwan Strait, western Taiwan Island, and off the Fujian (Fukian) coast. A few inactive volcanoes are found in the Korean Straits, on Cheju Island, and off the northern Taiwan coast. High concentrations of shallow to intermediate focus earthquakes and active submarine volcanism occur within the Taiwan–Ryukyu Island arc and the Okinawa Trough.

1. Tectonic Provinces

The tectonic provinces of the East China Sea are recognized through their magmatic activity, gravity, magnetic anomalies, heat flow, crustal movement fracture patterns, and sedimentary facies distribution (Emery *et al.*, 1969; Wageman *et al.*, 1970; Jin and Yu, 1982a; Emery, 1983). The general pattern of the seafloor is one of alternating northeast-trending uplifts and depressions; these are described successively (see Fig. 4).

(a) Fujian and Reinan Massif. The massif extends from the southeastern coast of China across the East China Sea to the southern Korean Peninsula and includes Cheju Island (Fig. 4-I). It is at the boundary between the East China Sea and the Yellow Sea. Over the massif, gravity values of about +20 mgal and magnetic anomalies in the range −100 to +300 gamma prevail.

Geologically, the massif consists of metamorphosed Proterozoic and late Paleozoic geosynclinal deposits. During Jurassic and Cretaceous time, the region was part of a volcanically active coastal foldbelt active during the Yanshanian orogeny (Jin and Yu, 1982a). It lay across the entrance to the present Yellow Sea, and by cutting off Pacific marine influences permitted the accumulation of terrigenous sediments over the Precambrian massif (Emery, 1983). Subsequently, in late Paleogene or early Neogene time, the barrier was breached by erosion with the consequent ingress of marine conditions into the Yellow Sea region. A Cenozoic fault basin formed in which about 1000 m of sedimentary and volcanic deposits accumulated.

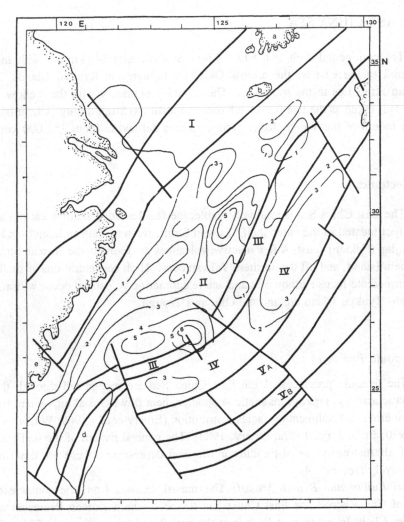

Fig. 4. Tectonic provinces with Cenozoic isopach (km) of the East China Sea. Provinces are bounded by fracture zones. I, Fujian and Reinan Massif; II, East China Sea Basin (Taiwan Basin); III, Taiwan–Sinzi Foldbelt; IV, Okinawa Trough; VA, Ryukyu Foldbelt of Island Arc; VB, Ryukyu Trench. Locations: a, Korean Peninsula; b, Cheju Island; c, Japan Island; d, Taiwan Island; e, Fujian; f, Yangzi estuary. $\bigcirc$, isopach; $\diagdown$, fracture. (After Zhang, 1983.)

(b) East China Sea Basin (Taiwan Basin). This basin underlies the entire continental shelf of the East China Sea with the exception of the outer edge (Fig. 4-II). The basement is formed of metamorphosed, eugeosynclinal, Mesozoic flysch and volcanic rocks which typically have seismic velocities of 5.3–6.2 km/sec. Resting unconformably upon the basement are Cenozoic sediments about 4 km thick.

The Cenozoic section is made up of a folded Paleogene sequence, with seismic velocities of 4.6 to 5.3 km/sec, unconformably overlain by Neogene sediments with seismic velocities ranging between 1.8 and 4.4 km/sec. There is a regional discon- formity separating the youngest Neogene (Pliocene) sediments from the Pleistocene deposits.

As a result of Early Paleogene spreading, three depocenters can be recognized within the basin. The northernmost depression, south of Cheju Island, contains a 1.5-km-thick sequence with the basement lying at depths of 3 to 4 km as indicated by the magnetic anomalies. In the central and southern depressions, however, the basement lies about 2 to 3 km deeper, with the Cenozoic sequence in both being about 5 km thick. The central depression lies off the Yangzi Estuary and the southern depocenter lies northwest of Taiwan (Fig. 4).

(c) *Taiwan–Sinzi Foldbelt.* The foldbelt lies at the edge of the continental shelf, extending from northeastern Taiwan to the southern Korean Strait. It has trapped sediments being transported oceanward. The basement is formed of meta- morphosed Paleozoic and Mesozoic rocks, over which is a Paleogene sequence with acidic volcanics in which the characteristic seismic velocity is 4.6 km/sec. The Paleogene is capped by a layer of unconsolidated Neogene sediments with velocities of approximately 2 km/sec. The large Lonjiang anticline, a roll-over structure associated with growth faults, is located in the eastern part of the foldbelt (Meyerhoff *et al.*, 1984).

During the Miocene, an andesitic and basaltic island arc formed which became inactive during the late Tertiary. It is characterized by linear positive magnetic anomalies separated from the Fujian–Reinan Massif by an area of predominantly negative anomalies. The positive anomaly following this foldbelt ends abruptly at about 26.5°N (Wageman *et al.*, 1970).

(d) *Okinawa Trough.* This trough, a Neogene depositional trap, is bounded by the continental slope to the west and by the Ryukyu Ridge to the east. At its deepest in the south near Taiwan, it is about 2270 m deep, but the floor shoals northward toward Japan. The basement is assumed to consist of Paleogene meta- morphic rocks and volcanics. The trough is filled by Quaternary turbidites and mafic volcanic ash more than 1.2 km thick. There are many active faults, indicated by the numerous shallow to intermediate focus earthquakes. A basement depth of about 3 km is calculated from the predominantly negative magnetic anomalies that average about −100 gamma. The Bouguer anomaly shows a large range, from +160 mgal in the south to +80 to 100 mgal over the northern trough. The highest heat flow values in the East China Sea, about 4 HFU, also occur in the southern trough, indicating a dynamic imbalance (Jin and Yu, 1982a). Li *et al.* (1982) proposed that the trough gradually developed in response to isostatic adjustment to the disequilibrium in the region.

(e) *Ryukyu Foldbelt, Island Arc, and Trench* (see Kobayashi, Volume 7A of this series). There are two arcs: the inner arc between the Okinawa Trough and the Ryukyu Islands, and the outer arc forming the islands of the foldbelt. This province is characterized by alternating positive and negative magnetic anomalies, low heat flow, high seismicity, and a Bouguer anomaly pattern marked by steep gradients. The inner arc, with magnetic anomalies of about 100 gamma, is thought to consist of Miocene and Pliocene andesites. The outer arc, with negative anomalies of zero to −100 gamma, consists of Paleozoic and Mesozoic metamorphic rocks and folded Tertiary sediments into which have been intruded granite and gabbro and which have been cut by recent volcanics.

Gravity anomalies, which are on the order of 20 mgal between the Ryukyu Island Arc and Trench, rise to about 300 mgal on the eastern side of the trench, indicating the transition from continental (27−30 km) crust to thin (6−9 km) oceanic crust.

2. *Tectonic History*

Within the tectonic framework of the East China Sea, there are three subduction zones (two extinct) separated by two secondary spreading axes, one of which is extinct. All of them trend northeastward and young eastward. The tectonic history is illustrated in a series of schematic diagrams by Jin and Yu (1982a).

During early Paleozoic time, a deep Cathaysian Sea lay between the North China Platform to the northeast and the Cathaysian Massif to the southwest (Fig. 5A). This sea extended from present-day northern Vietnam through southeastern China and the southern Yellow Sea to the southern Korean Peninsula. The northern part of this sea was obliterated during the Caledonian orogeny. From the Upper Paleozoic to the Triassic, a subduction zone existed between the Pacific and Asian plates and extended from the present southeast coast of China to southwestern Japan, passing through the western East China Sea (Fig. 5B).

The Yanshanian orogeny created the coastal ramp extending from present southeastern China through the northwestern East China Sea to southwestern Korea. It was exposed by regression during the late Cretaceous phase of this orogeny. A new Benioff zone developed (Fig. 5C) west of the coastal range and extended from the present eastern side of the Central Mountain range of Taiwan through the central East China Sea to southern Japan. Between these last two subduction zones, a Paleogene spreading axis (Fig. 5D) developed and remained active until the Late Miocene (Fig. 5E). The result of this spreading was the formation of a marginal sea and the gradual eastward migration and breakup of the coastal range (Fig. 5D'–F').

In late Miocene or early Pliocene time, a subduction zone forming the Ryukyu Trench and a backarc spreading center in the marginal basin became active (Fig. 5F).

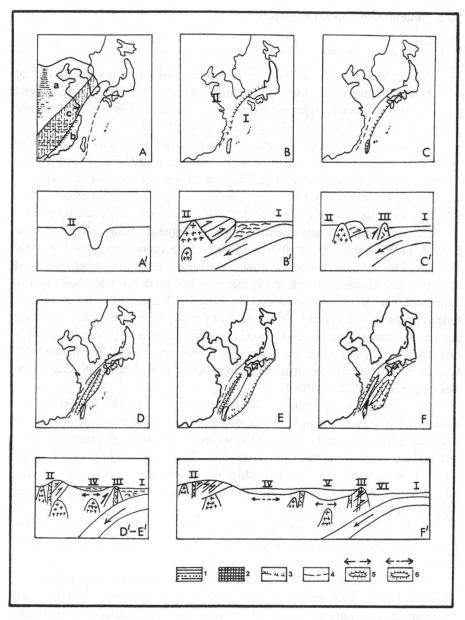

Fig. 5. Tectonic history of the East China Sea. A, Early Paleozoic; B, from Hercynian to Indosinian; C, Yanshanian Movement (Jurassic to Cretaceous); D, Paleogene; E, Late Miocene; F, Quaternary. Legend: 1, local transgression; 2, Cathaysian Sea Transgression; 3, active subduction zone of Benioff zone; 4, inactive subduction zone; 5, active secondary spreading axis; 6, inactive secondary spreading axis. Locations: a, North China Platform; b, Cathaysian Massif; c, Cathaysian Sea. Profile of plate tectonics (A′–F′): I, Pacific Plate; II, Asian Plate; III, Island Arc or coastal range; IV, East China Sea spreading axis; V, Okinawa Trough; VI, Ryuky Trench. (After Jin and Yu, 1982a.)

B. Sedimentation and Stratigraphy

1. *Distribution of Surface Sediments*

Three types of surface sediment are recognized based upon grain size, mineral and chemical composition, fossil or volcanic content, source area, currents, and depositional environments (Qin and Zhen, 1982; Chen *et al.*, 1980, 1982; Niino and Emery, 1961; Wageman *et al.*, 1970; Wang *et al.*, 1984).

(a) Inner Neritic Sediments. Such sediments are predominantly of terrigenous and deltaic origin from the Yangzi Estuary, and are distributed by a southerly, nearshore current to the Taiwan Strait. They spread from the shoreline to an approximate water depth of 50–60 m. They consist of silty clay and silt-sized quartz, feldspar, and clay minerals mixed with benthic organisms and plant fragments.

The suspended sediment discharge of the Changjiang (Yangzi) River into the East China Sea is about 500×10^6 tons annually, with a concentration of 100–1000 mg/liter. The sediment is derived from a drainage area lying in a warm, humid climate where chemical weathering is intensive and produces acidic soils rich in aluminum, thus forming kaolinite and halloysite. The Yangzi sediments are characterized by garnet and epidote. The Huanghe (Yellow) River is the other important source of sediment supplies, and produces reworked eolian loess with a high concentration of calcium which was formed in a cold dry climate. The marked difference in sediment type facilitates the recognition of provenance in the East China Sea (Yang and Milliman, 1983).

(b) Outer Neritic Sediments. The outer neritic sediments consist of terrigenous, generally well-sorted, fine sands and minor organic-rich muds. Some are found near the shelf break. The terrigenous sediments, although primarily quartz and feldspar, include hornblende and epidote. Some sediments are clearly relict deposits (Wageman *et al.*, 1970). Authigenic components are glauconite, pyrite, and carbonate. Most calcite is organic, pyrite is inorganic in origin, and glauconites may have either origin.

South of 30°N, glauconite may comprise 20% of the total sediment; lower percentages occur farther north. Glauconites of organic origin are particularly abundant in the Taiwan Shoal and South China Sea.

(c) Abyssal Sediments. These occur on the continental slope and in the Okinawa Trough; they mostly comprise pelagic muds and silts, turbidites, and mafic volcanics. Wang and Liang (1984) have reported spherical magnetite and ilmenite grains enclosed within volcanic glass in these deposits.

2. *Stratigraphy*

A composite stratigraphic sequence in the area of the outer shelf can be built up as shown in Table II.

TABLE II
Composite Stratigraphy of the Outer Shelf, East China Sea

Age	Thickness (m)	Lithology	Fauna
Quaternary	About 400	Neritic and deltaic clay and sand	Benthic and planktonic foraminifera and mollusks
		Local unconformity	
Pliocene	600–1000	Deltaic clastics, sand and clay with conglomerates	Benthic and planktonic foraminifera and mollusks
		Local unconformity	
Miocene	1000–4000	Continental clastics for the most part, sandstone, siltstone, mudstone, with conglomerates and extrusives	Benthic and planktonic foraminifera and mollusks
Upper Paleogene	1000–6000	Transitional to continental clastics with oil and gas	
Mesozoic	?	Neritic, bathyal, or abyssal flysch with volcanics	
Paleozoic	?	Metamorphic and metavolcanics	

3. *Paleoenvironment*

During the early Paleozoic, the old and deep Cathaysian Sea existed approximately along the present boundary of the southern Yellow Sea and the East China Sea, with a land area lying to the south. That sea was reduced in area and depth following Caledonian earth movements in the Devonian, leaving a limited shallow sea over southwest China–north Vietnam. Subsequently, in the early Carboniferous a marine transgression began in the south, and during the late Paleozoic a shallow sea came to occupy the site of the present Yellow Sea and western East China Sea and southeast China until well into Mesozoic time. In the area of the present eastern part of the East China Sea, however, bathyal or abyssal conditions prevailed. Late Mesozoic tectonism, the Yanshanian movements, resulted in the formation of the coastal range and, associated with it, marine regression.

During the Paleogene, subsidence of the East China Sea floor resulted in the formation of a marginal sea. Contemporaneous with this, backarc spreading was occurring and, as a result of plate collision, the coastal range was disrupted. During the Miocene a volcanic arc was generated that is still active today.

Reduced sea level during the Ice Age resulted in the subaerial exposure of about two-thirds of the East China Sea floor. During the Wurm (Wisconsin) glacial stage, this low coastal plain was covered by fluvial, estuarine, and lacustrine swamp deposits. With gradually rising sea level, littoral deposits were laid down over, or mixed with, sediments of terrestrial origin. Through most of the Holocene, deltaic deposits have been swept southward by coastal currents and therefore do not cover the outer shelf. This explains the occurrence of relict and reworked sediments and volcanics with their authigenic minerals in the outer shelf.

IV. YELLOW SEA

Geographically, the Yellow Sea is surrounded on three sides by land, the Chinese mainland to the west, Korea to the east, with the Bo Hai Gulf, treated in the next section, to the north (Fig. 6). Only to the south does the Yellow Sea open up to the East China Sea. The Yellow Sea is broad and shallow, with an average depth of 38 m and a total area of about 400,000 km². Geologically, the Yellow Sea is divided into two parts, the northern, which is part of the Sino-Korean Paraplatform with its thick sequence of Mesozoic sediments, and the southern, an extension of the Yangzi Paraplatform with its Cenozoic basinal sequence.

A. Tectonics

The southern Yellow Sea is tectonically the more active. A shallow seismic zone extends northeastward from the Subei Plains into the central part of the southern Yellow Sea. Shallow earthquake foci are found along the southern coast of the northern Yellow Sea, extending from the North China Plain through Bo Hai and trending northwestward (Fan *et al.*, 1982) (Fig. 6).

1. *Tectonic Provinces*

(a) North Yellow Sea Basin. The northern Yellow Sea is underlain by Precambrian metamorphic rocks which crop out in some islands and the surrounding land mass, but are normally at depths ranging from zero to 2.7 km, with a sedimentary cover which has a maximum thickness of 2.0–2.3 km. Magnetic anomalies over the basement are on the order of zero to −200 gamma, but uplift along fractures can lead to the recording of positive anomalies. The basin is essentially a fault graben formed in Jurassic time (Jin and Yu, 1982b). It can be divided into two subbasins.

(b) Central Yellow Sea Uplift. This zone, which forms the boundary between the northern and southern Yellow Sea, extends from the Shandong Peninsula across the central Yellow Sea to northern Korea. The boundary with the northern Yellow Sea is a major northeast-striking fault. The uplifted rocks are predominantly Sinian

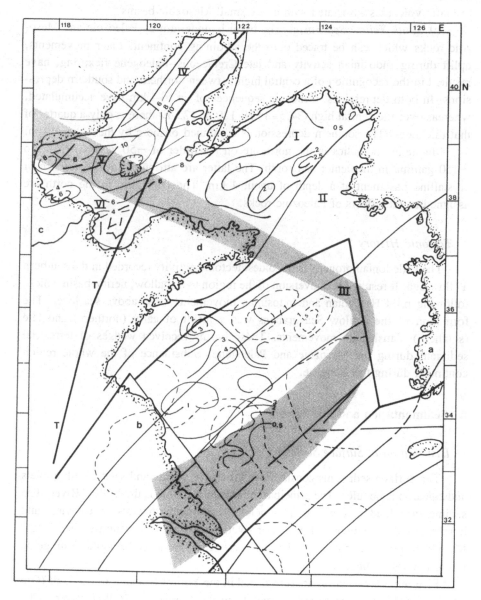

Fig. 6. Tectonic provinces with Mesozoic and Cenozoic isopach (km) of the Yellow Sea and the Bohai Gulf. Yellow Sea: I, North Yellow Sea Basin; II, Central Uplift; III, South Yellow Sea Basin. Bohai Gulf: IV, Panhandle Province; V, Central Gulf Province; VI, Southern Gulf Province. Locations: a, Korean Peninsula; b, Subei Plain; c, North China Plain; d, Shandong Peninsula; e, Liaodong Peninsula; f, Bohai Strait. Legend: ⅀, major fractures; /, fault; ⌒, isopach; J, triple junction; T, Tancheng–Lujiang Fracture System; ▪, earthquake zone.

and pre-Sinian and are cut by Yanshanian acidic intrusives. There are intermediate to mafic volcanics associated with a few small Mesozoic basins.

(c) South Yellow Sea Basin. The basin is underlain by Paleozoic and Mesozoic rocks which can be traced onto the adjoining continent. Later movements, uplift during Indosinian activity and late Cretaceous–Paleogene flexuring, have resulted in the recognition of a central high between northern and southern depressions. In both depressions, Cenozoic sequences of up to 4 km have accumulated, whereas over the central high (34.5–35°N, 120.5–123°E) there is only a quarter of that thickness. The southern depression is continued onshore by the Subei Basin.

Magnetic anomalies in the north are on the order of -50 gamma, rising to $+150$ gamma in the center and south. The latter are attributed to the presence of crystalline basement at a depth of about 4 km. The anomalies in the north are assigned to the effects of Paleozoic rocks.

2. Tectonic History

The Caledonian orogeny is the oldest tectonic activity recorded in the southern Yellow Sea. It resulted in conversion of the region to a shallow, neritic basin which, following mild Hercynian and Indosinian movements, rose above sea level. The formation of the Yellow Sea grabens was a result of early (northern) and late (southern) Yanshanian movements. These basins received wedges of terrestrial sediment during the Mesozoic and Paleogene. Subsidence of the whole region continued during the Neogene.

B. Sedimentation and Stratigraphy

1. Distribution of Surface Sediments

The surface sediments consist mainly of silts, muds, and sands, with gravels and calcareous nodules. The principal sedimentary source is the Yellow River. The sedimentary load is richer in plagioclase, carbonate minerals, muscovite, and hornblende than sediments from other rivers in the area; in contrast, the Yangzi River sediments are characterized by the presence of garnet and epidote. Authigenic pyrite is rarely found.

Five mineral provinces are recognized in the Yellow Sea and Bo Hai Gulf (Qin and Li, 1983), based upon components from the Yangzi and Yellow Rivers. The modern Yellow River flows into the Bo Hai Gulf with an average annual discharge of 1069×10^6 tons/yr. The old Yellow River flowed into the Yellow Sea; the Yangzi estuary was located between the East China and Yellow Seas.

Sediments of mineral province I (Fig. 7) consist of silt and mud with sand, gravel, shells, and calcareous nodules. They are characterized by a high percentage of hornblende and muscovite (40–50%), with 5–10% calcium carbonate—although

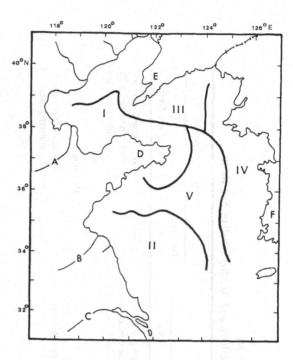

Fig. 7. Mineral provinces of the surface sediments in the Yellow Sea and the Bohai Gulf. Provinces: I, Southern Bohai Gulf and Northwestern Yellow Sea; II, Southwestern Yellow Sea; III, Northern Bohai Gulf and Northeastern Yellow Sea; IV, Eastern Yellow Sea; V, Central Yellow Sea. Locations: A, Modern Yellow River; B, Old Yellow River; C, Yangzi River; D, Shandong Peninsula; E, Liaodong Peninsula; F, Korean Peninsula. (After Qin and Li, 1983.)

the percentage is higher in the estuary of the Yellow River. The sediments are discharged from the modern Yellow River into the southwestern Bo Hai Gulf and are transported eastward by a longshore current parallel to the Shandong Peninsula, then pour southwestward along the coast into the northwestern Yellow Sea.

From its location between the estuaries of the old Yellow and Yangzi Rivers in the southwestern Yellow Sea, the sediments of province II are very heterogeneous—a mixture from both rivers. The hornblende content is reduced to 10–30%, and the calcium carbonate content remains about the same as in province I. The percentage of garnet, epidote, titanite, and orthoclase contributed by the Yangzi River generally remains low. The sediments form part of a delta complex, with submerged bars of silt and clay with calcareous nodules and shells.

Province III, in the northern Bo Hai Gulf and northeastern Yellow Sea, contrasts with provinces I and II, as sediments are derived from the rivers of the Liaodong Peninsula and North Korea. These sediments are generally coarser grained, with a lower, 1–3% carbonate content. The effect of the Yellow River is minimal.

Province IV, off the Korean coast, is almost free of the influence of sediment discharge from the Yellow River.

Province V, in the central Yellow Sea, shows a mineralogy transitional with all the other provinces. The major constituents are clay and silty clay, with a 3–5% carbonate content. A few relict sediments are found.

TABLE III

Comparison of the Successions in the Southern and Northern Yellow Sea Basin

Age	Southern Yellow Sea		Northern Yellow Sea	
	Thickness (m)	Lithology and fauna	Thickness (m)	Lithology and fauna
Quaternary	266–317	Clay, silt, sand, and gravel with intercalated marine horizons with foraminifera and mollusks	300	Clay, silt, and sand with marine horizons with foraminifera and mollusks
		Disconformity		Unconformity
Pliocene	About 700	Variegated clay, gray silt, sand, and gravel		Missing or represented by scattered thin lenses
		Local disconformity		
Miocene	About 750	Mostly shale, sandstone, with conglomerate and volcanics		
		Local angular unconformity or disconformity		Unconformity
Oligocene	Southern depression 1000	Mudstone, sandstone, siltstone, with carbonaceous shale and coal with ostracods, pollen and spores		Missing or represented by scattered thin lenses
	Northern depression 500	Siltstones, gypsiferous mudstones, pollen, and spores		

		Local disconformity or unconformity		
Eocene	Southern depression 160–1200	Sandstone, black shale, calcareous sandstone and limestone; ostracods and fish		Unconformity
	Northern depression 1000–2500	Gypsum, sandstones, siltstones, and oil shale		
		Unconformity		
Paleocene	0–400	Mainly mudstone with volcanics and basal conglomerate		Unconformity
		Unconformity		
Jurassic–Cretaceous	0–3000	Reddish-purple mudstone and argillaceous siltstone, metamorphics, and volcanics	1500–2000	Terrestrial clastics with coal, acidic intrusives, and local intermediate to basic extrusives
Triassic		Very thin		None
		Unconformity		
Paleozoic	1000–2000	Carbonates with clastics, coal, metamorphics		
		Unconformity		
Precambrian		Unknown	Metamorphic rocks	

2. *Stratigraphy*

It is possible to compare the stratigraphy of the northern and the southern Yellow Sea (Xu, 1982) as in Table III, since the depositional environment and tectonic history of these two regions are different.

3. *Paleoenvironment*

The northern Yellow Sea, part of the Sino-Korean Paraplatform, has been less subject to marine influences than the southern part. It was actively eroded during the Paleozoic, in marked contrast to the history of the southern Yellow Sea which, during early Paleozoic time, was a deep sea. As a result of Caledonian and Indosinian movements, it gradually became consolidated as part of the Yangzi Paraplatform and part of the Asian continent.

The more recent history of the Yellow Sea dates from the formation of a Mesozoic basin in the northern Yellow Sea during the early phase of the Yanshanian orogeny. This basin received a thick clastic wedge of Jurassic and Cretaceous terrestrial deposits and igneous products. The southern part of the Yellow Sea did not form as a basin until a late Yanshanian phase when it acquired a thick Cenozoic sequence, though at this time there was little or no sedimentation in the north. The northern and southern parts, however, do have a common Quaternary history.

The Cenozoic sequence in the southern Yellow Sea shows progressive environmental changes from lacustrine (Paleocene and Eocene) to swamp (Oligocene) to fluvial (Miocene and early Pliocene) (Yu, 1981), followed by marine transgression in the late Pliocene and early Quaternary after erosion breached the Fujian–Reinan coastal range. During the last glaciation, the Yellow Sea became a coastal plain, but with the rising sea levels accompanying deglaciation, the sea once more transgressed through the East China and Yellow Seas, flooding the Bo Hai Gulf.

V. BO HAI GULF

The shallow Bo Hai Gulf, which has an average depth of only 20 m, covers an area of about 84,000 km^2. It is enclosed on three sides and opens into the Yellow Sea to the east through the Bohai Strait between the Shandong and Liaodong Peninsulas (Fig. 6). Geologically, the Bo Hai Gulf belongs to the Sino-Korean Paraplatform, as does the northern Yellow Sea.

A. Tectonics

The basin occupying the Bo Hai Gulf is a graben complex with secondary and tertiary uplifts and depressions. As in all Tertiary grabens in eastern China, there are associated step faults and rollover anticlines, commonly referred to in the Chinese literature as "buried hills" (see Li, 1980, 1981a,b). The formation and development of the basin is attributed to tensional forces which resulted from upwarping of the lithosphere and upper mantle (Li, 1980, 1981a,b, 1984). The central Gulf region lies above up-domed mantle; here, the Mohorovicic discontinuity is at a depth of 29 km. At this point, the Cenozoic deposits are about 7 km thick (Fig. 8) and a gravity value of −5 mgal has been recorded.

The basement rocks are mostly Mesozoic clastics and volcanics, and Paleozoic carbonates and clastics. Surrounding the Bo Hai Gulf, the exposed rocks are mainly Precambrian carbonates or metamorphic rocks including carbonates.

1. *Tectonic Provinces*

The major tectonic feature of the Bo Hai Gulf region is the Tanchang–Lujiang fracture zone, which extends from northeast China through the Gulf to North China (Liu and Wang, 1981; see also Fig. 4 and Table III of Chapter 4 in this volume). Three tectonic provinces can be distinguished: the "Panhandle," named from its morphological appearance, and the central and northern subbasins. These converge

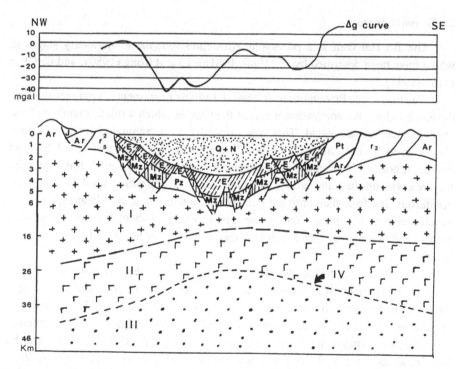

Fig. 8. A profile of Bouguer gravity anomaly and crustal structures of the Bohai Gulf. I, granitoid layer; II, basaltoid layer; III, upper mantle; IV, Moho discontinuity. Q, Quaternary; N, Neogene; E, Paleogene; J, Jurassic; Mz, Mesozoic; Pz, Paleozoic; Pt, Proterozoic; Ar, Archeozoic; r_2, Precambrian igneous intrusives; r_5, Mesozoic igneous intrusives. (After Li, 1981a.)

at the deepest depression, which has been called a triple junction by Li and Cong (1980) (Fig. 6).

(a) Panhandle Province. The elongated Panhandle region in the northeastern part of the Bo Hai Gulf is cut by the deep, lithospheric Tanchang–Lujiang Fault and other northeast-trending crustal faults, dividing it into three shallow central depressions bordered by marginal uplifts. The Tanchang–Lujiang Fault is associated with linear magnetic and gravity anomalies and also appears to be the locus of Mesozoic and Cenozoic magmatic activities (Liu and Wang, 1981).

(b) Central Gulf Province. In the central Gulf, interaction of the Tanchang–Lujiang fracture system with Paleozoic fractures results in a change in trend to a north-northeasterly direction. Most of the shallow crustal faults either follow this direction or trend east–west. The gravity and magnetic isolines generally coincide with the fault directions.

(c) Southern Gulf Province. The changed, north-northeasterly, trend of the Tanchang–Lujiang Fault persists in the southern Gulf; however, other crustal faults trend northwest, west-northwest, east-northeast, or east–west. Gravity and magnetic anomalies reflect several small depressions and uplifts.

2. Tectonic History

The Bo Hai Gulf is a polycyclic rift-depression, the evolutionary stages of which have been described by Li (1980, 1981a), Li and Cong (1980), and C. Chen *et al.* (1981).

During the late Precambrian, following Luliang movements, a parageosyncline developed along the northwestern rim of the Gulf in which a thick, mainly carbonate sequence was deposited. This was succeeded by a cratonic phase from Lower Cambrian to Middle Ordovician time, during which a sequence of 1.0–1.5 km of neritic carbonates interbedded with evaporites and argillites formed. Subsequently, the area was uplifted and eroded with no record of deposition until the mid-Carboniferous, but from then until the Permian, an intercratonic coal-bearing sequence developed in paralic basins. This phase is represented by about 300 m of alternating marine and nonmarine beds which become progressively more continental in character upwards.

There was uplift during the Hercynian orogeny, with the result that the Triassic is poorly represented. It is during the later Mesozoic and Cenozoic that taphrogenic rifting and subsidence became important and produced a series of horsts and grabens following the folding of Sinian to Triassic beds during Indosinian movements.

The beginnings of mantle upwarping in the central Bo Hai Gulf date from the late Cretaceous to early Paleogene, based upon a basalt dated at 70 Ma (Fig. 9A), with the initial rifting during the Paleocene (Fig. 9B) being more intense in the southern than in the northern province. This phase was associated with extrusion of a few alkali basalts and early Eocene tholeiites (C. Chen *et al.*, 1981). A more intense rifting phase followed (Fig. 9C), accompanied by marine transgression in

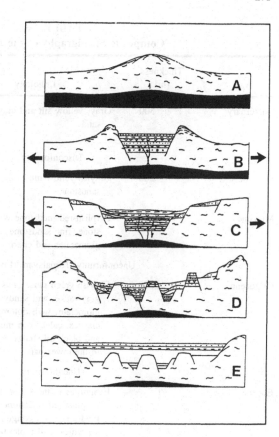

Fig. 9. Evolutionary stages of the rift basin of the Gulf of Bo Hai. For explanations see text. (After Li and Cong, 1980.)

the late Eocene or early Oligocene. In some of the deeper depressions created, deltaic and flysch deposits are found marking the beginning of the formation of the restricted Bo Hai Basin.

Intensive rifting with uplift and subsidence continued during the middle Oligocene (Fig. 9D), accompanied by transgression. Molasse deposits are found in several depressions and may extend from Oligocene into Miocene times (C. Chen *et al.*, 1981). The basin stabilized in the late Oligocene and reached a mature stage with diminishing tensional activity (Fig. 9E). Since the early Miocene (C. Chen *et al.*, 1981), the basin has been subjected to mild compression and uplift, and Neogene clastics are found unconformably overlying Paleogene strata.

B. Sedimentation and Stratigraphy

1. *Distribution of Surface Sediments*

The pattern of sediment distribution in the Bo Hai Gulf has been covered in the Yellow Sea section (see pp. 264–265 and Fig. 7).

TABLE IV
Composite Stratigraphy of the Bo Hai Gulf

Age	Thickness (m)	Lithology	Fauna
Quaternary	300	Gray, yellow silt and clay with sand	Marine pelecypods, foraminifera, pollen and spores
		Disconformity	
Pliocene	200	Grayish-yellow mudstone and fine sandstone	Freshwater ostracods, pollen, spores, and foraminifera
Miocene	1800	Purplish-red mudstone with silt-stone, fine sandstone, and con-glomerates and chert	Freshwater and marine ostracods
		Unconformity (Himalayan Movements)	
Oligocene	2000	Grayish-green mudstones, graywacke, and sandy con-glomerate, with dolomitic bio-clastics, calcareous mudstones, and black oil shales. Evaporites in the lower part	Brackish to freshwater gastropods, os-tracods, pollen, and spores
		Unconformity	
Eocene	?	Evaporites in the upper part. Dark purplish-red sandstones and mudstone with conglomerate, volcanics, and tholeiitic basalt	Freshwater ostracods and gastropods
		Unconformity (Yanshanian Movements)	
Cretaceous	2000–4000	Sandstone, conglomerate, siltstone, and shale	
	1000–5000	Tuff, andesite, basalt, and clastics	
Jurassic	1000–5000	Clastics, volcanics, agglomerates, with oil shale and coal, tuffaeous sandstone. Basalt at base	
Triassic		Generally absent	
		Unconformity	
Permian	500	Clastics with coal seams	
Carboniferous	300	Alternating marine and nonmarine sediments, the latter with coal	
		Unconformity	
Ordovician	1000–1500	Carbonate with evaporite and mudstone	
Cambrian	?	Carbonate	
		Unconformity	
Sinian	9000	Carbonate or metamorphic rocks	

2. *Stratigraphy*

The composite stratigraphic column is given in Table IV.

3. *Paleoenvironment*

Throughout the Paleozoic and Precambrian, the Bo Hai Gulf region experienced the history characteristic of a cratonic region, for it formed part of the Sino-Korean Paraplatform. There was little or no deposition from the mid-Ordovician to mid-Carboniferous or during the Triassic. Following Yanshanian movements, however, the region experienced tensional forces, and as a result a period of horst and graben formation prevailed during the Paleogene, with faulting renewed on several occasions. During the earliest Paleogene, subaerial volcanicity occurred, and mafic volcanics are interbedded with red beds. Turbidity deposits can be recognized in the deeper depressions, while deltaic deposits developed under shallower conditions. Evaporites formed occasionally when marine ingress was restricted, as for example in the middle Paleogene. There was a transition from fluvial and lacustrine conditions in the Neogene to the brackish water of the late Quaternary.

VI. CONCLUSIONS

The China Seas formed as backarc basins behind the converging margins of the Eurasian and Pacific plates. According to D. Li (1984), they form two types of tectonic rift basins: polyphase intraplate depressions, and epicontinental rift depressions. The Gulf of Bo Hai, the South Yellow Sea, and the Beibu Wan (northern Gulf of Tonkin), which developed on continental crust 29–37 km thick, fall into the first category. Here, the Sino-Korean (northern Yellow Sea) and Yangzi (southern Yellow Sea) Paraplatforms were subjected to tension and were broken up with many depressions following the intense Indosinian orogeny. These Neocathaysian basins have since been subjected to five episodes of Yanshanian movement and three of Himalayan activity.

The epicontinental rift depressions, which include the East China Sea, the mouth of the Pearl River, and the Yinggehai Basin of the South China Sea, formed on thinner (28–30 km) continental crust and extend close to oceanic crust. The South and East China Sea basins are thus bordered on one side by passive continental margins and on the other by active subduction zones (Manila and Ryukyu Trenches).

Two triple junctions occur, formed by the intersection, or interference, of two tectonic regimes: one is located at the northern margin of the South China Sea where the Pacific domain intersects the Tethyan–Himalayan domain; the other lies in the Central Bo Hai Gulf where the Mesozoic Neocathaysian system interacts with the Paleozoic Cathaysian system.

The East China Sea has an alternation of spreading axes and subduction zones younging eastward and striking northeastward.

The coastal features and surface sediment distribution reflect the balance of sediment discharge and distribution, tectonic movement, lithology, and so forth. In the China Seas north of Hangzhou Bay, coastal sediment deposition is prominent. Much of the sediment in the Bo Hai Gulf derives from the modern Yellow River, for the old channel discharged into the Yellow Sea. The Yangzi River sediment is deposited in the East China Sea. Erosional processes characterize the coast south of Hangzhou Bay. The basic pattern, however, is of Tertiary submergence–Quaternary emergence modified by glacially controlled sea level fluctuations.

The surface sediments fall into three distinct zones, with the provenance of the sediments defined by their composition and physical parameters. In the inner neritic zone, the Yellow River sediments show a predominance of reworked loess with a high concentration of calcium carbonate, whereas the Yangzi discharge consists mainly of fine clastics with a higher percentage of heavy minerals. Distribution is effected by a south-flowing coastal current out of the Bo Hai Gulf. Relict sediments and fine to medium clastics with authigenic glauconite characterize the outer neritic zone. The pelagic zone in the South and East China Seas contains hemipelagic turbidites and volcanics. Four turbidity layers attributed to glacial low sea level stands of late Pleistocene age are identified in the bathyal region of the South China Sea.

ACKNOWLEDGMENTS

I am grateful to Dr. A. Nairn and E. Samodai for editing the manuscript and to N. Humphreys for drafting all figures.

REFERENCES

Acta Oceanologica Sinica (ed.), 1983, *Proceedings of the International Symposium on Sedimentation on the Continental Shelf, with Special Reference to the East China Sea*, China Ocean Press, Beijing.

Chen, C., Huang, J., Chen, J., Tian, X., Chen, R., and Li, L., 1981, Evolution of sedimentary tectonics of Bohai rift system and its bearing on hydrocarbon accumulation, *Sci. Sin.* **24**:521–529.

Chen, J. C., and Chen, C., 1971, Mineralogy, geochemistry, and paleontology of shelf sediments of the South China Sea and Taiwan Strait, *Acta Oceanogr. Taiwan.* **1**:33–53.

Chen, L., Shi, Y., Shen, S., Xu, W., and Li, K., 1982, Glauconite in the sediments off the southern Fujian province coast to the continental shelf of Taiwan province [in Chinese], *Oceanol. Limnol. Sin.* **13**:35–47.

Chen, L., Yu, X., and Shi, Y., 1980, Glauconite in the sediments of the East China Sea [in Chinese], *Sci. Geol. Sin.* **3**:205–217.

Chen, P. Y., 1973, Clay minerals distribution in the sea-bottom sediments neighboring Taiwan Island and northern South China Sea, *Acta Oceanogr. Taiwan.* **3**:25–64.

Chen, P. Y., 1978, Minerals in bottom sediments of the South China Sea, *Geol. Soc. Am. Bull.* **89**:211–222.

Chen, S. Q., Liu, Z. H., Liu, Z. S., He, S. M., Yuan, H., and Zang, Y., 1981, Features of gravity and magnetic anomalies in central and northern parts of the South China Sea and their geological interpretation, *Sci. Sin.* **24:**1271–1284.

Cheng, T. C., and Cheng, S. Y., 1964, Planktonic foraminifera of the northern South China Sea [in Chinese], *Oceanol. Limnol. Sin.* **6:**38–77.

Chou, J. T., 1971, Recent sediments in the shallow part of the South China Sea, *Proc. Geol. Soc. China* **14:**99–117.

Damuth, J. E., 1980, Quaternary sedimentation processes in the South China Basin as revealed by echo-character mapping and piston-core studies, *Geophys. Monogr.* **23:**105–125.

Emery, K. O., 1983, Tectonic evolution of the East China Sea, in: *Proceedings of the International Symposium on Sedimentation on the Continental Shelf, with Special Reference to the East China Sea* (*Acta Oceanologica Sinica*, ed.), China Ocean Press, Beijing, Vol. 1, pp. 80–90.

Emery, K. O., Hayoshi, Y., Hilde, T., Kobayashi, K., Koo, J., Meng, C., Niino, H., Osterhagen, J., Renolds, L., Wageman, J., Wang, C., and Yang, S., 1969, Geological structure and some water characteristics of the East China Sea and the Yellow Sea, *ECAFE Tech. Bull.* **2:**3–43.

Fan, S., Chen, G., and Lin, Y., 1982, A study of seismicity and its mechanisms at eastern seas of China, in: *The Geology of the Yellow Sea and the East China Sea* [in Chinese] (Y. S. Qin, ed.), Science Press, Beijing, pp. 30–38.

Fan, X. K., 1982, The Tertiary system of the Pearl River mouth basin and its petroleum prospectives [in Chinese], *Mar. Geol. Res.* **2:**25–35.

Holloway, N. H., 1982, North Palawan block, Philippines—Its relation to Asian mainland and role in evolution of South China Sea, *Am. Assoc. Pet. Geol. Bull.* **66:**1355–1383.

Hsieh, C. M., 1973, *Atlas of China*, McGraw–Hill, New York.

Jin, X., and Yu, P., 1982a, Tectonics of the Yellow Sea and the East China Sea, in: *The Geology of the Yellow Sea and the East China Sea* [in Chinese] (Y. S. Qin, ed.), Science Press, Beijing, pp. 1–22.

Jin, X., and Yu, P., 1982b, Structural outline in the northern part of the Yellow Sea, in: *The Geology of the Yellow Sea and the East China Sea* [in Chinese] (Y. S. Qin, ed.), Science Press, Beijing, pp. 23–29.

Li, C. Z., Li, Y. Z., and Cai, H. M., 1980, Characteristics of the surface sediments on the north shelf of the South China Sea [in Chinese], *Nanhai Stud. Mar. Sin.* **1:**35–50.

Li, D., 1980, Geology and structural characteristics of Bohai Bay, China [in Chinese], *Acta Pet. Sin.* **1:**6–20.

Li, D., 1981a, Geological structural and hydrocarbon occurrence of Bohai Gulf oil and gas basin, China [in Chinese], *Mar. Geol. Res.* **1:**3–20.

Li, D., 1981b, Geological structure and hydrocarbon occurrence of the Bohai Gulf oil and gas basin (China), in: *Petroleum Geology in China* (J. F. Mason, ed.), Penn Well Publishing Co., Tulsa, Okla., pp. 180–192.

Li, D., 1984, Geologic evolution of petroliferous basins on the continental shelf of China, *Am. Assoc. Pet. Geol. Bull.* **68:**993–1003.

Li, J., and Cong, B., 1980, A preliminary research on the origin and evolution of the Bohai Basin, in: *Formation and Development of the North China Fault Block Region* [in Chinese] (W. Y. Zhang, ed.), Science Press, Beijing, pp. 206–220.

Li, Q., Jiang, J., Yan, Q., Xu, D., and Lu, W., 1982, The origin of the Okinawa Trough based on the data of gravity and magnetic profile [in Chinese], *Acta Oceanol. Sin.* **4:**324–334.

Li, Z. W., 1984, A discussion on the crustal nature of the central and northern parts of the South China Sea [in Chinese], *Acta Geophys. Sin.* **27:**153–166.

Liu, Z., and Wang, Z., 1981, Geological structural features of the Tanlu fracture zone in Bohai [in Chinese], *Mar. Geol. Res.* **1:**68–76.

Ludwig, W. J., 1970, The Manila Trench and West Luzon Trough—3. Seismic refraction measurement, *Deep Sea Res.* **17:**553–571.

Ludwig, W. J., Hayes, D. E., and Ewing, J. I., 1967, The Manila Trench and West Luzon Trough—1. Bathymetry and sediment distribution, *Deep Sea Res.* **14:**533–544.

Meyerhoff, A. A., Chen, C., Kamen-Kaye, M., Wang, K. K., and Willums, J., 1984, Energy in China: Summary, in: *Transactions of the 3rd Circum-Pacific Energy and Mineral Resource Conference* (S. Watson, ed.), American Association of Petroleum Geologists, Tulsa, Oklahoma. pp. 199–219.

Niino, H., and Emery, K. O., 1961, Sediments of shallow portions of East China and South China Seas, *Geol. Soc. Am. Bul.* **72:**731–762.

Polski, W., 1959, Foraminiferal biofacies off the North Asiatic coast, *J. Paleontol.* **33:**569–587.

Qin, G., 1980, Stratigraphic significance of the coiling direction in *Globorotalia* for the Late Cenozoic strata of the northern part of the South China Sea [in Chinese], in: *Papers on Marine Micropaleontology* (P. X. Wang, ed.), Oceanic Press, Beijing.

Qin, Y. S. (ed.), 1982, *The Geology of the Yellow Sea and the East China Sea* [in Chinese], Science Press, Beijing.

Qin, Y. S., and Li, F., 1983, Study of the influence of sediment loads discharged from Huanghe River on sedimentation in Bohai Sea and Huanghai Sea, in: *Proceedings of the International Symposium on Sedimentation on the Continental Shelf, with Special Reference to the East China Sea* (*Acta Oceanologica Sinica*, ed.), China Ocean Press, Beijing, Vol. 1, pp. 91–101.

Qin, Y. S., and Zhen, T., 1982, A study of distribution pattern of sediments on the continental shelf of the East China Sea, in: *The Geology of the Yellow Sea and the East China Sea* [in Chinese] (Y. S. Qin, ed.), Science Press, Beijing, pp. 39–51.

Ren, J. S., Jiang, C. F., Zhang, Z. K., and Qin, D. Y., 1980, *The Geotectonic Evolution in China* [in Chinese], Science Press, Beijing.

Rottman, M., 1977, Euthecosomatous pteropods and planktonic foraminifera in southeast Asian marine waters: Species associations and distribution in sediments, Ph.D. dissertation. University of Colorado.

Rottman, M., 1978, Species associations of planktonic foraminifera and zooplankton in the South China and Java Seas, *J. Foraminiferal Res.* **8:**350–359.

Samodai, E., Thompson, P., and Chen, C., 1986, Foraminiferal analysis of South China core with paleoenvironmental implications, *Proc. Geol. Soc. China* (Taipei) **29:**118–137.

Second Marine Geological Team, 1975, Geological Map of China Seas and Vicinity, 1 : 3,000,000, and the Explanatory Text [in Chinese], Marine Geological Survey of China.

Taylor, B., and Hayes, D. E., 1980, The tectonic evolution of the South China Basin, *Geophys. Monogr.* **23:**89–104.

Taylor, B., and Hayes, D. E., 1982, Origin and history of the South China Sea Basin, *Geophys. Monogr.* **27:**23–56.

Wageman, J. M., Hilde, T., and Emery, K. O., 1970, Structural framework of East China Sea and Yellow Sea, *Am. Assoc. Pet. Geol. Bull.* **54:**1611–1643.

Waller, H. O., 1960, Foraminiferal biofacies off the South China coast, *J. Paleontol.* **34:**1164–1182.

Wang, X., and Liang, J., 1984, On the origin of metal spherical minerals in the Okinawa Trough sediments [in Chinese], *Geochemica* **1:**31–35.

Wang, X., Ma, K., Chen, J., and Li, Z., 1984, Detrital minerals in the surface sediments of East China Sea shelf and their geological significance [in Chinese], *Mar. Geol. Quaternary Geol.* **4:**43–55.

Xu, W., 1982, Discussion on the two Cenozoic basins of the South Yellow Sea [in Chinese], *Mar. Geol. Res.* **2:**66–78.

Yang, Z., and Milliman, J. D., 1983, Fine-grained sediments of Changjiang and Huanghe Rivers and sediment sources of East China Sea, in: *Proceedings of the International Symposium on Sedimentation on the Continental Shelf, with Special Reference to the East China Sea* (*Acta Oceanologica Sinica*, ed.), China Ocean Press, Beijing, Vol. 1, pp. 436–446.

Yu, X., 1981, The Tertiary sedimentary features of the South Yellow Sea basin and its oil prospects [in Chinese], *Mar. Geol. Res.* **1:**77–82.

Zhang, W. Y. (ed.), 1983, *Tectonic Map of China and Adjacent Seas, 1 : 5,000,000* [in Chinese], Science Press, Beijing.

Zhao, X., 1984, Studies on the evolution of China's coast [in Chinese], Fujian, Technology Press.

Chapter 6

COMPLEXITIES IN THE DEVELOPMENT OF THE CAROLINE PLATE REGION, WESTERN EQUATORIAL PACIFIC

Kerry A. Hegarty

Department of Geology
University of Melbourne
Victoria 3182, Australia

and

Jeffrey K. Weissel

Lamont-Doherty Geological Observatory of Columbia University
Palisades, New York 10964

I. INTRODUCTION

Johnson and Molnar (1972) and Bracey and Andrews (1974) speculated that the region north of New Guinea might consist of a lithospheric plate separate and distinct from the adjacent Pacific, Indo-Australian, and Philippine plates. The boundaries of the Caroline plate (Fig. 1) were first identified by Weissel and Anderson (1978) using marine geophysical and seismological evidence. They concluded that an additional plate (the Caroline plate) currently exists, using arguments based on (1) plate motion models and (2) tectonic characteristics of the Sorol Trough, the Mussau Trench, and the curiously disrupted seafloor between these features (Fig. 2). Several models have been proposed to explain the formation of these and other structural features associated with the Caroline plate and its boundaries (Den *et al.*, 1971; Winterer *et al.*, 1971; Moberly, 1972; Bracey and Andrews,

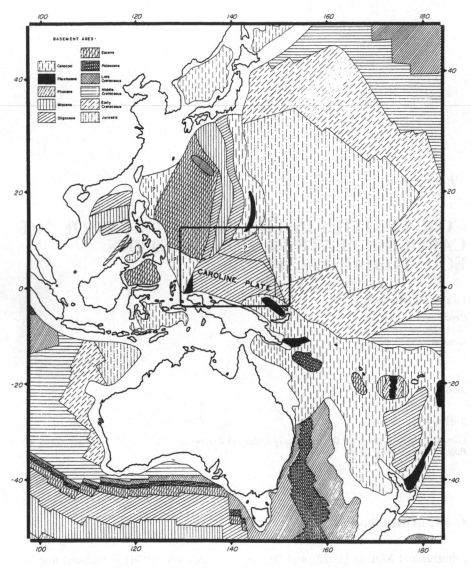

Fig. 1.   Map of oceanic basement ages of the western Pacific and eastern Indian Oceans. Ages are primarily based on magnetic anomaly lineations. (Modified after Larson *et al.*, 1984.)

1974; Bracey, 1975; Erlandson *et al.*, 1976; Mammerickx, 1978; Weissel and Anderson, 1978; Fornari *et al.*, 1979; Keating *et al.*, 1981; Hegarty *et al.*, 1983). However, the regional implications from several of these models have generally not been fully considered. In this chapter, we describe the prominent structural features using geological and geophysical data, and present a tectonic model for the evolution of the Caroline plate region that is consistent with these observations.

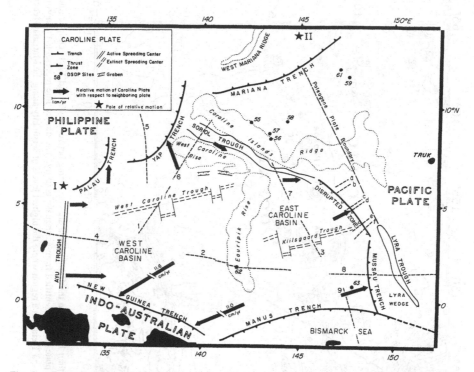

Fig. 2. Tectonic setting of the Caroline plate with respect to neighboring lithospheric plates (Pacific, Indo-Australian, Philippine, and the northern portion of the Bismarck). Relative motion vectors, referred to in the text, are shown. Locations of seismic reflection records, reproduced in Figs. 6, 7, and 10, are indicated. Dotted boundaries represents the 3000-m isobath. Stars I and II show the location of the pole describing relative plate motion for the Caroline–Philippine plates and Caroline–Pacific plates, respectively.

II. MAIN STRUCTURAL FEATURES

The Palau Trench, the Yap Trench, the Sorol Trough, the Mussau Trench (and its extension to the northwest), the Manus Trench, the New Guinea Trench, and the Ayu Trough form the present boundaries of the Caroline plate (Fig. 2). Teleseismic activity, though sparse, delineates each of these plate boundaries except the Mussau Trench (Fig. 3). Relative motion across each boundary (Fig. 2) was determined by adding the angular velocity vectors describing Philippine–Pacific (Karig, 1975), Philippine–Caroline (Weissel and Anderson, 1978), and Pacific–Indo-Australian (Minster and Jordan, 1978) relative motion.

The Caroline plate itself is divided structurally by the Eauripik Rise into two roughly equidimensional basins, the West Caroline Basin and the East Caroline Basin (Fig. 2). Water depths within the basins are typically 4000–5000 m, while the Eauripik Rise shoals to less than 1800 m. Small ENE–WSW-trending grabens or

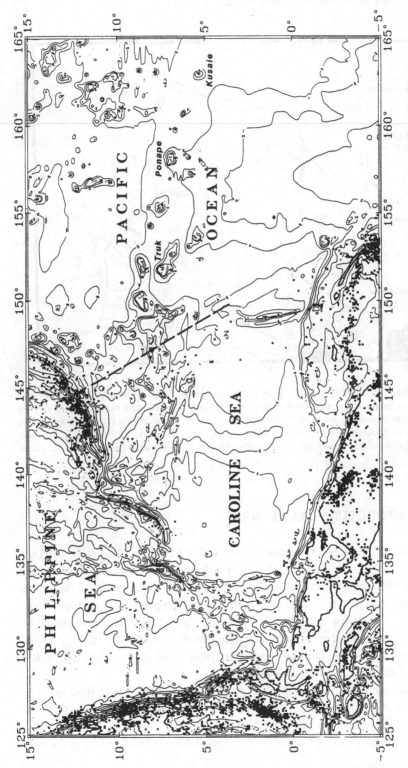

Fig. 3. Regional epicenter distribution pattern for shallow earthquakes (depths less than 100 km) in the Caroline Sea area occurring between 1962 and 1981. Despite low levels of teleseismicity along some boundaries, most of the perimeter of the Caroline plate can be identified from earthquake activity. The dashed line represents the northeastern Paleogene boundary of the Caroline Plate. This boundary intersects the Mariana Trench in a region where the highest levels of seismicity along the Mariana Trench are observed.

troughs occur in each basin, and these are referred to as the West Caroline Trough and the Kiilsgaard Trough. The Sorol Trough, a structural feature which disrupts the West Caroline Rise and Caroline Islands Ridge, constitutes the northern limit of the Caroline plate. The origin and structural characteristics and construction of all of these features will be discussed following a brief review of some important results from earlier work.

III. PREVIOUS WORK

Various explanations for the formation of major features associated with the Caroline plate have been suggested from marine magnetic and gravity anomaly data, seismic reflection and refraction profiles, earthquake focal mechanism solutions, bathymetric, petrological, and deep sea drilling data. Based on drilling at DSDP sites 62 and 63 (Fig. 2), Winterer *et al.* (1971) concluded that the West and East Caroline Basins formed by seafloor spreading along the Eauripik Rise during the Oligocene. Den *et al.* (1971), combining seismic reflection and nine two-ship refraction profiles, showed that the Eauripik Rise is a raised and thickened section of oceanic crust blanketed by a uniform thickness of calcareous sediments.

Moberly (1972) and Vogt and Bracey (1972) recognized east–west-striking magnetic lineations. They suggested that the Caroline Basins were generated along east–west-trending spreading centers, though they were unable to identify the location of the old spreading axes. The marine magnetic anomaly pattern of the West and East Caroline Basins has since been investigated by Bracey (1975), Erlandson *et al.* (1976), Mammerickx (1978), and Weissel and Anderson (1978). The simplest spreading history (discussed below) includes seafloor spreading at the Kiilsgaard Trough and West Caroline Trough during the Oligocene from magnetic anomaly 13 to anomaly 9 time (Weissel and Anderson, 1978).

The Caroline Islands Ridge (Fig. 2) has been modeled as a relict island arc (Bracey and Andrews, 1974). Using Karig's (1971) model of plate convergence and backarc basin evolution, Bracey and Andrews speculated that the Sorol Trough formed as an interarc basin north of an old northward-dipping subduction zone that now lies along the southern perimeter of the West Caroline Rise. Analysis of dredge samples from the Sorol Trough led Fornari *et al.* (1979) to agree that this feature is accompanied by significant amounts of shear, but that faulting is presently active. Further petrological analysis of the samples (Perfit and Fornari, 1982) suggests that both MORB and "hot-spot" type sources have been involved in the generation of magmas beneath the Sorol Trough.

IV. THE KIILSGAARD TROUGH AND THE WEST CAROLINE TROUGH

Marine magnetic lineations reflecting Oligocene seafloor spreading have been mapped as symmetrically disposed about the Kiilsgaard Trough (Weissel and Anderson, 1978) and about the West Caroline Trough (Bracey, 1983) which served as spreading centers during the genesis of the East and West Caroline Basins. Figure 4 compares the observed magnetic anomalies across the Kiilsgaard Trough to magnetic anomalies modeled using the time scale of LaBrecque et al. (1977). Seafloor spreading within the East Caroline Basin, which occurred at a half-rate of about 6 cm/yr, was active by at least 36 m.y.b.p. (anomaly 13 time) and terminated approximately 28 m.y.b.p. (anomaly 9 time).

Magnetic lineations can be identified in both basins as shown in Fig. 5. In general, the amplitudes of the magnetic anomalies in each of the basins are large, often exceeding 500 gammas peak-to-trough (Fig. 4). However, the fidelity of the seafloor spreading magnetic signal deteriorates near the flanks of both the Eauripik Rise and the West Caroline Rise, preventing identification and correlation of the anomalies across these features. Fracture zones also disrupt the lineation pattern, especially within the West Caroline Basin where anomaly identification is more difficult.

Basement structure and sediment thickness near the West Caroline and Kiilsgaard Troughs are relatively well known from single-channel seismic reflection records (Bracey, 1975; Erlandson et al., 1976; Weissel and Anderson, 1978; Bracey, 1983). Profiles 1 and 3 (Fig. 6) across these extinct spreading centers show small, sediment-filled (about 0.5 sec) grabens about 25–30 km wide and with about 500–1000 m of seafloor relief. Although sediment thickness increases away from the trough axes, basement depths do not always appear to increase away from the trough axes, as would be expected for cooling oceanic lithosphere (Sclater et al., 1971). This deviation from the commonly observed crustal age/depth relationship may be related to tectonism and volcanism that postdates seafloor spreading.

V. THE AYU TROUGH AND THE SOROL TROUGH

There are two boundaries along the perimeter of the Caroline plate where extensional tectonics have been suggested to occur (Weissel and Anderson, 1978). Crustal extension at the Ayu Trough is suggested from observations of strike-slip and extensional earthquake focal mechanisms, thin sediment cover, and rugged basement relief. However, no recognizable magnetic anomalies have been identified. A seismic reflection profile across the Ayu Trough (profile 4, Fig. 6) shows a 50-km-wide axial graben with over 2 km of seafloor relief. The initiation of spreading at the Ayu Trough was estimated to be late Miocene from observed sediment thicknesses combined with sedimentation rates measured at DSDP site 62 (Weissel and Anderson, 1978). Weissel and Anderson also showed that the observed subsidence away from the Ayu Trough, when compared with the empirical global depth–age curve (Sclater et al., 1971), is consistent with a Miocene age. Although these two methods allow only rough estimates of the age of the Ayu Trough, the results

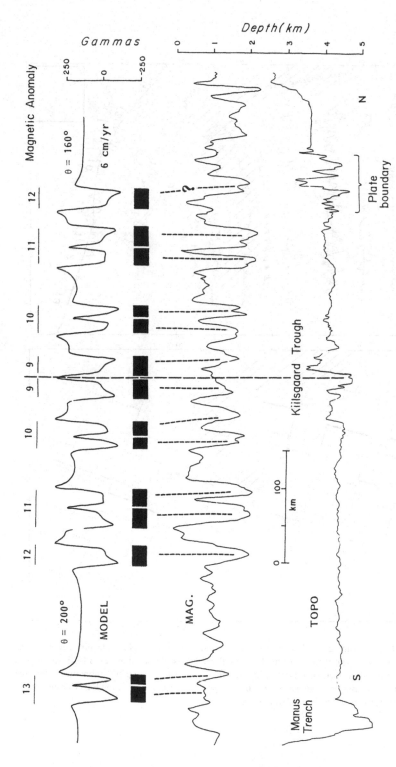

Fig. 4. Projected topographic and magnetic anomaly profiles (onto 000°) across the Kiilsgaard Trough within the East Caroline Basin compared to a theoretical magnetic anomaly profile based on the geomagnetic reversal time scale of LaBrecque et al. (1977). Profile location is shown by a bold track line in Fig. 5A. The magnetic layer was modeled as 0.5 km thick and 4.5 km deep with a half-spreading rate of 6.0 cm/yr. (From Weissel and Anderson, 1978.)

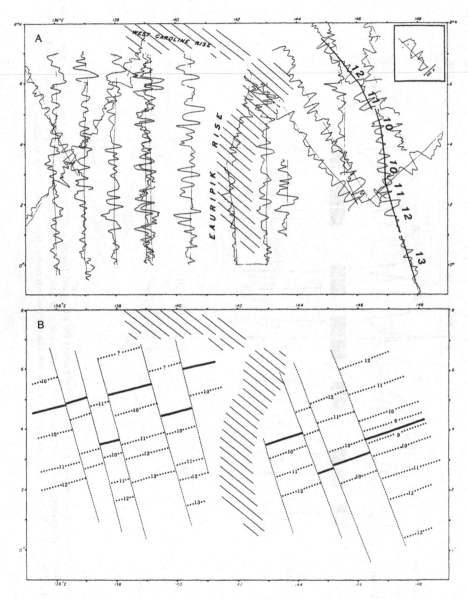

Fig. 5. (A) Magnetic anomalies along nearly north–south ship tracks. Note the diminution of the magnetic anomaly signal near the crest of the Eauripik Rise. The bold track line near 146°E was used in comparison to modeled magnetic anomalies (Fig. 4). (B) Correlation of magnetic anomalies within the West and East Caroline Basins. Reliable correlations are possible within most of the East Caroline Basin and the southern portions of the West Caroline Basin.

are compatible with one another. The boundary between Miocene crust generated at the Ayu Trough and Oligocene crust created at the West Caroline Trough is shown by the arrow in Fig. 6. This location was selected on the basis of variations in sediment thickness and basement roughness.

The Sorol Trough (profiles 6 and 7, Fig. 6) separates the Caroline and Pacific plates at about 8°N. Like the Ayu Trough, diffuse seismicity is associated with this

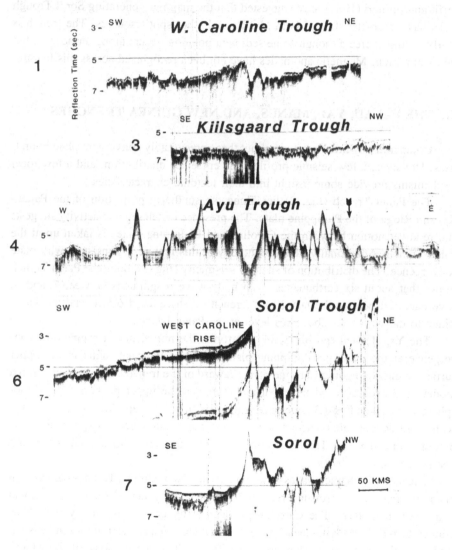

Fig. 6.   Seismic reflection profiles across the West Caroline (1), Kiilsgaard (3), Ayu (4), and Sorol (6 and 7) Troughs. Profile locations are shown in Fig. 2. The arrow (profile 4) shows the inferred boundary between Oligocene crust created at the West Caroline Trough (east) and Miocene crust generated at the Ayu Trough.

segment of plate boundary. However, to date there has not been a large enough earthquake to determine a first motion focal mechanism solution. Large basement scarps, sometimes exceeding 3 km in throw, have been interpreted as normal faults (Bracey and Andrews, 1974; Weissel and Anderson, 1978; Fornari et al., 1979), and thus indicative of an extensional environment. In addition, Fornari et al. (1979) suggested that extension is accompanied by a large component of strike-slip motion. Based on geochemical analysis of fresh samples dredged from the Sorol Trough, Perfit and Fornari (1982) have suggested that the magmas generating Sorol Trough rocks have characteristics of both MORB and "hot-spot" sources. The trough is nearly sediment-free although some sediment ponding occurs along terraces within the axial region. Magnetic anomalies have not been recognized across this feature.

VI. THE PALAU, YAP, MANUS, AND NEW GUINEA TRENCHES

Comparatively little is known about these presumably convergent plate boundaries. However, a few seismic profiles, the epicenter distribution, and a few focal mechanisms provide some insight into their tectonic characteristics.

The Palau Trench (Fig. 2) lies along the southerly projection of the Palau–Kyushu Ridge of the Philippine plate. The presence of teleseismic activity suggests that relative motion between the Caroline and Philippine plates is taken up at the Palau Trench. Additionally, the morphology of this feature is consistent with plate convergence. The distribution of shallow seismicity (Fig. 3) along the Palau Trench shows that about six earthquakes, with body-wave magnitudes between 4 and 5, have occurred here in the last 20 yr. Trench depths exceed 6 km, and the Palau Ridge to the west rises above sea level at the Palau Islands.

The Yap Trench (profile 5, Fig. 7) forms the northernmost segment of convergence at the Caroline–Philippine plate boundary. Normal faults in the upper surface at the downgoing Caroline plate seaward of the trench, are consistent with a model of convergence where extension occurs in the upper portion of the lithosphere during bending (Bodine and Watts, 1979). The Yap Trench is associated with more concentrated seismic activity than the Palau Trench (Fig. 3). Relative motion along the Yap Trench is dominated by thrusting with a small strike-slip component (Fitch, 1972).

The tectonic characteristics of the Manus and New Guinea Trenches are known primarily from a few seismic reflection profiles, earthquake studies, and global plate motion models. The Caroline plate probably converges rapidly with New Guinea across the New Guinea Trench, though significant uncertainties are associated with the direction of convergence (Fig. 2). The morphology of the Manus Trench (profile 9, Fig. 7) is similar to that of an active trench, and interplate convergence would normally be assumed (e.g., Mammerickx, 1978; Weissel and Anderson, 1978). However, relative motion across this boundary appears to be mainly transcurrent (Fig. 2), if the northern part of the Bismarck Sea region is

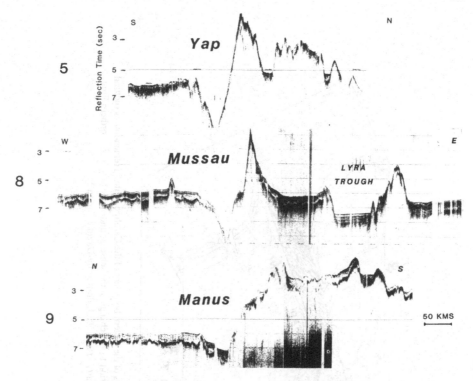

Fig. 7. Seismic reflection profiles across the Yap Trench (5), Mussau Trench and Lyra Trough (8), and Manus Trench (9). Profile locations are shown in Fig. 2.

treated as part of the Pacific plate. The low level of seismicity at both the New Guinea and Manus Trenches (Fig. 3) is surprising in view of the possibility of rapid relative motion (> 11 cm/yr) across these features. For the New Guinea Trench, we may argue that most of the relative motion is taken up on a broad zone of deformation through northern New Guinea marked by the wide belt of earthquake activity (Fig. 3).

VII. THE MUSSAU TRENCH AND THE DISRUPTED ZONE

The Mussau Trench and the disrupted zone to the north mark the easternmost boundary of the Caroline plate (Figs. 2 and 8). The Mussau Trench has been proposed as a site of former subduction (Erlandson *et al.*, 1976), present subduction (Weissel and Anderson, 1978; Hegarty *et al.*, 1983), and future subduction (Bracey, 1975).

The morphology of the Mussau Trench and Ridge is shown in Figs. 7 (profile 8) and 9. Both the trench depths and peak-to-trough free-air gravity anomaly amplitudes (Fig. 9) increase toward the south. The Mussau Trench is associated with

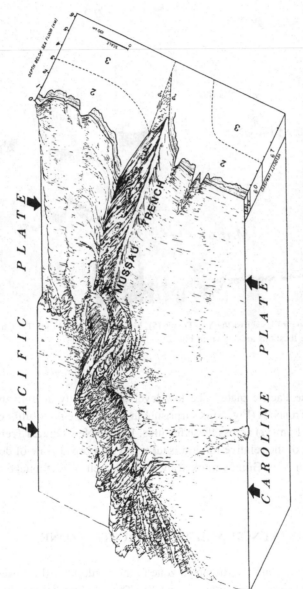

Fig. 8. Physiographic interpretation based on over 20 crossings of the eastern boundary of the Caroline plate. Tremendous structural variation is exhibited along the length of the boundary in addition to the change in the sense of underthrusting at the northern limit of the Mussau Trench, south of the disrupted zone. (From Hegarty et al., 1983; drafted by David Johnson.)

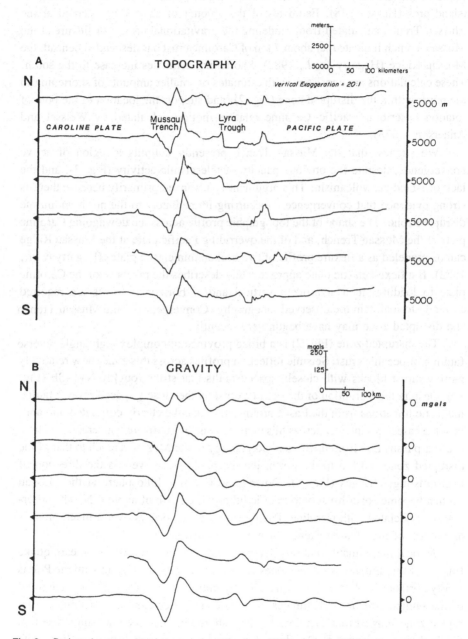

Fig. 9. Projected topographic and gravity profiles (onto 090°) across the Mussau Trench and Ridge. Topography (A) indicates a systematic steepening of the downgoing Caroline plate at the Mussau Trench from north to south. Gravity (B) profiles show the development of a slight outer bulge about 100 km west of the Mussau Trench. Spacing between profiles is about 60 km.

some of the greatest depths in the world's oceans, apart from trenches related to island arcs (Hess, 1948). Estimates of the amount of shortening accrued at the Mussau Trench calculated from modeling the gravitational effects of flexure at the Mussau Trench indicate that about 7 km of Caroline crust has descended beneath the Mussau Ridge (Hegarty et al., 1983). Shortening estimates increase to the south. These calculations are consistent with estimates of smaller amounts of shortening to the north within the disrupted zone (Fig. 2), and support the location of the pole of rotation describing Pacific–Caroline relative motion calculated by Weissel and Anderson (1978).

We suggest that the Mussau Trench presently delimits a region of active convergence, despite the obvious paucity of teleseismic activity (Fig. 3), and the lack of island arc volcanism. This argument is advanced primarily because there is strong evidence that convergence is occurring immediately to the north within the disrupted zone. The shape of the topographic profile across the downgoing Caroline plate at the Mussau Trench, and of the overriding Pacific plate at the Mussau Ridge can be modeled as a flexure profile of two broken thin elastic plates (Hegarty et al., 1983). If a flexed elastic plate approximately describes the response of the Caroline plate to loading, then significant vertical and/or horizontal forces are required currently to maintain the observed topography. Convergence at the Mussau Trench and disrupted zone may have begun very recently.

The disrupted zone (Fig. 2) is a broad province of complex high-angle reverse faulting of oceanic crust. Seismic reflection profiles across this zone show relatively narrow thrust blocks with closely spaced faults that strike roughly NW–SE (Fig. 10). Despite the disruption of the oceanic crust by the closely spaced reverse faults, magnetic lineations from the East Caroline Basin can be clearly correlated into and, in some cases, completely across this unusual zone of interplate convergence. There is a sharp transition from the trench morphology of the Mussau Trench to that of the disrupted zone, 50 km to the north; the morphology, as well as the direction of overthrusting, changes (Fig. 8). Downgoing Caroline lithosphere at the Mussau Trench becomes overthrust onto Pacific lithosphere north of about 4°N. The deformation associated with Caroline–Pacific convergence occurs over a much broader area north of the Mussau Trench than at the trench itself.

A focal mechanism solution (Bergman and Solomon, 1980) for an earthquake (m_b = 5.6) within the disrupted zones shows a compressional event with the P-axis nearly horizontal (plunge = 12°) and slightly oblique (azimuth = 272°) to the trend of the faulted structures. Estimates of the amount of convergence within the disrupted zone suggest that 2–5 km of crustal shortening has been accommodated on the high-angle reverse faults along this segment of the Pacific–Caroline plate boundary. The small amount of interplate convergence that has occurred within the disrupted zone as well as along the Mussau Trench to the south supports our suggestion that this is a very young plate boundary.

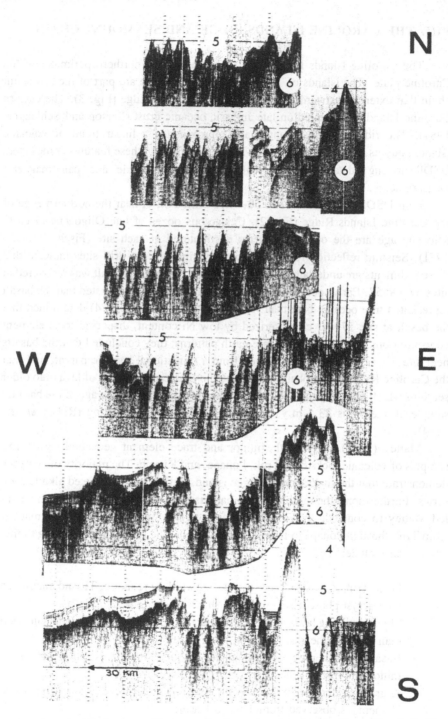

Fig. 10. Six seismic reflection profiles across the disrupted zone north of the Mussau Trench. Profile locations are shown in Fig. 2. Water depth is given in terms of seconds of two-way travel time (1 sec ≅ 750 m).

VIII. THE CAROLINE ISLANDS RIDGE AND SEAMOUNT CHAIN

The Caroline Islands Ridge (Fig. 2) lies along the northern perimeter of the Caroline plate. The islands of Truk, Ponape, and Kusaie are part of the seamount chain that extends eastward from the Caroline Islands Ridge (Fig. 3). The eastern Caroline Islands are located on late Jurassic oceanic crust (Larson and Schlanger, 1981). The ridge and seamount chain approximates a linear trend of volcanic islands, guyots, and atolls. Much of what is known about these features comes from DSDP drilling, detailed petrological analysis, radiometric and paleomagnetic measurements.

Four DSDP holes, sites 55 to 58, were drilled on or near the northern edge of the Caroline Islands Ridge (Fig. 2). Calcareous oozes of late Oligocene or early Miocene age are the oldest sediments encountered at each site (Fischer, et al., 1971). Seismic reflection records from the vicinity of the drill sites indicate that these sediments are underlain by smooth acoustic basement. Basalt was recovered at sites 57 and 58. Early speculations by Fischer et al. (1971) suggested that the basalt represented true oceanic basement. However, Ridley et al. (1974) determined that the basalt at site 57 was characterized by low Ni content, dispersed trace element concentrations, and fractionated rare-earth patterns; they concluded that the basalts near site 57 were slightly fractionated from a transitional basaltic parent, and that the Caroline Ridge was probably constructed by the emplacement of lava onto older seafloor which probably included a thin veneer of sediments. The age of the basaltic sample at site 57 is 23.5 m.y. as determined from K–Ar dating (Ridley et al., 1974).

Mattey (1982) analyzed the minor and trace element geochemistry of 150 samples of volcanic rocks from Truk, Ponape, and Kusaie. The majority of analyses demonstrate that the rocks from all three islands define a differentiated alkalic rock series. Furthermore, the more subtle variations in petrochemistry from each island led Mattey to conclude that a general shift toward more "alkaline" magmatism from Truk through Ponape to Kusaie was required. The observations leading to this conclusion include:

1. From Truk to Kusaie, there is a shift toward more mafic, Ti- and alkali-rich phenocryst phase assemblages in shield-building basaltic lavas.
2. Shield-building lavas to the west (Truk) tend to be less silica-undersaturated than those to the east (Kusaie).
3. Basalts possess progressively greater amounts of alkalies and incompatible minor and trace elements from Truk to Kusaie.
4. A gradual increase in the ratio of more incompatible to less incompatible elements is observed from Truk to Kasaie.

Mattey (1982) concluded that the general chemical properties of basalts from Truk,

Ponape, and Kusaie are typical of intraplate island alkalic basalts, and that the systematic variations are due to variable degrees of partial melting.

K–Ar ages and paleomagnetic measurements have been reported for Truk (Fuller *et al.*, 1980; Keating *et al.*, 1984), Ponape, and Kusaie (Keating *et al.*, 1981). These eastern islands of the seamount chain display an eastward progression from Truk (about 12 m.y.) through Ponape (about 6 m.y.) to Kusaie (about 1 m.y.). Paleomagnetic inclinations from 15 samples suggest that Truk formed about 4.4°N of the equator (Keating *et al.*, 1984). Fuller *et al.* (1980) arrived at a similar conclusion and showed that Truk formed at an ancient latitude of 3°N with a ±3° confidence interval. Further paleomagnetic results from Ponape and Kusaie suggest that all three islands formed at the same paleolatitude (Mattey, 1982). The radiometric and paleomagnetic results together suggest that the eastern, and possibly the western, Caroline Islands were constructed progressively from west to east along a line of latitude at approximately 4°N.

IX. SPECULATION ON THE ORIGIN OF THE EAURIPIK RISE, THE WEST CAROLINE RISE, AND THE CAROLINE ISLANDS RIDGE AND SEAMOUNT CHAIN

The Eauripik Rise, the West Caroline Rise, and the Caroline Islands Ridge were first recognized as major structural features by Hess (1948). Their formation and evolution remain somewhat enigmatic, despite the geophysical and geological data now available for study.

The Eauripik Rise, separating the East and West Caroline Basins, has been postulated to have formed as a result of excessive volcanism along the trace of a mantle plume (Bracey, 1975), at a spreading center (Winterer *et al.*, 1971; Erlandson *et al.*, 1976; Mammerickx, 1978), and along a leaky transform (Weissel and Anderson, 1978). The southern segment of the Eauripik Rise strikes nearly perpendicular to the relict spreading centers, and subparallel to mapped fracture zones within the basins (Fig. 5B). The Eauripik Rise is located at the position of the longest offset of the spreading centers within the Caroline basins.

A suite of geophysical profiles across the Eauripik Rise (Fig. 11) shows that this feature is associated with relatively long-wavelength seafloor magnetic anomalies, sediments of very uniform thickness (about 0.6 sec), relatively smooth and broad basement relief, thickened oceanic layers 2 and 3 (compared to the adjacent basins), 1 m of relief in the geoid, and a small free-air gravity anomaly. The refraction results (Den *et al.*, 1971), which demonstrate thickened crust beneath the crest of the Eauripik Rise, are consistent with the geoid and gravity measurements which indicate that the rise is nearly completely compensated. This evidence suggests that the Eauripik Rise may have formed as a result of excessive igneous

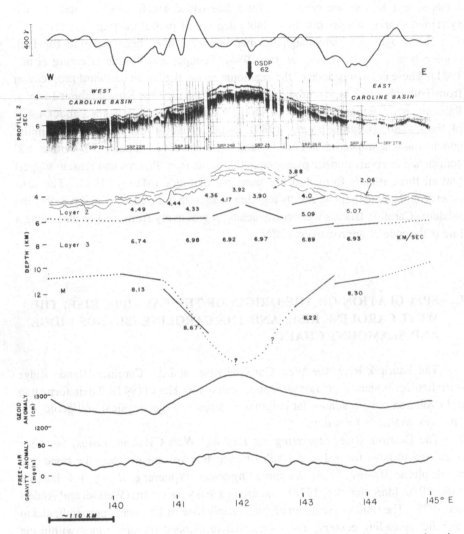

Fig. 11. Profile of several geophysical parameters across the Eauripik Rise—magnetic anomaly, seismic reflection and refraction, geoid and free-air gravity anomalies. Approximate location of these profiles is shown in Fig. 2 (profile 2). The arrow shows the position of DSDP site 62.

activity on young weak oceanic lithosphere during the early stages of seafloor spreading at the West Caroline and Kiilsgaard Troughs.

The West Caroline Rise (Fig. 2) is situated to the southwest of the Sorol Trough and, like the Eauripik Rise, has smooth basement characteristics (profile 6, Fig. 6). A series of E–W-trending troughs with about 400 m of seafloor relief is present along portions of the southern limit of the West Caroline Rise (Fig. 2). Bracey and Andrews (1974) postulated that these troughs represent the remnants of a subduction zone. If this is true, the basement and sedimentary cover of the West

Caroline Rise (which would represent a relict island arc or forearc) remain curiously undeformed.

The Caroline Islands Ridge (Fig. 2) and seamount chain (including Truk, Ponape and Kusaie; Fig. 3) may have formed during the passage of the Pacific plate over a melting anomaly or "hot spot" in the mantle (Clague and Jarrard, 1973) as discussed above. We suggest that the origins of the Eauripik Rise and the West Caroline Rise are related to the same melting anomaly as that required to explain the Caroline Islands Ridge and seamount chain.

Several workers (e.g., Jarrard and Clague, 1977; Epp, 1978) have noted that the general trend of portions of the Caroline Islands Ridge and seamount chain forms a small circle which can be described by the same pole of rotation that describes Pacific plate/mantle relative motion deduced from the trend of the Hawaiian seamounts. Such a relationship between the two seamount chains implies that ages along the Caroline chain should be less than about 42 m.y. (the age of the Hawaiian–Emperor bend), with the ages decreasing to the east. Predicted hot-spot traces generated as small circles about calculated Hawaiian poles (Minster *et al.*, 1974; Epp, 1978; Turner *et al.*, 1980) are shown in Fig. 12. Each trace can account for the recognized age progression for Truk, Ponape, and Kusaie. The difference between predicted island ages and observed K–Ar ages may be explained by the significant scatter and poor reproducibility of the age data (Keating *et al.*, 1984).

We suggest that the Caroline Islands Ridge and West Caroline Rise were constructed on top of Oligocene or late Eocene Caroline oceanic crust during the eastward passage of a hot spot from about 36 to 18 m.y.b.p. The petrochemistry (Ridley *et al.*, 1974) and age of the basalt and sediments encountered at DSDP sites 55–58 (Fischer *et al.*, 1971; Ridley *et al.*, 1974) are consistent with this hypothesis. The same melting anomaly responsible for the Caroline Islands Ridge and West

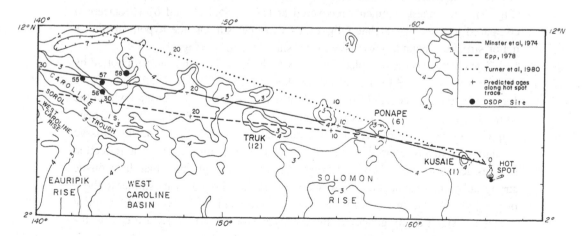

Fig. 12. Predicted hot-spot tracks generated as small circles to proposed Hawaiian poles by Minster *et al.* (1974), Epp (1978), and Turner *et al.* (1980). The hot spot is assumed to be currently at about 5°N, 164°E, southeast of Kusaie. Numbers in parentheses indicate the approximate age in m.y.b.p. of the islands. Bathymetric contours are in thousands of meters.

Caroline Rise may have also created the Eauripik Rise. The strongly oblique trend of parts of the Eauripik Rise to the nearly east–west fabric of the Caroline Islands Ridge and West Caroline Rise may have been influenced by the orientation of preexisting fracture zones within the Caroline basins.

X. TECTONIC EVOLUTION OF THE CAROLINE PLATE REGION

The following speculations on the evolution of the Caroline plate region attempt to account for the geological and geophysical observations outlined above.

We propose that the present Caroline plate and the lithosphere north of the Sorol Trough (south of the Mariana Trench) formed as a backarc basin to northward-dipping subduction of the Indo-Australian plate beneath the Central Highlands of northern New Guinea (Fig. 13A). Backarc spreading probably began prior to anomaly 13 time (early Oligocene, late Eocene) at the West Caroline Trough and Kiilsgaard Trough. The geology within the orogenic belt of the Central Highlands is consistent with this hypothesis. Davies (1978) and Hamilton (1979) report Cretaceous, Paleogene, and early Miocene island-arc volcanic rocks, mélange complexes, and masses of ophiolite along the northern half of New Guinea. During extension within the backarc, the present Caroline plate consisted of two plates, northern and southern "Caroline platelets," analogous to the present-day Mariana platelet and Philippine plate.

During Oligocene time, the eastern boundary of both of these relatively small plates was probably located at the Lyra Trough and along its NNW projection (Fig. 2), a trend now delineated by low levels of teleseismic activity extending SSE from the Mariana arc (Fig. 3; Katsumata and Sykes, 1969). The Lyra Trough trend intersects the Mariana Trench near a region of unusually high earthquake activity (Fig. 3). The oldest sediments recovered at DSDP sites 59 and 61 (Cretaceous) compared to sites 55–58 (late Oligocene–Miocene) support the hypothesis that a major structural boundary exists between sites 58 and 59 (Fig. 2).

The existence of an old plate boundary at the Lyra Trough is also supported by magnetic anomaly and seismic reflection data. Mesozoic magnetic anomalies (late Jurassic–early Cretaceous) have been identified east of the Lyra Trough within the Lyra Basin (Taylor, 1978). No recognizable magnetic anomalies have yet been mapped on· the wedge of oceanic crust between the Lyra Trough and the Mussau Trough; this area is referred to here as the Lyra Wedge (Fig. 2). Seismic reflection profiles across the East Caroline Basin and the Lyra Trough show that the acoustic stratigraphy and basement depths within the East Caroline Basin are similar to those of the Lyra Wedge (e.g., profile 8, Fig. 7). Sediment thicknesses and basement depths are significantly greater within the Lyra Trough and the Lyra Basin (to the east) than within the East Caroline Basin and Lyra Wedge (profile 8, Fig. 7).

Thus, the Lyra Wedge was possibly generated as part of the Caroline spreading system during the Oligocene, and at that time, the Caroline–Pacific plate boundary

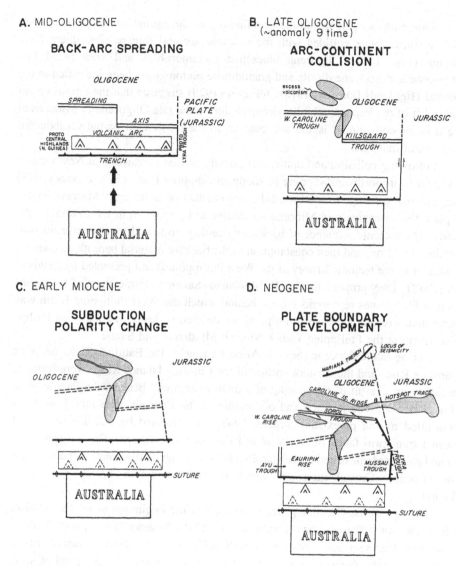

Fig. 13. Schematic illustrations of the tectonic evolution of the Caroline plate region from mid-Oligocene time (A) to the present (D).

was probably along the Lyra Trough and its NNW projection. We speculate that the Lyra Trough served as a transform fault as there is no structural or geophysical evidence to support subduction or seafloor spreading activity along this extinct plate boundary. If the Lyra Trough was a transform fault, then relative motion must have ceased across it when spreading stopped in the present Caroline basins near anomaly 9 time. Near this time, the present Caroline plate began moving with the Pacific plate (Fig. 13B).

Spreading stopped within the Caroline basins during the late Oligocene when the Australian plate collided with the volcanic arc and forearc of northern New Guinea (Fig. 13B). Oligocene blueschist metamorphics and widespread late Oligocene and Miocene diorite and granodiorite plutons have been identified in the Central Highlands (Davies, 1978). Johnson (1979) suggests that the emplacement and uplift of the Papuan ophiolite occurred during the late Oligocene or Miocene at the time of collision of island arc material to the north with partially subducted continental material.

Following collision and uplift, it is postulated that subduction at New Guinea reversed from northward-dipping to southward-dipping (Fig. 13C). Davies (1978) suggests that a change to southward-dipping subduction in the early Miocene would explain the extensive mid-Miocene volcanism and plutonism in the Central Highlands. The tectonic sequence of backarc spreading, followed by a reversal in subduction direction, and then consumption of the backarc material beneath its own arc is similar to the tectonic history of the West Philippine Basin presented by Lewis *et al.* (1982). They propose that the E. Mindanao–Samar volcanic chain of the southeastern Philippines represents the arc behind which the West Philippine Basin was generated. Present-day eastward-dipping subduction is destroying the West Philippine Basin at the Philippine Trench beneath Mindanao and Samar.

During late Oligocene (or early Miocene time?), the Eauripik Rise, the West Caroline Rise, and the western portion of the Caroline Islands Ridge were formed, probably as a result of the passing of a melting anomaly beneath the Pacific plate (Fig. 13B–D). The position and orientation of the Eauripik Rise may have been controlled by the preexisting structural fabric represented by fracture zones and extinct transform faults of the West and East Caroline Basins. The hot spot continued eastward constructing the Caroline Islands Ridge and seamount chain during the Miocene. Crustal extension, which is presently active, began at the Ayu Trough.

The age of the Caroline plate, as defined by the beginning of relative motion across the Sorol Trough, the disrupted zone, and the Mussau Trench, is difficult to constrain. Thin sediment cover within the Sorol Trough, the small amount of crustal shortening at the disrupted zone and at the Mussau Trench, the low level of seismicity at each of these features, and the youthful morphology of the Mussau Trench all suggest that Caroline–Pacific boundaries are young, perhaps younger than a few million years. Alternatively, the age of the Caroline–Pacific plate boundaries may be as old as the early Miocene (following the formation of the West Caroline Rise and westernmost Caroline Islands Ridge). In this case, some of the above observations may be attributed to very slow Caroline–Pacific relative motion. The lack of sediments within the Sorol Trough may be explained by deep water current scouring (Bracey and Andrews, 1974). The uncertainties associated with the Caroline–Pacific relative motion vectors shown in Fig. 2 are too large to help resolve the question of the age of the Caroline plate.

Present-day relative motion on each boundary of the Caroline plate (Fig. 2) is based on interpretations of teleseismic, seismic reflection, and gravity data. The exact age of the initiation of motion between the Caroline plate and the Pacific plate remains unresolved (a possible range of 1–20 m.y.). Another remaining question is: Does the boundary, defined by the trend in seismic activity and DSDP drilling results along the NNW projection of the Lyra Trough, represent a new or an old plate boundary, or both? We believe that there is strong evidence to suggest that this trend represents, at least, an old plate boundary and that the crust north of the Sorol Trough is of Caroline plate affinity. That is, the crust north of the Sorol Trough and south of the Mariana Trench (identified by a question mark in Fig. 1) is of Oligocene or possibly Eocene age. If the Lyra Trough trend (Figs. 2 and 3) north of 4°N represents a young plate boundary currently developing along a relict plate boundary, then the Caroline plate region actually consists of two small plates separated by the Sorol Trough and each one bounded on the east by Pacific plate. This hypothesis requires further supporting observations; we presently prefer the simpler one-plate model to describe the complex tectonic patterns in the Caroline Sea region.

ACKNOWLEDGMENTS

We thank Steven Cande, Dennis Hayes, Robyn Leslie, and Stephen Lewis for helpful reviews and criticisms. Discussions with S. Lewis were particularly useful. Technical assistance was provided by Anita Morgan and Norma Iturrino. Drafting by David Johnson is greatly appreciated. Carol Elevitch patiently and quickly typed each iteration of this manuscript. The National Science Foundation and the Office of Naval Research supported our research under grants OCE78-19830 and N00014-75-C-0210.

REFERENCES

Bergman, E. A., and Solomon, S. C., 1980, Oceanic intraplate earthquakes: Implications for local and regional intraplate stress, *J. Geophys. Res.* **85**:5389–5411.

Bodine, J. H., and Watts, A. B., 1979, On lithospheric flexure seaward of the Bonin and Mariana Trenches, *Earth Planet. Sci. Lett.* **43**:132–140.

Bracey, D. R., 1975, Reconnaissance geophysical survey of the Caroline Basin, *Geol. Soc. Am. Bull.* **86**:775–784.

Bracey, D. R., 1983, Geophysics and tectonic development of the Caroline Basin, Technical Report, TR-283, Naval Oceanographic Office.

Bracey, D. R., and Andrews, J. E., 1974, Western Caroline Ridge: Relic island arc? *Mar. Geophys. Res.* **2**:111–125.

Clague, D. A., and Jarrard, R. D., 1973, Tertiary Pacific plate motion deduced from the Hawaiian–Emperor Chain, *Geol. Soc. Am. Bull.* **84**:1135–1154.

Davies, H. L., 1978, Geology and mineral resources of Papua New Guinea, Third Regional Conference on Geology and Mineral Resources of Southeast Asia, pp. 685–689.

Den, N., Ludwig, W. J., Murauchi, S., Ewing, M., Hotta, H., Asanuma, T., Yoshii, T., Kubotera, A., and Hagiwara, K., 1971, Sediments and structure of the Eauripik–New Guinea Rise, *J. Geophys. Res.* **76**:4711–4723.

Epp, D., 1978, Age and tectonic relationships among volcanic chains on the Pacific plate, Ph.D. dissertation, University of Hawaii.

Erlandson, D. L., Orwig, T. L., Kiilsgaard, G., Mussells, J. H., and Kroenke, L. W., 1976, Tectonic interpretations of the East Caroline and Lyra Basins from reflection-profiling investigations, *Geol. Soc. Am. Bull.* **87**:453–462.

Fisher, A. G., Heezen, B. C., Boyce, R. E., Bukry, D., Douglas, R. G., Garrison, R. E., Kling, S. A., Krasheninnikov, V., Lisitzin, A. P., and Pimm, A. C., 1971, *Initial Reports of the Deep Sea Drilling Project*, Vol. 6, U.S. Government Printing Office, Washington, D.C.

Fitch, T. J., 1972, Plate convergence, transcurrent faults, and internal deformation adjacent to Southeast Asia and the western Pacific, *J. Geophys. Res.* **77**:4432–4460.

Fornari, D. J., Weissel, J. K., Perfit, M. R., and Anderson, R. N., 1979, Petrochemistry of the Sorol and Ayu Troughs: Implications for crustal accretion at the northern and western boundaries of the Caroline plate, *Earth Planet. Sci. Lett.* **45**:1–15.

Fuller, M., Dunn, J. R., Green, G., Lin, J.-L., McCabe, R., Torey, K., and Williams, I., 1980, Paleomagnetism of Truk Islands, eastern Carolines and of Saipan, Marianas, *Geophys. Monogr. Am. Geophys. Union* **22**:235–245.

Hamilton, W., 1979 *Tectonics of the Indonesian region*, Geological Survey Professional Paper 1078, U.S. Government Printing Office, Washington, D.C.

Hegarty, K. A., Weissel, J. K., and Hayes, D. E., 1983, Convergence at the Caroline–Pacific plate boundary: Collision and subduction, in: *The Tectonic and Geologic Evolution of Southeast Asian Seas and Islands*, Part 2 (D. E. Hayes, ed.), *Geophys. Monogr.* 27.

Hess, H. H., 1948, Major structural features of the western North Pacific, and interpretation of H. O. 5485, bathymetric chart, Korea to New Guinea, *Bull. Geol. Soc. Am.* **59**:417–446.

Jarrard, R. D., and Clague, D. A., 1977, Implications of Pacific Island and Seamount ages for the origin of volcanic chains, *Rev. Geophys. Space Phys.* **15**(1):57–76.

Johnson, T. L., 1979, Alternative model for emplacement of the Papuan ophiolite, Papua New Guinea, *Geology* **7**:495–498.

Johnson, T., and Molnar, P., 1972, Focal mechanisms and plate tectonics of the southwest Pacific, *J. Geophys. Res.* **77**:5000–5032.

Karig, D. E., 1971, Structural history of the Mariana Island arc system, *Geol. Sco. Am. Bull.* **82**:323–344.

Karig, D. E., 1975, Basin genesis in the Philippine Sea, in: *Initial Reports of the Deep Sea Drilling Project*, Vol. 31, U.S. Government Printing Office, Washington, D.C., pp. 857–879.

Katsumata, M., and Sykes, L. R., 1969, Seismicity and tectonics of the western Pacific: Izu–Mariana–Caroline and Ryukyu–Taiwan regions, *J. Geophys. Res.* **74**:5923.

Keating, B., Mattey, D., Naughton, J., Epp, D., and Helsley, C. E., 1981, Evidence for a new Pacific hotspot, *EOS Trans. Am. Geophys. Union* **62**:381–382.

Keating, B. H., Mattey, D. P., Naughton, J., and Helsley, C. E., 1984, Age and origin of Truk Atoll, eastern Caroline Islands: Geochemical, radiometric-age and paleomagnetic evidence, *Geol. Soc. Am. Bull.* **95**:350–356.

LaBrecque, J. L., Kent, D. V., and Cande, S. C., 1977, Revised magnetic polarity time scale for Late Cretaceous and Cenozoic time, *Geology* **5**:330–335.

Larson, R. L., and Schlanger, S. O., 1981, Geological evolution of the Nauru Basin and regional implications, in: *Initial Reports of the Deep Sea Drilling Project*, Vol. 61, U.S. Government Printing Office, Washington, D.C., pp. 841–862.

Lewis, S. D., Hayes, D. E., and Mrozowski, C. L., 1982, The origin of the West Philippine basin by inter-arc spreading, in: *Geology and Tectonics of the Luzon–Marianas Region* CCOP Technical Publication, (G. R. Balce and A. S. Zanoria, eds.), pp. 31–51.

Mammerickx, J., 1978, Re-evaluation of some geophysical observations in the Caroline Basins, *Geol. Soc. Am. Bull.* **89**:192–196.

Mattey, D. P., 1982, The minor and trace element geochemistry of volcanic rocks from Truk, Ponape and

Kusie, Eastern Caroline Islands; the evolution of a young hot spot trace across old Pacific Ocean crust, *Contrib. Mineral. Petrol.* **80:**1–13.

Minster, J. B., and Jordan, T. H., 1978, Present-day plate motions, *J. Geophys. Res.* **83:**5331–5354.

Minster, J. B., Jordan, T. H., Molnar, P., and Haines, E., 1974, Numerical modeling of instantaneous plate tectonics, *Geophys. J. R. Astron. Soc.* **36:**541–576.

Moberly, R., 1972, Origin of the lithosphere behind island arcs, with reference to the western Pacific, *Geol. Soc. Am. Mem.* **132:**35–55.

Perfit, M. R., and Fornari, D. J., 1982, Mineralogy and geochemistry of volcanic and plutonic rocks from the boundaries of the Caroline Plate: Tectonic implications, *Tectonophysics* **87:**279–313.

Ridley, W. I., Rhodes, J. M., Reid, A. M., Jakes, P., Shih, C., and Bass, M. N., 1974, Basalts from Leg 6 of the Deep-Sea Drilling Project, *J. Petrol.* **15**(1):140–159.

Sclater, J. G., Anderson, R. N., and Bell, M. L., 1971, Elevation of ridges and evolution of the central eastern Pacific, *J. Geophys. Res.* **76:**7888–7915.

Taylor, B., 1978, Mesozoic magnetic anomalies in the Lyra Basin, *EOS Trans. Am. Geophys. Union* **59**(4):320.

Turner, D. L., Jarrard, R. D., and Forbes, R. B., 1980, Geochronology and origin of the Pratt–Welker seamount chain, Gulf of Alaska: A new pole of rotation for the Pacific plate, *J. Geophys. Res.* **85:**6547–6556.

Vogt, P. R., and Bracey, D. R., 1972, Geophysical investigation of the Caroline Basin, Abstracts of the International Symposium on the Oceanography of the South Pacific, New Zealand National Committee for UNESCO and the Royal Society of New Zealand.

Weissel, J. K., and Anderson, R. N., 1978, Is there a Caroline plate? *Earth Planet. Sci. Lett.* **41:**143–158.

Winterer, E. L., Riedel, W. R., Moberly, R. M. Jr., Resig, J. M., Kroenke, L. W., Gealy, E. L., Heath, G. R., Bröhniman, P., Martini, E., and Worsley, T. R., 1971, *Initial Reports of the Deep Sea Drilling Project,* Vol. 7, U.S. Government Printing Office, Washington, D.C., pp. 323–339.

Chapter 7

THE NORFOLK RIDGE SYSTEM AND ITS MARGINS

James V. Eade*

New Zealand Oceanographic Institute
Division of Marine and Freshwater Science, DSIR
Wellington, New Zealand

I. INTRODUCTION

New Caledonia, Norfolk Island, and northernmost New Zealand lie on a long, narrow, ridge system, the Norfolk Ridge System (Fig. 1), which is situated in the center of a geologically complex region of ridges and basins between Pacific Basin oceanic crust and continental crust of Australia.

The geology of the Norfolk Ridge System in the northern and the southern parts is known mainly from land studies on New Caledonia and Northland, New Zealand. Marine geophysical studies (seismic reflection, seismic refraction, magnetics, and gravity) have provided information on the formation and tectonic history of only the Norfolk Ridge, the central part of the Norfolk Ridge System. New data, most notably seismic reflection profiles collected by the research vessels *Fred H. Moore* (Mobile Oil Corporation) and *Sonne* (Bundesanstalt für Geowissenschaften und Rohstoffe, West Germany) across the West Norfolk and Wanganella Ridges, and southern part of the Norfolk Ridge, have been used here to reinterpret the structure of the southern part of the study area. A new bathymetric map (Fig. 2), constructed from new data and a reinterpretation of old data, is also presented. From these data and a review of previous work, a new tectonic history of the Norfolk Ridge System and its margins is suggested.

*Present address: CCOP/SOPAC, % Mineral Resources Department, Suva, Fiji.

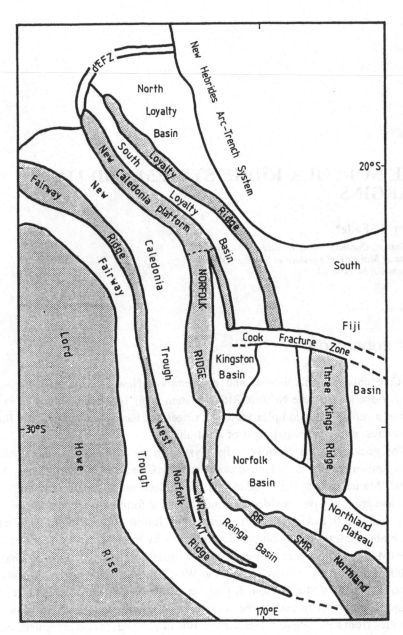

Fig. 1.   Physiographic map showing main morphological units in the study area. Ridges are stippled.

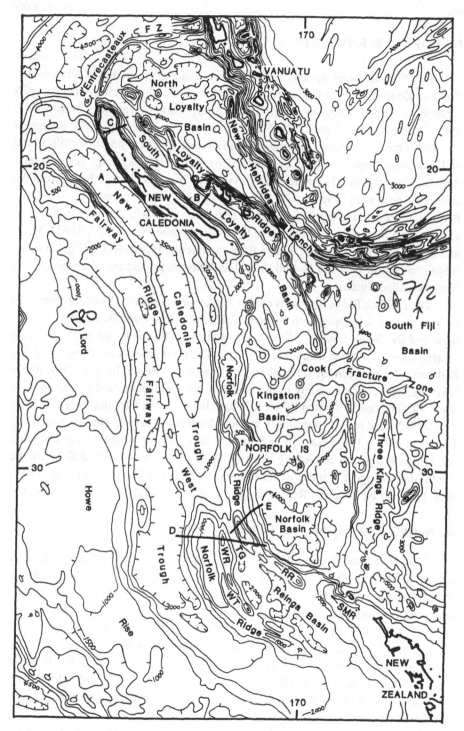

Fig. 2.   Bathymetric map. Location of seismic reflection profiles A–E are indicated. Abbreviations: WR, Wanganella Ridge; WT, Wanganella Trough; TG, Taranui Gap; RR, Reinga Ridge; SMR, South Maria Ridge.

II. NORFOLK RIDGE SYSTEM

The Norfolk Ridge System, as defined here, extends from the d'Entrecasteaux Fracture Zone (Fig. 1) in the north (at latitude 18°S) south to New Zealand at 34°S (Dubois *et al.*, 1974; Dupont *et al.*, 1975). Many workers (van der Linden, 1967, 1968; Woodward and Hunt, 1971; Dupont *et al.*, 1975; Lapouille, 1977; Launay *et al.*, 1982) have considered the ridge system in the south to connect Norfolk Island and Northland (northern New Zealand) along West Norfolk Ridge (Fig. 2) as a double-ridge system. However, a gap physically separates Norfolk Ridge from the double-ridge system (Eade, 1972; Dupont *et al.*, 1975). C. Ravenne and C.-E. de Broin have concluded that West Norfolk Ridge connects with Fairway Ridge (Fig. 2) and is not part of the Norfolk Ridge System (Ravenne *et al.*, 1977, 1982). They, and the present writer, regard Reinga Ridge and South Marina Ridge (Fig. 2) as the ridge connection between Norfolk Island and Northland and therefore part of the Norfolk Ridge System.

The Norfolk Ridge System is mostly flat-topped at depths shallower than 1500 m with steep sides sloping into the New Caledonia Trough and Reinga Basin to the west and the South Loyalty, Kingston, and Norfolk Basins to the east. The ridge system consists of three distinct sections. (1) The northern section (New Caledonia Platform) lies between the d'Entrecasteaux Fracture Zone and latitude 23°S, trends NW–SE, and consists mainly of the island of New Caledonia and its surrounding shelf. (2) The central section (Norfolk Ridge) lies between latitudes 23°S and 32°S, trends N–S, and has its crest mostly at about 1000 m but rises to Norfolk and Philip Islands at 29°S. (3) The southern section consists of a more subdued ridge centering on Reinga Ridge and connecting Norfolk Ridge to the New Zealand shelf via the South Maria Ridge. Crest depths here are mostly less than 1200 m but reach 1750 m deep between Norfolk and Reinga Ridges, and Reinga and South Maria Ridges.

A. Northern Section (between Latitudes 18°S–23°S)

New Caledonia lies on a long (750 km), narrow (70 km) platform which lies mostly at shelf depths of less than 70 m (Dubois *et al.*, 1974). The southwest margin falls steeply to the floor of the New Caledonia Trough at more than 3500 m deep and the northeast margin falls less steeply to the floor of the South Loyalty Basin at 2500 m deep. At its northwest end (lat. 18°S) the platform terminates against the d'Entrecasteaux Fracture Zone where the seafloor falls to more than 3500 m deep. At its southeastern end (lat. 23°S) the New Caledonia Platform bifurcates and deepens, connecting with the prominent N–S-trending Norfolk Ridge and also extending into an unnamed, less prominent ridge east of the northernmost part of Norfolk Ridge (Figs. 1 and 2).

The submerged part of the New Caledonia Platform consists of an irregular basement covered by a complete sedimentary sequence which is more than 1 sec (two-way time) thick in discrete basins (Ravenne *et al.*, 1982). In the north, the platform consists of outcropping basement ridges along both margins with a central sediment-filled depression (Fig. 3) (Bitoun and Recy, 1982). Average crustal thickness beneath New Caledonia is 22 km (Dubois *et al.*, 1973).

Most of our knowledge of the geology and geological history of the platform is based on the geology of New Caledonia which has been summarized by Paris and Lille (1977), Paris (1981), Kroenke (1984), and Brothers and Lillie (this volume). The dominant geological elements of New Caledonia are as follows:

1. Pre-late Cretaceous rocks with continental affinities that form the core of the island. These are thought to have been formed on an active Australian continental margin (Packham, 1982) and then metamorphosed either during uplift and emergence of the New Caledonia insular core (Paris and Lille, 1977) or during subduction to the southwest beneath Australia (Blake *et al.*, 1977) in late Jurassic–early Cretaceous time.

2. Late Cretaceous rhyolites and dolerites which are thought to have formed during initial breakup stages of the east coast of Australia.

3. Tholeiitic basalts of the Formation of Basalts (= Poya Formation in Kroenke, 1984) (late Cretaceous–early Tertiary) along the west coast of New Caledonia. These are thought to have been formed in the New Caledonia Trough along a spreading axis during the late Cretaceous and uplifted and incorporated into the accretionary wedge associated with a period of collision along a northeast-dipping subduction zone beneath New Caledonia in Paleocene and Eocene times (Kroenke, 1984). Volcaniclastic debris from arc volcanism associated with this period of collision was

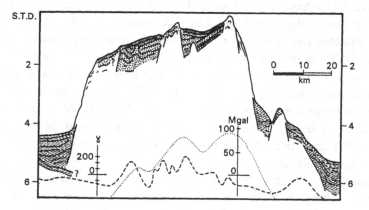

Fig. 3.   Interpreted seismic reflection profile across New Caledonia Platform, north of New Caledonia (Line C). For location see Fig. 2. (From Bitoun and Recy, 1982, with permission.)

deposited north of the arc and may have been redeposited farther north in the North Loyalty Basin to form the thick Eocene volcaniclastic sequence seen at DSDP site 286 (Kroenke, 1984).

4. Peridotites and metamorphosed basalts of northern New Caledonia associated with the major obductive event from the northeast at the end of the Eocene (Kroenke, 1984) or during the Oligocene (Paris and Lille, 1977).

5. Miocene volcanism on Loyalty Ridge and northeast-dipping subduction of South Loyalty Basin crust beneath Loyalty Ridge.

B. Central Section (between Latitudes 23°S–32°S)

The central part of the Norfolk Ridge System consists of the Norfolk Ridge proper. This is a N–S-trending, flat-topped, long (1000 km), narrow (70 km), steep-sided feature entirely submerged except for Norfolk Island and Philip Island at latitude 29°S. Apart from a slight offset at Norfolk Island, the ridge is distinctively straight. Along most of its length the ridge crest lies at 1000–1500 m deep (Fig. 2), shallowing to a major 75-m-deep platform around Norfolk Island (Main and McKnight, 1981). At its margins the ridge slopes steeply west to the floor of the New Caledonia Trough (at more than 3500 m deep) and east into the Kingston Basin (at approximately 4000 m deep) and Norfolk Basin (more than 4000 m deep). At its southern end (lat. 32°S), the ridge bends to the southeast to join the Reinga Ridge (Fig. 2). It does not connect with Wanganella Ridge or West Norfolk Ridge as shown on many bathymetric maps but is separated from it by a deep, narrow gap, the Taranui Gap (Eade, 1972) which has also been named "Bassin Intermediaire" by Dupont et al. (1975).

The structure of the ridge is similar to the New Caledonia Platform. Basement ridges along both margins outcrop or are thinly covered by sediment. The basement ridge along the western margin is wider than that along the eastern margin and is thought to consist of aligned volcanic centers (Ravenne et al., 1982). In between the margins, reflectors form a shallow syncline with an eastward asymmetry (Ravenne et al., 1982). The sedimentary sequence is at least 3 km thick (Shor et al., 1971; Dupont et al., 1975) and overlies a basement which could be either metamorphosed volcanics or continental intrusives (Shor et al., 1971). Three major reflectors have been identified in profiles of this ridge by Dupont et al. (1975). Reflector A, at 0.2–0.5 sec below the seabed, is thought to correspond to the base of the Pliocene or a change in lithification in the Miocene; reflector B is thought to represent the Upper Eocene–Oligocene Regional Unconformity recognized at DSDP sites 206 and 208; and reflector C is thought to be either at the Cretaceous–Paleocene boundary, or more likely a change in facies within the Cretaceous. Reflector C has frequently been called acoustic basement but on some profiles other reflectors are visible below it (Dupont et al., 1975).

There is no clearly recognizable single basement reflector beneath the sedimentary sequence (Dupont *et al.*, 1975). Ravenne *et al.* (1982) thought this to be due to the probable volcaniclastic nature of the older sediments which would mask basement reflectors.

Dupont *et al.* (1975) note the presence of two types of faulting on Norfolk Ridge. Old faults with significant throws have determined the shape of the ridge and young faults with small vertical displacement except at the margins of the ridge where they downthrow into the adjacent basins are also present.

Basement features interpreted as intrusions are seen on some profiles and appear to be scattered along the central part of the ridge. They are thought to be related to the Pliocene volcanic activity seen on Norfolk and Philip Islands (Dupont *et al.*, 1975) dated at 3.1–2.3 Ma (Jones and McDougall, 1973; Aziz-Ur Rahman and McDougall, 1973).

The ridge is characterized by a low-amplitude magnetic signature except for local, moderately strong, positive anomalies (> 600 nT) associated with the intrusions around Norfolk Island (Fig. 4) (van der Linden, 1971; Lapouille, 1977). Magnetic variations, where they do occur, are not continuous and cannot be correlated over long distances (Lapouille, 1977).

Crustal thickness beneath Norfolk Ridge is 21 km (Shor *et al.*, 1971) which is greater than for a volcanic arc. Although somewhat thinner than typical continental crust, the crustal thickness, seismic character, and magnetic signature lead to the assumption that Norfolk Ridge is of continental or continental margin origin.

C. Southern Section (between Latitudes 32°S–34°S)

The southern section of the Norfolk Ridge System is 450 km long and connects the southern end of the Norfolk Ridge at 32°S, to the New Zealand continental shelf at Three Kings Islands at 34°S. It consists of three separate ridge segments arranged *en echelon,* and trending NW–SE. The northern segment is a subdued extension of the southern part of Norfolk Ridge, the central segment is Reinga Ridge rising to a peak less than 500 m deep, and the southern segment is South Maria Ridge rising to shelf depths west of Three Kings Islands off northern New Zealand. The ridge crest falls to approximately 1750 m deep between each segment. The southern section is bounded to the southwest by Reinga Basin and to the northeast by Norfolk Basin. The steep slope from depths less than 500 m on the ridge to over 4000 m in the Norfolk Basin has been described as a fracture zone, the Vening Meinesz Fracture Zone (van der Linden, 1967). New bathymetric and structural data show that the evidence for a fracture zone existing here—offsets in Lord Howe Rise and Norfolk Ridge, and linear northeastern slopes of Reinga Ridge, Northland, and Coromandel continental shelves—does not exist. The character of this margin is the same as that for the Norfolk Ridge to the north and like Norfolk Ridge margins it appears to be an old (pre-Cenozoic) margin of unknown origin separating continental crust of the ridge from oceanic crust of Norfolk Basin.

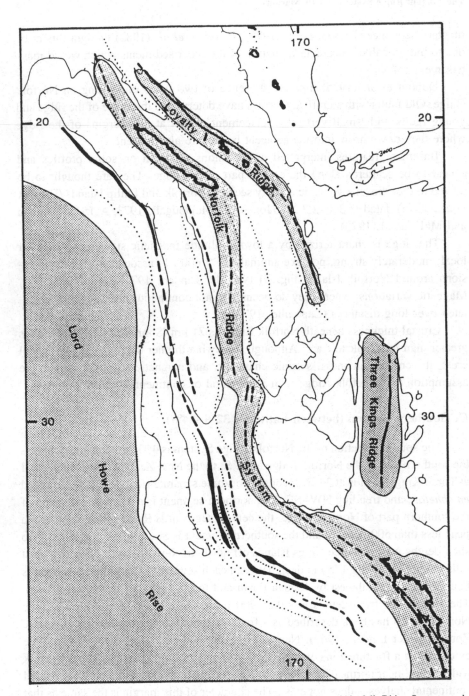

Fig. 4. Magnetic anomaly map showing principal magnetic anomalies of Norfolk Ridge System, and troughs and ridges along its margin (magnetic anomalies on Lord Howe Rise and South Fiji Basin are not shown). Solid anomalies: thick > 1000 nT; thin 500–1000 nT; dashed anomalies: < 500 nT; dotted anomalies: negative nT. (Adapted from van der Linden, 1971; Hunt, 1978; Lapouille, 1977; Ravenne *et al.*, 1977.)

Pockets of sediment partially cover basement on Reinga Ridge (Davey, 1977). The structure and geology of South Maria Ridge, the southern third of this section, are a continuation of the same structure and geology of Northland, New Zealand (Summerhayes, 1969). Fine-grained basic volcanics, augite basalts, and ker-atophyres of the Cretaceous Whangakea Volcanic Series (= Tangihua Volcanic), indurated graywackes (probably pre-Cretaceous), and Tertiary sandstone and silt-stone have been dredged from South Maria Ridge (Summerhayes, 1969).

The magnetic signature of Northland also continues northwest offshore (Hunt, 1978) (Fig. 4). The Eastern Belt of the Stokes Magnetic System, 100–500 nT in Northland, can be traced along South Maria Ridge and Reinga Ridge as a subdued positive anomaly (250–500 nT), apparently slightly offset between each ridge seg-ment and the shelf (Fig. 4).

It is concluded here that the southern section of the Norfolk Ridge System, as described above, is part of the continental sliver which connects New Zealand to Norfolk Ridge and New Caledonia.

III. TROUGHS AND BASINS WEST OF NORFOLK RIDGE

The whole Norfolk Ridge System from d'Entrecasteaux Fracture Zone to New Zealand parallels very closely the eastern margin of the Lord Howe Rise and like most of the Rise the System is thought to be continental in origin. The intervening troughs and ridges are considered here to be oceanic in origin and originally part of an extensional basin which formed between the Rise and Ridge System.

It is evident from the work of Dubois *et al.* (1974), Dupont *et al.* (1975), and Ravenne *et al.* (1977), and from the new bathymetry presented here (Fig. 2) that two major troughs lie between Lord Howe Rise and the Norfolk Ridge System (Figs. 1 and 2). New Caledonia Trough lies west of the northern and central parts of the ridge system between latitudes 19°S and 31°S. Fairway Trough lies east of Lord Howe Rise along its entire length, i.e., between 22°S and 38°S. Previously, many authors (e.g., van der Linden, 1967; Woodward and Hunt, 1971; Burns *et al.*, 1973; Dubois *et al.*, 1974; Dupont *et al.*, 1975; Lapouille, 1977; Kroenke, 1984) have described both of these troughs as a single feature under the name "New Caledonia Basin" (or "Trough").

The two troughs are separated by the Fairway Ridge and the West Norfolk Ridge which are the northern and southern parts of the same feature, the Fairway–West Norfolk Ridge (Dubois *et al.*, 1974; Ravenne *et al.*, 1977). In the north at 20–23°S and in the south at 32–34°S, these ridges trend NW–SE and are major struc-tures, with their crests rising to within 500 m of the sea surface. In contrast, between 25°S and 30°S the Fairway–West Norfolk Ridge trends N–S and is a subdued feature with crest depths at 2500–3200 m deep. Although subdued, it does appear, from the limited amount of data available, to form a topographic barrier which separates the two troughs lying at 3400–3500 m on either side (Fig. 2).

The name "Wanganella Ridge" is a new name proposed here for the ridge which lies immediately east of West Norfolk Ridge, is shallowest (80 m) at Wanganella Bank (32°30'S, 167°30'E), extends north from Wanganella Bank to at least 31°S and possibly to 28°S, and southeast from the bank to about 35°S, 170°E. On many bathymetric charts (e.g., van der Linden, 1968), Wanganella Ridge is named "Norfolk Ridge" and is shown connected at 32°S to Norfolk Ridge as its southern extension. This connection does not exist. The West Norfolk and Wanganella Ridges (Fig. 2) diminish in size to the southeast and disappear altogether both as basement and morphological features before reaching the foot of the New Zealand continental slope off Northland.

The New Caledonia Trough, southwest of New Caledonia, is flat-floored at 3600 m. Sediment is generally 2000–4000 m thick and overlies a strong basement reflector (Ravenne et al., 1977). Maximum sediment thickness (more than 4000 m) lies along the southwest coast of New Caledonia where basement dips deep beneath New Caledonia (Fig. 5) (Dubois et al., 1974). This has been interpreted as indicating the existence of a fossil subduction zone dipping northeast under New Caledonia (Dubois et al., 1974; Kroenke, 1984). The sedimentary sequence on reflection profiles off New Caledonia has a well-bedded appearance with many parallel reflectors. These reflectors have been interpreted as pulses of terrigenous sediments derived from New Caledonia by turbidity currents following its uplift at the time of obduction in the early Oligocene (Dubois et al., 1974; Ravenne et al., 1977). The sedimentary sequence thins to the south where on profiles it is less well-bedded. Consequently, it is assumed that terrigenous input was much less to the south where pelagic sedimentation apparently dominated.

The Fairway Trough lies between Lord Howe Rise and the Fairway–West Norfolk Ridge. It slopes gently from less than 2000 m deep in the north at latitude 22°S and in the south off the west coast of Northland, New Zealand, to a maximum depth of 3550 m in the center of the Trough at 29°S. Sediments are generally 2000 m thick, and thicken toward New Zealand. On seismic reflection profiles in the southern part of the trough, uppermost sediments are well-bedded and therefore assumed to be turbidites (Fig. 6, unit R in Fairway Trough). These can be seen on profiles as far north as 32°S, 1100 km from New Zealand, their probable source. Some may have traveled a further 300 km north to the deepest part of the trough. Unbedded, supposedly pelagic sediments dominate in the northern part of the trough. The turbidites and pelagic sediments overlie a sequence of disturbed sediments (unit S) where slumping has been common, especially at the margins of the trough. These in turn unconformably overlie gently folded sediments (Fig. 6, unit T in Fairway Trough). On the basis of calculated sedimentation rates from site 206, assumed pelagic and turbidite sedimentation rates, plus sediment thickness on profiles from Fairway Trough near site 206, this unconformity appears to be the

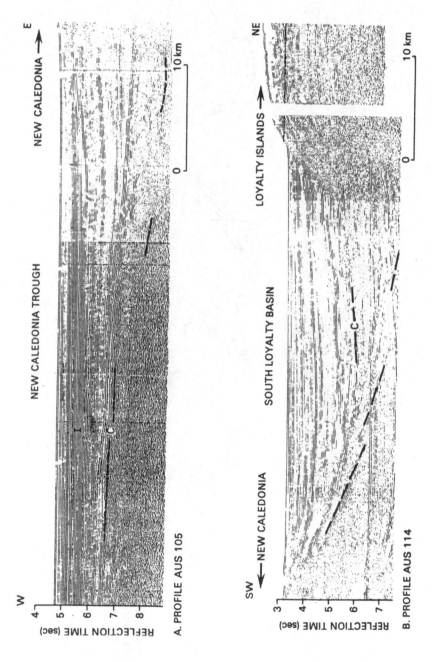

Fig. 5. Multichannel seismic reflection lines off New Caledonia. For location see Fig. 2. Line A (Profile AUS105) shows the sediment filling the New Caledonia Trough and thickening toward New Caledonia. Line B (Profile AUS114) shows sediment filling the South Loyalty Basin and thickening toward Loyalty Islands. Both profiles suggest northeastward-dipping subduction. (From Dubois et al., 1974, with permission.)

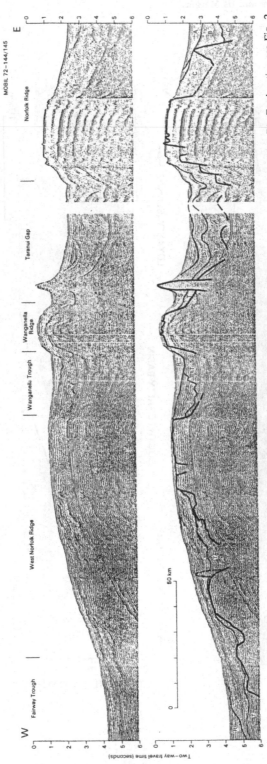

Fig. 6. Seismic reflection profile (Mobil 72-144/145) across the southern Norfolk Ridge System and ridges and troughs to the west (Line D). For location see Fig. 2. Upper profile uninterpreted. Lower profile interpreted: R, Miocene to Recent unit (turbidites in Fairway Trough); S, Oligocene "slumped" unit; T, late Cretaceous–Eocene "folded" unit overlying basement. S/T is the Regional Unconformity. A, B, and C are probable correlatives of R, S, and T, respectively.

Regional Unconformity, late Eocene to early Oligocene in age, seen at site 206 (Burns *et al.*, 1973). This is in contrast to Willcox *et al.* (1980) who interpret this unconformity to be the Maastrichtian breakup unconformity.

DSDP site 206 was drilled in Fairway Trough west of West Norfolk Ridge on a structural high which has been interpreted as a compressional feature formed about midway in the sedimentary history of the trough (Burns *et al.*, 1973). Only the pelagic sediments which cover the high were penetrated, the turbidite sequence only occurring beneath the flat floor of the trough surrounding the high. The Regional Unconformity, representing a hiatus in the late Eocene and early Oligocene, was penetrated. The entire sequence beneath the unconformity is characterized by disturbed bedding. Slump structures and disturbed bedding also occur just above the unconformity but most of the upper sequence is undisturbed.

A sedimentary sequence similar to that seen in seismic profiles in Fairway Trough is also recognized on West Norfolk Ridge (Fig. 6). The topmost sediments (unit R) are not well-bedded (and therefore assumed to be pelagic), and very thin. Beneath this unit, slumped sediments (unit S) are present mainly in depressions. The lower unit (unit T) appears folded on profiles, usually much more so than in Fairway Trough, and is often highly folded. It also occurs mainly in depressions. The folded sediments (unit T) lie directly on a well-defined basement reflector both in Fairway Trough and on West Norfolk Ridge. As a similar sedimentary sequence overlies basement both in Fairway Trough and on West Norfolk Ridge, it is suggested that they have a similar geological history including similar origin for their basements.

Basement is very close to the surface on both West Norfolk and Wanganella Ridges and may crop out in places. West Norfolk Ridge and its western flank consists of two or three longitudinal basement ridges with thin sediment cover (< 0.5 sec) separated by long, narrow, sediment-filled troughs (> 1.5 sec thick) (Fig. 6). Wanganella Trough, at its northern shallowest end, is filled with more than 2.0 sec of sediment. Magnetic anomalies reflect this structure, and strong positive magnetic anomalies (> 1000 nT) are associated with each basement ridge except the northern end of Wanganella Ridge where they are less than 500 nT (Fig. 4). These connect to the north with the major positive anomalies associated with Fairway Ridge (Lapouille, 1977).

Reinga Basin lies between Wanganella Ridge and Reinga Ridge at a depth of about 2000 m. Seismic profiles reveal more than 1 sec of sediment overlying basement (Davey, 1977) which dips eastward toward the New Zealand continental slope west of South Maria Ridge where sediments are more than 3 sec thick. This basement morphology is similar to that off the southwest coast of New Caledonia where asymmetric trench morphology has been recognized. The same three sedimentary units (R, S, and T) seen in reflection profiles in Fairway Trough and West Norfolk Ridge are also tentatively recognized for Reinga Basin. An upper largely undisturbed and essentially flat-lying unit unconformably overlies a folded and disturbed lower unit which overlies a strong basement reflector. Some disturbed bedding occurs in the lowest part of the upper unit especially in depressions on the unconformable surface. At its northern end, Reinga Basin narrows toward Taranui Gap (Fig. 2). At the Gap the Norfolk Ridge is less than 20 km from Wanganella Ridge. Beneath Taranui Gap there is 1.0–1.5 sec or more of highly folded and faulted sediments (Figs. 6, 7). Folding and faulting beneath the Gap appear to be most intense at the foot of the Norfolk Ridge where sediments are thickest and basement deepest. North of the Gap the sediments are less folded and the Gap broadens into the New Caledonia Trough. At the southern end of Reinga Basin the sedimentary sequence grades into that beneath the New Zealand continental rise in the Fairway Trough.

Crustal thickness has been measured at 10–16 km in New Caledonia Trough (Shor et al., 1971) and 9–16 km in Fairway Trough (Woodward and Hunt, 1971), and both studies conclude that these troughs are floored by oceanic crust. Hochstein (1973) has shown that oceanic-type basaltic rocks form the bulk of the West Norfolk Ridge and Wanganella Ridge, and that these rocks were extruded from extensive linear rifts similar to those of active midoceanic rises. Dubois et al. (1974) note the difference between acoustic basement on Lord Howe Rise and Fairway Ridge, and Ravenne et al. (1977) conclude that Fairway Ridge is of volcanic origin, possibly oceanic, in contrast to the continental origin for Lord Howe Rise.

All of the basement and the older sedimentary cover between the continental slivers of Lord Howe Rise and the Norfolk Ridge System are undulating and deformed in a manner similar to intraplate deformation of oceanic crust in the northeast Indian Ocean (Weissel et al., 1980). In the Indian Ocean the lithospheric deformation consists of broad undulations 100–300 km across with 1- to 3-km relief, high-angle faulting of acoustic basement, and folding of the sedimentary section. The scale of this deformation is very similar to that seen between Lord Howe Rise and the Norfolk Ridge System.

The work of Shor et al. (1971), Woodward and Hunt (1971), Hochstein (1973), Dubois et al. (1974), and Ravenne et al. (1977) and the similarity of Indian Ocean deformation and the character of basement and older sediments between Norfolk Ridge and Lord Howe Rise lead to the speculation that all crust between Lord Howe Rise and Norfolk Ridge is oceanic, including that beneath both Fairway–West Norfolk and possibly Wanganella Ridges. The origin of Wanganella Ridge is less clear, as at least in part it does not have a strong positive magnetic character like Fairway-West Norfolk Ridge but rather a weak positive anomaly similar to that of Norfolk Ridge. If this speculation is correct, then most of this crust was probably formed during the late Cretaceous and Paleocene extensional phase (Kroenke, 1984) by seafloor spreading to form a single basin, the proto-New Caledonia Basin. This protobasin was later deformed into the broad undulations which form Fairway–West Norfolk and possibly Wanganella Ridges, and other elongate structural highs in the New Caledonia and Fairway Troughs. This hypothesis is similar to that first suggested by Dubois *et al.* (1974) and later offered as an alternative by Ravenne *et al.* (1977) who thought the Fairway–West Norfolk Ridge to be a frontal bulge of oceanic crust in front of a subduction zone located to the east.

Weissel *et al.* (1980) speculate that in the northeast Indian Ocean a major tectonic event, the Himalayan Stage of collision of India and Asia, as recorded by a prominent unconformity in the sedimentary section, could have triggered intraplate activity in late Miocene time and produced the deformation. The major tectonic event which could have deformed the proto-New Caledonia Basin may be represented by the Late Eocene to Late Oligocene Regional Unconformity, which is seen in DSDP holes drilled in Fairway Trough and on Lord Howe Rise. The tectonism was most likely related to oblique collision of Australian and Pacific Plates at the Eocene plate boundary which lay along the western side of Norfolk Ridge (Kroenke, 1984).

IV. BASINS EAST OF NORFOLK RIDGE

Morphological and structural elements east of the Norfolk Ridge System generally parallel the ridge system (Figs. 1 and 2). In the north the Loyalty Ridge and intervening narrow trough, the South Loyalty Basin, closely parallel the ridge system. The South Loyalty Basin has a maximum depth of more than 3500 m in the north, shallows to almost 1500 m deep at 23°S off the southeastern tip of New Caledonia, and deepens to 3000 m at its southern end. Sediments in the basin are very thick, reaching 6 km between emerged land areas of New Caledonia and Loyalty Islands (Dubois *et al.*, 1974). Maximum sediment thickness occurs at the foot of the Loyalty Ridge (Fig. 5), suggesting asymmetric trench morphology and implying northeastward subduction beneath the Loyalty Ridge (Kroenke, 1984). Basement in the South Loyalty Basin is thought to be oceanic and Eocene in age, formed during the same event that formed the North Loyalty Basin where anomalies 23–18 (52–42 Ma) have been identified (Weissel, 1981; Kroenke, 1984). However, it is possible that South Loyalty Basin may have formed earlier as part of another, larger basin, a proto-Norfolk Basin which may have existed along the eastern margin of the Norfolk Ridge System from New Caledonia to Northland, New Zealand, and included the Kingston and Norfolk Basins (see below).

South of South Loyalty Basin the eastern margin of the Norfolk Ridge is bounded by Kingston and Norfolk Basins (van der Linden, 1967) (Fig. 2). Both basins are similar in size, shape, depth, subbottom character, and magnetic signature. Both are flat-floored at depths of 4000–4250 m. making them the deepest basins in the Southwest Pacific marginal basin complex, apart from Tasman Basin. Sediments, up to 1.0 sec thick in the center of each basin and thicker around the margins, overlie a distinct basement reflector (Fig. 7). The crust in Kingston Basin is 12–15 thick (Shor *et al.*, 1971) which although thicker than normal oceanic crust

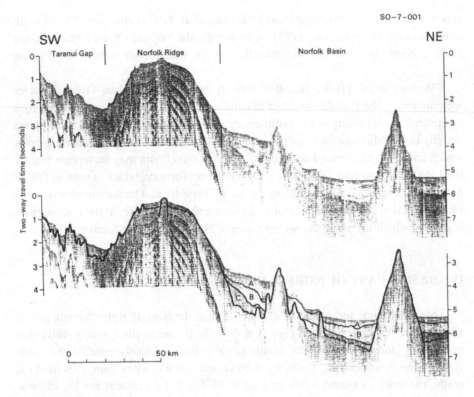

Fig. 7. Seismic reflection profile (SO-7-001) across the southern Norfolk Ridge System and Norfolk Basin to the east (Line E). For location see Fig. 2. Upper profile uninterpreted. Lower profile interpreted: A, B, and C are probable correlatives of R, S, and T, respectively, in Fig. 6.

is nevertheless considered to be oceanic. The magnetic anomalies in both basins are very subdued (less than 200 nT) (Hochstein and Reilly, 1967) and generally negative with respect to the International Reference Field (Launay et al., 1982). Launay et al. (1982) recognize anomalies 34 and 33 in Norfolk and Kingston Basins giving a 85–75 Ma (late Cretaceous) age for formation of these basins.

A late Cretaceous age is supported by the depth of the Norfolk Basin and by the age of abducted oceanic crust and sediments in Northland, New Zealand. Sclater et al. (1971) show that there is a direct relationship between water depth in a basin and the age of the underlying oceanic crust; i.e., the deeper the basin, the greater the age. Subsidence of the seafloor is accounted for by the thermal contraction of a cooling lithosphere as it moves away from a center of spreading or other areas of high heat flow. Using observations for basin depths in the South Pacific (Sclater et al., 1971, Fig. 2a; Molnar et al., 1975, Fig. 3), the Norfolk Basin at 4250 m would appear to have been formed at approximately 32 Ma. However, sediment in Norfolk Basin is approximately 0.75 sec thicker than that on the Pacific Plate from which Sclater et al. (1971) and Molnar et al. (1975) used data to construct their depth/age curves. If this extra sediment is taken into account, then the age increases to 65–70 Ma. Also, close proximity to plate boundaries active in the Cenozoic in the vicinity of Norfolk Ridge (Kroenke, 1984) and Three Kings Ridge (Kroenke and Eade, 1982) and the high heat flow probably associated with them, could have prevented Norfolk Basin from subsiding in a regular predictable manner and resulted in less subsidence than otherwise might be expected. Therefore, the basin could have been deeper than it is now and consequently could be older than the 65–70 Ma deduced.

A Cretaceous age is also supported by the existence of obducted oceanic crust and deep-sea sediments as old as 102 Ma in Northland, New Zealand (Brothers and Delaloye, 1982). These Cretaceous rocks were obducted from the northeast presumably from a basin Cretaceous in age. This basin, originally the southeast extension of Norfolk Basin, has largely disappeared but remnants can be found along the inner margin of Northland Plateau (Fig. 1) (Eade, 1983).

The original Cretaceous basin, which includes South Loyalty Basin, Kingston Basin, Norfolk Basin, and other basins to the southeast off Northland, could be remnants of a piece of Pacific Plate trapped between the Norfolk Ridge System and Three Kings Ridge. Three Kings Ridge is a west-facing arc (Kroenke and Eade, 1982) with seafloor being subducted from the west beneath the ridge and not the east as suggested by some authors (Watts *et al.*, 1976; Davey, 1982; Malahoff *et al.*, 1982). Three Kings Ridge was possibly the active arc behind which the South Fiji Basin formed as a backarc basin in the Oligocene. If so, then this Oligocene event could have isolated a piece of Pacific Plate between Three Kings Ridge and the Norfolk Ridge System. Much of this isolated plate appears to have been subducted beneath Three Kings Ridge, but remnants are possibly those preserved as Norfolk Basin and other basins along the eastern margin of the Norfolk Ridge System.

Between Kingston and Norfolk Basins, and Three Kings Ridge, the seafloor is much shallower (2000–3000 m) and more irregular than in the basins, with thin (< 0.5 sec) sediment cover and many basement peaks which are not buried. This area appears from seismic reflection records to be oceanic in origin but may be much younger (probably mid-Cenozoic) than the deeper basins (Norfolk and Kingston Basins) to the west. This area is complex and at present poorly understood. Much work remains to be done here.

V. TECTONIC HISTORY

The tectonic history of the Norfolk Ridge System consists of three major phases. The late Cretaceous to early Eocene was a time of extension and breakup of the eastern margin of Gondwanaland. The early Eocene to early Oligocene was dominated by oblique collision at a plate boundary which lay along the western margin of the Norfolk Ridge System. The early Oligocene to Recent was a time of inactivity along the Norfolk Ridge System except for sporadic intraplate volcanic activity.

At the time when rifting of the eastern margin of Gondwanaland began, the seafloor east of this continent was floored by Cretaceous oceanic crust. This seafloor was formed at the Pacific–Phoenix spreading ridge as it migrated rapidly to the southeast (Hilde *et al.*, 1977) and swept along what is now the eastern margin of the Norfolk Ridge and New Zealand at about 100 Ma (Bradshaw, 1983). Once the ridge had reached a point a little to the east of Chatham Island, it then propagated into the Tasman Basin (Bradshaw, 1983).

Extension and rifting of the eastern margin of Gondwanaland began about 89 Ma (Kroenke, 1984) and was followed later by seafloor spreading in Tasman Basin and in proto-New Caledonia Basin. Spreading began in both basins at about the same time (approximately 75 Ma, Kroenke, 1984) and continued in proto-New Caledonia Basin into the Paleocene and in Tasman Basin into the early Eocene displacing Lord Howe Rise and the Norfolk Ridge System to the northeast. By the early Eocene (approximately 53 Ma) all spreading has ceased and the Norfolk Ridge System, the easternmost sliver of continental Gondwanaland, had reached its maximum distance from Australia.

In the early Eocene the proto-New Caledonia Basin was a large essentially flat-floored feature, thinly covered by pelagic sediments, extending from the north near New Caledonia to New Zealand and from the eastern rifted margin of Lord Howe Rise to the western rifted margin of the Norfolk Ridge System. The remnants of this proto-basin are now represented by all seafloor, the ridges as well as the troughs, that lie between the Norfolk Ridge System and the Lord Howe Rise.

By middle Eocene time, a convergent/strike-slip boundary between Pacific and Australia Plates had been formed along the western margin of the Norfolk Ridge System (Kroenke, 1984). In the north near New Caledonia where the Ridge System trends NW-SE the plates collided and the Australia Plate was subducted beneath the Ridge System. During this collision tholeiitic basalts from the New Caledonia Trough were incorporated on to the southwest margin of New Caledonia to form part of the Formation of Basalts (Kroenke, 1984). Northeast dipping subduction may also have occurred beneath the southern section of the ridge system at Reinga Ridge where the ridge trends NW-SE and where asymmetric trench morphology is recognized. However, subduction may have been minimal here and much of the compression appears to have resulted in crustal shortening by a major crumpling of the floor of the proto-New Caledonia Basin. During this collision phase the central part of the basin floor was pushed up to form the Fairway–West Norfolk and Wanganella Ridges and produced considerable distortion of the overlying sediments (Fig. 8). This crumpling appears to have also produced other minor features such as the structural high in Fairway Trough on which DSDP site 206 was drilled (Burns *et al.*, 1973).

In the Eocene the pole of rotation of the Pacific Plate relative to the Australia

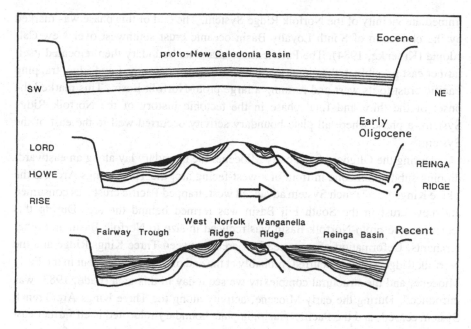

Fig. 8.   Diagrammatic representation of formation of West Norfolk and Wanganella Ridges. Upper figure: Eocene—seafloor spreading ceased, pelagic sediments accumulating on oceanic crust in proto-New Caledonia Basin. Middle figure: Early Oligocene—plate collision along western margin of Reinga Ridge, crustal shortening by folding and possibly thrusting to form midbasin structural ridges, slumping of sediment from ridges to troughs. Lower figure: Recent—irregular seafloor morphology smoothed by nondeposition of sediment on ridges and sediment deposition in troughs.

Plate lay at approximately 27.2°S, 153.6°W (Packham and Terrill, 1975). As a consequence, collision along the plate boundary was greatest where the Norfolk Ridge System trends NW–SE and least (mainly strike-slip) where it trends N–S. It was not surprising therefore that the ridges produced by the crumpling of floor of the basin are largest where collision was greatest in the northern and southern parts of the protobasin. In the central part of the area where ridges and troughs trend N–S, relative plate movement along the boundary was mainly strike-slip with some minor convergence (Kroenke, 1984) and therefore deformation of the floor of the proto-New Caledonia Basin was less.

   Throughout the Eocene tectonic event, both sediment and basement were deformed and as the ridges grew, sediments on their crests and side slopes were removed by slumping and redeposited in the adjacent troughs and basins. The resulting unconformity between these slump deposits and the underlying folded and faulted sediments, is the Regional Unconformity of Kennett *et al.* (1972), which was formed as a result of tectonism and not erosion of bottom waters as proposed by Kennett *et al.* (1972).

   At the beginning of the Oligocene, all plate boundary activity ceased in the immediate vicinity of the Norfolk Ridge System. The end of this phase was marked by the obduction of South Loyalty Basin oceanic crust southwest over New Caledonia (Kroenke, 1984). The Pacific/Australia plate boundary then relocated itself farther east possibly in the vicinity of the present position of the Lau Ridge, trapping Pacific crust to its west and forming a large proto-Norfolk Basin. This marked the onset of the third and final phase in the tectonic history of the Norfolk Ridge System, a phase where all plate boundary activity occurred well to the east of the System.

During the Oligocene and early Miocene, the boundary lay along an eastward-dipping subduction zone in front of a west-facing arc, the Three Kings Arc. As the Three Kings Arc/Trench System advanced west, trapped Pacific crust was consumed and new crust in the South Fiji Basin was formed behind the arc. During this tectonism the proto-Norfolk Basin was reduced in size to Norfolk Basin and other remnants. Deformation of some seafloor areas between Three Kings Ridge and the Norfolk Ridge System occurred, probably at the end of the tectonic event in the Early Miocene, and the structural complexity we see today in this area (Eade, 1983) was introduced. During the early Miocene, activity along the Three Kings Arc/Trench System ceased, and the Pacific/Australia plate boundary relocated itself farther east as the Tonga/Lau–Kermadec System.

The scattered Miocene to Recent basaltic volcanism on Norfolk Ridge and in the northern part of the Tasman appears to be intraplate activity not directly related to any plate boundary.

ACKNOWLEDGMENTS

I gratefully acknowledge useful discussion and comments from Dr. L. W. Kroenke, Hawaii Institute of Geophysics, Professor R. N. Brothers, University of Auckland, and Dr. K. B. Lewis, N. Z. Oceanographic Institute. Dr. J. Recy, ORSTOM, Noumea, kindly granted permission to use the ORSTOM data represented in Figs. 3 and 5. Use of seismic reflection data collected by Mobil Oil Corporation, USA, and Bundesanstalt für Geowissenschaften und Rohstoffe, West Germany, held on open file by DSIR, Wellington, is also gratefully acknowledged. Thanks go to Rose-Marie Thompson and Karen Jackson for typing the manuscript, and Chris Heaton for drafting the figures.

REFERENCES

Aziz-Ur Rahman, and McDougall, I., 1983, Paleomagnetism and paleosecular variation in lavas from Norfolk and Phillip Islands, Southwest Pacific Ocean, *Geophys. J. R. Astron. Soc.* **33**:141–145.
Bitoun, G., and Recy, J., 1982, Origine et Evolution du Bassin des Loyaute et de ses Bordures apres la Mise en Place de la Serie Ophiolitique de Nouvelle-Caledonie, *Trav. Doc. ORSTOM* **147**:505–539.

Blake, M. C., Brothers, R. N., and Lanphere, M. A., 1977, Radiometric ages of blueschists in New Caledonia, International Symposium Geodynamics South-west Pacific, Nouméa, 1976, Technip Editions, pp. 279–288.

Bradshaw, J. D., 1983, Cretaceous tectonics of New Zealand, *Geol. Soc. N.Z. Annu. Conf. 1983, Misc. Publ.* **30A**:38.

Brothers, R. N., and Delaloye, M., 1982, Obducted ophiolites of North Island New Zealand: Origin, age emplacement and tectonic implications for Tertiary and Quaternary volcanicity, *N.Z. J. Geol. Geophys.* **25**:257–274.

Burns, R. E., Andrews, J. E., van der Lingen, G. J., Churkin, M., Galehouse, J. S., Packham, G. H., Davies, T. A., Kennett, J. P., Dumitrica, P., Edwards, A. R., and von Herzen, R. P., 1973, Site 206, in: *Initial Reports of the Deep Sea Drilling Project, Volume 21*, U.S. Government Printing Office, Washington, D.C., pp. 2103–2195.

Davey, F. J., 1977, Marine seismic measurements in the New Zealand region, *N.Z. J. Geol. Geophys.* **20**:719–777.

Davey, F. J., 1982, The structure of the South Fiji Basin, *Tectonophysics* **87**:185–241.

Dubois, J., Guillon, J., Launay, J., Recy, J., and Trescases, J. J., 1973, Structural and other aspects of the New Caledonia-Norfolk area, in: *The Western Pacific: Island arcs, marginal seas, geochemistry* (P. J. Coleman, ed.), University West Australia Press, Perth. pp. 223–235.

Dubois, J., Ravenne, C., Aubertin, A., Louis, J., Guillaume, R., Launay, J., and Montadert, L., 1974, Continental margins near New Caledonia, in: *The Geology of Continental Margins* (C. A. Burk and C. L. Drake, eds.), Springer-Verlag, Berlin, pp. 521–535.

Dupont, J., Launay, J., Ravenne, C., and de Broin, C. E., 1975, Donnees nouvelles sur la ride de Norfolk, Sud-Ouest Pacifique, *C.R. Acad. Sci. Ser. D* **281**:605–608.

Eade, J. V., 1972, Wanganella Bank bathymetry, 1 : 200,000, *N.Z. Oceanogr. Inst. Chart, Misc. Ser.* **20**.

Eade, J. V., 1983, Morphology and structure of the seafloor off northern New Zealand, and a suggested tectonic evolution, *Geol. Soc. N.Z. Annu. Conf. 1983, Misc. Publ.* **30A**:38.

Hilde, T. W. C., Uyeda, S., and Kroenke, L. W., 1977, Evolution of the Western Pacific and its margin, *Tectonophysics* **38**:145–165.

Hochstein, M. P., 1973, Interpretation of magnetic anomalies across Norfolk Ridge, Southwest Pacific, in: *The Western Pacific: Island Arcs, Marginal Seas, Geochemistry* (P. J. Coleman, ed.), University of Western Australia Press, Nedlands, pp. 65–76.

Hochstein, M. P., and Reilly, W. I., 1967, Magnetic measurements in the South-west Pacific Ocean, *N.Z. J. Geol. Geophys.* **10**:1527–1562.

Hunt, T., 1978, Stokes magnetic anomaly system, *N.Z. J. Geol. Geophys.* **21**:595–606.

Jones, J. G., and McDougall, I., 1973, Geological history of Norfolk and Philip Islands, Southwest Pacific Ocean, *J. Geol. Soc. Aust.* **20**:239–254.

Kennett, J. P., Burns, R. E., Andrews, J. E., Churkin, M., Davies, T. A., Dumitrica, P., Edwards, A. R., Galehouse, J. S., Packham, G. H., and van der Lingen, G. J., 1972, Australian–Antarctic continental drift, palaeocirculation changes and Oligocene deep-sea erosion, *Nature Phys. Sci.* **239**:51–55.

Kroenke, L. W., 1984, Cenozoic tectonic development of the Southwest Pacific, *U.N. ESCAP, CCOP/SOPAC Tech. Bull.* **6**.

Kroenke, L. W., and Eade, J. V., 1982, Three Kings Ridge: A west facing arc, *Geo-Mar. Lett.* **2**:5–10.

Lapouille, A., 1977, Magnetic surveys over the rises and basins in the Southwest Pacific, International Symposium Geodynamics South-west Pacific, Nouméa, 1976, Technip Editions, Paris, pp. 15–28.

Launay, J., Dupont, J., and Lapouille, A., 1982, The Three Kingsridge and the Norfolk Basin (Southwest Pacific): An attempt at structural interpretation, *South Pac. Mar. Geol. Notes* **2**:121–130.

Main, W., and McKnight, D. G., 1981, Norfolk Island Bathymetry, 1 : 200,000, *N.Z. Oceanogr. Inst. Chart, Island Ser.*

Malahoff, A., Feden, R. H., and Fleming, H., 1982, Magnetic anomalies and tectonic fabric of marginal basins north of New Zealand, *J. Geophys. Res.* **87**:4109–4125.

Molnar, P., Atwater, T., Mammerickx, J., and Smith, S. M., 1975, Magnetic anomalies, bathymetry and late tectonic evolution of the South Pacific since the Late Cretaceous, *Geophys. J. R. Astron. Soc.* **40**:383–400.

Packham, G. H., 1982, Foreward to paper on tectonics of the Southwest Pacific Region, in: *The Evolution of the India–Pacific Plate Boundaries, Tectonophysics* **87**:(1–4):1–10.

Packham, G. H., and Terrill, A., 1975, Submarine geology of the South Fiji Basin, in: *Initial Reports of the Deep Sea Drilling Project, Volume 30*, U.S. Government Printing Office, Washington, D.C., pp. 617–645.

Paris, J. P., 1981, Geologie de la Nouvelle Caledonie: un essai de synthese, *Mem. Bur. Rech. Geol. Minieres* **113**:1–279.

Paris, J. P., and Lille, R., 1977, New Caledonia: Evolution from Permian to Miocene, mapping data and hypotheses about geotectonics, International Symposium Geodynamics South-west Pacific, Nouméa, 1976, Technip Editions, Paris, pp. 195–208.

Ravenne, C., de Broin, C.-E., Dupont, J., Lapouille, A., and Launay, J., 1977, New Caledonia Basin–Fairway Ridge: Structural and sedimentary study, International Symposium Geodynamics South-west Pacific, Nouméa, 1976, Technip Editions, Paris, pp. 145–154.

Ravenne, C., Dunand, J. P., de Broin, C.-E., and Aubertin, F., 1982, Les bassins sedimentaires du Sud-Ouest Pacifique, *Trav. Doc. ORSTOM* **147**:461–477.

Sclater, J. G., Anderson, R. N., and Bell, M. L., 1971, Elevation of ridges and evolution of the Central Eastern Pacific, *J. Geophys. Res.* **76**:7888–7915.

Shor, G. G., Kirk, H. K., and Menard, H. W., 1971, Crustal structure of the Melanesian area, *J. Geophys. Res.* **76**:2562–2586.

Summerhayes, C. P., 1969, Submarine geology and geomorphology off Northern New Zealand, *N.Z. Jl. Geol. Geophys.* **12**:507–525.

van der Linden, W. J. M., 1967, Structural relationships in the Tasman Sea and South-west Pacific Ocean, *N.Z. J. Geol. Geophys.* **10**:1280–1301.

van der Linden, W. J. M., 1968, Lord Howe Bathymetry, 1 : 1,000,000, *N.Z. Oceanogr. Inst. Chart, Oceanic Ser.*

van der Linden, W. J. M., 1971, Western Tasman Sea, geomagnetic anomalies, 1 : 200,000, *N.Z. Oceanogr. Inst. Chart, Misc. Ser. 19.*

Watts, A. B., Weissel, J. K., and Davey, F. J., 1976, Tectonic evolution of the South Fiji marginal basins, in: *Island Arcs, Deep Sea Trenches and Back Arc Basins* (M. Talwani and W. C. Pitman III, eds.), Maurice Ewing Series I, American Geophysical Union, pp. 419 427.

Weissel, J. K., 1981, Magnetic lineations in marginal basins of the Western Pacific, *Philos. Trans. R. Soc. London Ser. A* **300**:223 247.

Weissel, J. K., Anderson, R. N., and Geller, C. A., 1980, Deformation of the Indo-Australian plate, *Nature* **287**:284–291.

Willcox, J. B., Symonds, P. A., Hinz, K., and Bennett, D., 1980, Lord Howe Rise, Tasman Sea—Preliminary geophysical results and petroleum prospects, *BMR J. Aust. Geol. Geophys.* **3**:225–236.

Woodward, D. J., and Hunt, T. M., 1971, Crustal structure across the Tasman Sea, *N.Z. J. Geol. Geophys.* **14**:39–45.

Chapter 8

REGIONAL GEOLOGY OF NEW CALEDONIA

R. N. Brothers and A. R. Lillie

Department of Geology
University of Auckland
Auckland, New Zealand

I. INTRODUCTION

The main island of New Caledonia is elongated in a northwest direction (bearing 305°) and is some 400 km long and only 50 km wide, between latitudes 20 and 22°S and longitudes 164 and 167°E. A few lesser islands to the northwest and southeast belong to the same structural belt and the island group is surrounded by a barrier reef enclosing a shallow lagoon of some 15 km maximum width. The main island is very mountainous with plains restricted to valleys and to piedmont slopes on its western side, but the highest summits (Mont Panié 1628 m, Mont Colnett 1514 m) are very close to the east coast. The Loyalty Islands, lying some 100 km east of and aligned parallel to the main island, consist largely of uplifted coral reefs built on a foundation of Tertiary olivine basalt and dolerite and separated from the main island by a narrow marine trench exceeding 2000 m depth; thus, they seem to form a geologically distinct group (Guillon, 1974; Dubois *et al.*, 1974b).

Although very distant from any large land masses (about 1500 km from Australia, New Zealand, and New Guinea), the narrow island of New Caledonia has a very full range of thick marine and terrigenous strata ranging in age from Permian to Quaternary and some of these strata are highly metamorphosed to greenschists, blueschists, and eclogites. Basalt flows are extensive and huge masses of peridotite have yielded deposits of nickel for which New Caledonia was once the world's leading producer. The emplacement of the peridotites as a nappe and much of the metamorphism took place during the Late Eocene. These features and the geological relations of the territory to neighboring lands obviously require explanations, particularly the connection to New Zealand along the sinuous Norfolk Ridge and the close resemblances between the Permian and Mesozoic strata and faunas of New Caledonia and New Zealand.

II. BRIEF HISTORY OF RESEARCH

The first comprehensive account of the geology is that of Piroutet (1917) but other geological observations in New Caledonia date back to 1862 (Deslongchamps, 1863, 1864). Garnier's initial geological explorations (1865, 1868) are notable, and a number of scattered papers on minerals and mineral prospections published between 1862 and 1917 refer to the regional geology but none are based on very extensive field research. Lacroix worked intermittently on New Caledonian minerals and rocks from 1888 onwards but his classic memoirs on the glaucophanites (1941) and the peridotites (1942) describe materials collected by other workers. Davis (1925) was almost entirely concerned with geomorphology.

After 1917 a long period followed during which attention was mostly concentrated on mineral prospection with only a few reports on field mapping or fossil collection. Extensive field studies comparable to those of Piroutet were only resumed in 1946 when Routhier, Avias, and Arnould arrived on a geological mission sent by ORSOM (later ORSTOM) to map the country systematically and several colored maps on the scale of 1/100,000 were published. Subsequently, earth scientists attached to BRGM, CNRS, and ORSTOM published a great range of articles on New Caledonia, leading to a geological map series on the 1/50,000 scale which now covers almost the whole island. An index map (Fig. 1) lists the 1/50,000 map numbers and titles, and these are quoted in our text so that the positions of particular localities can be more easily found within the general geological map of New Caledonia (Fig. 2).

Lillie and Brothers (1970) summarized information and publications up to that date, but since then a considerable volume of new research has been recorded. Among this recent literature the most notable contribution has been the geological synthesis and discussion published by Paris (1981) as a memoir with a colored 1/200,000 map of the whole territory and a very extensive bibliography. The present authors have drawn extensively on the monograph by Paris (1981) for a description of the on-land geology, as well as using research data produced by the Geology Department at Auckland University particularly in metamorphic studies and Mesozoic stratigraphy.

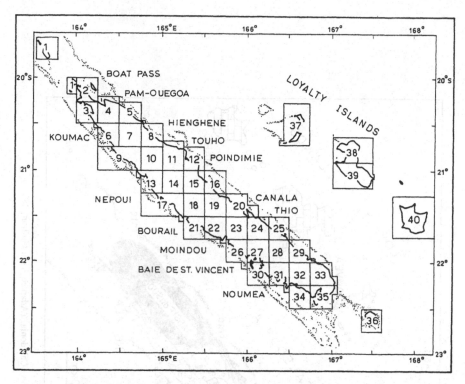

Fig. 1.   Index map of New Caledonia for 1/50,000 map numbers and titles. 1, Iles Belep; 2, Poum–Isle Yandé; 3, Paagoumène; 4, Pam–Ouégoa; 5, Pouébo; 6, Koumac; 7, Paimboas; 8, Hienghène; 9, Ouaco–Voh; 10, Goyéta-Pana; 11, Touho; 12, Poindimié; 13, Pouembout; 14, Paéoua; 15, Ponérihouen; 16, Baie Lebris; 17, Poya–Plaine des Gaiacs; 18, Mé Maoya; 19, Houailou; 20, Kouaoua; 21, Bourail; 22, Moindou; 23, Canala–la-Foa; 24, Thio; 25, Port Bouquet; 26, Oua-Tom; 27, Bouloupari; 28, Humboldt; 29, Kouakoué; 30, La Tontouta; 31, Nouméa; 32, St Louis; 33, Yaté; 34, Mont Dore; 35, Prony; 36, I. des Pins; 37, Ouvéa; 38, Lifou N; 39, Lifou S; 40, Maré.

III. STRUCTURAL OUTLINE

The total thickness of crust in New Caledonia is established by geophysics as 32–35 km (Crenn, 1952, 1953; Dubois, 1969; Collot *et al.*, 1979). It has been suggested by Dubois *et al.* (1974b) that a crust 35 km thick underlies the Central Chain and thins to 18–20 km below the present coasts. These estimates are entirely consistent with the great thicknesses of strata measured by field geologists.

Most of the large folds and faults follow the elongation of the island. This general strike of 110–130° appears to be the result of several orogenies producing structures mostly of similar trends, but of ages dated chiefly as pre-Permian, early Cretaceous, and late Eocene. Many of the folds are overturned toward the southwest and cut by thrust faults of parallel trend. Major folds and faults that depart from the

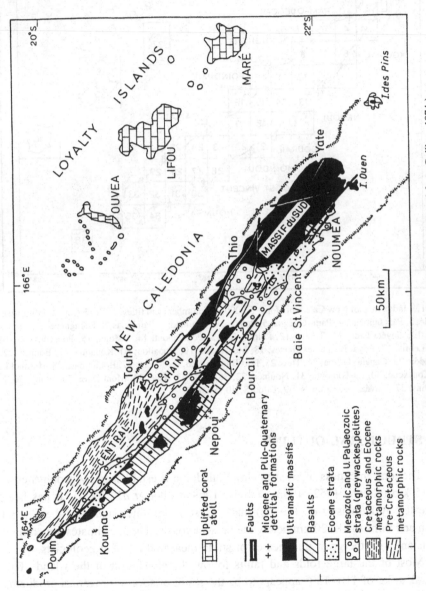

Fig. 2. Generalized geological map of New Caledonia. (Adapted from Guillon, 1974.)

general trend are found, however, in the complex ground west-southwest of Touho (map 11) and southwest of Houailou. In Late Miocene time a period of faulting and lesser differentiated tiltings profoundly affected the physiography. The main structural units (Paris, 1981), corresponding approximately with regional distribution, are as follows:

1. The Ultramafic Nappe, most clearly exposed in the Massif du Sud
2. The Central Chain
3. The Western Coastal Belt
4. The Northern Region
5. The Eastern Coastal Belt

A. The Ultramafic Nappe

In addition to the great southern mass northeast of Nouméa, for most writers the other peridotite masses along both western and eastern sides of the island belong to this once continuous nappe which formerly stretched as an overthrust sheet over most of the island before becoming disrupted by late faulting and erosion (Fig. 3). The relations of the Massif du Sud to the rocks of surrounding units, generally lower-lying, are unusually clear. On its western edge the peridotite sheet east of Nouméa (sheet 31) rests with overthrust contact on Upper Cretaceous or Lower Tertiary strata of the Western Coastal Belt. Cross sections through the Western Belt near Nouméa (Fig. 4) show all these later sedimentary formations as strongly overpushed southwestwards (Gonord, 1977; Paris, 1981) and within this coastal belt along the regional strike to Tontouta (sheet 30), overthrust slices of Tertiary and Cretaceous strata can be mapped in detail.

The general regional overpush toward the southwest is certain. Along the base of the peridotite sheet toward the east near Koua (sheet 27), about 28 km north-northwest of Tontouta, the serpentinized and sheared base of the ultramafic sheet dips at 20° eastwards, resting on Cretaceous and Tertiary beds which form the abbreviated unconformable stratigraphic cover to older Triassic strata of the Central Chain. Toward the east coast, the basal contact between peridotites and other beds steepens and near Nakety (sheet 24) and the promontory of Bogota (sheet 20), immediately north of Nakety Bay, the contact dips steeply north at 70°; the strata immediately south of the contact include Cretaceous sandstones and Tertiary lavas which are sheared along with serpentinite layers. Foliation and mineral banding within the peridotite are parallel to the dip of the contact (Fig. 5; Fromager, 1968). To the southwest this shear zone is flanked by Triassic and Liassic beds outcropping widely as part of the basement of the Central Chain. This is striking evidence that the peridotites at Bogota represent a steep root zone to the great nappe (Rod, 1974; Lillie, 1975) and is fully supported by Crenn's (1952, 1953) gravimetric and magnetic determinations which indicate that the zone of peridotites extends almost

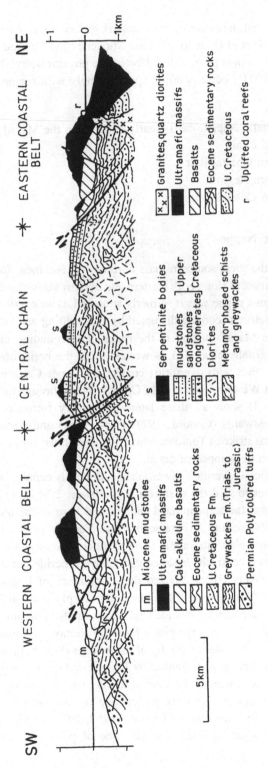

Fig. 3. Geological cross section of New Caledonia. (Following Guillon, 1974.)

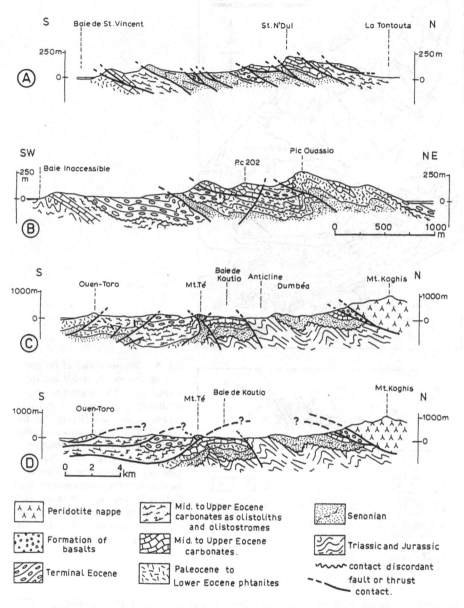

Fig. 4. Cross sections of the Western Coastal Belt. A and B: thrust sheets in the region of Tontouta–Baie de St Vincent (sheet 30); C and D: structure in the peninsula of Nouméa (sheet 31), west of the peridotite nappe, with C showing the autochthonist interpretation (Paris, 1981) and D showing the allochthonist interpretation (Gonord, 1977). (Adapted from Paris, 1981.)

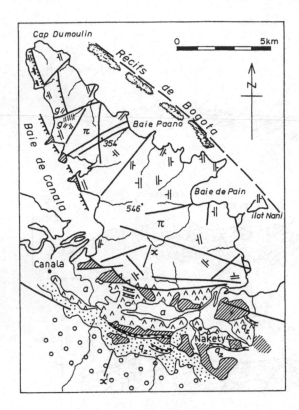

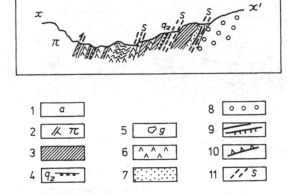

Fig. 5. Structural map of the peninsula of Bogota (sheet 20) and cross-section $x-x'$ of the contact of the peridotites against the sedimentary terrain between Canala and Nakety. 1a: alluvium; 2–5 are the ultrabasic association (with 2: peridotite mineral foliation, 3: serpentinites, 4: quartz "dikes," 5: gabbros); 6: Paleogene volcano-sedimentaries; 7: Cretaceous sandy pelites; 8: Jurassic graywackes; 9: normal fault with downthrow symbol; 10: thrust contact with teeth on upper slice; 11: schistosity plane. (Adapted from Fromager, 1968.)

vertically to some 8–9 km depth, confirmed later by Collot and Missègue (1977) and Collot *et al.* (1987). This root zone stretches southeast toward Thio and beyond to become continuous on the map with the overthrust sheet; it also follows the east coast northwest almost to Ponérihouen (sheet 15), but in this distance it is always separated by outcrops of Cretaceous or/and Tertiary basalts from the older Mesozoic, Permian, or pre-Permian strata of the Central Chain. Although the great peridotite mass does not extend further northwestwards, the Tertiary basalts and some Tertiary sediments can be traced to Hienghène, in places with small exposures of Cretaceous. It is likely that these east coast strata have a slightly different Cretaceo-Tertiary stratigraphy from those of the Western Belt so we have tentatively separated the stretch as a distinct structural belt; in this, we differ slightly from Paris (1981). It is agreed by all field workers that the great nappe was emplaced by a general southwesterly overthrust profoundly affecting all the structures west of its original position, in the Eastern Belt, Central Chain, and Western Belt. As yet the amount of repetition resulting from overthrusting has not been calculable. The age of overthrusting of the peridotite mass is fixed as late Eocene on precise stratigraphic evidence (see later).

The ultramafics in the southern massif are cut by major faults (Fig. 2) some of which extend beyond the massif into rocks of the Central Chain. Part of the Plaine des Lacs is due to subsidence along fault lines as extension followed nappe emplacement. One of many such faults extends from the Massif du Sud parallel to the regional trend into rocks of the Central Chain, dislocating the wide outcrop of Triassic and Jurassic beds and marked by a line of serpentinite lenses at Koindé, 10 km southwest of Canala (sheet 23). Such late faults are typical and follow many of the boundaries of formations and structural units; indeed, it is the multiplicity of such faults, diverging or in echelon, that makes mapping difficult. Following nappe emplacement, fault movements continued into post-Miocene time and affected the whole island, leading to widespread disruption of the ultramafics; subsequent erosion developed outliers or klippe with steep slopes and flat tops such as those nearest the Massif du Sud. Thus, the peridotite massifs of the east coast seem to represent a continuation of the root zone of Bogota. In addition, all the large peridotite masses of the west coast as far as Tiébaghi (sheet 3), and even the Isle Yandé (sheet 2), are regarded by most writers as also remnants of this one nappe.

Along the west coast the basal junctions of all the peridotites against older rocks, very commonly Tertiary basalts (see later), are everywhere sheared and faulted contacts which in places are mapped as steep, in others as flat, but mostly are very obscure due to downhill creep of serpentinites. Basalts are not abundant at the Massif du Sud, but other peridotites in both Western and Eastern Belts are so closely associated with the Tertiary basalts that these have been interpreted (Avias, 1967, 1977) as the former cover, inverted and allochthonous, of the whole ophiolite assemblage. But Guillon (1972, 1975), Gonord (1967), and Coudray (1976) after detailed research in the Massif du Sud and the western basalts, agree with Routhier (1953) that the Paleogene basalts represent flows formed within the Western Belt, along the "sillon" of Routhier (1953) (see Western Belt).

B. The Central Chain

To the south, the rocks of the Central Chain evidently plunge below the ultramafics of the Southern Mass. To the east they are bounded by the Senonian beds of the Eastern Belt, in many places in fault contact but also locally in true

stratigraphic discordant contact; nearby, in the Central Chain, Senonian conglomerates and sandstones cover stratigraphically the pre-Senonian volcanic–sedimentary
group. To the west, the boundary of the Central Chain is a complex fault zone
termed by Paris (1981) the West Caledonian Accident which is more or less coincident with the margin of the "sillon" of Routhier (1953) who seems to have used
this term to include, in addition to the fault zone, those Cretaceous sediments and
Tertiary sediments and basalts lying east of the Permian and Jurassic outcrops of the
west coast. To the north, the boundary of the Central Chain is described by Paris
(1981) as the zone of inverse faulting of Pouépai–Kokengoné which brings older
pre-Permian rocks against younger undifferentiated Mesozoic sediments of the
Northern Region. (Pouépai is in the interior sheet 10 and Kokengoné lies in sheet 12
on the east coast, south of Touho sheet 11.) The fault zone between these localities
runs very irregularly in an east–west direction, but to the northwest it turns into a
series of echelon faults dislocating and roughly parallel to the strike of the Cretaceous beds of the Northern Region. Likewise, some Cretaceous strata continue for
a long way down the western edge of the Central Chain, so the boundary is
indistinct in the northwest corner.

The Central Chain was identified as having a structural identity from the early
ORSOM mapping of Routhier (1953) and Avias (1953). The rocks of this belt were
identified as sediments, commonly volcanogenic, with occasional fossils identified
as Permian or Triassic or Jurassic, but many beds were and are still mapped
comprehensively as undifferentiated Permian–Lower Triassic or Triassic–Jurassic.
Metamorphic grade varies from prehnite–pumpellyite metagraywackes to greenschists, with local occurrences of lawsonite and glaucophane suggesting superposed
metamorphisms of different ages (Guérangé et al., 1973, 1975, 1976, 1977; Paris
and Lille, 1976, 1977).

Within the Central Chain two large areas of schist have been regarded as pre-
Permian by Bard and Gonord (1971): the region of Ouango-Netchaot stretching
east–west and bordering the Northern Belt south of Hienghène (sheets 9–14); and
the extensive massifs of Boghen, east and northeast of Bourail (sheets 18–20, 22,
23). The pre-Permian age was assigned because nearby Permian rocks are not
metamorphosed and have simple structure. However, according to Paris (1981)
identifiable Permian strata do not cover the schists stratigraphically and all known
contacts between the schists and neighboring rocks are faulted. Paris (1981) suggested that these schists may well be of Permo-Triassic age, but mapped them
provisionally as pre-Permian. Fossiliferous beds of Permian and Lower Triassic age
in the Central Chain are confined mostly to the southeast end and the bulk of the
strata elsewhere are mapped as Triassic–Liassic. Jurassic beds later than Liassic
have not been extensively found in the Central Chain although they occur northeast
of Poya (sheet 17) and near the east coast at Poindimié (sheet 12) and in a few other
small outcrops. Several large overturned folds aligned on a bearing of 110°, with
amplitudes of about 1 km, affect the Permian, Triassic, and Jurassic sequence.

Complexes of gabbros, dolerites, and basalts within the Central Chain (Lozes, 1975; Gonord, 1977; Guérangé *et al.*, 1977) have a pre-Senonian age of emplacement indicated by Cretaceous sedimentary covers (Gonord, 1972) and outcrop on sheets 15, 18–20, 23. Their chemical analyses, summarized by Paris (1981), have a generally tholeiitic character compatible with an origin at an active continental margin.

The Permian–Jurassic rocks of the whole Central Chain were strongly folded and faulted, along trends striking 110°, immediately before being covered discordantly by Cretaceous sediments. Other faults also indicate pronounced post-Cretaceous deformation, presumably accompanying late Eocene emplacement of the ultramafic nappe and late Miocene to Pliocene fracturing of that nappe. Thin Tertiary covering beds within the Central Chain are very shallow-water facies, indicating that the region was largely emergent from Cretaceous into Tertiary time.

C. The Western Coastal Belt

Figure 4 summarizes the structure of the south end of the western belt between Nouméa and Baie de St Vincent. Folded Triassic–Jurassic strata are overlain unconformably by shallow-water Cretaceous beds succeeded by Eocene conglomerates, siliceous shales, quartzites, limestones, and flysch. Upper Eocene flysch with strong unconformity covers the Lower Eocene and contains interbedded basalt flows and debris. The Tertiary beds of shallow-water origin show many lateral variations. Both the Tertiary and the Mesozoic basement are involved in strongly overturned folds and minor nappes cut by reverse faults and thrusts; within this fold complex are olistoliths (Gonord, 1970a) of Tertiary limestone, evidence of gravity-glide and tectonic disturbance during sedimentation. Tectonic interpretations by Paris (1981) and Gonord (1970a) recognize that emplacement of the ultramafic nappe was later than the terminal Eocene, but for Gonord (1970a) the Upper Eocene flysch had been thrust at least 8 km westwards from a position originally close to the peridotites.

North of Baie de St Vincent overthrusting within the Tertiary strata steadily diminishes and wide outcrops of Permian, Triassic, and Jurassic strata, and also of Senonian–Eocene beds, can be conveniently subdivided in relation to the Synclinal of Oua-Tom which is composed of folded and infaulted Cretaceous and Tertiary rocks. The syncline extends northwest from the head of the bay through La Foa and Moindou toward Col de Boghen and the component shallow-water Senonian and Eocene beds are spread inland from the west coast and form the extensive sedimentary basin of Bouloupari. To the Western Belt are assigned the moderately folded and faulted strata ranging in age from Permian to Jurassic west of the Synclinal of Oua-Tom and extending to the coast (sheets 22, 26, 27 and Fig. 6). Their stratigraphy and paleontology are very similar to beds of the same age in New Zealand. They have always been regarded as representing the marginal deposits of the ancient

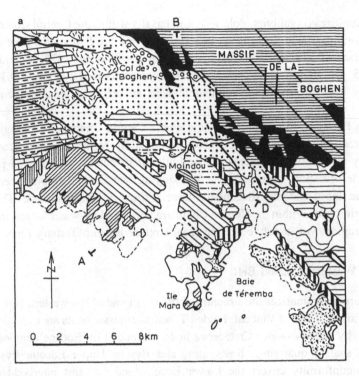

Fig. 6.   Geological map (a) of the region near Moindou (sheet 22) with (b) cross sections A–B, C–D. (From Paris, 1981.)

more extensive continent of Gondwanaland. All the beds—Permian to Jurassic— are of shallower-water facies than those of the Central Chain, but new evidence shows that the differences between the Permian–Jurassic strata of the two belts are not so great as formerly imagined. These older strata of the west coast are covered unconformably by the "Couches à Charbon" (Cretaceous) in a continuation of the Syncline of Oua-Tom. Near the Col de Boghen, Senonian and Tertiary strata discordantly cover the Cretaceous and west of the Col the Tertiary beds form the wide and thick flysch basin of Bourail. Immediately northwest of Bourail, Triassic– Liassic occupies the core of the Bourail–Cap Goulvain anticline and the flysch locally overlaps unconformably on these older beds (Fig. 7). The tightly folded, infaulted, and overthrust syncline of Oua-Tom continues northwestward without interruption along the whole length of the Central Chain, widening out to form stratigraphically significant outcrops northeast of Voh (sheet 9) where Senonian conglomerates rest unconformably on schists now mapped as pre-Permian. Further- more, the belt runs northwestwards without interruption into extensive outcrops of thick Cretaceous and Tertiary strata on the western side of the Northern Belt; thus, there is no sharp boundary between the Western and Northern Belts.

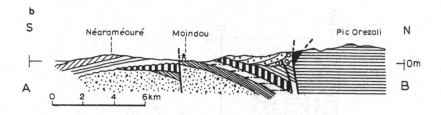

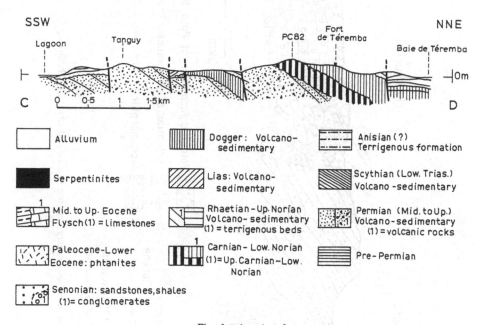

Fig. 6. (*continued*)

The Formation of Basalts

Near Nouméa some basalts are intercalated with the Middle–Upper Eocene sediments, but elsewhere other ages have been determined within the Western Coastal Belt for such volcanic flows which cover large areas from Bourail (sheet 21) to Koumac (sheet 6). At Koné (sheet 13) associated red and green jaspilites contain occasional *Inoceramus* (Carroué, 1972) suggesting a Senonian age, but nearby Coudray and Gonord (1966) described pillow lavas and basaltic breccia with a Paleocene microfauna. At Népoui (sheet 17), studies by Coudray (1969, 1976) showed that the basalts are overlain a short distance inland by ultramafics; however, to the west (Fig. 8) the coastal fringe of Upper Eocene flysch sequence covers serpentinites stratigraphically and consists of basal carbonates and clays, largely derived from weathered serpentinite, which are overlain by sandy limestones with

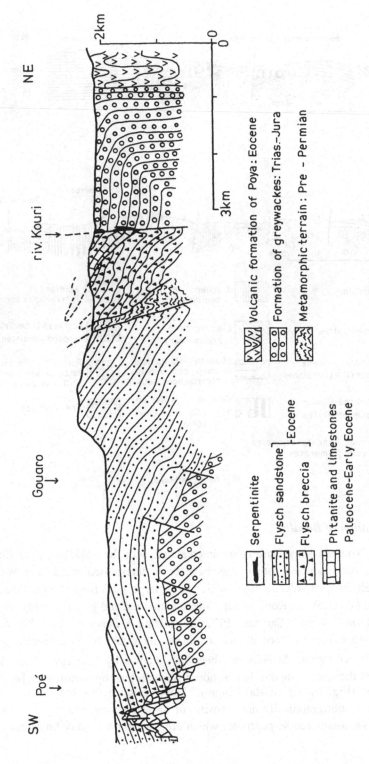

Fig. 7. Interpretative cross section in the sedimentary sequence near Bourail (sheet 21). (Following Espirat, 1971.)

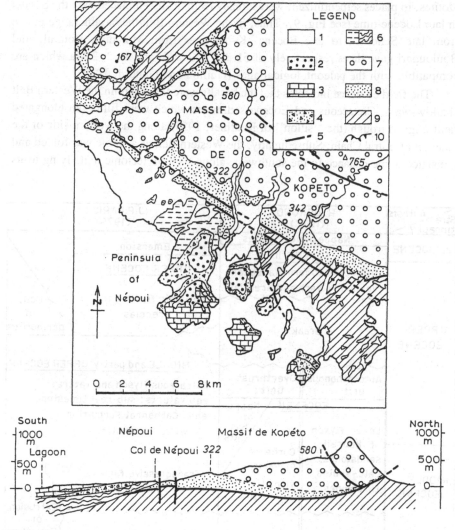

Fig. 8. Geological map and cross section for the region of Népoui (sheet 17). 1: Quaternary alluvium; 2: Upper Miocene–Lower Pliocene littoral sediment; 3: Lower Miocene reef carbonates; 4: Lower Miocene sandstones and conglomerates; 5: fault, or thrust contact (cross section); 6: Upper Eocene flysch; 7: peridotite nappe; 8: peridotite nappe serpentinite; 9: Formation of Basalts; 10: thrust contact (map) with symbol on upper unit. (From Paris, 1981.)

sandstone layers containing jasper and basalt debris; several of the beds have yielded quite rich late Eocene microfaunas. Marine Miocene conglomerates and sandstones overlying this Eocene flysch with very marked unconformity contain a basal conglomerate composed 99% of pebbles of peridotite. These observations show that the basalts of Népoui cannot be younger than Eocene and that the peridotites, in places serpentinized, were already tectonically emplaced over the basalts in late Eocene time (see Fig. 9). Thus, in the Western Belt the basalts range in age from late Senonian to late Eocene. K–Ar ages from Moindou, Bourail, and Bouloupari gave dates respectively of 51 ± 7, 59 ± 6, and 42 ± 2 Ma which are compatible with the paleontological ages (Guillon and Gonord, 1972).

The field evidence indicates that the basalts were flows within the Western Belt shallow-water Cretaceous and Eocene sediments which were filling an elongated fault-aligned trough (the "sillon" of Routhier, 1953) along the western side of the emergent Central Chain. Subsequently, the Western Belt as a whole was folded and translated some distance to the southwest, along with all tectonic units lying to its

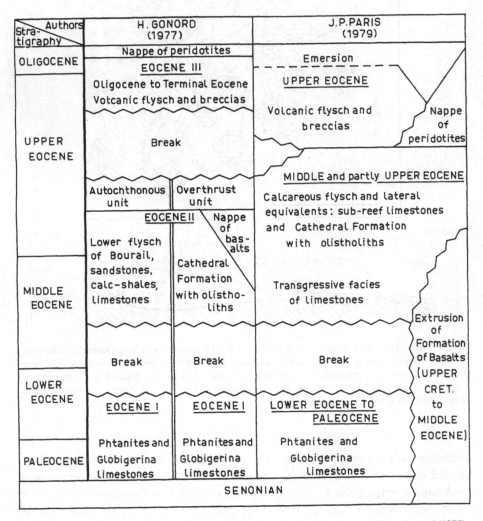

Fig. 9. Stratigraphic subdivision of Tertiary formations in New Caledonia, according to Gonord (1977) and Paris (1979). (Adapted from Paris, 1981.)

east, but the basalts are not allochthonous in relation to other rocks of the Western Belt. The peridotite massifs of the Western Belt sit above the basalt sheets from which they are separated by thick layers of mylonitized serpentinites which were developed at the base of the ultramafic nappe during its emplacement from northeast to southwest. Boninites near this tectonic contact have been described by Sameshima *et al.* (1983).

D. The Northern Region

The Northern Region lies within sheets 2–8 and following Paris (1981) this wide area can be described as two distinct sectors.

1. *The Western Sector (Sheets 3–7)*

The common rocks are phtanites (compact, cryptocrystalline silicified, fine shales or arenites) and fine limestones of the Paleocene and Lower Eocene, which east of Koumac form a wide outcrop of synclinal disposition between strips of Cretaceous shale, all steeply dipping. Southeastwards the Tertiary outcrop narrows considerably and passes without break within the zone of faulted Cretaceous beds flanking the western side of the Central Chain. Furthermore, at Koumac the Eocene beds of the Western Belt are continuous with those of the Northern Region, so that lithostratigraphic distinction between the three belts has become vague. Northwest of Koumac the Tertiary strata continue to the top of the island, surrounding some complexly faulted long inliers of Cretaceous rocks; similar outcrops of serpentinite occur along fault lines.

The eastern limit of this sector can be taken as a zone of faults followed from the Baie de Banaré, north of Poum (sheet 2) on the west coast, then inland SW of Ouégoa (sheet 4). The line continues southeastwards but becomes indistinct and undulous with the appearance of several irregular serpentinite outcrops among Cretaceous shales (sheet 7). Irregular sigmoidal boundaries of outcrop are found in this field where the mesoscopic folds are generally steeply plunging and irregular in trend. Many thrust faults, dipping to the southwest, are aligned along the masses of serpentinite and metabasalts regarded as dismembered ophiolites injected along a regional mélange zone which reaches the east coast south of Touho on sheet 11 (Brothers and Blake, 1973; Black and Brothers, 1977). In most of this structural block the major structures have more or less horizontal fold axes but near serpentinites and faults, and generally approaching the schists to the east, mesoscopic folds of steep plunge appear, as well as cleavage planes oblique to the bedding.

2. *The Eastern Sector (Sheets 4, 5, 7, 8)*

A well-developed belt of Tertiary high-pressure metamorphic rocks forms an axial range parallel to the eastern shoreline, with a maximum elevation of 1628 m at Mont Panié. The western boundary of this belt is the line of faults already described as trending southeast from Baie de Banaré and swinging east as a northern limit to the mass of older low-grade schists mapped as pre-Permian in the Central Chain. This fault system marks a tectonic junction between the high-pressure schists and the overlying mélange zone which formed a suprametamorphic cover (Brothers and Blake, 1973). Within the schist belt there is a progression, from southwest to northeast, from blueschists to eclogites so that high-grade polarity is toward the Pacific Ocean, as described later. As shown by Paris (1981), faults within the metamorphic sequence have a general northwest alignment, a trend which parallels the tectonic boundary along the western side of the whole schist belt.

The regional structure of the metamorphic complex has been described as anticlinal (Espirat, 1963; Lillie, 1975) from a variety of field evidence along the northeast high-grade coastal margin, including a Tertiary limestone lens within coarse gneisses (Arnould, 1958) and a downfaulted strip of low-grade lawsonite schists (Bell and Brothers, 1985). All of the paraschists show a foliation parallel to original bedding and this can be mapped along the belt for a distance of some 170 km with strike roughly between 120 and 130°, parallel to the length of the island. The dip is seldom less than 40°, quite commonly greater than 60°, and in places beds are overturned. From Tiabet in the far north (sheet 2) to Touho in the southeast (sheet 11), a feature in all the schists is the abundance of mesoscopic folds plunging steeply down the dip, or at slight angle to the dip, of the schistosity plane. The amplitudes vary from 1 cm to 5 m, or even much larger, and lineations commonly follow the fold axes. In many places throughout the northern region a second steeply dipping schistosity (S_2) cuts the other at some 15 to 25° dihedral angle and the intersection of the two forms the lineation which may locally become very close-spaced (Lillie, 1975). Near certain faults only, vergence patterns of the steep folds suggest that they are related to transcurrent movements on the planes of nearby faults but no regional vergence pattern has yet been detected for the whole mass of schists. The steep plunges of the abundant smaller folds may result from a general very pervasive wrench movement (Lillie, 1970, 1975) but alternatively they could have originated from the reorientation of folds and lineations, originally only rough-ly horizontal, as a result of crustal shortening with simple shear. Superimposed over all other folds and lineations at some places is a set of very late angular kink folds, with roughly horizontal axes and amplitudes ranging from 0.5 cm to 40 cm. They can be found over distances of a few kilometers with their axial planes parallel to nearly vertical joints, themselves cut by other joints roughly normal to them; the orthogonal pattern is conspicuously different from that of the older structures within the schists.

E. The Eastern Coastal Belt

The Eastern Belt contains the eastern margin of the Central Chain and the root zone of the Ultramafic Nappe which can be followed as a steeply dipping peridotite sheet thrust over Cretaceous–Eocene sediments and basalt flows. It seems likely that all of the separate peridotite masses along the east coast, from Canala (sheet 23) northwest to Poindimié (sheet 12), are outliers of this ultramafic root zone. The east coast basalts rest on Senonian beds and they are covered by the overthrust peridotite in the same manner as that in the separate Western Belt. Thus, the parallel Eastern and Western Belts have similar rock sequences and tectonic history.

IV. STRATIGRAPHY

A. Pre-Permian

The pre-Permian greenschists cover two large areas: the Ouango–Netchaot massifs in the north of the Central Chain (sheets 7, 10, 14), and the Boghen massifs in the middle of the chain (sheets 19, 22, 24). The schists are very constant in character, largely quartzo-feldspathic, often sericitic or micaceous, and contain scattered layers of calcschists. They seem to have been originally very fine tuffaceous sediments of acid and/or basic composition, and they enclose locally interbedded green volcanic rocks with pillow structures, partly altered to chlorite–epidote metabasalts. Bard and Gonord (1971) first suggested that the schists were pre-Permian, largely on the evidence of structures within the schists. Later writers in adopting this view (Guérangé *et al.*, 1973, 1975, 1976, 1977; Paris and Lille, 1976) noted also that the schists of Boghen were polymetamorphic, an idea also applied to the schists of Ouango–Netchaot by Carroué (1971, 1972); a first green-schist metamorphism had been shortly succeeded by a high-pressure metamorphism yielding lawsonite and glaucophane, these new minerals lying along the plane of foliation but showing no common alignment. The metamorphic effects were attributed to a Hercynian, perhaps Permian orogeny. Paris (1981) in critically reviewing the evidence stresses that it is entirely based on structural data. Apart from doubtful ovoid traces that may be radiolarian, no fossils are known and he refutes the view of Avias and Gonord (1973) that the schists are stratigraphically overlain by Permian beds near La Foa (sheet 23). However, he stresses that near the outcrops of Permian schist the adjacent fossiliferous Permo-Triassic strata show no signs of the polymetamorphism visible in the schist. Thus, the pre-Permian age is accepted provisionally. Nevertheless, in places the post-Jurassic and pre-Senonian green-schist metamorphism yielding pumpellyite and prehnite, first noted by Routhier (1953), has been recognized by later writers (see Paris, 1981, pp. 182–183) in the Central Chain and even in the Northern Region; but the Mesozoic rocks showing these later metamorphisms are said to differ distinctly in structural style from the pre-Permian schists.

B. Permian and Lower Triassic

These strata comprise two main assemblages. The lower part consists of the volcanic and volcanogenic sedimentary rocks to which Avias (1953) assigned the name *tufs polycolores* because of their varied conspicuous colors—green, violet, and red. These rocks are found both in the Central Chain, especially its southeast end (sheets 22–24, 27) where they are largely undifferentiated, and in the Western Belt, stretching from the Baie de St Vincent to the vicinity of Moindou, where they form exposures which have been mapped in detail allowing clear differentiation. Our descriptions are based largely on the latter outcrops (Fig. 6). The bands of tuff, ranging from tens of centimeters to a few meters in thickness, are interstratified with volcanic agglomerates and occasional pillow-lavas. The volcanic material varies in composition from andesitic to dacitic, including keratophyres, and the tuffs contain abundant phenocrysts of altered plagioclase. In some places, green calcareous silt-stones have yielded plant remains as well as other numerous fossils. The tuffs show locally signs of metamorphism with albite–oligoclase replacing more calcic pla-gioclase; chlorite and epidote have also formed in place (Paris, 1981). Campbell (1979) has found that this local metamorphism ranges from prehnite–pumpellyite to greenschist facies.

This lower part of the Permian–Triassic in the Western Belt has been studied by a number of authors including Piroutet (1917), Avias (1953), Avias and Guérin (1958), Pharo (1967), Lille and Paris (1976), Paris and Lille (1976), Waterhouse (1976), Grant-Mackie (1977), Campbell (1979), Campbell and Grant-Mackie (1984), Campbell and Bando (1985), Campbell *et al.* (1985), and Grant-Mackie (1985). The outcrops appear on the following maps on the 1/50,000 scale: Moindou (sheet 22), Oua-Tom (sheet 26), and La Tontouta (sheet 30) where the beds are shown as striking generally northwest with dips to the southwest ranging from 10° to 60°. But the exposures are discontinuous and cut by numerous faults; hence, although their aggregate thickness must be in the order of a few hundred meters, no attempt has been made to estimate the total. The map and sections (Fig. 6) of the Moindou region (from Paris, 1981) are characteristic of the regional sequence. A dike of rhyodacite cutting the Permian beds, described by Campbell (1979), was radiometrically dated as 214.3 ± 1.5 Ma (Lower Triassic). The names of the abundant fossils are summarized in Paris (1981); they consist mostly of brachiopods, bivalves, and gastropods, all identical or allied to New Zealand spe-cies so that very close correlation with the New Zealand stages is possible. Al-though it seems that only some of these stages have yet been recognized, namely Middle Permian and an incomplete Lower Triassic (Fig. 10), eventually many gaps will be filled in. The *tufs polycolores* strongly resemble the Permian Te Anau Group of New Zealand, but in New Caledonia the Permian has not yet been lithologically distinguished from the Lower Triassic. These volcanogenic sedimentary beds also form extensive outcrops in the Central Chain where massive fine tuffites are inter-bedded with subordinate phyllitic shales, the sequence being locally meta-morphosed to prehnite–pumpellyite schistose rocks. Also interbedded are meta-basalts and metadolerites, and this basic volcanism contrasts with the more acid lavas of the Western Belt. Fossils are very rare and consist only of *Atomodesma trechmanni.* The assemblage in the Central Chain is likely to include Triassic as well as Permian beds.

The upper part of the Permo-Triassic consists largely of siltites and arenites, rich in white mica and plant debris. It overlies the Lower Triassic (Scythian) volcanosedimentary beds which form the core of the anticline of Moindou (see Fig. 6) and it is covered by volcanosedimentary fossiliferous beds of Ladino-Carnian age; it is of inferred Anisian age although the only fossil recorded is a broken *Daonella* (?).

C. Middle Triassic to Upper Jurassic

These beds correspond approximately to the comprehensive formation of graywackes recognized by Avias (1953) and Routhier (1953) in the Central Chain and in the Western Belt alongside the Permo-Triassic. In the Central Chain few fossils were found and the sequence remained largely undifferentiated, but along the west coast between Nouméa and Moindou (sheet 22) locally abundant fossils made it possible to establish more complete successions. In the early literature, the beds of the Western Belt were regarded as shelf deposits and those of the Central Chain as mostly fine-grained deeper-water axial sediments. This model is now discarded as a result of work in the Central Chain by Guérangé *et al.* (1973, 1975, 1976, 1977), Lille and Paris (1976), Paris and Lille (1976), and Paris and Bradshaw (1976).

In the Central Chain, the Middle Triassic–Upper Jurassic beds consist of interdigitations in variable proportions of two main lithologies, namely beds of volcanoclastic origin, dominantly andesitic, and fine terrigenous beds containing much plant debris; the latter are like the plant beds of the English Jurassic and were previously, and mostly rather erroneously, designated as *couches à charbon.* One or the other type of lithology dominates at different stratigraphic levels but lateral passages of one facies to the other have been seen. These two facies are described briefly. The volcanosedimentary beds are always well stratified, consisting essentially of volcanics with little terrigenous material. Reworked tuffs dominate, ranging in texture from very fine to conglomeratic. Clastic minerals include plagioclase (An_{30-50}), quartz of rhyolitic aspect, pyroxene, epidote, and chlorite. The lithic fragments are largely dacitic and andesitic, rarely basaltic or quartz-feldspathic. Interbedded among the coarse beds are extremely fine tuff beds (cinerites). The terrigenous beds, or *couches à charbon,* although containing plant remains, do not include coal seams. They are commonly very fine, black, richly micaceous silts and interbedded sandstones (arenites), some beds of conglomerate and occasionally calcarenites. Both of these facies occur in the Central Chain and the Western Belt.

Based on these two main lithostratigraphic types, three megasequences are observed in the Western Belt (Paris, 1981, p. 56): a first sequence contains the Ladino-Carnian to Rhaetian; a second begins with the Liassic and ranges into the

Stratigraphic divisions			New Zealand stages	Faunal zones	Zones identified in New Caledonia	
Periods and dating	Stages	Substages Tozer (1967)		Ammonoid zones (Tozer, 1967)	Avias and Guerin (1958)	Campbell (1979)
Middle triassic — 220 Ma	Anisian		Etalian			
Lower Triassic	Scythian (werfenian)	Spathian	Malakovian	*Keyserlingites subrobustus* / *Olenikites pilaticus*	— ⌉ 1	— ⌉ 1
		Smithian		*Wasatchites tardus* / *Euflemingites romunderi*	⌉ 1	
		Dienerian		*Paranorites sverdrupi* / *Proptychites candidus*		
		Griesbachian		*Pachyproptychites strigatus* / *Ophiceras commune* / *Otoceras boreale* / *Otoceras concavum*	⌉ ? ⌋	

— 230 Ma

		Waterhouse (1976)	Waterhouse (1976)	Atomodesma zones (Waterhouse, 1976)	Waterhouse (1970)	Campbell (1979)
Upper Permian	Dorashamian	Ogbinian	Makarewan	Atomodesma sp. B		
		Vedian	Waiitian			
	Djulfian	Baisalian		Atomodesma trabeculum	Mt. Melpai	Téremba south of Moindou
240 Ma		Urushtenian	Puruhauan	Atomodesma trechmanni		
Middle Permian	Pendjabian	Chidderuan		Atomodesma woodi		
		Kalabagian				
	Kazanian	Sosnovian		Atomodesma aff. mitchelli		
250 Ma		Kalinovian	Braxtonian	Atomodesma obliquatum		
	Kungurian	Irenian				
		Filippovian	Mangapirian	Atomodesma marwicki, At. sp. A		
Lower Permian	Baigendzinian					
	Sakmarian		Telfordian			
280 Ma	Asselian					

Fig. 10. Stratigraphic subdivisions of the Permian–Lower Triassic and faunal zones identified in New Caledonia. (From Paris, 1981.)

Dogger; an important break occurs for Callovian to Kimmeridgian time and is succeeded by the third sequence entirely Upper Jurassic. In the Central Chain two sequences are recognized: a Triassic–Liassic and an Upper Jurassic ensemble, each formed of a volcanosedimentary part and a terrigenous plant bed sequence, passing laterally one into the other. On the whole the megasequences tend to terminate with terrigenous beds. Although these two main lithostratigraphic types are now recognized on both the West Coast and in the Central Chain the rocks of the two belts differ significantly. In the Western Belt a calcareous cement is common in the reworked tuff beds, stratigraphic thinnings and overlaps are common, as well as evidences of ancient ravining; fossils are in places very abundant. All these features are evidence of a littoral margin. In the Central Chain, the beds are thicker, fossils are rare, but above all the volcanosedimentary beds (including volcanic conglomerates) give evidence of a nearby volcanic source which must have lain to the east of the Central Chain.

1. The Western Belt

The Middle Triassic to Upper Jurassic stages have been studied mainly in the Western Belt where the fossiliferous Permian, Triassic, and Jurassic are commonly found in close juxtaposition, often faulted (Fig. 3). As a result, descriptions are of very local areas and few thicknesses are quoted. Because of considerable lateral facies-variation, any attempt to present a comprehensive Middle Triassic–Upper Jurassic lithological column would be misleading, but a list of key fossils is given in Fig. 11.

Near Moindou (sheet 22), the Ladino-Carnian, with a thickness of 800 m, is locally transgressive on the Permian and overlaps the Lower Triassic beds (Fig. 6) as a sequence of dacitic tuffs containing conglomerate bands. The fauna consists almost entirely of New Zealand species of brachiopods and mollusks (Campbell, 1979). The succeeding and lithologically similar Norian–Rhaetian is at least 700 m thick and is also richly fossiliferous. The Liassic, over 500 m thick, commences with conglomerates and again can be closely correlated with the equivalent New Zealand strata. Further south, in the bays of St Vincent and Dumbéa (sheets 27, 30, 31), the Upper Carnian rests directly on the Permian and is succeeded by the Lower Norian. Among many plant fossils a new dicotyledonous species, *Homoxylon neocaledonicum* Boureau, was discovered by Avias (1953), indicating that angiosperms had appeared as early as the Triassic.

In the Middle Jurassic on the west side of sheet 27, Pharo (1967) described fossiliferous tuffs and silts with a fauna similar to the New Zealand Temaikan stage (Dogger and perhaps Lower Callovian; Stevens, 1971), but an overlying set of beds commenced with a conglomerate and give a fauna similar to the Ohauan (Tithonic) of New Zealand (Paris, 1981). It would seem that a stratigraphic break lasting from the Callovian to the Tithonian is present in both New Caledonia and New Zealand.

2. The Central Chain

A very wide outcrop of Middle Triassic to Upper Jurassic beds is mapped as two megasequences. The Triassic–Liassic megasequence consists largely of the volcanosedimentary facies in which reworked tuffs contain remarkably little cement. Slumps and strongly graded beds, said to be characteristic of true graywackes, are completely absent. A number of new fossil sites have been discovered in recent years (Lozes, 1975; Paris *et al.*, 1979) particularly in the Thio region (sheet 24) and the continuous sedimentary sequence from Carnian to Liassic is clear. The shells of *Monotis*, very abundant in the Norian, are found in slabs where all are intact and lie parallel to the bedding plane (Lozes *et al.*, 1976); none of the features of the beds are characteristic of deep-water sedimentation such as was formerly suggested. Near Canala (sheet 23) the volcanosedimentary beds overlie an Anisian facies of terrigenous micaceous silts, arenites and black shales with abundant plant remains, and even very thin coal seams, which outcrops most widely west of Ponérihouen (sheet 15) where richly fossiliferous bands occur within very extensive outcrops of the Triassic–Jurassic (Guérangé *et al.*, 1975). Again, in no part of the whole Triassic–Jurassic sequence is there any evidence of deep-water sedimentation.

The younger megasequence, of Middle (?) to Upper Jurassic age, is present on both sides of the Central Chain and includes shales and terrigenous beds with *Malayamaorica*, *Inoceramus*, and plant remains (Lozes *et al.*, 1977; Arène *et al.*, 1979; Guérangé *et al.*, 1975).

D. Pre-Senonian Volcanoplutonic Complexes

In describing the Central Chain as a structural unit it was shown that conspicuous volcanoplutonic masses, of generally tholeiitic character, were covered unconformably by Cretaceous conglomerates. These bodies are evidence of Late Jurassic intrusion and volcanicity, immediately preceding, or contemporaneous with, the pre-Senonian folding (the Rangitata orogeny of New Zealand).

E. Upper Cretaceous

The Upper Cretaceous (Senonian) is always transgressive and discordant on older formations which are commonly folded by the late Jurassic pre-Senonian orogeny. As in New Zealand, no Lower Cretaceous fossils have been found, and the earliest Cretaceous beds appear to be Senonian. In places the sequences contain a transition from Cretaceous to Tertiary without any apparent break, so the existence of Maastrichtian must be assumed. The principal lithological ensemble consists of terrigenous conglomerates, sandstones, and coaly siltstones which have led to the common name "*formation à charbon*" but in only a few places could the beds truly be called coal measures. In the Northern Region and near Nouméa are interstratified

International stages	New Zealand stages		Key fossils for New Caledonia and New Zealand	Correspondence with characteristic international fossils
140 Ma				
Tithonic (?)	Oteke	Puaroan	Hibolithes arkelli grantmackiei	Aulacosphinctoides Uhligites
				Paraboliceras
				Kossmatia
Kimmeridgian	Kawhia	Ohauan	Inoceramus haasti	
Oxfordian		Heterian	Inoceramus galoi	Idoceras Epicephalites
Callovian				
Bathonian		Temaikan	Meleagrinella (cf. echinata, Inoceramus inconditus)	Macrocephalites
Bajocian				Brachybelus, Belemnopsis mackayi
Aalenian		Ururoan	Pseudaucella marshalli	
Toarcian				Dactylioceras, etc.

JURASSIC

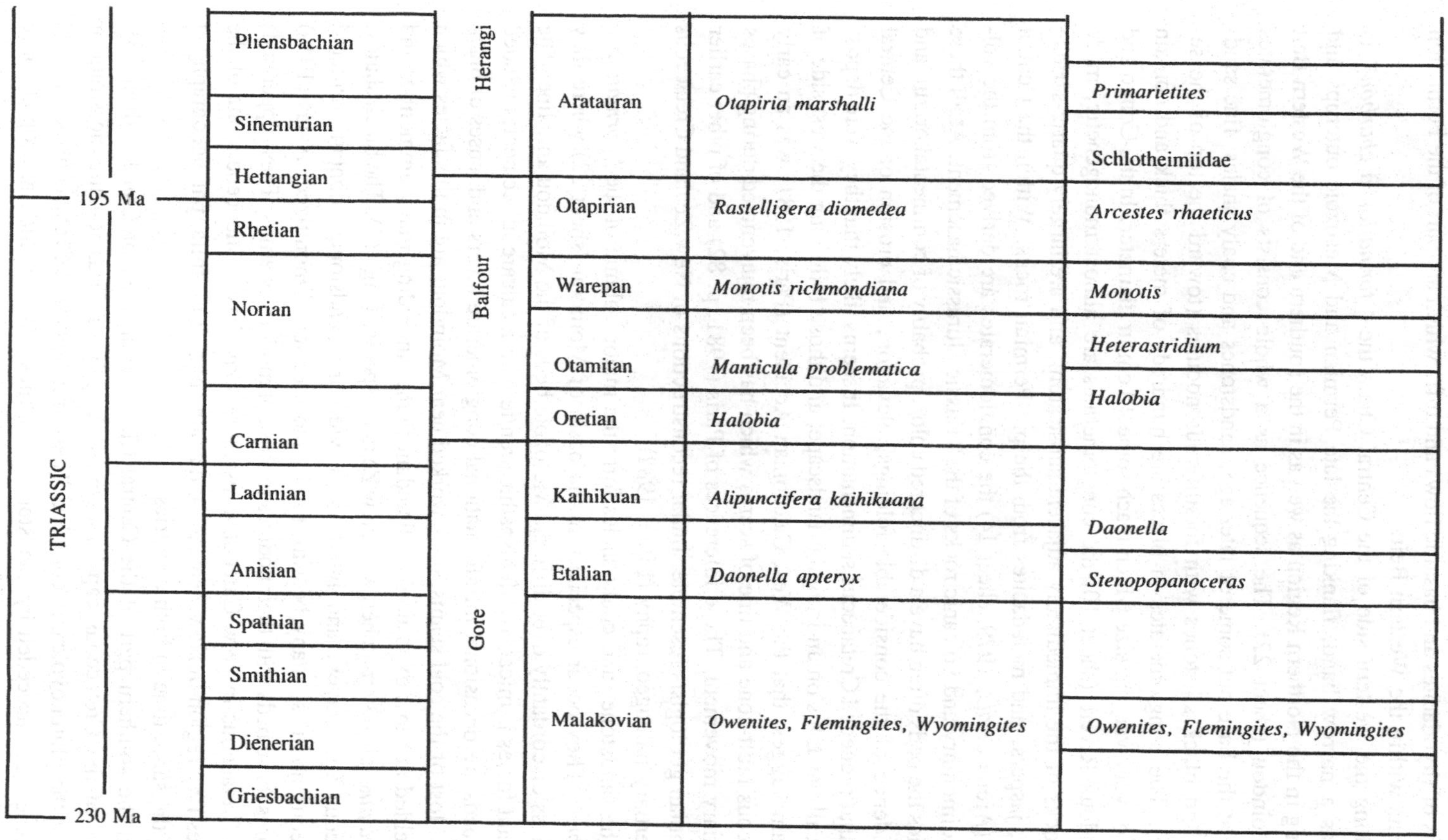

Period	Ma	International stage	NZ stage	NZ group	New Zealand key fossil	International key fossil
		Pliensbachian		Herangi		
		Sinemurian				
	195 Ma	Hettangian	Aratauran		*Otapiria marshalli*	*Primarietties*
TRIASSIC		Rhetian	Otapirian	Balfour	*Rastelligera diomedea*	Schlotheimiidae
		Norian	Warepan		*Monotis richmondiana*	*Arcestes rhaeticus*
			Otamitan		*Manticula problematica*	*Monotis*
		Carnian	Oretian		*Halobia*	*Heterastridium*
		Ladinian	Kaihikuan	Gore	*Alipunctifera kaihikuana*	*Halobia*
		Anisian	Etalian		*Daonella apteryx*	*Daonella*
		Spathian	Malakovian		*Owenites, Flemingites, Wyomingites*	*Stenopopanoceras*
		Smithian				
		Dienerian				*Owenites, Flemingites, Wyomingites*
	230 Ma	Griesbachian				

Fig. 11. Correspondence between the international stratigraphic scale of the Triassic–Jurassic and the New Zealand stages, with key fossils for New Caledonia–New Zealand and characteristic international fossils. (From Paris, 1981.)

horizons of rhyodacites and basaltic flows and tuffs which are part of the Formation of Basalts within the Western Belt.

Along the western side of the Central Chain the *"formation à charbon"* is found as a narrow band, flanking the late Permian and Mesozoic outcrops and widening in the Northern Region as well as in the southern end of the Western Belt near Moindou (sheet 22). The sequence as a whole consists of conglomerates mostly at the base but some of later age, sandstones and coaly shales, fine sandstones and black siltstones with fossiliferous nodules (toward the top of the sequence). The conglomerates in places are hundreds of meters thick and contain blocks of very varying size which reach some 40 cm in diameter. In the Congo and Kamendoua Rivers (sheet 10) the conglomerates are almost monogenetic, made from rocks of the immediately adjacent substratum, and are largely quartz-sericite schists, jaspers, and metadacites from the pre-Permian rocks. Within the Central Chain (Arène *et al.*, 1979; sheet 14) the conglomerates are derived from the volcanosedimentary and volcanic rocks of the Triassic–Jurassic basement. At all these localities the beds often have a distinct red color, probably of continental origin, and are evidence of the considerable foldings, elevation, and erosion of the Central Chain that preceded Cretaceous sedimentation. It seems likely that they were deposited locally as talus on an irregular landscape; and those following the west side of the chain suggest that the West Caledonian Accident (Paris, 1981) was an early Cretaceous fault-zone and line of scarps which has been rejuvenated during phases of Tertiary movement. These inferences of Paris (1981, p. 82) and of other earlier Caledonian geologists resemble similar reconstructions of New Zealand Cretaceous and Tertiary paleogeography (Lillie, 1982).

The sandstones and coaly shales form the major volume of the *"formation à charbon."* They occur especially in the basin of Moindou (sheet 22) where they transgress discordantly over different Mesozoic beds of the Moindou anticline. The dominant facies is micaceous feldspathic arenite with a calcareous cement; included in the beds are occasional intraformational conglomerate layers and lenses of black shale. Most of the coal seams, once worked near Moindou, are in this facies which has yielded at most exposures an abundant fauna, including many ammonites and *Inoceramus,* closely matched with New Zealand species (Fig. 12). The black shales, micaceous and clayey, and interbedded with fine sandstones, contain scattered calcareous nodules, many bearing fossils more or less limonitized; Avias (1949) compares these beds with their nodules to those of modern swamps. These upper beds of the Cretaceous in New Caledonia resemble some basal Paleocene strata of the Hawkes Bay region in New Zealand, and again suggest a similar paleogeography at nearly the same time in both countries.

In the southern part of the Central Chain, southwest of Canala (sheet 23), outliers of the Cretaceous cap buttes made largely of Triassic–Jurassic strata or rocks of the plutonovolcanic complex. These thin scattered Cretaceous beds, basal conglomerates succeeded by sandstones or siltites, have yielded gastropods once

International stages	New Zealand stages		Characteristic *Inoceramus* faunas
	H. W. Wellman (1959)	I. G. Speden (1975)	
Maestrichtian	Haumurian	Haumurian	*Inoceramus matatorus*
Campanian	Piripauan	Piripauan	*I. pacificus, I. australis*
	Teratan		
Santonian		Teratan	*I. opetius, I. nukeus*
	Mangaotanean		
Coniacian			
	Arowhanan	Mangaotanean	*I. bicorrugatus*
Turonian	Ngaterian	Arowhanan	*I. rangitira*
	Motuan		
Cenomanian			
	Urutawan	Ngaterian pp	*I. hakarius, I. fyfei*

Fig. 12. Stratigraphic distribution of *Inoceramus* faunas in the Upper Cretaceous of New Zealand. (From Paris, 1981.)

attributed to the Portlandian but now regarded as Senonian. Typical of such outcrops is Table Unio where Cretaceous beds are capped by Eocene limestone (western edge of sheet 23). These outliers within the Central Chain show that, although the early conglomerates accumulated as talus in a faulted and irregular landscape, the later shallow-water Cretaceous and Tertiary sediments overlapped the early beds to cover some of the higher ground.

In the Nouméa region the Cretaceous strata, overlying the Triassic–Liassic discordantly and overthrust by the nappe of peridotites, contain many tuff and lava beds, both basic and rhyolite, intercalated within fine sandstones, coaly shales, and conglomerates in a sequence topped by siltites with nodules. This early evidence of volcanicity resembles that seen in the Koné Formation farther north in the Western Belt (see Section III.C) and also in the Northern Region. In the Eastern Belt and below the peridotite thrust masses, the Cretaceous strata form thin undulous outcrops separating those masses from the pre-Permian or Triassic–Liassic beds that outcrop widely in the Central Chain. The strata, of the same lithologies as those on the western side of the Chain, have yielded less numerous but similar faunas, and rest unconformably with a basal conglomerate on the older Permian and Mesozoic strata. In the upper parts of the beds, siliceous shales appear as lenses and the succession seems to pass up without a break into the Eocene. In the Northern Region the Cretaceous sequence is some thousands of meters thick, is repeated by folding, and passes eastwards into the high-pressure schist belt (Briggs *et al.*, 1978). The beds are arenites and siltites or shales, with rare bands of dolomite and conglomerate, and show less varying lithologies than elsewhere. Within these beds are sills of dolerites and rhyolitic tuffs and lavas, as well as trachyandesites, showing volcanism analogous to˙ that seen in the Nouméa region. The least metamorphosed lawsonite schists contain a few *Inoceramus*. Westwards, the Cretaceous beds pass upwards without a break into Paleocene shales with interbedded siliceous argillites and limestones.

1. *The Koné Formation*

Strata of the Koné Formation extend along the west coast and are closely associated with the Formation of Basalts. They abut, always with a fault contact, against the other Senonian or Paleocene beds that flank the western side of the Central Chain. In several places *Inoceramus* have been found in the Koné Formation which consists largely of fine-grained tuffs, tuffaceous sandstones, and red-green claystones. Near Koné (sheet 13) similar beds are interstratified within the overlying basalts which, nevertheless, are mainly of Tertiary age (Carroué, 1972). Probably sedimentation continued from Cretaceous into early Tertiary time, as already noted in the Northern Region and the Eastern Belt, and the Koné Formation belongs to the highest Cretaceous stages.

2. *The Formation of Basalts*

The basalts have very extensive outcrops, especially along the Western Belt between Bourail (sheet 21) and Koumac (sheet 6), and as far south as Nouméa (sheet 31). In the Eastern Belt, the outcrops are not so wide but the rocks can be traced from Thio (sheet 24) northwest to beyond Touho (sheet 11). The age varies from place to place, ranging from Senonian to Eocene; basalt extrusion ceased before the late Eocene at Koné (sheet 13) but near Nouméa (sheet 31) flows are interstratified with late Eocene calcareous flysch (Gonord, 1967). As suggested by Routhier (1953) and Guillon (1975), the basalts of the Western Belt formed a great series of flows along a slight downward slope on the edge of a land that emerged in late Eocene time. Besides the scattered marine microfossils and macrofossils, pillow lavas occur at many places, supporting evidence of a marine origin. The basalts of the Eastern Belt probably originated in a similar but distinct groove to the east of the rock masses now forming the Central Chain which was emergent in early Tertiary time. The relations of the basalts to underlying Eocene and Senonian beds are clear, and it seems that peridotites always overlie the basalts but the thrust contacts are commonly not very evident. On the other hand, at the peridotite masses closely adjacent to the great Massif du Sud, contacts are exposed at a number of places and peridotites are thrust over the basalts as well as over adjacent rocks.

The Tertiary sedimentary rocks associated with the basalts are commonly red or green jaspers, often showing manganese stains, as well as green clays, tuffs, and fine-grained limestones containing Paleocene microfaunas; these beds are often folded. The basalts are mostly massive flows, but some show pillow structures and a few bombs have been found. The rocks are mostly augite basalts, much altered and weathered. Texture is coarse, generally doleritic, more rarely microlitic, and some of the rocks are even gabbroic. The plagioclase is usually near labradorite and may be locally altered to albite. The clinopyroxene is often partly replaced by green hornblende although primary amphiboles also occur. Olivine is very rare and a little interstitial quartz is always found. The petrology described by Challis and Guillon (1971), Rodgers (1975), and Gonord (1977) shows that rocks with a low content of K_2O fall in the category of oceanic tholeiites, but in some other respects the basalts are calc-alkaline; thus, two magmatic lines seem to be indicated. Paris (1981) concluded that they formed in an island-arc environment on the edge of an active continental margin.

F. Eocene

A new classification for the New Caledonian Eocene, involving three subdivisions, was proposed by Gonord (1967, 1970a,b, 1977) and Paris (1981) adopted an analogous scheme, as follows:

1. Paleocene–Lower Eocene: siliceous argillites (phtanites) and *Globigerina* limestones; in places the Lower Eocene is concordant with the Senonian and elsewhere is discordant on all older beds.
2. Middle and part of the Upper Eocene: transgressive and discordant, essentially calcareous and includes several facies of reef limestones, calc-shales, and flysch.
3. Terminal Eocene: transgressive and discordant on all earlier beds; characterized by flysch and breccia containing reworked volcanic rocks.

1. Paleocene–Lower Eocene

The strata occupy a wide area in the Northern Region and continue sporadically along the west coast to Bourail (sheet 21) and to the Nouméa district (sheet 31) where they are present in tectonic slices (Fig. 4). In the Central Chain, the beds form thin outliers and on the east coast they reappear commonly under basalts or peridotite. The dominant lithology, phtanite, is fine-grained, smooth, and hard, consisting of small quartz grains in a siliceous cement; the coarser rocks are fine-grained arenites. The phtanites are more than 2000 m thick in the Northern Region where they contain interbedded black shales and lensoid limestone beds, 100–200 m thick and 1–2 km long. The limestones are very fine-grained, almost sub-lithographic, and essentially microlites cut by thin veins of sparite; they form karst features and ridges especially at Koumac (sheet 6) and Hienghène (sheet 8). Their planktonic microfaunas range in age from Paleocene to Eocene. The character of these sediments indicates that the shallow-water lagoonal conditions of the late

Senonian were followed by a deepening and a marked decrease in detrital contributions, and the silica content suggests a source area of very mature relief (Routhier, 1953). It is likely that most of the present land was covered by water, probably deepest in the Northern Region and in the Eastern Belt and becoming shallower in the southwest where the coarser lithological types are found.

2. Middle and Upper Eocene (in Part)

This formation is essentially calcareous, in contrast to the Paleocene–Lower Eocene sediments, and shows very localized facies variations.

On the west coast, the principal outcrops are in the basins of Nouméa–Bouloupari (sheets 31, 27) and Bourail (sheet 21), and small outliers lie on the older rocks of the Central Chain. The shallow-water limestones contain algae, bryozoans, echinoid spines, debris of bivalves, globigerines, and other foraminifera including *Discocyclina, Asterocyclina, Amphistegina,* and *Chapmanina,* a key foraminifer of the Upper Eocene.

Near the basin of Bourail the beds occur in two different facies (Paris and Lille, 1976, 1977). To the north, sandy marls and calcareous sandstones rest on either Senonian or the Lower Eocene phtanites and contain abundant benthic foraminifera. To the south, the beds either cover the Senonian or overlap it to transgress directly onto pre-Senonian rocks; sandy glauconitic limestones with benthic foraminifera pass upwards into a flysch of calcarenites and microbreccias containing fragments of phtanite, tuff, basalt, and rhyodacite. The uppermost beds are marls with planktonic foraminifera. The northern facies records the margin of emergent land; the southern facies shows differential sinking after the original emergence and has a thickness of about 2000 m (Espirat, 1971).

The wide basin of Nouméa–Bouloupari has a Middle to Upper Eocene sequence of variable thickness which reaches a maximum of 200 m on the north side. Lithologies include basal glauconitic limestone of marginal-reef facies, sandy shales and marls, all strongly folded and with a rich fauna described by Grekoff and Gubler (1951), Tissot and Noesmoen (1958), Pharo (1967), and Gonord (1977). The sequence is interrupted by layers of conglomerate which, together with other evidence of intraformational erosion, indicate that these epicontinental beds were affected by vertical movements causing displacement of coastlines (Gonord, 1977).

In the Nouméa district (sheet 31) two facies of flysch form the Middle to Upper Eocene. The lower unit consists of alternating bands of calcarenite, microbreccia, and sandy marl with debris from phtanites and *Globigerina* limestone; it is similar to lithologies elsewhere in the Western Belt. Higher in the sequence, these beds appear more folded and broken and pass into the upper unit (the Cathedral Formation of Tissot and Noesmoen, 1958; Gonord, 1970a, 1977). This contains largely marls and shales which are strongly folded and enclose olistoliths of various dimensions as well as exotic blocks, many meters long, from neighboring paleoreliefs including deformed phtanite, micrites with *Globigerina,* bioclastic limestone, and arenite. The Cathedral Formation forms lenses separated by tectonic contacts; it rests transgressively and discordantly on Lower Eocene phtanites, and is covered by subsequent stratigraphic deposition of the terminal Eocene. Paris (1981) favors an autochthonist origin for the Cathedral Formation by synsedimentary subaqueous sliding of an original more coherent sequence, but Gonord (1977) regards the Cathedral flysch and its overlying terminal Eocene cover as emplaced together by tectonic translation from a root in front of the southwest-moving ultramafic nappe.

3. *Terminal Eocene*

The basins of Bourail (sheet 21) and Nouméa–Bouloupari (sheets 31, 27) contain extensive thin outcrops of the terminal Eocene and in both regions the dominant facies above basal conglomerate is a flysch consisting of rhythmic alterations of sandstone and microbreccia or breccia, plus minor calcarenites and siltstones. Clastic fragments are representative of older rocks (Triassic–Jurassic, Paleocene, phtanites, basalts) which the beds cover across an irregular eroded surface. Within the formation, sedimentary features include grading, slumping, currentbedding, and ravining. In the Nouméa district, the discordant contact between the terminal Eocene and the underlying folded carbonates of the Middle and Upper Eocene is pronounced. In the Bourail basin the carbonate flysch of the Middle to Upper Eocene shows a transitional passage into the flysch of the terminal Eocene which, farther north, transgresses beyond the other Tertiary beds to overlap on the paleostructure of the developing Bourail–Cap Goulvain anticline.

At Népoui (sheet 17) serpentinite at the base of the Kopéto peridotite massif is covered by several meters of arenite containing serpentinite debris and these beds pass laterally to a conglomerate with a benthic microfauna including *Discocyclina, Heterostegina,* and *Amphistegina.* The basal littoral strata are succeeded by fossiliferous limestones, dolomites, sandstones, altered volcanic ash, and claystones; the rich planktonic microfauna has been determined as Upper Eocene and the beds are mapped as terminal Eocene (Coudray, 1976; Paris *et al.,* 1979). Farther north on the coast at Koumac (sheet 6) are breccias containing serpentinite, jasper, and dolerite, and resting on basalts and on serpentinites; covering strata are alternating arenites and dolomites with chalky layers representing zeolitized ash beds. This sequence is recorded by Paris (1981) as terminal Eocene.

In general, during late Eocene sedimentation most of the island was emergent with littoral deposition confined mostly to the area of visible outcrops of the terminal Eocene. The basalts were being eroded and parts of the peridotite were already in place, with their lower flanks covered by the late Eocene sediments. It is not clear if the Massif du Sud was fully emplaced at this stage and most authors (e.g., Paris, 1981) believe that these peridotites moved southwestwards after the deposition of the terminal Eocene in the Nouméa basin, thus providing an explanation for the highly thrust nature of both the Middle and Upper Eocene sequence and the terminal Eocene, all of which are closely involved in the same pattern of deformation.

G. Oligocene

Sediments of this age are entirely absent and it is inferred that all of the land had risen and that peridotite emplacement was completed during this time.

H. Miocene

Outcrops of this age are few and small in the north of the main island and at the Belep Islands (sheet 1) and the Loyalty Islands, but they are important in fixing a lower age limit for peridotite emplacement. Near Népoui (sheet 17) conglomerates and fossiliferous sandstones, composed almost entirely of peridotite debris, cover the Eocene Flysch with angular discordance. Overlying, gently inclined calcareous sediments contain interdigitations of continental and marine beds, the latter being lagoon facies or coralline with corals in position of growth. Vertebrate bones have been found, and the fauna gives a Lower Miocene age, Aquitanian–Burdigalian ranging into Vindobonian. These beds, with paleosols showing recurrent and slight retreats of the sea, indicate a general marine incursion of approximately 100–150 m above present sea level which followed the post-Eocene emergence.

I. Neogene and Quaternary Movements

The four phases of post-Oligocene history follow the original interpretation of Davis (1925) with modifications based on Routhier (1953) and Coudray (1976), and including the ideas of Dubois et al. (1974a) from geophysics.

During Phase I the whole island was peneplained, the process starting in early Miocene time when the land was at a low elevation and continuing probably well into or beyond mid-Miocene time. The evidence is the existence of flat erosion surfaces, now high, cut across different rocks but best seen on the ultramafic masses which had become tectonically emplaced. The concentration of nickel at the base of the laterite horizons began in this phase (Avias, 1953) and may have been largely completed then.

In Phase II active erosion of the peneplain took place following great uplift of the old surface to heights reaching at least 1300 m (Dubois et al., 1974b) in the middle of the island, but less on the west side. Davis thought that the peneplain was tilted down along its southwest edge and up along its northeast edge. This upheaval was later than the Lower Miocene beds and probably contemporaneous with deposition of lateritic breccias which unconformably cover the Miocene of Népoui and which are dated as Mio-Pliocene (Coudray, 1976). Following the uplift, and contemporaneous with its later stages, was faulting that produced blocks carrying the dislocated peneplain remnants at different heights; in the fault-angle depressions terrestrial breccias accumulated (Orloff and Gonord, 1968; Gonord and Trescases, 1970; Coudray, 1976). This differential uplifting in places has continued until recently as shown by some river profiles. According to Dubois *et al.* (1974b) the differences in river profiles render it difficult to accept with confidence the idea of a first simple tilting of the whole peneplain (Davis, 1925).

Phase III, represented by subsidence with submergence of some valleys and the beginning development of the great barrier reef, especially along the west coast, was fairly late and the amount of movement differed from place to place. Good evidence of such subsidence is the Miocene limestone dredged from a depth of greater than 450 m, some 100 km south of the Ile des Pins (Daniel *et al.*, 1976). Even more striking evidence of subsidence and of coral reef growth is a drill hole at Tenia on the western barrier reef off the Baie de St Vincent, studied by Coudray (1976). This hole penetrated coral rock, associated reef limestones and sands, and lagoon sediments to a depth of 226 m before meeting phtanites, presumably Eocene (Fig. 13). These types of rock form four sequences or reef complexes, each type of sediment representing a particular reef environment within a complex (see Fig. 13). Discontinuities at the top of sequences, at -11, -40, and -105 m, represent marked periods of emergence due to eustatic oscillations interrupting a general subsidence of the reef basement. Radioactive dating of corals above the level of -11 m gives 125,000 $\pm$ 2000 yr, but at 18.3 m depth a date of 235,000 yr b.p.;

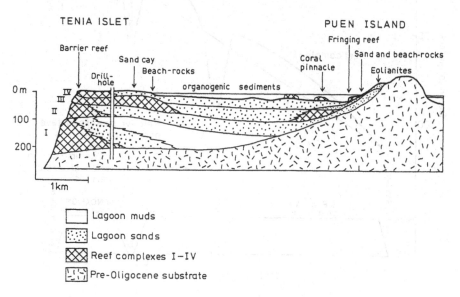

Fig. 13.   Interpretative section through the outer part of Baie de St Vincent, showing the four superposed reef complexes. (Adapted from Coudray, 1976.)

these dates, respectively within the Holocene and the Pleistocene, are confirmed by
the clear evidence of microfaunas (Coudray and Margerel, 1974). Microfaunas
from the base of the series are clearly Pleistocene; in consequence, with the upper
reef complex being certainly postglacial, the lower three complexes are interpreted
by Coudray (1976) as representing the warm interglacial periods, and the interrupting emergences correspond with the three glacial periods Mindel, Riss and Wurm
(Fig. 14).

Phase IV covers the Late Quaternary history concerned with recent rises and
falls of sea level, some certainly eustatic, leaving undercut cliffs, high beaches, and
some flat reefs. But certain raised reefs near Ile des Pins and other parts of the
southeast coast, reaching heights of 20 m (Fontes *et al.*, 1977) and dated radiometrically as between 120,000 and 200,000 yr b.p., are probably tectonically
upraised. Dubois *et al.* (1974a, 1975, 1977) calculate that isostatic uprise as a result
of unloading by erosion would not account for the amount of uplift of the New
Caledonian peneplain which must be the result of tectonics. In the southernmost of
the Loyalty Islands (Maré) the top of a coral reef is 130 m above sea level but
proceeding northwestwards the tops of raised reefs are seen to be lower until, in the

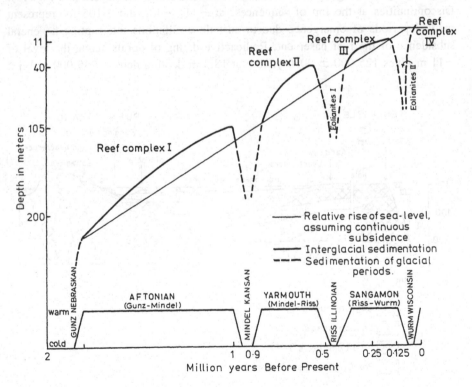

Fig. 14.   Climatic and chronological interpretation of reef and near-reef sedimentation peripheral to
New Caledonia during the Quaternary. (From Coudray, 1976.)

extreme northern atolls of the group, the reef is not yet uplifted. Those vertical differences can be interpreted to represent tiltings of the Loyalty Group which have, as yet, only slightly affected the Ile des Pins in southern New Caledonia.

From calculations based on such data plus geophysics, the authors argue that an epeirogenic wave started in the Quaternary affecting first Maré in the Loyalties, then the more northerly islands of the group, and now the southernmost end of New Caledonia; the amount of uplift to be seen varies with the age of movement. This wave is due to a lithospheric bulge of the Indian plate before its subduction at the very active east-dipping New Hebrides trench. The obliquity of this trench, trending 340°, in relation to the Loyalty Islands and New Caledonia, both trending generally 310°, partly explains why the greatest movement has been concentrated in the southern ends of both island groups. That this subduction zone dips toward the Pacific Plate and toward massive peridotite blocks on New Hebrides is a fact possibly very relevant in attempting to explain New Caledonian tectonics.

In addition to stratigraphic facts concerned with Late Tertiary and Quaternary tectonics just outlined, a great deal of precise sedimentological study is concerned with the Pleistocene and Holocene, including very recent sedimentation in lagoons and marshes. Some of this work describes the present processes of chemical erosion, and some the chemical alteration of rock and detritus in marine environments. These studies are outlined in the memoir by Paris (1981), but are omitted here.

V. REGIONAL METAMORPHISM

Paris (1981) has summarized the grades of metamorphism that have been recognized within the major age-lithological divisions in New Caledonia and on this basis three periods of regional metamorphism were defined: pre-Permian, post-Jurassic to pre-Senonian, and Tertiary. The first two events, in the older protoliths, are difficult to distinguish separately because they were dominantly low-pressure low-temperature in character (prehnite–pumpellyite to greenschist) and the metamorphic petrology of those regions has not been studied systematically, but Tertiary metamorphism of blueschist–eclogite grade in the northern region has been well documented.

A. Pre-Permian Metamorphism

The pre-Permian massifs of Ouango–Netchaot (sheets 7, 10, 14) and Boghen (sheets 19, 23, 24) contain foliated quartzofeldspathic schists, phyllites, micaceous quartzites with interleaved calc-schists, metavolcanic basic flows and tuffs, and acid tuffs. Greenschist assemblages are common, with quartz–albite–chlorite–muscovite–epidote, and local gradation into amphibolites with oligoclase–green hornblende has been described by Lozes (1975). The high-pressure phases lawsonite and glaucophane developed at a late stage in both para- and orthoschists, and do not show the preferred lineation that is common in the greenschist mineralogy. From structural evidence a pre-Permian age for both types of assemblage was deduced by Bard and Gonord (1971) and Paris (1981), whereas a post-Jurassic and pre-Senonian age was preferred by Lozes (1975), Guérangé *et al.* (1973), and Paris and Lille (1976). No additional data are available to determine the age of the greenschist event, but samples from the Ouango–Netchaot massif have provided whole-rock K–Ar dates of 128 Ma for glaucophane–epidote metabasalts and lawsonite–albite metagraywackes, and $^{40}Ar/^{39}Ar$ dates of 132–159 Ma for lawsonite–albite metagraywackes (Blake *et al.*, 1977). However, these ages may date only an Upper Jurassic greenschist assemblage rather than the later blueschist mineralogy (lawsonite–glaucophane) which is notably poor in potassium.

B. Post-Jurassic to Pre-Senonian Metamorphism

In contrast to the pre-Permian schists, metamorphites of this age lack pronounced foliation or isoclinal folding and dominantly are prehnite–pumpellyite metagraywacke in grade. In the Central Chain and along its northern borders, in Triassic–Jurassic volcanosedimentary terrains, tuffs and basic volcanics carry assemblages of quartz–albite–prehnite–pumpellyite ± muscovite ± chlorite ± calcite ± sphene. Associated terrigenous sediments are only weakly metamorphosed although locally some pelites and arenites reach low-grade epidote-bearing greenschist grade. Sporadic occurrences of lawsonite are always close to fault zones and appear to be the result of localized augmentation of tectonic pressure. In the Western Coastal Belt (Campbell, 1979) the pre-Senonian rocks similarly contain prehnite–pumpellyite associations with rare appearances of epidote.

C. Metamorphism of the Formation of Basalts

The Formation of Basalts (Senonian to Eocene) has a low-grade metamorphic mineralogy which variously contains quartz, albite, chlorite, prehnite, pumpellyite, epidote, and sometimes sphene or stilpnomelane. These assemblages are regarded by Paris (1981) as the products of hydrothermal submarine alteration at the time of extrusion, and thus not related directly to regional metamorphism of Tertiary age.

D. Tertiary Metamorphism

In the Northern Region, high-pressure schists cover an area of about 4500 km^2 in a coastal belt, 175 km long and 25 km wide, extending from Boat Pass to Touho and reaching elevations up to 1600 m (Figs. 15 and 16). Metamorphic polarity is toward the Pacific Ocean so that the highest grade rocks are located along the northeastern shoreline. The southwestern margin of the metamorphic belt is a regional mélange zone (Brothers and Blake, 1973) with a maximum width of 30 km,

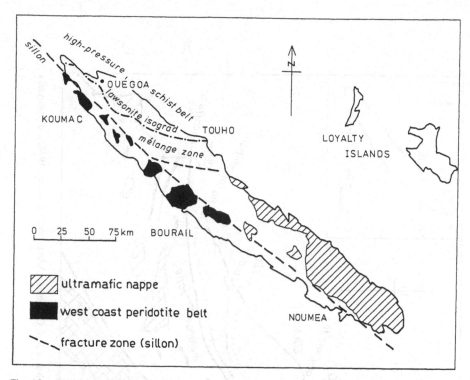

Fig. 15. Regional setting for the mélange zone and the Tertiary high-pressure schist belt within New Caledonia.

containing large schuppen of pre-Permian and Permian–Jurassic metagraywackes with overlying Cretaceous–Eocene sediments which were thrust northeastwards as a tectonic cover on the high-pressure schists during the mid-Tertiary metamorphism. The thrust slices range in metamorphic grade from prehnite–pumpellyite to low-grade greenschist and they are often separated by thin seams of serpentinite below which there is local development of narrow bands (up to 4 m) of lawsonite–albite schists. In places the mélange zone has been penetrated from below by large tectonically injected segments of ophiolitic ocean crust which have a high-pressure mineralogy of higher temperature and pressure origin, and of slightly older age (41 Ma), than the regional schist belt (Black and Brothers, 1977).

1. Metamorphic Protolith

The parent rocks were a contrasted association of Senonian–Eocene sedimentary and igneous lithologies, and the northeastern side of the belt is dominated by thick recrystallized units of shallow-water Cretaceous sandstones, siltstones, and carbonaceous claystones with minor calcarenites, cherts, and conglomerates. Igneous members were dolerite sills and basalt to rhyolite flows and tuffs, the acid

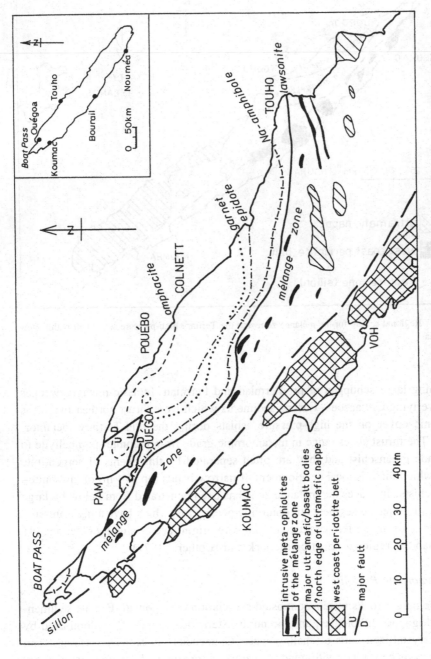

Fig. 16. Regional distribution of isograds within the high-pressure schist belt of New Caledonia. (Adapted from Brothers, 1974, and Yokoyama et al., 1986.)

volcanism being accompanied by stratiform Cu-Pb-Zn (-Ag-Au) mineralization which was later metamorphosed during the high-pressure event (Briggs *et al.*, 1977, 1978). Toward the southwest, on the low-grade side of the metamorphic belt, there is increase in the carbonate content of the sediments which pass upwards into conformable Eocene calcareous and siliceous lithologies; this level contains a number of fossil localities, including Cretaceous *Inoceramus* within lawsonite–albite metasediments. Arnould (1958) identified Eocene foraminifera in limestone within high-grade epidote gneisses on the east coast, indicating a broad closure in the regional structure. From stratigraphic evidence the age of metamorphism must be at least as young as the Eocene, and K–Ar dates within the schist belt at 39–36 Ma are the uppermost Eocene (Blake *et al.*, 1977).

2. *Metamorphic Zones*

Along the length of the metamorphic belt there is a progression, from SW to NE, from fully recrystallized lawsonite–albite schists through blueschists to eclogites that can be traced in sedimentary and igneous lithologies. The sequence of main metamorphic zones is defined in metasediments by regional isograds for lawsonite, Mn-rich garnet, epidote, and omphacite (Fig. 16). In the northern part of the belt, additional isograds have been described for ferroglaucophane, fully or-dered graphite, lawsonite-out, Mn-poor garnet, and various sliding equilibria com-positions in almandine and phengite (Brothers, 1974; Briggs *et al.*, 1978; Diessel *et al.*, 1978; Brothers and Yokoyama, 1982; Yokoyama *et al.*, 1986). The silicate mineralogy has been studied in considerable detail by Black (1970a,b,c, 1973a,b, 1974a, 1975, 1977), Briggs (1975), and Ghent *et al.* (1987a,b) as well as the paragenesis of matrix oxides and sulfides (Itaya *et al.*, 1985), in all lithologies from the fine-grained lawsonite–albite schists to the coarse eclogites on the east coast.

Oxygen isotope temperatures (Black, 1974b) increase across the isograds from 250°C for lawsonite to 340°C for Mn-poor garnet, 390°C for epidote, 430°C for omphacite, and 550°C in the deepest-exposed portion of the omphacite zone. Re-constructions of the metamorphic thermal profile (Brothers, 1974; Diessel *et al.*, 1978) identified low gradients of 7–10°C/km across the lawsonite and Mn-poor garnet zones, steepening to more than 30°C/km across the epidote zone. Subsequent mapping of almost flat-lying isogradic surfaces for epidote and omphacite in the deeper part of the belt, within the elevated coastal mountain chain, allowed accurate measurement of these high-grade zone thicknesses which gave a thermal gradient near 50°C/km (Yokoyama *et al.*, 1986). However, from structural evidence Bell and Brothers (1985) show that the lower-grade lawsonite and Mn-garnet isogradic surfaces are flexed and stand at high angles.

3. *Deformation and Metamorphic Crystallization*

Microstructural and textural studies (Bell and Brothers, 1985) clearly define two stages of deformation–crystallization within which the sequence for critical mineral associations shows the metamorphic development of the schist belt. The first two phases of deformation (D_1-D_2) were noncoaxial with a large thrust component and affected the whole metamorphic sequence under conditions of temperature and pressure prograde crystallization (Fig. 17); this is correlated with northeastward overthrusting of the mélange zone as a regional suprametamorphic cover. Subsequent D_3-D_4 folding involved bulk horizontal shortening, with concurrent thickening of the metamorphic pile, but this deformation was overprinted on D_1-D_2 microstructures only within the epidote and omphacite zones.

D_1-D_2 mineralogy includes assemblages in the lawsonite and Mn-garnet zones, and at deeper levels the cores of albite porphyroblasts and the MnO-rich centers of almandine garnets; most importantly, crystallization of omphacite was confined to the D_2 phase. D_3-D_4 was a period of pressure-retrograde temperature-prograde conditions (Fig. 17), with depressurization initiated by upfolding elevation of the schist belt, flexing of the D_1-D_2 lawsonite and Mn-garnet zones, and erosion of the metamorphic overburden. Concurrently, within the deeper part of the pile,

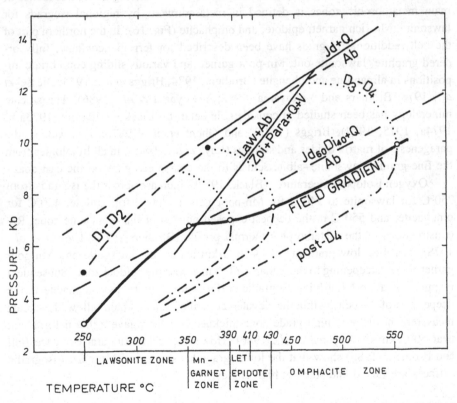

Fig. 17. Metamorphic paths during deformation–crystallization in the time sequence D_1-D_2, D_3-D_4, and post-D_4 relative to the preserved field gradient. LET is the lawsonite–epidote transition subzone. (From Bell and Brothers, 1985.)

the D_1-D_2 lawsonite eclogites began pressure-retrogressive recrystallization with the conversion of lawsonite to epidote and of omphacite to porphyroblastic albite, so that the two zones indexed by these two minerals (epidote-in and omphacite-out) were developed at this late stage. Two additional processes took place in D_3-D_4 time across the developing epidote and omphacite zones. All rock units penetrated by D_3-D_4 microstructures underwent annealing recrystallization and matrix decarbonization, such that their assemblages lack evidence of strain and mineral rims do not contain graphite inclusions. Additional independent evidence from oxide and sulfide mineral chemistry within the epidote and omphacite zones (Itaya *et al.*, 1985) indicates concomitant changes in the metamorphic fluid (gas) phase by decrease in the mole fraction of H_2O and increases in CO_2, CH_4, and H_2S, suggesting a link between the three concurrent processes of depressurization, volatilization, and decarbonization which were accompanied by an absolute increase in metamorphic temperature. Detailed discussion of the metamorphic path and field gradient is given by Bell and Brothers (1985) and Brothers (1985).

4. *Regional Tectonics*

Plate boundary tectonics during the mid-Tertiary (Fig. 18) seem to correlate most satisfactorily field relationships for the coeval processes of high-pressure metamorphism and ultramafic obduction (Brothers, 1974; Briggs *et al.*, 1978; Paris, 1981). The interface for collision between the Indian and Pacific plates in the

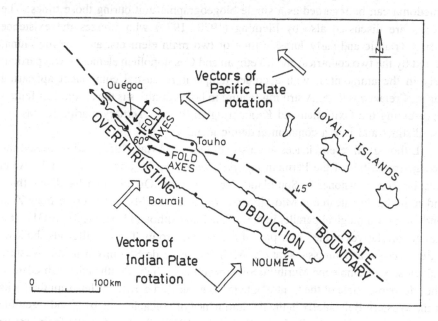

Fig. 18. Suggested vectors for contrasted plate rotations during evolution of the mid-Tertiary plate junction, or intraplate suture, in New Caledonia. (From Briggs *et al.*, 1978.)

late Eocene behaved like a transform system and the suture line, which traversed the island as the mélange zone, appears not to have been a subduction–convergence site because Cenozoic volcanic arcs and regional plutonism are absent. In the northern part of the island, northeastward overthrusting of basement sialic crust onto a fringing Cretaceous–Eocene marine basin created the location and conditions for high-pressure metamorphism and this tectonic phase corresponded to D_1–D_2 deformation within the schist belt. In the southern part of the island, southwestward obduction of Pacific ocean crust and upper mantle emplaced the ultramafic nappe, but there is no evidence of metamorphism below the base of this peridotite sheet. It is possible that the ultramafic cover originally extended across the north end of the island as an additional high-density unit in the metamorphic overburden of the high-pressure schist belt, but field evidence to support this concept is lacking since perched peridotite erosional remnants on the mélange zone are absent north of Touho.

VI. GEOLOGICAL RESEMBLANCES BETWEEN NEW CALEDONIA AND NEW ZEALAND

All recent research in New Caledonia has confirmed the early impressions of close resemblances of the late Paleozoic and Mesozoic faunas to those of New Zealand. Paleontologists agree with Stevens (1977) that New Zealand and New Caledonia can be regarded as a single biogeographic unit during those times. The faunas are discussed also by Fleming (1970, 1979) who stresses the existence during Triassic and early Jurassic time of two main elements: an endemic fauna, shared by the two countries; and a Tethyan and Cosmopolitan element, seen particularly in the ammonites, with a colder temperature austral component appearing during Cretaceous time. A striking feature in both countries is the absence of faunas representing the Oxfordian and Kimmeridgian Stages. Tertiary marine faunas are less distinct and lack a common endemic identity.

Lithological resemblances are also very close. Avias (1953) early stressed the strong similarities of the Permian *tufs polycolores* to the Lower Permian Te Anau breccias and sandstones of Southland, New Zealand. On the other hand, the thick and remarkably calcareous Mid-Upper Permian of the Maitai Group in New Zealand has no lithological parallel in New Caledonia although Paris (1981, p. 31) cites faunal correlatives of the lower part of the Maitai Group. The fossiliferous shallow-water Triassic and Jurassic strata of Moindou–Baie de St Vincent in the Western Belt closely resemble the Murihiku Supergroup of both the South and North Islands. The Mesozoic strata of the Central Chain, as explained earlier, differ from the rocks of the Western Belt chiefly in the predominance of volcaniclastic beds and resemble most closely, as noted by Paris and Bradshaw (1977), the Morrinsville facies of the North Island, found east of the beds of the Murihiku Supergroup. In both countries shallow-water plant beds correspond to those levels above which Oxfordian and Kimmeridgian faunas are lacking. The absence of Neocomian strata in both countries, following on the Rangitata orogeny of circum-Pacific extent, is very striking. Although beds of Aptian–Albian age recorded in New Zealand are as yet unknown in New Caledonia, the discrepancy may be misleading because many of the basal beds of the Cretaceous in both regions are coarse conglomerates, locally talus breccias. The sandstones and shales which succeed the breccias in both countries have their parallels, especially in the "*couches à charbon*" containing poor seams in New Caledonia but represented by rich coalfields in the South Island of New Zealand. The equivalent of the marine Cretaceous shales of the Northern Region are the shales of the east coast geosyncline in New Zealand, and the passage of the marine Cretaceous to Tertiary strata, seen in the north of New Caledonia, recalls the transition from Cretaceous to Tertiary on the eastern side of New Zealand. On the other hand, in both countries considerable breaks occur locally with folded Senonian beds covered by Eocene or later beds.

Within the Tertiary sequence there are fewer direct resemblances, although certain similar lithologies were established in earlier ages in New Caledonia than in New Zealand. Thus, the Eocene calcareous strata of the Western Belt have some parallels with the widespread shallow-water Te Kuiti Group of New Zealand which, however, is of Oligocene age. In New Caledonia, the marine history ends abruptly with the Lower Miocene, recalling the early uprise of the North Auckland peninsula in the mid-Miocene which was followed by great andesite activity in mid-Miocene–Pliocene. Olistoliths were emplaced in the Tertiary of North Auckland, but at times later than the Eocene events known in New Caledonia. After Miocene time New Caledonia appears to have become consolidated as a landmass and to have reached a stage of comparative seismic quiescence. In contrast, within parts of New Zealand, thick marine Miocene, Pliocene, and even Pleistocene strata accumulated and strong seismic and volcanic activity continues at present. Thus, New Caledonia contains strata representing only part of the geological history seen in New Zealand which, as a whole, appears to have strata equivalent to those of New Caledonia and also of the major islands lying east and north of New Caledonia.

REFERENCES

Arène, J., Guérangé, B., Lozes, J., and Paris, J. P., 1979, Carte géologique de la Nouvelle-Calédonie à l'échelle du 1/50000: feuille Paéoua, Minute provisoire Bureau de Recherches Géologiques et Minières, Paris.

Arnould, A., 1958, Etude géologique de la partie nord-est de la Nouvelle-Calédonie, Thèse, Paris Université.

Avias, J., 1949, Note préliminaire sur quelques phénomènes actuels ou subactuels de pétrogénèse et autres dans les marais côtiers de Moindou et de Canala (Nouvelle-Calédonie). *C. R. Soc. Geol.* **13**:277–280.

Avias, J., 1953, Contribution à l'étude stratigraphique et paléontologique des formations antécrétacées de la Nouvelle-Calédonie centrale, *Sci. Terre* **1**:1–276.

Avias, J., 1967, Overthrust structure of the main ultrabasic New Caledonia massives, *Tectonophysics* 4:531–541.

Avias, J., 1977, About some features of allochtonous ophiolitic and vulcano-sedimentary suites and their contact zones in New Caledonia, International Symposium Geodynamics South-west Pacific, Nouméa, 1976, Technip Editions, Paris, pp. 245–264.

Avias, J., and Gonord, H., 1973, Existence dans la chaîne centrale de la Nouvelle-Calédonie (bassin de la Boghen et région du col d'Amieu) de horst de formations plissées à métamorphisme principal d'âge anté-permien et très probablement hercynien, *C. R. Acad. Sci.* 276:17–18.

Avias, J., and Guérin, S., 1958, Contribution à l'étude des faunes de Céphalopodes permotriasiques de Nouvelle-Calédonie. Nautiloidés et Ammonoidés du Permien et du Trias inférieur, *Bull. Geol. Nouvelle-Caledonie* 1:117–133.

Bard, J. P., and Gonord, H., 1971, Découverte d'associations antésénoniennes à lawsonite, pumpellyite et glaucophane dans les "masses cristallophylliennes" paléozoiques du centre de la Nouvelle-Calédonie, *C. R. Acad. Sci.* 273:280–283.

Bell, T. H., and Brothers, R. N., 1985, Development of P-T prograde and P-retrograde, T-prograde isogradic surfaces during blueschist to eclogite regional deformation metamorphism in New Caledonia as indicated by progressively developed porphyroblast microstructures, *J. Metamorphic Geol.* 3:59–78.

Black, P. M., 1970a, Ferroglaucophane from New Caledonia, *Am. Mineral.* 55:508–511.

Black, P. M., 1970b, P_2 omphacite intermediate in composition between jadeite and hedenbergite from metamorphosed acid volcanics, Bouehndep, New Caledonia, *Am. Mineral.* 55:512–515.

Black, P. M., 1970c, Coexisting glaucophane and riebeckite–arfvedsonite from New Caledonia, *Am. Mineral.* 55:1060–1064.

Black, P. M., 1973a, Mineralogy of New Caledonian metamorphic rocks. I. Garnets from the Ouégoa district, *Contrib. Mineral. Petrol.* 38:221–235.

Black, P. M., 1973b, Mineralogy of New Caledonian metamorphic rocks. II. Amphiboles from the Ouégoa district, *Contrib. Mineral. Petrol.* 39:55–64.

Black, P. M., 1974a, Mineralogy of New Caledonian metamorphic rocks. III. Pyroxenes and major element partitioning between coexisting pyroxenes, amphiboles and garnets from the Ouégoa district, *Contrib. Mineral. Petrol.* 45:281–288.

Black, P. M., 1974b, Oxygen isotope study of metamorphic rocks from the Ouégoa district, New Caledonia, *Contrib. Mineral. Petrol.* 47:197–206.

Black, P. M., 1975, Mineralogy of New Caledonian metamorphic rocks. IV. Sheet silicates from the Ouégoa district, *Contrib. Mineral. Petrol.* 49:269–284.

Black, P. M., 1977, Regional high-pressure metamorphism in New Caledonia, *Tectonophysics* 43:89–107.

Black, P. M., and Brothers, R. N., 1977, Blueschist ophiolites in the mélange zone, northern New Caledonia, *Contrib. Mineral. Petrol.* 65:69–78.

Blake, M. C., Brothers, R. N., and Lanphere, M. A., 1977, Radiometric ages of blueschists in New Caledonia, International Symposium Geodynamics South-west Pacific, Nouméa, 1976, Technip Editions, Paris, pp. 279–281.

Briggs, R. M., 1975, Structure, metamorphism and mineral deposits in the Diahot region, northern New Caledonia, Thesis, Auckland University.

Briggs, R. M., Kobe, H. W., and Black, P. M., 1977, High-pressure metamorphism of stratiform sulphide deposits from the Diahot region, New Caledonia, *Miner. Deposita* 12:263–279.

Briggs, R. M., Lillie, A. R., and Brothers, R. N., 1978, Structure and high-pressure metamorphism in the Diahot region, northern New Caledonia, *Bull. Bur. Rech. Géol. Minières Fr. Sect. 4* 3:171–189.

Brothers, R. N., 1974, High-pressure schists in northern New Caledonia, *Contrib. Mineral. Petrol.* 46:109–127.

Brothers, R. N., 1985, Regional mid-Tertiary blueschist–eclogite metamorphism in New Caledonia, *Bull. Bur. Rech. Géol. Minières Géol. Fr.* 1:37–44.

Brothers, R. N., and Blake, M. C., 1973, Tertiary plate tectonics and high-pressure metamorphism in New Caledonia, *Tectonophysics* 17:337–358.

Brothers, R. N., and Yokoyama, K., 1982, Comparison of the high-pressure schist belts of New Caledonia and Sanbagawa, Japan, *Contrib. Mineral. Petrol.* **79**:219–229.

Campbell, H. J., 1979, Permian–Mesozoic stratigraphy of the Moindou–Téremba area (west coast, New Caledonia), Thesis, Auckland University.

Campbell, H. J., and Bando, Y., 1985, Lower Triassic ammonoids of New Caledonia, *Bull. Bur. Rech. Géol. Minières Géol. Fr.* **1**:5–14.

Campbell, H. J., and Grant-Mackie, J. A., 1984, Biostratigraphy of the Mesozoic Baie de St Vincent Group, New Caledonia, *J. R. Soc. N.Z.* **14**:349–366.

Campbell, H. J., Grant-Mackie, J. A., and Paris, J. P., 1985, Geology of the Moindou–Téremba area, New Caledonia: Stratigraphy and structure of the Téremba Group (Permian–Lower Triassic) and Baie de St Vincent Group (Upper Triassic–Lower Jurassic), *Bull Bur. Rech. Géol. Minières Géol. Fr.* **1**:19–36.

Carroué, J. P., 1971, Carte et notice explicative de la carte géologique de la Nouvelle-Calédonie à l'échelle du 1/50000: feuille Pouébo, Bureau de Recherches Géologiques et Minières, Paris.

Carroué, J. P., 1972, Carte et notice explicative de la carte géologique de la Nouvelle-Calédonie à l'échelle du 1/50000: feuille Pouembout, Bureau de Recherches Géologiques et Minières, Paris.

Challis, G. A., and Guillon, J. H., 1971, Etude comparative à la microsonde électronique du clinopyroxène des basaltes et des péridotites de Nouvelle-Cáledonie. Possibilité d'une origine commune de ces roches, *Bull. Bur. Rech. Géol. Minières Sect. 4* **2**:39–45.

Collot, Y., and Missègue, F., 1977, Gravity measurements in Loyalty archipelago, southern New Caledonia and Pines Island, International Symposium Geodynamics South-west Pacific, Nouméa, 1976, Technip Editions, Paris, pp. 125–134.

Collot, Y., Missègue, F., and Malahoff, A., 1979, Anomalies gravimétriques et structure de la croûte dans la région de Nouvelle-Calédonie, in: *Contribution à l'étude du Sud-Ouest Pacifique, Trav. Doc. ORSTOM* (à paraître).

Collot, J. Y., Malahoff, A., Recy, J., Latham, G., and Missègue, F., 1987, Overthrust emplacement of New Caledonia ophiolite: geophysical evidence, *Tectonics* **6**:215–232.

Coudray, J., 1969, Observations nouvelles sur les formations miocènes et post-miocènes de la région de Népoui (Nouvelle-Calédonie): précisions lithologiques et preuves d'une tectonique "recente" sur la côte sud-ouest de ce Territoire, *C. R. Acad. Sci.* **269**:1599–1602.

Coudray, J., 1976, Recherches sur le Néogène et le Quaternaire marins de la Nouvelle-Calédonie. Contribution à l'étude sédimentologique à la connaissance de l'histoire géologique post-éocène, Thèse, Montpellier Université 1975, Singer-Polignac Editions **8**:1–272.

Coudray, J., and Gonord, H., 1966, Extension de la paléosurface d'émersion intra-éocène dans les regions nord-ouest du bassin de Nouméa, *C. R. Soc. Géol.* **3**:105–107.

Coudray, J., and Margerel, J. P., 1974, Les Foraminifères de la série récifale traversée par le sondage Ténia (côte sud-ouest de la Nouvelle-Calédonie) apports stratigraphiques, paléo-écologiques et sédimentologiques, *C. R. Acad. Sci.* **279**:231–234.

Crenn, Y., 1952, Mesures gravimétriques en Nouvelle-Calédonie, *C. R. Acad. Sci.* **236**:105–106.

Crenn, Y., 1953, Anomalies gravimétriques et magnétiques liées aux roches basiques de Nouvelle-Calédonie, *Ann. Geophys.* **9**:291–299.

Daniel, J., Dugas, F., Dupont, J., Jouannic, C., Launay, J., and Monzier, M., 1976, La zone charnière Nouvelle-Calédonie ride de Norfolk (SW Pacifique). Résultats de dragages et interprétations, *Cah. ORSTOM Géol.* **8**:95–101.

Davis, W. M., 1925, Les côtes et les récifs coralliens de la Nouvelle-Calédonie, *Ann. Géogr.* **34**:244–269, 332–359, 423–441, 521–558.

Deslongchamps, E., 1863, Documents sur la géologie de la Nouvelle-Calédonie suivis du catalogue des roches recueillies dans cette île (fossiles triasiques de l'île Hugon), *Bull. Soc. Linn. Normandie* **8**:332–378.

Deslongchamps, E., 1864, On the geology of New Caledonia and on some Triassic fossils from the island of Hugon, *Q. J. Geol. Soc. London* **20**:31–32.

Diessel, C. F. K., Brothers, R. N., and Black, P. M., 1978, Coalification and graphitization in high-pressure schists in New Caledonia, *Contrib. Mineral. Petrol.* **68**:63–78.

Dubois, J., 1969, Contribution à l'étude structurale du sud-ouest du Pacifique d'après les ondes sismiques observées en Nouvelle-Calédonie et aux Nouvelles-Hébrides, *Ann. Géophys.* **25**:923–972.

Dubois, J., Launay, J., and Recy, J., 1974a, Uplift movements in New Caledonia–Loyalty Islands area and their plate tectonics interpretation, *Tectonophysics* **24**:133–150.

Dubois, J., Ravenne, C., Aubertin, A., Louis, J., Guillaume, R., Launay, J., and Montadert, L., 1974b, Continental margins near New Caledonia, in: *The Geology of Continental Margins* (C. A. Burk and C. L. Drake, eds.), pp. 521–535, Springer-Verlag, New York.

Dubois, J., Launay, J., Marshall, J., and Recy, J., 1975, Some new evidence on lithospheric bulges close to island arcs, *Tectonophysics* **26**:189–196.

Dubois, J., Launay, J., Marshall, J., and Recy, J., 1977, Subduction rate from associated lithospheric bulge, *Can. J. Earth Sci.* **14**:250–255.

Espirat, J. J., 1963, Etude géologique de régions de la Nouvelle-Calédonie septentrionale (extrémité nord et versant est), Thèse, Clermont Université.

Espirat, J. J., 1971, Carte et notice explicative de la carte géologique de la Nouvelle-Calédonie à l'échelle du 1/50000: feuille Bourail, Bureau de Recherches Géologiques et Minières, Paris.

Fleming, C. A., 1970, The Mesozoic of New Zealand: Chapters in the history of the circum-Pacific mobile belt, *Q. J. Geol. Soc.* **125**:125–170.

Fleming, C. A., 1979, *The Geological History of New Zealand and Its Life*, Auckland University Press, Auckland.

Fontes, J. C., Launay, J., Monzier, M., and Recy, J., 1977, Genetic hypothesis on the ancient and recent reef complexes in New Caledonia, International Symposium Geodynamics South-west Pacific, Nouméa, 1976, Technip Editions, Paris, pp. 289–300.

Fromager, D., 1968, Nouvelles données pétrographiques et structurales sur les massifs d'ultrabasites et leur contact avec les terrains sédimentaires entre la baie de Canala et la rivière Comboui (Nouvelle-Calédonie), *C. R. Soc. Géol.* **3**:83–84.

Garnier, J., 1865, Coup d'oeil sur la géologie de la Nouvelle-Calédonie dans la traversée de Port de France (côte ouest) à Kanala (cote est), *Nouv. Ann. Voyages* **4**:351–373.

Garnier, J., 1868, Note sur la Nouvelle-Calédonie, *Bull. Soc. Géogr.* **15**:453–468.

Ghent, E. D., Black, P. M., Brothers, R. N., and Stout, M. Z., 1987a, Eclogites and associated albite–epidote–garnet paragneisses between Yambé and Cape Colnett, New Caledonia, *J. Petrol.* **28**:627–643.

Ghent, E. D., Stout, M. Z., Black, P. M., and Brothers, R. N., 1987b, Chloritoid-bearing rocks associated with blueschists and eclogites, northern New Caledonia, *J. Metamorphic Geol.* **5**:239–254.

Gonord, H., 1967, Note sur les quelques observations nouvelles précisant âge et mode de formation du flysch volcanique sur la côte sud-occidentale de la Nouvelle-Calédonie, *C. R. Soc. Géol.* **7**:287–289.

Gonord, H., 1970a, Sur la présence d'olistolites et sur la mise en place probable de nappes de glissement dans le flysch éocène du bassin tertiaire de Nouméa–Bouloupari (Nouvelle-Calédonie), *C. R. Acad. Sci.* **270**:3010–3013.

Gonord, H., 1970b, Découverte de formations sédimentaires d'age éocène (éocène moyen à supérieur) dans la chaîne centrale de Nouvelle-Calédonie, *C. R. Acad. Sci.* **271**:1953–1955.

Gonord, H., 1972, Sur la présence de roches éruptives post-liasiques et anté-sénoniennes dans les regions centrales de Nouvelle-Calédonie, *C. R. Acad. Sci.* **274**:17–20.

Gonord, H., 1977, Recherches sur la géologie de la Nouvelle-Calédonie, sa place dans l'ensemble structural du Pacifique sud-ouest, Thèse, Montpellier Université.

Gonord, H., and Trescases, J. J., 1970, Observations nouvelles sur la formation post-miocène du Muéo (côte de la Nouvelle-Calédonie), *C. R. Acad. Sci.* **270**:584–587.

Grant-Mackie, J. A., 1977, Mesozoic chronostratigraphy of New Zealand and New Caledonia, *Lethaia* **10**:302.

Grant-Mackie, J. A., 1985, New Zealand–New Caledonian Permian–Jurassic faunas, biogeography and terranes, *N.Z. Geol. Surv. Record* **9**:50–52.

Grekoff, N., and Gubler, Y., 1951, Données complémentaires sur les terrains tertiaires de la Nouvelle-Calédonie, *Rev. Inst. Fr. Pét.* **6**:283–293.

Guérangé, B., Lille, R., and Lozes, J., 1973, Données nouvelles concernant la stratigraphie, la sédimentologie, la pétrologie et la structure de la chaîne centrale de la Nouvelle-Calédonie, *Bull. Bur. Rech. Géol. Minières Sect. 2* 1:24–25.

Guérangé, B., Lille, R., and Lozes, J., 1975, Etude géologique des terrains anté-oligocènes de la chaîne centrale néo-calédonienne: stratigraphie, régimes de sédimentation, évolution structurale et métamorphique, *Bull. Bur. Rech. Géol. Minières Sect. 4* 2:127–137.

Guérangé, B., Lozes, J., and Autran, A., 1976, Mesozoic metamorphism in the New Caledonia central chain and its geodynamic implications in relation to the evolution of the Cretaceous Rangitata orogeny, International Symposium Geodynamics South-west Pacific, Nouméa, 1976, Technip Editions, Paris, pp. 265–278.

Guérangé, B., Lozes, J., and Autran, A., 1977, Le métamorphisme mésozoique dans la chaîne centrale de Nouvelle-Calédonie et ses implications géodynamiques dans l'évolution de l'orogenèse Rangitata au Crétacé, *Bull. Bur. Rech. Géol. Minières Sect. 4* 1:53–68.

Guillon, J. H., 1972, Essai de résolution structurale d'un appareil ultramafique d'âge alpin: les massifs de Nouvelle-Calédonie. Implications concernant la structure de l'arc mélanésien, *C. R. Acad. Sci.* 274:3069–3072.

Guillon, J. H., 1974, New Caledonia, in: *Mesozoic–Cenozoic Orogenic Belts* (A. M. Spencer, ed.), pp. 445–452, Scottish Academic Press, Edinburgh.

Guillon, J. H., 1975, Les massifs péridotitiques de Nouvelle-Calédonie. Type d'appareil ultrabasique stratiforme de chaîne récente, *Mem. ORSTOM* 76:11–120.

Guillon, J. H., and Gonord, H., 1972, Premières données radiométriques concernant les basaltes de Nouvelle-Calédonie. Leurs relations avec les grands événements de l'histoire géologique de l'arc mélanésien interne au Cénozoique. *C. R. Acad. Sci. Fr.*, 275:309–312.

Itaya, T., Brothers, R. N., and Black, P. M., 1985, Sulfides, oxides and sphene in high-pressure schists from New Caledonia, *Contrib. Mineral. Petrol.* 91:151–162.

Lacroix, A., 1941, Les glaucophanites de la Nouvelle-Calédonie et les roches qui les accompagnent, leur composition et leur genèse, *Mem. Acad. Sci. Fr.* 65:1–103.

Lacroix, A., 1942, Les péridotites de la Nouvelle-Calédonie, leurs serpentinites et leurs gîtes de nickel et de cobalt, les gabbros qui les accompagnent, *Mem. Acad. Sci. Fr.* 66:1–143.

Lille, R., and Paris, J. P., 1976, Révision stratigraphique des terrains anté-éocènes de Nouvelle-Calédonie, *C. R. Acad. Sci.* 282:965–968.

Lillie, A. R., 1970, The structural geology of lawsonite and glaucophane schists of the Ouégoa district, New Caledonia, *N.Z. J. Geol. Geophys.* 13:72–116.

Lillie, A. R., 1975, Structures in the lawsonite–glaucophane schists of New Caledonia, *Geol. Mag.* 112:225–234.

Lillie, A. R., 1982, *Strata and Structure in New Zealand*, Tohunga Press, Auckland.

Lillie, A. R., and Brothers, R. N., 1970, The geology of New Caledonia, *N.Z. J. Geol. Geophys.* 13:145–183.

Lozes, J., 1975, Etude géologique de la chaîne centrale de Nouvelle-Calédonie entre Thio et Ponérihouen, Thèse, Toulouse Université.

Lozes, J., Yerle, J. J., Schmid, M., Lajoinie, J. P., and Vogt, J., 1976, Carte et notice explicative de la carte géologique de la Nouvelle-Calédonie à l'échelle du 1/50000: feuille Thio, Bureau de Recherches Géologiques et Minières, Paris.

Lozes, J., Guérangé, B., Fromager, D., Doumenge, J. P., and Schmid, M., 1977, Carte et notice explicative de la carte géologique de la Nouvelle-Calédonie à l'échelle du 1/50000: feuille Ponérihouen, Bureau de Recherches Géologiques et Minières, Paris.

Orloff, O., and Gonord, H., 1968, Note préliminaire sur un nouveau complexe sédimentaire continental situé sur les massifs du Goa N'Doro et de Kadjitra (régions côtières à l'est de la Nouvelle-Calédonie), définition de la formation et conséquences de cette découverte sur l'âge des fracturations majeure récemment mises en évidence dans les mêmes régions, *C. R. Acad. Sci.* 267:5–8.

Paris, J. P., 1979, Les recherches pétrolières en Nouvelle–Calédonie, Note dactylo, Colloque sur le développement des ressources de la mer dans les territoires français du Pacifique, *Nouméa* 1979.

Paris, J. P., 1981, Géologie de la Nouvelle-Calédonie: un essai de synthèse, *Mem. Bur. Rech. Géol. Minières* **113**:1–278.

Paris, J. P., and Bradshaw, J. D., 1977, Triassic and Jurassic paleogeography and geotectonics of New Zealand and New Caledonia, International Symposium Geodynamics South-west Pacific, Nouméa, 1976, Technip Editions, Paris, pp. 209–215.

Paris, J. P., and Lille, R., 1976, New Caledonia: Evolution from Permian to Miocene, mapping data and hypotheses about geotectonics, International Symposium Geodynamics South-west Pacific, Nouméa, 1976, Technip Editions, Paris, pp. 195–208.

Paris, J. P., and Lille, R., 1977, La Nouvelle-Calédonie du Permien au Miocène: données cartographiques, hypothèses géotectoniques, *Bull. Bur. Rech. Géol. Minières* **1**:79–95.

Paris, J. P., Guy, B., Doumenge, J. P., and Jegat, P., 1979, Carte et notice explicative de la carte géologique de la Nouvelle-Calédonie à l'échelle du 1/50000: feuille Canala-La Foa, Bureau de Recherches Géologiques et Minières, Paris.

Pharo, C. H., 1967, The geology of some islands in Baie de Pritzbauer (New Caledonia), Thesis, Auckland University.

Piroutet, M., 1917, Etude stratigraphique sur la Nouvelle-Calédonie, Thèse, Protat frères, Macon.

Rod, E., 1974, Geology of eastern Papua: Discussion, *Geol. Soc. Am. Bull.* **85**:653–658.

Rodgers, K. A., 1975, Lower Tertiary tholeiitic basalts from southern New Caledonia, *Mineral. Mag.* **40**:25–32.

Routhier, P., 1953, Etude géologique du versant occidental de la Nouvelle-Calédonie entre le col de Boghen et la pointe d'Arama, *Mem. Soc. Géol.* **32**:1–127.

Sameshima, T., Paris, J. P., Black, P. M., and Heming, R. F., 1983, Clinoenstatite-bearing lava from Népoui, New Caledonia, *Am. Mineral.* **68**:1076–1082.

Speden, I. G., 1975, Cretaceous stratigraphy of Raukumara peninsula, *Bull. N.Z. Geol. Surv.* 91.

Stevens, G. R., 1971, The Jurassic system in New Zealand: Colloque du Jurassique Luxembourg 1967, *Mem. Bur. Rech. Géol. Minières* **75**: 739–751.

Stevens, G. R., 1977, Mesozoic biogeography of the south-west Pacific and its relationship to plate tectonics, International Symposium Geodynamics South-west Pacific, Nouméa, 1976, Technip Editions, Paris, pp. 309–326.

Tissot, B., and Noesmoen, A., 1958, Les bassins de Nouméa et de Bourail (Nouvelle-Calédonie), *Rev. Inst. Fr. Pét.* **13**:739–760.

Tozer, E. T., 1967, A standard for Triassic time, *Bull. Geol. Surv. Canada* 156.

Waterhouse, J. B., 1970, A new Permian fauna from New Caledonia and its relationships to Gondwana and the Tethys, IUGS Symposium Gondwana Stratigraphy, Buenos Aires, 1967, Unesco, pp. 249–272.

Waterhouse, J. B., 1976, The Permian system: World correlations for Permian marine faunas, *Univ. Queensl. Pap. Dept. Geol.* **7**(2):1–232.

Wellman, H. W., 1959, Division of the New Zealand Cretaceous, *Trans. R. Soc. N.Z.* **87**:99–163.

Yokoyama, K., Brothers, R. N., and Black, P. M., 1986, Regional eclogite facies in the high-pressure metamorphic belt of New Caledonia, *Geol. Soc. Am. Mem.* **164**:407–423.

Chapter 9

THE TONGA AND KERMADEC RIDGES

J. Dupont
ORSTOM
Paris, France

I. INTRODUCTION

A. History

The 150 islands forming the Kingdom of Tonga seem to have been sought unconsciously by daring navigators. Uninhabited, they were first visited and settled during Polynesian ocean migrations four or five centuries before the Christian era, and it was only 2000 years later that they were discovered afresh by the first European voyagers.

Now, as the 20th century comes to a close, a flight of about 48 hours separates old Europe from the soil of Nuku'alofa, the capital of the Kingdom, where can be seen rows of colonial pines mounting guard over the Royal Palace. It is therefore difficult to imagine the courage of the first adventurous navigators and their joy on seeing on the horizon the peaks of the highest volcanoes or the green fringe of coconut palms bordering the coral islands. Nevertheless, the 17th century had scarcely begun when, in 1616, a Dutch vessel commanded by Lemaire and Schouten was the first to sight part of the Tonga archipelago. Following that, ships regularly visited these shores. Tasman stopped there during his voyage from New Zealand to Fiji. The great English navigator, James Cook, made two stops there, in 1773 and 1774, and, delighted by the welcome of the inhabitants, named them the Friendly Islands or archipelago. Toward the end of the 18th century, the French navigator d'Entrecasteaux discovered the Kermadec archipelago, naming it after his second officer, Chevalier Huon de Kermadec, who, already sick, was destined to die shortly afterwards (1793) in New Caledonia.

Other travelers have perhaps seen or landed on the islands, for the archipelago is vast. Can this have been the case for Wallis, Malaspina, or Dumont d'Urville?

Then gradually, beginning in the 19th century, the Kingdom of Tonga entered contemporary history. Now, the ships docking there range from small interisland coasters to passenger ships discharging their loads of tourists and to oceanographic vessels flying the French, American, Japanese, or German flags.

B. Geographic Location

1. *The Tonga and Kermadec Islands*

The islands lie in the South Pacific between 14°30′ and 36°S and between 172°W and 179°E (Fig. 1). The structures from which the Tonga islands emerge are more than 1300 km long and quasilinear, striking north-northeast to south-south-west for about 1000 km. Only at the extremities can a change in orientation be observed. The ridge which forms the base of the Kermadec Islands is also linear for about 1000 km, and is a prolongation of the Tonga ridge (Fig. 2).

The Tonga Kermadec archipelago is only the emergent part of a vast structural

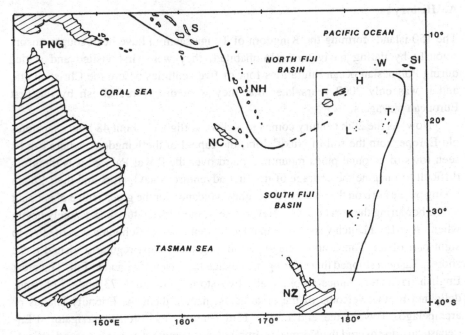

Fig. 1.　The geographic location of the Tonga–Kermadec zone in the southwest Pacific. A, Australia; PNG, Papua–New Guinea; S, Solomon; NH, New Hebrides; NC, New Caledonia; F, Fiji; H, Horn (or Futuna); W, Wallis; SI, Samoa; T, Tonga; L, Lau; K, Kermadec; NZ, New Zealand.

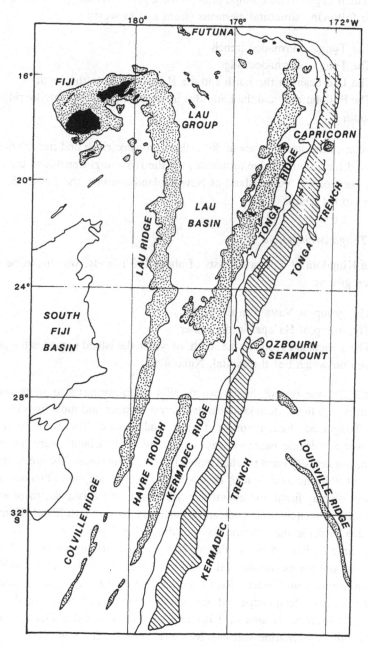

Fig. 2.   General map of the Tonga–Kermadec area. The simplified bathymetry of Tonga is drawn from the map of Hawkins (1974), that of Fiji and Kermadec from Mammerickx *et al.* (1971).

edifice which began to take shape prior to the Upper Eocene. This edifice is made up of the following structural elements (from east to west):

- The Tonga–Kermadec trench
- The Tonga–Kermadec ridge
- The Lau basin in the north and the Havre trough to the south
- The Fiji platform and the Lau ridge in the north, and the Colville ridge to the south

Farther to the south, at about 36°S, the Kermadec ridge and trench, the Havre trough, and the Colville ridge terminate, but the Hikurangi trench and the Rotorua volcanic zone of the North Island of New Zealand continue the Tonga–Kermadec structures to the south.

2. *The Tonga Archipelago*

The Kingdom of Tonga consists of about 150 islands, which can be divided into three groups:

- The group of Vava'u in the north
- The group of Ha'apai in the center
- The Tongatapu group in the south, of which the island named is the principal and on which lies the capital, Nuku'alofa

Schematically, the islands lie in two parallel north-northeast to south-southwest trends, of which the eastern contains the largest number and most important of the islands: Tongatapu, Eua, Nomuka, Lifuka, and Vava'u. These are coral islands grown over a volcanic basement, as on Eua, the only island where the volcanic basement has actually been reached below the coral limestone. The western trend is made up of volcanic islands such as Ata, Tofua, Kao, Late, and Fonulaei, some of which still display fumarolic activity. In addition to these islands, there are some reefs formed of ancient eroded volcanoes, such as Falcon, Metis Shoal, Home Reef, and Curacao Reef at the extreme north of the archipelago.

Volcanic activity persists; Falcon Island, since its discovery in 1781, has disappeared and reappeared several times—disappearing most recently in February 1949. An eruption on Curacao Reef occurred in July 1973 and, more recently, in June 1979 a new volcano appeared between Kao and Late. It was named Lateiki in August, and by then its area of 3 hectares was already under attack by the sea. Farther south, a submarine volcano, Monowai, erupted in 1979.

The highest point in the archipelago, with an elevation of 1109 m, is the extinct volcano on Kao. There is, in addition, the volcanic island of Niua Fo'ou which belongs to neither alignment.

3. The Kermadec Archipelago

The Kermadec archipelago is formed by a small number of volcanic islands of which the most important, from north to south, are: Raoul, Macaulay, Curtis, Herald, and Esperance. The islands are administered by New Zealand but, with the exception of Raoul on which a meteorological station requires the presence of ten persons, they are uninhabited.

C. Geological Framework

The Tonga–Kermadec region has been studied for many years, for it is a tectonically important zone. This can be well illustrated by reporting the earlier geological and geophysical work and completing it by reference to more recently acquired data (Dupont, 1982a).

Prior to 1800, the only information on the Tonga Islands was contained in navigational reports. Between 1800 and 1900, in addition to descriptions of voyages, there were also descriptions of the volcanoes and some geological observations, of which one dates back to 1811. Between 1900 and 1950, scientific studies multiplied, in geology (Hoffmeister, 1932), in petrography, and in volcanology—all enriching knowledge of these volcanic or coral islands. More specialized studies date from 1950, with works on seismicity (Gutenberg and Richter, 1954), refraction (Raitt *et al.*, 1955), gravity (Talwani *et al.*, 1961), and petrography (Ewart and Bryan, 1973). It was the seismological study on the Tonga–Kermadec island arc which enabled Sykes (1966) and Isacks *et al.* (1968, 1969) to demonstrate the disappearance of oceanic lithosphere in subduction zones.

Geological studies on the whole zone linked tectonically to the Tonga–Kermadec island arc are numerous and can be grouped according to the structure studied.

1. The Fiji Platform and Lau Ridge

From the work of Rodda (1967), which was concerned principally with the petrography and tectonics of Fiji (Viti Levu), three broad conclusions can be drawn:

- The oldest volcanic rocks date from the middle Eocene.
- The alternation of volcanic and sedimentary rocks with foraminifera implies alternate periods of emergence and immersion.
- Volcanic activity ended between 5 and 4.7 Ma.

Magnetic and paleomagnetic studies indicate that the Fiji platform has undergone an anticlockwise rotation, although the magnitude of the rotation varies according to author (James and Falvey, 1978; Malahoff *et al.*, 1979). Whatever the magnitude of the rotation of the platform, it can be regarded as an ancient element of the Lau ridge and formerly a continuation of it.

Between 1880 and 1930, the islands on the Lau ridge were studied by Agassiz, Dana, and Davis, who sought in them a perfect example of the evolution of atolls. In 1945, Ladd and Hoffmeister studied the morphology and petrography of the islands which have a volcanic origin and which have undergone numerous vertical movements. Periods of stability have permitted the development of marine terraces now found at a variety of altitudes.

The oldest rocks found on the Lau islands date to between 9 and 6.4 Ma, which does not exclude, for Gill (1976), the possibility of finding Upper Eocene rocks as on Fiji (Viti Levu), for, according to him, the petrographic character of the Fiji platform is the same as that of the Lau ridge.

2. *The Tonga Ridge*

The first important geological work was that of Hoffmeister (1932) on Eua, who showed the existence of a volcanic basement covered by Upper Eocene limestones. It was this work which established a pre-Upper Eocene age for the Tonga island arc.

Later, following the Nova expedition in 1967, basalts were dredged from more than 7000 m, and fresh peridotites and dunites from depths in excess of 9000 m. According to Fisher and Engel (1969), these ultramafic rocks formed the internal wall of the trench. In 1972, Bryan *et al.* pointed out that the activity of the present volcanic line of Tonga was recent, and in the synthesis by Ewart *et al.* (1977) it was recognized that:

- The lavas of Tonga and Kermadec are typical island arc tholeiites.
- The tholeiitic lavas of the Niua Fo'ou volcano, as those of the Lau basin, approximate to those of the oceanic plates.

The first indications of the sedimentary thickness on the Tonga ridge were derived from a combination of drilling and seismic reflection profiling. On Tongatapu, two wells exceeding 1680 m passed through volcaniclastic sediments almost exclusively. These sediments had the character of marine deposits in medium to abyssal depths, with the basal sediments dating from the Lower Miocene (Tongilava and Kroenke, 1975; Katz, 1976). The change in sedimentation marked by the transition to pelagic and reef limestones occurred at a depth of about 300 m and in time between the Upper and Lower Pliocene (about 3.5 Ma). Elsewhere, multichannel reflection profiling indicates a thickness in excess of 3000 m (Kroenke and Tongilava, 1975; Katz, 1976), although the profiles measured by *R/V Lee* in 1982 showed from 2 to 3 km over the Tonga platform.

Data collected on the Lau and Tonga ridges make it possible to retrace the history of the zone:

1. Eocene—a shallow marine fauna of Upper Eocene age is found on both ridges. The fauna is the oldest in the zone and overlies undated volcanics on

Eua. This volcanic basement is made up of rocks of tholeiitic island arc type.

2. Miocene—the Lower and Middle Miocene are unchanged on the two ridges. According to Gill (1976), the petrographic data imply the two ridges were still contiguous (Fig. 3). In the schema of a single Lau–Tonga arc, Lau would represent the forearc and the volcanic line, while Tonga comprises the area between the forearc and the trench under which the Pacific plate is descending from east to west.

3. Pliocene—the Miocene island arc was disrupted at about the Upper Miocene–Lower Pliocene limit. One consequence of this rupture of the Lau–Tonga ridge was the formation of the Lau basin which developed along a zone of structural weakness such as those followed by lines of volcanos.

After the opening of the Lau basin, volcanism continued along the Lau ridge (2.8 Ma) and until even later in Fiji (Taveuni, 2000 to 700 yr).

The petrographic data of Gill (1976) confirm Karig's (1970) hypothesis in which he regarded the Lau and Tonga ridges as two parts of the same arc presently separated by the opening of the Lau interarc basin. The hypothesis finds further support from morphology and sedimentology. There exists, in fact, a symmetry between the ridges with respect to the Lau basin—a morphological symmetry, for the facing flanks are more abrupt than the external slopes, and a sedimentological symmetry, in the distribution and thickness of the sediments.

3. *The Lau and Havre Basins*

The study of the Tonga–Kermadec arc is incomplete without consideration of the Lau basin and the Havre trough.

Karig (1970) was the first to regard the Lau and Havre basins as interarc basins due to spreading. During the opening of the Lau basin, the tholeiitic basalts emitted were fairly similar to those of midoceanic ridges, but showing an island-arc tendency in their isotopic character and in the presence of certain trace elements (Gill, 1976).

Following Karig, Chase (1971) and Sclater *et al.* (1972) studied the opening of the Lau basin in which the Peggy Ridge played the role, by turns, of axis of spreading or transform fault. However, according to Hawkins (1974), the ancient arc was not split, rather a new one was created by the migration eastwards of the seismic zone.

Weissel (1977) proposed a two-stage opening of the Lau basin, the first between 6–5 and 3.5 Ma was the opening corresponding to the present Havre trough, and from 3.5 Ma to the present day, consequent upon a ridge jump, the present opening forms a triple junction formed by the north–south spreading axis and two transform faults (the Peggy and Roger ridges). It must be noted that, according to

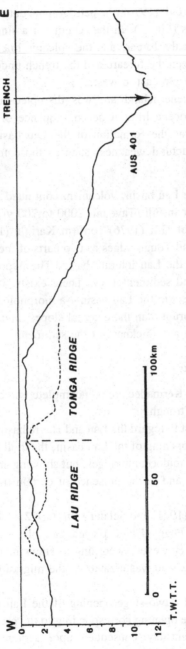

Fig. 3.   Hypothetical reconstruction of the Lau–Tonga protoarc from the juxtaposition of bathymetric profiles of the Lau ridge and the Tonga arc (AUS 401). The reconstruction relies only on the morphology of abrupt opposing flanks. The present volcanic line, the Tofua trough, and the eastern flank of the Lau ridge are dashed.

Malahoff *et al.* (1982), the opening of the Lau and Havre basins is more recent—between 2.5 and 1.8 Ma.

II. MORPHOLOGY OF THE TONGA–KERMADEC ISLAND ARC

The existing bathymetric maps (Mammerickx *et al.*, 1971; Eade, 1971; Hawkins, 1974) show morphological differences between the Tonga and Kermadec ridges. More recent work by the *N/O Noroit* and *Coriolis* and by the *R/V Lee* allows a precise definition of the morphological character of the oceanic plate and the arc-trench system. Each of these zones will be studied in turn; certain of the morphological characteristics are summarized in Table I.

A. The Pacific Plate

The oceanic crust generated at the East Pacific Rise forms the Pacific plate which fronts the Tonga–Kermadec arc. The depth of the plate is relatively uniform, between 5100 and 5800 m. Beyond the 5800-m isobath, depth increases rapidly down the outer flank of the trench (Fig. 4).

On the plate, seamounts or isolated guyots provide significant relief, but in the majority of cases, they do not break the surface. There are, from north to south:

- In the north Tonga zone, at 15°S, 172°10′W, a peak rising from the Tonga Trench that reaches to within 1460 m of the surface from a depth of 4400 m
- At 18°30′S and 172°15′W, borders and deforms the Tonga Trench, the Capricorn seamount which rises about 4400 m to a peak 730 m below the surface
- South of the Kermadec, at 35°30′S and 176°W, a seamount whose base is at −5120 m and whose peak is less than 365 m below the surface

In addition to these isolated seamounts, the oceanic plate is cut by a northwest–southeast-oriented ridge which ends in the trench in the Ozbourn guyot (26°S, 175°W). This topographic relief, in part in the trench, delimits the Tonga and

TABLE I
Principal Morphological Parameters of the Tonga and Kermadec Island Arc

Arc	Length of the trench	Maximum depth of the trench	Volcano–trench distance	Art crest–trench distance
Tonga	1.550 km[a]	10.882 m[b]	170–206 km[c]	116–155 km[c]
Kermadec	1.250 km[a]	10.047 m[b]	150–180 km[c]	150–180 km[c]

[a]After Mammerickx *et al.* (1971).
[b]After Faleyev *et al.* (1977).
[c]After Dupont (1982b).

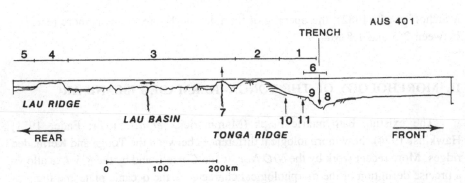

Fig. 4.   The principal structures of the island arc (after Karig and Sharman, 1975; Dupont, 1982b) from profile AUS 401. 1, accretionary prism; 2, frontal arc; 3, active marginal basin; 4, remnant arc; 5, inactive marginal basin; 6, subduction zone; 7, volcanic chain; 8, outer wall of the trench; 9, inner wall of the trench; 10, upper slope discontinuity; 11, trench slope break.

Kermadec trenches. The ridge, called the Louisville Ridge, in reality consists of a sequence of isolated eminences still poorly known. Some authors (Hayes and Ewing, 1971) consider it as a possible prolongation of the Eltanin fracture zone, although this is not firmly established.

Although there are no certain data on the age of the oceanic crust bordering the Tonga–Kermadec arc, there is, nevertheless, a general convergence from several lines of research to indicate an age of about 130 Ma:

- The relationship of age as a function of depth (Parsons and Sclater, 1977), giving 120 Ma for depths of 5900 m
- DSDP holes in the Pacific at depths of 5969 and 5981 m, which indicated ages between 130 and 140 Ma without reaching basement (Sclater and De-trick, 1973)
- DSDP hole 204 (Burns *et al.*, 1973) at 5354 m indicated a probable Lower Cretaceous age (130–140 Ma) after penetrating 160 m of oceanic crust

According to Dubois *et al.* (1977), the thickness of the oceanic lithosphere facing the Tonga–Kermadec is between 90 and 110 km.

B. The Tonga–Kermadec Trench

The trench is oriented north-northeast to south-southwest, but at the northern end it bends toward the west and has an east–west trend before attenuating and dying out. To the south, the Kermadec trench, without direction change, becomes progressively less well defined morphologically, and is no longer identifiable at the latitude of North Island, New Zealand.

A longitudinal profile following the axis of the trench (Fig. 5) shows that:

- The deepest part of the Tonga Trench, whose depths exceed 10,800 m, faces the deepest parts of the oceanic plate.

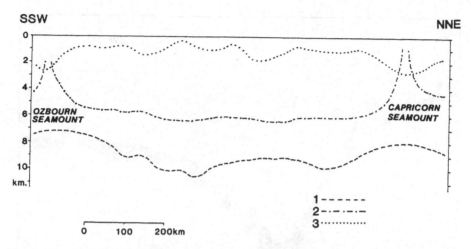

Fig. 5. Longitudinal profiles of the Tonga arc, from the Capricorn seamount to the Ozbourn seamount, based upon 18 bathymetric profiles normal to the arc. The regularity of the downgoing plate between the points of relief (seamounts), and the deepening of the accretionary prism facing the same relief, should be noted. 1, longitudinal section along the axis of the trench; 2, section through the downgoing plate parallel to the trench and intersecting the Capricorn and Ozbourn seamounts; 3, section through the accretionary prism, parallel to the trench.

• The sills in the trench, corresponding to the Capricorn and Ozbourn guyots, still lie on the descending oceanic plate as the guyots gradually descend into the trench.

However, if this morphological comparison is extended to the ridge, then it must be recorded that the seamounts face the deepest parts of the island arc (Fig. 5). According to Dupont (1982b), the relief on a plate plunging toward the base of the trench has an influence on the morphology of that part of the island arc facing it.

The bathymetric profiles and reflection profiles indicate fracturing of the descending plate, which facilitates the adjustment of its curvature and final disappearance under the trench. The network of the present profiles is too wide to determine whether the fractures remain parallel to the trench. Nor, in general, can the reflection profiles show the thin sediment cover over each fractured segment (0.2 to 0.4 sec two-way travel time) (Fig. 6).

C. The Tonga–Kermadec Frontal Arc

Morphologically, the frontal arc is the most elevated part of the island arc. It is crowned by either volcanic or coral islands. In the case of Tonga, it forms a more or less flattened swell 90 to 120 km wide. It lies at variable depths, for it is composed of highs or plateaus with an average depth of 500 m, from which rise the coral islands, separated by deeper water zones which may reach a depth of 1000 m. The arc remains linear and, toward 15°S, abuts the trench. A series of massifs, more or

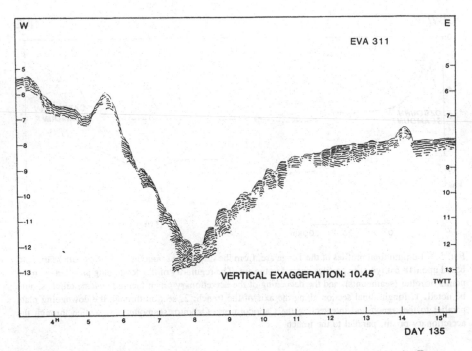

Fig. 6. The fracturing of the Pacific plate due to its curvature prior to descent under the Tonga arc (Australo-Indian plate) (profile EVA 311). Fractures can be distinguished cutting the plate into blocks. In general, the sedimentary cover is too thin and so does not show on the single-channel reflection seismic traces.

less well individualized, mark its continuation to the west along the line of the trench.

The frontal arc can be subdivided into the following three important blocks, delimited by a grid of west-southwest to east-northeast and north-northwest to south-southeast faults (Dupont and Herzer, 1985):

- To the south, a submerged block with an average depth of 500 to 1000 m
- In the center, the most important block, with an average depth of 500 m, on which lie all the coral islands
- A more broken-up northern zone, over which the depth varies from 1000 to 1500 m

There are two discontinuities in the last zone. The first discontinuity, occurring toward 18°30′S, displaces the northern part some 35 km west with respect to the part farther to the south. The distance between the axis of the frontal arc and the trench is 170 km to the north of Vava'u, but is only 140 km to the south. This displacement occurs at about the latitude of the Capricorn seamount. The second displacement, found at about 17°S, seems to be correlated with northwest–southeast structures in the Lau basin, along which is found the Niua Fo'ou volcano (Fig. 7).

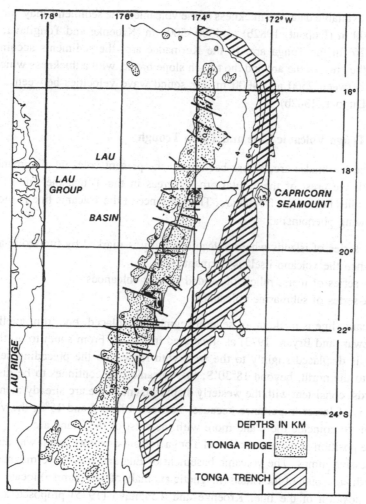

Fig. 7. Network of fractures which cut the Tonga arc into blocks of different size. There is a first system oriented E–W to WSW–ENE approximately normal to the trench, and a second system N–S to NNW–SSE which cuts the first set from the blocks as mentioned (see Dupont and Herzer, 1985). There is, in addition, a third SE–NW fault set. This latter trend is most evident where it cuts the volcanic line, in particular in north Tonga where these faults are responsible for the successive westerly displacements of the crest of the arc.

The frontal arc of the Kermadec has the form of a continuous ridge, the crest of which is, at the same time, the crest of the island arc and the line of the volcanic arc, emergent at several points. The depth between the islands and reefs is highly variable, but there are no large plateaus.

In the Tonga as well as the Kermadec arc, the arc itself is formed of volcaniclastic sediments, as shown by borings on Tongatapu. Volcanic intrusions pierce the sediment layers, which may be broken up by faulting into tilted blocks (Tonga).

From the available data, the thickness of the volcaniclastic sediments may vary from 2000–3000 m (Dupont, 1982b) to 4000–5000 m (Kroenke and Tongilava, 1975; Katz, 1976) in the Tonga arc. In the Kermadec arc, the sediments accumulated between the crest of the arc and the trench slope break, with a thickness which may reach to between 2500 and 4000 m, for sound wave velocities between 2 and 3 km/sec (Dupont, 1982b).

D. The Tonga Volcanic Line and Tofua Trough

In the Kermadec arc, as has been seen, the present volcanic alignment coincides with the crest of the frontal arc, whereas in the Tonga the line of active volcanoes lies behind the frontal arc. The existence of the volcanic line is shown by the following phenomena:

- A series of islands where volcanic activity is marked by fumarolic activity when the volcano itself is dormant
- A series of reefs, relicts of eroded former volcanoes
- A series of submarine volcanoes

The volcanic line is made up of several segments displaced, one from another, by faults (Ewart and Bryan, 1973) as illustrated in Fig. 7. From south to north, each segment is displaced slightly to the west with respect to the preceding segment. Farther to the north, beyond 18°30'S, the volcanic line continues to be displaced westwards, consistent with the westerly displacement of the arc already mentioned; but this is of lesser importance because, between 18°30'S and 15°S, the volcanic line tends to coincide more and more with the crest of the island arc.

The position of the present line of Tonga volcanoes shows that it was displaced by successive jumps. The volcanic basement of Eua and transverse magnetic profiles north and south of the island (magnetic anomaly of 480 gamma) locate the pre-Eocene position of the line. Kroenke and Tongilava (1975) proposed a second alignment parallel to this and to the present alignment, through Tongatapu, Nomuka, and Ha'apai, assigning it a Mio-Pliocene age. Actually, the third alignment is morphologically to the rear of the arc, at least between 23 and 18°30'S, and refraction data even seem to indicate that the volcanoes rest on the crust of the Lau basin or at least at the junction between the Lau basin and the Tonga arc (Figs. 7 and 12).

Refraction lines show that the Tofua trough occurs at the crustal boundary between the Lau basin and the Tonga arc in a tectonically weak zone. The depression, 300 km long and 25–30 km wide, runs parallel to the arc, with a morphology which has been accentuated by movement along faults lying along the western flank margin of the frontal arc. Sediments emitted by the volcanoes along the western margin of the depression are therefore trapped within the depression and have accumulated to a considerable thickness (2500 m) (see Figs. 8 and 12).

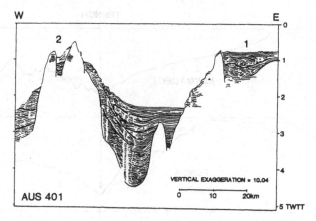

Fig. 8. Transverse section of the Tofua trough (reflection profile AUS 401). 1, Crest of the arc north of the Ha'apai island group; 2, submarine volcano. (After Dupont, 1982b.)

E. Morphological Evolution of the Tonga and Kermadec Arcs

Following the classification of island arcs based upon morphological characteristics (Karig and Sharman, 1975), Dupont (1979, 1982b) defined the morphology of the Tonga–Kermadec arc from a study of more than 30 transverse profiles. He classified the profiles into three families (Fig. 9) which are, in their order of evolution:

- Kermadec, or original morphology stage
- Ozbourn–Capricorn, or transitional morphology
- Tonga, or the morphology of the final stage

In a geographical sense, these morphological types can be ordered from south to north as follows (see Fig. 10):

- Kermadec arc, original stage
- Area of Ozbourn seamount, transitional stage
- Southern part of the Tonga arc, final stage
- Area of Capricorn seamount, transitional stage
- Northern part of the Tonga arc (18–15°S), original stage

According to Dupont (1979, 1982b), subduction of the Louisville ridge is the principal cause of the morphological changes recorded in the arc, but the geographical distribution of the three families of profiles poses a problem. Although the morphology of the Kermadec, Ozbourn, and southern part of Tonga may easily be explained by the subduction of the Louisville ridge beneath the Tonga–Kermadec arc as a consequence of Pacific plate motion, the northern part, in contrast, is not involved. Only by using the seismic data from the North Tonga region (Louat and

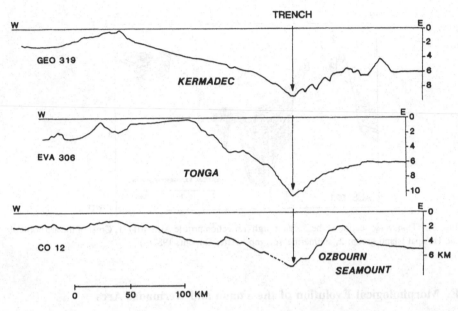

Fig. 9.   Three profiles which characterize the morphology of the Kermadec (GEO 319), Tonga (EVA 306), and the zone opposite the Capricorn or Ozbourn seamounts (CO 12). The profiles are typical of the three morphological families defined by Dupont (1982b) and illustrate that author's hypothesis on the morphological evolution of the Kermadec and Tonga.

Dupont, 1982), which show that the Benioff zone is shallower in the north than in the south, can an explanation be found for the morphology of the northern zone. The entire North Tonga region is of recent formation (3–4 Ma) and, hence, its morphology has in no way been affected by the subduction of the ridge. Only the occurrence of the Capricorn seamount locally modifies the morphology of the arc.

If the effect of a subducted ridge on an island arc is to be admitted, then it must be established that the morphology of all arcs depends upon:

- The type of topographic relief existing or having existed on the subducted plate
- The orientation of the relief on the plate
- The direction of motion of the plate with respect to the subduction zone

The effect of isolated topographic relief, such as the Capricorn guyot, can only have a local influence on arc morphology, in contrast to the effect of a ridge such as the Louisville Ridge. Yet, a ridge perpendicular to an arc and trench, and moving approximately perpendicular to them, would have the same effect as an isolated seamount. The effects on the morphology of the arc are most apparent in the frontal arc and the elevation of this zone, and not only explain the numerous evidences of vertical uplift in this part of the island arc, but more particularly the uplift of the

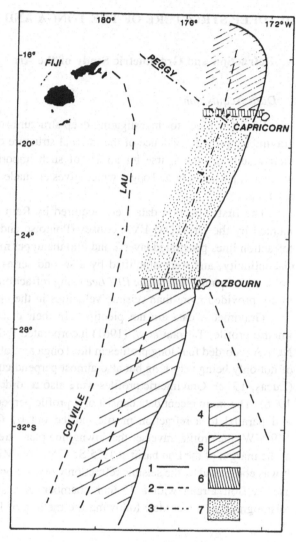

Fig. 10. Geographic distribution of the three morphological families. 1, Tonga and Kermadec trench; 2, axes of structures; 3, limit of the fossil plate which passes into the Peggy ridge; 4, location of the Kermadec morphological family; 5, same morphological type as the preceding, but reflecting recent subduction in north Tonga; 6, Ozbourn and Capricorn morphological family; 7, Tonga morphological family.

great thickness of sediments which accumulated in the deep waters over the flank of the arc between the crest and slope break.

At the present time, the motion of the Louisville ridge on the Pacific plate is toward the Kermadec. If the Louisville Ridge were a continuous chain instead of being a series of isolated prominences (Hayes and Ewing, 1971; Mammerickx et al., 1971), vertical uplift ought to have been found of the frontal arc south of Tonga, and a trench either absent or much shallower than normal, as is the case in the New Hebrides opposite the d'Entrecasteaux ridge (Daniel et al., 1982; Jouannic et al., 1982).

III. DEEP STRUCTURE OF THE TONGA AND KERMADEC RIDGES

A. Refraction and Gravimetric Study of the Arc

1. *Data Acquisition*

While methods for investigating deep structure such as seismic refraction and gravimetry provide evidence of the internal structure of the Tonga–Kermadec arc, their results are too sparse for an arc of such importance. They also tend to be localized, particularly in Tonga, which gives an inadequate view of the structure as a whole.

The first refraction data were acquired by Raitt *et al.* (1955) and were augmented by the ORSTOM EVA cruises (Pontoise and Latham, 1982). These first refraction lines penetrated layer 3 and into the upper mantle below the Mohorovicic discontinuity, and were amplified by a second series that was carried out by petroleum companies and by the *R/V Lee* using refraction buoys. The latter, however, rarely provided more than seismic velocities in the upper layers of sediment.

Gravimetric data are not plentiful. In their discussion of a transverse gravimetric profile, Talwani *et al.* (1961) incorporated Raitt's refraction data. In 1971, NOAA recorded four long profiles in the Tonga arc, of which 2D had the advantage of not only being very long but also almost perpendicular to the structure of the arc (Lucas, 1972). Gravimetric profiles were also recorded by the *R/V Vitiaz* (Kogan, 1976). The more recent data consist of a profile perpendicular to the arc and trench and parallel to a refraction profile carried out by ORSTOM. Starting at 20°S, 169°5'W, the profile traverses the downgoing plate, trench, and island arc, and ends at the margin of the Lau basin at 19°5'S, 175°5'W (Missègue and Malahoff, 1982). It was completed by the seismic refraction work of Pontoise and Latham (1982) who used a Bolt airgun with a 15-liter chamber as a source and an ocean bottom seismograph as a recorder, following the method of Lathan *et al.* (1978).

2. *Structure of the Ridge*

The refraction and gravimetric data are recorded in Fig. 11, and the principal features are discussed passing from east to west.

(a) The Downgoing Plate. This is formed of three layers which can be described in the following manner:

- A thin sedimentary layer which cannot be recognized in refraction profiles, but which is clearly visible in reflection profiles. The relative thinness, less than 400 m, is essentially due to the distance from terrestrial or volcanic sources and the depth of deposition, which is always greater than the carbon-

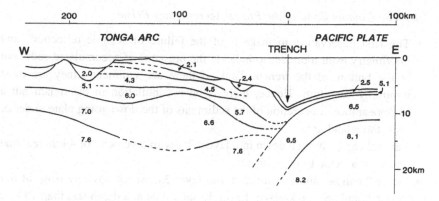

Fig. 11. Sketch of the deep structure of the Tonga arc, based on seismic refraction (Pontoise and Latham, 1982), amplified by gravimetric data (Missègue and Malahoff, 1982), with respect to the ascent of dense material to 14.5 km along the outer surface of the subducted lithosphere.

ate compensation depth. The gravimetric data provide evidence of a thin sedimentary layer with a density of 1.97, which agrees remarkably well with the first horizons visible in the reflection profile.

- Layer 2 is also thin (about 600 m) with a velocity of 5.3 km/sec. It thickens toward the trench, which may be the result of the onset of accretion. The gravimetric model of Missègue and Malahoff (1982) gives a density of 2.55 for the layer corresponding to layer 2, but this layer seems already to thicken 30 km east of the trench. The gravimetric as well as the refraction data show that the zone of accumulation, classically situated at the contact of the plates, may begin well before reaching the trench by the thickening of layer 2.

- Layer 3 has a thickness of 4.5 km and a velocity of 6.5 km/sec. The thickness is abnormal for layer 3 of an oceanic plate. At the trench, it dips under the arc at an angle of 56°. The best-fit gravimetric model requires a density of 2.81 and a thickness of 5 km—that is, be even thicker than indicated by refraction and dip at no more than 50°, which approximates to the angle indicated by refraction.

The velocities of 8.1 and 8.2 km/sec seen on the refraction profiles of Raitt *et al.* (1955) are characteristic of upper mantle velocities immediately below the Mohorovicic discontinuity. In the trench, the discontinuity plunges to a depth which varies according to the method used—to about 20 km from refraction, but only 17.5 km by gravimetric calculation.

Finally, the characteristic bulge of the oceanic lithosphere before dipping under the arc is particularly clearly seen on NOAA long profile 2D. The maximum upwarping lies 220 km east of the Tonga trench (Lucas, 1972).

(b) The Contact Zone of the Plates/Accretionary Prism:

- The first observation to make is of the failure of seismic reflection, and gravimetry even more so, to indicate the presence of accumulated sediment at the bottom of the trench. Thus, if there are sediments, they are most assuredly very thin. Hence, there is no accumulation in the trench but a disappearance or accretion of the sediments of the downgoing plate under or onto the island arc.
- Layer 2 can be broken down into layers 2a, with a velocity of 4 km/sec, and 2b, with a velocity of 6 km/sec.
- Layer 3 can be similarly divided into layer 3a, which has a velocity of 6.9 km/sec, and 3b, 7.6 km/sec. Layer 3b should be at a depth less than 15 km.

The gravimetric data for the upper layers are no better than the refraction data, but they do confirm that, at the point of contact of the plates under the accretionary prism, the Mohorovicic discontinuity is at a depth of 14.5 km. However, the upper mantle velocities below the Tonga arc are not above 8 km/sec as under the downgoing slab, but are only on the order of 7.6 km/sec.

(c) The Island Arc. The refraction profiles detail the arc structure and, in particular, point to a marked thickening of layers 2 and 3. In these lower layers, the velocities recorded are always below the velocities normally measured in the upper mantle—7.6 km/sec instead of 8.1 or 8.2 km/sec.

If velocities of 7.6 km/sec can be accepted as characteristic of the type of upper mantle found below island arcs, then the thickness of the crust below the Tonga arc is about 16 km. The incompleteness of the gravimetric data for this segment of the arc permitted Missègue and Malahoff (1982) to only estimate a possible Mohorovicic depth of about 35 km, a figure which does not agree with that calculated from the refraction results.

(d) The Backarc Zone. In the backarc area, i.e., the zone of the Tofua trough and the volcanic line, the refraction data bring out the different characteristics of the sediments in the Tofua trough and those of the arc *sensu stricto*. A great thickness of unconsolidated sediments (see Fig. 8), thus of relatively recent age, is found in the trough, whereas on the arc, below a thin layer of recent unconsolidated sediments, there are thicker layers with velocities in the range 2.7 to 3.5 km/sec which are formed by older sediments in the process of consolidation.

The variety of the layers, as much in seismic velocity as in thickness, seems to delimit two distinct zones of different age and origin. The Tofua trough thus lies at the junction of these two zones, as is underlined by Fig. 12. As for the crust of the Lau basin, it has all the characteristics of young oceanic crust, and the great thickness of unconsolidated sediments shown in Fig. 11 finds an explanation in its proximity to the volcanic line.

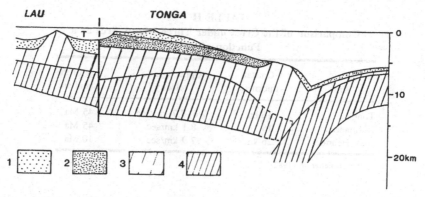

Fig. 12. Schematic representation of the structure of the Tonga arc, illustrating the differences between the crust of the Lau basin and that of the island arc. The boundary between the two corresponds to the break of the protoarc Lau–Tonga. T, Tofua; 1, 2.0–2.5 km/sec layer; 2, 2.7–3.5 km/sec layer; 3, layer with 3.8–6.0 km/sec velocity; 4, layer with 6.4–7.2 km/sec velocity. Under layer 4, velocities range from 7.6 to 7.7 km/sec under both the Lau basin and the Tonga arc. Under the downgoing slab, velocities are on the order of 8.2 km/sec. (After Dupont *et al.*, 1982.)

3. Conclusions

Gravimetric data show that the Mohorovicic discontinuity rises below the accretionary prism, with dense material (d = 3.33) rising to 14.5 km. The modeling of Missègue and Malahoff (1982) demonstrates that it is necessary to take this fact into account if the theoretical curve of the calculated anomaly is to most closely approximate the observed curve. This rise of dense material along the side external to the downgoing plate was interpreted by Elsasser (1971) in terms of a shear at the plate contact resulting from the convergence of the plates and the density of the downgoing slab, a shear which facilitated the rise of denser material. This phenomenon, too little studied, has still to be verified and proved.

Below the Tonga arc, a crustal thickness of less than 16 km is recorded in which seismic velocities do not exceed 7.6–7.7 km/sec, while Shor *et al.* (1971) found velocities in the upper mantle of 8.1 km/sec below a crustal thickness of 18.4 km in the Kermadec arc. The velocity of 7.6 km/sec in the upper mantle below the Tonga arc approaches that found under the New Hebrides arc, where velocities of 7.7 to 7.9 km/sec were recorded by Pontoise *et al.* (1980, 1982) and Daniel *et al.* (1982).

As Table II indicates, comparisons of the three island arcs in the southwest Pacific are difficult insofar as age, crustal thickness, and upper mantle velocities are concerned.

Should the current data be doubted or should they be regarded as indicating that the island arcs may have, for reasons as yet unknown, a crust of variable

TABLE II
Comparison of the Crust under Three Island Arcs as a
Function of Age[a]

Arc	Crustal thickness	Upper mantle velocity	Age of the island arc
Tonga	16 km	7.6–7.7 km/sec	45 Ma
Kermadec	18 km	8.1 km/sec	45 Ma
New Hebrides	26 km	7.9 km/sec	10 Ma

[a]After Daniel *et al.* (1982).

thickness over an upper mantle in which velocities range from 7.6 to 8.1 km/sec? All this goes to underline the need to augment refraction and gravimetric data from the Tonga–Kermadec ridge.

B. Seismological Study of the Subduction of the Pacific Plate below the Tonga–Kermadec Arc

1. Introduction

The existence of subduction in the Tonga–Kermadec region was demonstrated by Oliver and Isacks (1967) and Isacks *et al.* (1968, 1969) in their interpretation of seismological data. The seismic foci they found defined a continuous narrow zone which extended from zero to 700 km in depth. In this zone, seismic waves were little attenuated and had high velocities, whereas in the enclosing asthenosphere, particularly in the backarc region, only the low-frequency waves were recorded at local seismological stations. The focal mechanisms of shallow earthquakes were thrust displacements in the majority of cases, leading the authors to claim the descent of the Pacific plate below the Tonga–Kermadec arc—thereby adding the phenomenon of subduction in the context of plate tectonics.

In addition to the descent of the Pacific plate below the Australo-Indian plate, seismology also shows the transition of this subduction zone to a sinistral east–west transform fault north of Tonga (Isacks *et al.*, 1968, 1969) and the continuity of the Benioff zone between Tonga and Kermadec (Isacks and Barazangi, 1977).

The geographic distribution of seismicity is not uniform; there is a minimum of activity at the junction of the Tonga–Kermadec trench and the Louisville ridge, and to the north, a zone of strong activity marking the passage of the subduction into an east–west transform fault (Louat, 1977).

Finally, west of the Tonga and Kermadec ridges there are found the Lau and Havre basins, respectively. The upper mantle below the Lau basin is characterized by a marked attenuation of P and S wave velocities, with the eastern boundary of this zone lying near the active volcanic line of Tonga (Fig. 13). These characteristics do not seem to extend to the Havre trough (Barazangi and Isacks, 1971; Aggarwal *et al.*, 1972).

The morphology of the Benioff zone varies according to author. According to Isacks and Barazangi (1977), it can be represented by three sections drawn through Kermadec, northern and southern Tonga (Fig. 14). Despite the difference in the dip of the Benioff zone between south Tonga and Kermadec, the authors demonstrate the continuity of the Benioff zone between Tonga and Kermadec. Hanuš and Vaněk (1979) grouped their seismic data along lines perpendicular to the trench. From the distribution of earthquakes, they argued for the existence of an earlier subduction zone preceding the present Tonga–Kermadec subduction (Fig. 15). To these two phases of subduction separated in time, there may be added two further fossil subduction zones—the so-called Lau zone lying west of the present Tonga–Kermadec trench, and the Horn (or Futuna Island) zone north of Tonga. Hanuš and Vaněk (1979) further concluded that the rate of subduction of the Pacific plate is variable and that, in particular, the subduction of relief, a chain, or seamount (Louisville ridge or Capricorn seamount) may slow down the rate of disappearance of the downgoing plate.

A study by Louat and Dupont (1982) was based upon a similar principle. In interpreting the thickness of the seismic zone and its dip, it is usual to consider sections normal to the trench on which foci are positioned as a function of depth, the technique used by Hanuš and Vaněk whose method on deep earthquakes (400 to 700 km) in the Tonga–Kermadec area led them to conclude that the direction of subduction has not varied since its inception 45–50 Ma according to geological estimates. Louat and Dupont (1982) concluded, from a study of an east–west section, that it was possible to discern a coherent pattern in the deep Benioff zone between 35 and 22°S (sections A–D, Fig. 16). Farther north, between 22 and 18°S, the interpreta-

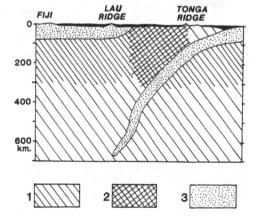

Fig. 13. Schematic section across the Tonga arc, Lau basin, and Lau ridge, showing the zones of low, extremely low, and high attenuation. 1, low attenuation; 2, extremely low attenuation; 3, high attenuation. (After Barazangi and Isacks, 1971.)

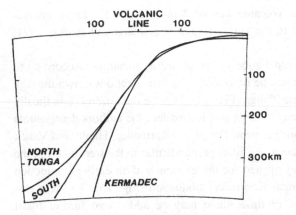

Fig. 14. The morphology of the Benioff zone of Tonga and Kermadec, according to Isacks and Barazangi (1977), showing the high dip of the Benioff zone in the Kermadec and its decrease in the South Tonga and North Tonga.

tion of the sections (E–H in Fig. 16) is only possible by adopting the pattern seen farther to the south. The paucity of seismic stations, due to the distribution of the islands, can induce considerable uncertainty in the localization of earthquakes recorded by only a small number of stations, and for this reason, Louat and Dupont (1982) selected only the better established events, retaining only those earthquakes recorded at more than 100 stations.

From the study of earthquakes from successive depths—deep, intermediate, and shallow—different hypotheses have been formulated, as below.

2. Deep Earthquakes

Since about 30 Ma (Packham and Andrews, 1975), the rotation pole between the Pacific and Australo-Indian plates, assuming the latter fixed, has lain south of

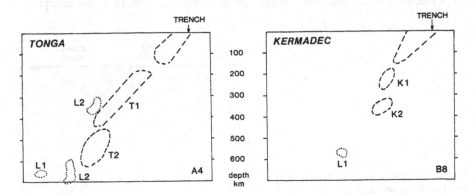

Fig. 15.   The morphology of the Benioff zone of Tonga and Kermadec, according to Hanuš and Vaněk (1979). On the section through Tonga (A4 near 20°S), the interpretation of the different subductions is shown: L2 and L1, fossil subduction of Lau; T2, former subduction of Tonga; T1, present subduction. On the Kermadec section (B8, about 30°S): L1, fossil subduction of Lau; K2, former subduction of Kermadec; K1, present-day subduction.

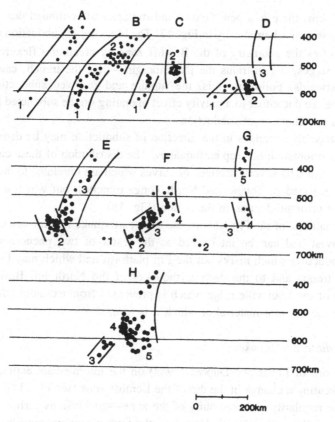

Fig. 16. The interpretation of seismic sections of Tonga and Kermadec, based on deep earthquakes (400–700 km), according to Louat and Dupont (1982). Section A groups earthquakes between 33 and 28°S; B, between 27 and 24°S; C, between 24 and 23°S; D, between 23 and 22°S; E, between 22 and 21°S; F, between 21 and 20°S; G, between 20 and 19°S; H, between 19 and 18°S. The numbers 1 to 5 represent the different segments of lithosphere subducted after rupture.

the Kermadec, with the present pole lying at approximately 60°S, 180°. Consequently, the velocity of convergence of the plates is lower in the Kermadec than it is in Tonga, and it is therefore to be expected that the subducted plate is shorter and thus, *a priori,* simpler. Figure 16 shows a Benioff zone becoming more complex from south to north.

Louat and Dupont (1982) suggested that the maximum depth of the Benioff zone is 700 km and, having attained that depth, it cannot descend farther. This depth, therefore, occupies the role of a buffer zone or anchor point. The Benioff zone begins to flex at depths between 400 and 500 km (section B, Fig. 16); then, as subduction continues, a break develops in the subducted slab at about 500 km, and the slab continues to descend behind the break (the beginning of rupture is shown in section B and subsequent stages in C and D). When the newly fractured edge itself reaches 700 km, there is a new flexure and rupture and continued descent of the plate; this is shown schematically in Fig. 17. The blocking of subduction at 700 km at first modifies the geometry of the Benioff zone by creating a flexure; then, in succeeding stages, it constrains the plate margin—i.e., the trench, causing it to migrate eastwards. Following this, the rupture and renewed subduction of the Benioff zone are due either to a gravity effect operating on the subducted plate or to a change in the direction of subduction.

The successive changes in the direction of subduction may be deduced from the changed orientation of deep earthquakes. The distribution of these earthquakes may be illustrated by several sections of curves which are unrelated to the trends of the present Kermadec, Tonga, and North Tonga trenches, but which indicate the evolution of prior subduction in the region (Fig. 18).

At the latitude of the Fiji Islands, the deep earthquake foci are oriented southeast–northwest and can be interpreted as the result of two phenomena: (1) an ancient subduction which marks out the Fiji platform and which may be linked to the Vitiaz trench and to the deep earthquakes of the North Fiji Basin; (2) the subduction of the Louisville ridge which is prevented from extending deeper than 600 km by the buoyant material of which it is formed.

3. *Intermediate Earthquakes*

The work of Louat and Dupont (1982) on the intermediate earthquakes, although indicating a change in the dip of the Benioff zone (see Fig. 14), in no way changes the regularity and continuity of the zone—conclusions earlier underlined by Isacks and Barazangi (1977). However, the former authors considered that the change in dip was not solely a consequence of the subduction of the Louisville ridge, but that it was much more a response to the arresting of subduction at 700 km which leads to a flexing of the subducted lithosphere when movement continues.

(a) The Kermadec. The intermediate focus earthquakes in the Kermadec, as the deep earthquakes, show that the Benioff zone in this region has not been altered by the movement of plate boundaries or by changes in the orientation of the con-

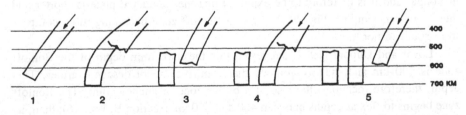

Fig. 17. Schematic interpretation of the mechanism represented in Fig. 16. 1, dip of downgoing lithosphere to 700 km; 2, flexing and rupture after motion is arrested at 700 km; 3, dip behind the first slab of the newly fractured slab; 4 and 5, continuation of the mechanism. The arrows indicate the direction of motion of the subducted lithosphere.

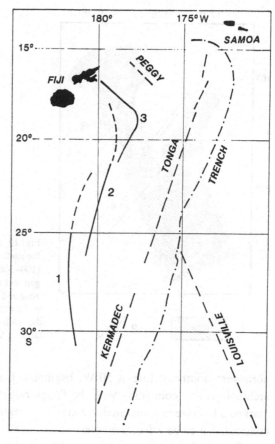

Fig. 18. Distribution of deep earthquakes of the Kermadec–Tonga showing the change in orientation and illustrating the migration of the Benioff zone. 1, original direction, little different from present subduction. 2, first direction change; the change in the south is very little different from the present zone but, in contrast, the northern part shows an inflection of the subduction zone around the Fiji platform. 3, orientation of subduction prior to the formation of the north Tonga zone. (After Louat and Dupont, 1982.)

verging Pacific and Australo-Indian plates. The northern part of the Benioff zone in the Kermadec area may still represent original subduction.

In the southern part of the Kermadec, the surface trace of the intermediate earthquakes curves westwards, and this seismicity ceases abruptly between 32 and 33°S. These features may be linked to the existence of an ancient major fault trending southeast–northwest through this area, and can be correlated with changes in the morphology of the Havre trough and the hypothetical Cook fracture zone (Mammerickx *et al.*, 1971).

(b) The North Tonga. In the north Tonga region, the principal seismic discontinuity in the intermediate earthquakes is not correlated with the trace of the trench between the Tonga and Samoa archipelagos, and the same is true of shallow earthquakes, especially in the Lau basin where they have a southeast–northwest trend (Fig. 19). This distribution of intermediate and shallow earthquakes was interpreted by Louat and Dupont (1982) as indicating the existence of an old boundary between the Pacific and Australo-Indian plates. This boundary was formerly situated along a

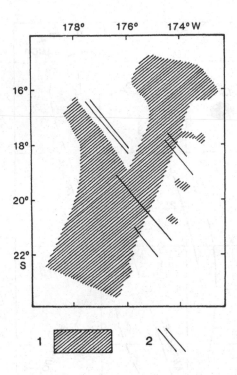

Fig. 19. Shallow and intermediate seismicity of the north Tonga region. Intermediate earthquakes (100–500 km) in hachures (1) show a seismicity gap and a turn to the northwest of the subducted zone and the shallower part of the subduction zone in the extreme north. Only in the Lau basin does the shallow seismicity (2) retain a SE–NW orientation. (After Louat and Dupont, 1982.)

northwest–southeast line at 50°W, beginning from a region close to the Vava'u archipelago and coinciding with the Peggy ridge; they did not specify whether it marked a fracture or a subduction zone, in contrast to Hanuš and Vaněk (1979) who considered it to be a fracture zone.

Present-day subduction between 19 and 15°S may be understood in terms of a northward migration of Tonga–Kermadec subduction, a phenomenon possessing, in its own right, extensional dynamics which lead to the conclusion that the Lau basin is formed of recent crust. The same phenomenon is also visible in the southern New Hebrides subduction where the same depth of intermediate earthquakes can be observed as in north Tonga and where the Benioff zone is quite as irregular as that of north Tonga. It thus seems that the extremities of subduction zones may display similar characteristics, whatever the region. The north Tonga region shows that the Tonga–Kermadec subduction has produced secondary movements, prolonging subduction to the north. The region is still evolving.

Nevertheless, it has to be reported that the study of the extreme northern Tonga by Louat and Dupont is based on the convergence of the trace of intermediate and shallow focus earthquakes recorded at less than 100 stations, and on the morphology of the arc. If an epicenter whose location is in no doubt can be found lying within the zone, which these authors consider as a permanent gap in intermediate seismicity, all their conclusions will require revision. However, using a similar method with some variations, Hanuš and Vaněk arrived at similar conclusions, in particular on the existence of an ancient boundary between the plates along the Peggy ridge.

4. Shallow Earthquakes

Shallow focus earthquakes in the Tonga–Kermadec island arc have been associated with subduction, principally through the interpretation of focal mechanisms. Johnson and Molnar (1972) distinguish three kinds of mechanism: thrust mechanisms (subduction), tear (hinge) faulting mechanisms, and mechanisms resulting from internal tension due to the flexuring of the descending Pacific plate.

Along the length of the Tonga–Kermadec arc, earthquakes which can be tied to tension at plate contacts can be grouped in a zone about 40 km wide at Tonga and 70 km in the Kermadec. The transition between the two zones is marked by a seismicity gap toward 26°S, opposite the Ozbourn guyot (Louisville ridge). It should also be noted that there are concentrations of shallow earthquakes in front of the arc where the trench is deformed, where the trench bends in north Tonga and north and south of the contact of the Louisville ridge and the island arc.

The shallow earthquakes north of Tonga interpreted by Isacks et al. (1969) as the east–west shearing of the Pacific plate form a group of earthquakes oriented about 30°W. The shear is marked by a network of shallow and intermediate faults spread over 100 km from the curvature of the trench. Farther to the west, the margin of the plates, which has always been considered as a fracture zone, is not, seismologically speaking, at all well marked by shallow earthquakes.

The reduction of shallow seismic activity toward 26°S seems to have been constant since 1900 (Meyers and Von Hake, 1976). Kelleher and McCann (1976) consider this seismic gap as due to the greater buoyancy of the material forming the Louisville ridge and carried down by subduction. Nevertheless, it can be noted that the frequency of shallow earthquakes does not decrease along the extension of the Louisville ridge, except opposite the contact of the ridge and the arc in the direction of subduction. The cause of the seismicity gap may thus be related to deformation of the downgoing plate in the ridge–arc contact zone, having as a consequence a strain reduction at the interface of the two plates (Louat and Dupont, 1982). According to this hypothesis, the zone of weak shallow seismicity should follow to the south the arc–Louisville ridge contact during the subduction of the latter. Confirmation or negation of this conclusion must wait a few million years!

5. Longitudinal Seismological Section of the Tonga–Kermadec

Both Hanuš and Vaněk (1979) and Louat and Dupont (1982) have presented longitudinal sections showing the distribution of earthquakes in Tonga–Kermadec.

On the simplified section presented by the latter authors, the shortening of the Benioff zone in north Tonga is particularly clearly shown. This corresponds to recent subduction (of 330 km) in this region. Hanuš and Vaněk (1979) figure in their section the influence of relief of the downgoing plate on the morphology of the Benioff zone; thus, opposite the Niue ridge, Louisville ridge (Ozbourn seamount), and the south and north Kermadec seamounts, the present Benioff zone attains depths of only 250, 200, 220, and 250 km, respectively. These shallow depths in regions where focal depths greater than 500 km are found support the concept of the fossil subduction zones (Lau and Horn), prior to the present subduction zone of Tonga and Kermadec proposed by these authors (see Fig. 15).

The effect of relief on the descending plate on subduction remains a difficult problem to resolve. In the present state of knowledge, two models which often have no common link can be compared: on the one hand, well-established relief may cause changes in arc morphology (Dupont, 1982b); on the other hand, there are irregularities in the Benioff zone which can be explained by the subduction of relief whose existence has never been established. Only hypotheses link one with the other. In the case of the Tonga–Kermadec arc system, if for the Louisville ridge it is reasonably certain that there is an elongate structure of which part has been subducted, what can be said of other relief? Is it isolated relief or a chain? The morphology of the Pacific plate makes it impossible to decide with certainty. If the hypothesis of Hanuš and Vaněk can be verified, there will be a means of going back in time and determining whether former relief existed and disappeared.

6. Conclusion

The study of the seismicity of the Tonga–Kermadec arc shows that subduction is geographically unstable along active oceanic margins, that subduction in the Tonga–Kermadec arc is not simple—particularly in the north—and that the relative linearity of the arc is unlikely to be a permanent feature of it. The morphology of the Benioff zone can be explained by the anchoring of subducted lithosphere at 700 km, and this plays a important role in the flexure and rupture of the subducted slab with the resultant eastward migration of the trench.

The deep focus earthquakes, where they are coordinated, provide an excellent indication of the geodynamic history of the region; it is by this means that several successive periods of subduction have been discerned. The latest of these modified the north of Tonga where the old limiting subduction–transform fault passes into the Peggy ridge.

Finally, it is worth recalling that the interpretation of seismic data is closely linked to the number and quality of the data (location and focal depth, number of available records).

IV. CONCLUSIONS

Studies carried out over the years on the Tonga–Kermadec ridge have led to an improvement in our knowledge of the area. It can be shown that a knowledge of morphological, refraction, and gravimetric data leads to a better understanding of the general form of the ridge and the superficial and internal structure of this subduction zone, one of the most important structural zones in the South Pacific. The new hypotheses born of the confrontation of new data sought by the scientific community have advanced our ideas on the origin, formation, and evolution of this active margin. We are at but a staging point and research must continue.

It can be stated that the morphology of an island arc is not immutable. An island arc is born, grows, evolves, dies, and becomes fossilized. The morphology not only changes as a result of the volcanic activity which is characteristic of island arcs, with the birth of new volcanoes, rejuvenation of volcanic activity, and the disappearance of volcanic islands by marine erosion, but also through seismic activity through earthquakes with the disappearance or uplifting of parts of an island, or by marine activity with the formation of terraces. Such morphological changes, very common in the Tonga–Kermadec island arc, are all local. In contrast, it appears the subduction of relief present on the subducting oceanic plate may cause, more slowly, morphological changes not detectable on a human time scale and less spectacular than the earlier-mentioned changes, but they are, nevertheless, much more important geographically.

Seismicity permits us, through the study of present-day earthquakes, to go back in time and hypothesize on the mechanisms which operated in the past. The paradox for the nonscientist is that the more earthquakes there are, the more the seismologist is satisfied, for in this way knowledge is improved. It is thus that the study of deep focus earthquakes permits the conclusion that subduction below the Tonga–Kermadec arc is not stable, but is being progressively displaced toward the east. The amount is very small in the Kermadec zone, but is much greater for Tonga—in particular, the northern part where subduction appears to be very recent (3 to 4 Ma). This latter conclusion is supported by intermediate focus earthquakes, or rather by their absence, for there is a seismicity gap in the extreme northern part of the Tonga arc where the Benioff zone only extends down to 330 km.

These regions, which have been called active margins, merit that designation for they are not only active seismically and volcanically but also morphologically. The scientific cruise of the R/V Lee in 1982 in the Tonga area, and the projected mission of the N/O Charcot in the Lau and Havre basins and over the junction of the arc with the Louisville ridge, can provide new data which may confirm or invalidate the different hypotheses presented here.

The work on the Tonga–Kermadec island arc discussed in this chapter represents only a resumé, an incomplete resumé, for the author was forced to make a selection out of the many papers published. The interested reader is encouraged to plunge into the abundant literature on this area of the southwest Pacific, using the basic list presented here.

V. ADDENDUM

Since the completion of this chapter on the Tonga–Kermadec arc, many new data have been published as a result of the cruises of the *R/V S. P. Lee* (Australia–New Zealand–United States Tripartite Agreement) during March and April 1982, the *R/V Sonne* (West German cruise S035) from December 1984 through February 1985, and the *N/O Jean Charcot* (SEAPSO cruise of GIS, with ORSTOM–IFREMER–UBO–CNRS and BRGM) from December 1985 through January 1986.

The cruise of the *R/V S. P. Lee* (Scholl and Vallier, 1985) provided much new information on the southern part of the Tonga arc platform from single- and multi-channel seismic and refraction data, drilling and dredging, accompanied by onshore studies of the Ata and Eua Islands and by studies of the Lau Ridge and Basin. In particular, Morton and Sleep (1985) demonstrated the existence of the top of the magma chamber crustal between 22 and 23°S (Valu Fa Ridge).

The cruise of the *R/V Sonne* provided precision on the trend and morphology of the Valu Fa ridge. Hydrothermal activity on this ridge was demonstrated by Stackelberg *et al.* (1985), and dredging in the northern Lau Basin by the *R/V Thomas Washington* in 1986 brought up sulfur.

Continuing the investigations begun in 1975, French researchers carried out a single-channel seismic and seabeam survey over the southern part of the Tonga Trench and over the zone of contact between the Louisville Ridge and the Tonga arc. These studies also showed the beginning of normal faulting of the Ozbourn guyot. This guyot has been preceded down the subduction zone by another, which still appears in the depths of the trench, and the internal trench wall may be a zone of tectonic erosion (Pontoise *et al.*, 1986).

The Tonga–Kermadec island arc—as well as its backarc basin, the Lau Basin—continues to generate scientific interest. The international scientific community (American, Australian, French, German, Japanese, and New Zealand) tends to combine in small groups, proposing deep drilling within the framework of the Ocean Drilling Program as well as deep dives with American, French, and Japanese submersibles.

ACKNOWLEDGMENTS

The author expresses his thanks to Dr. A. E. M. Nairn for the translation of the text, and to C. Heaton for completing the illustrations.

REFERENCES

Aggarwal, Y. P., Barazangi, M., and Isacks, B., 1972, P and S traveltimes in the Tonga–Fiji region: A zone of low velocity in the uppermost mantle behind the Tonga island arc, *J. Geophys. Res.* **77**:6427–6434.

Barazangi, M., and Isacks, B. L., 1971, Lateral variations of seismic-wave attenuation in the upper mantle above the inclined earthquake zone of the Tonga island arc: Deep anomaly in the upper mantle, *J. Geophys. Res.* **76**:8493–8516.

Bryan, W. B., Stice, G. D., and Ewart, A., 1972, Geology, petrography, and geochemistry of the volcanic islands of Tonga, *J. Geophys. Res.* **77**:1566–1585.

Burns, R. E., Andrews, J. E., van der Lingen, G. J., Churkin, M., Galehouse, J. S., Packham, G. H., Davies, T. A., Kennett, J. P., Dumitrica, P., Edwards, A. R., and von Herzen, R. P., 1973, Site 204, in: *Initial Reports of the Deep Sea Drilling Project*, Vol. 21, U.S. Government Printing Office, Washington, D.C., pp. 33–56.

Chase, C. G., 1971, Tectonic history of the Fiji plateau, *Geol. Soc. Am. Bull.* **82**:3087–3110.

Daniel, J., Collot, J. Y., Ibrahim, A. K., Isacks, B., Latham, G. V., Louat, R., Maillet, P., Malahoff, A., and Pontoise, B., 1982, La subduction aux Nouvelles-Hébrides, *Trav. Doc. ORSTOM* **147**:149–156.

Dubois, J., Dupont, J., Lapouille, A., and Recy, J., 1977, Lithospheric bulge and thickening of the lithosphere with age: Examples in the South-West Pacific, International Symposium Geodynamics South-west Pacific, Nouméa, 1976, Technip Editions, Paris, pp. 371–380.

Dupont, J., 1979, Le système d'arc insulaire des Tonga et Kermadec: deux morphologies différentes, une seule zone de subduction (Pacifique Sud), *C.R. Acad. Sci.* **289**:245–248.

Dupont, J., 1982a, Le cadre général et les traits essentiels de l'arc insulaire des Tonga–Kermadec, *Trav. Doc. ORSTOM* **147**:249–261.

Dupont, J., 1982b, Morphologie et structures superficielles de l'arc insulaire des Tonga–Kermadec, *Trav. Doc. ORSTOM* **147**:263–282.

Dupont, J., and Herzer, R. H., 1985, Effect of subduction of the Louisville Ridge on the structure and morphology of the Tonga Arc, in: *Geology and Offshore resources of Pacific island arcs–Tonga region* (D. W. Scholl and T. L. Vallier, eds.), Circum–Pacific Council for Energy and Mineral Resources, Houston, pp. 323–332.

Dupont, J., Louat, R., Pontoise, B., Missegue, F., Latham, G. V., and Malahoff, A., 1982, Aperçu morphologique, structural et sismologique de l'arc insulaire des Tonga–Kermadec, *Trav. Doc. ORSTOM* **147**:319–323.

Eade, J. V., 1971, Tonga bathymetry, 1/1,000,000, *N.Z. Oceanogr. Inst. Chart, Oceanic Ser.*

Elsasser, W. M., 1971, Sea-floor spreading as thermal convection, *J. Geophys. Res.* **76**:1101–1112.

Ewart, A., and Bryan, W. B., 1973, The petrology and geochemistry of the Tongan islands, in: *The Western Pacific: Island Arcs, Marginal Seas, Geochemistry* (P. J. Coleman, ed.), University of Western Australia Press, Nedlands, pp. 503–522.

Ewart, A., Brothers, R. N., and Mateen, A., 1977, An outline of the geology and geochemistry, and the possible petrogenetic evolution of the volcanic rocks of the Tonga–Kermadec–New Zealand island arc, *J. Volcanol. Geotherm. Res.* **2**:205–250.

Faleyev, V. I., Udintsev, G. B., Agapova, G. V., Domanitskiy, Y. A., and Marova, N. A., 1977, Data on the maximum depths of trenches in the world ocean, *Oceanology* **17**:311–313.

Fisher, R. L., and Engel, C. G., 1969, Ultramafic and basaltic rocks dredged from the nearshore flank of the Tonga trench, *Geol. Soc. Am. Bull.* **80**:1373–1378.

Gill, J. B., 1976, Composition and age of Lau basin and ridge volcanic rocks: Implications for evolution of an interarc basin and remnant arc, *Geol. Soc. Am. Bull.* **87**:1384–1395.

Gutenberg, B., and Richter, C. F., 1954, *Seismicity of the Earth*, 2nd ed., Princeton University Press, Princeton, N.J.

Hanuš, V., and Vaněk, J., 1979, Morphology and volcanism of the Wadati–Benioff zone in the Tonga–Kermadec system of recent subduction, *N.Z. J. Geol. Geophys.* **22**:659–671.

Hawkins, J. W., 1974, Geology of the Lau basin, a marginal sea behind the Tonga arc, in: *The Geology of Continental Margins* (C. A. Burk and C. L. Drake, eds.), Springer-Verlag, Berlin, pp. 505–520.

Hayes, D. E., and Ewing, M., 1971, The Louisville ridge—A possible extension of the Eltanin fracture zone, *Antarct. Res. Ser.* **15**:223–228.

Hoffmeister, J. E., 1932, Geology of Eua, Tonga, *Bull. Bernice P. Bishop Mus.* **96**:3–93.

Isacks, B. L., and Barazangi, M., 1977, Geometry of Benioff zones: Lateral segmentation and downwards bending of the subducted lithosphere, in: *Island Arcs, Deep Sea Trenches and Back-Arc Basins* (M. Talwani and W. C. Pitman, III, eds.), M. Ewing Series, 1, American Geophysical Union, pp. 99–114.

Isacks, B. L., Oliver, J., and Sykes, L. R., 1968, Seismology and the new global tectonics, *J. Geophys. Res.* **73**:5855–5899.

Isacks, B. L., Sykes, L. R., and Oliver, J., 1969, Focal mechanisms of deep and shallow earthquakes in the Tonga–Kermadec region and the tectonics of island arcs, *Geol. Soc. Am. Bull.* **80**:1443–1470.

James, A., and Falvey, D. A., 1978, Analysis of palaeomagnetic data from Viti Levu, Fiji, *Bull. Aust. Soc. Explor. Geophys.* **9**:115–117.

Johnson, T., and Molnar, P., 1972, Focal mechanisms and plate tectonics of the Southwest Pacific, *J. Geophys. Res.* **77**:5000–5032.

Jouannic, C., Taylor, F. W., and Bloom, A. L., 1982, Sur la surrection et la déformation d'un arc jeune: l'arc des Nouvelles-Hébrides, *Trav. Doc. ORSTOM* **147**:223–246.

Karig, D. E., 1970, Ridges and basins of the Tonga–Kermadec island arc system, *J. Geophys. Res.* **75**:239–254.

Karig, D. E., and Sharman, G. F., 1975, Subduction and accretion in trenches, *Geol. Soc. Am. Bull.* **86**:377–389.

Katz, H. R., 1976, Sediments and tectonic history of the Tonga ridge, and the problem of the Lau basin, *U.N. ESCAP, CCOP/SOPAC Tech. Bull.* 2 pp. 153–165.

Kelleher, J., and McCann, W., 1976, Buoyant zones, great earthquakes, and unstable boundaries of subduction, *J. Geophys. Res.* **81**:4885–4896.

Kogan, M. G., 1976, Gravity anomalies and main tectonic units of the southwest Pacific, *J. Geophys. Res.* **81**:5240–5248.

Kroenke, L. W., and Tongilava, S. L., 1975, A structural interpretation of two reflection profiles across the Tonga arc, *South Pac. Mar. Geol. Notes* **1**:9–15.

Ladd, H. S., and Hoffmeister, J. E., 1945, Geology of Lau, Fiji, *Bull. Bernice P. Bishop Mus.* **181**.

Latham, G. V., Donoho, P., Griffiths, K., Roberts, A., and Ibrahim, A. K., 1978, The Texas ocean bottom seismograph, Paper presented at the Society of Exploratory Geophysics, Houston, Offshore Technical Conference, p. 1467.

Louat, R., 1977, Relative seismic energy released in South Pacific area, International Symposium Geodynamics South-west Pacific, Nouméa, 1976, Technip Editions, Paris, pp. 29–36.

Louat, R., and Dupont, J., 1982, Sismicité de l'arc des Tonga–Kermadec, *Trav. Doc. ORSTOM* **147**:299–317.

Lucas, W. H., 1972, South Pacific traverse RP-7-SU-71. Pago-Pago to Callao to Seattle, NOAA Technical Report ERL 230-POL 8, Boulder, Colo.

Malahoff, A., Hammond, S., Feden, R., and Larue, B., 1979, Magnetic anomalies: The tectonic setting and evolution of the southwest Pacific marginal basins, Third Southwest Pacific Workshop Symposium, Sydney.

Malahoff, A., Feden, R., and Fleming, H., 1982, Magnetic anomalies and tectonic fabric of marginal basins north of New Zealand, *J. Geophys. Res.* **87**:4109–4125.

Mammerickx, J., Chase, T. E., Smith, S. M., and Taylor, I. L., 1971, Bathymetry of the South Pacific, Chart No. 12, Scripps Institution of Oceanography.

Meyers, H., and Von Hake, C. A., 1976, Earthquake data file summary-key to geophysical records documentation No. 5, NOAA, National Geophysical and Solar Terrestrial Data Center, Boulder, Colo.

Missègue, F., and Malahoff, A., 1982, Etude gravimétrique de l'arc des Tonga, *Trav. Doc. ORSTOM* **147**:293–298.

Morton, J. L., and Sleep, N. H., 1985, Seismic reflections from a Lau Basin magma chamber, in: *Geology and Offshore Resources of Pacific Island Arcs—Tonga Region* (D. W. Scholl and T. L. Vallier, eds.), Circum-Pacific Council for Energy and Mineral Resources, Houston, pp. 441–453.

Oliver, J., and Isacks, B. L., 1967, Deep earthquake zones, anomalous structures in the upper mantle, and the lithosphere, *J. Geophys. Res.* **72**:4259–4275.

Packham, G. H., and Andrews, J. E., 1975, Results of leg 30 and the geologic history of the southwest Pacific arc and marginal sea complex, in: *Initial Reports of the Deep Sea Drilling Project,* Vol. 30, U.S. Government Printing Office, Washington, D.C., pp. 691–705.

Parsons, B., and Sclater, J. G., 1977, An analysis of the variation of ocean floor bathymetry and heat flow with age, *J. Geophys. Res.* **82**:803–827.

Pontoise, B., and Latham, G. V., 1982, Etude par réfraction de la structure interne de l'arc des Tonga, *Trav. Doc. ORSTOM* **147**:283–291.

Pontoise, B., Latham, G. V., Daniel, J., Dupont, J., and Ibrahim, A. K., 1980, Seismic refraction studies in the New Hebrides and Tonga area, *U.N. ESCAP, CCOP/SOPAC Tech. Bull. 3* pp. 47–58.

Pontoise, B., Latham, G. V., and Ibrahim, A. K., 1982, Sismique réfraction: structure de la croûte aux Nouvelles-Hébrides, *Trav. Doc. ORSTOM* **147**:79–90.

Pontoise, B., Pelletier, B., Aubouin, J., Baudry, N., Blanchet, R., Butscher, J., Chotin, P., Diament, M., Dupont, J., Eissen, J. P., Ferriere, J., Herzer, R., Lapouille, A., Louat, R., d'Ozouville, L., Soakai, S., and Stevenson, A., 1986, La subduction de la ride de Louisville le long de la fosse des Tonga: premiers résultats de la campagne SEAPSO (leg V), *C.R. Acad. Sci.* **303**:911–918.

Raitt, R. W., Fisher, R. L., and Mason, R. G., 1955, Tonga trench, *Geol. Soc. Am. Spec. Pap.* **62**:237–254.

Rodda, P., 1967, Outline of the geology of Viti Levu, *N.Z. J. Geol. Geophys.* **10**:1260–1273.

Scholl, D. W., and Vallier, T. L. (compilers and editors), 1985, *Geology and Offshore Resources of Pacific Island Arcs—Tonga Region,* Circum-Pacific Council for Energy and Mineral Resources, Houston.

Sclater, J. G., and Detrick, R., 1973, Elevation of midocean ridges and the basement age of Joides deep sea drilling sites, *Geol. Soc. Am. Bull.* **84**:1547–1554.

Sclater, J. G., Hawkins, J. W., Mammerickx, J., and Chase, C. G., 1972, Crustal extension between the Tonga and Lau ridges: Petrologic and geophysical evidence, *Geol. Soc. Am. Bull.* **83**:505–518.

Shor, G. G., Kirk, H. K., and Menard, H. W., 1971, Crustal structure of the Melanesian area, *J. Geophys. Res.* **76**:2562–2586.

Stackelberg, U. Von, and The Shipboard Scientific Party, 1985, Hydrothermal Sulfide Deposits in Back-Arc Spreading Centers in the Southwest Pacific, *BGR Circular 2,* Hannover, West Germany, pp. 3–14.

Sykes, L. R., 1966, The seismicity and deep structure of island arcs, *J. Geophys. Res.* **71**:2981–3006.

Talwani, M., Worzel, J. L., and Ewing, M., 1961, Gravity anomalies and crustal section across the Tonga trench, *J. Geophys. Res.* **66**:1265–1278.

Tongilava, S. L., and Kroenke, L. W., 1975, Oil prospecting in Tonga 1968–1974, *South Pac. Mar. Geol. Notes* **1**:1–8.

Weissel, J. K., 1977, Evolution of the Lau basin by the growth of small plates, in: *Island Arcs, Deep Sea Trenches and Back-Arc Basins* (M. Talwani and W. C. Pitman, III, eds.), M. Ewing Series 1, American Geophysical Union, pp. 429–436.

Chapter 10

THE GEOLOGICAL EVOLUTION OF NEW ZEALAND AND THE NEW ZEALAND REGION

R. J. Korsch* and H. W. Wellman

Department of Geology
Victoria University of Wellington
Wellington, New Zealand

I. MAIN INTRODUCTION

The Geology of New Zealand, edited by R. P. Suggate, G. R. Stevens, and M. T. Te Punga (1978), contains a near-complete set of references up to 1970, and, to cover a printing delay, a less complete set up to 1976. For brevity and where appropriate, we use *"Geology NZ"* with page and figure numbers for references up to 1970. New Zealand stratigraphy is ordered into stage divisions that have unique two-letter symbols, and have been set out with a time scale by Stevens (1980). The letter symbols are used below in parentheses when appropriate.

In the following, "New Zealand Region" is used for the area shown by Fig. 1, and "New Zealand" (shortened to "NZ") for the present NZ land area, which is a minor part of the NZ Region.

First a regional account is given of the geology of NZ. There are 20 regions that are mostly determined by the pre-Cretaceous structure. Salient points that need explanation are stressed. The regional account makes it clear that as a country, NZ is unusual, because of the large tectonic movements that have taken place in the Cenozoic, and in particular in the late Cenozoic. The Cenozoic movements are part of more extensive movements that can be traced into the seafloor well beyond NZ

*Present address: Bureau of Mineral Resources, Canberra, Australian Capital Territory, Australia.

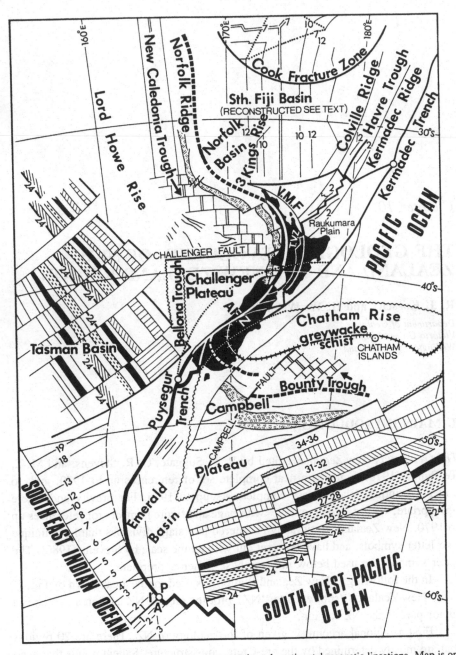

Fig. 1. Map of New Zealand Region showing oceanic and continental magnetic lineations. Map is on conical projection with standard parallels at 30°S and 50°S. Distortion is negligible and the map can be cup up and reconstructed by taking out the younger seafloor. VMF, Vening Meinesz Fault; AFZ, Alpine Fault Zone; IPA, Indian–Pacific–Antarctic Triple Junction. Triangle of dotted lines with northern apex truncated by northern margin of map represents the three vectors for the South Fiji Basin triple opening junction. Stokes Magnetic Lineations consist of Dun Mountain Lineation shown by heavy lines which is flanked to west and south by a pair of lines that enclose a "granite" patterned belt. Light lines on Campbell Plateau and Lord Howe Rise are Ordovician (?) magnetic lineations. Movement of the Endeavour and Three Kings slides is taken out.

(Fig. 2) and most of the data for the extension of the movements are from oceanic and continental magnetic lineations.

According to the oceanic magnetic lineations, most of the oceans of the NZ Region were created in the Cenozoic. If on a map the Cenozoic oceans are cut out and the older parts reconstructed, then the resultant movement through the Alpine Fault Zone is in fair agreement with the movement inferred from unbending and unfaulting NZ for the Cenozoic.

Oceanic sliding is a related phenomenon and a further complication. Within a restricted area to the south of the Campbell Plateau there is a large detached continental fragment named Bollons Seamount, an anomalous strike for a series of magnetic lineations, and an unexpected oceanic rise named the Endeavor Rise (see Fig. 5). The three features are thought to form the Endeavour Slide. Although reasonably well established as an oceanic slide, the sliding movement is relatively small and is almost smoothed out in seafloor spreading calculations.

To the north of NZ the Three Kings Rise is thought to be a slide related to the formation of the Northland Chaos. It is less well established than the Endeavour Slide but more important for reconstruction if correctly identified. It would explain the position of the otherwise anomalous Three Kings Rise, make it possible to restore a missing part of the Dun Mountain Ophiolite Line of the Stokes Magnetic Anomaly, and provide the missing force needed to trigger the well-established onshore slide known as the Northland Chaos.

The reconstructions discussed next and used to illustrate the changes in the geography of the NZ Region are based on data from a variety of sources. The

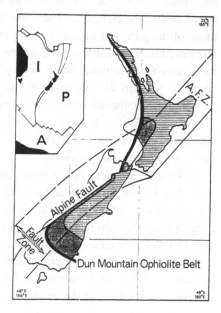

Fig. 2.   Sketch map on Mercator projection with a 5° grid showing main structural features of NZ. Graywacke shown by horizontal hatching and schist by heavy horizontal hatching. Dun Mountain Ophiolite and Alpine Fault limit graywacke and schist to east. Dun Mountain Ophiolite dextrally bent (c. 500 km) by NZ Recurved Arc and dextrally displaced 500 km by Alpine Fault. The Recurved Arc and the Alpine Fault make up the Alpine Fault Zone which extends beyond NZ as the Indian–Pacific plate boundary. The map inset shows the triple junction and the three boundaries of the Indian, Pacific, and Antarctic plates with relation to NZ.

amount of rotation is taken from seafloor spreading, the shiftpoles are also from seafloor spreading but are modified slightly so that their changes are geologically reasonable. Also taken into account are geological reference lines and dating from NZ geology. Gondwana in which several continental parts were once united provides an invaluable starting point for reconstruction.

A series of reconstructions and a geological history can be given either forward in time or backward in time. Each has advantages and disadvantages. The present is all that is known with certainty and this is the main reason for going backward. On the other hand, going backward reverses cause and effect and perhaps more important reverses the sense of tectonic movements. For instance, going backward changes the apparent sense of faulting, dextral for sinistral and so on and is confusing. The reconstructions and geological descriptions below, although constructed backward in time, are given forward in time, and consist of three main phases that relate to the separation of NZ from Gondwana.

In the first phase, Precambrian to Carboniferous, NZ is part of Gondwana, and a description is given of parts (terranes) that happened to be exposed. There are many undescribed terranes in the submerged continental parts of the NZ Region. The nature and ages of the terranes are discussed and an attempt made to determine when they came together.

The second phase, Carboniferous to mid-Cretaceous, deals with subduction on the Pacific coast of Gondwana and the source and accretion of the huge mass of graywacke that now makes up two-thirds of NZ.

The third and last phase, mid-Cretaceous to the present, deals with the separation of the NZ Region from Gondwana and summarizes the complex events that have taken place since. It ends with an account of the microplates that make up NZ today and in the Holocene. The plate tectonic data for the present greatly exceed those of the past, and include faulting and tilting rates from dated river and marine terraces, historical earthquake movements, and data for elastic movements within the plates from retriangulation surveys. The Holocene deformation rates which equal or perhaps exceed those of the past show that for NZ, deformation and orogeny are continuing processes and not events that happened in the distant past.

From 1959 to 1968 the geology of NZ was effectively described on 28 rectangular maps with a 1 : 250,000 scale. Each map included a concise account of geology and mineral resources.

Geology NZ is largely a restatement of the map data. Presentation is first according to time divisions and second according to districts, there being six main time divisions and seven districts which are political rather than geological and distinct from the map rectangles mentioned above. *Geology NZ* contains an excellent 1 : 1,000,000 colored geological map which we cannot attempt to reproduce.

In 1956 one of the present writers gave a brief account of the geology of NZ (Wellman, 1956). Presentation was first in terms of 20 regions that are essentially geological, and then for the whole of NZ in terms of time. The 20 regions proved satisfactory and are used below with updated geology largely from the map sheets and *Geology NZ*. The position of the 20 regions is shown by Fig. 3.

A. Introduction to New Zealand Structural Regions

There are four closely related structural features that extend over the whole of NZ and transcend the 20 regions (compare Fig. 2 and Fig. 3):

1. The Dun Mountain Ophiolite (eastern part of Stokes Magnetic Anomaly).
2. The recurved arc shape (mirror image "S") of the structural grain of both basement, including the Dun Mountain Ophiolite, and cover. Basement is pre-mid-Cretaceous and cover post-mid-Cretaceous (greater and less than 100 Ma).
3. The Alpine Fault and its main extension the Wairau Fault. The basement, including the Dun Mountain Ophiolite, is dextrally displaced by 500 km at the Alpine Fault.
4. The graywacke and metamorphosed graywacke that make up the two-thirds of the basement that lies to the east of the line of the Dun Mountain Ophiolite and its line of displacement on the Alpine Fault.

The Dun Mountain Ophiolite is a near-vertical slab of serpentinite and related rocks up to 1 km wide. It produces a moderately strong and unusually continuous positive magnetic lineation (= anomaly). Except where displaced by the Alpine Fault the ophiolite crops out almost continuously in South Island. The magnetic lineation extends through South Island and can be traced north, with minor displacements, from the north end of South Island along the west coast of North Island almost to North Cape, but in North Island the ophiolite, except for a single outcrop, is buried beneath younger rocks. Within NZ itself the Dun Mountain Ophiolite is a reference line of major importance. The interpretation of the structure of the NZ Region thus depends largely on correct identification of the magnetic lineation produced by the Dun Mountain Ophiolite outside NZ on the assumption that it is not confined to NZ itself.

The recurved arc was first described by Macpherson in 1946. He showed that the cover structures as well as the undermass structures have the mirror image "S" shape. It has a dextral sense of bending and is considered here to be drag on the Alpine Fault. Thus, the recurved arc together with the Alpine Fault make up the Alpine Fault Zone, for which the total dextral offset is about 1000 km.

The Alpine Fault is the most conspicuous structural feature in NZ. The total dextral displacement of about 500 km is generally accepted, but there is no general agreement as to when the faulting started. According to *Geology NZ*, almost all the faulting took place 100 Ma or more ago during the Rangitata Orogeny. On the other hand, reconstructions based on seafloor spreading, using the fault zone as the Indian–Pacific plate boundary, require a dextral displacement of about 1000 km

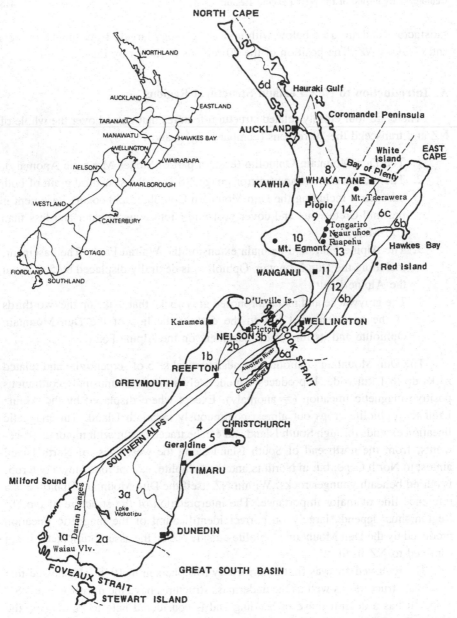

Fig. 3. Sketch map of NZ showing the 20 structural regions described in the text. Geological affinities indicated by letter suffixes. For instance, 1a, 2a, and 3a have been dextrally displaced by 1b, 2b, and 2c. Also 6a, 6b, 6c, and 6d contain Cretaceous and Tertiary trench deposits. The first three are close to the present subduction zone. The last (6d, Western Northland) is the Northland Chaos and was originally to the east of NZ. Major geographical regions are: 1a, Fiordland; 1b, West Coast of South Island; 2a, Southland Syncline; 2b, Nelson Syncline; 3a, Otago and Alpine Schist (= Haast Schist); 3b, Marlborough Schist; 4, Alpine Greywacke; 5, Canterbury Plains and Banks Peninsula; 6a, Eastern Marlborough; 6b, East Coast Ranges of North Island; 6c, Raukumara Peninsula; 6d, Western Northland; 7, Eastern Northland; 8, South-east Auckland; 9, South-west Auckland; 10, Taranaki & Egmont; 11, Rangitikei Basin; 12, Wairarapa and Hawke's Bay Depression; 13, Axial Range of North Island; 14, Volcanic Belt (Taupo Volcanic Zone).

together with a 60° rotation between the two sides of the fault zone during the last 60 Ma, i.e., during the Cenozoic, and after the Rangitata Orogeny. It so happens that reference lines with well-defined fault displacement are either undermass features or river-terrace-risers of Holocene age, i.e., of the last 10,000 yr only. For this reason the Holocene tectonics of NZ are discussed in some detail in order to show that the Holocene movement rates are consistent with those of the seafloor spreading.

Graywacke is the problem rock of NZ. The name itself is unsatisfactory, but Torlesse Supergroup, the alternative used in *Geology NZ*, is longer and no better. The graywacke is well-bedded, strongly and irregularly deformed, and consists of detrital fragments that range in size from clay to cobbles, and represent a wide range of compositions. Red beds generally associated with basaltic pillow lavas are widespread and appear to be interbedded but may have a different source from the graywacke. The graywacke is virtually nonmagnetic and directly adjoins the Dun Mountain Ophiolite which limits the graywacke to the east. Consequently, in NZ on moving from east to west there is a magnetic plain bounded by the positive lineation that marks the Dun Mountain Ophiolite. The same magnetic plain and positive lineation is assumed to continue throughout the NZ Region.

The ages, sources, site of deposition with special reference to the Dun Mountain Ophiolite, and mode of deformation of the graywacke are discussed later.

The main stratigraphic division is between undermass and cover. The division extends over about three-quarters of NZ, undermass being 100 m.y. old and older and the cover being younger. There is a well-defined peneplain at the base of the cover. Along the northern east coast of NZ there is uncertainty about the undermass–cover contact, and this is discussed later. In the west and south of South Island there are freshwater beds that span the 100 Ma age. They cover small areas only and are thought to mark a failed opening prior to the breakaway of NZ from Gondwana.

In NZ there is a generally accepted sequence of metamorphic grade using coal rank, zeolites, and minerals in regionally metamorphosed rocks that allow maximum depth of burial to be estimated. At almost all places there is a "metamorphic" unconformity between undermass and cover, showing that the thickness of undermass eroded prior to the formation of the peneplain was much greater than that deposited on the peneplain. Figure 4 shows the estimated thickness of cover in kilometers for South Island. Note that the boundary between subbituminous and bituminous coal is at about 3 km depth burial. The inferred increase in cover thickness toward the Alpine Fault and what is now the Southern Alps is the most striking feature of the map, and cover thickness of up to 4 km in the Oligocene along the line of the present South Alps is likely.

The strongly compacted Cenozoic strata infaulted in the Moonlight Fault near Lake Wakatipu mentioned later, and the Cenozoic strata infaulted in the Alpine Fault at Kaipo Slip 10 km NE of Milford Sound (*Geology NZ:* p. 503), are critical outcrops within the Southern Alps.

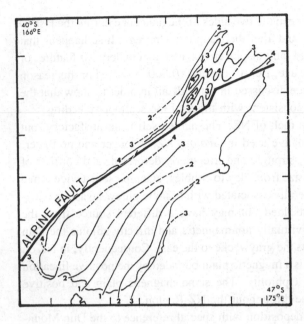

Fig. 4. Sketch map of South Island of NZ on Mercator projection with a 1° grid. Contours give estimated depth in kilometers of maximum burial of base of cover which ranges in age from Upper Cretaceous to Oligocene. Depth estimates from coal analyses, mostly mine samples. (Modified from Suggate, 1959.)

It is worth noting that no direct relation exists between coal rank and either age or degree of deformation. Jurassic coal of the south of South Island is slightly lower in rank than Upper Miocene coal (3 km burial) of Murchison 55 km NE of Reefton. The almost horizontal bituminous coal of Maui offshore 90 km SW of Mt. Egmont in North Island with a present and total depth of burial of 3.1 km is much higher in rank than the vertical subbituminous coal of Fletcher Creek 15 km N of Reefton with a 2.1 km infered depth of burial. However, it is certain that the crustal heat flow is not constant and the depth of maximum burial inferred from coal rank is probably up to 30% in error.

1. Fiordland

The region includes the Waiau Valley on the east and to the south the southern part of Stewart Island. The old rocks comprise gneisses, intrusives which are mostly granites, and Ordovician sediments. They are separated by a major unconformity from the next younger rocks (Lower Cretaceous freshwater sandstone and breccia similar to the Ohika beds and Hawks Crag Breccia of the West Coast Region) that crop out over a small area on the southwest coast (*Geology NZ:* p. 400).

Tertiary beds are almost confined to the southern and eastern part of the region. The Eocene consists of thin freshwater beds. The Lower Oligocene is thicker and consists of freshwater beds overlain by calcareous mudstone and limestone. The Upper Oligocene is thin, fairly uniform, calcareous sandstone and limestone. The

Lower Miocene consists of mudstone and sandstone which contain abundant fossils at a few places. The fossils are about the same age as a set of entirely different and colder water fossils that are restricted to Oamaru 90 km NNE of Dunedin and to the north on the east coast of South Island. Being different the two sets were thought to be of different ages and this resulted in an excessive number of NZ Miocene stages. The stages (Awamoan and Hutchinsonian) based on the South Island east coast fossils have since been discarded but were retained in *Geology NZ*. The Upper Miocene and Pliocene consist of freshwater sandstone. Except in the center of Waiau Valley they rest directly on older rocks and mark the coastline around a central Fiordland (*Geology NZ:* Fig. 7.90). The single area of marine beds on the west coast stands as a vertical infaulted strip on the inside of Five Finger Peninsula at Dusky Sound 70 km SSW of Milford Sound. It is Pliocene and represents an old coastline now turned on end (*Geology NZ:* p. 689). The overall history of Tertiary deformation is probably similar to that of the West Coast Region described next.

2. *West Coast of South Island*

The pre-Carboniferous rocks of NZ are confined to this region and region 1a. Fossiliferous rocks include Vendian, Cambrian, Ordovician, Silurian, and Devonian. They are discussed in greater detail later. Next comes freshwater and marine Permian which is restricted to Parapara Peak in the extreme north, and rests unconformably on Ordovician, but is equally as strongly metamorphosed. Mesozoic beds are entirely freshwater and rest unconformably on older rocks. The Topfer is Triassic and confined to a small area near Reefton, the Ohika is probably Lower Cretaceous and confined to the lower Buller Gorge 30 km north. The distinctive red-colored and locally thick Hawks Crag Breccia overlies. It was deposited in fault-angle depressions (half-grabens) that strike NW at a decided angle to the NNE grain of the country and are thought here to represent a failed opening. Paparoa coal measures, Senonian in age, overlie the breccia at Greymouth and in the southern part of this region, and at the extreme north end of the region the similar but Maastrichtian Pakawau beds rest directly on the old rocks (*Geology NZ:* pp. 375–380).

A peneplain was cut over the whole region before Tertiary sedimentation commenced. It now shows as a stripped surface on the flanks of the mountains in the less deformed part and provides an obvious structural reference surface. Tertiary beds are preserved in all the valleys and as outliers on many of the mountains and the Tertiary history is known in some detail. During the Eocene and Oligocene, fault-angle depressions (half-grabens) along the site of the present mountain ranges filled with coal measures, sandstone, mudstone, and then limestone. Metamorphism of the coal measures ranged from subbituminous to low volatile bituminous depending on the depth of burial which ranged from 1 km to 3 km over as little as 10 km horizontal distance. In the middle of the Oligocene the rate of subsidence slowed down, stopped, and then reversed. The deposits in the depression were then slowly elevated. At present the parts that had been most deeply buried and having the highest coal rank are the parts most uplifted, so that coal rank for each tectonic block increases with increasing altitude (*Geology NZ:* pp. 495 and 674). The nature of the faulting changed in the Oligocene from normal to reverse, and the reverse movement still continues as active folding and faulting. As in most other parts of South Island, the intensity of late Tertiary deformation increases fairly regularly toward the Alpine Fault which bounds the region to the southeast. The regional dip of the Oligocene limestone is about 10° in northwest Nelson and steepens to 60° to 90° near the fault.

The late Tertiary was deposited in the present-day fault-angle depressions (half-grabens) that formed by the reverse faulting. It consists of marine sandstone and mudstone with minor freshwater beds and ended with early Quaternary gravels that contain the first schist from the rising Southern Alps to the east of the region.

3. *Southland Syncline*

The old rocks range in age from basal Permian to Lower Jurassic. The major structure, the Southland Syncline, widens and youngs to the southeast in the direction of plunge and is well-defined down to the base of the Triassic. The synclinal axis curves through the region as a wide arc and divides it into two parts. On the northeast side the beds are arranged in regular belts that are steep near the axis and become overturned toward the schist to the NE and are interrupted by folds with axial planes that dip toward the schist. The beds, near-graywackes, sandstones, mudstones, and conglomerates, and the Dun Mountain Ophiolite, have a total thickness of over 15 km. As might be expected from the great thickness, the lower parts are considerably more metamorphosed than the upper. To the north of the Dun Mountain Ophiolite the rocks are subschistose and the boundary with the underlying schist is gradational and somewhat arbitrary. The lowest rocks on the southwest side of the syncline are thick, Lower Permian pillow lavas and andesite sills. The overlying sediments are entirely of shelf facies and are thinner than the more steeply dipping beds on the northeast side of the axis.

The next oldest rocks are small intrusive quartz porphyries and andesites of Jurassic age. Small intrusives of the same composition occur in the Nelson Syncline to be described next.

Except for coal measures at Ohai 140 km W of Dunedin, the Paleocene and Eocene are absent, the Lower Tertiary consisting of Lower Oligocene freshwater beds and marine limestones, and Upper Oligocene calcareous sandstone.

The Neogene is represented in the western part by small areas of Lower Miocene marine mudstone and Upper Miocene marine and freshwater sandstone, and in the eastern part by freshwater quartzose sands and conglomerates.

4. *Nelson Syncline*

The old rocks of the region closely match those of the Southland Syncline 500 km to the southwest. In addition to the strong folding in the early Cretaceous, they were strongly deformed in the Neogene but the original structure is synclinal and similar to that in Southland, with a strongly deformed and partly overturned east flank and a more gently dipping west flank. The synclinal structure is evident in the Permian down to the Dun Mountain Ophiolite. To the east of the ophiolite the Lower Permian Pelorus group is graywacke with bands of serpentinite and red beds. Its structure is complex and unknown. The west flank of the syncline consists of Permian andesite lavas and agglomerates, called the Brook Street Volcanics. There is a single thin bed of serpentinite interleaved in pillow lava in the andesite lavas. The volcanics are intruded by the uppermost part of the Rotorua Igneous Complex which is mostly covered by Moutere Gravels.

The Cretaceous is absent. Tertiary beds are strongly deformed and poorly exposed. They consist of Eocene coal measures and marine dark mudstone, and Oligocene marine sandstone and mudstone turbidites that are quite different from the other Oligocene rocks of New Zealand. The western side of the region is covered by the Moutere Gravels, a thick, deeply weathered graywacke outwash from a mid-Quaternary glaciation of the Spencer Mountains at the north end of the Southern Alps.

In terms of the overall structure of the NZ Region the Dun Mountain Ophiolite is the most important feature of the Nelson Syncline.

5. *Otago and Alpine Schist (= Haast Schist)*

Schist forms the undermass over the whole region. It is many kilometers thick and was formed by burial to 10 km or more from graywacke and red beds similar to those into which it passes laterally.

The Haast Schist is subdivided into the Alpine Schist and Otago Schist. The Alpine Schist lies between the Alpine Fault and graywacke along the crest of the Southern Alps. It widens southward to pass into the Otago Schist. Metamorphic grade increases uniformly away from the graywacke and toward the Alpine Fault.

According to the first appearance of schist fragments, the Otago Schist was first exposed about 100 Ma ago and the Alpine Schist much later and not until 2 Ma ago.

The folding of the graywacke and the unroofing of the Otago Schist is usually attributed to the Rangitata Orogeny. An alternative explanation in terms of subduction and isostatic uplift of an excessively thick crustal section held down by subduction is described later.

It is universally accepted that uplift is continuing on the Alpine side of the Alpine Fault, and that the Southern Alps are a direct result of the uplift, but there is no general agreement on whether the schist was passively uplifted or whether it was heated up or even metamorphosed by the uplift. That the schist and its schistosity was caused by the faulting can be disproved by lamprophyre dikes that intrude it. The dikes, about 80 Ma old, contain ocelli that have remained spherical. The metamorphism that caused the schistosity is thus older than the dikes.

The extreme opposite view, that the schist is the result of simple uplift, is adopted here. It is thought that the crust is being obducted so that the cross section from the crest of the Southern Alps to the Alpine Fault is effectively a section through part of the crust that has been turned on end. The model adopted (Wellman, 1979, Fig. 6) is one that would be in steady state provided that the Southern Alps were being eroded as fast as they were being uplifted. Adams (1980) found that most of the rock eroded from the Southern Alps was transported as the suspended load in the Alpine rivers. On estimating the suspended load from various parts of the Southern Alps, he found to within 10% or so that erosion equaled uplift. It is thus possible that the Southern Alps are, and have been for a few million years, in steady state with erosion equaling uplift. In order to provide perspective, the rate of Alpine erosion is compared with that required to produce the total volume of graywacke and schist. For the Southern Alps, detritus is being eroded away at the rate of about 10×10^6 km^3 in 140 Ma. By comparison the total required for the NZ graywacke and schist which were built up in 140 Ma is about 20 times as great. Thus, it would need 20 Southern Alps being uplifted at the present rate for 140 Ma to produce sufficient material to form the NZ graywacke and schist.

In the uppermost Cretaceous and Paleogene the sea advanced westward over the Otago Schist and the region to the north; in the Neogene the sea retreated. The stages of the advance are better known than those of the retreat. Each division of the marine beds can be traced westward to an old shoreline beyond which it passes into freshwater beds. From the uppermost Cretaceous to the end of the Oligocene, each shoreline is farther west than the preceding one; after the Oligocene the reverse is the case. The complete sequence is similar at all places (lower freshwater beds, marine beds, upper freshwater beds), the marine rocks wedging out inland and at most places containing Oligocene limestone in the middle. The freshwater beds, mostly quartzose and in some places auriferous, contain thick seams of lignite at many places. As mentioned, according to the structure of the West Coast Region the sea advanced during a period of normal faulting and retreated during a period of reverse faulting which still continues.

Deformation of the Tertiary beds increases inland toward the Alpine Fault. Near the coast, where inliers are most numerous, the dip seldom exceeds 10°. The most spectacular inlier is an infaulted strip near Lake Wakatipu. The beds here are vertical and almost parallel to the schistosity planes in the enclosing schist. The schists were gently dipping when the Tertiary beds of the outlier were laid down, and must have been folded with them later. Such late folding is in agreement with the mode of formation of the Southern Alps by upturning as mentioned above.

6. *Marlborough Schist*

The old rocks of the region are entirely schist and essentially anticlinal in structure. They pass westward by decreasing metamorphism into the Permian Pelorus graywacke to the end of the Dun Mountain Ophiolite, and eastward by fault contacts and gradation into graywacke at the north end of the Alpine Greywacke region. To the south they are cut off by the Wairau Fault, which is the main continuation of the Alpine Fault, and are thought to have once joined with the Otago Schist on the opposite side of the Alpine Fault 500 km to the southwest. To the north they probably extend continuously at depth to the schist of the Kaimanawa Range 150 km NE in North Island, having been found about halfway to the Kaimanawa Range in a borehole at a depth of 2.3 km. Their most remarkable feature is a narrow band of steeply dipping red beds that extends continuously SSW for 40 km near the top of the schist. The band is parallel to the strike of the schistosity and its continuity proves the structural simplicity of this part of the schist.

Tertiary beds are confined to small infaulted areas near Picton. They consist of strongly deformed low volatile bituminous coal measures probably of Eocene age, and Oligocene calcareous mudstone.

7. *Alpine Greywacke (Torlesse Complex)*

The oldest rocks in this region are New Zealand's only fossiliferous Carboniferous: marble, quartzite, and basic volcanics are exposed over a very small area west of Geraldine in South Canterbury. They are faulted against Triassic graywacke on the west and covered by Tertiary beds on the east. Schist like that of the Otago belt covers a small area to the south. The bulk of the old rocks are graywackes with the usual associated red beds and range in age from Permian to Upper Triassic.

The uniformity of the sediments and the rarity of fossils along with at least three periods of deformation (Bradshaw, 1972; Spörli, 1979; Findlay, 1979) make it difficult to decipher the structure.

The graywacke at several places is overlain possibly conformably by freshwater beds with plant fossils of Jurassic age that appear to be equally as strongly deformed as the graywacke. The close association between graywacke and freshwater beds raises difficult problems regarding depth of deposition of the graywacke and this is discussed below.

The next younger rocks are acid and basic volcanics with interbedded freshwater beds at Mt. Somers 80 km W of Christchurch and Malvern Hills 55 km WNW of Christchurch. They rest unconformably on older rocks and are early Upper Cretaceous in age.

Upper Cretaceous and Tertiary beds extend along the eastern side of the region and have been preserved in downfaulted areas within the foothills of the Southern

Alps. They rest unconformably on all underlying rocks and provide much of the data on the advance and retreat of the sea discussed under the Otago and Alpine Schist region (3a).

8. *Canterbury Plains and Banks Peninsula*

The Canterbury Plains consist of huge fans of gravel and sand built up by the main rivers at a time when glaciers extended well out from the Southern Alps, probably during the culmination of the last main stages of the last glaciation. Surprisingly, although the gravels reach the coast they are completely absent offshore.

The fans have since been trenched by the main rivers and cut back by the sea, but the underlying rocks which are the continuation of those in the regions to the west, are exposed only near the alpine foothills. Below the gravels, the Tertiary strata, as found in boreholes, are from about 1 km to 2 km thick and consist of mudstone, sandstone, and limestone, thinner but otherwise similar to those in the adjoining regions.

Two late Tertiary volcanoes composed of interbedded flows of basalt and basic andesite form Banks Peninsula. Magnetic anomalies show that similar basic rocks underlie the gravel plains to the southwest. Still farther southwest on the edge of the plain at Geraldine and Timaru, slightly younger basalts show above the gravels. There is an isolated outcrop of graywacke, unfortunately unfossiliferous, within the eastern volcano of Banks Peninsula.

9. *Eastern Marlborough*

During the late Mesozoic and early Tertiary a marine trench extended along the east coast of North Island and south to Marlborough where it shallowed out. The oldest rocks of the region are graywackes that are younger but otherwise similar to the Alpine graywackes of region 4. At the trench axis, which lies close to the coast, they pass up abruptly but conformably into sandstone–siltstone turbidites with bands of turbidite conglomerate. Westwards the graywacke grades laterally through transitional beds into thin shelf deposits that rest unconformably on Jurassic graywacke. Fossils are rare in the graywacke. The oldest are Upper Jurassic, and the youngest, near the trench axis, Albian. The transitional and shelf deposits are more fossiliferous, all Cretaceous stages from Albian upwards being represented.

The Cenomanian in the middle Awatere and middle Clarence valleys is noteworthy for its thick marine basalts and for the numerous feeding dikes that cut through the Albian below.

Throughout NZ the rate of deformation slowed down toward the end of the Cretaceous and reached its lowest value in the period of tectonic calm during the Upper Maastrichtian and Paleocene. Elevated regions were peneplained, and strongly leached clays and quartzose sands deposited. In the Eastern Marlborough

Region the trench sediments became less distinctive with time. The upper Campanian is decidedly quartzose and is overlain by sulfurous quartz sand or clays of Maastrichtian age that pass up through Paleocene bentonitic mudstone and bentonitic limestone and basalt. The Oligocene as usual is represented by limestone and the Lower Miocene by calcareous sandstones and mudstones. In the Upper Miocene the rate of deformation increased and turbidites including a thick turbidite conglomerate (Marlborough Conglomerate) composed of clasts of all the older rocks were deposited. Continued deformation uplifted and tilted the conglomerate which now dips at 30° to 90°.

10. *East Coast Ranges*

The region is the northern continuation of eastern Marlborough and is composed of similar rocks. The oldest are Upper Jurassic graywackes with apparently interbedded basalt and limestone at Ekatahuna 110 km NE of Wellington and to the south near Tinui 100 km ENE of Wellington. Other basalts at Red Island and to the south are of early to mid-Cretaceous age.

The Lower Cretaceous, confined to the southern part of the region, consists of crushed dark mudstone and massive tuffaceous-looking graywacke sandstone with bands of redeposited conglomerate.

All the Middle and Upper Cretaceous stages are present and show the same upward change of facies from graywacke through transitional facies to shelf sediments as in eastern Marlborough but the lateral facies changes are less clear and the trench axis is ill-defined and probably offshore.

From the Middle Cretaceous to Upper Miocene the strata closely match those of eastern Marlborough, the main difference being that the Paleocene flint beds and flinty limestone are thinner and that the Oligocene limestone is replaced by calcareous mudstone. A Miocene turbidite conglomerate is present but less well known than the one in eastern Marlborough. The Pliocene consists of coquina limestone and freshwater beds that have a fairly uniform lithology over the whole region.

The region was strongly folded along NNE axes in the Cenozoic including the Quaternary, and folding is still continuing. At Cook Strait the present rate of tilting of the more gentle (ENE) flanks of the folds decreases from about 30°/Ma on the west to 15°/Ma on the east. Small diapirs similar to those of the Raukumara Peninsula (region 6c) occur in the northern part of the region.

11. *Raukumara Peninsula*

The oldest rocks are Jurassic but only a few fossil localities are known so the full extent of the beds is uncertain. The Lower Jurassic fossils are in limestone lenses within basalt apparently interbedded with graywacke near Whakatane. The Upper Jurassic fossils are in siltstone beds interbedded with graywacke near Opotiki, 25 km E of Whakatane. Freshwater beds probably of the same age occur nearby. Jurassic to Lower Cretaceous fossils occur in a small limestone lens within the thick Matakaoa basalts and probably Lower Cretaceous fossils occur in a limestone lens with basalt in Jurassic-looking graywacke on the coast 30 km SE.

The main structural feature is the south end of a north-plunging Cretaceous syncline along the central highlands of the region. The southeast, south, and southwest margins are fairly well defined, Cretaceous beds thinning and shallowing in facies in much the same way as in eastern Marlborough.

The uppermost Cretaceous and Eocene rocks are similar to those in Eastern Marlborough and East Coast Ranges, and consist of Maastrichtian sulfurous flinty siltstones and turbidites, Paleocene bentonite and flinty beds, and Eocene flinty limestone, red beds, and minor basalts. In the axial part, some Paleocene beds are thick and interbedded with turbidites. This suggests local mobility but the overall stability is shown by the absence of fresh detritus in the uppermost Cretaceous and Paleocene.

The Oligocene contains flinty limestone and calcareous mudstone. The comparative stability that had continued since the end of the Cretaceous ended at the end of the Oligocene. The Miocene is extremely thick and contains a high proportion of turbidites. It seems to have been deposited in a deep basin along the western side of the region. Marine Pliocene and marine and freshwater lower Quaternary rocks have local distribution only.

The pattern of folding is far more complex than in the two coastal regions to the southwest. Many of the Cretaceous and Eocene beds have been folded along northwest-striking axes at a decided angle to folds in the younger Tertiary sediments which strike northeast parallel to the coast. In the north the Miocene rests with a 90° unconformity on the Upper Cretaceous. Elsewhere the only evidence for a break is the apparent absence of the Upper Oligocene in sections that otherwise appear to be conformable.

There are about 20 diapiric structures. They contain bentonite and other Eocene beds plus fossiliferous Cretaceous concretions. Bentonite on burial compacts less than other sediments and is less dense at all depths. Accordingly, the diapirs may be bentonite domes analogous to salt domes and may have risen through the more dense rocks.

12. *Western Northland*

Structure is chaotic (see next section) and it is now generally accepted that the whole region is a huge slide.

The Tangihua basalts are the oldest rocks and form the highest hills. They are without interbedded sediment other than layers of tuff and small limestone lenses. They closely resemble the Matakaoa basalts of Raukumara Peninsula. Fossils from one of the limestone lenses are Upper Jurassic to Lower Cretaceous in age.

The next younger set of rocks when sorted into age order by fossils provide an almost complete sequence from Albian to Oligocene with lithologies and fossil species that closely match those of the east coast trench regions. The Miocene is different from, and younger than, the slide. It consists largely of marine andesites which form a fairly continuous belt along the western margin of the region.

13. *Eastern Northland*

The pre-Miocene sequence is quite different from that of western Northland. The oldest rocks are steeply dipping graywackes with associated basalts and lenses of limestone from which Permian foraminifera and corals have been collected at several places on the coast over a length of 80 km. Except that the fossil localities lie on a northwest-trending line, nothing is known of the structure of the graywacke beyond that it is complex. Mesozoic beds are represented by graywacke with *Inoceramus*. Next come Eocene coal measures and marine mudstone, and Oligocene limestones which rest on a peneplained graywacke surface and are now preserved along the western margin of the region. As in the other regions with a clear unconformity between undermass and cover, the Tertiary structure is essentially simple and determined by tilting of undermass blocks.

Boreholes that have progressively reached farther and farther down have traced the rocks of eastern Northland progressively farther and farther westward below the structurally chaotic rocks of western Northland. The contact dips gently westwards and is a sinuous line at the ground surface.

Rocks that postdate the slide are Miocene basalts and agglomerates near the coast and more extensive Quaternary basalts that lie farther inland.

14. *Southeast Auckland*

The northern part of the region consists of the Coromandel Range to the east and lowlands to the west of the dashed line in Fig. 3. The lowlands have a low central ridge that separates the middle Waikato Basin from the Hauraki Depression. They differ from other NZ lowlands in the apparent absence of either Upper or Lower Tertiary marine or freshwater sediments.

The southern part is mostly covered by ignimbrite from the Volcanic Belt, and slopes north. Graywacke projects through the ignimbrite hills in the west. The most easterly graywacke is near the Bay of Plenty coast. It is well to the east and narrows the Volcanic Belt to a width of 20 km only. In the east there is a narrow belt of Pliocene and Lower Pleistocene freshwater beds that interfinger with marine beds at the coastal end of the Volcanic Belt.

The old rocks are graywackes in which Jurassic fossils have been found in place and in concretionary pebbles in conglomerates interbedded in the graywacke. There are red beds that are extensive enough to have been used for correlation. Manganese from the red bed jaspilites was mined in this region and in eastern Northland. Manganiferous jaspilites occur with the red beds in all the graywacke and schist regions but concentrations of manganese occurred only where the relief was gentle and the erosion rate slow. Dips are steep throughout and the structure is far more complex than in the syncline to the west.

Tertiary beds cover small areas in the Coromandel Peninsula. They consist of Lower Oligocene coal measures and limestone. The upper Tertiary is represented by thick andesitic and rhyolitic volcanics with thin interbedded freshwater beds of Upper Miocene age. Silver–gold bullion to the value of 41 million pounds was won from lodes in thermally altered andesite.

15. *Southwest Auckland*

The general structure of the old rocks of the region is synclinal. The synclinal axis extends south beneath Upper Miocene marine sediments probably joining that through Nelson. The rocks range in age from the base of the Upper Triassic to the top of the Jurassic. The Triassic part consists of tuffaceous sandstone and siltstone similar to rocks of the same age in the synclines of Southland and Nelson. The strata are up to 2 km thick and most were deposited at shallow to moderate depths. There is conglomerate with 2-m granite boulders at the base in the west and some interbedded freshwater beds near the top of the Triassic. The upper part of the sequence is not known elsewhere in NZ. It consists of shallow-water marine and freshwater deposits that extend up to the top of the Jurassic and are about 1 km thick.

Cretaceous and Eocene beds are absent. Lower Oligocene beds rest on a fairly even surface cut across the older rocks. The oldest are coal measures with subbituminous coal of considerable economic importance. The usual calcareous Lower Oligocene beds overlie the coal, and are followed by relatively thick Upper Oligocene mudstone and sandstones. The Miocene contains graded beds and passes into thick turbidites to the north and south. The Pliocene is represented by richly fossiliferous marine beds at Otahuhu 15 km SE of Auckland and Kaawa Creek 60 km S of Auckland in the northern part of the region; elsewhere it consists of freshwater beds and volcanics. Large areas are covered by Quaternary basalts and andesites. The large andesite cones of Pirongia and Karioi near Kawhia are conspicuous in the south, as are the small but younger and less dissected basalt cones near Auckland in the north.

Deformation of Tertiary strata is mostly by block faulting and tilting, but the rate of deformation is relatively slow and no active faults are known.

16. *Taranaki and Egmont*

The eastern side of the region is entirely covered with Upper Tertiary beds and the western side with Quaternary andesite flows, lahars, and ash from the 2.5-km-high Mt. Egmont volcano. The regional dip of the Upper Tertiary beds is to the south and southwest. They rest on the Lower Tertiary on the west side of the

Volcanic Belt and directly on the graywacke of the axial range on the east side. To the south they dip beneath Quaternary marine beds.

Mt. Egmont is the youngest of three andesite volcanoes that lie on a line and young (500,000 yr, 250,000 yr, and active) to the SSE.

The strike of the younging line is tectonically unexplained and different in character from the line of persistent volcanic activity that defines the Volcanic Belt (region 14). It may be analogous to the several lines of Pliocene and Upper Miocene andesite volcanoes that extend along and to a large extent determine the west coast of North Island to the north of Mt. Egmont.

The Tertiary beds thicken abruptly west of the Taranaki Fault which extends NNE through the region 40 km E of Mt. Egmont, and the old rocks are 3 km deep below Mt. Egmont and to the west offshore. There is a useful gas and condensate field at Kapuni 20 km south of Mt. Egmont and a giant gas field at Maui offshore and 90 km SW of the mountain.

The geological conditions are unusual. Thick coal of Eocene age was buried from 2 to 3 km in the Miocene, and was metamorphosed from subbituminous to low volatile bituminous. The methane that was produced from the coal was held in the coal measure sands.

The Tertiary structure is simple and mostly determined by faulting. Dips are mostly low, and the folds broad and gentle.

17. *Rangitikei Basin*

The entire region is covered by Quaternary beds, and it contains an unusually complete Quaternary sequence. The beds form a wide SSE-plunging syncline with the two limbs at an angle of 120°. At the north limb, which extends to the coast at Wanganui, the beds dip gently south and rest conformably on Upper Pliocene beds. At the east limb the beds rest directly on the graywacke of the Axial Range and are locally fairly steep. Small outliers and stripped surfaces show that they recently extended east across what is now the range to connect with similar rocks on the eastern side. Deposition and uplift was controlled by a migrating zero-isobase line that during the mid- and late Quaternary has moved across the basin to its present position near the present-day coastline. Faults and drag folds strike NNE and further complicate the structure.

The Quaternary is underlain by fairly thick Pliocene beds that rest directly on the undermass. The undermass consists of graywacke on the east, and schist on the west. The schist, found in a single borehole, lies directly between that of Marlborough to the SW and that of the Axial Range to the NE.

The main feature of geophysical interest is the culmination of a negative isostatic gravity anomaly that extends southwest from East Cape, bends around at the culmination, and then extends south through Cook Strait. At the culmination the intensity is −150 mgal. About a quarter of the anomaly is caused by the relatively thick and only slightly compacted Pliocene and Quaternary beds and the remainder by thick undermass actively held down by tectonic forces. With a release of the holding force the undermass would rise, presumably in the same way as the Otago schist is thought to have risen in the early Cretaceous.

18. *Wairarapa and Hawke's Bay Depression*

Early Quaternary and Pliocene marine and freshwater sediments similar to those of the Rangitikei Basin cover most of the region. The basal beds contain thick bands of Pliocene conglomerate derived from the Axial Range and serve to date its uplift. To the east the sediments rest on the Upper Tertiary beds of the Wairarapa Ranges and to the west and north either directly on the graywacke of the Axial Range or on Upper Tertiary strata. Although Cretaceous rocks are well developed in the Wairarapa ranges and to the north in Raukumara Peninsula, they are absent here and the underlying graywackes are probably Jurassic on the east and Triassic on the west.

In the southern part of the region the sediments are strongly tilted along NE-striking fold axes. In the northern part around Hawke's Bay they are more gently tilted and dip toward the bay. The sediments have been uplifted as much as 1 km to produce relief that is remarkably steep for such young sediments.

The Wairarapa Fault (last movement 1855) extends through the region from Hawke's Bay to Cook Strait, and displaces fold axes and late Quaternary lake deposits and river terraces. The fault as traced south changes from low-dipping reverse-dextral to steep-dipping essentially dextral and then at Cook Strait back to low-dipping reverse-dextral.

19. *Axial Range of North Island*

The region is composed entirely of graywacke, except for one small outlier of Lower Tertiary, several outliers of Upper Tertiary, and a thin covering of Quaternary pumice and ignimbrite from the volcanic belt. It represents the least known area of graywacke in NZ. Except for ubiquitous but unidentifiable plant fragments, the only fossils are a few *Monotis* (Upper Triassic) localities in the southern part, a few Jurassic localities in the north and east, and the annelid *Torlessia* on the coast near Wellington. Increase in degree of metamorphism to the northwest reaches subschist rank in the Kaimanawa Range southeast of Mt. Ruapehu.

The structure, probably complex throughout, has been deciphered only in relatively few areas (e.g., Spörli and Bell, 1976; Spörli, 1978) including the coast south of Wellington where graded bedding shows that an original syncline about 2 km wide has been inverted and turned into an antiform.

Lower Tertiary beds are preserved in a small faulted outlier about 50 km NNE of Wellington. They consist of compacted quartz sand and greensand of Upper Eocene and Lower Oligocene age and are different from most NZ strata of these ages. They are the only remnant of a once thick and extensive sequence.

20. *Volcanic Belt (Taupo Volcanic Zone)*

The historically active (less than 150 yr old) volcanoes of New Zealand, Ruapehu, Ngauruhoe, Tongariro, Tarawera, and White Island, are on a line that strikes toward the active volcanoes of the Kermadec and Tonga Islands. The line is of major importance but if the historical period extended as far back in NZ as it does in Europe, then three other volcanoes that lie well off the Volcanic Belt would have to be added to the "active" list. The three, Egmont in Taranaki, Rangitoto near Auckland, and an unnamed volcano in Eastern Northland, were all active less than 500 yr ago. The volcanoes outside the belt were probably active for short periods only, and the Volcanic Belt is best considered as one with persistent volcanism.

There are Upper Pliocene marine beds without volcanic debris at both ends of the Volcanic Belt and it is likely that the volcanism of the belt is less than 2 Ma old.

Within the central part of the belt the volcanic rocks are several kilometers thick and consist of a variety of rhyolitic, andesitic, and basaltic volcanics but there is no clear progressive change in composition with time.

Most of the surface rocks within the volcanic belt and for a considerable distance outside are rhyolite ignimbrite sheets and rhyolite ash deposits including pumice. They represent the larger eruptions with volumes of 100 km^3 or more, the inner part being the ignimbrite and the outer part, now mostly eroded, ash. Smaller eruptions were entirely of ash. [The term *ignimbrite* was coined by Marshall (1932) to describe rocks from this belt.]

Ignimbrite sheets form a broad arch that is downfaulted at the two margins of the belt and either buried or absent in the central part. Within the belt there are many well-defined faults that strike NNE on the strike of the belt and at an appreciable angle to the 15-km-long ENE linear rift of the Tarawera Eruption of 1886. The faults are generally assumed to be normal but they dip too steeply (70°–90°) to have caused appreciable widening.

Retriangulation of approximately 80-yr-old surveys indicates that the belt is widening at about 7 mm/yr (Sissons, 1979). A basalt dike up to 3 m wide was intruded during the Tarawera Eruption, and the retriangulation is probably record-ing extensional strain related to dike intrusion. Releveling records subsidence (negative uplift) of several millimeters per year and confirms the surface geology which suggests that the center of the belt remains at the same level, gradual subsi-dence being counterbalanced by the intermittent deposition of volcanic ash.

The general topic of andesite volcanism in NZ is discussed in terms of subduc-tion rates at the end of this chapter.

B. Magnetic Lineations for New Zealand Region

Table I sets out the tectonic divisions of the NZ Region. The basic division is between oceanic and continental crust, with an intermediate division where oceanic crust is being changed to continental. A description has already been given of the geology of NZ in terms of 20 regions, in which the salient points needing explanation are emphasized. Most of the salient points can be explained in terms of reconstructions based on plate tectonics and are discussed in the following sections.

A knowledge of the age of the oceans in and around the NZ Region is essential for understanding the geology of NZ itself. It is accepted here that the oceans are formed by seafloor spreading and that magnetic lineations define how and when they spread. As mentioned for the NZ Region, continental magnetic anomalies, mainly the Stokes, and mostly caused by ophiolites or gabbros and diorites are almost as important for reconstruction as are magnetic linations in the oceans. As shown in Table I the greater part of the continental portion of the NZ Region is submerged and outcrops hidden. The continental margin anomalies provide the main link between the submerged and emergent parts.

1. *Magnetic Lineations and Oceanic Slides* (Figs. 5 and 6)

It has been suggested here that slides have to be considered in addition to seafloor spreading and subduction (plus obduction) in the interpretation of seafloor and continental magnetic lineations. The importance of continental slides, frequently termed "olistostromes," is commonly accepted and the Northland Chaos has already been mentioned. The slides to be discussed lie one to the south and one to the north of NZ and each is over 100 times as large as the Northland Chaos. Slides differ from microplates in that they are superficial features that have slid downslope independently of the movement of the tectonic plates. The low value of friction for serpentinite and the presence of serpentinite in the seafloor may explain the large size of the postulated slides. It may be significant that there are many small pockets of serpentinite in part of the Northland Chaos (*Geology NZ:* p. 449).

The Endeavour Slide to the SE of the Campbell Plateau provides a simple explanation for three otherwise unrelated anomalous features. Bollons Seamount is a continental fragment that has slid south for 200 km from the nearest continental crust. The main part of the slide consists of magnetic lineations that appear to have been rotated 10° in a clockwise sense. The Endeavour Rise lies to the west of the slide and contains a set of magnetic lineations which have already been mapped by Christoffel and Ross (1965) as though bent around to the north. The slide is shown by Fig. 5, first as it is thought to have been before sliding, and second as it is now.

The Three Kings Slide lies just north of NZ. It is less well established than the Endeavour Slide and is suggested as an alternative to obducting the Three Kings Ridge over the southwest corner of the South Fiji Basin and its magnetic lineations. The magnetic lineations span the Oligocene and the slide (or the obduction) would

TABLE I
Main Tectonic Divisions for New Zealand Region

OCEANS (ocean crust)

Pacific (old), then New Caledonia, Bellona, and Bounty troughs and then Tasman Sea and southwest Pacific Ocean; and now southeast Indian Ocean and southwest Pacific Ocean. In addition, backarc spreading in South Fiji and now in Havre Trough. Ocean crust being worked over in Emerald Basin.

NEW ZEALAND TRENCH AND SLOPE (transformation of sediment plus ocean crust into continental crust)

North slope Chatham Rise [became inactive 82 (?) Ma ago]. Off Northland, now represented by Northland Chaos (became inactive 25 (?) Ma ago). Eastland, Hawkes Bay, Wairarapa, and Marlborough (youngest slope deposits about 7 Ma old). Present-day deposition is in Hikurangi Trench.

PLATFORM (continental crust)

Now submerged: Lord Howe Rise, Challenger Rise, Chatham Rise, and Campbell Plateau.

Now emergent: New Zealand.

Single southern rigid part which includes parts of Otago, Southland, and Canterbury. Alpine Fault Zone plus Fiordland.

Two northern rigid parts separated by the North Island Shear Belt. Eastern part now faulted south and no longer rigid.

---

have taken place probably in the early Miocene and at about the same time as the emplacement of the Northland Chaos and the sinistral displacement of the Glenroy Fault (see below). According to the slide explanation the Three Kings Ridge was once on the west side of what is now the Norfolk Basin and connected the east side of the Norfolk Ridge with the submarine peninsula NE of North Cape thus completing the line of the Dun Mountain Ophiolite from NZ to the Norfolk Ridge. The slide is thought to have been triggered by movement of the Glenroy and Vening Meinesz faults. The north end of the slide moved eastward to fill the west side of the South Fiji Basin. The southern end of the slide is thought to consist of the axial (= trench) deposits that moved SE as the Northland Chaos and now cover western Northland.

The slide is shown by Fig. 6, first as it is thought to have been before sliding, and second as it is now. For the Three Kings Slide, plate tectonics as well as the sliding have to be taken into account, and the map outline of the "before sliding" map has been displaced on the "as now" map to show the amount and sense of the plate movement.

The postulated movement of the two slides is taken out in Fig. 1.

2. Continental Magnetic Lineations (Fig. 1)

There are two sets of continental magnetic lineations: an older set and the "Stokes Magnetic Anomaly System" (Wellman, 1973). The older set is best de-

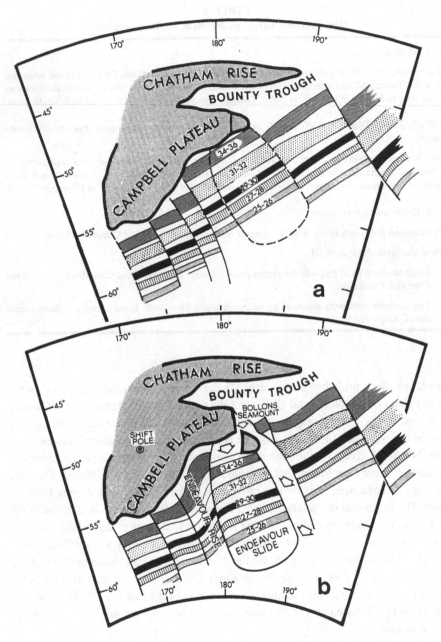

Fig. 5.   Maps showing the development of the Endeavour Slide. (b) shows Bollons Seamount and the
oceanic magnetic lineations as they are today. (a) shows the Endeavour Slide moved back and Bollons
Seamount joined to the east end of Campbell Plateau. In (b) note how the slide is supposed to have
created a gap in the anomalies to the east of the slide and a rucking up of the seafloor into the Endeavour
Rise on the western side, already mapped by Christoffel and Ross (1965).

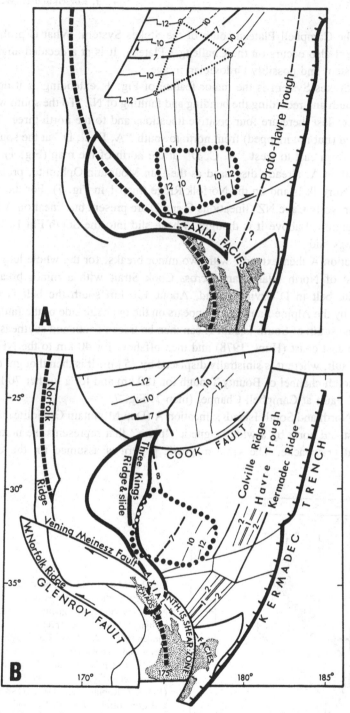

Fig. 6. Maps showing the development of the Three Kings Slide. (B) shows magnetic lineations and NZ in present-day position. (A) shows supposed original position. Note that the margin of (A) is unbroken, and that the margin is broken in (B) to show inferred nature of movement. The latitude and longitude grid is that of the present day.

fined on the Campbell Plateau south of the Stokes System. What is probably the same system also occurs on the Challenger Plateau. It is at a decided angle to the Stokes System and possibly Ordovician in age.

The Stokes System is the major feature of Fig. 1, extending as it does from north to south and recording the bending and faulting of NZ. To the south where the system is widest there are four positive lineations and to the north three. They are here termed (but not mapped) from north to south "A, B, C, D" at the south of the map, and from east to west "A, C, D" at the north of the map (Fig. 1).

Lineation A, already discussed as the Dun Mountain Ophiolite, probably occurs near Norfolk Island on the Norfolk Ridge (map 1 in Fig. 7). For the next 650 km to near North Cape NZ, lineations C and D are present but lineation A is absent and as mentioned above it is thought to have slid into the South Fiji Basin as the Three Kings slide.

Lineation A then extends, with two minor breaks, for the whole length of the west coast of North Island, and across Cook Strait with a minor break to the serpentinite belt in D'Urville Island. About 150 km south the belt is dextrally displaced by the Alpine Fault and reappears on the opposite side of the fault 485 km SSW in the south of South Island. It can then be traced south and southeast for 250 km to the east coast (Hunt, 1978) and then offshore for 40 km to the NE-striking Mernoo Fault, where it is sinistrally displaced by 25 km. It is then thought to extend near the south channel of Bounty Trough for 180 km and for a further 70 km to the Campbell Fault at Campbell Channel (map 7, Fig. 7).

For North and South Islands, lineation A (Dun Mountain Ophiolite) is flanked to the east and north by a wide magnetic "plain" that represents the nonmagnetic schist and graywacke. The same magnetic pattern is assumed for the Campbell

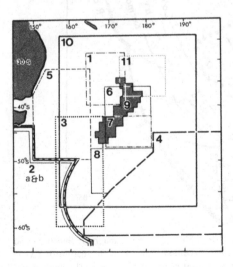

Fig. 7. Map of NZ Region giving a key to the coverage of the main maps of the ocean around NZ. Maps showing magnetics and magnetics plus bathymetry are given preference. 1, van der Linden (1971); 2a, Weissel and Hayes (1972); 2b, Hayes and Conolly (1972); 3, Hayes and Talwani (1972); 4, Christoffel and Falconer (1972); 5, Weissel *et al.* (1977); 6, Davey and Robinson (1978); 7, Robinson and Davey (1981); 8, Davey and Christoffel (1978); 9, NZ land area (Numerous maps); 10, Carter (1980); 11, Malahoff *et al.* (1982).

Plateau, the first magnetic lineation south of the plain being taken as lineation A. The lineation is a continuous feature with an intensity of about 500 ± 150 nT. It strikes almost E–W along 48°S and lies just to the north of Bounty Island which is composed of 189 Ma old granite.

The Campbell Fault dextrally offsets the Stokes System and extends southwest across the Campbell Plateau (Cook, 1981) from the western end of the Bounty Trough to dextrally offset the continental margin. Existing maps (maps 7 and 10, Fig. 7) show the western end of the trough and the offset continental margin, and the fault is considered to extend from 45.2°S, 176°E through 50°S, 170°E to the continental margin at 55°S, 165.5°E.

In 1978 Davey and Christoffel, in mapping the southern part of the Stokes System, disregarded our lineation "A" and considered our lineation "B" to be the most northerly part of the system. Lineation "B" lies to the south of the Bounty Granite.

In 1978 Christoffel placed the Campbell Fault about 100 km west of the position adopted here, probably because the continental shelf was then wrongly mapped (compare first and second editions of map 10, Fig. 7).

Off the west coast of North Island, lineations C and D constitute a pair that are offset by faults (transforms?) that do not extend east to lineation A. The linear pair comes ashore at the north end of South Island. It is now generally accepted that lineation C marks the Rotoroa Igneous Complex, and it is also generally accepted that the complex after being cut off by the Alpine Fault reappears on the opposite side some 500 km south-southwest as the Darran Complex.

Lineation D, the other member of the pair, comes ashore on the northern end of South Island at Separation Point (40.7°S, 173°E) and is thought here to be caused by the Riwaka Complex. Like the other lineations of the system it extends south to be cut off by the Alpine Fault. On the opposite side of the fault the lineation is probably represented by the N–S-striking belt of gabbro through the center of Fiordland. The gabbro is associated with sulfide deposits of economic interest and was mapped by Williams (1974, p. 394) but it is not shown in *Geology NZ* (colored map of Fiordland, Fig. 2.29). The gabbro contains magnetic bands and produces a positive magnetic lineation (Hunt, 1978). At the north end of South Island the Separation Point Granite lies between lineations C and D, and in Fig. 1 a granite symbol is mapped between the two lineations for their whole extent.

A well-defined positive magnetic lineation extends just inland of the Fiordland coast for 200 km from 44.5°S (Milford Sound) to 45.8°S (Hunt, 1978, Fig. 3). The anomaly is attributed but with some uncertainty to ultramafics that are exposed at Anita Bay in Milford Sound. The ultramafics are correlated with the Cobb Ultramafics of North West Nelson which extended south from near the northern part of South Island to near the NE end of the Alpine Fault proper at 42°S, 172.3°E. Their dextral displacement by the Alpine Fault is 440 km.

3. *Oceanic Magnetic Lineations*

Lineations of probable Jurassic age that cover too small an area to be shown on the map (Fig. 1) occur off the east coast of North Island. Transforms are closely spaced (about 30 km apart) and strike southeast. Thus, the magnetic lineations strike northeast and are parallel to the magnetic lineations already sketched in (Pitman *et al.*, 1974) near 25°S, 190°E and at right angles to the southeast-striking Louisville Ridge (26°S, 185°E to 38°S, 192°E).

(a) Basal Cretaceous. The transforms for the supposed Jurassic seafloor extend landwards to about the −2 km contour line. A fairly continuous positive magnetic anomaly lies between the −2 km contour and the coast, and between East Cape and Hicks Bay, 25 km NW of East Cape, there is a positive anomaly just north of the coast and a strong negative one extending 15 km inland. The 1 : 1,000,000 geological map (*Geology NZ*) shows that the Matakaoa Basalt causes the anomaly, and the strong negative value suggests a low-angle thrust below the basalt.

Figure 8 shows the outcrops of the Matakaoa Basalt and its correlatives together with the magnetic anomalies and their supposed seaward extensions. From its composition (*Geology NZ:* p. 632) the Matakaoa Basalt was once part of the

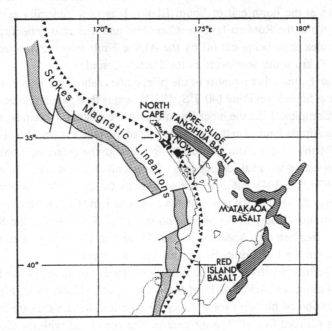

Fig. 8. Map of North Island of NZ showing outcrops of Matakaoa Basalt and associated rocks and inferred seaward extension according to recorded positive magnetic anomalies. The map also shows the several lineations of the Stokes Magnetic Anomaly System. The position shown for the pre-slide Tangihua Basalt is inferred.

ocean floor, but is now stranded within lower (?) Cretaceous turbidites and is part of the continental crust.

(b) Mid-Cretaceous. For the Bounty Trough from 175°E to 181°E, i.e., the part of the trough included in the Bounty Sheet (map 7, Fig. 7), five oceanic transforms were mapped from the magnetics and found to span the trough axis. To the NE the Campbell Fault transfers the opening movement of the Bounty Basin to the continental margin. The movement (see Fig. 1 and imagine young seafloor to be cut out) then crossed the line of the present Alpine Fault to the Bellona Trough and then to the New Caledonia Trough. The strike adopted for the transforms to the west of North Island depend largely on the strike of the major transform (Challenger Fault, Fig. 1) that connects the Bellona Trough with the New Caledonia Trough, and is represented by the single magnetic discontinuity across the Challenger Plateau which is east–west and at latitude 39°S. The fault provides a southern limit to the New Caledonia Trough and a northern limit to the Bellona Trough. For the New Caledonia Trough the limit agrees with the depth of the trough when allowance is made for some infilling from the NZ land area. Agreement is less good for the Bellona Trough which is deeper than would be expected to the north of the line of the Challenger Fault. The spacing of the transforms is based on magnetics (mostly from map 1, Fig. 7) but the observation lines are widely spaced and the mapped positions are approximate. As far as can be judged from the contoured maps the magnetic signature is the same for the two troughs, but as yet the age is not known.

A possible date for the opening of the three troughs is about 85 Ma ago (late Cretaceous) and is based on the age of the basal cover beds in wells drilled to 3 km in the Great South Basin which lies on the landward extension of the Bounty Trough. A less direct age is from the Hawks Crag Breccia and its correlatives. The three troughs represent a failed opening. The Hawks Crag Breccia lies around the edge of the failed opening and its age is about 100 Ma (*Geology NZ:* p. 330). It is thus likely that the three troughs started opening when the Hawks Crag Breccia grabens stopped opening and when the Tasman Sea and the Campbell Plateau–West Antarctic Sea started to open.

(c) Late Cretaceous and Paleocene. LOM 36 to LOM 21. For the Senonian there are two sets of numbers for identical oceanic magnetic lineations. If the first set by Christoffel and Falconer (1972) are here prefixed "LOM" (linear oceanic magnetics) and the other set as compiled by Pitman *et al.* (1974) prefixed "P" then LOM 32 = P 32 and LOM 36 = P 33. On priority we here adopt the first set and have renumbered accordingly.

The lineations are partly to the south and east of the Campbell Plateau and partly in the Tasman Sea. The Campbell Plateau part is taken from Falconer (1974), with the Endeavour Slide (Fig. 5) restored. The true pattern is doubtless a regular and simple one, but a minor problem remains. At 176°E there is an offset in the lineations. Well before "transforms" were understood, Christoffel and Ross (1965)

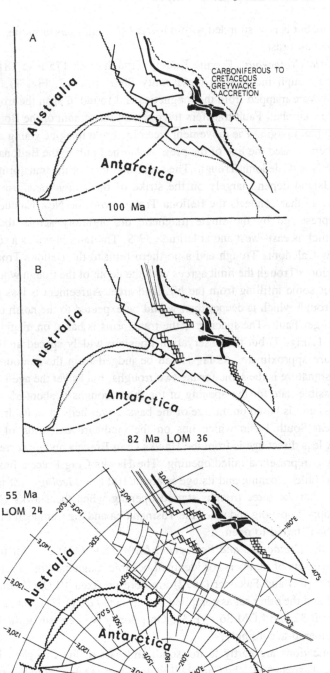

Fig. 9.   Three sketch maps showing the development of the NZ Region from 100 Ma ago (mid-Cretaceous) to 55 Ma ago (beginning of Eocene). Projection is Polar Equidistant. The next three maps

mapped the offset as a horizontal fold (see Fig. 5) but Falconer (1974) replaced the fold by two transforms. It would seem that fashion has played some part in the interpretations.

The Tasman Sea part is from Weissel *et al.* (1977). Figure 1 shows that the two sets of lineations appear to have been separated by the growth of new seafloor. In order to unify the two parts, the Campbell Plateau part has to be rotated 64° clockwise relative to the Tasman Sea part. The rotation has taken place since LOM 24 formed (−55 Ma ago) and more probably since LOM 21 formed (−50 Ma ago).

(d) Oligocene, South Fiji Basin. LOM 7 to LOM 12. Lineations are from Malahoff *et al.*, (1982) with the Three Kings Slide slid back (see Fig. 6). A triple junction centered at 25.7°S, 175.5°E is the critical part of the lineation system. Two large transforms determined the triple junction opening. A northeast-striking sinistral fault limits the system to the west, and an east-northeast-striking sinistral fault limits the system to the south. It should be noted that the Vening Meinesz Fault is mapped 1° too far to the north at 170°E in Fig. 16 of Malahoff *et al.* (1982).

(e) Eocene to Present Day. LOM 21(?) to LOM 0. For the ocean between Australia and Antarctica the lineations are well-defined except that the age of the oldest lineations is uncertain. At Greymouth in South Island (42.3°S, 171.5°E) a coalfield has been surveyed in some detail and, after a long period of standstill and peneplanation, faulting commenced about 50 Ma ago. In order to conform with the Greymouth data, LOM 21 is taken as being the oldest Australian–Antarctic LOM lineation. The most easterly of the Australian–Antarctic (Indian–Antarctic) lineations are at about 160°E. The present-day lineation LOM 0 is of particular importance because being at the Indian–Antarctic–Pacific triple junction its strike must pass through the present-day shiftpole for the Indian–Pacific plate boundary. Lineations LOM 24 to LOM 3 are poorly defined southeast of Campbell Plateau and much better defined northeast of Louisville Ridge. Thus, it is uncertain if there has been movement (leakage) on the ridge itself.

Lineations LOM 18 to LOM 13 on 160°E immediately to the south of the

---

(Fig. 10) and these three were constructed backward in time, but in order to give the correct sense of fault displacement the maps are set out forward in time. (A) About 100 Ma ago (mid-Cretaceous). The NZ Region is united with Australia and Antarctica as eastern Gondwanaland. There is subduction on the Pacific coast, with graywacke continuing to accrete. (B) 82 Ma ago. LOM 36. Mid-late Cretaceous. The main events since about 100 Ma ago have been continuing subduction along the Pacific coast plus the formation of an abortive rift now represented by the Bounty Trough, the Bellona Trough, and the New Caledonia Trough. The western end of the Bounty Trough is connected with the southern extension of the Bellona Trough by the dextral Campbell Fault. The northern end of the Bellona Trough is connected with the southern end of the New Caledonia Trough by the sinistral Challenger Fault. (C) 55 Ma ago. LOM 24. The main event since 82 Ma ago has been the opening of the Tasman Sea and the southwest Pacific Ocean. The angle of opening is about 25° for the Tasman Sea, and about 21° for the southwest Pacific. The difference, about 200 km, is attributed to sinistral faulting through Greater Antarctica on what is here termed the Vostok Fault, and which is supposed to end at the triple junction (Australia, Antarctica, unknown continent or ocean) at the southwest corner of Australia.

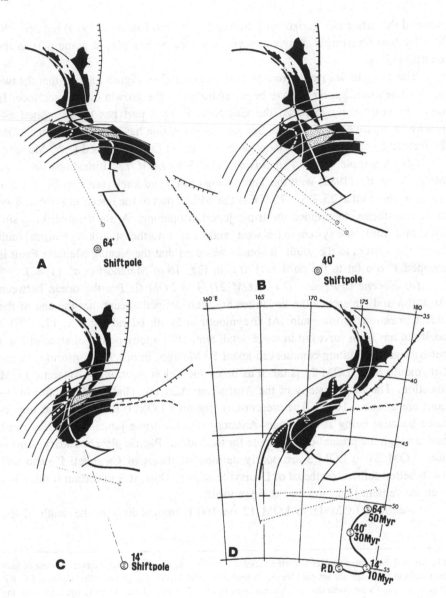

Fig. 10.   Four maps on conical projection with standard parallels at 35°S and 50°S to show progressive change in shape of NZ. Map A is on a larger scale but is otherwise identical with map C of Fig. 9. Map D is for the present day and maps B and C are interpolations between map A and map D. The lines with ticks are the two sides of the present Kermadec Trench, and the amount of subduction is shown by the progressive decrease in the width between them. The line with the scalloped edge is the present northern margin of the Chatham Rise. The stippled area is land that has been obducted, eroded, and destroyed. It adjoins the subducted part of the Pacific. The Alpine Fault Zone is all-important. In maps A, B, and C, it is divided evenly by seven arcs. The part to the north of the outer arc is the rigid part of the Indian Plate, and the part southeast of the inner arc is the rigid part of the Pacific Plate. Dextral shear on the fault zone causes the Indian Plate to rotate clockwise relative to the Pacific Plate. During the shearing the lengths along the arcs remain constant, but the width between the arcs changes. The heavy curved line is the Dun

Tasman Sea lineations (map 5, Fig. 7) mark spreading between the Indian and Pacific plates (not shown in Fig. 1). According to the shiftpole used here, the rotation rate was about 2°/Ma and more rapid than the average rate for the last 50 Ma.

C. Reconstructions (Figs. 9 and 10)

The gradual development of NZ from 100 Ma ago when it was part of Gondwana to the present day is illustrated by seven sketch maps. The first three (Fig. 9) show the breakup of Gondwana, and the second four, (Fig. 10) which are on a larger scale, the gradual change in the area and shape of NZ from the breakup of Gondwana to the present day. The maps are based on the fit of the now separated rigid fragments of Gondwana, on the magnetic lineations of the oceans that are shown in Fig. 1, and on the structure and faults within NZ. A further controlling factor is that the reconstructions have to be geologically reasonable at all times. For instance the change of area for NZ is limited to what is known from geology, a few tens of kilometers of extension for a transgression, and a few tens of kilometers of narrowing for mountain building. But there is virtually no direct geological control for change of shape, and the inferred change is so large that the NZ of Gondwana is almost unrecognizable in the NZ of today, and this in spite of the fact that slightly over half of NZ remains rigid, distortion being confined to the Alpine Fault Zone. Needless to say, the present-day coastline on the reconstructions has no meaning other than for location.

The first map is for about 100 Ma ago (Fig. 9A). It unites the now separated parts of western Gondwana, the only innovation being a hypothetical Vostok Fault through East Antarctica. The second map (Fig. 9B) is for about 83 Ma ago (LOM 36). The three basins, Bounty, Bellona, and New Caledonia, have opened, and the

---

Mountain Ophiolite (DMO), either exposed or defined by its magnetic lineation. Its length is held constant from map A to map C. From map C to map D it is displaced by dextral-reverse movement of the Alpine Fault. The amount of displacement, 328 km (360 km if parallel drag is included), is appreciably less than the present offset, 480 km, because allowance has to be made for the part of the ophiolite that has been uplifted and eroded away.

Map A: 50 Ma ago. LOM 21. Within the Alpine Fault Zone the Dun Mountain Ophiolite has a long "S" shape, a length of 522 km, and an end-to-end angle of about 45° to the arcs of the shear zone. The width of the fault zone is 422 km.

Map B: 30 Ma ago. LOM 10. Within the Alpine Fault Zone the Dun Mountain Ophiolite is almost straight, its length remains at 522 km, and it is at right angles to the arcs of the shear zone. The width of the shear zone has increased from 422 km to 522 km.

Map C: 10 Ma ago. LOM 5. Within the Alpine Fault Zone the Dun Mountain Ophiolite has a mirror image long "S" shape, its length remains at 522 km, and its end-to-end angle is again about 45° but in the opposite sense to that in map A. The width of the shear zone has decreased from 522 km to 422 km, i.e., to its Fig. 10A value.

Map D: Present day. LOM 0. From map C to map D there is no bending of the shear zone, but there is dextral displacement on the North Island Shear Belt, dextral displacement on the Marlborough Faults, dextral-reverse movement on the Alpine Fault, and sinistral movement on the Fiordland Fault.

time of opening is thought to be mid-late Cretaceous from the age of the oldest strata in the Great South Basin which lies on the extension of the Bounty Trough. The third map is for about 55 Ma ago at the end of the opening of the Tasman Sea and the then directly adjoining part of the southwest Pacific (Fig. 9C). The timing is from LOM 24. Reconstruction is based on the magnetic map of the NZ Region (Fig. 1). The two lines that now limit LOM 24 to LOM 36 and enclose younger oceanic lineations are made to coincide, and the two lines then slid against each other so as to give the Dun Mountain Ophiolite (DMO) the correct length for the set of reconstructions that extend on to the present day. It is the adjustment to the length and position of DMO that requires a Vostok Fault with a 200-km dextral displacement.

The next four maps are on a larger scale and show NZ only. The first (Fig. 10A) for 50 Ma ago is the same as that for 55 Ma ago, and this in spite of the formation of lineations LOM 23 to 21 in the southwest Pacific. From 55 Ma to 50 Ma ago there was peneplanation and decreasing uplift within NZ, and the LOM 23 to 21 opening is thought to have been transmitted to Antarctica, and to be possibly represented by the uplift of the Transantarctic Mountains. The next three maps (Fig. 10B–D) are based entirely on data from within NZ. As already mentioned, the period of peneplanation ended about 50 Ma ago when the tilting of the crustal blocks within the Alpine Fault Zone started. Until 30 Ma ago the tilting was by normal faulting, and from then until the present day by reverse faulting (see below, Fig. 19B). The net result is that the blocks have now overshot and the parts that were most deeply buried are now the parts most uplifted. The timing and amount of block tilting determines the reconstructions for 30 Ma and 10 Ma ago. The two steps 50 Ma to 30 Ma ago and 30 Ma to 10 Ma ago consist of dextral bending within the Alpine Fault Zone. In the first there is widening and marine transgression, and in the second narrowing and marine regression. The final step from 10 Ma ago to the present day is entirely by faulting. Because the 10-Ma-old map has to fit the pattern of earlier reconstructions, the amount of inferred fault movement from 10 Ma ago to the present day is largely independent of the rate of present-day (Holocene) movement. That the movements agree well provides a check on the validity of the reconstructions.

II. PRECAMBRIAN TO CARBONIFEROUS

In terms of the pre-Mid-Cretaceous, NZ consists of an old part (Western Province) and a young part (Eastern Province), the two parts being separated by the Median Tectonic Line (Landis and Coombs, 1967). The rocks of the Eastern Province are nowhere older than Carboniferous, whereas those of the Western are as old as Precambrian. The main difference is in the age of the crust. For the Western Province the crust is Ordovician or even older, but for the Eastern it ranges in age from no older than Upper Carboniferous to Middle Cretaceous. Both provinces are or became part of Gondwana, but the eastern was built onto it, partly while it was breaking up.

Precambrian to Devonian rocks occur only in the Western Province, which consists of regions 1a and 1b in Fig. 3. Because of dextral faulting by the Alpine Fault the province is now in two parts some 500 km apart: a NW Nelson part, and a Fiordland part. The Fiordland part has been uplifted and eroded more than the NW Nelson part and the metamorphic grade of the rocks at the ground surface is higher in Fiordland than in NW Nelson. The grade difference and the lack of detailed surveys in Fiordland has hindered correlation across the Alpine Fault.

The division adopted here depends largely on correlation of ultramafic and mafic rocks and their magnetic lineations (Hunt, 1978; this chapter). Three terranes are postulated, the Greenland, Cobb, and Arthur terranes; they match the three sedimentary belts of Cooper (1979) but also contain crystalline rocks (Fig. 11).

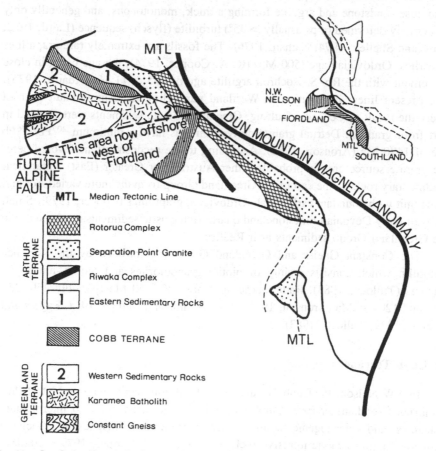

Fig. 11. Pre-Carboniferous distribution of terranes in the Western Province. Terrane and unit names are from Northwest Nelson; inferred correlatives in Fiordland are discussed in text. Refer to inset and Fig. 7 for position of Western Province in reconstruction of NZ.

The only NZ Carboniferous known is a small outcrop of Upper Carboniferous limestone and volcanics at Kakahu 30 km NNW of Timaru in the Torlesse complex that is interpreted here as being infaulted ocean floor material.

A. Greenland Terrane

The Greenland Terrane is known only from NW Nelson and Westland but is inferred to extend to the offshore continental fragment to the west of Fiordland and the Alpine Fault (Fig. 11). The terrane consists of at least three units.

The Constant Gneiss of the Charleston Metamorphic Group (Shelley, 1970, 1972) is supposed to unconformably underlie the Greenland Group and has a Rb–Sr isochron of 680 ± 21 Ma (Adams, 1975) and other than the Vendian of NW Nelson is the only Precambrian group known from NZ. It extends over 100 km^2 only.

The Greenland Group, which here includes the Waiuta Group, consists of quartzose sandstone and argillite forming a thick, monotonous, and generally only moderately deformed (dips usually > 45°) turbidite (flysch) sequence (Laird, 1972; Laird and Shelley, 1974; Nathan, 1976). The fossils are extremely rare graptolites of earliest Ordovician age (500 Ma) (R. A. Cooper, 1974), the age being in close agreement with the Rb–Sr isochron argillite age of 495 ± 11 Ma (Adams, 1975). The present linear distribution in Westland paralleling the Alpine Fault is most likely the result of Cenozoic faulting. Greenland Group sediments were derived in part from granites. Detrital granite-derived zircons have a minimum $^{207}Pb/^{206}Pb$ age of 1470 Ma (Aronson, 1968). According to the reconstructions favored here, the granite source area is probably either Australia or Greater (East) Antarctica. Sedimentary rocks to the east of the Greenland Group form the more varied Aorangi Mine unit with abundant early to late Ordovician graptolites (Cooper, 1979). Small areas of early Devonian limestone and quartz-rich clastic sediments are faulted into the Greenland Group sediments near Reefton.

The Constant Gneiss and Greenland Group are intruded by the Karamea Batholith which consists mainly of biotite granodiorites and biotite–muscovite granites (Tulloch, 1983). Radiometric ages are 370–350 Ma (possibly 430–280 Ma) and 120–95 Ma (Aronson, 1965, 1968; Adams and Nathan, 1978; Eggers and Adams, 1979; Tulloch, 1983).

B. Cobb Terrane

In NW Nelson, the Cobb Terrane consists of complexly deformed latest Pre-cambrian (Vendian) or early Cambrian to late Ordovician mafic to intermediate volcanics, and volcanogenic sediments and quartz and talc magnesites associated with ultramafic to mafic intrusive rocks (Hunter, 1977; Cooper, 1979; Grindley, 1980). The Anita Ultramafics from the Milford Sound area of Fiordland (Wood, 1972) are correlated with ultramafic rocks of the Cobb Terrane in NW Nelson, and associated metamorphic rocks (Wood, 1972; Blattner, 1978; Oliver, 1980) are here correlated with the remainder of the Cobb Terrane. Also correlated are the Western Fiordland metagabbroic diorite basement and cover rocks of Oliver and Coggan (1979).

Grindley (1971) mapped at least four major nappes in NW Nelson, and a major nappe has since been inferred in the Fiordland correlatives of the Cobb Terrane by Oliver and Coggan (1979). Hunter (1977) considered that the mafic–ultramafic intrusions in NW Nelson are stratiform and just possibly part of an ophiolite. Grindley (*Geology NZ:* p. 130) interpreted the NW Nelson segment of the Cobb Terrane as a volcanic island arc complex resting on thin continental or thin oceanic crust.

C. Arthur Terrane

The sedimentary part of the Arthur Terrane consists of quartz-rich sandstone, siltstone, and thick limestone (Arthur Marble) ranging in age from early Ordovician to early Devonian (Cooper, 1979) and, although in part coeval with Cobb and Greenland sedimentary rocks, it is lithologically different, being mostly limestone. Correlatives in Fiordland, now amphibolite to granulite grade metamorphics, are the Central Fiordland metasediments of Oliver and Coggan (1979), the Kellard Point and Deep Cover Gneisses of Oliver (1980), and the Doubtful Sound Province rocks of Gibson (1982) all of which contain a considerable amount of marble.

There are three distinct belts of igneous rocks (Fig. 11): Riwaka Complex; the Rotorua Complex; and the Separation Point Granite with which is included the Mackay Intrusives. The Riwaka Complex consists of mafic to ultramafic intrusive rocks (see Grindley, 1980; Bates, 1980) considered to be late Paleozoic in age by Aronson (1968) but an $^{40}Ar/^{39}Ar$ analysis by Harrison and McDougall (1980) gave an age for the gabbro part of the complex of 367 ± 5 Ma. In Fiordland a linear belt of mafic and ultramafic rocks with a linear magnetic anomaly from Mt. Soaker to Mt. Aitken that includes Mt. George Gabbro and associated layered intrusions (Williams, 1974, p. 395; Gibson, 1982) was correlated with the Riwaka Complex by Bates (1980), and is so correlated here.

The Rotorua Complex (with a strong magnetic anomaly) in the eastern part of NW Nelson consists of intermediate to basic intrusives (Tasman Intrusives) and amphibolite facies metamorphic rocks. The Tasman plutons intrude the Permian Brook Street Volcanics, are Permian to Cretaceous in age, and are thought by Grindley (*Geology NZ:* p. 99) to be part of the Rotorua Complex. However, little dating has been done and the age and the full content of the Rotorua Complex are uncertain. Fiordland correlatives include the Eastern Fiordland gabbros and diorites of Oliver and Coggan (1979) and the western sector of the Darran Complex (Pirajno, 1981; Blattner, 1978; Williams and Harper, 1978), but ages and relations with granites to the east are uncertain.

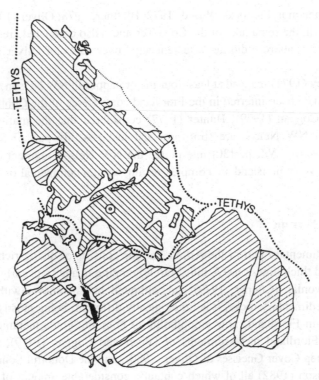

Fig. 12.   Central equal distance projection of a Pacific-free world that possibly existed in the middle Paleozoic.

In NW Nelson, the Cretaceous Separation Point Granite, in keeping with its name, is mainly biotite granite (Tulloch, 1983) and has an age of 114 Ma (Harrison and McDougall, 1980) which is about the same as the K–Ar and Rb–Sr ages of Aronson (1965, 1968) and Hulston and McCabe (1972). Fiordland correlatives are the Eastern Fiordland granodiorites and granites (Oliver and Coggan, 1979). It is worth noting that in NW Nelson the Separation Point Granite and the Mackay Intrusives lie to the *west* of the Rotorua Complex but that their correlatives in Fiordland lie to the *east* of the correlative of the Rotorua Complex.

D. Evolution and Tectonic Events

If the several parts of the Western Province were once widely separated as is commonly thought, then the time when they first came together will be the time of the first tectonic event to affect all parts simultaneously. The tectonic events reported by Adams *et al.* (1975), Cooper (1979), Grindley (1980), *Geology NZ* (pp. 117–135) for the Western Province are: unnamed event (early to mid-Cambrian);

Haupiri Disturbance (c. 525 Ma ago); Greenland Event (c. 440 Ma ago); First Tuhua Orogeny (c. 400 Ma ago); Second Tuhua Orogeny (c. 380–360 Ma ago); Rangitata Orogeny (c. 120–100 Ma ago); Kaikoura Orogeny (present day). The first to be simultaneous in all parts of the Western Province was the Second Tuhua at 380–350 Ma ago (Aronson, 1968; Hunter, 1977; Oliver, 1980; Harrison and McDougall, 1980).

Grindley (*Geology NZ:* p. 124) considered that during the Silurian the nappes of the Cobb terrane were thrust north or northwestwards over autochthonous Greenland and Arthur terranes. Cooper (1979) on the other hand interpreted the Cobb terrane as basement to the Arthur terrane, with the Cobb terrane being later upthrust to its present position. With the existing uncertainties in Western Province geology, several other possibilities can be envisaged.

The initial independent development of the three terranes, the mutual obliquity of the Cobb and Arthur terranes (Fig. 11), the formation of nappes in the Cobb terrane, and the several deformational, metamorphic, and plutonic events indicate a complex history. The three terranes may represent separate allochthonous terranes which were amalgamated during the Silurian or Devonian by mechanisms such as complex suturing (Dewey, 1976, 1977) or possibly by obduction of the Cobb island-arc ophiolitic material. Closure of the ocean basins in which sediments of the Greenland and Arthur terranes were accumulating resulted in a complex contact between the two terranes. After closure the terranes were possibly part of a Pacific-free world (Fig. 12).

III. PERMIAN TO MID-CRETACEOUS

A. Western Province

The Permian Parapara Group and the Triassic Topfer Coal Measures were deposited and granitic plutons were intruded, but on the whole the Western Province was relatively stable between the Devonian and the Jurassic.

1. *Parapara Group*

The small outcrop of Permian rocks at Parapara Peak in NW Nelson (Clark *et al.*, 1967) is a shallow-water deposit and markedly different from the other NZ Permian (Fig. 13). They are mainly quartzose conglomerate, sandstone, and shale that rest unconformably upon Ordovician rocks. Permian and Ordovician rocks are both strongly deformed, and both have suffered the same regional high T/P metamorphism (Landis, 1981). The fauna is a Middle Permian one, similar to that of Tasmania (Waterhouse and Vella, 1965).

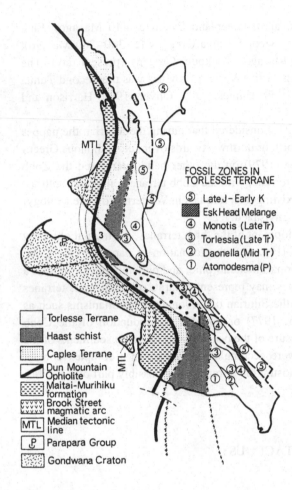

FOSSIL ZONES IN
TORLESSE TERRANE

⑤ Late J – Early K
▨ Esk Head Melange
④ Monotis (Late Tr)
③ Torlessia (Late Tr)
② Daonella (Mid Tr)
① Atomodesma (P)

☐ Torlesse Terrane
▨ Haast schist
▦ Caples Terrane
◧ Dun Mountain
Ophiolite
▨ Maitai–Murihiku
formation
⩔ Brook Street
magmatic arc
MTL Median tectonic
line
ℙ Parapara Group
▦ Gondwana Craton

Fig. 13. Pre-130 Ma reconstruction of NZ showing distribution of units in the Eastern Province. The Dun Mountain Magnetic Lineation is the reference line and represents the Dun Mountain Ophiolite. Fossil zones in the Torlesse terrane are based on zones defined by Campbell and Warren (1965), Andrews *et al.* (1976), Speden (1976), and MacKinnon (1983).

B. Eastern Province

Six tectonostratigraphic units, set out below from west to east, and grouped according to composition, make up the Eastern Province (Fig. 13):

WEST
1. Brook Street Magmatic Arc
2. Maitai–Murihiku forearc basin   Hokonui Association (or Western or Marginal)—regions 2a and 2b, Fig. 3
3. Dun Mountain Ophiolite
4. Caples terrane
5. Haast Schist   transitional—regions 3a and 3b, Fig. 3
6. Torlesse terrane   Torlesse Association (or Eastern or Axial
EAST   or Alpine or Trench)—regions 4, 12,
                                          13, and parts of regions 5, 6a, 6b, and
                                          6c, Fig. 3

The following points have to be taken into account in interpreting the Eastern Province (see Landis and Bishop, 1972):

a. The difference in composition between the quartzofeldspathic Torlesse association and the volcanic-derived Hokonui association
b. The location(s) of the source area(s) for the two associations
c. The mechanisms of transportation of detritus and the nature of the depositional environments
d. The nature of the contact, now located somewhere within the Haast Schist, between the Caples and Torlesse units
e. The origin of basalts and associated cherts and red and green argillites within units 4 to 6
f. The contrast in deformational style between unit 2 and units 4 to 6 inclusive
g. The mechanism of formation of the tectonic mélanges of the Caples and Torlesse units

Plate tectonic models have been proposed by Fleming (1970), Landis and Bishop (1972), Blake et al. (1974), Coombs et al. (1976), Carter et al. (1978), Spörli (1978), Wood (1978), Howell (1980), Bradshaw et al. (1981), Dickinson (1982), and MacKinnon (1983). The only point of general agreement is a convergent plate margin along the east side of NZ from the Carboniferous to early Cretaceous, which implies a trench and subduction.

1. Brook Street Magmatic Arc

The Brook Street magmatic arc (Figs. 13 and 14) abuts the Median Tectonic Line and consists of a once continuous suite of volcanic rocks and related intrusives that until 1979 were considered entirely early Permian. In the Eglington Valley 40 km SSE of Milford Sound, Williams and Smith (1979) found two parallel belts of extrusives, an andesite belt and a belt of Permian primitive island arc tholeiites to the east. They inferred that the andesites are possibly post-Permian. In Nelson, recently found plant fossils show that the Brook Street arc volcanoes were active until the Mesozoic, probably until the Jurassic, and possibly until the Cretaceous (Johnston, 1981).

Calcalkaline granitic intrusives (Tasman and Mackay Intrusives) with isotopic ages of 250 Ma to 130 Ma (Aronson, 1968; Devereux et al., 1968; Williams and Harper, 1978) and minor ultramafics intrude the Brook Street volcanics to form part of the arc complex.

2. Maitai–Murihiku Forearc Basin

The Maitai part is mid to late Permian and is known only in South Island. The Murihiku part is early Triassic to late Jurassic and is best exposed in Southland and Kawhia (west Auckland). Both parts are thought to have been deposited in a forearc basin.

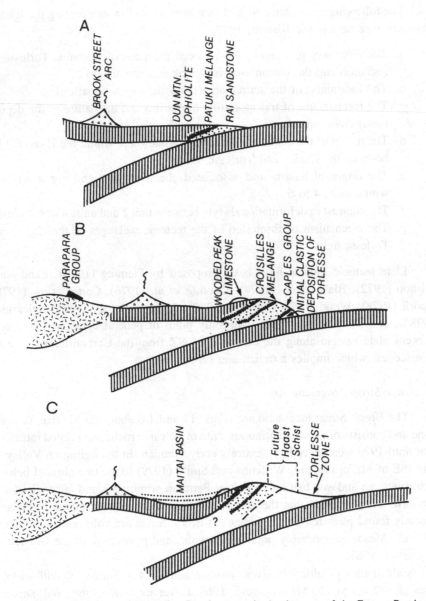

Fig. 14. Sketch cross sections illustrating Permian tectonic development of the Eastern Province, South Island. (A) Early Permian, failure of oceanic crust, initiation of subduction, formation of Patuki ophiolite mélange, and commencement of volcanism in the Brook Street magmatic arc. (B) Mid-early late Permian, elevation of the accretionary prism at the line of the trench–slope break, erosion of pelagic sediment and uppermost part of the Dun Mountain Ophiolite followed by deposition of the Wooded Peak Limestone (Maitai), formation of Croisilles mélange and incorporation of Caples and Pelorus groups into the accretionary prism, initial deposition of Torlesse terrane. (C) Late late Permian, subsidence in forearc basin with deposition of Maitai turbidites, continued deposition and accretion of Torlesse fossil zone 1 rocks.

Sediments of the Maitai sequence rest unconformably (see later) upon the Dun Mountain Ophiolite (Figs. 1 and 15) and consist of about 4 km of mainly volcano-derived siltstone with subordinate turbidite sandstone, conglomerate, breccia, limestone, and nonvolcanogenic quartzofeldspathic sandstone. The Murihiku sequence, which is up to 10 km thick, is also largely volcano-derived. It has been suggested by Carter *et al.* (1978), Landis (1980), and MacKinnon (1983) that the Maitai and Murihiku sequences were both derived from the Brook Street volcanic arc and that the progressive changes in sediment composition reflect the progressive change in the chemical composition of the volcanic material (Boles, 1974).

According to Coombs *et al.* (1976), the metamorphic grade of the Maitai sequence is mainly lawsonite–albite–chlorite facies whereas that of the younger Murihiku sequence is zeolite facies only. Deformation produced open to close and near-isoclinal folds with regular axial planes dipping steeply (greater than 45°) to the east and slaty cleavage in the siltstone that contrasts with the more complex and less organized deformation of the Caples and Torlesse terranes.

3. *Dun Mountain Ophiolite*

The basal units of the Maitai–Murihiku forearc basin are either faulted against or rest disconformably on the Dun Mountain Ophiolite. The ophiolite consists of pillow lavas, and rare mafic dikes that grade down through gabbros to ultramafic rocks that are now mostly serpentinites (Blake and Landis, 1973; Coombs *et al.,* 1976; Sinton, 1980; Davis *et al.,* 1980). Two rocks, dunite (after Dun Mountain) and rodingite (after Roding River), derive their names from localities in Nelson which occur within this unit. Compilation and detailed mapping in Nelson by Johnston (1981) shows that the ophiolite has been tilted strongly to the west (now near-vertical) and a geochemical study by Davis *et al.* (1980) indicates an ocean

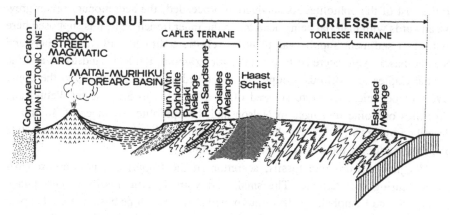

Fig. 15.   Sketch cross section showing arrangement of Eastern Province units in South Island at about 100 Ma ago.

ridge origin. The absence of pelagic sediment above the pillow lava makes the Dun Mountain Ophiolite sequence incomplete [see Coleman (1977) for a description of complete and incomplete ophiolites]. The pelagic sediments are thought to have been eroded and the ophiolite is thought to be significantly older than the overlying limestone as is shown in Fig. 14.

The basal (eastern) side of the ophiolite is faulted by or faulted against the Patuki Melange (Johnston, 1981).

4. *Caples Terrane*

The Caples Terrane (Figs. 13 and 14) includes three groups and consists mainly of unfossiliferous volcano-derived graded bedded graywacke and argillite with minor metabasalt, chert, and limestone. It includes the Caples Group (Western Otago), the Pelorus Group (Nelson), and the Waipapa Group (Northland). Also included are the Patuki Melange, the Rai Sandstone, the Croisilles Melange (Johnston, 1981), and some other formations. Derivation is mainly from the Brook Street volcanics (e.g., Carter *et al.*, 1978; Turnbull, 1979; MacKinnon, 1983). Transport was from the volcanic arc, across the forearc basin, down submarine canyons through the inner trench slopes to the trench floor.

The terrane is complexly deformed (e.g., Turnbull, 1980) and weakly metamorphosed (Kawachi, 1974), and is here interpreted (Fig. 14) as an accretionary prism oceanward of the Brook Street arc and Maitai–Murihiku forearc basin. All three groups that comprise the Caples Terrane were produced by a west-dipping subduction zone (Coombs *et al.*, 1976; Carter *et al.*, 1978; Spörli, 1978).

At some time possibly in latest Carboniferous or earliest Permian the oceanic crust and lithosphere failed, with the eastern portion being subducted, and the western portion (the Dun Mountain Ophiolite, etc., Fig. 15) being much later turned on edge. The initial subduction line is marked by the Patuki Melange immediately to the east of the ophiolite. As subduction proceeded, the accretionary prism grew oceanwards with material being accreted to the inner trench wall. It is supposed here that when sediment supply was plentiful, clastic sediment with minor chert and oceanic basalt were accreted to produce, for example, the Rai Sandstone, and the normal Caples and Pelorus groups, but when sediment supply was low the upper layers of oceanic crust were sheared off and accreted to produce the Patuki and Croisilles ophiolite mélanges in Nelson and the ophiolite mélanges of Southland described by Craw (1979). No detailed descriptions exist of the deformational style or of the nature of the contacts between the units.

As judged from rare fossils, accretion of the Caples terrane was probably mostly during the Permian. The small part with Triassic fossils in east Otago (Campbell and Campbell, 1970) is interpreted as trench–slope basin material deposited on top of the accretionary prism and subsequently kneaded into the prism.

The Waipapa Group contains Jurassic fossils (Speden, 1976) as well as the well-known Permian fossils (Spörli, 1978; Spörli and Gregory, 1981) in the lime-stone associated with basalt. We suggest that the Waipapa graywackes are of Late Jurassic age, were derived mainly from the Brook Street arc, and were deposited in trench–slope basins above portions of the accretionary prism containing the Permian fossils from seafloor that had already been accreted. The Jurassic deposits were then kneaded into the prism and have been deformed in the same way as the rest.

5. *Haast Schist*

The boundaries between the Haast Schist and the less metamorphosed Caples and Torlesse terranes on either side are taken as being at the base of the Chlorite zone of the greenschist facies (e.g., Wood, 1978) and are transitional (e.g., Bishop, 1972; Kawachi, 1974).

The schist is regionally metamorphosed up to amphibolite facies grade but there are relatively few recent studies of metamorphism and deformation (e.g., Bishop, 1972, 1974; A. F. Cooper, 1972, 1974; Findlay, 1979). The Haast Schist is here interpreted as a deeply eroded part of the accretionary prism that was once more deeply buried and more strongly deformed by flowage than the rest.

From 100 or so K–Ar "ages," Sheppard *et al.* (1975) and Adams (1979) thought that the schist was uplifted first in the late Jurassic or early Cretaceous and later in the late Cenozoic by throw on the Alpine Fault as a narrow strip along the west side of the Southern Alps. Adams (1979) considered that the minimum age of metamorphism of the Haast Schist and Torlesse terrane decreased from 200 Ma to 170 Ma with decreasing distance from the Alpine Fault.

It is generally accepted that the contact between the Caples and Torlesse terranes lies within the Haast Schist. According to the models proposed by Coombs *et al.* (1976), Spörli (1978), Bradshaw *et al.* (1981), Landis (1981), and MacKinnon (1983) the contact is an abrupt break and either a strike-slip fault-contact or a collision zone, but the abrupt break has not been found. On the other hand, according to our model, the contact within the schist is gradational. For the schist of Marlborough, Vitaliano (1968) presented chemical analyses and petrological descriptions that favor gradation. The contact between the Waipapa and Torlesse units in North Island being without schist is an unsolved problem. Mayer (1969) and Skinner (1972) have shown that the sandstones within Waipapa Group sequences have affinities with Hokonui and with Torlesse terranes (cf. Dickinson, 1971, 1982; MacKinnon, 1983).

6. *Torlesse Terrane*

The Torlesse terrane consists of quartzofeldspathic graywacke and argillite (e.g., Dickinson, 1971, 1982; Beggs, 1980; MacKinnon, 1983) with minor amounts of basalt, chert, red and green argillite, conglomerate, and limestone. The terrane has suffered low-grade (prehnite–pumpellyite to pumpellyite–actinolite) metamorphism. It is intensely and complexly deformed and contains zones of tectonic mélange with argillite matrix (e.g., Bradshaw, 1972, 1973; Spörli and Bell, 1976).

Bradshaw (1972) and Bradshaw and Andrews (1973) considered that the major part of the Torlesse consists of shallow-water marine with some freshwater sediments. It is now generally thought, in agreement with earlier (1950–1970) views, that some 95% is deep-water marine and some 5% shallow marine and freshwater sediments.

Most of the shallow-water marine and freshwater sediments occur in Canterbury well to the east (oceanward) of deep-water marine rocks of the same age and this led Bradshaw and Andrews (1973) and Andrews *et al.* (1976) to propose a Torlesse source area to the east. From the reconstruction (Fig. 13) it can be seen that the terrestrial deposits were in fact deposited landward of the accreting margin. They pass conformably up into graywacke–argillite turbidites (Retallack, 1979; Campbell and Force, 1972).

The Torlesse terrane is now generally accepted as being an accretionary prism (e.g., Spörli, 1978; MacKinnon, 1983) although Bradshaw *et al.* (1981) regard it as a giant submarine fan unit that was deposited on the Pacific plate from an unknown source and was gradually added to the active Gondwana margin. We accept the accretionary prism interpretation but consider that the Torlesse terrane grew oceanward from the Caples terrane and is not an accretionary prism that grew somewhere else and was later attached to the Caples terrane.

We assume that the graywacke–argillite was deposited in an ocean trench and that subduction scraped it from the seafloor and stacked it up in sheets each up to 5 km thick, subduction migrating oceanward to keep pace with the stacking. In theory the assumption is easily tested, because within the sheets, which should young oceanward, the graywacke layers should young landwards. When seafloor basalt is scraped off, it will form the base of a sheet and will be faulted against the graywacke of the next younger sheet. Fossils associated with the basalt may be many millions of years older than the fossils in the graywacke itself, the exact age difference depending on the nature of the basalt, ocean plate boundary (midocean ridge) basalt being probably much older, and ocean plate (intraplate) basalt need be only slightly older. The stacking mechanism is fairly satisfactory for the older graywackes which are complexly folded and sparsely fossiliferous (few ages) and less satisfactory for the Cretaceous graywackes which are fairly fossiliferous (many ages) and have a simpler structure. The shallow-water marine and freshwater sediments in Canterbury are interpreted as trench–slope basin deposits that were ponded above an older portion of the accretionary prism (see Spörli, 1978).

Five fossil zones (1, *Atomadesma*, Permian; 2, *Daonella*, mid-Triassic; 3, *Torlessia*, late Triassic; 4, *Monotis*, late Triassic; 5, late Jurassic, early Cretaceous) are recognized within the Torlesse terrane (Fig. 13, after Campbell and Warren, 1965; Andrews *et al.*, 1976; Speden, 1976; MacKinnon, 1983). The original and simple map pattern was bent and faulted during the Cenozoic and is now fairly complex (see Andrews *et al.*, 1976). With a few exceptions the zones become progressively younger eastwards toward the trench, but within each zone the dominant (> 75%) younging direction is westward. Agreement with the accretionary prism model is good.

The volume of the Torlesse graywacke belt is enormous. It is about 3500 km long, with an average width of about 300 km and a normal crustal thickness of say 30 km and any postulated source must be able to provide such a volume.

The Western Province has the right composition, and was the first choice as a source (e.g., Landis and Bishop, 1972) but the Brook Street volcanics and the volcano-derived Hokonui and Maitai strata lie between and appear to block transport from the Western Province to the Torlesse terrane. Other possible sources are an eastern unidentified continent (e.g., Bradshaw and Andrews, 1973; Andrews *et al.*, 1976; Retallack, 1979) and Marie Byrd Land in Lesser (West) Antarctica (e.g., Bradshaw *et al.*, 1981; Cooper *et al.*, 1982; inferred in Griffiths, 1975). According to the reconstructions, we suggest a localized source area in Lesser Antarctica nearer the Antarctic Peninsula than is Marie Byrd Land (Fig. 16). The shiftpole for Carboniferous to Mesozoic seafloor spreading was probably located in Lesser Antarctica and there was oblique subduction along the Australian and South American margins of Gondwana (Fig. 16). Clockwise rotation of part of Lesser Antarctica about this shiftpole led to obduction which provided the source material for Torlesse detritus. The Jones Mountains and Eights Coast of Lesser Antarctica are the closest area to the part obducted and eroded. Also, the clockwise rotation accounts for the present position of the Ellsworth Orogen (see Craddock, 1975, Fig. 41.2).

Rocks of the younger fossil zones within the Torlesse terrane contain an increasing sediment contribution from older Torlesse rocks. The conglomerates in the younger fossil zones contain up to 90% of older Torlesse graywacke (e.g., MacKinnon, 1983). The only likely source is at the line of the trench–slope break to the west of the trench. To allow for accretion there would have to have been relative eastward migration of the trench–slope break and the trench itself (Karig and Sharman, 1975).

Basalts and associated cherts, limestones and red and green argillites, are widespread and make up about 5% of the Torlesse terrane, of the Caples terrane, and of the Haast Schist. An oceanic origin either midocean ridge (ocean plate boundary), or intraplate (ocean plate) has been proposed (e.g., Spörli, 1975, 1978) but was not favored by Coombs *et al.* (1976). New chemical analyses indicate that the basalts in the Torlesse, in the Waipapa, and in the Haast units are entirely oceanic (Roser, 1983; Grapes and Palmer, 1984) and not alkaline (island arc). The basalts and associated rocks are assumed to represent oceanic material that was sheared off the seafloor and incorporated into the accretionary prism.

According to Feary and Pessagno (1980), radiolaria from chert within early Cretaceous (zone 5) Torlesse rocks of North Island are early Jurassic and about 50 Ma older than the fossils in the graywackes and argillites.

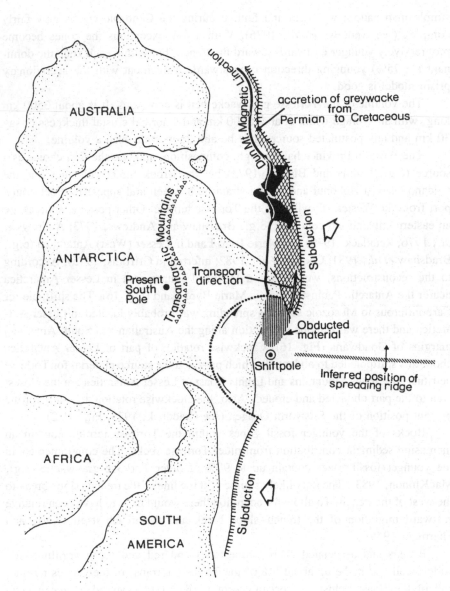

Fig. 16.   Three-plate model for c. 130 Ma ago showing the source of the Torlesse sediments. Gondwana is assumed fixed, Pacific is in two parts with spreading axis in middle and two lines of subduction.
Obduction produced by sinistral faulting provides the material for the Torlesse terrane.

Thus, the Torlesse accretionary prism consists of four different kinds of material that were mixed and then subjected to the same sequence of deformations:

1. Oceanic basalt and associated chert and red and green argillite and limestone from the oceanic plate

2. Sandstone and siltstone derived from Antarctica and deposited as turbidites in the trench

3. Conglomerate, sandstone and siltstone, derived from Torlesse that had been compacted and then exposed, and deposited as turbidites in the trench

4. Freshwater to shallow marine deposits, derived from Torlesse graywacke and deposited in small basins above older, already accreted Torlesse material

7. *Discussion*

The following Eastern Province features are of major importance:

1. The Eastern Province consists of a number of subparallel geological units that are consistent with their formation in association with a subduction zone dipping westward.

2. Present-day geological maps and, even better, reconstructions show that the Torlesse terrane progressively widens southwards from New Caledonia, through New Zealand to the Chatham Rise (Figs. 9, 13, and 16). Thus, the source area for the Torlesse terrane is likely to be near the point on Antarctica from which the Chatham Rise split off (Fig. 16). The major transportational system was from south to north, parallel to the trench axis (see also Dickinson, 1982). This contrasts with the eastward (transverse) transport direction for the Hokonui association from its Brook Street source.

3. In Fig. 13 the accretionary prism units (Caples, Waipapa, and the Torlesse fossil zones) are shown as overlapping sheets that are younger northwards. The anomalous Permian fossils in the Waipapa unit in Northland are regarded as being infaulted ocean floor and are disregarded. The northward narrowing is attributed to increasing distance from a southern source. Possibly the line of subduction changed in the same way as today, the trench being full of sediment at the south (Hikurangi Trough) and empty to the north (Kermadec Trench).

IV. LAST HUNDRED MILLION YEARS (Tables I and II)

A. Introduction

From the Permian to the mid-Cretaceous a distinction is made between moderately deformed platform deposits (Hokonui) with thickness that although large can be estimated with some degree of certainty, and trench deposits (Torlesse) that have been stacked up into complex assemblages. Torlesse stratigraphic thickness is uncertain but almost certainly greater than that of the platform deposits. The division between platform and trench deposits continues through to the present day, with the trench deposits being largely turbidites and generally at least three times as thick as the platform deposits. In the reconstruction (Figs. 9 and 10) the trench deposits lie on a near-straight line. One of the main uncertainties is the degree of trench activity particularly from mid-Cretaceous (100 Ma ago) to 50 Ma ago. From Permian to mid-Cretaceous the trench is generally accepted as being a line of subduction. From 50 Ma ago to the present day, subduction on the line of the trench is basic to the narrowing in our reconstruction. For the present day the trench is an evident topographic feature (the Kermadec Trench and its filled-up southern end, the Hikurangi Trough). Associated in their appropriate positions are the negative gravity anomaly, the intermediate and deep focus earthquakes, and the andesite volcanoes.

B. Cretaceous and Lower Cenozoic

1. New Zealand Trench

(a) Cretaceous. The trench rocks and fossils are remarkably uniform throughout NZ and a single basin that once extended from Marlborough to Northland has been proposed for them (Geology NZ: p. 472, Eastern Basin). Simple two-plate reconstructions for NZ that use the Wairau Fault as the Indian–Pacific plate boundary dismember the trench rocks by separating off the Marlborough part. We avoid this by distributing the plate movement over all five Marlborough faults.

The trench rocks are least disturbed in Marlborough. In Wairarapa, Hawke's Bay, and Eastland they are jumbled by folding, faulting, and sliding. In Northland they were brought in 25 Ma ago by a huge slide and are now spread over the platform undermass and cover as the Northland Chaos.

For correlating the Cretaceous we assume roughly equal lengths for the NZ stages, and differ slightly from the NZ Geological Society time chart (Stevens, 1980) and considerably from Geology NZ. Internal correlation and that with S. America is by the fairly common and facies tolerant species of Inoceramus. Correlation with Europe is by rare to very rare ammonites.

Does the trench contain Lower Cretaceous? This is the major problem. The all too easy answer is no. With no Lower Cretaceous, the undermass–cover unconformity can be extended out from the platform to the trench, and terrane mappers can map all the graywacke and then stop, or so they generally suppose. Because of the anomalous twofold division commonly used for the Cretaceous, we first have to define our Cretaceous divisions. The following letter symbols for NZ stages are from Wellman (1959). If Maastrichtian to Turonian (Mh, Mp, Rt, Rm, and Ra) are Upper Cretaceous, Cenomanian to Aptian (Cn, Cc, Cm, Cu, and Uk) are Middle Cretaceous, and Neocomian (no NZ stages) is Lower Cretaceous, then age-diagnostic Lower Cretaceous fossils are unknown from the trench of NZ. The absence of the key fossils is generally taken to mean that the Lower Cretaceous is missing. And this in spite of the absence of the expected unconformity and the presence of several kilometers of strata covering some 2500 km² in North Island that cannot be mapped better than "Upper Jurassic to Lower Cretaceous" (1 : 1,000,000 map in *Geology NZ*).

The rock types are thought to solve the problem because the Upper Cretaceous is entirely flysch suite (sandstone–siltstone turbidites plus some siltstone) and the Upper Jurassic entirely graywacke suite (graywacke–argillite turbidites) and the contact between the two suites, although not specially sought after, is considered to mark the undermass–cover unconformity. The actual contact (Wellman, 1959) is well exposed within the Albian at several places. It is gradational and there is *no* unconformity. Within a few tens of meters the flysch suite rocks pass down into graywacke suite rocks, and the fossils show that there is no appreciable age difference between the two. Conglomerates enhance the confusion. There are, at several places, conglomerates 10 m to 30 m thick at or just above the lithological break. They have been taken, not once, but repeatedly since 1917, as being the basal conglomerate of the cover but are actually turbidite conglomerates in no way different from the numerous conglomerates in the underlying graywacke suite that have already been mentioned in the discussion on the origin of the Torlesse.

In Marlborough, in South Island, over a kilometer of effectively unfossiliferous graywacke–argillite is mapped as Lower Cretaceous and a convincing cross section (*Geology NZ*: Fig. 6.28) drawn for the end of the Cretaceous. It is shown here without vertical exaggeration and extended down into the Upper Jurassic as Fig. 17. It should be noted that strong Cenozoic deformation is not shown by the section. The main features are the progressive east-to-west change from shelf sandstone through silt-

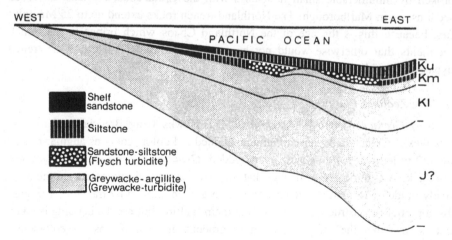

Fig. 17.   Cross section modified from *Geology NZ* (Fig. 6.28) illustrating the facies changes for the Cretaceous strata from platform to trench at the north end of South Island. The Early Cretaceous rocks at the base of the section are entirely of graywacke and argillite. The Mid-Cretaceous rocks are partly graywacke and argillite and partly sandstone–siltstone flysch. The Late Cretaceous is entirely sandstone and siltstone and changes regularly from shallow-water sandstone on the shelf through massive siltstone to flysch in the trench. Folding during the overall subsidence is shown by an eroded fold in Mid-Cretaceous strata at the east end of the section.

stone to sandstone–siltstone turbidite for the uppermost Cretaceous, and the already-described downward change from flysch to graywacke suite. Analogous changes take place in the more disturbed North Island east coast sections (Wellman, 1959). Even more important as a subduction model is the progressive eastward migration of the zero isobase (often incorrectly termed ''hinge line'') as shown by the lines of intersections of successive time planes on the western side of the section; and the fold that grew within the trench while the overall movement was one of subsidence. The present-day Hikurangi Trough east of North Island contains not one but several growing folds (Lewis, 1971, Fig. 4) that appear to be analogous to that in the cross section (Fig. 17).

(b) *Lower Cenozoic*. The Paleogene trench sediments are highly distinctive and extend from Marlborough to Northland. Supply of terrigenous sediments was reduced suddenly at the end of the Cretaceous and the strata changed from sandstone–siltstone to flint, limestone, glauconite, and bentonite, the last two in thin layers mixed with sandstone and siltstone. Turbidites, some with unusual composition, are common and the thicknesses considerably greater than for most of the platform sections. The flint, and the overlying limestone into which it grades, are rhythmically bedded and probably turbidites. The flints are generally accepted as being metasomatized limestone. The bentonite is commonly supposed to be a product of the peneplanation which is described next, but, although glass shards have yet to be found, it is more probably weathered acid ash. The most unusual sediment is probably a turbidite of glauconite, sandstone, and bentonite in thin layers, and now broken by innumerable small faults. As with the Cretaceous, the least disturbed sections are in Marlborough. The Northland trench rocks extend up to 25 Ma only; first because this is the age of the Northland Chaos which brings in the trench sediments that otherwise would not be exposed and second because the trench bypassed Northland shortly after.

2. New Zealand Platform

(a) *Cretaceous (Failed Openings)*. The Hawks Crag Breccia (and its correlatives), distinctive because of an unexplained red color, contains angular blocks up to 2 m long. It is remarkably even-bedded (*Geology NZ:* Fig. 6.19) and up to 1 km thick. Air-lubricated landslides that traveled for several kilometers are the most likely mode of formation. Deposition was in grabens linked by transforms. Figure 18, a partial reconstruction, in which the main faulting but not the bending is taken out, shows that the inferred pattern of grabens and transforms is geologically reasonable.

At places the breccia grades down into, and up into, fairly thick freshwater beds. Angiosperms are present in the upper but not in the lower freshwater beds and the top of the breccia is probably about 100 Ma old. The upper freshwater beds are best developed as the Paparoa coal measures of Greymouth and contain the only worthwhile deposits of low-sulfur bituminous coal in NZ.

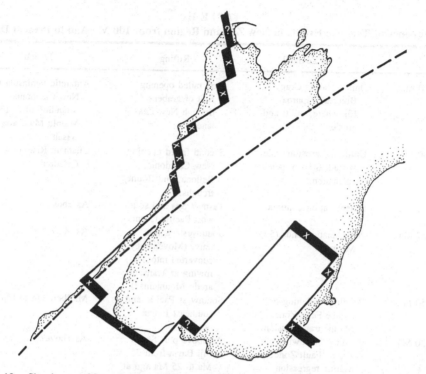

Fig. 18.   Sketch map of South Island of NZ showing the line of the 100-Ma-old Hawks Crag Breccia as a "failed" opening. The map is partly unfaulted but not unbent. Crosses mark outcrops of Hawks Crag Breccia and associated rocks, and question marks outcrops that may be associated. The double lines that contain the outcrops are lines of supposed grabens, and the joining lines the transforms that are supposed to have taken the widening from graben to graben. In the rigid southeast part, transforms and grabens should be at right angles. Elsewhere the angle between them will in general be changed from a right angle by the bending.

   The Hawks Crag graben formation, probably sharp and brief, is taken as being the earliest *failed* attempt at ocean opening. The second *failed* attempt is thought to be represented by the three troughs, New Caledonia, Bellona, and Bounty, already discussed under Reconstructions. There are no lineation ages for the troughs, but 2-km-thick freshwater beds at the base of the Great South Basin are early Upper Cretaceous, and make an 85 to 95 Ma age likely. The *successful* opening started about 82 Ma ago with the formation of LOM 36 in the Tasman Sea and Southwest Pacific Ocean and continued until 57 Ma ago.

   The sequence of openings is related to other events in Table II.

   (b) Lower Cenozoic (Peneplanation and Marine Transgression). After the opening of the Hawks Crag grabens, progressive peneplanation was followed by freshwater deposition and then by marine transgression and deposition. Five transgression shorelines are shown in Fig. 20. From 100 Ma to 50 Ma ago the transgression was caused by the subsidence that is an unexplained but general feature when

TABLE II
Sequence of Tectonic Events in New Zealand Region from 100 Ma Ago to Present Day

	New Zealand	Rifting	Trench
100 Ma	Ohika, Hawks Crag Breccia, Paparoa. Elsewhere, uplift and erosion.	First failed opening. Belt of grabens through New Zealand.	Antarctic Peninsula to New Caledonia. Stranding of c. 130-Ma-old Matakaoa Basalt.
90 Ma	Uplift and erosion widespread. Start of peneplanation.	Second failed opening. New Caledonia, Bellona, and Bounty troughs.	Chatham Rise to New Caledonia.
82 Ma	Peneplanation continues.	Tasman Sea and southwest Pacific Ocean.	As above.
57 Ma	Peneplanation ends (50 Ma ago)	Southwest Pacific only. (Movement converted into narrowing at Transantarctic Mountains?)	As above?
50 Ma	Bending–widening of Alpine Fault Zone. Marine transgression.	Southwest Pacific and southeast Indian oceans.	Marlborough to Fiji?
30 Ma	Bending–narrowing of Alpine Fault Zone. Marine regression.	As above plus South Fiji Basin from 35 Ma to 25 Ma ago at least. Three Kings Slide and Northland Chaos 25 Ma ago.	As above?
10 Ma	Mountain building at accelerating rate. Dextral-reverse movement on Alpine Fault. Dextral on North Island Shear Belt. Sinistral on Fiordland Fault.	As above plus Havre Trough in last 2 Ma at least.	Now from Cook Strait along Kermadec and Tonga trenches to Fiji.
0 Ma	Present day.		

fragments are rifted from their mother continent. After 50 Ma ago the subsidence was caused by the widening of the Alpine Fault Zone (A.F.Z.) and is directly related to the widening of the blocks that make up the Alpine fault zone (Fig. 19 and Tables III and IV).

Subsidence was counterbalanced, but progressively less and less, by the uplift of the schist belt and perhaps Fiordland. The schist belt had been deeply buried through being held down by subduction, and when subduction slowed down or stopped, it rose and continued to rise as its mountains were worn down. The net result was an upward and then a downward movement that was much the same for all shorelines but progressively younger for the inner ones.

Fig. 19. Map diagram and cross sections showing how the bending on the Alpine Fault Zone (AFZ) is thought to have resulted in the faulting and tilting of earth blocks that is recorded by the geology of the Greymouth and other South Island coalfields.

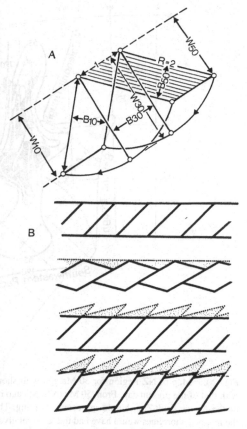

(A) Map diagram showing mechanism of bending proposed for the AFZ. Hachured part with ten lines is a segment of the fault zone 50 Ma ago. Line marked "1" is a base line, supposed to be fixed, and part of the Indian Plate. "R = 2" is a radius across the AFZ taken as being twice as long as the base line. "B_{50}" is breadth of segment 50 Ma ago and "W_{50}" is width of AFZ 50 Ma ago. The segment is supposed to rotate about the base line in the direction shown by the arrows on the arcs, the ten hachured slices moving against each by sinistral faulting. While they are rotating the slices are tilting to retain a constant volume, the amount and sense of tilt being shown by the cross sections (B) and discussed later. The breadth of the segment (maximum width for coalfields) B_{30} and the width W_{30} of AFZ had maximum value 30 Ma ago. The final step for the map diagram (but not for the sections) is for 10 Ma ago when continued rotation, which had amounted to 90°, resulted in a mirror image situation with respect to 50 Ma ago; W10 being equal to W50 and B10 to B50.

(B) Cross sections (horizontal-vertical scale) showing faulting and associated breadth changes according to data from the Greymouth Coalfield that was caused by the bending of the Alpine Fault Zone. The sections are taken in the directions of B50, B30, and B10 in the map diagram "A." The top (first) section is for the end of peneplanation 50 Ma ago, and just before faulting started. There are no cover beds, the top of the undermass is level, and the fault planes dip west at 45°. The second section is for the time of greatest extension, greatest breadth, 30 Ma ago. The earth blocks have rotated by normal faulting, and fault angle depressions have filled with covering sediments, with coal measures at the base. The faults dip west at 33°. The third section is for 10 Ma ago. Reverse faulting has brought the top of the undermass back to level, but the covering sediments form block mountains that are being eroded. The fourth and last section is for the present day. Continued reverse faulting has caused the surface of the undermass to overshoot so that it now dips west and not east as it had done previously. Faults dip west at 52°. Most of the cover above sea level has been eroded. The coal increases in rank updip in the direction of greatest burial. In the above the faulting is described as being first normal and then reverse, but as mentioned in (A) there was continuous sinistral faulting, but this cannot be shown in the cross sections.

The general transgression sequence is first sandstone, then carbonaceous siltstone, then calcareous mudstone, and finally limestone. The sequence of layers is synchronous along each shoreline, but progressively younger and less complete from shoreline to shoreline toward the schist.

The existence of Gondwana species such as tuataras, moas (until 150 yr ago),

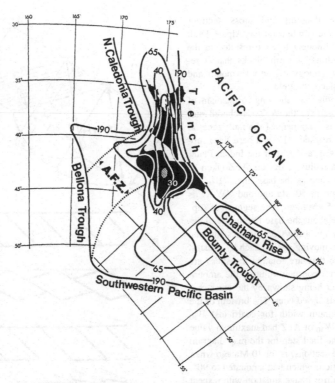

Fig. 20.   Map of NZ Region for 30 Ma ago with shorelines for 190, 65, 40, and 30 Ma ago together with that of the present day. From 50 Ma to 10 Ma ago it is supposed that there was considerable bending on the Alpine Fault Zone but no appreciable faulting. There would have been a rubber sheet situation and the mapped shorelines would have had the same relative positions on their appropriate reconstructions as they have on the 30 Ma ago reconstruction.

kiwis, Southern beech trees, and conifers shows that at least one small island remained unsubmerged. It is shown in stipple in Fig. 20 and placed in the center of the schist.

The stripped peneplain surface is a scenic feature over much of NZ. The best examples are in South Island, particularly in Northwest Nelson where the difference in resistance to erosion between cover and undermass is greatest.

The economic deposits of the transgression are varied and have the greatest importance for NZ. Coal is widespread and of direct and indirect use. Some 10 Ma ago, thick coal measures were buried to about 3 km in Taranaki, metamorphosed, and a large gas field created. The transgression and early regression limestones are widespread and essential for agriculture in a country with an adequate or more than adequate rainfall.

At Nelson there are two anomalies in the otherwise simple sedimentation pattern. There is a dextral jog in the 40-Ma-old shoreline, and some 30 Ma ago

TABLE III
Cumulative Shiftpoles for Reconstructing New Zealand and Movement Rates and Widths of Alpine Fault Zone at 42°S

| | | | | | | | At 42°S | | | |
| | | | Cumulative shiftpole | | | | | | | |
Age (ma)	Rotation	Rotation rate (degrees/Ma)	Lat. S	Long. E	Rotation	Radius (degrees lat.)	Net displacement (km)	Net rate (mm/yr)	Width Alpine Fault Zone (km)	Rate of widening (mm/yr)
c. 50	0°		50°	178°	64°	8.3°	0		370	
c. 30	24°	1.2°	52°	176°	40°	10.0°	254	13	490	+6
c. 10	50°	1.3°	55°	179°	14°	13.0°	678	21	370	−6
0	64°	1.4°	56°	175°	—	13.0°	1030	35	300	−7

instead of the expected limestone there are breccia and turbidites. Both anomalies are explained by movement on the Glenroy Fault. The 20 km of sinistral displacement shown in Fig. 21 is not taken out in the shoreline map (Fig. 20) as it should have been, and earlier dip-slip displacement on the fault will explain the anomalous rocks.

TABLE IV
Relative Widths of Alpine Fault Zone Calculated from Dip of Peneplain at Greymouth Compared with Relative Reconstruction Widths

Age (Ma)	Rate of tilt of peneplain (degree/Ma)	NZ stage symbols	P = eastward dip of peneplain (Gage, 1952, Fig. 12)	D = westward dip of fault planes $(45° − P)$	sin 45°/sin D	W/370 km	W = width of Alpine Fault Zone from shiftpole (Table III)
c. 50	0.6	Dm	0°	45°	1.00	1.00	370
c. 30	0.6	Lwh/Ld	12°	33°	1.30	1.32	490
c. 10	0.6	Sw/Tt	0°	45°	1.00	1.00	370
0	0.7	H	−7°	52°	0.90	0.81	300

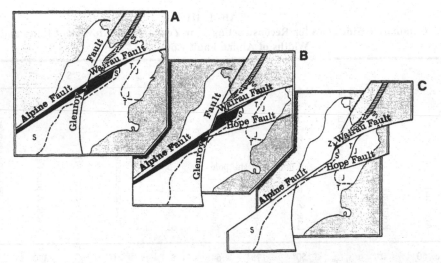

Fig. 21.   Three maps showing the development of the 20-km-long sinistral jog in the Alpine–Wairau fault system at the Glenroy River. (A) About 25 Ma ago. For convenience of identification the reconstruction is made by simple unfaulting, the appreciable bending being not taken out. The black represents the part to be obducted. The schist of the Southern Alps is shown by ''S'' and the Maitai rocks and the Dun Mountain Ophiolite by ''Z.'' The Alpine and Wairau are a single unbroken fault. (B) About 25 Ma ago. There has been 20-km sinistral displacement on the Glenroy Fault but no other change. The Glenroy Fault is supposed to extend north along the west side of the Waimea Depression to Northland and south to about Dunedin. The Hope Fault appears but is without displacement. (C) Present day. There has been dextral movement on the Wairau Fault and on the Hope Fault and the direct line of the Glenroy Fault has been broken in the same way as the direct line of the Alpine–Wairau Fault was broken by the Glenroy Fault. Note the small area of ''Z'' which is now at the north end of the jog and was separated from the main belt by the dextral displacement of the Wairau Fault. Note also that the nature of the faulting is shown by the breaks in the margins of maps B and C. Map A should be copied and then cut out and shifted in order to understand the faulting sequence.

C. Upper Cenozoic (Marine Regression): 30 Ma Ago to Present Day

The reconstructions (Fig. 10B,C) slightly oversimplify the tectonics. They show no faulting but bending by reverse-sinistral block faulting in north-striking faults within the Alpine Fault Zone from 30 Ma to 10 Ma ago and faulting without bending for the last 10 Ma. They neglect the 20 km of sinistral displacement on the Glenroy Fault that began about 25 Ma ago, and also neglect the present-day bending at an unknown rate shown by the faulting during the 1929 West Nelson Earthquake and by the faulting inferred from nodal-plane solutions. The faulting of the last 10 Ma is almost the same as that of the Holocene described at the end of this chapter.

After the culmination of the transgression the deposition of limestone continued for 5 to 10 Ma and the pattern of sedimentation remained fairly simple. With continuing block faulting the number of small basins increased to produce a complex than can be explained here only in the most general way. The overall effect was that NZ grew in the last 30 Ma from being a tiny island to its present size. Growth was the result, not of uniform overall uplift, but of uplift eventually prevailing over the subsidence that is recorded by the Upper Cenozoic strata.

Growth took place earlier in South Island than in North Island, the southern half of North Island being under the sea a mere 3 Ma ago. Shorelines up to about half a million years old extend around most of the coast and provide basic data for estimating the rate of uplift of South Island (*Geology NZ:* Fig. 7.90; Wellman, 1979) and the southern part of North Island (Ghani, 1978).

D. Holocene Tectonics

Alexander McKay was the first to recognize that more had happened geologically in NZ in the last 10 Ma than in the last 100 Ma in most of the world. He described the 2-m dextral displacement on the Hope Fault from the 1888 Earthquake, and spotted other active faults. The steep Cenozoic dips of the Kaikouras were obvious to all, but only McKay faced up to the young age from the fossils and accepted the reality of an ongoing deformation.

McKay's discoveries are now commonplace and part of plate tectonics. The faulting and much of the folding seen by McKay took place during the Holocene (late 10,000 yr) and the following account depends on the kind of features seen by McKay plus data from resurveys, from seismology, and not surprisingly from soil science.

Holocene faulted terrace flights that extend down to river level with fault displacement that progressively increases with age and height of terrace are universally accepted as proving ongoing faulting. There are about five such flights in NZ and a few more overseas.

Consequently, the rate accepted for Holocene faulting in NZ depends almost entirely on the age accepted for the conspicuous, widespread, and generally loess-free river aggradation surface in which the faulted terraces are cut. Those who assume that river aggradation coincides with glacial advances favor an age of 16,000 yr of even more, whereas those who study loess stratigraphy favor 10,000 yr or even less. For example, for the Waiohine Surface which is a typical example, Lensen (in Lensen and Vella, 1971) accepted 35,000 yr ("second to last stadial of Last Glaciation") and Suggate and Lensen in 1973 accepted 18,000 yr ("principal aggradation surface for latest glaciation"). On the other hand, Vella who had accepted 20,000 yr in 1971 and who has since studied loess stratigraphy now favors about 10,000 yr (Dr. P. Vella, personal communication, 1979).

An indirect but independent dating method (Wellman, 1972) is by correlating the youngest five of the eight river terraces at Waiohine River 60 km NE of Wellington which are cut in the aggradation surface with five fault-uplifted marine beach ridges at Turakirae Head 20 km SE of Wellington. The beach ridges are a few kilometers NW of the Wairarapa Fault and 60 km SSW of the Waiohine River terraces which are faulted by the same fault.

The amount of dextral displacement and throw for the river terraces and the uplift of the beach ridges are set out in Table V. Correlation is by pairing off in upward order each of the five lowest faulted terraces with each of the five uplifted marine beach ridges. The uniformity of the ratios of Turakirae uplift to Waiohine throw makes the odds over 120 to 1 against a chance result. By simple proportion, using the now generally accepted 6.5 ± 0.2 k.y. age (Ota et al., 1982) for the oldest Holocene beach ridge, the age of the aggradation surface is 11 ± 1 k.y., the dextral faulting rate for the Wairarapa Fault 11 ± 1 mm/yr, and the Turakirae uplift rate 4.2 ± 0.1 mm/yr.

It is universally accepted that the latest uplift at Turakirae Head took place during the 1855 Earthquake (New Zealand's largest historical earthquake, M 8.0 ± 0.1) and generally accepted by geologists that earlier earthquakes caused the earlier uplifts. The river-terrace/beach-ridge correlation makes it likely that the Waiohine river terraces were formed by the same series of earthquakes that uplifted the Turakirae beach ridges. On the other hand, the correlation spans a distance of 60 km and can be true only if each of the postulated earthquakes had a faulting length of considerably more than 60 km.

The mean net fault displacement for the eight Waiohine River terraces is 15 ± 6 m. Assuming from Gibowicz and Hatherton (1975) that $M - 7 = \log D$ ($D =$ meters net fault displacement), then the mean magnitude of the postulated earth-

TABLE V
Correlation of Waiohine River Terraces Faulted by the Wairarapa Fault with Beach Ridges Uplifted by the Wairarapa Fault at Turakirae Head; 65 km NE and 20 km SE of Wellington City, Respectively

River terrace	WAIOHINE RIVER TERRACES (Lensen and Vella, 1971) Cumulative dextral displacement meters	Cumulative throw displacement meters (T)	TURAKIRAE MARINE RIDGES (Wellman, 1972) Beach ridge	Cumulative uplift meters (U)	U/T
a	$(123.2)^a$	18.3		Obliterated by	
b	99.0	14.7		the postglacial	
c	85.5	12.3		rise of sea level	
d	66.9	12.0	F	27.0	2.3
e	n.d.^{b}	10.2	E	24.0	2.3
f	n.d.^{c}	n.d.^{c}	D	18.0	n.d.
g	32.1	3.6	C	9.0	2.5
h	12.0	0.9	B	2.5	2.8

^{a}The "a" terrace, being the top terrace, i.e., the aggradation surface, is without a riser from which to measure dextral displacement, and the "123.2" is by proportion. ($99.0 \times 18.3/14.7$.)
^{b}The "e" terrace is eroded and the dextral displacement is unmeasurable.
^{c}The "f" terrace is eroded and the throw and dextral displacement are both unmeasurable.

quakes is M 8.15 ± 0.15. According to historical earthquakes of similar magnitude the faulting length would have been about 300 ± 100 km. The length is considerably greater than the 60 km, and the earthquake explanation can be accepted.

River terraces similar to those of the Waiohine are a striking feature of the NZ landscape particularly within the Alpine Fault Zone. They are considered here to define the amount and extent of the uplift caused by large earthquakes. Mostly there are up to about eight terraces that are cut in the main aggradation surface giving an earthquake recurrence interval of about 1200 yr. The estimated earthquake magnitudes range from M 7.8 ± 0.4 to M 8.0 ± 0.25 depending on the size of the faults.

The Holocene tectonics of NZ are modeled by Fig. 22 in terms of four plates, five shiftpoles (with rotation rates), and eight cross sections. For simplicity of description, fault belts with parallel faults having similar sense of displacement are consolidated into a single fault. For example, there are five main Marlborough faults, but only the most active, the Hope, is mapped.

The North Island Shear Belt and its northern extension as the Havre Trough is shown in sections 1, 2, and 3. Movement on the Taupo Volcanic Zone and on the Galpin Fault (Fig. 23) are consolidated into the shear belt. At the Havre Trough at section 1 (35°S) the LOM 2 magnetic lineation pair with an age of 1.8 Ma are 63 km apart (Malahoff *et al.*, 1982). The calculated separation from the model is 54 km (6.2° lat. × 2.5°/Ma for 1.8 Ma). Agreement is reasonable.

The shear belt itself is shown as an arc of the EI shiftpole. Calculated movement is dextral at 19 mm/yr (4.0° lat. × 2.5°/Ma). The retriangulation rate at section 2 (Sissons, 1979) is 21 mm/yr and is made up of 14 mm/yr for the shear belt and 7 mm/yr for the Taupo Volcanic Zone. Agreement with the 19 mm/yr calculated value is good.

At section 3 the shear belt consists of four faults that have a consolidated dextral rate of 20 mm/yr: Galpin, the transform for the Taupo Volcanic Zone (Fig. 23), 7 mm/yr as mentioned above; Shepherds Gully possibly 3 mm/yr; Ohariu 6 mm/yr; and Wellington 4 mm/yr. Data for the last three faults are from Ota *et al.* (1982). Agreement with the calculated value of 19 mm/yr is excellent and probably accidental.

Also on section 3 there is the line of subduction which is shown on the map in consolidated form as the extension of the Kermadec Trench. The calculated rate of 38 mm/yr is almost pure narrowing. The surface expression is by reverse-faulting and the tilting of earth blocks between the North Island Shear Belt and the line of the Kermadec Trench. To account for the 38 mm/yr narrowing, the earth blocks would need to be tilting at 22°/Ma over a width of about 100 km if the fault planes dipped at 45°. For the landward part of the narrowing belt, Ghani (1978) surveyed a width of 60 km and found that the tilting rates decreased from 30°/Ma in the west to 15°/Ma in the east. With a 45° fault dip, about 24 mm/yr of the required narrowing is accounted for. The remaining 14 mm/yr is probably off the coast where growing folds are shown by bathymetry.

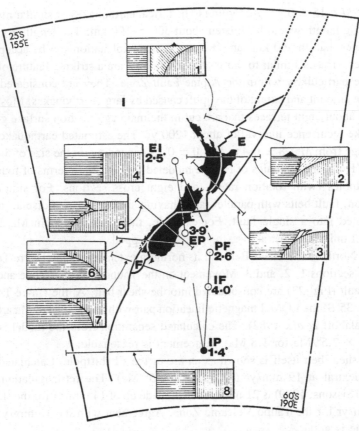

Fig. 22. Sketch map with eight cross sections illustrating Holocene tectonics of NZ. Parallel and similar faults are consolidated to delimit four plates: the Indian (I), the Pacific (P), and two microplates, Eastland (E) and Fiordland (F). The order of the letter pair on the shiftpole gives sense of plate rotation and can be used to give sense of movement on the consolidated faults. For instance, "EI" means that Eastland Plate is rotating clockwise on the EI shiftpole relative to the Indian Plate. The numbers on the shiftpole are the relative rotation rates in degrees per million years and are used in the text to calculate faulting rates. In the cross sections the faults with ticks on their left-hand side are dextral, and those with ticks on their right sinistral. All faults are dextral except the Fiordland Fault (arc to PF shiftpole). Note that the poles for the two pairs of three-plate combinations are collinear. Also because the map area is small and effectively flat, the rotation rates are proportional to the length of the "opposite" intercepts. For example, if IP:PF is the length from IP to PF, and so on, then

$$\frac{4.0}{\text{IP:PF}} = \frac{1.4}{\text{PF:IF}} = \frac{2.6}{\text{IF:IP}}.$$

Surprisingly, a major feature—the Wairarapa Fault—is surplus to requirements. From near Wellington it extends NE for 250 km and changes from a steeply dipping essentially dextral fault to a series of eastdipping thrusts. The dextral movement has to be balanced out, and sinistral faults with an equal displacement rate (11 mm/yr) are proposed to the east of the Wairarapa Fault. The blocks

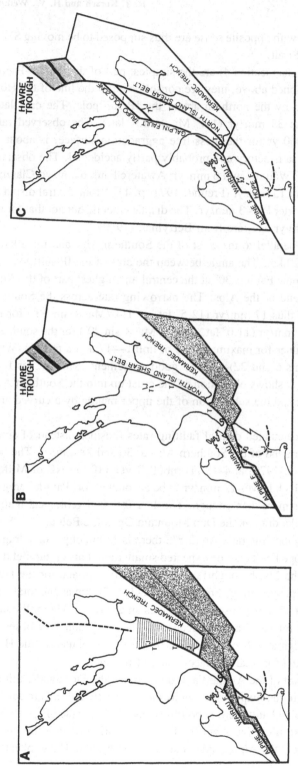

Fig. 23. Three maps showing development of the North Island Shear Belt, the Taupo Volcanic Zone, and the Galpin Fault. (A) About 10 Ma ago. The boundary between the Triassic (T) and Jurassic (J) graywacke is an unbroken line with a long "S" shape analogous to the shape of the Dun Mountain Ophiolite. The stippling represents the area to be subducted. (B) About 2 Ma ago. There has been 90 km of dextral displacement on the North Island Shear Belt and the Triassic–Jurassic boundary is displaced accordingly. There is a corresponding opening of the Havre Trough, and a reduction in the area to be subducted. (C) Present day. There has been a total of 170 km of dextral displacement on the North Island Shear Belt and the same displacement on the Triassic–Jurassic boundary, and within the Havre Trough. The area to be subducted is eliminated but subduction is still continuing. The Taupo Volcanic Zone has opened and the same displacement taken place by dextral movement on the Galpin Fault. For the Coromandel volcanoes which are up to 10 Ma old, note that opening similar to that of the Taupo Volcanic Zone can be envisaged provided that the Galpin Fault is extended north to the coast of the Firth of Thames.

between the pair of faults with opposite sense are thus supposed to be moving SW to be subducted into Cook Strait.

The Marlborough faults (section 4) are the simplest part of the Indian–Pacific plate boundary. As mentioned above, they are consolidated to the line of the Hope Fault and are represented by the northern arc of the IP shiftpole. The calculated displacement is dextral at 35 mm/yr (1.4° Ma × 13° lat.). The observed rate, assuming an age of 10,000 yr for the loess-free aggradation surface, is about 36 mm/yr dextral, which is satisfactory and probably partly accidental. The observed 36 mm/yr rate consists of: Wairau Fault 6 mm/yr; Awatere Fault 6 mm/yr; Clarence Fault 6 mm/yr; Hope Fault 15 mm/yr (Freund, 1971, p. 11: "four dextral displacements of 490 ft."); and Porter Pass 3 mm/yr. The displacements, but not that for the Hope Fault, or the 10,000 yr age, are from Berryman (1979).

The Alpine Fault is parallel to the crest of the Southern Alps and remarkably straight for a length of 500 km. The angle between the arcs of the IP shiftpole and the straight line of the Alpine Fault is 30° at the central and highest part of the Alps, and 20° at the southern end of the Alps. The narrowing rate across the Southern Alps and Alpine Fault is thus 17 mm/yr (12.5° lat. × 1.4°/Ma × sin 30°) for the center of the Alps and 10 mm/yr (11.0° lat. × 1.4°/Ma × sin 20°) for the south end of the Alps. Observed values for maximum rate of uplift—17 and 8 mm/yr (Wellman, 1979, Table I, items 6 and 22)—are in good agreement. Section 5 is from Wellman (1979, Fig. 4). It shows obduction of the crust up into the Southern Alps by a curved Alpine Fault and the subduction of the upper mantle by a curved anti-Alpine Fault.

The corresponding calculated dextral faulting rates (cosines instead of sines) for the center and southern end of the Southern Alps at 30 and 28 mm/yr. The best observed value (Berryman, 1979) at 44°S (Turnbull River 110 km NE of Milford Sound) is at 15 mm/yr which leaves 15 mm/yr to be accounted for. Parallel drag on the Alpine Fault provides more than enough. The 1 : 1,000,000 geological map of NZ shows 35 km of parallel drag on the Dun Mountain Ophiolite Belt at 44.1°S but there is no way of finding the drag rate. At 42.8°S there is 33 km of parallel drag on the western end of the Hope Fault and the expected smaller amounts of parallel drag on the western ends of the Marlborough faults to the north. An age for the Hope Fault will give a rate for parallel drag. Freund (1971) gives 20 km as the total fault displacement. The rate of 15 mm/yr used above gives an age of 1.3 Ma for the fault and a rate of 25 mm/yr for the drag displacement. The total indirectly observed value is 10 mm/yr too high and the 10 mm/yr difference is attributed to the Hope Fault being underestimated for total displacement and for age.

The southernmost part of the Alpine Fault on land and the Fiordland Fault are shown in section 6, and in the map are shown as the arc for IF and the arc for PF, respectively. For the Alpine Fault there is no throw displacement (see Fig. 10.3 of *Geology NZ*), strong dextral displacement, and perfect continuity with the line of the fault to the north where the Alpine side is strongly upthrown. The explanation adopted, and possibly the only one available, is that the Fiordland microplate (F) is being squeezed south like a pip, thus increasing the dextral rate for the Alpine Fault and causing sinistral displacement on the Fiordland Fault which lies on the east of the Fiordland microplate. The calculated rates are 54 mm/yr for the Alpine Fault (7° lat. × 4°/Ma) and 25 mm/yr for the Fiordland Fault (5° lat. × 2.6°/Ma). There are no rate observations for this part of the Alpine Fault. For the Fiordland Fault, observations (Wellman, 1952, Items S303 to S305) indicate sinistral displacement at about 12 mm/yr but they are poor and inconclusive.

The two southernmost cross sections (7 and 8) are offshore and the calculated rates cannot be checked by field observations.

Bibby (1981) has used the angular differences found by retriangulation across the Marlborough Faults from 1888 to 1977 to infer the Indian–Pacific plate movement. He assumed that lengths parallel to the faults have not changed and further assumed that the rate of angular change for the observation period which did not include major earthquakes equaled the long-term rate which included major earthquakes. The difference between his rate of 53 ± 9 mm/yr and our rate of 35 mm/yr is attributed to the rate changing with major earthquakes. The rate was slow before the 1848 and 1855 earthquakes and more rapid after them.

Walcott (1978) adopted a shiftpole from Chase (1978) at 62.0°S, 174.3°E with a relative rotation rate of 1.3°/Ma which gives a movement rate of 50 mm/yr across the Marlborough Faults. He accepted the older ages and slower displacement rates for the active faults and considered that some two-thirds of the plate movement was taking place by "pervasive shear" which was not specified further except that it was assumed to be aseismic. It should be noted that the Earth Deformation Branch of the New Zealand Geological Survey has made and repeated surveys across the main active faults for the last 10 yr and found no indication of movement by creep. Also it must be supposed that "pervasive shear" unless defined otherwise will change the lengths of lines parallel to the strike of the Marlborough and other faults and thus nullify the rates inferred from retriangulation for plate movement.

As outlined above we consider the present day and past plate movement through NZ to have been on the kind of faults and folds that are already known and mapped and see no need to postulate additional kinds.

Backarc Spreading, Subduction, and Andesite Volcanism

The knowledge of subduction rates for the Holocene, and of past minimum subduction rates from the reconstructions now makes it possible to consider andesite volcanism in terms of subduction rates. However, backarc spreading significantly increases the calculated minimum rates and has to be described first. Backarc spreading is, and has been, of minor importance within NZ itself, but it is of major importance immediately north of NZ.

In the simplest case "backarc spreading" as used here can be explained in terms of three plates. In cross section a large plate, say the Indian, on the left; a large plate, say the Pacific, on the right; and a small plate in the middle. With respect to the small plate there is a left- and a right-hand boundary. For NZ north of 45°S the line between the two large plates is almost a radius from the Indian–Pacific shiftpole and as far north as 25°S, i.e., 30° from the shiftpole the rate of narrowing between the two large plates (I–P) narrowing) is almost proportional to distance from shiftpole. The natural expectation would be for the I–P narrowing to be shared between the left-hand and right-hand plate boundaries. Surprisingly, what is actually happening is widening (backarc spreading) on the left-hand boundary and a corresponding increase in the amount of narrowing (subduction north of 45°S) on the right-hand boundary. At present on moving northward the first backarc spreading is at the Taupo Volcanic Zone where it increases the rate of subduction from 43 mm/yr to 55 mm/yr. At the Havre Trough backarc spreading is at 21 mm/yr and increases the subduction rate from 53 mm/yr to 74 mm/yr. Farther north at 32°S the backarc spreading rate increases to 35 mm/yr from LOM 2–LOM 1–LOM 2 (last 2 Ma) and to 46 mm/yr from LOM 12–LOM 7–LOM 12 (33 Ma to 24 Ma ago). By comparison, if it is assumed that the distance from the Norfolk Ridge to the Kermadec Trench is backarc spreading that has taken place in the last 53 Ma (LOM 22), then the average rate is 31 mm/yr.

For NZ at present there is a successive dropout of convergent plate features to the south with decreasing I–P narrowing and decreasing subduction. Backarc spreading and andesite volcanism drop out at 39.5°S where the I–P narrowing rate is 43 mm/yr and the backarc spreading rate 7 mm/yr. Subduction plus deep-focus earthquakes plus the negative gravity anomaly drop out at 41.5°S where the I–P narrowing rate is 38 mm/yr. Obduction then takes over and continues south to 45°S where the I–P narrowing rate is 20 mm/yr (Fig. 22).

An I–P narrowing rate of 43 mm/yr is the present minimum rate for andesite volcanoes, and the minimum rate may well have been the same as in the past, at least for the Cenozoic. According to the reconstructions used here, the I–P narrowing rate for the north of NZ was 35 mm/yr from 50 Ma to 39 Ma ago, and 42 mm/yr from 30 Ma to 10 Ma ago, and at 45 mm/yr, and increasing, from 10 Ma ago to the present day. For these rates, and taking into account the high values for backarc spreading immediately north of NZ, there should have been continuing andesite volcanism north of NZ, with the volcanism just reaching NZ 30 Ma ago and then moving south at an accelerating-rate to reach 39.5°S at the present day. The forecast agrees well with geology. The earliest NZ Cenozoic andesites are confined to the north of NZ and are about 25 Ma old. As mentioned under Region 14 (Taupo Volcanic Zone), there has been migration to the south at 15 km/Ma in the last 10 Ma from Coromandel to the Taupo Volcanic Zone in the east, and migration to the SE at 60 km/Ma in the last 0.5 Ma at the Mt. Egmont volcano line in the west.

ACKNOWLEDGMENTS

We would like to thank Hugh Bibby for sharpening the last section, Paul Vella for looking over the Cretaceous part, and Barry P. Roser and Rodney Grapes for comments on the terrane mapping. Ted Hardy drafted the maps for the space available, and Val Hibbert typed the manuscript in the time available. To both our sincere thanks. We have to thank Andy Duncan and Alan Nairn for suggestions for improving the present and somewhat longer draft copy.

This manuscript, submitted in 1982, revised in 1983 for Volume 7A, was deferred to Volume 7B with the authors' permission.

REFERENCES

Adams, C. J. D., 1975, Discovery of Precambrian rocks in New Zealand: Age relations of the Greenland Group and Constant Gneiss, West Coast, South Island, *Earth Plant. Sci. Lett.* **28**:98–104.

Adams, C. J. D., 1979, Age and origin of the Southern Alps, *R. Soc. N.Z. Bull.* **18**:73–78.

Adams, C. J. D., and Nathan, S., 1978, Cretaceous chronology of the Lower Buller Valley, South Island, New Zealand, *N.Z. J. Geol. Geophys.* **21**:455–462.

Adams, C. T., Harper, C. T., and Laird, M. G., 1975, K–Ar ages of low grade metasediments of the Greenland and Waiuta Groups in Westland and Buller, New Zealand, *N.Z. J. Geol. Geophys.* **18**:39–48.

Adams, J., 1980, Contemporary uplift and erosion of the Southern Alps, New Zealand, *Geol. Soc. Am. Bull.* **Part II, 91**:1–114.

Andrews, P. B., Speden, I. G., and Bradshaw, J. D., 1976, Lithological and paleontological content of the Carboniferous–Jurassic Canterbury Suite, South Island, New Zealand, *N.Z. J. Geol. Geophys.* **19**:791–819.

Aronson, J. L., 1965, Reconnaissance rubidium–strontium geochronology of New Zealand plutonic and metamorphic rocks, *N.Z. J. Geol. Geophys.* **8**:401–423.

Aronson, J. L., 1968, Regional geochronology of New Zealand, *Geochim. Cosmochim. Acta* **32**:669–697.

Bates, T. E., 1980, The origin, distribution and geological setting of copper and nickel sulphides in the Riwaka Complex, North West Nelson, New Zealand, *Australas, Inst. Min. Metall. Annu. Conf. Proc.* **35**:51.

Beggs, J. M., 1980, Sedimentology and paleogeography of some Kaihikuan Torlesse rocks in mid Canterbury, *N.Z. J. Geol. Geophys.* **23**:439–445.

Berryman, K., 1979, Active faulting and derived PHS directions in the South Island NZ, *R. Soc. N.Z. Bull.* **18**:29–34.

Bibby, H. M., 1981, Geodetically determined strain across the southern end of the Tonga–Kermadec–Hikurangi subduction zone, *Geophys. J. R. Astron. Soc.* **66**:513–533.

Bishop, D. G., 1972, Progressive metamorphism from prehnite–pumpellyite to greenschist facies in the Dansey Pass area, Otago, New Zealand, *Geol. Soc. Am. Bull.* **83**:3177–3197.

Bishop, D. G., 1974, Stratigraphic, structural, and metamorphic relationships in the Dansey Pass area, Otago, New Zealand, *N.Z. J. Geol. Geophys.* **17**:302–335.

Blake, M. C., Jr., and Landis, C. A., 1973, The Dun Mountain ultramafic belt: Permian oceanic crust and upper mantle in New Zealand, *U.S. Geol. Surv. J. Res.* **1**:529–534.

Blake, M. C., Jr., Jones, D. L., and Landis, C. A., 1974, Active continental margins: Contrasts between California and New Zealand, in: *The Geology of Continental Margins* (C. A. Burk and C. L. Drake, eds.), pp. 853–872, Springer-Verlag, Berlin.

Blattner, P., 1978, Geology of the crystalline basement between Milford Sound and the Hollyford Valley, New Zealand, *N.Z. J. Geol. Geophys.* **21**:33–47.

Boles, J. R., 1974, Structure, stratigraphy, and petrology of mainly Triassic rocks, Hokonui Hills, Southland, New Zealand, *N.Z. J. Geol. Geophys.* **17**:337–374.

Bradshaw, J. D., 1972, Stratigraphy and structure of the Torlesse Supergroup (Triassic–Jurassic) in the foothills of the Southern Alps near Hawarden (S60–61), Canterbury, *N.Z. J. Geol. Geophys.* **15**:71–87.

Bradshaw, J. D., 1973, Allochthonous Mesozoic fossil localities in melange within the Torlesse rocks of North Canterbury, *J. R. Soc. N.Z.* **3**:161–167.

Bradshaw, J. D., and Andrews, P. B., 1973, Geotectonics and the New Zealand Geosyncline, *Nature* **241**:14–16.

Bradshaw, J. D., Adams, C. J., and Andrews, P. B., 1981, Carboniferous to Cretaceous on the Pacific margin of Gondwana: The Rangitata phase of New Zealand, in: *Gondwana Five* (M. M. Cresswell and P. Vella, eds.), pp. 217–221, A. A. Balkema, Rotterdam.

Campbell, J. D., and Force, E. R., 1972, Stratigraphy of Mount Potts Group at Rocky Gully, Rangitata Valley, Canterbury, *N.Z. J. Geol. Geophys.* **15**:157–167.

Campbell, J. D., and Warren, G., 1965, Fossil localities of the Torlesse Group in the South Island, *Trans. R. Soc. N.Z.* **3**:99–137.

Campbell, J. K., and Campbell, J. D., 1970, Triassic tube fossils from Tuapeka rocks, Akatore, South Otago, *N.Z. J. Geol. Geophys.* **13**: 392–399.

Carter, L., 1980, New Zealand Region Bathymetry (2nd ed.) 1 : 6,000,000, Department of Scientific and Industrial Research, Wellington.

Carter, R. M., Hicks, M. D., Norris, R. J., and Turnbull, I. M., 1978, Sedimentation patterns in an ancient arc–trench–ocean basin complex: Carboniferous to Jurassic Rangitata Orogen, New Zealand, in: *Sedimentation in Submarine Canyons, Fans, and Trenches* (D. J. Stanley and G. Kelling, eds.), pp. 340–361, Dowden, Hutchinson & Ross, Stroudsburg, Pa.

Chase, C. G., 1978, Plate kinematics: The Americas, East Africa, and the rest of the World, *Earth Planet. Sci. Lett.* **37**:353–368.

Christoffel, D. A., 1978, Interpretation of magnetic anomalies across the Campbell Plateau, South of New Zealand, *Aust. Soc. Exp. Geophys. Bull.* **9**:143–145.

Christoffel, D. A., and Falconer, R. K. H., 1972, Marine magnetic measurements in the south-west Pacific Ocean and the identification of new tectonic features, *Antarct. Res. Ser.* **19**:197–209.

Christoffel, D. A., and Ross, D. I., 1965, Magnetic anomalies south of the New Zealand Plateau, *J. Geophys. Res.* **70**:2857–2861.

Clark, R. H., Vella, P., and Waterhouse, J. B., 1967, The Permian at Parapara Peak, North-west Nelson, *N.Z. J. Geol. Geophys.* **10**:232–246.

Coleman, R. G., 1977, *Ophiolites*, Springer-Verlag, Berlin.

Cook, R. A., 1981, Geology and bibliography of the Campbell Plateau, New Zealand, *N.Z. Geol. Surv. Rep.* **97**.

Coombs, D. S., Landis, C. A., Norris, R. J., Sinton, J. A., Borns, D. J., and Craw, D., 1976, The Dun Mountain Ophiolite Belt, New Zealand, its tectonic setting, construction, and origin, with special reference to the southern portion, *Am. J. Sci.* **276**:561–603.

Cooper, A. F., 1972, Progressive metamorphism of metabasic rocks from the Haast Schist Group of southern New Zealand, *J. Petrol.* **13**:457–492.

Cooper, A. F., 1974, Multiphase deformation and its relationship to metamorphic crystallisation at Haast River, South Westland, New Zealand, *N.Z. J. Geol. Geophys.* **17**:855–880.

Cooper, R. A., 1974, Age of the Greenland and Waiuta groups, South Island, New Zealand [Note], *N.Z. J. Geol. Geophys.* **17**:955–962.

Cooper, R. A., 1979, Lower Paleozoic rocks of New Zealand, *J. R. Soc. N.Z.* **9**:29–84.

Cooper, R. A., Landis, C. A., Le Masurier, W. E., and Speden, I. G., 1982, Geologic history and regional patterns in New Zealand and West Antarctica—Their paleotectonic and paleogeographic significance, in: *Antarctic Geoscience* (C. Craddock, ed.), pp. 43–53, University of Wisconsin Press, Madison.

Craddock, C., 1975, Tectonic evolution of the Pacific margin of Gondwanaland, in: *Gondwana Geology* (K. S. W. Campbell, ed.), pp. 609–618, Australian National University Press, Canberra.

Craw, D., 1979, Melanges and associated rocks, Livingstone Mountains, Southland, New Zealand, *N.Z. J. Geol. Geophys.* **22**:443–454.

Davey, F. J., and Christoffel, D. A., 1978, Magnetic anomalies across Campbell Plateau, New Zealand, *Earth Planet. Sci. Lett.* **41**:14–20.

Davey, F. J., and Robinson, A. G., 1978, Cook Magnetic anomaly map 1 : 1,000,000 Oceanic Series, Department of Scientific and Industrial Research, Wellington.

Davis, T. E., Johnston, M. R., Rankin, P. C., and Stull, R. J., 1980, The Dun Mountain ophiolite belt in East Nelson, New Zealand, in: *Ophiolites* (A. Panayiotou, ed.), pp. 480–496, Cyprus Geological Survey Department, Nicosia.

Devereux, I., McDougall, I., and Watters, W. A., 1968, *Potassium–argon mineral dates on intrusive rocks from the Foveaux Strait area, N.Z. J. Geol. Geophys.* **11**:1230–1235.

Dewey, J. F., 1976, Ophiolite obduction, *Tectonophysics* **31**:93–120.

Dewey, J. F., 1977, Suture zone complexities: A review, *Tectonophysics* **40**:53–67.

Dickinson, W. R., 1971, Detrital modes of New Zealand graywackes, *Sediment. Geol.* **5**:37–56.

Dickinson, W. R., 1982, Compositions of sandstone in Circum-Pacific Subduction Complexes and fore-arc basins, *Am. Assoc. Petrol. Geol. Bull.* **66**:121–137.

Eggers, A. J., and Adams, C. J., 1979, Potassium–argon ages of molybdenum mineralization and associated granites at Bald Hill and correlation with other molybdenum occurrences in the South Island, New Zealand, *Econ. Geol.* **74**:628–637.

Falconer, R. K. H., 1974, Geophysical studies in the Southwest Pacific, Unpublished Ph.D. thesis, Victoria University of Wellington, New Zealand.

Feary, D. A., and Pessagno, E. A., Jr., 1980, An Early Jurassic age for chert within the Early Cretaceous Oponae Melange (Torlesse Supergroup), Raukumara Peninsula, New Zealand, *N.Z. J. Geol. Geophys.* **23**:623–628.

Findlay, R. H., 1979, Summary of structural geology of Haast Schist Terrain, Central Southern Alps, N.Z.: Implications of structures for uplift of and deformation within Southern Alps, *R. Soc. N.Z. Bull.* **18**:113–120.

Fleming, C. A., 1970, The Mesozoic of New Zealand: Chapters in the history of the Circum-Pacific Mobile Belt, *Q. J. Geol. Soc. London* **125**:125–170.

Freund, R., 1971, The Hope Fault, a strike-slip fault in New Zealand, *N.Z. Geol. Surv. Bull.* **86**.

Gage, M., 1952, The Greymouth Coalfield, *N.Z. Geol. Surv. Bull.* n.s. **45**.

Geology of New Zealand, 1978, (R. P. Suggate, G. R. Stevens, and M. T. Te Punga, eds.), DSIR, Wellington, New Zealand.

Ghani, M. A., 1978, Late Cenozoic vertical crustal movement in the southern North Island, New Zealand, *N.Z. J. Geol. Geophys.* **21**:117–125.

Gibowicz, S. J., and Hatherton, T., 1975, Source properties of shallow earthquakes in New Zealand and their tectonic associations, *Geophys. J. R. Astron. Soc.* **43**:589–605.

Gibson, G. M., 1982, Stratigraphy and petrography of some metasediments and associated intrusive rocks from central Fiordland, New Zealand, *N.Z. J. Geol. Geophys.* **25**:21–43.

Grapes, R. H., and Palmer, K., 1984, Magma type and tectonic setting of metabasites, Southern Alps, New Zealand, using immobile elements, *N.Z. J. Geol. Geophys.* **27**:21–25.

Griffiths, J. R., 1975, New Zealand and the southwest Pacific Margin of Gondwanaland, in: *Gondwana Geology* (K. S. W. Campbell, ed.), pp. 619–637, Australian National University Press, Canberra.

Grindley, G. W., 1971, Sheet S8 Takaka, Geological Map of New Zealand 1 : 63,360, Department of Scientific and Industrial Research, Wellington.

Grindley, G. W., 1980, Sheet S13 Cobb, Geological Map of New Zealand 1 : 63,360, Department of Scientific and Industrial Research, Wellington.

Harrison, T. M., and McDougall, I., 1980, Investigations of an intrusive contact, northwest Nelson, New Zealand—II. Diffusion of radiogenic and excess ^{40}Ar in hornblende revealed by ^{40}Ar/^{39}Ar age spectrum analysis, *Geochim. Cosmochim. Acta* **44**:2005–2020.

Hayes, D. E., and Conolly, J. R., 1972, Morphology of the Southeast Indian Ocean, *Antarct. Res. Ser.* **19:**125–146.

Hayes, D. E., and Talwani, M., 1972, Geophysical investigations of the Macquarie Ridge Complex, *Antarct. Res. Ser.* **19:**211–234.

Hayward, B. W., 1982, A lobe of "Onerahi Allochthon" within Otaian Waitematas, *Geol. Soc. N.Z. Newsl.* **58:**13–19.

Howell, D. G., 1980, Mesozoic accretion of exotic terranes along the New Zealand segment of Gondwanaland, *Geology* **8:**487–491.

Hulston, J. R., and McCabe, W. J., 1972, New Zealand potassium–argon age list No. 1, *N.Z. J. Geol. Geophys.* **15:**406–432.

Hunt, T. M., 1978, Stokes Magnetic Anomaly System, *N.Z. J. Geol. Geophys.* **21:**595–606.

Hunter, H. W., 1977, Geology of the Cobb Intrusives, Takaka Valley, northwest Nelson, New Zealand, *N.Z. J. Geol. Geophys.* **20:**469–502.

Johnston, M. R., 1981, Sheet 027AC Dun Mountain, Geological Map of New Zealand 1 : 50,000, Department of Scientific and Industrial Research, Wellington [Metric Series].

Karig, D. E., and Sharman, G. F., III, 1975, Subduction and accretion in trenches, *Geol. Soc. Am. Bull.* **86:**377–389.

Kawachi, Y., 1974, Geology and petrochemistry of weakly metamorphosed rocks in the Upper Wakatipu District, southern New Zealand, *N.Z. J. Geol. Geophys.* **17:**169–208.

Laird, M. G., 1972, Sedimentology of the Greenland Group in the Paparoa Range, West Coast, South Island, *N.Z. J. Geol. Geophys.* **15:**372–393.

Laird, M. G., and Shelley, D., 1974, Sedimentation and early tectonic history of the Greenland Group, Reefton, New Zealand, *N.Z. J. Geol. Geophys.* **17:**839–854.

Landis, C. A., 1980, Little Ben Sandstone, Maitai Group (Permian): Nature and extent in the Hollyford–Eglington region, South Island, New Zealand, *N.Z. J. Geol. Geophys.* **23:**551–567.

Landis, C. A., 1981, The Permian Collage, *Geol. Soc. N.Z., Hamilton Conf. Progr. Abstr.* p. 56.

Landis, C. A., and Bishop, D. G., 1972, Plate tectonics and regional stratigraphic–metamorphic relations in the southern part of the New Zealand Geosyncline, *Geol. Soc. Am. Bull.* **83:**2267–2284.

Landis, C. A., and Coombs, D. S., 1967, Metamorphic belts and orogenesis in southern New Zealand, *Tectonophysics* **4:**501–518.

Lensen, G., and Vella, P., 1971, The Waiohine River faulted terrace sequence, *R. Soc. N.Z. Bull.* **9:**117–119.

Lewis, K. B., 1971, Growth rate of folds using tilted wave-planed surfaces: Coast and continental shelf, Hawke's Bay, New Zealand, *R. Soc. N.Z. Bull.* **9:**225–231.

MacKinnon, T. C., 1983, Origin of Torlesse and related rocks, South Island, New Zealand, *Geol. Soc. Am. Bull.* **94:**967–985.

Macpherson, E. L., 1946, An outline of Late Cretaceous and Tertiary diastrophism in New Zealand, *N.Z. Geol. Surv. Mem.* **6.**

Malahoff, A., Feden, R. H., and Fleming, H. S., 1982, Magnetic anomalies and tectonic fabric of marginal basins north of New Zealand, *J. Geophys. Res.* **87:**4109–4125.

Marshall, P., 1932, Notes on some volcanic rocks of the North Island of New Zealand, *N.Z. J. Sci. Technol.* B **13:**198–202.

Mayer, W., 1969, Petrology of the Waipapa Group near Auckland, New Zealand, *N.Z. J. Geol. Geophys.* **12:**412–435.

Nathan, S., 1976, Geochemistry of the Greenland Group (Early Ordovician), New Zealand, *N.Z. J. Geol. Geophys.* **19:**683–706.

Oliver, G. J. H., 1980, Geology of the granulite and amphibolite facies gneisses of Doubtful Sound, Fiordland, New Zealand, *N.Z. J. Geol. Geophys.* **23:**27–42.

Oliver, G. J. H., and Coggan, J. H., 1979, Crustal structure of Fiordland, New Zealand, *Tectonophysics* **54:**253–292.

Ota. Y., William, D. N., and Berryman, K. R., 1982, Late Quaternary tectonic map of Wellington 1 : 50,000 (with notes), Department of Scientific and Industrial Research, Wellington.

Pirajno, F., 1981, Geochemistry and mineralisation of the southern part of the Darran Complex, Fiordland, New Zealand, *N.Z. J. Geol. Geophys.* **24**:491–513.

Pitman, W. C., Larson, R. L., and Herron, E. M., 1974, Map of Magnetic Lineations of the Oceans, Lamont–Doherty Geological Observatory of Columbia University, Palisades, N.Y.

Retallack, G. J., 1979, Middle Triassic Coastal Outwash Plain Deposits in Tank Gully, Canterbury, New Zealand, *J. R. Soc. N.Z.* **9**:397–414.

Robinson, A. G., and Davey, F. J., 1981, Bounty magnetic anomaly map 1 : 1,000,000 Oceanic Series, Department of Scientific and Industrial Research, Wellington.

Roser, B. P., 1983, Comparative studies of copper and manganese mineralisation in the Torlesse, Waipapa and Haast Schist terrances, Unpublished Ph.D. thesis, Victoria University of Wellington, New Zealand.

Shelley, D., 1970, The structure and petrography of the Constant Gneiss near Charleston, southwest Nelson, *N.Z. J. Geol. Geophys.* **13**:370–391.

Shelley, D., 1972, Structure of the Constant Gneiss near Cape Foulwind, southwest Nelson, and its bearing on the regional tectonics of the West Coast, *N.Z. J. Geol. Geophys.* **15**:33–48.

Sheppard, D. S., Adams, C. J., and Bird, G. W., 1975, Age of metamorphism and uplift in the Alpine Schist Belt, New Zealand, *Geol. Soc. Am. Bull.* **86**:1147–1153.

Sinton, J. M., 1980, Petrology and re-evaluation of the Red Mountain Ophiolite Complex, New Zealand, *Am. J. Sci.* **280A**:296–328.

Sissons, B. A., 1979, The horizontal kinematics of the North Island of New Zealand, Unpublished Ph.D. thesis, Victoria University of Wellington, New Zealand.

Skinner, D. N. B., 1972, Subdivision and petrology of the Mesozoic rocks of Coromandel (Manaia Hill Group), *N.Z. J. Geol. Geophys.* **15**:203–227.

Speden, I. G., 1976, Fossil localities in Torlesse rocks of the North Island, New Zealand, *J. R. Soc. N.Z.* **6**:73–91.

Spörli, K. B., 1975, Waiheke and Manaia Hill Groups, East Auckland, Comment, *N.Z. J. Geol. Geophys.* **18**:757–762.

Spörli, K. B., 1978, Mesozoic tectonics, North Island, New Zealand, *Geol. Soc. Am. Bull.* **89**:415–425.

Spörli, K. B., 1979, Structure of South Island Torlesse in relation to the origin of the Southern Alps, *R. Soc. N.Z. Bull.* 99–104.

Spörlie, K. B., and Bell, A. B., 1976, Torlesse melange and coherent sequences, eastern Ruahine Range, North Island, New Zealand, *N.Z. J. Geol. Geophys.* **19**:427–447.

Spörli, K. B., and Gregory, M. R., 1981, Significance of Tethyan fusulinid limestones of New Zealand, in: *Gondwana Five* (M. M. Cresswell and P. Vella, eds.), pp. 223–229, A. A. Balkema, Rotterdam.

Stevens, G. R., 1980, Geological Time Scale showing NZ divisions. Compiled for Geological Society of New Zealand.

Suggate, R. P., 1959, New Zealand coals: Their geological setting and its influence on their properties, *N.Z. Dep. Sci. Ind. Res. Bull.* **134**.

Suggate, R. P., and Lensen, G. J., 1973, Rate of horizontal fault displacement in New Zealand, *Nature* **242**:518–519 [reply to Wellman, 1972].

Suggate, R. P., Stevens, G. R., and Te Punga, M. T. (eds.), 1978, *The Geology of New Zealand*, Government Printer, Wellington.

Tulloch, A. J., 1983, Granitoid rocks of New Zealand—A brief review, *Geol. Soc. Am. Mem.* **159**:5–20.

Turnbull, I. M., 1979, Petrography of the Caples terrane of the Thomson Mountains, northern Southland, New Zealand, *N.Z. J. Geol. Geophys.* **22**:709–727.

Turnbull, I. M., 1980, Structure and interpretation of the Caples terrane of the Thomson Mountains, northern Southland, New Zealand, *N.Z. J. Geol. Geophys.* **23**:43–62.

van der Linden, W. J. M., 1971, Magnetic anomaly map 1 : 2,000,000 Western Tasman Sea, Department of Scientific and Industrial Research, Wellington.

Vitaliano, C. J., 1968, Petrology and structure of the south-eastern Marlborough Sounds, New Zealand, *N.Z. Geol. Surv. Bull.* n.s. **74**.

Walcott, R. I., 1978, Present tectonic and Late Cenozoic evolution of New Zealand, *Geophys. J. R. Astron. Soc.* **52**:137–164.

Waterhouse, J. B., and Vella, P., 1965, A Permian fauna from North-west Nelson, New Zealand, *Trans. R. Soc. N.Z. Geol.* **3**:57–84.

Weissel, J. K., and Hayes, D. E., 1972, Magnetic anomalies in the southeast Indian Ocean, *Antarct. Res. Ser.* **19**:165–196.

Weissel, J. K., Hayes, D. E., and Herron, E. M., 1977, Plate tectonic synthesis: The displacements between Australia, New Zealand, and Antarctica since the late Cretaceous, *Mar. Geol.* **25**:231–277.

Wellman, H. W., 1952, Data for the study of Recent and late Pleistocene faulting in the South Island of New Zealand, *N.Z. J. Sci. Technol.* **34B**:270–288.

Wellman, H. W., 1956, Structural outline of New Zealand, *N.Z. Dep. Sci. Ind. Res. Bull.* **121**.

Wellman, H. W., 1959, Divisions of the New Zealand Cretaceous, *Trans. R. Soc. N.Z.* **87**:99–163.

Wellman, H. W., 1972, Rate of horizontal fault displacement in New Zealand, *Nature* **237**:275–277.

Wellman, H. W., 1973, The Stokes Magnetic Anomaly, *Geol. Mag.* **110**:419–429.

Wellman, H. W., 1979, An uplift map for the South Island of New Zealand, and a model for uplift of the Southern Alps, *R. Soc. N.Z. Bull.* **18**:13–20.

Williams, G. J., 1974, Economic geology of New Zealand, *Australas. Inst. Min. Metall. Monogr.* **4**.

Williams, J. G., and Harper, C. T., 1978, Age and status of the Mackay Intrusives in the Eglinton–upper Hollyford area, *N.Z. J. Geol. Geophys.* **21**:733–742.

Williams, J. G., and Smith, I. E. M., 1979, Geochemical evidence for paired arcs in the Permian volcanics of southern New Zealand, *Contrib. Mineral. Petrol.* **68**:285–291.

Wood, B. L., 1972, Metamorphosed ultramafics and associated formations near Milford Sound, New Zealand, *N.Z. J. Geol. Geophys.* **15**:88–128.

Wood, B. L., 1978, The Otago Schist Megaculmination: Its possible origins and tectonic significance in the Rangitata Orogen of New Zealand, *Tectonophysics* **47**:339–368.

Chapter 11

GEOPHYSICS OF THE PACIFIC BASIN

Chapter 11A

GRAVITY

Hiromi Fujimoto and Yoshibumi Tomoda

Ocean Research Institute
University of Tokyo
Tokyo 164, Japan

I. INTRODUCTION

Gravity anomalies show the effects of both the topography of the solid earth and its compensation. Free-air gravity anomalies, therefore, approximately represent isostatic anomalies, and the small range of free-air anomalies of about 800 mgal (1 mgal = 10^{-5} m sec^{-2}), or 0.08% of the earth's gravity field, shows that the earth is nearly in isostatic equilibrium (Bowin *et al.*, 1982). Large-amplitude free-air anomalies, or deviations from isostasy, are maintained by tectonic activities (Vening Meinesz, 1932). Isostasy is fairly well achieved in wavelengths longer than about 500 km in the Pacific basin, and the corresponding free-air anomalies are as small as ±20 mgal (McKenzie *et al.*, 1980). The value of 20 mgal is 0.002% of the earth's gravity field, and the small value indicates that gravity measurements need high precision. It was difficult for the prealtimeter gravity field to discuss such gravity anomalies, because the spatial resolution of the gravity data obtained from satellite orbit perturbations was not sufficient, and because surface observations were sparse and not free from measurement errors of about ±10 mgal.

The geoid approximately coincides with the topography of the ocean surface, and represents the earth's gravity field with its longer-wavelength components enhanced (see Section II). Another interpretation of geoid anomaly is that it expresses density anomaly with its deeper part enhanced (Ockendon and Turcotte,

1977). The precision and dense coverage of the satellite altimetry data obtained in the last decade by GEOS-3 and SEASAT have greatly improved our information of the gravity field over the world oceans. Improved sea gravimeters and the satellite navigation both of which became popular around 1970 have made sea gravimetry accurate to several milligals. Surface gravity data have been the base of the analyses of the satellite altimeter data.

Depth of principal mass anomalies contributing to the earth's gravity field has recently been estimated. Jordan (1978) analyzed the observed gravity field assuming, *a priori,* five horizons of density contrast. According to his results, 80% of the energy (variance) in free-air gravity anomalies can be explained by terrain and Moho variations, and 11% by deep sources (at 350 and 2880 km). Concerning the geoid anomalies, 7% can be explained by the shallow sources and the rest by the deep sources. Bowin (1983) proceeded with Jordan's analyses assuming four main sources: the core–mantle boundary, lower mantle, upper mantle in the outer 600 km, and the crust. His results show that about 90% of the observed free-air anomalies can be attributed to mass anomalies in the crust and upper mantle in the outer 600 km of the earth.

There has been considerable progress in the interpretation of the earth's gravity field. The thermal and mechanical properties of the oceanic lithosphere have been studied by many investigators. Global mapping of mantle convection is now one of the important targets of the interpretations of the gravity field. Seismic tomography and numerical simulation have shown their potential ability in the study of the earth's interior. These studies give us hope that a significant insight into the physical processes of plate motions and mantle convections can be obtained in the near future.

The terrestrial and satellite gravimetry in the Pacific are reviewed in Section II. Several approaches to the geophysical interpretations of gravity data are discussed in Section III, gravity anomalies in the Pacific and their geophysical interpretations in Section IV, and future gravity work in Section V.

II. GRAVITY MEASUREMENTS IN THE PACIFIC

A. Pendulum Gravity Observation at Sea

Since gravity measurement requires high precision, large and complicated motions of a surface ship stand in the way of gravity measurement at sea. Upon the advice of a director of the Dutch Government Mines, Vening Meinesz initiated pendulum gravity measurement at sea on board a submarine in 1923 (Vening Meinesz, 1929). Besides reducing the rolling and pitching, a submarine is not subject to vibrations during submergence. He carried out a gravity expedition from Holland to Java via the Panama Canal in 1926–1927. Combined with the gravity expedition in 1923 along the route from Holland to Java by way of the Suez Canal, a continuous ring of gravity values encircling the earth was obtained for the first time (Vening Meinesz, 1932).

Convinced that the clarification of the relation between gravity and tectonic activity would be of primary importance, he conducted a gravity survey around the Java Trench in 1923 and 1927. He remarked that tectonic areas showed large gravity anomalies and strong deviations from isostasy, and that the principal tectonic phenomenon taking place in the deeper layers of the crust might be a downbuckling of the crust (Vening Meinesz, 1932).

A gravity survey across the Japan Trench was carried out in 1934 by Matuyama (1934) with a Vening Meinesz pendulum apparatus. After gravity measurements around the Izu–Bonin area the next year, a map of free-air anomalies in the subduction zones around Japan was compiled (Kumagai, 1940).

After World War II, Worzel conducted extensive pendulum gravity expeditions at sea using many (= 26) submarines in 1947–1959 (Worzel, 1965). These results are estimated to be reliable to about ±3 mgal (Worzel, 1965).

With the beginning of satellite gravimetry, Jeffreys (1959, Fig. 21) tried spherical expansions of the gravity field by using gravity data obtained by land observations and pendulum gravity measurements obtained on submarines. But the results were restricted to the poorly determined first few terms of the spherical harmonic expansions (Lambeck and Coleman, 1983).

B. Ship Gravity Measurements

After many unsuccessful trials, surface ship gravity measurements started in the early 1960s. The developments in 1960s have been reviewed by Worzel and Harrison (1963), LaCoste (1967), and Talwani (1970).

The Eötvös correction has been one of the most serious problems in gravity measurements at sea. Most of the shipborne gravity measurements in the 1970s were carried out with the aid of the U.S. Navy Navigation Satellite System (NNSS), which was developed in the mid 1960s. Eötvös correction by the NNSS is usually accurate to better than a few milligals (e.g., Talwani *et al.*, 1966), but the NNSS is not so reliable in strong ocean currents. The errors in Eötvös correction, sometimes exceeding 10 mgal with the celestial navigation adopted in the 1960s, can be about 1 mgal with the Loran C in good condition, and is smaller than 1 mgal with the Global Positioning System (GPS).

Both a Graf–Askania Gss-2 sea gravimeter (Graf and Schulze, 1961; Schulze, 1962) and a LaCoste & Romberg sea gravimeter mounted on a gyrostabilized platform (LaCoste, 1967) have been widely used for scientific researches. Many LaCoste & Romberg meters are also used on seismic vessels for commercial explorations. The two instruments are highly damped spring-type gravimeters and their serious problems were cross-coupling errors. In order to reduce the errors, the gravimeters were mounted on stable platforms with sophisticated gyro erection loops (Talwani, 1970). Cross-coupling errors were corrected by an analogue computer (Gss-2) or in the postcruise cross-correlation analysis (LaCoste, 1973). In both gravimeters, the effect of heave accelerations was reduced by analogue filters, which had phase delays.

LaCoste & Romberg sea gravimeters prior to 1965 were mounted on gimbals and were only suitable for use in relatively calm seas (e.g., Chiburis et al., 1972). Watts et al. (1985) have evaluated sea surface gravity data obtained in the 1960s and 1970s from analysis of the discrepancies at intersecting ship tracks. They estimated measurements by Gss-2 to be accurate to 5–10 mgal, and those by the LaCoste & Romberg meter on a stable platform to be accurate to 3–5 mgal.

LaFehr (1980) summarizes the accuracy of shipborne gravity data in commercial explorations to be usually better than 2 mgal and quite often better than 1 mgal. Except in high sea-state conditions, the limiting source of error is introduced by errors in the Eötvös correction.

In order to reduce cross-coupling errors, straight-line models of the two spring-type sea gravimeters have been introduced. The weight is suspended by several strings to move only vertically. The Bodenseewerk KSS 30 sea gravimeter succeeded the Graf–Askania Gss-2 meter. The new straight-line LaCoste & Romberg sea gravimeter is described by LaCoste (1983). Both models show an accuracy of better than 1 mgal in field trials (e.g., Valliant, 1983).

A vibrating string sea gravimeter was first built by Gilbert (1949) for use on submarines. Lozinskaya (1959) made gravity measurements in the Caspian Sea with an instrument very similar to Gilbert's. The Tokyo Surface Ship Gravity Meter (TSSG) developed by Tomoda and Kanamori (1962) is also a vibrating string gravimeter. The TSSG has been used for more than 20 years in gravity measurements at the Ocean Research Institute, University of Tokyo, and at the Hydrographic Department, Japan. It is free from cross-coupling errors because two pairs of cross strings suspend the weight. The gravity sensor is so simple and compact that it can be directly mounted on a vertical gyroscope with a simple erection loop. The quantity being measured is the period of the string vibration. It is not linearly related to vertical acceleration, so that gravity values are obtained through a digital processing. The most serious problem for the string gravimeter, i.e., the nonlinear rectification error, was completely solved by using a high-speed minicomputer (Fujimoto, 1976; Fujimoto and Tomoda, 1985a).

Wing (1969) has solved the nonlinear rectification problem alternatively by using a double-string accelerometer, and Bowin et al. (1972) assembled the Vibrating-String Accelerometer (VSA) of the same principle mounted on a Sperry MK-19 gyrostabilized platform and showed its reliability to be better than a LaCoste & Romberg meter. Gravity data obtained by string gravimeters are estimated to be accurate to several milligals (Fujimoto, 1976; Watts et al., 1985).

A few sea gravimeters with an electromagnetic servo accelerometer developed by the Bell Aerospace Company (Anonymous, 1967) are now in commercial explorations (LaFehr, 1980). Segawa et al. (1981) have also developed a similar sea gravimeter for gravity measurements in the Antarctic Sea on board an icebreaker.

A helicopter gravity measuring system has been operational in North America since 1977, and gravity measurements have also been attempted in an airplane (LaCoste *et al.*, 1982). Airborne gravimetry requires precise data of the aircraft's altitude and will be useful in detailed continuous gravity surveys on land and at sea in coastal areas.

Mapping of free-air anomalies by use of surface ship gravity data has been actively carried out in the Pacific Ocean: in the northeast Pacific (Dehlinger *et al.*, 1970), around the Macquarie Ridge complex (Hayes and Talwani, 1972), around Japan (Tomoda, 1973a), in Hawaii and vicinity (Watts, 1975a), in the Aleutian arc-trench system (Watts, 1975b), around Australia (Bureau of Mineral Resources, 1976), in the Philippine Sea (Watts, 1976a), in the Taiwan–Luzon region (Bowin *et al.*, 1978), in the east and southeast Asian seas (Watts *et al.*, 1978a), in the Kurile arc-trench system (Watts *et al.*, 1978b), in the Banda Sea region (Bowin *et al.*, 1980), in the southwest Pacific (Watts *et al.*, 1981), in the northwest Pacific (Tomoda and Fujimoto, 1982), near the coasts of Japan by the Hydrographic Department, Japan (Ganeko and Harada, 1982), and in the central Pacific by the Geological Survey of Japan (Ishihara, 1982). The U.S. National Oceanic and Atmospheric Administration (NOAA) published Pacific SEAMAP 1961–70 data of bathymetry, magnetics, and gravity in the central north Pacific (e.g., Chiburis *et al.*, 1972). A detailed bathymetric and gravimetric survey was carried out over the subduction zones around Japan during the Franco-Japanese cooperative Project Kaiko I in 1984 (KAIKO I Research Group, 1986).

Bowin *et al.* (1981, 1982) have performed the global mapping of free-air anomalies using terrestrial gravity data. Figure 1 shows the distribution of gravity data compiled by them. It shows sparse gravity data in the continents of Asia, Africa, and South America as well as in the south Pacific. Watts *et al.* (1985) compiled maps of bathymetry, gravity, and geoid in the Pacific Ocean, by computing the values on 90 × 90-km square grids. Watts and Leeds (1977) computed 1° × 1° free-air gravity anomaly averages in the northwest Pacific from sea gravity data. Rapp (1983b) has compiled an improved set of terrestrial mean gravity anomaly data on 1° × 1° blocks over the world.

C. Satellite Gravimetry from Satellite Orbits

The use of artificial satellites to determine the earth's gravity field has been followed in two ways. One is to analyze the perturbation of the satellite motion (Kaula, 1966), and is discussed in this section. The other way is called satellite altimetry and will be discussed in the next section.

In the following discussion, the earth's gravity potential is expanded in spherical harmonics as

$$U(r,\theta,\varphi) = \frac{GM}{r} \left\{ 1 + \sum_{n=2}^{\infty} \sum_{m=0}^{\infty} \left(\frac{R}{r}\right)^n P_{nm} (\sin\theta) \left[C_{nm} \cos m\varphi + S_{nm} \sin m\varphi\right] \right\}$$

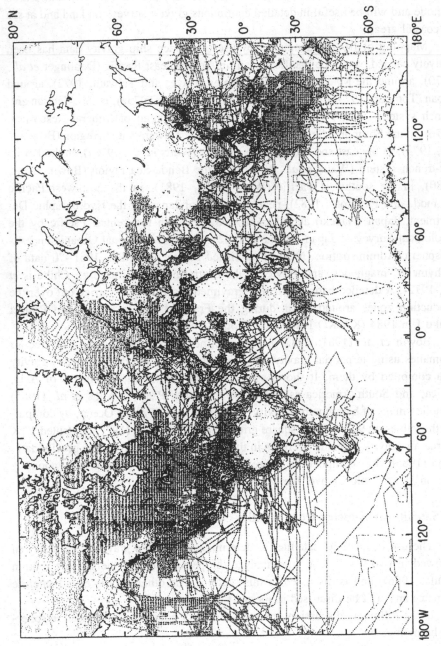

Fig. 1.   Distribution of terrestrial gravity data compiled by Bowin et al. (1982).

where r, θ, φ are the geocentric distance, latitude, and longitude, and R is the mean equatorial radius. The $P_{nm}(\sin \theta)$ are the Legendre functions of degree n and order m, and the C_{nm}, S_{nm} are the Stokes coefficients.

Geoid height is obtained by dividing anomalous gravity potential by normal gravity. A coefficient of gravity anomalies (G_{nm}) is simply related to that of the geoid (N_{nm}) as

$$G_{nm} = (\gamma/R)(n - 1)N_{nm}$$

where γ is the normal gravity (Heiskanen and Moritz, 1967). The geoid, therefore, approximately shows a kind of gravity anomaly with its longer-wavelength components enhanced.

Analyses of satellite orbit perturbations deduced from ground-based tracking data have become a principal method to study the earth's gravity field since the Russian Sputnik 1 launched in 1957 and the succeeding satellites. The progress of the study of the earth's gravity field from 1958 to 1982 has been reviewed and evaluated in detail by Lambeck and Coleman (1983). Some of the recent geopotential models have been evaluated by Lerch et al. (1985b).

The Smithsonian Astrophysical Observatory (SAO) carried out pioneering works in the field of satellite gravimetry in the 1960s and its geopotential models were called "Standard Earth" models. Smithsonian standard earth (I) was complete to degree and order 8 (Lundquist and Veis, 1966). The solution was based on Baker–Nunn Camera observations. Talwani (1970) pointed out the solution showed a gravity high associated with the Pacific margin, especially with the island arc areas along the western margin. He suggested that the gravity high over the island arc areas might well be caused by the higher densities associated with a descending lithospheric plate in the mantle. Runcorn (1967) interpreted the long-wavelength gravity anomalies in terms of the fluid motion in the mantle. Smithsonian standard earth (II) was expanded to degree and order 16 (Gaposchkin and Lambeck, 1970, 1971) based on laser range measurements in addition to Baker–Nunn observation. Surface gravity data were introduced to improve the solution. Kaula (1972) showed that most of the large gravity anomalies in the solution were related to plate tectonics or post-glacial uplift. Smithsonian standard earth (III) was expanded to degree and order 18 (Gaposchkin, 1973, 1974).

Several Goddard Earth Models, or GEM solutions, were presented by NASA. An odd-numbered GEM solution is generally based on satellite tracking data only and an even-numbered one includes terrestrial information. By 1978 the three principal prealtimeter solutions of the global gravity field were the Smithsonian model (SAO 77) through degree 24 (Gaposchkin, 1977), GEM solutions (GEM9 and GEM10) through degree 22 (Lerch et al., 1979), and GRIM2 model through degree 30 by a French–German cooperative program (Balmino et al., 1978). GEM9 is based on satellite tracking data only, and was designed principally to benefit the Geos3 altimetry (Lerch et al., 1985b).

Lambeck and Coleman (1983) evaluated these solutions. According to them, Gaposchkin (1977) solved for the harmonics without assuming any *a priori* information on the power spectrum of the potential. The powers of GEM9 and GEM10 are too small at high degrees, and those of the GRIM2 solution are too small at low degrees, because in their inversions it is assumed that the potential coefficients decay according to some *a priori* rule, say Kaula's (1966) rule, and much of the contribution to the estimated error comes from the lower-degree harmonics (Lambeck and Coleman, 1983).

Lerch *et al.* (1985b) have shown that this is not the case by comparing these solutions with independent data. One of their results is shown in Fig. 2, which gives the results of comparison in 5° × 5° blocks of marine gravity anomalies. The values from SEASAT altimetry were obtained from a set of 1° × 1° blocks (Rapp, 1983a). According to Lerch *et al.* (1985b), the poor result of SAO 77 is due principally to the high weight of land gravimetry in the SAO solution. The even-numbered GEMs (GEM3 versus GEM4, GEM5 versus GEM6, GEM7 versus GEM8) also suffer somewhat from this data weighting problem.

The Laser Geodynamics Satellite (LAGEOS), which is one of the first artificial

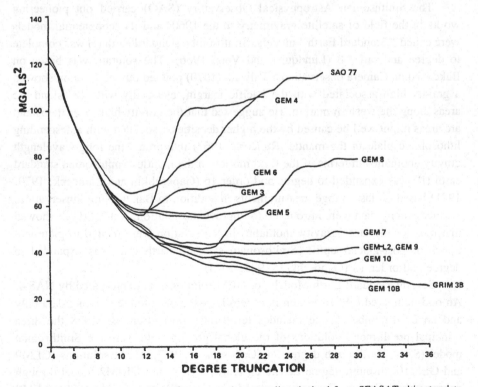

Fig. 2.   Gravity field versus 5° × 5° mean gravity anomalies obtained from SEASAT altimeter data evaluated at 1114 points. (From Lerch *et al.*, 1985b.)

satellites developed for geodynamic measurements using satellite laser-ranging techniques, has contributed to improve the lower-degree harmonics of the earth's gravity field, because it is on a high altitude and thus sensitive only to harmonics of relatively low degrees through 8 (Cohen and Smith, 1985). The GEM9 solution has been improved to become the GEM-L2 solution expanded to degree 20 by using LAGEOS observations (Lerch *et al.*, 1982, 1983, 1985a). The GEM-L2 solution was used in the SEASAT altimetry. The evaluation of this solution will be discussed later.

Satellite-to-satellite tracking (SST) in low altitude was first proposed by Wolff (1969) and its analysis has been reviewed in detail by Colombo (1984). This method has already been applied to make a local map of gravity anomalies near the East Pacific Rise by Marsh and Marsh (1977), in the Central Pacific (Marsh *et al.*, 1981), or over the continents (Kahn *et al.*, 1982). The gravity map obtained from 90 passes of SST data (Marsh *et al.*, 1984) shows good agreement with the conventional GEM solutions and with the results of SEASAT altimetry. Future gravity work in this field is introduced in Section V.

D. Satellite Altimetry and Gravity Anomalies

The determination of the earth's geoid has been considerably improved with the direct measurements of sea surface topography by a radar altimeter from three artificial satellites: SKYLAB launched in 1973 (Leitao and McGoogan, 1974), GEOS-3 in 1975 (Stanley, 1979), and SEASAT in 1978 (Born *et al.*, 1979, 1981). The accuracy of each altimetry datum is estimated to be 1 m for SKYLAB, 20–50 cm for GEOS-3, and about 10 cm for SEASAT. The footprint of the altimetry data on the sea surface is reported to be in the range of 2–12 km for SEASAT (Born *et al.*, 1981). The altimetry data are, therefore, capable of resolving anomalous features with wavelengths as short as 14 km. Disturbing oceanographic effects rarely exceed ±1 m, while geoidal undulations exceed ±50 m, so that the mean ocean surface obtained represents the marine geoid. A number of ground truth studies showed that after adjustment for tilt and bias, discrepancies are typically 50 cm for GEOS-3 (Rapp, 1979) and 30 cm for SEASAT (Rowlands, 1981).

The resolution and dense coverage of the satellite altimetry over the world's oceans have improved the knowledge of not only the global gravity field but also detailed regional gravity anomalies. Lambeck and Coleman (1983) emphasize that no other geophysical observations, not even bathymetry, are better known than the regional geoid over the oceans. Many uncharted seamounts have been founded by means of satellite altimetry (Lazarewicz and Schwank, 1982; Lambeck and Coleman, 1982; Sandwell, 1984). Geoid maps of the Pacific Ocean basin have been prepared based on the adjusted SEASAT altimetry data: the 1° × 1° mean geoid anomaly map (Rowlands, 1981), and the image-processed geoid map by the Jet Propulsion Laboratory (Parke and Dixon, 1982). Rapp (1983a) has compiled an improved set of $1° \times 1°$ mean gravity anomaly data from SEASAT altimetry data. Segawa and Matsumoto (1987) have compiled detailed maps of free-air anomalies over the world's oceans from satellite altimetry data.

The global gravity models mentioned above have also been improved by using GEOS-3 altimetry data: the SAO 80 model expanded to degree 30 by Gaposchkin (1980), GEM10B and GEM10C by Lerch *et al.* (1981), and GRIM3 (Reigber *et al.*, 1983a) and GRIM3B (Reigber *et al.*, 1983b) expanded to degree 36. GEM10B is complete to degree and order 36. GEM10C is expanded to degree 180 with the GEM10B coefficients fixed. Lambeck and Coleman (1983) point out that GEM10C is based on gravity field involving some predictions on land areas. Rapp (1981) also expanded the earth's gravity field to degree and order 180 combining satellite altimetry data, terrestrial gravity data, and satellite orbit data. The GRIM3 model has further been improved to GRIM3-L1 solution (Reigber *et al.*, 1985) by using LAGEOS satellite laser-ranging data and $1° \times 1°$ mean free-air anomaly data from SEASAT (Rapp, 1983a) and from surface observations (Rapp, 1983b).

Some of these solutions as well as the GEM-L2 solution have been evaluated by Lambeck and Coleman (1983), and recently by Lerch *et al.* (1985b). Comparing three solutions of SAO 80, GEM10B, and GRIM3, Lambeck and Coleman (1983) have pointed out that the solutions do not model the geoid with the accuracies claimed by some of the authors, because the differences among the solutions are considerable.

Lerch *et al.* (1985b) have evaluated the GEM-L2 and GEM10B solutions, by comparing them with the GRIM3B solution and other independent data. Part of the result is shown in Fig. 2. They remark that GRIM3B has considerably been improved from GRIM3 since Lambeck and Coleman's paper, and that GEM-L2 and GRIM3B show an overall agreement of 17.7 cm through degree and order 4. According to their evaluation, the geoid errors in the GEM-L2 may be as low as 20 cm (rms worldwide) through degree and order 6 and 110 cm through degree and order 20. The GEM10B geoid appears to be accurate to better than 150 cm rms through degree 36 (Lerch *et al.*, 1985b). NASA has a plan of satellite altimetry of the next generation (see Section V).

III. APPROACHES TO THE GEOPHYSICAL INTERPRETATIONS OF GRAVITY ANOMALIES

A geoid anomaly for isostatically compensated topography under the long-wavelength, flat earth approximation is

$$\Delta N(x) = \frac{-2\pi G}{g} \int_0^\infty \Delta\rho(x,z) \cdot z \cdot dz$$

where $\Delta N(x)$ is the geoid height anomaly, G is the gravity constant, g is the gravity acceleration, z is the depth, and $\Delta\rho(x,z)$ is the density anomaly (Ockendon and Turcotte, 1977). Seafloor depth is proportional to $\int\Delta\rho(x,z)dz$. These equations well express the enhanced sensitivity of the geoid slope to density anomaly in the lower portion of the mantle (Sandwell and Schubert, 1982). Free-air anomaly reflects the effect of the topography and the reduced effect of its compensation. Therefore, the relationship among the topography, gravity, and geoid has useful information on the spatial distribution of the density anomaly.

A. Gravity Anomalies and Topography

It has been shown in the United States (Bowie, 1917) and in the Atlantic (McKenzie and Bowin, 1976) that isostatic compensation begins when the wavelength exceeds about 100 km. Such a study is based on the analysis of isostasy, i.e., the relationship between gravity anomalies and topography.

Tsuboi (1938) founded the basis of Fourier analysis of gravity anomalies by showing that, if the observed surface gravity distribution is expressed by

$$g(x,0) = \sum_m B_m \frac{\cos}{\sin} mx$$

the responsible mass distribution at depth d is given by

$$M(x,d) = \frac{1}{2\pi G} \sum_m e^{md} \cdot B_m \frac{\cos}{\sin} mx$$

Dorman and Lewis (1970) formulated an analytical method to interpret isostatic compensation, and McKenzie and Bowin (1976) proceeded with the method. This method utilizes time series techniques. In the wavenumber domain, the spectrum of gravity anomaly is related to that of topography by a simple multiplication of a transfer function $Z(k)$:

$$F(\Delta g(x))_k = Z(k) \cdot F(h(x))_k$$
$$\Delta g(x) = F^{-1}\{Z(k) \cdot F(h(x))\}$$

where $\Delta g(x)$ is gravity anomaly due to the topography and its compensation, and $h(x)$ is topography. F and F^{-1} denote Fourier transform and its inverse, and k is the wavenumber, $k = \lambda/2$. The transfer function $Z(k)$ is defined according to the assumed isostatic compensation model, and is called admittance. McKenzie and Bowin (1976) introduced complex functions in order to show the effects of noisy data as deviations in the phase of the admittance. In this case the admittance is given by the cross spectrum of gravity and bathymetry divided by the power spectrum of the bathymetry.

Isostatic compensation has been analyzed in two ways. One is to compare the admittance obtained from the Fourier transforms of the observed gravity anomalies and topography with the theoretically obtained admittance (e.g., McKenzie and Bowin, 1976; Watts, 1978; Cochran, 1979). The other is to compare the observed gravity anomalies with the synthetic gravity anomalies computed from the topography by using the theoretically obtained admittance (e.g., Lewis, 1982).

In order to discuss the relationship between intermediate-wavelength gravity anomalies and ocean bottom topography, the effects of lithospheric age and of sediment loading should be subtracted from the depth data (e.g., Watts et al., 1985), and the residuals obtained after the corrections are called residual depth anomalies. The effect of lithospheric age can be corrected by using the parameters for the ocean depth (e.g., Parsons and Sclater, 1977) and an isochron map (e.g., Sclater et al., 1980). According to the sediment thickness data over the Pacific (Ludwig and Houtz, 1979), the correction for sediment loading is unlikely to exceed about 0.03 km and is negligible over much of the Pacific Ocean (Watts et al., 1985).

Three basic models to interpret isostatic compensation are introduced here following the reviews by Lewis (1982). Discussions here are on free-air anomalies and not on the geoid, but the discussions can be applied to the geoid with only minor changes, because the geoid is in a simple relationship with gravity anomalies in the wavenumber domain.

1. Density Contrast Function

Density variation in the topography is obtained from gravity anomalies by this analysis. If the result is compared with the Pratt isostatic compensation model, the depth of compensation can be estimated from the result. Parker (1972) proposed a method to calculate the gravitational attraction by using Fourier transforms of the topography, and Oldenburg (1974) proposed an inverse method to solve for the density variation in a body with a flat bottom and a top surface defined by the topography that fits the observed data. Lewis (1982) modified Oldenburg's method to give

$$\rho_i(x) = \frac{1}{h(x)} F^{-1} \left\{ F[\Delta g(x)] \cdot \exp(kZ_0) / (2\pi G) - \sum_{n=2}^{\infty} \frac{(k)^{n-1}}{n!} F[\rho_{i-1}(x) \cdot h^n(x)] \right\}$$

where $\rho_i(x)$ is the density on the ith iteration, $\Delta g(x)$ is the observed gravity anomaly, $h(x)$ is the height of the top surface above Z_0. Some filtering of $\rho(x)$ is required to achieve convergence of this iterative method (Lewis, 1982). This method is easily modified into the two-dimensional Fourier transforms, which will be applied not only to detailed geophysical surveys in a small area (see Ishihara, 1986) but also to interpretations of regional gravity anomalies.

2. Airy Local Compensation

Suppose that an axial topography is locally compensated at some depth by low-density material having a mass of $(\rho_c - \rho_w) \cdot h(x)$, where ρ_c is density of the crust, ρ_w density of seawater, and $h(x)$ topography. If we use the linear approximation, we get a theoretical admittance:

$$Z(k) = 2\pi G(\rho_c - \rho_w) \, [\exp(-kZ_1) - \exp(-kZ_2)]$$

where Z_1 is mean depth of the topography and Z_2 is mean depth of the compensation (Lewis, 1982).

3. Elastic Plate Regional Compensation

The regional compensation model of Vening Meinesz (1941) has become one of the basic models for investigating the structure of the lithosphere (e.g., Walcott, 1970; McKenzie and Bowin, 1976). The outer shell of the earth (lithosphere) is treated as an elastic plate floating on the surface of a fluid (asthenosphere).

The equation for the gravity field from a topographic load and its compensation has been derived by McKenzie and Bowin (1976) and is modified by Lewis (1982) as

$$Z(k) = 2\pi G(\rho_c - \rho_w) \left[\exp(-kz_1) - \left(1 + \frac{4\mu t k^2 P_1}{g P_2 (\rho_m - \rho_c)} \right)^{-1} \cdot \exp(-kz_2) \right]$$

where $t = T/2$, T is the plate thickness, $P_1 = [\sinh(2kt)/(2kt)]^2 - 1$, $P_2 = [\sinh(4kt)/(4kt)]+1$, g is the gravitational acceleration, μ is the rigidity modulus, and ρ_m is the mantle density (Lewis, 1982).

Banks et al. (1977) used a thin plate approximation to obtain

$$Z(k) = 2\pi G(\rho_c - \rho_w) \left[\exp(-kz_1) - \left(1 + \frac{k^4 D}{g(\rho_m - \rho_c)} \right)^{-1} \cdot \exp(-kz_2) \right]$$

where D is the flexural rigidity defined by

$$D = \mu T^3 / [6(1-\sigma)]$$

where σ is the Poisson ratio (Lewis, 1982). The flexural rigidity D can be used as the variable parameter in fitting the data, and the value is used to estimate the effective elastic thickness of the lithospheric plate. Judging from the analysis by Madsen *et al.* (1984), there are only minor differences in the gravity field between the Airy model with $Z = 10$ km and the plate model with a plate thickness of about 3 km ($D = 10^{20}$ N m). If the relaxation time is the problem, the viscosity of the asthenosphere must be taken into account (Watts and Daly, 1981).

B. Gravity Anomalies and Velocity Structures

Recognizing the systematic discrepancy between observed gravity anomalies and crustal structures obtained by explosion seismology, Yoshii (1972) introduced "residual gravity anomaly" (RGA), which is defined as the gravity anomaly corrected for the structure above the Moho. In other words, RGA is a kind of Bouguer gravity anomaly reduced to the depth of the Moho, and represents the mass anomaly below the Moho. The ocean depth has essentially the same meaning as RGA only if there is no change in the crustal structure; RGA is generally a more direct indication of the upper-mantle structure than the ocean depth (Yoshii, 1973). Yoshii (1972) has also shown that RGA has a clear relation to the terrestrial heat flow of the Pacific and that the relation is explained by the change in thickness of the lithosphere. According to the lithospheric model by Yoshii *et al.* (1976), the density contrast between the lithosphere and asthenosphere is 0.1 g cm^{-3}, and then thickness of the lithosphere is proportional to RGA.

The actual calculation of RGA in the simple one-dimensional case is

$$\text{RGA} = \Delta g + 2\pi G \sum_i H_i(\rho_m - \rho_i)$$

where Δg is the free-air gravity anomaly, G is the universal constant of attraction, and ρ_m is 3.3 g cm^{-3} corresponding to the mantle seismic wave velocity of 8.1 km sec^{-1}. H_i is the thickness of the ith crustal layer determined by explosion seismology. The density of this layer, ρ_i, is assumed from the seismic velocity by use of the empirical law compiled by Ludwig *et al.* (1970). Even if a slightly different velocity–density relationship (e.g., Nafe and Drake, 1957) is used, the change in relative values of RGA is negligibly small (Yoshii, 1973).

Recent advances in seismic tomography (e.g., Dziewonski, 1984; Woodhouse and Dzienwonski, 1984) have resulted in images of long-wavelength lateral heterogeneity in seismic velocity. These velocity variations presumably are proportional to density variations, both resulting from temperature differences associated with mantle convection (Hager *et al.*, 1985).

C. Gravity Anomalies as a Constraint to Numerical Simulations

McKenzie *et al.* (1974) carried out numerical experiments on several simple two-dimensional models for constant-viscosity convection to estimate the characteristics of the topography, heat flow, and gravity anomaly for the models. According to their results, positive gravity anomaly of small amplitude (10–20 mgal) is predicted over rising parts of the flow, which is compatible with the observed data. The total gravity anomaly is produced partly by the upward deformation of the surface over a rising current, and partly by the smaller density due to the higher temperature of the rising material, both effects being of the same order of magnitude in opposite directions (McKenzie et al., 1974). McKenzie (1977) has studied the effect of viscosity variations by analyzing the admittance between topography and gravity anomaly as a function of wavelength, and suggested that variations of viscosity with temperature are dominant only in the vicinity of island arcs.

Elaborate simulations on the global mantle convection have been performed by means of the spherical finite element method (Baumgardner, 1985), and simplified simulations on the lithospheric subduction have been carried out by means of the finite difference method (Matsumoto and Tomoda, 1983a).

IV. GRAVITY ANOMALIES IN THE PACIFIC AND THEIR GEOPHYSICAL INTERPRETATIONS

A. Overview of Gravity Anomalies in the Pacific

Overview of gravity anomalies in the Pacific has been presented by Bowin et al. (1982), Marsh et al. (1984), and Watts et al. (1985). Figure 3 is a map of free-air anomalies in the Pacific compiled by Segawa and Matsumoto (1987).

Although the East Pacific Rise intersected by many fracture zones is the most striking topographic feature in the Pacific Ocean, it shows only a slightly positive free-air anomaly (Bowin et al., 1982). Gravity anomalies associated with some of the large fracture zones are well recognized in Fig. 3. Marsh et al. (1984) remark that gravity (or geoid) anomalies in the central and eastern Pacific basin are dominated by a pattern of roughly east–west anomalies with a transverse wavelength of about 2000 km. A further comparison with regional bathymetric data shows a good correlation of gravity and geoid with plate age. Each anomaly band is framed by major fracture zones whose regular spacing (1000 km) seems to account for the fabric in the field (Marsh et al., 1984).

Striking gravity anomalies are associated with the island arcs and adjacent trenches of the western Pacific and of the Tonga–Kermadec region. Two branches of the western Pacific arc nearly completely circle the Philippine Basin (Bowin et al., 1982). A band of positive anomalies trends about S70°E over Melanesia in the western South Pacific. Within the positive band lie narrow zones of strong positive and negative anomalies associated with convergent zones, volcanic arcs, and trenches within Melanesia (Bowin et al., 1982). The Tonga–Kermadec arc–trench system extends onto land at eastern North Island, New Zealand. The anomaly

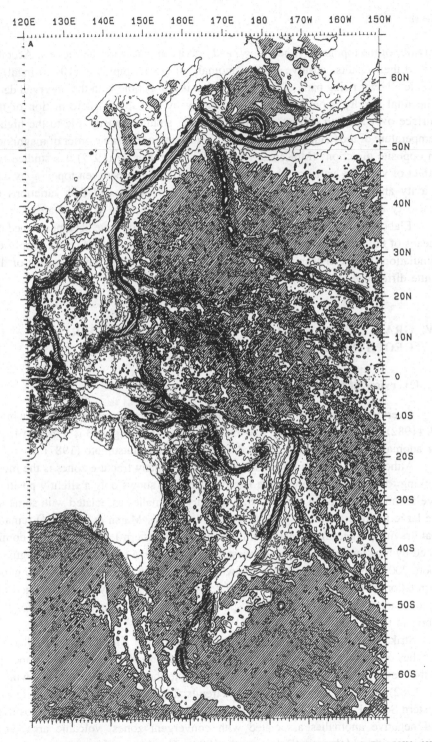

Fig. 3. A map of free-air gravity anomalies in the Pacific, (A) 120E–150W and (B) 150W–60W, obtained by satellite altimetry data (modified slightly from Segawa and Matsumoto, 1987, original maps drafted in color). Contours are drawn at values from −100 mgal to +100 mgal at 20-mgal intervals. Hatched areas show negative free-air anomalies. Compared with the free-air anomaly map compiled by Watts *et al.* (1985), gravity anomalies in this map tend to be rather negative.

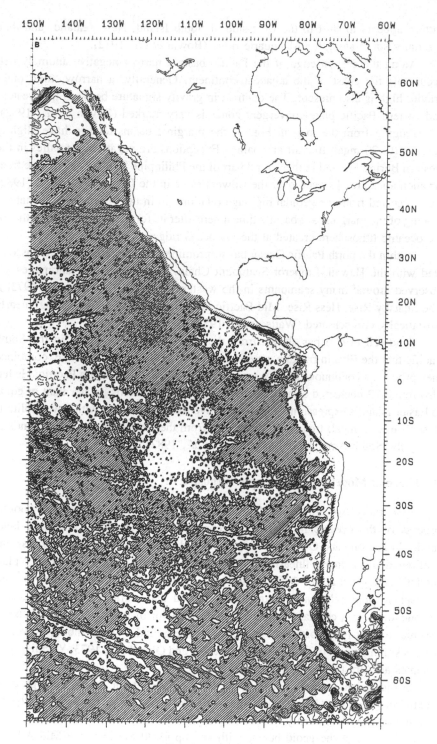

Fig. 3. (*continued*)

pattern appears to shift to the west side of South Island, New Zealand, and then continues south along the Macquarie ridge (Bowin et al., 1982).

Along the eastern edge of the Pacific basin a narrow negative anomaly belt occurs off the coast of the adjacent continent. Generally, a narrow, mild outer gravity high is also present. The contrast in gravity signature between the western and eastern Pacific plate-convergent zones is very marked (Bowin et al., 1982).

Judging from the map in Fig. 3, the marginal basins tend to have slightly positive (0–20 mgal) free-air anomalies. Exceptions are negative anomalies in the western Bering Sea and in the central part of the Philippine Sea. Large positive free-air anomalies are observed over the Bowers ridge in the Bering Sea. Horai (1982) has suggested from the geoidal high-age relationship in the Philippine Sea that the cooling of the marginal sea basin's lithosphere after its formation is similar to that of the oceanic lithosphere created at the midocean ridges.

Within the north Pacific Ocean basin, prominent gravity anomalies are associated with the Hawaii–Emperor Seamount Chain. Negative free-air anomalies are observed around many seamounts in the western Pacific basin (Tomoda, 1973b). The Shatsky Rise, Hess Rise, Mid-Pacific Islands, and Tuamotu Islands are nearly isostatically compensated (Watts et al., 1985).

Sandwell (1984) has shown from the satellite altimetric geoid in the South Pacific that the Eltanin fracture zone is connected to the Louisville ridge; combined they produce a continuous gravity anomaly across most of the South Pacific. He has also found 72 uncharted seamounts having geoid expressions greater than or equal to Easter Island's expression. A broad positive free-air anomaly with low amplitude (less than +25 mgal) trends southeastward from the vicinity of the Line Islands to nearly the South American coast (Bowin et al., 1982).

B. Thermal Models of the Oceanic Lithosphere

Essentially two simple models have been proposed to explain the cooling process of the oceanic lithosphere. According to the thermal "boundary layer model" (Turcotte and Oxburg, 1969; Davis and Lister, 1974), plate thickness and seafloor depth increase linearly with the square root of age. The cooling "plate model" (Langseth et al., 1966; McKenzie, 1967) predicts for large ages that the lithosphere approaches a constant thickness and that seafloor depth also decays exponentially to a constant value. From a viewpoint of isostasy, the plate model assumes a Pratt-type isostatic compensation. The boundary layer model as well as the thickening model of the lithosphere (Parker and Oldenburg, 1973; Yoshii, 1973) assumes an Airy-type isostatic compensation.

Geoid height–age relationship can be a constraint for the cooling process. Haxby and Turcotte (1978) pointed out that the geoid height should linearly decrease with age (0.16 m/m.y.) in the case of the boundary layer model, and found a linear decrease of the geoid height with age up to 50 Ma over the Mid-Atlantic ridge. Investigation of the relationship over four symmetrically spreading midocean ridges (Sandwell and Schubert, 1980) also showed that the geoid height–age data were, except for the southeast Pacific, consistent with constant slopes up to 80 Ma.

Crough (1979) has pointed out, however, that geoid height–age relationship is considerably disturbed by long-wavelength geoid undulations, so that the offset in geoid height across a large fracture zone gives a more accurate relationship. Several studies have been carried out over the Mendocino fracture zone (Crough, 1979; Detrick, 1981; Sandwell and Schubert, 1982). According to the results of Sandwell and Schubert (1982), the thermal structure begins to deviate from the boundary layer model at the thickness of 20–40 km, and a cooling plate model with a thickness of 125 km is most compatible with the geoid offset and with the depth–age relation. According to Cazenave et al. (1983), the observed data in the South Pacific and southeast Indian Ocean are most compatible with a cooling plate model with a plate thickness of 50–70 km for ages less than 30 Ma and 70–90 km for older ages. The results of these two studies show significant deviations, which are presumably caused by local and regional variations in the thermal properties of the lithosphere. Further studies are necessary in this field.

Lithospheric structure has also been studied using seismic results. Yoshii (1973) has shown, on the basis of the crustal structures in the Pacific, that the thickness of the lithosphere inferred from the RGA is approximately proportional to the square root of the age, and that the RGA in the Hawaii region, which is smaller than the value predicted from the standard RGA versus age curve by about 100 mgal, implies hot material beneath the region. His model is compatible with other geophysical data, such as regionalized group velocities of Rayleigh waves (Yoshii et al., 1976).

Figure 4 shows profiles of crustal structure and RGA across the northern Pacific from the Japan Sea to the Gorda Ridge off the west coast of the United States (Tomoda and Fujimoto, 1981). RGA increases with increasing distance from the midocean ridge. The change of RGA by 300 mgal corresponds to that of lithospheric thickness by about 70 km under the density contrast of 0.1 g cm^{-3}.

RGA versus square root of seafloor age in the Pacific (compiled by Fujimoto and Tomoda, 1985b) is shown in Fig. 5. Panel a shows the observed RGA and panel b shows the RGA obtained after corrections for the effects of lithospheric thinning in hot-spot swells and thickening near trenches. The curve in panel b indicates the lithospheric thickening rate expected from the thermal model of lithospheric thickening which Kono and Yoshii (1975) proposed on the basis of numerical experiments under unsteady conditions. These results show that the largest variation in the upper-mantle structure is caused by the cooling effect of the lithosphere, and that reasonable thickness of the lithosphere can be inferred from RGA.

C. Midocean Ridge

Interpretations of the long-wavelength gravity and topographic features of the Mid-Atlantic Ridge were the focus of the early studies of midocean ridges (e.g.,

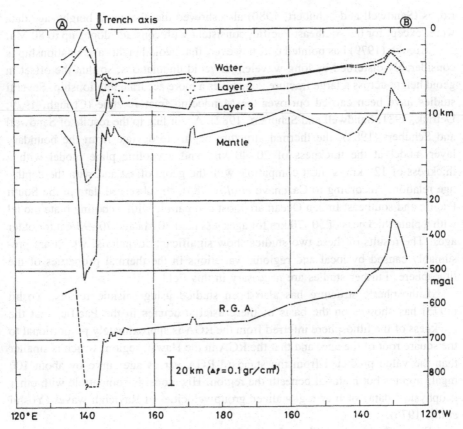

Fig. 4. Profiles of subterranean structure across the northern Pacific. The line on the bottom indicates the lithosphere–asthenosphere boundary estimated on the basis of residual gravity anomalies. (From Tomoda and Fujimoto, 1981.)

Talwani *et al.*, 1965). Since it was recognized that simple thermal models of a cooling lithosphere could account for the long-wavelength geophysical features, the focus has been on the short-wavelength free-air anomalies over ridge crests.

Cochran (1979) has analyzed isostasy of the world's midocean ridge crests. He found that the East Pacific Rise crest is characterized by a positive horstlike topographic feature about 20 km across and a few hundred meters high. The characteristic gravity anomaly associated with the central high is a relatively small-amplitude (15–25 mgal) high which is 20–30 km wide and is flanked by smaller-amplitude gravity lows. The anomaly pattern is similar to that of the Reykjanes Ridge and is in contrast to that of the Mid-Atlantic Ridge, where negative and positive free-air anomalies are associated with the rift valley and the rift mountain, respectively.

Cochran (1979) has estimated that gravity and bathymetry of the East Pacific

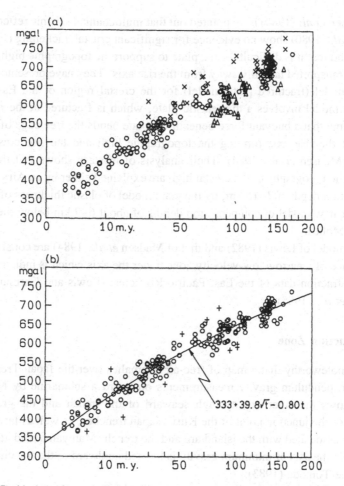

Fig. 5. (a) Residual gravity anomalies versus square root of seafloor age in the Pacific. Crosses are near a trench and triangles are in a hot-spot swell. (Modified slightly from Fujimoto and Tomoda, 1985b.) (b) Residual gravity anomalies versus square root of seafloor age obtained after corrections of the effects of thickening of the lithosphere near trenches and its thinning in hot-spot swells. The curve indicates lithospheric thickening rate expected by the lithospheric thickening model proposed by Kono and Yoshii (1975).

Rise crest are best explained by a plate model of regional compensation with elastic thickness of 2–6 km. Louden (1981) has estimated that the Juan de Fuca ridge is compensated either regionally by an elastic plate 5 to 10 km thick or locally by subcrustal low densities at depths of 15–20 km. Considering that it is unlikely that the plate is strong enough to support the axial topography, Lewis (1982) has suggested from an analysis of the gravity and topography that the East Pacific Rise crest is supported dynamically by other forces, such as plate driving forces.

Madsen *et al.* (1984) have pointed out that multichannel seismic reflection data (Stoffa *et al.*, 1980) show no evidence for significant crustal thickening beneath the rise axis and that it is difficult for the plate to support the topographic high, because it will be transported laterally away from the rise axis. They have presented another isostatic model (fractured plate model) for the crestal region of the East Pacific Rise. The model involves a thin elastic plate, which is fractured at the rise crest. They assume that a buoyant force beneath the plate bends the free edge of the plate upward at the rise axis forming the topographic high and the rise crest gravity anomaly (Madsen *et al.*, 1984). Their analysis of isostasy shows that the gravity anomaly and topography of the crestal high are explained either by an Airy model of compensation depth of 7–15 km, by the plate model of elastic thickness of 2–5 km, or by their new model of compensation depths of about 6–7 km below the seafloor (near the Moho).

The model of Lewis (1982) and that of Madsen *et al.* (1984) are consistent with the presence of a narrow low-velocity zone under the axis which is indicated by the seismic refraction data of the East Pacific Rise crest (Lewis and Garmany, 1982; McClain *et al.*, 1985).

D. Subduction Zone

It is noteworthy that a map of free-air anomalies over the Japan Trench compiled from pendulum gravity measurements on board a submarine by Matuyama (1934) shows the outer gravity high seaward of the trench and the gravity low landward of the junction point of the Kuril–Japan trenches, as well as large gravity anomalies associated with the island arc and the trench. Matuyama paid attention to the fact that the axis of minimum gravity is located landward of the trench axis. For details, see Tomoda (1983).

The varied amplitude of the outer gravity high seaward of oceanic trenches described by Watts and Talwani (1974) is related to the magnitude of horizontal compressive stress, and is consistent with the classification of subduction zones into two types according to their present-day backarc activities (Uyeda and Kanamori, 1979). The seismic results near the Japan–Bonin trenches (Houtz *et al.*, 1980) suggest positive density anomalies in the upper mantle, or thickening of the lithosphere near the trench (Tomoda and Fujimoto, 1981). Other geophysical data are needed to settle whether the outer gravity high is caused by the subducting lithosphere (Morgan, 1965; Grow and Bowin, 1975; Sager, 1980) or not (Watts and Talwani, 1975; Chapman and Talwani, 1982).

Geoidal swell and apparent discrepancy between the gravity and seismic results in the arc–trench system are also important observational facts (e.g., Chapman and Talwani, 1982). Karig (1971) noticed that seismic results in the marginal basins showed standard oceanic thickness of the crust with the seafloor shallower than in deep ocean basins, while free-air anomalies were near zero in both regions. These results indicate lower-density material below the Moho of the marginal basins. Yoshii (1972) also recognized this discrepancy across the Japan Trench and proposed the RGA to represent the density anomaly below the Moho. He explained the discrepancy by the thinning of the lithosphere under the island arc.

Chapman and Talwani (1982) have obtained similar results by using gravity and geoid anomalies as well as seismic results, i.e., topographic high landward of the trench is partially compensated with the mantle by a thinning of the lithosphere. They also remark that such a dipolar layered mass distributed landward of the trench is the fundamental cause of the gradual slope of the geoid, which generally starts 1000–3000 km seaward of trenches. Density contrast of the descending lithosphere cannot be 0.05 g cm^{-3} or larger, and its effect is only of secondary importance in explaining this geoid anomaly (Chapman and Talwani, 1982).

This conclusion is in contrast to the result obtained by Sager (1980), who investigated the structure of the Mariana arc using gravity and seismic results. According to his results, the low-density mantle, the frontal arc root, and the subducted oceanic crust all provide negative anomalies of similar amplitude. The downgoing slab contributes a large ($\sim$ 130 mgal), broad, positive anomaly over the arc and trough (Sager, 1980).

Chapman and Talwani (1982) emphasized that it is quite difficult to match both gravity and geoid values; it is a severe constraint. Because gravity anomaly is sensitive to the shallower density anomaly and geoid anomaly reflects density anomaly enhanced by its depth (see Section III), the geoid is more sensitive to changes in the mantle structure. It will be necessary, therefore, to interpret the observed geoid anomaly globally and to check whether the result by Sager (1980) is compatible with the geoid anomaly or not.

Matsumoto and Tomoda (1983a) carried out simplified numerical simulations on the evolution of lithospheric subduction on an assumption that both the lithosphere and asthenosphere behave as Newtonian viscous incompressible fluids with constant viscosities and densities. An initiation of lithospheric subduction and its evolution presented by them indicate the potential ability of this approach.

Figure 6 shows topography and gravity anomaly obtained by the author through a two-dimensional simulation. No force is assumed except for gravity. Density contrast between the lithosphere and asthenosphere is, therefore, the only source of the driving force of lithospheric subduction. Because isostasy is assumed at the initial condition, a horizontal pressure gradient is absent at the lower part of the asthenosphere, but it exists at the boundary of the two lithospheres. This horizontal pressure gradient drives the initial motion. The subducting angle and the spatial distribution of gravity anomalies change according to the viscosities and to the thicknesses of the two lithospheres. The density contrast of the subducting lithosphere is assumed to be 0.1 g cm^{-3} (Yoshii *et al.*, 1976). This value seems to be too large considering the amplitudes of the positive and negative gravity anomalies.

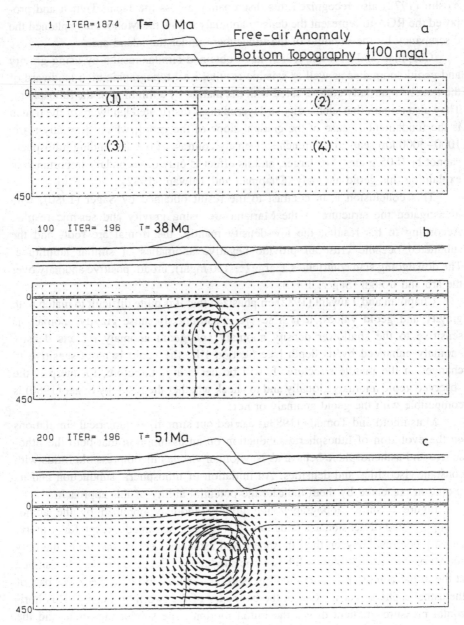

Fig. 6. A result of numerical simulation on the evolution of a lithospheric subduction. (1, 2) Lithosphere of density 3.3 g cm^{-3} and viscosity 10^{23} poise. (3, 4) Asthenosphere of density 3.2 g cm^{-3} and viscosity 10^{21} poise.

E. Seamount and Rise

The structure of the Hawaiian–Emperor seamount chain has been studied by many investigators. Watts (1975a) compiled a map of gravity anomalies in this region and later (Watts, 1976b) summarized the features of the gravity anomalies: the large-amplitude positive anomalies (up to +700 mgal) over the Hawaiian ridge, large-amplitude negative anomalies (up to −136 mgal) flanking the ridge, and a broad belt (> 250 km) of positive anomalies (+25 to +50 mgal) bordering the negative anomalies. The long-wavelength (2200 km) positive anomaly (up to +15 mgal) over the ridge recognized in a 5° × 5° mean free-air anomaly map correlates closely with the Hawaiian swell upon which the Hawaiian ridge is superimposed (Watts, 1976b).

Vening Meinesz (1941) presented a regional compensation model for the isostatic structure of the Hawaiian ridge. Watts (1978) and Watts and Ribe (1984) show that the elastic thickness of the plate increases with the age of the lithosphere at the time of loading. But some results show considerable deviations (Lambeck, 1981). Watts *et al.* (1980) classify the seamounts in the Pacific basin into two groups: nearly isostatic seamounts which probably were formed on younger and more flexible lithosphere (e.g., Mid-Pacific mountains and Line Islands ridge) and nearly uncompensated seamounts which probably were formed on old lithosphere (e.g., Hawaiian ridge and Magellan seamounts). Judging from the gravity map in Fig. 3, however, considerable negative gravity anomalies are associated with Hawaiian ridge and Magellan seamounts.

Detrick and Crough (1978) and Crough (1978) have shown that the midplate swells are probably caused by thinning of the lithosphere over a mantle hot spot.

Tomoda *et al.* (1982) have compiled the data of explosion seismology and shown that the residual gravity anomaly is locally negative by 50–70 mgal over the Hawaiian swell (Fig. 7). The amplitude of the locally negative residual gravity anomaly decreases to about 20 mgal across the Emperor seamount chain. These results can be explained by the reheating model of the lithosphere presented by Detrick and Crough (1978).

Tomoda (1973b) paid attention to the negative free-air anomalies associated with many seamounts in the western Pacific basin (see Fig. 3), and Fujimoto (1976) proposed a model wherein the lithosphere is thinner under a seamount due to the remaining effect of the hot magma. The seamount is supported by the buoyancy of the asthenosphere and will be subducted at a trench when the buoyancy fades away. Tomoda and Fujimoto (1981, 1983) have tried to explain the complicated accretion processes at trench axes by classifying seamounts or rises into three types as is shown in Fig. 8. Delayed subduction shown in the figure may cause a node of the trench (Fujimoto, 1976). A rise with a thick crust may accrete to the continent at the trench axis and, if the subduction force is large enough, a new subduction zone would be formed seaward of the original trench so that a jump of the plate boundary

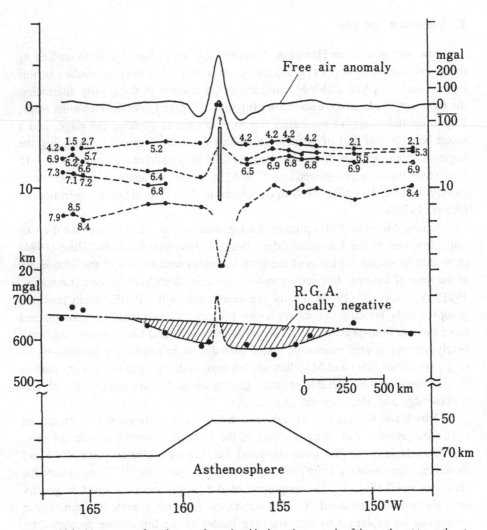

Fig. 7.   Velocity structure, free air anomaly, and residual gravity anomaly of the southwest to northeast section across Oahu Island. (From Tomoda *et al.*, 1982.)

would take place (Tomoda and Fujimoto, 1981, 1983). Jumping of the trench axis caused by a collision of a rise is also presented by a numerical simulation carried out by Matsumoto and Tomoda (1983b).

F. Mantle Convection

According to McKenzie (1985), the mantle contains at least three scales of circulation. One is comparable to the plate dimensions, known as the large scale circulation, which returns material from the island arcs to the ridges. A second of an

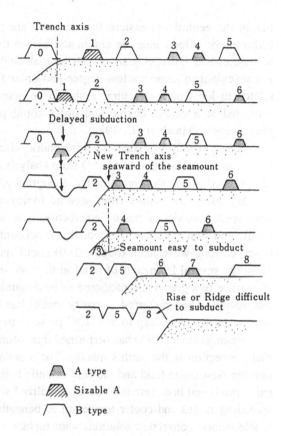

Fig. 8. Seamounts or rises can be classified into three types. (1) A type: seamounts supported by buoyancy of the asthenosphere which is easily subducted. (2) Sizable A type: large seamounts supported by buoyancy of the asthenosphere which will take a long time to be subducted. (3) B type: rise supported by buoyancy of the crust, which is difficult to be subducted. These various structures of seamounts or rises would result in complicated accretion processes at trench axes. A rise with a thick crust may accrete to the landward plate at the trench axis and if the subduction forces are large enough, a new subduction zone would be formed seaward of the original trench so that a jump of the plate boundary would take place. (From Tomoda and Fujimoto, 1983.)

intermediate scale is comparable to the depth of the upper mantle and a third with a small scale is comparable to the thickness of the asthenosphere.

Parsons and McKenzie (1978) suggested the possibility of the small-scale mantle convection. McKenzie *et al.* (1980) postulate that the correlation between variations of the bathymetry and of the geoid in the Pacific Ocean suggests that both are the surface expression of mantle convection. The plan form of the convection which is required to generate the observed anomalies does not consist of rolls but is three-dimensional, with rising and sinking jets elongated in the direction of the absolute motion of the Pacific plate (McKenzie *et al.*, 1980). These interpretations have been supported by the more extensive investigation carried out in the Pacific basin by Watts *et al.* (1985). They point out that one of the characteristic features of the gravity anomaly over a topographic load dynamically supported by a convection is that it is much smaller ($\sim$ 30 mgal km^{-1}) than would be expected for just an uncompensated warping of the seafloor ($\sim$ 90 mgal km^{-1}).

On the other hand, Marsh *et al.* (1984) have pointed out that because of the significant effect of age-dependent plate structure on the geoid and gravity fields, especially over young (< 80 Ma) seafloor, intermediate gravity (or geoid) anomalies in the central and eastern Pacific basin are framed by major fracture zones. Lithospheric effects must be eliminated before these anomalies can be systematically ascribed to deeper sources such as small-scale mantle convection. Although it is convenient to remove a low-degree-and-order field model to express anomalies related to local plate structure, it should be noted that removal of a field model truncated at a certain degree and order could produce ripples through Gibbs's phenomenon (Marsh *et al.*, 1984).

Kaula (1980) estimated the lithospheric viscosity to be about 10^{23} poise (g $cm^{-1} sec^{-1} = 0.1 \ kg \ m^{-1} sec^{-1}$) from analysis of the spherical harmonic spectra of observed plate velocities, gravity, topography, and heat flow.

By using the results from seismic tomography, Hager *et al.* (1985) have developed a model of mantle convection with subducting slabs and postglacial uplift, and remarked that this model can account for 90% of the variance in the observed long-wavelength (degree 2–9) geoid anomalies. According to their conclusions, several kilometers of relief at the core–mantle boundary and a kilometer of surface topography are predicted to be dynamically maintained by the flows in the mantle. Their preferred viscosity model has an asthenosphere of 10^{21} poise, with viscosity increasing to 3×10^{23} poise in the lower mantle.

Baumgardner (1985) has performed three-dimensional simulations on the thermal convection in the earth's mantle. The mantle is treated as an infinite Prandtl number Newtonian fluid and the core–mantle boundary is assumed to be free-slip, isothermal, and undeformable. Cases initialized with warmer temperatures beneath spreading ridges and cooler temperatures beneath present subduction zones yield whole-mantle convection solutions with surface velocities that correlate well with currently observed plate velocities. His conclusions are that a substantial portion of the heating of the earth's mantle is from the core, and that the flow pattern in the mantle is dominated by a relatively small number (about seven) of localized upwelling plumes. It is surprising since the locations of these plumes correlate well with volcanic activity in Iceland, the Horn of Africa, the Kerguelen Islands, and Hawaii. For the interpretation of gravity anomalies, it is necessary to assume that the mantle–core boundary is deformable.

V. FUTURE WORK

According to Edelson (1985), although satellite altimetry has brought revolutionary advances in gravity measurements over the world's oceans, it cannot measure gravity over the continents of great tectonic interest, such as the collision belts of the Himalayas and the Andes, and rift zones in East Africa. New gravity field information is also necessary over the oceans, since oceanographers must have an independently determined gravitational geoid in order to derive the ocean currents from the next generation of altimeter data expected from the ocean topography experiment (TOPEX) mission. The improved gravity field model will result in significant improvements in tracking the TOPEX and other satellites (Edelson, 1985). According to NASA Conference Publication 2305 (Wells, 1984), NASA has two programs for the improvement of the gravity field model. One is the Geopotential Research Mission (GRM) by use of satellite-to-satellite tracking, and the other is the Gravity Gradiometer Mission (GGM).

According to the conference report, it is planned that the GRM will be approved in fiscal year 1988 for a 1992 launch.* The mission concept is to make satellite-to-satellite measurements between two drag-free spacecraft in the same nearly circular polar orbit at a low enough altitude (160 km) to obtain gravity data with an accuracy of about 2 mgal at a resolution (half-wavelength) of 100 km. Global magnetic field measurements would be made by one of the satellites throughout the mission. The pair satellites would stay in orbit for some 6 months. A Disturbance Compensation System (DISCOS) will be included on each spacecraft to eliminate the effects of atmospheric drag and other nongravitational forces. The resulting improvement in knowledge of the earth's gravity field, compared with GEM10B, is shown in Fig. 9 (Wells, 1984). The geoid undulation uncertainty for all degrees from 2 through 180 would be about 1 cm (Breakwell, 1980; Colombo, 1981). This compares with a present uncertainty of about 70 cm for short wavelengths and 8 cm for long wavelength (Wells, 1984). GRM will have a sensitivity to detect time variations in the earth's gravity field caused by tides, postglacial re-

---

*After the space shuttle accident, NASA decided to delay the schedule.

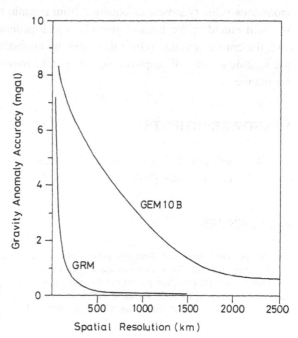

Fig. 9. Expected accuracy of the gravity field before and after the Geopotential Research Mission. (From Wells, 1984.)

bound, great earthquakes, and so on (Wagner and McAdoo, 1986). Various technical problems of this method have been reported in detail by Colombo (1984).

NASA has a plan of gravity gradiometry in the 1990s as a follow-on to GRM. The long-term objective beyond GRM is to obtain an accuracy of 0.5 to 1.0 mgal at a spatial resolution (half-wavelength) of 50 km. The status and prospects of gravity gradiometer instruments now under development are described by Wells (1984). Two types of cryogenic detectors, the superconducting quantum interference device (SQUID) and superconducting cavity oscillator (SCO), have been suggested; each has the sensitivity required for the gravity gradiometry.

Because bathymetric data are principally important for the interpretation of gravity data, precise shipborne gravimetry will still be important to investigate local structures combined with a detailed topographic map and other geophysical observations. Ocean bottom gravimeters in combination with ocean bottom pressure gauges will be important to detect vertical motion of the lithosphere and preearthquake crustal movements.

In interpretations of the gravity field, the following statement in the report of the Geopotential Research Mission Science Steering Group (1985) is noteworthy: "GRM will give profound insight into the deep interior of the earth through gravity and magnetic measurements over the continents as well as the oceans. The gravity field will elucidate the pattern and energetics of the thermal convection in the mantle which drives the plate motion." In order to make such advances in the study of the earth's gravity field, seismic works and numerical simulations in a regional as well as global scale will be principally important. Gravity anomalies associated with anomalous velocity structures obtained from seismic tomography of higher resolution will elucidate the thermal convections. Residual gravity anomalies obtained from the crustal structure reflect these density anomalies in the mantle. The gravity and seismic results will improve the numerical simulations of global circulations in the mantle.

ACKNOWLEDGMENTS

We thank Drs. J. Segawa and T. Matsumoto for providing a free-air anomaly map. We wish to thank Professor S. Uyeda for his comments.

REFERENCES

Anonymous, 1967, Report on prototype gravity measuring system, constructed by Bell Aerosystems Company for U.S. Naval Oceanographic Office and U.S. Coast and Geodetic Survey, in: Proc. First Marine Geodesy Symposium, U.S. Government Printing Office, pp. 187–194.
Balmino, G., Reigber, C., and Moynot, B., 1978, Le modele de potential gravitationnel terrestre, GRIM 2. Determination, evaluation. *Ann. Geophys.* **34**:55–78.

Banks, R. J., Parker, R. L., and Huestis, S. P., 1977, Isostatic compensation on a continental scale: Local versus regional mechanisms, *Geophys. J. R. Astron. Soc.* **51**:431–452.

Baumgardner, J. R., 1985, Three-dimensional treatment of convective flow in the earth's mantle, *J. Stat. Phys.* **39**:501–511.

Born, G. H., Dunne, J. A., and Lame, D. B., 1979, SEASAT mission overview, *Science* **204**:1405–1406.

Born, G. H., Lame, D. B., and Rygh, P. J., 1981, A survey of the goals and accomplishments of the SEASAT mission, in: *Oceanography from Space* (J. F. R. Gower, ed.), *Mar. Sci.* **13**:3–14, Plenum Press, New York.

Bowie, W., 1917, Investigation of Gravity and Isostasy, Spec. Publ. USCGS 40.

Bowin, C., 1983, Depth of principal mass anomalies contributing to the Earth's geoidal undulations and gravity anomalies, *Mar. Geod.*, **7**:61–100.

Bowin, C., Aldrich, T. C., and Folinsbee, R. A., 1972, VSA gravity meter system: Tests and recent developments, *J. Geophys. Res.* **77**:2018–2033.

Bowin, C., Lu, R. S., Lee, C., and Schouten, H., 1978, Plate convergence and accretion in Taiwan–Luzon region, *Am. Assoc. Pet. Geol. Bull.* **62**:1653.

Bowin, C., Purdy, G. M., Johnston, C., Shor, G., Lawver, L., Hartono, H. M. S., and Jezek, P., 1980, Arc–continent collision in Banda Sea region, *Am. Assoc. Pet. Geol. Bull.* **64**:876.

Bowin, C., Warsi, W., and Milligan, J., 1981, Free-air gravity anomaly map of the world, Geological Society of America Map and Chart Series, No. MC-45.

Bowin, C., Warsi, W., and Milligan, J., 1982, Free-air gravity anomaly atlas of the world, Geological Society of America Map and Chart Series, No. MC-46.

Breakwell, J. V., 1980, More about satellite determination of short wavelength gravity variations, Gravsat User Working Group.

Bureau of Mineral Resources (BMR), 1976, Gravity map of Australia, BMR, Geology and Geophysics, Australia.

Cazenave, A., Lago, B., and Dominh, K., 1983, Thermal parameters of the oceanic lithosphere estimated from geoid height data, *J. Geophys. Res.* **88B**:1105–1118.

Chapman, M. E., and Talwani, M., 1982, Geoid anomalies over deep sea trenches, *Geophys. J. R. Astron. Soc.* **68**:349–369.

Chiburis, E. F., Dowling, J. J., Dehlinger, P., and Yellin, M. J., 1972, Pacific SEAMAP 1961–70 data for areas 15636-12, 15642-12, 16836-12, and 16842-12, NOAA Tech. Rep. NOS51, U.S. Government Printing Office, Washington, D.C.

Cochran, J. R., 1979, An analysis of isostasy in the world's oceans. 2. Mid-ocean ridge crests, *J. Geophys. Res.* **84**:4713–4729.

Cohen, S. C., and Smith, D. E., 1985, LAGEOS scientific results: Introduction, *J. Geophys. Res.* **90B**:9217–9220.

Colombo, O. L., 1981, Global geopotential modeling from satellite-to-satellite tracking, Ohio State University Report No. 317.

Colombo, O. L., 1984, The global mapping of gravity with two satellites, Netherlands Geodetic Commission Publications on Geodesy, New Series Vol. 7.

Crough, S. T., 1978, Thermal origin of mid-plate hot-spot swells, *Geophys. J. R. Astron. Soc.* **55**:451–469.

Crough, S. T., 1979, Geoid anomalies across fracture zones and the thickness of the lithosphere, *Earth Planet. Sci. Lett.* **44**:224–230.

Davis, E. E., and Lister, C. R. B., 1974, Fundamentals of ridge crest topography, *Earth Planet. Sci. Lett.* **21**:405–413.

Dehlinger, P., Couch, R. W., McManus, D. A., and Gemperle, M., 1970, Northeast Pacific structure, in: *The Sea* (A. E. Maxwell, ed.), Vol. 4, Part II, Wiley, New York, pp. 133–189.

Detrick, R. S., 1981, An analysis of geoid anomalies across the Mendocino fracture zone: Implications for thermal models of the lithosphere, *J. Geophys. Res.* **86B**:11751–11762.

Detrick, R. S., and Crough, S. T., 1978, Island subsidence, hot spots, and lithospheric thinning, *J. Geophys. Res.* **83B**:1236–1244.

Dorman, L. M., and Lewis, B. T. R., 1970, Experimental isostasy. 1. Theory of the determination of the earth's isostatic response to a concentrated load, *J. Geophys. Res.* **75**:3357–3365.

Dziewonski, A. M., 1984, Mapping the lower mantle: Determination of lateral heterogeneity in P velocity up to degree and order 6, *J. Geophys. Res.* **89B**:5929–5952.

Edelson, B. I., 1985, Keynote address, in: Geopotential Research Mission (GRM), NASA Conference Publication 2390, NASA, Washington, D.C., pp. iii–vi.

Fujimoto, H., 1976, Processing of gravity data at sea and their geophysical interpretation in the region of the western Pacific, *Bull. Ocean* Res. Inst., Univ. Tokyo, **8**.

Fujimoto, H., and Tomoda, Y., 1985a, A compact on-line data processing system for the Tokyo Surface Ship Gravity Meter, *J. Phys. Earth* **33**:45–58.

Fujimoto, H., and Tomoda, Y., 1985b, Lithospheric thickness anomaly near the trench and possible driving force of subduction, *Tectonophysics* **112**:103–110.

Ganeko, Y., and Harada, Y., 1982, Gravity anomalies around Japan, Report of Hydrographic Researches, Hydrographic Department, Japan, *17:*163–180.

Gaposchkin, E. M. (ed.), 1973, Smithsonian standard earth (III), Spec. Rep. 353, Smithson. Astrophys. Obs., Cambridge, Mass., pp. 262–276.

Gaposchkin, E. M., 1974, Earth's gravity field to the eighteenth degree and geocentric coordinates for 104 stations from satellite and terrestrial data, *J. Geophys. Res.* **79**:5377–5411.

Gaposchkin, E. M., 1977, Gravity-field determination using laser observations, *Philos. Trans. R. Soc. London Ser. A* **284**:515–527.

Gaposchkin, E. M., 1980, Global gravity field to degree and order 30 from GEOS 3 satellite altimetry and other data, *J. Geophys. Res.* **85B**:7221–7234.

Gaposchkin, E. M., and Lambeck, K., 1970, 1969 Smithsonian standard earth (II), Spec. Rep. 315, Smithson. Astrophys. Obs., Cambridge, Mass.

Gaposchkin, E. M., and Lambeck, K., 1971, The earth's gravity field to sixteenth degree and station coordinates from satellite and terrestrial data, *J. Geophys. Res.* **76**:4855–4883.

Geopotential Research Mission Science Steering Group, 1985, Geopotential Research Mission—Scientific Rationale, NASA, Washington, D.C.

Gilbert, R. L. G., 1949, A dynamic gravimeter of novel design, *Proc. Phys. Soc. London Sect. B* **62**:445–454.

Graf, A., and Schulze, R., 1961, Improvements on the sea gravimeter Gss2, *J. Geophys. Res.* **66**:1813–1821.

Grow, J. A., and Bowin, C. O., 1975, Evidence for high-density crust and mantle beneath the Chile Trench due to the descending lithosphere, *J. Geophys. Res.* **80**:1449–1458.

Hager, B. H., Richards, M. A., and O'Connel, R. J., 1985, The source of the earth's long wavelength geoid anomalies: Implications for mantle and core dynamics, in: Geopotential Research Mission (GRM), NASA Conference Publication 2390, NASA, Washington, D.C., pp. 57–59.

Haxby, W. F., and Turcotte, D. L., 1978, On isostatic geoid anomalies, *J. Geophys. Res.* **83**:5473–5478.

Hayes, D. E., and Talwani, M., 1972, Geophysical investigation of the Macquarie Ridge complex, American Geophysical Union, Antarctic Research Series, Vol. 19, Fig. 3.

Heiskanen, W. H., and Moritz, H., 1967, *Physical Geodesy,* Freeman, San Francisco.

Horai, K., 1982, A satellite altimetric geoid in the Philippine Sea, *Nature* **299**:117–121.

Houtz, R., Windisch, C., and Murauchi, S., 1980, Changes in the crust and upper mantle near the Japan–Bonin Trench, *J. Geophys. Res.* **85**:267–274.

Ishihara, T., 1982, Free air gravity anomaly map of the central Pacific, Geological Survey of Japan Marine Geology Map Series 19.

Ishihara, T., 1986, Density determination of seamounts along the Bonin–Mariana Arc by the gravity versus effective depth analysis, Dr. thesis, Geophys. Inst., Univ. Tokyo, Tokyo.

Jeffreys, H., 1959, *The Earth,* 4th ed., Cambridge University Press, London.

Jordan, S. K., 1978, Statistical model for gravity, topography, and density contrasts in the earth, *J. Geophys. Res.* **83**:1815–1824.

Kahn, W. D., Klosko, S. M., and Wells, W. T., 1982, Mean gravity anomalies from a combination of Apollo/ATS 6 and GEOS 3/ATS 6 SST tracking campaigns, *J. Geophys. Res.* **87B**:2904–2918.

KAIKO I Research Group, 1986, Topography and Structure of trenches around Japan—Data atlas of Franco-Japanese KAIKO Project, Phase I. Univ. Tokyo, Tokyo.

Karig, D. E., 1971, Structural history of the Mariana island arc system, *Geol. Soc. Am. Bull.* **82**:323–344.

Kaula, W. M., 1966, *Theory of Satellite Geodesy,* Ginn (Blaisdell), Boston.

Kaula, W. M., 1972, Global gravity and mantle convection, *Tectonophysics* **13**:341–359.

Kaula, W. M., 1980, Material properties for mantle convection consistent with observed surface fields, *J. Geophys. Res.* **85B**:7031–7044.

Kono, Y., and Yoshii, T., 1975, Numerical experiment on the thickening plate model, *J. Phys. Earth* **23**:63–75.

Kumagai, N., 1940, Studies in the distribution of gravity anomalies in the northeast Honsyu and the central part of the Nippon Trench, Japan, *Jpn. J. Astron. Geophys.* **17**:477–551.

LaCoste, L. J. B., 1967, Measurement of gravity at sea and in the air, *Rev. Geophys.* **5**:477–526.

LaCoste, L. J. B., 1973, Crosscorrelation method for evaluating and correcting shipboard gravity data, *Geophysics* **38**:701–709.

LaCoste, L. J. B., 1983, LaCoste and Romberg straight-line gravimeter, *Geophysics* **48**:606–610.

LaCoste, L. J. B., Ford, J., Bowles, R., and Archer, K., 1982, Gravity measurements in an airplane using state-of-the-art navigation and altimetry, *Geophysics* **47**:832–838.

LaFehr, T. R., 1980, Gravity method, *Geophysics* **45**:1634–1639.

Lambeck, K., 1981, Lithospheric response to volcanic loading in the Southern Cook Islands, *Earth Planet. Sci. Lett.* **55**:482–496.

Lambeck, K., and Coleman, R., 1982, A search for seamounts in the Southern Cook and Austral region, *Geophys. Res. Lett.* **9**:389–392.

Lambeck, K., and Coleman, R., 1983, The earth's shape and gravity field: A report of progress from 1958 to 1982, *Geophys. J. R. Astron. Soc.* **74**:25–54.

Langseth, M. G., Le Pichon, X., and Ewing, M., 1966, Crustal structure of the mid-ocean ridges. 5. Heat flow through the Atlantic ocean floor and convection currents, *J. Geophys. Res.* **71**:5321–5355.

Lazarewicz, A. P., and Schwank, D. C., 1982, Detection of uncharted seamounts using satellite altimetry, *Geophys. Res. Lett.* **9**:385–388.

Leitao, C. D., and McGoogan, J. T., 1974, Skylab radar altimeter: Short-wavelength perturbations detected in ocean surface profiles, *Science* **186**:1208–1209.

Lerch, F. J., Klosko, S. M., Laubscher, R. E., and Wagner, C. A., 1979, Gravity model improvement using Geos 3 (GEM9 and 10), *J. Geophys. Res.* **84B**:3897–3916.

Lerch, F. J., Putney, B. H., Wagner, C. A., and Klosko, S. M., 1981, Goddard earth models for oceanographic applications (GEM10B and 10C), *Mar. Geod.* **5**:145–187.

Lerch, F. J., Klosko, S. M., and Patel, G. B., 1982, Gravity model development using LAGEOS, *Geophys. Res. Lett.* **9**:1263–1266.

Lerch, F. J., Klosko, S. M., and Patel, G. B., 1983, A refined gravity model from LAGEOS (GEM-L2), NASA Tech. Memo. TM84986.

Lerch, F. J., Klosko, S. M., Patel, G. B., and Wagner, C. A., 1985a, A gravity model for crustal dynamics (GEM-L2), *J. Geophys. Res.* **90B**:9301–9311.

Lerch, F. J., Klosko, S. M., Wagner, C. A., and Patel, G. B., 1985b, On the accuracy of recent Goddard gravity models, *J. Geophys. Res.* **90B**:9312–9334.

Lewis, B. T. R., 1982, Constraints on the structure of the East Pacific Rise from gravity, *J. Geophys. Res.* **87B**:8491–8500.

Lewis, B. T. R., and Garmany, J. D., 1982, Constraints on the structure of the East Pacific Rise from seismic refraction data, *J. Geophys. Res.* **87B**:8417–8425.

Louden, K. E., 1981, A comparison of the isostatic response of bathymetric features in the north Pacific Ocean and Philippine Sea, *Geophys. J. R. Astron. Soc.* **64**:393–424.

Lozinskaya, A. M., 1959, The string gravimeter for the measurement of gravity at sea, *Bull. Acad. Sci. USSR Engl. Transl. Geophys. Ser.* **3**:398–409.

Ludwig, W. J., and Houtz, R. E., 1979, Isopach map of sediments in the Pacific Ocean basin and marginal sea basins, American Association for Petroleum Geologists.

Ludwig, W. J., Nafe, J. E., and Drake, C. L., 1970, Seismic refraction, in: *The Sea* (A. E. Maxwell, ed.), Vol. 4, Part I, Wiley, New York, pp. 53–84.

Lundquist, C. A., and Veis, G. (eds.), 1966, Geodetic parameters for a 1966 Smithsonian Institution standard earth, Vol. 1, Spec. Rep. 200, Smithson. Astrophys. Obs., Cambridge, Mass.

McClain, J. S., Orcutt, J. A., and Burnett, M., 1985, The East Pacific Rise in cross section: A seismic model, *J. Geophys. Res.* **90B:**8627–8639.

McKenzie, D. P., 1967, Some remarks on heat flow and gravity anomalies, *J. Geophys. Res.* **72:**6261–6273.

McKenzie, D. P., 1977, Surface deformation, gravity anomalies and convection, *Geophys. J. R. Astron. Soc.* **48:**211–238.

McKenzie, D. P., 1985, The importance of GRM gravity observations in continental regions, in: Geopotential Research Mission (GRM), NASA Conference Publication 2390, NASA, Washington, D.C., p. 56.

McKenzie, D. P., and Bowin, C., 1976, The relationship between bathymetry and gravity in the Atlantic Ocean, *J. Geophys. Res.* **81:**1903–1915.

McKenzie, D. P., Roberts, J. M., and Weiss, N. O., 1974, Convection in the earth's mantle: Towards a numerical simulation, *J. Fluid Mech.* **62:**465–538.

McKenzie, D. P., Watts, A., Parsons, B., and Roufosse, M., 1980, Planform of mantle convection beneath the Pacific Ocean, *Nature* **288:**442–446.

Madsen, J. A., Forsyth, D. W., and Detrick, R. S., 1984, A new isostatic model for the East Pacific Rise crest, *J. Geophys. Res.* **89B:**9997–10015.

Marsh, J. G., and Marsh, B. D., 1977, Gravity anomalies near the East Pacific Rise with wavelengths shorter than 3300 km recorded from GEOS-3/ATS-6 satellite-to-satellite Doppler tracking data, NASA Technical Memorandum 79533.

Marsh, J. G., Marsh, B. D., Williamson, R. G., and Wells, W. T., 1981, The gravity field in the Central Pacific from satellite-to-satellite tracking, *J. Geophys. Res.* **86B:**3979–3997.

Marsh, B. D., Marsh, J. G., and Williamson, R. G., 1984, On gravity from SST, geoid, from Seasat, and plate age and fracture zones in the Pacific, *J. Geophys. Res.* **89B:**6070–6078.

Matsumoto, T., and Tomoda, Y., 1983a, Numerical simulation of the initiation of subduction at the fracture zone, *J. Phys. Earth* **31:**183–194.

Matsumoto, T., and Tomoda, Y., 1983b, Numerical simulation of the mutual interaction between a trench and a seamount, *J. Phys. Earth* **31:**281–297.

Matuyama, M., 1934, Measurements of gravity over the Nippon Trench on board the I. J. Submarine Ro-57, Preliminary Report, *Proc. Imp. Acad. Jpn.* **10:**626–628.

Morgan, W. J., 1965, Gravity anomalies and convection currents, *J. Geophys. Res.* **70:**6175–6204.

Nafe, J. E., and Drake, C. L., 1957, Variation with depth in shallow and deep water marine sediments of porosity, density and the velocity of compressional and shear waves, *Geophysics* **22:**523–552.

Ockendon, J. R., and Turcotte, D. L., 1977, On the gravitational potential and field anomalies due to thin mass layers, *Geophys. J. R. Astron. Soc.* **48:**479–492.

Oldenburg, D. W., 1974, The inversion and interpretation of gravity anomalies, *Geophysics* **39:**526–536.

Parke, M. E., and Dixon, T. H., 1982, Topography relief from Seasat altimeter mean sea-surface, July 7–December 10, 1978, *Nature* **300:**317.

Parker, R. L., 1972, The rapid calculation of potential anomalies, *Geophys. J. R. Astron. Soc.* **31:**447–455.

Parker, R. L., and Oldenburg, D. W., 1973, Thermal model of ocean ridges, *Nature Phys. Sci.* **242:**137–139.

Parsons, B., and McKenzie, D. P., 1978, Mantle convection and the thermal structure of the plates, *J. Geophys. Res.* **83:**4485–4496.

Parsons, B., and Sclater. J. G., 1977, An analysis of the variation of ocean floor bathymetry and heat flow with age, *J. Geophys. Res.* **82:**803–827.

Rapp, R. H., 1979, Geos 3 data processing for the recovery of geoid undulations and gravity anomalies, *J. Geophys. Res.* **84B:**3784–3792.

Rapp, R. H., 1981, The earth's gravity field to degree and order 180 using Seasat altimeter data, terrestrial data, and other data, Rep. 322, Dep. Geod. Sci., Ohio State Univ., Columbus.

Rapp, R. H., 1983a, The determination of geoid undulations and gravity anomalies from Seasat altimeter data, *J. Geophys. Res.* **88**:1552–1562.

Rapp, R. H., 1983b, The development of the January 1983 1° × 1° mean free-air anomaly data tape, Internal report, Dep. Geod. Sci., Ohio State Univ., Columbus.

Reigber, C., Balmino, G., Moynot, B., and Müller, H., 1983a, The GRIM3 earth gravity field model, *Manuscr. Geod.* **8**:93–138.

Reigber, C., Balmino, G., Moynot, B., Müller, H., Rizos, C., and Bosch, W., 1983b, An improved GRIM3 earth gravity model (GRIM3B), Paper presented at XVIII IUGG–IAG General Assembly, Hamburg.

Reigber, C., Balmino, G., Müller, H., Bosch, W., and Moynot, B., 1985, GRIM gravity model improvement using LAGEOS (GRIM3-L1), *J. Geophys. Res.* **90B**:9285–9299.

Rowlands, D., 1981, The adjustment of SEASAT altimeter data on a global basis for geoid and sea surface height determinations, Dep. Geod. Sci. Rep. No. 325, Ohio State Univ., Columbus.

Runcorn, S. K., 1967, Flow in the mantle inferred from the low degree harmonics of the geopotential, *Geophys. J.* **14**:375–384.

Sager, W. W., 1980, Mariana arc structure inferred from gravity and seismic data, *J. Geophys. Res.* **85B**:5382–5388.

Sandwell, D. T., 1984, A detailed view of the South Pacific geoid from satellite altimetry, *J. Geophys. Res.* **89B**:1089–1104.

Sandwell, D. T., and Schubert, G., 1980, Geoid height versus age for symmetric spreading ridges, *J. Geophys. Res.* **85B**:7235–7241.

Sandwell, D. T., and Schubert, G., 1982, Geoid height–age relation from SEASAT altimeter profiles across the Mendocino fracture zone, *J. Geophys. Res.* **87B**:3949–3958.

Schulze, R., 1962, Automation of the sea gravimeter Gss2, *J. Geophys. Res.* **67**:3397–3401.

Sclater, J. G., Jaupart, C., and Galson, D., 1980, The heat flow through oceanic and continental crust and the heat loss of the Earth, *Rev. Geophys. Space Phys.* **18**:269–311.

Segawa, J., and Matsumoto, T., 1987, Free air gravity anomaly of the world ocean as derived from satellite altimeter data, *Bull. Ocean Res. Inst., Univ. Tokyo* **25**.

Segawa, J., Kaminuma, K., and Kasuga, T., 1981, A new surface ship gravity meter "NIPRORI-1" with a servo accelerometer, *J. Geod. Soc. Jpn.* **27**:102–130.

Stanley, H. R., 1979, The GEOS 3 project, *J. Geophys. Res.* **84**:3779–3783.

Stoffa, P. L., Buhl, P., Herro, T. J., Kan, T. K., and Ludwig, W. J., 1980, Mantle reflections beneath the crustal zone of the East Pacific Rise from multichannel seismic data, *Mar. Geol.* **35**:83–97.

Talwani, M., 1970, Gravity, in: *The Sea* (A. E. Maxwell, ed.), Vol. 4, Part I, Wiley, New York, pp. 251–297.

Talwani, M., Le Pichon, X., and Ewing, M., 1965, Crustal structure of the mid-ocean ridges. 2. Computed model from gravity and seismic refraction data, *J. Geophys. Res.* **70**:341–352.

Talwani, M., Dorman, J., Worzel, J. L., and Bryan, G. M., 1966, Navigation at sea by satellite, *J. Geophys. Res.* **71**:5891–5902.

Tomoda, Y., 1973a, *Maps of Free-air and Bouguer Gravity Anomalies in and around Japan*, University of Tokyo Press, Tokyo.

Tomoda, Y., 1973b, Gravity anomalies in the Pacific Ocean, in: *The Western Pacific: Island Arcs, Marginal Seas, Geochemistry* (P. J. Coleman, ed.), University of Western Australia Press, pp. 5–20.

Tomoda, Y., 1983, Marine geodesy of the northwestern Pacific, *Mar. Geod.* **7**:11–38.

Tomoda, Y., and Fujimoto, H., 1981, Gravity anomalies in the western Pacific and their geophysical interpretation of their origin, *J. Phys. Earth* **29**:387–419.

Tomoda, Y., and Fujimoto, H., 1982, Maps of gravity anomalies and bottom topography in the western Pacific and reference book for gravity and bathymetric data, Bull. Ocean Res. Inst., Univ. Tokyo, 14.

Tomoda, Y., and Fujimoto, H., 1983, Roles of seamount, rise, and ridge in lithospheric subduction, in:

Accretion Tectonics in the Circum-Pacific Regions (M. Hashimoto and S. Uyeda, eds.), Terra
Scientific Publishing Company, Tokyo, pp. 319–331.

Tomoda, Y., and Kanamori, H., 1962, Tokyo Surface Ship Gravity Meter—1, *J. Geod. Soc. Jpn.*
7:116–145.

Tomoda, Y., Fujimoto, H., Matsumoto, T., and Kono, Y., 1982, Residual gravity anomalies of the
Hawaiian seamounts and anomalies of the thickness of lithosphere, *Zishin* 35:293–301 (in
Japanese).

Tsuboi, C., 1938, Gravity anomalies and the corresponding subterranean mass distribution, *Proc. Imp.
Acad. Jpn.* 14:170.

Turcotte, D. L., and Oxburg, E. R., 1969, Convection in a mantle with variable physical properties, *J.
Geophys. Res.* 74:1458–1474.

Uyeda, S., and Kanamori, H., 1979, Back-arc opening and the mode of subduction, *J. Geophys. Res.*
84:1049–1061.

Valliant, H. D., 1983, Field trials with the LaCoste and Romberg strait-line gravimeter, *Geophysics*
48:611–617.

Vening Meinesz, F. A., 1929, Theory and practice of pendulum observations at sea, Netherlands Geod.
Comm., Delft.

Vening Meinesz, F. A., 1932, Gravity expeditions at sea 1923–1930, Vol. 1, Netherlands Geod.
Comm., Delft.

Vening Meinesz, F. A., 1941, Gravity over the Hawaiian Archipelago and over the Madiera area:
Conclusions about the earth's crust, Proc. Kon. Ned. Akad. Wetensch.

Wagner, C. A., and McAdoo, D. C., 1986, Time variations in the earth's gravity field detectable with
Geopotential Research Mission intersatellite tracking, *J. Geophys. Res.* 91:8373–8386.

Walcott, R. I., 1970, Flexural rigidity, thickness, and viscosity of the lithosphere, *J. Geophys. Res.*
75:3941–3954.

Watts, A. B., 1975a, Gravity field of the northwest Pacific Ocean basin and its margin: Hawaii and
vicinity, Geological Society of America Map and Chart Series, No. MC-9.

Watts, A. B., 1975b, Gravity field of the northwest Pacific Ocean basin and its margin: Aleutian island
arc–trench system, Geological Society of America Map and Chart Series, No. MC-10.

Watts, A. B., 1976a, Gravity field of the northwest Pacific Ocean basin and its margin: Philippine Sea,
Geological Society of America Map and Chart Series, No. MC-12.

Watts, A. B., 1976b, Gravity and bathymetry in the central Pacific Ocean, *J. Geophys. Res.* 81:1533–
1553.

Watts, A. B., 1978, An analysis of isostasy in the world's oceans. 1. Hawaiian–Emperor seamount
chain, *J. Geophys. Res.* 83:5989–6004.

Watts, A. B., and Daly, S. F., 1981, Long wavelength gravity and topography anomalies, *Annu. Rev.
Earth Planet. Sci.* 9:415–448.

Watts, A. B., and Leeds, A. R., 1977, Gravimetric geoid in the Northwest Pacific Ocean, *Geophys. J.
R. Astron. Soc.* 50:249–277.

Watts, A. B., and Ribe, N. M., 1984, On geoid heights and flexure of the lithosphere at seamounts, *J.
Geophys. Res.* 89B:11152–11170.

Watts, A. B., and Talwani, M., 1974, Gravity anomalies of deep sea trenches and their tectonic
implications, *Geophys. J. R. Astron. Soc.* 36:57–90.

Watts, A. B., and Talwani, M., 1975, Gravity effect of downgoing lithospheric slabs beneath island
arcs, *Geol. Soc. Am. Bull.* 86:1–4.

Watts, A. B., Bodine, J. H., and Bowin, C. O., 1978a, Free-air gravity field, in: A geophysical atlas of
east and southeast Asia seas (D. E. Hayes, ed.), Geological Society of America Map and Chart
Series, No. MC-25.

Watts, A. B., Kogan, M. G., and Bodine, J. H., 1978b, Gravity field of the northwest Pacific Ocean
basin and its margin: Kuril island arc–trench system, Geological Society of America Map and Chart
Series, No. MC-27.

Watts, A. B., Bodine, J. H., and Ribe, N. M., 1980, Observation of flexure and geological evolution of
the Pacific Ocean basin, *Nature* 283:532–537.

Watts, A. B., Kogan, M. G., Mutter, J., Karner, G. D., and Davey, F. J., 1981, Free-air gravity field of the southwest Pacific Ocean, Geological Society of America Map and Chart Series, No. MC-42.

Watts, A. B., McKenzie, D. P., Parsons, B. E., and Roufosse, M., 1985, The relationship between gravity and bathymetry in the Pacific Ocean, *Geophys. J. R. Astron. Soc.* **83:**263–298.

Wells, W. C. (ed.), 1984, Spaceborne gravity gradiometers, NASA Conference Publication 2305, Science Applications, Inc., Mean, Virginia.

Wing, C. G., 1969, M.I.T. vibrating string sea gravimeter, *J. Geophys. Res.* **74:**5882–5894.

Wolff, M., 1969, Direct measurements of the earth's gravity potential using a satellite pair, *J. Geophys. Res.* **14:**5295–5300.

Woodhouse, J. H., and Dziewonski, A. M., 1984, Mapping the upper mantle: Three-dimensional modeling of earth structure by inversion of seismic waveforms, *J. Geophys. Res.* **89B:**5953–5986.

Worzel, J. L., 1965, *Pendulum Gravity Measurements at Sea, 1936–1959,* Wiley, New York.

Worzel, J. L., and Harrison, J. C., 1963, Gravity at sea, in: *The Sea* (M. N. Hill, ed.), Vol. 3, Wiley, New York, pp. 134–174.

Yoshii, T., 1972, Terrestrial heat flow and features of the upper mantle beneath the Pacific and the Sea of Japan, *J. Phys. Earth* **20:**271–285.

Yoshii, T., 1973, Upper mantle structure beneath the North Pacific and the marginal seas, *J. Phys. Earth* **21:**313–328.

Yoshii, T., Kono, Y., and Ito, K., 1976, Thickening of the oceanic lithosphere, *Geophys. Monogr. Am. Geophys. Union* **19:**423–430.

Chapter 11B

HEAT FLOW

Makoto Yamano and Seiya Uyeda
Earthquake Research Institute
University of Tokyo
Tokyo 113, Japan

I. INTRODUCTION

In the early 1950s, the first measurements of heat flow through the ocean floor were made in the Pacific Ocean (Revelle and Maxwell, 1952). Thereafter, heat flow data in the Pacific steadily accumulated (e.g., von Herzen, 1959). The large number of measurements made by von Herzen and Uyeda (1963) on the Risepac expedition greatly contributed to our understanding of heat flow distribution in the Pacific. Particularly, they showed that high heat flow prevails on the crest of the East Pacific Rise, supporting the ideas of seafloor spreading that midocean ridges are the centers of spreading. Many measurements were also made in the western Pacific and clarified the broad features of heat flow distribution in trenches and backarc basins (e.g., Uyeda and Vacquier, 1968; Watanabe *et al.*, 1970; Sclater *et al.*, 1972).

During the 1960s, the global heat flow pattern was described by Lee and Uyeda (1965) and Simmons and Horai (1968). Figure 1 is a heat flow profile across the Pacific from the East Pacific Rise to Japan (Langseth and von Herzen, 1970). A serious problem in heat flow distribution at this early stage was that the mean heat flow values in young ocean basins are much lower than the values expected from thermal models of oceanic plates (e.g., Sclater and Francheteau, 1970). This problem was settled by the proposition of hydrothermal circulation by Lister (1972) based on observations in the eastern Pacific. Now it is widely accepted that the major features of heat flow distribution in normal ocean basins can be explained by

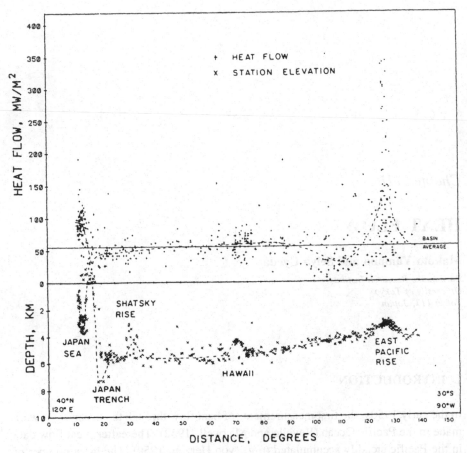

Fig. 1.   Profile of heat flow (top) and station elevation (bottom) along a great circle across the Pacific Ocean from the Japan Sea to the East Pacific Rise (Langseth and von Herzen, 1970).

thermal models of oceanic lithosphere (e.g., Sclater *et al.*, 1980). Consequently, current heat flow studies are more or less concentrated on the detailed thermal structures of midocean ridges, subduction zones, and backarc basins, although there have been some attempts to make a detailed heat flow survey in the old Pacific to check the validity of different plate cooling models (E. Davis, personal communication).

     To clarify the nature of the hydrothermal circulation, detailed heat flow surveys have been made on spreading centers in the eastern Pacific. Multiple penetration-type heat flow probes, which were developed in the 1970s, have proved to be extremely useful in making closely spaced measurements. These instruments have the ability to record temperature for many hours and transmit the data to the surface ship by acoustic signals for on-line monitoring. Thus, one can obtain multiple heat flow values by one lowering. Some of the instruments of this type can also measure the thermal conductivity of sediments *in situ* (e.g., Hyndman *et al.*, 1979). Recently, closely spaced heat flow measurements using these probes have been made in backarc basins also (e.g., the Mariana Trough and the Okinawa Trough) and furnished extremely interesting results.

Some new methods of marine heat flow measurement have been developed in the last decade. They include temperature measurement in sediments at the bottom of DSDP holes (e.g., Erickson *et al.*, 1975; Yokota *et al.*, 1980; Horai and von Herzen, 1985), development of a shallow sea heat flow probe (Matsubara *et al.*, 1982), and estimates of heat flow from seismic profiling records on gas hydrate layers in seafloor sediments (Yamano *et al.*, 1982).

In this chapter, we will show the distributions of presently available heat flow data in the Pacific area, review the possible implications of heat flow distributions and the results of recent measurements in ocean basins, midocean ridges, and subduction zones.

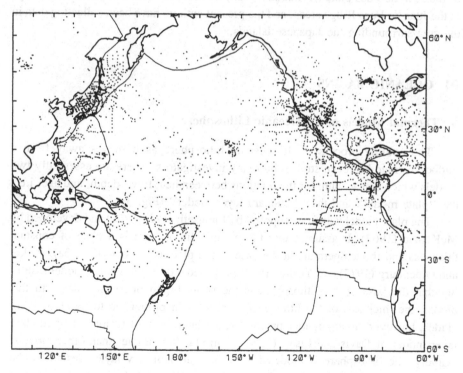

Fig. 2.   Distribution of heat flow stations as of 1982 according to the compilation by IHFC (Chapman, 1983).

II. HEAT FLOW DATA IN THE PACIFIC

Worldwide compilations of heat flow data have been made by the International Heat Flow Commission (IHFC) of the International Association of Seismology and Physics of the Earth's Interior (e.g., Jessop *et al.*, 1976). Up to 1982, they had collected over 10,000 heat flow data (Chapman, 1983). The locations of these data in and around the Pacific are plotted in Fig. 2. Although a considerable number of data, published or unpublished, were acquired subsequent to this compilation, especially surveys using multiple penetration-type devices, this map still shows the overall aspect of heat flow data distribution. Most of the Pacific area has been covered by heat flow measurements, though the density of data changes from place to place and present-day measurements tend to be concentrated on the geophysically more active regions.

In addition to the worldwide compilation, heat flow maps for smaller areas have been constructed. The geothermal maps of the east and southeast Asian seas by Anderson *et al.* (1978a) and of the southwestern Pacific by Shen (1982) are useful for they show the data quality and the sedimentary environment. The heat flow map of the Juan de Fuca plate (Hyndman. 1983) includes the results of detailed surveys in the Juan de Fuca Ridge areas, and the present authors have been collecting data in the area surrounding the Japanese Islands.

III. OCEANIC BASINS

A. Thermal Models of the Oceanic Lithosphere

The cooling process of the oceanic lithosphere as it moves away from a midocean ridge has been studied by many investigators. There are essentially two models which can account for the relations between heat flow, depth, and age, i.e., the "plate model" and the "boundary layer model" (Fig. 3).

The plate model was first proposed by Langseth *et al.* (1966) and developed by McKenzie (1967). It assumes that the oceanic lithosphere is a plate of constant thickness and the temperature at the base of the plate is constant (Fig. 3a). Parker and Oldenburg (1973) and Yoshii (1973) proposed a thickening plate model assuming that the base of the lithosphere is the solid–liquid phase boundary. In this model, the thickness of the lithosphere increases in proportion to $(age)^{1/2}$. It includes, however, an ambiguous parameter, the latent heat of fusion. The boundary layer model of Davis and Lister (1974) is similar but simpler than other models. Namely, the lithosphere is modeled as a thermal boundary layer created by the instantaneous cooling of a semi-infinite half-space (Fig. 3b). In this model, supposing that the base of the lithosphere is an isotherm, the thickness of the lithosphere

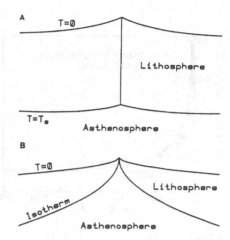

Fig. 3. Schematic diagram of thermal models for oceanic lithosphere. A, Plate model. T_s, Solidus temperature. B, Boundary layer model.

and the water depth are proportional to (age)$^{1/2}$ and surface heat flow is proportional to (age)$^{-1/2}$. Until the cooling of the lithosphere proceeds to some extent, the plate model is indistinguishable from the boundary layer model. Actually, Parsons and Sclater (1977) showed that the (age)$^{1/2}$ dependence of depth and heat flow holds in the plate model as well for sufficiently young ocean floor. For larger ages, both heat flow and depth were shown to approach constant asymptotic values (Fig. 4).

These models could be tested by the observed heat flow and bathymetry. Sclater and Francheteau (1970) examined heat flow data in the north Pacific and south Atlantic and demonstrated that heat flow decreases systematically with age. The heat flow values in young ocean basins were significantly lower than the values expected from the plate model, while the topography was consistent with the model. This discrepancy was explained by Lister (1972) as resulting from the hydrothermal circulations in the oceanic crust. Then Sclater *et al.* (1976) showed that the "reliable" heat flow values, measured in well-sedimented areas where water circulation was sealed, were close to the values predicted by the thermal models. However, the difference between the heat flow values predicted by the plate model and the boundary layer model is so small that it is difficult to distinguish between the two models by the existing heat flow data (Fig. 4).

The mean depth of the north Pacific and north Atlantic departs from the (age)$^{1/2}$ law around 70 m.y.b.p. in agreement with the plate model (Parsons and Sclater, 1977). As a plate of constant thickness near the spreading center is physically unrealistic, Parsons and McKenzie (1978) proposed a revised model. In this model, the thickness of the lithosphere increases while it is young, but small-scale convection develops at the lower boundary beyond some age and this convection supplies the excess heat flux to keep the temperature at the base of the lithosphere constant. Davies (1980) criticized that even this model is physically unrealistic. On

A

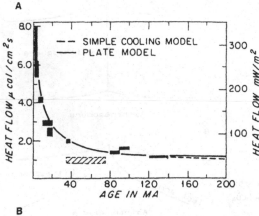

B

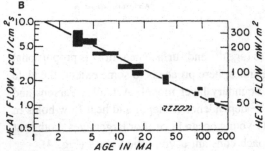

Fig. 4. Reliable heat flow plotted against age for the well-sedimented areas in the north Pacific and the north Atlantic (Sclater *et al.*, 1980). Solid curve is the heat flow versus age relation from the plate model, and dashed curve that from the boundary layer model. The hatched rectangle represents an anomalously low heat flow area in the equatorial Pacific. A, Linear scale. B, Logarithmic scale.

the other hand, Heestand and Crough (1981) showed that depth in the north Atlantic increases in proportion to (age)$^{1/2}$ beyond 150 m.y.b.p. in the areas far from hot-spot traces where reheating of the lithosphere may have occurred. Davis *et al.* (1984) carried out precise heat flow measurements in the areas of the northwest Atlantic with ages of 140 to 160 m.y. Their results showed that the heat flow is higher and the depth is shallower than expected from both of the thermal models, though there is a possibility that the discrepancy from the models is attributed to the effect of hot spots. Similar results were obtained off the Bermuda Rise in the northwest Atlantic (Detrick *et al.*, 1986) and in the northwest Pacific basin older than 150 m.y. (E. Davis, personal communication). To solve these problems, it seems necessary to make more heat flow measurements in old ocean basins.

These results indicate that it is still uncertain which model best describes the thermal structure of oceanic lithosphere. In any case, heat flow, Q, can be represented by the relation,

$$Q(t) = Ct^{-1/2} \tag{1}$$

where t is the age and C is a constant, at least up to about 120 m.y. (Fig. 4). The value of C is 473 or 502 mW m^{-2} (m.y.)$^{1/2}$ according to Parsons and Sclater (1977) and Lister (1977), respectively. We will use these constants and equation (1) for calculating the theoretical heat flow values in the following.

B. Hawaiian Swell

There are anomalously shallow areas called swells or rises around many hot-spot traces. The most representative case is the Hawaiian Swell, elongated along the younger part of the Hawaiian–Emperor seamount chain. These swells are thought to be caused by reheating of the lithosphere by hot spots, resulting in thinning of the lithosphere (Detrick and Crough, 1978; Crough, 1978; Menard and McNutt, 1982). The main basis for this interpretation is the fact that each swell subsides at a rate expected for ocean basins of the age corresponding to the depth of the swell, not the age of the surrounding seafloor.

If hot-spot swells are due to reheating of the lithosphere, anomalously high heat flow may be observed there. von Herzen et al. (1982) made a detailed heat flow survey along the Hawaiian Swell and calculated heat flow anomaly defined as the difference between the observed value and the theoretical value for the crustal age (Fig. 5). The anomalous heat flow is near zero around Hawaii and gradually increases to about 12 mW m^{-2} near Midway. This observation can be explained as a result of reheating of the lower lithosphere, because it takes a certain time for the temperature anomaly to be conducted up to the surface. It is consistent with the gravity and geoid anomalies requiring the existence of a low-density portion produced by the reheating of the lower lithosphere. They also made a numerical simulation of heat flow and bathymetry anomalies, assuming that the temperature below a depth L increased to the asthenospheric temperature owing to reheating. As

Fig. 5. Seafloor subsidence and anomalous heat flow on the Hawaiian Swell as a function of time since reheating the base of the lithosphere (Detrick et al., 1981). (A) Dots are observed depth, and solid curves are theoretical subsidence predicted from reheating model where L is the lithospheric thickness after reheating. (B) Observed anomalous heat flow compared with predicted anomalous heat flow for $L = 37$ km (upper solid curve) and $L = 45$ km (lower solid curve) when lithosphere with an age of 90 m.y. was reheated. Dashed curve shows heat flow anomaly for reheating of lithosphere with an age of 80 m.y.

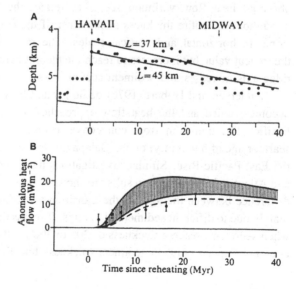

shown in Fig. 5, heating of the lithosphere below 40 to 50 km can explain both the heat flow and bathymetry profiles.

IV. MIDOCEAN RIDGES

A. Hydrothermal Circulation

As mentioned above, a problem in the theory of plate tectonics is that heat flow values around midocean ridges are highly scattered and the average is significantly lower than the values predicted by thermal models of oceanic lithosphere. Lister (1972) suggested that such variable and apparently low heat flow is caused by hydrothermal circulation in the oceanic crust, based on observations on the East Pacific Rise and Juan de Fuca Ridge areas. Since the oceanic crust formed at the spreading center has many cracks and, therefore, high permeability, convection of high-temperature water occurs. Most of the heat is convectively carried out into the seawater, so that the conductive heat flow measured by usual means should be lowered. The observed heat flow should be high in the areas of upwelling and low in the areas of downwelling, causing the large scatter in heat flow data. As the sediment cover with much lower permeability thickens, the circulation becomes sealed and the average heat flow approaches the theoretical value. While convection continues within the basement, spatial variation of heat flow remains high.

The existence of such a hydrothermal circulation was clearly shown by Williams et al. (1974) on the Galapagos Spreading Center. They made a detailed heat flow profile perpendicular to the ridge axis and found that the heat flow oscillates with a wavelength of about 6 km (Fig. 6). The spatial scale of the observed heat flow variation seems to reflect the pattern of water circulation throughout the entire thickness of the crust, if the convective cells are equidimensional in horizontal and vertical scales. The mean heat flow is lower than the theoretical value but increases to reach the theoretical value as the distance from the ridge axis increases and sediment accumulates.

Anderson and Hobart (1976) demonstrated how the hydrothermal circulation becomes sealed and the heat flow approaches the predicted value in the eastern Pacific. The transition from convective to conductive heat transfer occurs at a seafloor age of 5 to 6 m.y. on the Galapagos Spreading Center and 10 to 15 m.y. on the East Pacific Rise. Similar investigations have been made in the Indian Ocean and Atlantic Ocean and the results are shown in Fig. 7 (Anderson et al., 1977). The difference between the ages of the transition for different oceans is believed to be mainly due to different sedimentation rates. It seems that circulation becomes sealed when sediments reach a thickness of 200 to 300 m, though the condition of sealing does not depend only on sediment thickness but also on the permeability of the

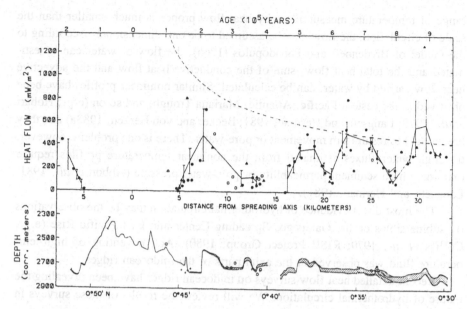

Fig. 6. Heat flow, topography, and sediment thickness (stippled area) profiles across the Galapagos Spreading Center around 86°W (Williams *et al.*, 1974). Dashed line is the theoretical heat flow and solid lines connect running averages of the measured heat flow values over 2-km intervals every 1 km along a profile.

oceanic crust and sediments and the wave numbers of convection (Anderson and Skilbeck, 1981).

When heat flow measurements are made in an area where hydrothermal circulation is occurring, it is expected that nonlinear temperature profiles in the sediments are observed together with an oscillatory pattern of heat flow variation. Anderson *et al.* (1979) obtained many nonlinear profiles in the Indian Ocean and interpreted them as caused by water flow through the sediments. Because the depth

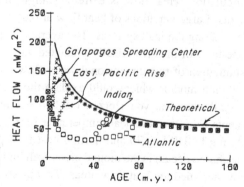

Fig. 7. Observed mean heat flow plotted versus age for several midocean ridges (Anderson *et al.*, 1977). Solid curve represents the theoretical heat flow.

range of temperature measurement by heat flow probes is much smaller than the scale of circulation, the flow may be assumed to be one-dimensional. According to the model of Bredehoeft and Papadopulos (1965), the flow of water can be estimated and the total heat flow, sum of the conductive heat flow and the advective heat flow carried by water, can be calculated. Similar nonlinear profiles have been observed in the eastern Pacific, Atlantic, Mariana Trough, and so on (e.g., Hobart et al., 1979; Langseth and Herman, 1981; Becker and von Herzen, 1983a) and they are thought to result from movement or pore-water. There is one problem, however: the high water fluxes estimated from the nonlinear temperature profiles require extremely high sediment permeability or pore-water pressure (Abbott et al., 1981; Langseth and Herman, 1981).

The most direct evidence of hydrothermal circulation may be the observations by submersibles on the Galapagos Spreading Center and East Pacific Rise (e.g., Corliss et al., 1979; RISE Project Group, 1980). Active venting of high-temperature fluid was observed in the axial parts of the midocean ridges.

Recent detailed heat flow surveys on midocean ridges have been revealing the nature of hydrothermal circulation. We will review the results of these surveys in the eastern Pacific.

B. Galapagos Spreading Center

Active thermal vents were found on the axial ridge of the Galapagos Spreading Center around 86°W by deep submersible Alvin (Corliss et al., 1979). The maximum temperatures of rising fluid were 7 to 17°C, lower than those of the vents found on the East Pacific Rise.

Green et al. (1981) conducted a detailed heat flow survey on the flank just south of the ridge axis where hydrothermal vents were observed. They obtained a two-dimensional heat flow distribution in an area about 30 km long and 20 km wide (Fig. 8). The pattern of heat flow is lineated parallel to the ridge axis with a wavelength of 5 to 15 km and appears to represent the pattern of hydrothermal circulation. Heat flow is extremely high, higher than 400 mW m^{-2}, in mound areas. Large variations of heat flow at fault scarps indicate a concentration of water flow along the fault systems. The average conductive heat flow in this area is about one-third of the theoretically predicted value. Fehn et al. (1983) made a numerical simulation of two-dimensional water flow through porous media and obtained circulation models which could explain the heat flow profile normal to the ridge.

Becker and von Herzen (1983a) estimated vertical advection through sediments from the nonlinear temperature profiles obtained by downhole measurement during DSDP Leg 70 and normal heat flow probes. Results indicate that localized discharge occurs in the mound areas with high heat flow and recharge prevails in the low and medium heat flow areas. This pattern is consistent with the inference from

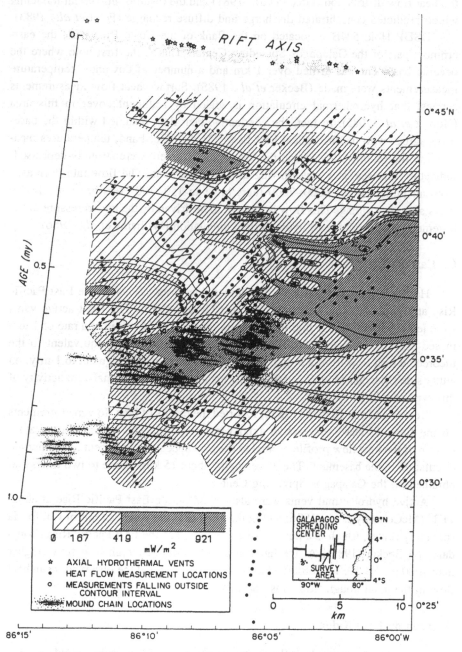

Fig. 8. Heat flow distribution on the south flank of the Galapagos Spreading Center (Green *et al.*, 1981). Contour values are in HFU (10^{-6} cal cm^{-2} sec^{-1}).

the heat flow distribution (Green *et al.*, 1981) and the result of numerical modeling which predicted concentrated discharge and diffuse recharge (Fehn *et al.*, 1983).

DSDP Hole 504B is located on the flank of the Costa Rica Rift, the easternmost part of the Galapagos Spreading Center. This is the first hole where the oceanic basement was drilled over 1 km and a number of downhole temperature measurements were made (Becker *et al.*, 1985). Surface heat flow measurements suggest that hydrothermal circulation is sealed by sediment cover in this area (Hobart *et al.*, 1985). In fact, the conductive heat flow obtained within the basement agrees with the theoretical heat flow. On the other hand, temperatures measured in the sediment layers and the top of the basement were strongly depressed, indicating that seawater was flowing into the basement. The flow rate decreased from an initial rate of 6000–7000 liters hr^{-1} to 150–200 liters hr^{-1} nearly 3½ years after the drilling. This flow is thought to be driven by the underpressure of the basement pore fluid due to hydrothermal circulation (Williams *et al.*, 1986).

C. East Pacific Rise

Hot springs were discovered by submersible observations on the East Pacific Rise at around 21°N as well (RISE Project Group, 1980). The most active vents were jetting out extremely high-temperature, at least 350°C, fluid at a rate of 1 to 5 m sec^{-1}. The estimated heat flux from such a single vent is equivalent to the theoretical heat flow from a 3- to 6-km segment of midocean ridge out to 1 m.y. on either side (Macdonald *et al.*, 1980). This means that hydrothermal vent activity of this type should be episodic.

Becker and von Herzen (1983b) reported the results of heat flow measurements on the flank of these very active axial vents (Fig. 9). The observed oscillatory pattern of the heat flow profile normal to the axis appears to represent hydrothermal circulation in the basement. The wavelength, about 15 km, is one to two times that obtained on the Galapagos Spreading Center.

Active hydrothermal vents were also found on the East Pacific Rise at about 11°N (McConachy *et al.*, 1986) and in the axial rift valley of the Gulf of California spreading center, which is a northward continuation of the East Pacific Rise (Lonsdale and Becker, 1985). At the latter site, extraordinarily high heat flow, higher than 2000 mW m^{-2}, was observed near the hot springs and the observed mean heat flow in the axial trough floor was 660 mW m^{-2}.

D. Juan de Fuca Ridge

Many closely spaced heat flow measurements have been made around the Juan de Fuca Ridge and Explorer Ridge (e.g., Davis and Lister, 1977; Davis *et al.*, 1980). The obtained heat flow is lower than the theoretical value and shows oscillatory pattern probably due to hydrothermal circulation.

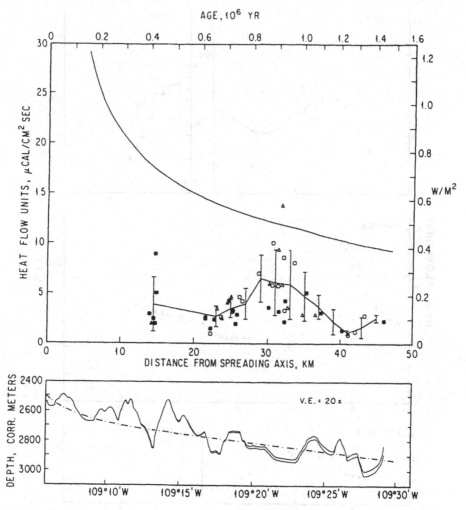

Fig. 9. Heat flow and bathymetry projected onto a line normal to the axis of the East Pacific Rise around 21°N (Becker and von Herzen, 1983b). Running averages of 4-km intervals are plotted every 2 km. Solid curve represents the theoretical heat flow.

Davis *et al.* (1980) found two scales of heat flow variations, i.e., 10 km to 20 km and 1 km to a few kilometers (Fig. 10). They supposed that large-scale variations reflect horizontal heat transport in the basement through water recharge at outcrops such as seamounts. The amplitude of small-scale variations decreases with increasing crustal age. It may be caused by thickening of the sediment cover.

A detailed survey in the Middle Valley of the northern Juan de Fuca Ridge revealed that heat flow is especially high, higher than 1000 mW m^{-2} on a mound, which is believed to be of hydrothermal origin (Davis *et al.*, 1987). In the Escanaba

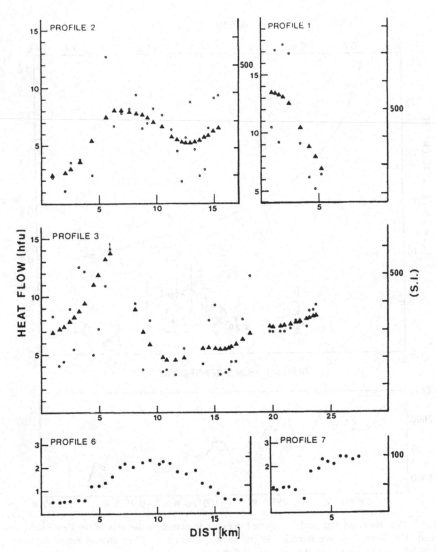

Fig. 10.   Heat flow profiles in the Juan de Fuca Ridge area (Davis *et al.*, 1980). Profiles 1–3 were taken over young crust with thin sediment cover, while profiles 6 and 7 were taken over older and more heavily sedimented crust. Open and closed circles are observed values. Triangles represent profiles smoothed with a Gaussian filter of 1 km width, which demonstrate large-scale variations in heat flow.

Trough of the Gorda Ridge, measurements of maximum heat flow exceeding 1200 mW m^{-2} suggest the existence of hydrothermal vent activity (Abbott *et al.*, 1986).

Direct observations using submersibles found active hydrothermal vents on a seamount at the axis of the central Juan de Fuca Ridge (Canadian American Seamount Expedition, 1985) and in the axial valley of the southern Juan de Fuca Ridge (U.S. Geological Survey Juan de Fuca Study Group, 1986).

V. SUBDUCTION ZONE

A. Heat Flow Distribution

There are many subduction zones and marginal basins on the margins of the Pacific, where a large number of heat flow measurements have been made. The Japanese Islands and the surrounding area are one of the best investigated subduc-

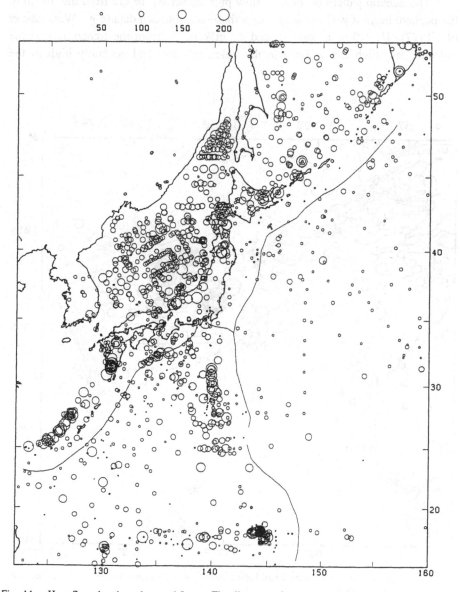

Fig. 11. Heat flow data in and around Japan. The diameter of each circle is proportional to the heat flow value except for values higher than 200 mW m^{-2} or lower than 20 mW m^{-2}.

tion zones. Figure 11 is a heat flow map in and around Japan compiled by the authors, updating and revising the compilation by Yoshii (1979). The diameter of each circle is proportional to the heat flow value except for values higher than 200 mW m^{-2} and lower than 20 mW m^{-2}. As another example of heat flow distribution in marginal basins, heat flow values in the southwestern Pacific are shown in Fig. 12. The data sources are the compilations by IHFC (Chapman, 1983) and Halunen (1979).

The general pattern of the heat flow profiles across the arc from the trench to the backarc basin is well known through these accumulated data (e.g., Watanabe *et al.*, 1977). Heat flow is low, around 40 mW m^{-2}, from the trench axis to the volcanic arc, high and variable in the volcanic zone, and generally high in the

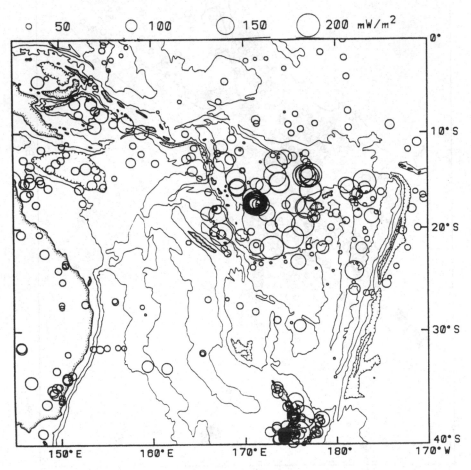

Fig. 12.   Heat flow data in the southwestern Pacific area shown in the same way as in Fig. 11. Intervals of depth contours are 2000 m.

backarc basin. The modes of subduction, however, differ subduction zone by sub-duction zone (e.g., Uyeda, 1982). For example, Uyeda and Kanamori (1979) classified the subduction zones into Mariana type and Chilean type based on whether or not backarc basins are actively spreading and showed that differences in major characteristic features such as seismicity, stress state, volcanism, crustal movement among different subduction zones are concordant with such a classifica-tion. The detailed thermal structures of individual subduction zones should also depend on various factors such as the rate and angle of subduction, the ages of subducting plate and backarc basin, and the flows in the asthenosphere. Surface heat flow distribution may reflect the differences in such factors. Naturally, the thermal state depends not only on the present state of subduction but also on its history.

It is first necessary to know what is different and what is common in the features of the heat flow distribution of subduction zones. We have plotted heat flow versus distance from the trench axis in several arcs on the margins of the Pacific where a reasonable amount of data has been obtained (Fig. 13). The general features stated above (Watanabe *et al.*, 1977) can be seen with the exception of high heat flow in the Nankai Trough (Southwest Japan arc) and around the trench axis off Oregon.

In the following, we will discuss the heat flow in backarc basins and trench–arc gap zones.

B. Heat Flow versus Age in Backarc Basins

The major problem of heat flow in backarc basins is whether or not the relation between heat flow and age of the basin is different from that for ocean basins. Two questions may be concerned with this problem: (1) Is the process of backarc spread-ing basically the same as that of spreading of midocean ridges? (2) Does subduction of the slab underneath the backarc basin affect the thermal evolution of the backarc basin? If backarc basins were formed through spreading similar to midocean ridges, the heat flow versus age relation might be similar. Even in such a case, however, the existence of convective flow beneath the backarc basin induced by the subduct-ing slab could make the thermal evolution different from that of oceanic lithosphere.

The heat flow versus age relationship for backarc basins was examined in detail by Watanabe *et al.* (1977). They found that heat flow in backarc basins is higher than the theoretical value for the oceans and proposed a two-stage cooling model. After their study, the ages of many marginal basins were determined to a higher accuracy. Sclater *et al.* (1980) showed that heat flow in six marginal basins with reliable heat flow values and well-determined ages is close to the prediction from the thermal models for normal oceanic lithosphere. Anderson (1980) also concluded that the heat flow versus age relation for oceans holds for several margin-al seas of southeast Asia.

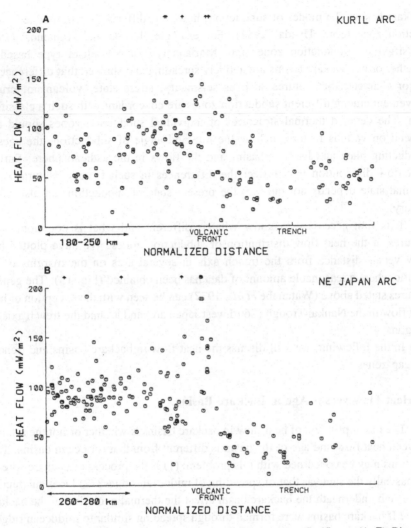

Fig. 13. Heat flow versus distance across subduction zones in the circum-Pacific area (A–E). Distance is normalized to volcanic front–trench distance except in the Southwest Japan arc (D). Small arrows at the top of each profile represent values higher than 210 mW m^{-2}.

Here we present a plot of heat flow versus age of marginal basins adding newly obtained heat flow and age data to the former compilations (Fig. 14). The data sources for age and heat flow are listed in Table I. It should be noted that the plot includes the basins which apparently were not formed by backarc spreading and/or which are not located adjacent to the active subduction zones. For example, the Coral Sea (CO) was believed to have been generated by branches of midocean ridges (Weissel and Watts, 1979) and the Aleutian Basin (AL) is a trapped part of the Kula plate (Cooper et al., 1976). The West Philippine Basin (WP) may also be a

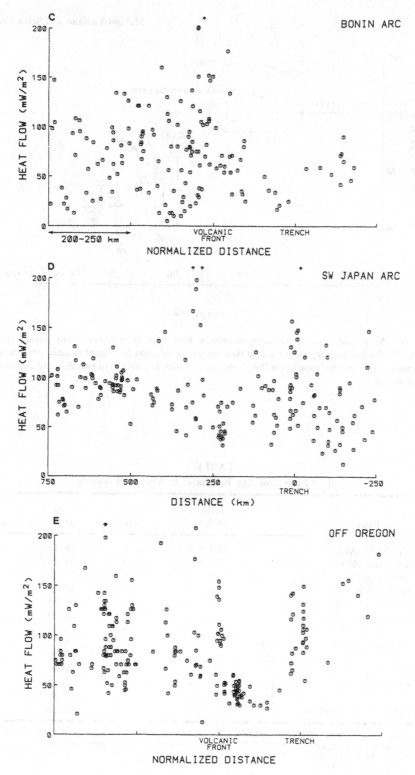

Fig. 13. (continued)

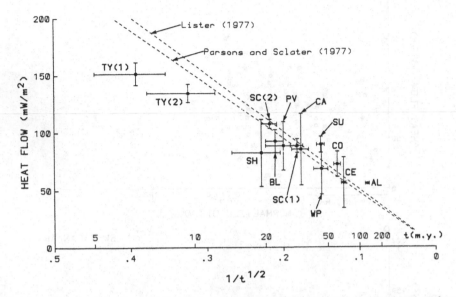

Fig. 14. Mean and standard deviation of reliable heat flow in marginal basins plotted against $(age)^{-1/2}$. Dashed lines represent the heat flow versus age relations for normal oceanic lithosphere by Parsons and Sclater (1977) and Lister (1977). Names of basins and sources of heat flow and age data are listed in Table I.

TABLE I
Heat Flow and Age Estimates in Marginal Basins

Basin		Heat flow (mW m^{-2})	Ref.[a]	Age (m.y.)	Ref.[a]
Aleutian Basin (AL)		55.2 ± 0.4	1	117–132	2
Balearic Basin (BL)		92 ± 10	3	20–25	3
Caroline Basin (CA)		85 ± 31	4	28–36	5
Celebes Sea (CE)		56 ± 22	4	65–72	6
Coral Sea Basin (CO)		72 ± 11	7, 8	56–64	9
Parece Vela Basin (PV)		88 ± 21	10	20–30	10
Shikoku Basin (SH)		82 ± 29	4	14–24	11
South China Sea (SC)	(1)	88 ± 6	12	27.5–33	12
	(2)	107 ± 4	12	19–23	12
Sulu Sea (SU)		89 ± 7	4	41–47	6
Tyrrhenian Sea (TY)	(1)	134 ± 8	3	7–12	3
	(2)	151 ± 10	3	5–8	3
West Philippine Basin (WP)		68 ± 22	4	39–50	13

[a]1, Langseth et al. (1980); 2, Cooper et al. (1976); 3, Hutchison et al. (1985); 4, Anderson et al. (1978a); 5, Weissel and Anderson (1978); 6, Lee and McCabe (1986); 7, Shen (1982); 8, M. G. Langseth (personal communication); 9, Weissel and Watts (1979); 10, Mrozowski and Hayes (1979); 11, Shih (1980); 12, Taylor and Hayes (1983); 13, Mrozowski et al. (1982).

trapped part of the Pacific Ocean (Uyeda and Ben-Avraham, 1972; Hilde and Lee, 1984).

It seems that the heat flow versus age relation for marginal basins does not significantly differ from the relation for oceanic basins, though there are some exceptions. The mean heat flow seems to be lower than the theoretical value in the Balearic Basin (BL), Tyrrhenian Basin (TY) and Shikoku Basin (SH), and higher in the Aleutian Basin. In the case of the Balearic Basin and Tyrrhenian Basin, heat flow values would be increased through correction for the sedimentation effect and may become consistent with the thermal models within the range of errors (Hutchison et al., 1985). The scatter of heat flow data in the Shikoku Basin is very high, suggesting that some advective heat transfer by water flow is operative. The high heat flow in the Aleutian Basin might be explained as a result of reheating by subduction of the Kula–Pacific ridge (Langseth et al., 1980). Another possible exception is the Japan Sea, which is not shown in Fig. 14 as its age has not been well determined. Heat flow in the Japan Sea is rather uniform and the average of the reliable heat flow is 95 and 97 mW m^{-2} in the Japan Basin and Yamato Basin, respectively (Tamaki, 1986). If the Japan Sea opened about 15 m.y.b.p. as suggested by the rotation of southwest Japan (e.g., Otofuji and Matsuda, 1984), heat flow is lower than expected even after correction for the effect of sedimentation. There remains, however, a large uncertainty in the age of the Japan Sea. If we assume that the heat flow versus age relation for backarc basins is similar to that for oceans, the age of the Japan Sea would be about 25 m.y.

Therefore, at the present stage, we do not have strong evidence that the heat flow versus age relation for oceanic lithosphere is not applicable to marginal basins. This means that subduction and induced convection in the mantle wedge may not have much influence upon the thermal evolution of the marginal basin lithosphere.

Heat flow in the very young marginal basins, such as the Mariana Trough, is highly scattered and the mean is lower than the theoretical value. This is most probably due to hydrothermal circulation in the crust as in young oceanic basins (Watanabe et al., 1977). The results of heat flow studies in two well-investigated young backarc basins, the Mariana Trough and Okinawa Trough, will be given below.

C. Mariana Trough

The Mariana Trough is one of the actively spreading backarc basins. Karig (1971) first suggested that the Mariana Trough was formed by recent backarc spreading based on geological information. An analysis of magnetic anomalies and the drilling on DSDP Leg 60 revealed that the spreading began around 6 m.y.b.p. and has continued to the present (Bibee et al., 1980; Hussong and Uyeda, 1981).

Hydrothermal circulation is expected to be occurring in this young and active backarc basin. Actually, Anderson (1975) found that heat flow is high, 181 and 354 mW m^{-2}, in the axial part of the trough and low heat flow, lower than 40 mW m^{-2}, prevails in other areas. The result was thought to indicate high magmatic activity on the axis and convective heat transfer by water circulation. Uyeda and Horai (1981) measured heat flow on DSDP Leg 60 using a new temperature probe attached to the nose of the water sampler (Yokota et al., 1980). They obtained nearly zero heat flow at one site and variable heat flow within a small area at another site. It also appears to be due to hydrothermal activity.

After DSDP Leg 60, heat flow measurements by the multiple penetration-type probe were made and extremely high heat flow was observed on about 3-m.y.-old crust. The high heat flow, up to about 2000 mW m^{-2}, was concentrated around small (about 20 m high) mounds (Hobart et al., 1983). In other places, most of the measurements were lower than the theoretical values (Fig. 15). These observations, as well as the large variation of heat flow and nonlinear temperature gradients, assured the existence of hydrothermal circulation. Thus, the thermal state of the Mariana Trough is very similar to that of the midocean spreading centers. Actually, Alvin dives at the axial part of the Trough at around 18°N found "clear smokers" and the maximum temperature of the vent fluid was 285°C (Craig et al., 1987).

D. Okinawa Trough

The Okinawa Trough is also thought to be formed by recent backarc extensional tectonics. In the southwestern part of the trough, spreading appears to have started during Pliocene time (e.g., Lee et al., 1980; Letouzey and Kimura, 1985). The oceanic crust may have developed along the trough axis. The middle part of the trough has apparently been under extension, but seems to be still in the stage of rifting, for the crustal structure is continental rather than oceanic (Nagumo et al., 1986) and the sampled volcanic rocks are island arc type (Kimura et al., 1986).

Heat flow in the southwestern Okinawa Trough is quite variable, from 9 mW m^{-2} to 235 mW m^{-2} (Lu et al., 1981; Yamano et al., 1988). This is possibly caused by advective heat transfer. However, the Okinawa Trough has very thick sediment layers, thus water may be flowing through outcrops and/or faults, not sediment cover.

Figure 16 shows heat flow distribution in the middle Okinawa Trough. Heat flow is less variable than in the southwestern part and rather high. Especially, there is a high heat flow area around 126°E as suggested by Vander Zouwen (1984). Point A in Fig. 16 represents the location of a small depression called the Natsushima-84 Deep, around 3 km long and 1 km wide. It lies in the central rift region of the trough and is surrounded by seamounts. Recently, closely spaced heat flow measurements were made in this depression and very high and variable heat flows (maximum value about 1600 mW m^{-2}) were obtained (Yamano et al., 1986a,b). The mean and standard deviation are about 600 and 400 mW m^{-2}, respectively. Some of the temperature profiles are nonlinear, which implies the existence of hydrothermal

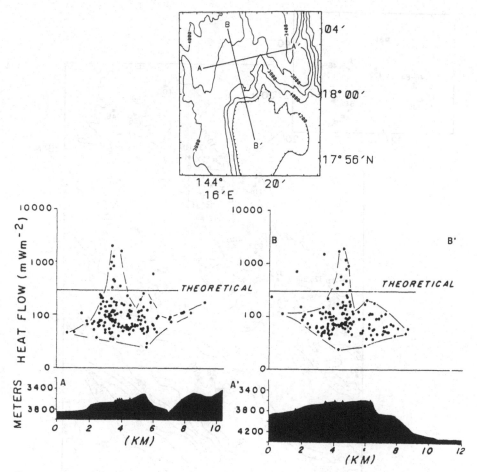

Fig. 15. Heat flow and topography profiles along the lines A–A' and B–B' on the floor with an age of
3 m.y. in the Mariana Trough. Theoretical heat flow is estimated from the heat flow versus age relation
for oceanic lithosphere. Note that heat flow is in a logarithmic scale.

circulation. A hydrothermal mound with shimmering water was discovered on a
seamount directly south of the Deep by a submersible (Kimura *et al.*, 1988).

Measurements were also made in the vicinity of the depression (Yamano *et al.*,
1986b, 1988). Heat flow is projected along a north–south line (line X–Y in Fig. 16)
passing through the Natsushima-84 Deep in Fig. 17. The heat flow anomaly appears
to be concentrated in the central rift region. This anomaly may be caused by magma
reservoirs beneath the depression, since an anomalous reflector found on the multichannel seismic reflection record and peculiar phases of seismic waves detected by
OBSs indicate the existence of magma reservoirs in this region (Nagumo *et al.*,
1986). The young age of the sampled rocks, 0.2 to 0.3 m.y. or younger, also
suggests active volcanism near the depression (Kimura *et al.*, 1986).

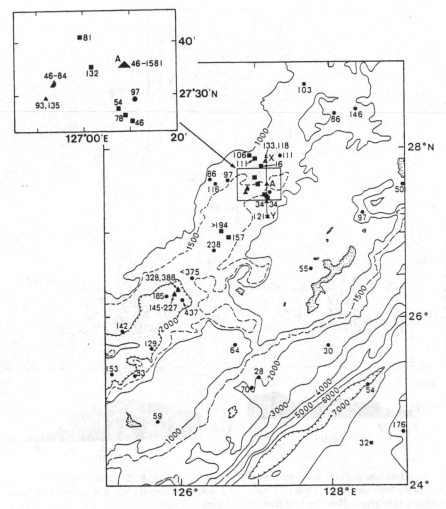

Fig. 16.   Heat flow data in the middle Okinawa Trough area (in mW m⁻²). Circles represent existing data, triangles are from Yamano *et al*. (1986a,b), and squares are from Yamano *et al*. (1988). Point A is the Natsushima–84 Deep.

E. Heat Flow in the Trench–Arc Gap

Heat flow is generally subnormal in the trench–arc gap as previously stated (Watanabe *et al.*, 1977). This fact can be explained as a result of heat absorption by the subducting cold oceanic lithosphere. However, the thermal structure of subduction zones should be determined by not only the cooling effect of the descending oceanic plate but also by other factors such as frictional heating along the plate boundary, convection in the mantle wedge above the subducted slab, dehydration of the oceanic crust. Although it was shown that the frictional heating between the slab

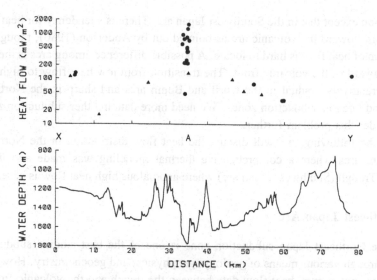

Fig. 17. Heat flow and topography profiles along the line X–Y in Fig. 16. Scale for heat flow is logarithmic. Circles represent data on the line and triangles are data in the vicinity, projected onto the line. (A) is the location of the Natsushima–84 Deep.

and the overlying mantle is not a plausible mechanism to supply heat for arc volcanism (e.g., Yuen *et al.*, 1978), it may be an important heat source in the trench–arc gap. It is widely accepted that subduction of the oceanic plate induces convection in the upper mantle. The spatial extent of this convection can control the thermal structure beneath the trench–arc gap (Honda, 1985). Dehydration of hydrous minerals in the subducted oceanic crust acts as a heat sink (e.g., Anderson *et al.*, 1978b). Accretion of sediments will affect the thermal structure in the vicinity of the trench. Wang and Shi (1984) proposed a thermal model for subduction complexes considering the flow within the sedimentary prism driven by the descending slab.

Although other geophysical and geological data are necessary to estimate the effects of these factors, detailed observation of surface heat flow is also a source of useful information. Anderson (1980) examined the heat flow profiles in the Japan Trench and Bonin Trench areas. He concluded that heat flow in the trench–arc gap is not uniformly low but increases toward the volcanic arc from a minimum about equidistant between the trench and the arc. He also maintained that heat flow in the forearc basin is normal, not low.

The heat flow profiles in Fig. 13 include newly obtained data after Anderson (1980). Heat flow is anomalously high around the trench axis in the Southwest Japan and Oregon subduction zones probably because the subducting plates are young. This point will be discussed later. The profiles between the trenches and the volcanic arcs are generally similar to each other in spite of various modes of subduction except that in the Southwest Japan arc. There is a tendency for heat flow to increase toward the volcanic arc as pointed out by Anderson (1980), though the minimum of heat flow is hard to locate. A possible difference among arcs is the heat flow seaward of the volcanic front. The transition from low heat flow to high heat flow is relatively gradual in the Kuril and Bonin arcs and sharp in the Northeast Japan and Oregon subduction zones. We need more data in other subduction zones to consider this problem further.

In the following, we will discuss the heat flow distribution in the Northeast Japan arc area where a comprehensive thermal modeling was made and in the Nankai Trough (Southwest Japan arc) where anomalous high heat flow is observed.

F. Northeast Japan Arc

The Northeast Japan subduction zone is one of the best studied subduction zones through various means of geology, geophysics, and geochemistry. However, there were not enough heat flow data between the trench and the volcanic front in the 1970s (Yoshii, 1979). Recently, heat flow data in this area have much increased through measurements using DSDP holes (Burch and Langseth, 1981), shallow holes on land (Nagao, 1987), oil/gas wells (Sekiguchi, unpublished), and normal seafloor heat flow probes (Burch and Langseth, 1981; Yamano, 1986). The results of these measurements are included in the heat flow profile in Fig. 13.

Many thermal models for this subduction zone have been proposed since Hasebe *et al.* (1970). Burch and Langseth (1981) stated that heat flow in the forearc area is almost uniformly low and consistent with the thermal model for subduction by DeLong *et al.* (1979). But this model is basically conductive and cannot explain heat flow increase around the volcanic front.

Honda (1985) conducted a more detailed numerical simulation that included the effects of the induced flow in the mantle wedge, taking account of petrological and seismological information. Figure 18 shows his model and the calculated heat flow distribution together with the observed heat flow values. His main conclusions are as follows: To explain the low heat flow around the aseismic front (Yoshii, 1975), a triangular region seaward of the aseismic front within the mantle wedge (stippled area in Fig. 18) should be rigid and not be involved in convection; to fit the heat flow values from the trench to the aseismic front, heating by shear stress of 500 to 1000 bars is required between the subducting slab and the overlying mantle seaward of the aseismic front; to reproduce the high heat flow around the volcanic front, higher than 120 mW m^{-2}, advective heat transfer by ascending magma is necessary. Open circles and triangles in Fig. 18 represent the data obtained after his simulation. Heat flow seems to increase from the aseismic front to the volcanic front as predicted by the model.

An apparent defect of this model is that it does not include radiogenic heat generation in the crust. Estimation of crustal radiogenic heat, including measure-

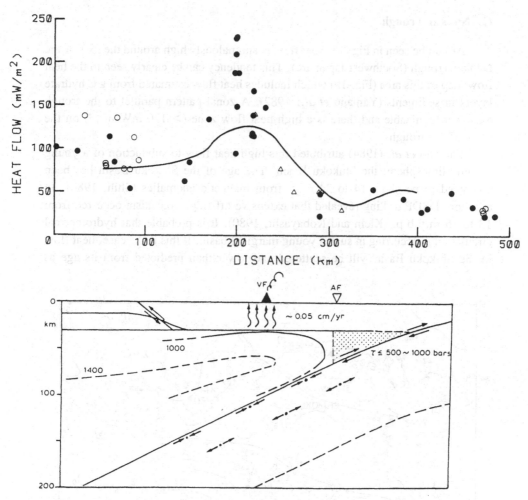

Fig. 18.   (Top) Heat flow profile (solid curve) calculated by Honda (1985) through numerical simulation of the thermal structure of the Japan Trench subduction zone. Closed circles are heat flow data used as constraints, open circles data obtained after his simulation, and triangles preliminary values obtained using shallow holes (Nagao, 1987). (Bottom) Schematic model of thermal structure of the Northeast Japan arc which corresponds to the calculated heat flow profile presented above and demonstrates isotherms in the mantle wedge, shear stress along the slab boundary, and upward migration of magma beneath the volcanic front (Honda, 1985). Stippled area represents the rigid part of the mantle wedge. Arrows indicate earthquake source mechanisms.

ments of radioisotope contents of representative crustal rocks, and research on its implications in the thermal process under the Japanese area, have just been started by the authors' group. Preliminary results indicate that the crustal heat reduces the temperature at the base of the crust by about 250°C compared to Honda's model, and the stress needed under the forearc is also reduced to 100 to 200 bars (Furukawa and Uyeda, 1986).

G. Nankai Trough

As can be seen in Fig. 13, heat flow is anomalously high around the axis of the Nankai Trough (Southwest Japan arc). This tendency can be clearly seen in the heat flow map of this area (Fig. 19) which includes heat flow estimated from gas hydrate layers in sediments (Yamano *et al.*, 1982). A zonal pattern parallel to the trough axis is recognizable and there is a high heat flow zone (> 130 mW m^{-2}) on the floor of the trough.

Yamano *et al.* (1984) attributed this high heat flow to subduction of a young oceanic lithosphere, the Shikoku Basin. The age of the Shikoku Basin has been estimated to be about 14 to 24 m.y. from magnetic anomalies (Shih, 1980). In addition, DSDP drilling revealed that extensive off-ridge volcanism occurred from 13 to 15 m.y.b.p. (Klein and Kobayashi, 1980). It is probable that hydrothermal circulation is occurring in such a young marginal basin. If this is the case, heat flow in the Shikoku Basin will be scattered and lower than predicted from its age as

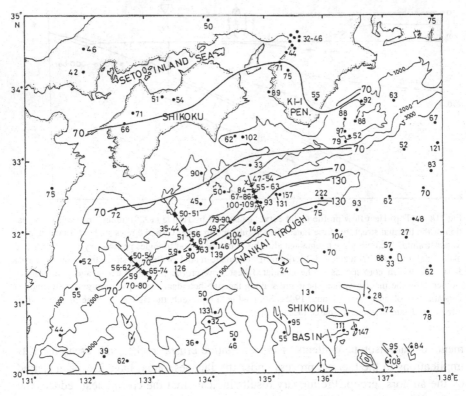

Fig. 19. Heat flow data in the Nankai Trough area (in mW m^{-2}). Stars are the values measured on DSDP Leg 87. Data along three lines are estimated from BSRs (bottom simulating reflectors) due to gas hydrates.

actually observed and the theoretical value will be observed only in the thickly sedimented Nankai Trough. The theoretical heat flow for an age of 13 to 15 m.y. is about 130 mW m^{-2}. This value is not sufficiently high to explain the observed high values in the Nankai Trough, especially when the effect of sedimentation is taken into account. Seepage of heated pore-water from the trench bottom sediments may also contribute to the high heat flow. This inference might be supported by high heat flow and fluid movement along faults found in the Barbados subduction zone (ODP Leg 110 Scientific Party, 1987). But, at this stage, it seems reasonable to assume that the high heat flow is basically due to subduction of a hot plate. Off Oregon, the age of subducting plate is about 8 m.y. and the corresponding theoretical heat flow is about 170 mW m^{-2}. The observed values are lower than those expected (Fig. 13), which may be due to the sedimentation effect.

The heat flow distribution landward of the trough is rather complicated and difficult to explain by a simple thermal model of subduction. It might be influenced by the various factors mentioned above, such as movement of sediments, dehydration, frictional heating, and time variation of the mode of subduction.

VI. SUMMARY

1. Although the data density is far from uniform, most of the Pacific Ocean, except high-latitude regions, has been covered by heat flow data and the present-day studies tend to be focused on detailed surveys of tectonically active regions, e.g., midocean ridges, subduction zones, and backarc basins.

2. Heat flow and basement depth of oceans can be explained by cooling of a plate of constant thickness or a semi-infinite half-space. In very old ocean basins, however, some discrepancies between theory and observation are found.

3. Precise heat flow measurements on the Hawaiian Swell revealed that the observed heat flow is higher than the theoretical heat flow for the basement age by up to 12 mW m^{-2}. This anomaly is probably a result of reheating of the lower lithosphere by the Hawaiian hot spot.

4. Hydrothermal circulation in the young oceanic crust was proposed as an explanation for the fact that heat flow is variable and lower than that expected from thermal models near midocean ridges. Detailed heat flow distributions obtained by closely spaced measurements on the flanks of ridges and the observed nonlinear temperature–depth profiles support the existence of such a circulation. Direct observation by submersibles discovered active hydrothermal vents and hydrothermal deposits in the axial parts of the spreading centers in the Pacific.

5. In subduction zones, heat flow is generally low seaward of the arcs and high in the volcanic zones and backarc basins. Differences in the mode of subduction among arcs, however, may cause differences in heat flow distributions. We con-

structed heat flow profiles across several subduction zones. The transition from low heat flow to high heat flow in the arc–trench gap is more gradual in the Kuril and Bonin arcs than in the Northeast Japan arc and Oregon subduction zone.

6. Reliable heat flow data in marginal basins were plotted versus age. The heat flow versus age relation for oceans seems to hold for marginal basins. It suggests that subduction of the cold slab and the induced convection beneath the basin may not significantly affect the thermal evolution of marginal basins.

7. High and variable heat flows were observed in very young backarc basins. The thermal state of the Mariana Trough seems to be similar to that of midocean ridges. In the middle Okinawa Trough, an extraordinarily high and concentrated heat flow anomaly was found in the central rift. This anomaly is probably caused by present or recent volcanic activity.

8. Heat flow distribution in the arc–trench gap of the Northeast Japan arc has been determined fairly well. It is consistent with the results of numerical simulation which took account of the effects of asthenospheric convection induced by subduction and shear heating along the slab surface.

9. There is a high heat flow zone on the floor of the Nankai Trough, higher than 130 mW m^{-2}, which is anomalous for a trench. The high heat flow may result from subduction of young and hot Shikoku Basin plate.

REFERENCES

Abbott, D., Menke, W., Hobart, M., and Anderson, R., 1981, Evidence for excess pore pressures in southwest Indian Ocean sediments, *J. Geophys. Res.* **86**:1813–1827.

Abbott, D. H., Morton, J. N., and Holmes, M. L., 1986, Heat flow measurements on a hydrothermally-active slow-spreading ridge: The Escanaba Trough, *Geophys. Res. Lett.* **13**:678–680.

Anderson, R. N., 1975, Heat flow in the Mariana marginal basin, *J. Geophys. Res.* **80**:4043–4048.

Anderson, R. N., 1980, 1980 update of heat flow in the east and southeast Asian seas, *Geophys. Monogr. Am. Geophys. Union* **23**:319–326.

Anderson, R. N., and Hobart, M. A., 1976, The relation between heat flow, sediment thickness, and age in the eastern Pacific, *J. Geophys. Res.* **81**:2968–2989.

Anderson, R. N., and Skilbeck, J. N., 1981, Oceanic heat flow, in: *The Sea*, Vol. 7 (C. Emiliani, ed.), Wiley–Interscience, New York, pp. 489–523.

Anderson, R. N., Langseth, M. G., and Sclater, J. G., 1977, The mechanisms of heat transfer through the floor of the Indian Ocean, *J. Geophys. Res.* **82**:3391–3409.

Anderson, R. N., Langseth, M. G., Hayes, D. E., Watanabe, T., and Yasui, M., 1978a, Heat flow, thermal conductivity, thermal gradient, in: *Geophysical Atlas of the East and Southeast Asian Seas* (D. E. Hayes, ed.), Map and Chart Series MC-25, Geological Society of America, Boulder, Colo.

Anderson, R. N., DeLong, S. E., and Schwarz, W. M., 1978b, Thermal model for subduction with dehydration in the downgoing slab, *J. Geol.* **86**:731–739.

Anderson, R. N., Hobart, M. A., and Langseth, M. G., 1979, Geothermal convection through oceanic crust and sediments in the Indian Ocean, *Science* **204**:828–832.

Becker, K., and von Herzen, R. P., 1983a, Heat transfer through the sediments of the mounds hydrothermal area, Galapagos Spreading Center at 86°W, *J. Geophys. Res.* **88**:995–1008.

Becker, K., and von Herzen, R. P., 1983b, Heat flow on the western flank of the East Pacific Rise at 21°N, *J. Geophys. Res.* **88**:1057–1066.

Becker, K., Langseth, M. G., von Herzen, R. P., and Anderson, R. N., 1985, Deep crustal geothermal measurements, Hole 504B, Deep Sea Drilling Project Legs 69, 70, 83, and 92, *Initial Reports of the Deep Sea Drilling Project* **83**:405–418.

Bibee, L. D., Shor, G. G., Jr., and Lu, R. S., 1980, Inter-arc spreading in the Mariana Trough, *Mar. Geol.* **35**:183–197.

Bredehoeft, J. D., and Papadopulos, I. S., 1965, Rates of vertical groundwater movement estimated from the earth's thermal profile, *Water Resour. Res.* **2**:325–328.

Burch, T. K., and Langseth, M. G., 1981, Heat-flow determination in three DSDP boreholes near the Japan Trench, *J. Geophys. Res.* **86**:9411–9419.

Canadian American Seamount Expedition, 1985, Hydrothermal vents on an axis seamount of the Juan de Fuca Ridge, *Nature* **313**:212–214.

Chapman, D. S., 1983, "International Heat Flow Commission Global Heat Flow Density Compilation," World Data Center A For Solid Earth Geophysics, Boulder, Colo.

Cooper, A. K., Marlow, M. S., and Scholl, D. W., 1976, Methozoic magnetic lineations in the Bering Sea marginal basin, *J. Geophys. Res.* **81**:1916–1932.

Corliss, J. B., Dymond, J., Gordon, L. I., Edmond, J. M., von Herzen, R. P., Ballard, R. D., Green, K., Williams, D., Bainbridge, A., Crane, K., and van Andel, T. H., 1979, Submarine thermal springs on the Galapagos Rift, *Science* **203**:1073–1083.

Craig, H., Horibe, Y., Farley, K. A., Welhan, J. A., Kim, K.–R., and Hey, R. N., 1987, Hydrothermal vents in the Mariana Trough: Results of the first ALVIN dives, *Eos* **68**:1531.

Crough, S. T., 1978, Thermal origin of mid-plate hot-spot swells, *Geophys. J. R. Astron. Soc.* **55**:451–469.

Davies, G. F., 1980, Review of oceanic and global heat flow estimates, *Rev. Geophys. Space Phys.* **18**:718–722.

Davis, E. E., and Lister, C. R. B., 1974, Fundamentals of ridge crest topography, *Earth Planet. Sci. Lett.* **21**:405–413.

Davis, E. E., and Lister, C. R. B., 1977, Heat flow measured over the Juan de Fuca Ridge: Evidence for widespread hydrothermal circulation in a highly heat transportive crust, *J. Geophys. Res.* **82**:4845–4860.

Davis, E. E., Lister, C. R. B., Wade, U. S., and Hyndman, R. D., 1980, Detailed heat flow measurements over the Juan de Fuca Ridge system, *J. Geophys. Res.* **85**:299–310.

Davis, E. E., Lister, C. R. B., and Sclater, J. G., 1984, Towards determining the thermal state of old ocean lithosphere: Heat-flow measurements from the Blake–Bahama outer ridge, northwestern Atlantic, *Geophys. J. R. Astron. Soc.* **78**:507–545.

Davis, E. E., Goodfellow, W. D., Bornhold, B. D., Adshead, J., Blaise, B., Villinger, H., and Le Cheminant, G. M., 1987, Massive sulfides in a sedimented rift valley, northern Juan de Fuca Ridge, *Earth Planet. Sci. Lett.* **82**:49–61.

DeLong, S. E., Schwarz, W. M., and Anderson, R. N., 1979, Thermal effects of ridge subduction, *Earth Planet. Sci. Lett.* **44**:239–246.

Detrick, R. S., and Crough, S. T., 1978, Island subsidence, hot spots, and lithospheric thinning, *J. Geophys. Res.* **83**:1236–1244.

Detrick, R. S., von Herzen, R. P., Crough, S. T., Epp, D., and Fehn, U., 1981, Heat flow on the Hawaiian Swell and lithospheric reheating, *Nature* **292**:142–143.

Detrick, R. S., von Herzen, R. P., Parsons, B., Sandwell, D., and Dougherty, M., 1986, Heat flow observations on the Bermuda Rise and thermal models of midplate swells, *J. Geophys. Res.* **91**:3701–3723.

Erickson, A. J., von Herzen, R. P., Sclater, J. G., Girdler, R. W., Marshall, B. V., and Hyndman, R., 1975, Geothermal measurements in deep-sea drill holes, *J. Geophys. Res.* **80**:2515–2528.

Fehn, U., Green, K. E., von Herzen, R. P., and Cathles, L. M., 1983, Numerical models for the hydrothermal field at the Galapagos Spreading Center, *J. Geophys. Res.* **88**:1033–1048.

Furukawa, Y., and Uyeda, S., 1986, Thermal state under Tohoku Arc with consideration of crustal heat generation, *Bull. Volcanol. Soc. Jpn.* **31**:15–28(in Japanese with English abstract).

Green, K. E., von Herzen, R. P., and Williams, D. L., 1981, The Galapagos Spreading Center at 86°W: A detailed geothermal field study, *J. Geophys. Res.* **86**:979–986.

Halunen, A. J., Jr., 1979, Tectonic history of the Fiji Plateau, Ph.D. thesis, University of Hawaii, Honolulu.

Hasebe, K., Fujii, N., and Uyeda, S., 1970, Thermal processes under island arcs, *Tectonophysics* **10**:335–355.

Heestand, R. L., and Crough, S. T., 1981, The effect of hot spots on the oceanic age–depth relation, *J. Geophys. Res.* **86**:6107–6114.

Hilde, T. W. C., and Lee, C.-S., 1984, Origin and evolution of the West Philippine Basin: A new interpretation, *Tectonophysics* **102**:85–104.

Hobart, M. A., Anderson, R. N., Fujii, N., and Uyeda, S., 1983, Heat flow from hydrothermal mounds in two million year old crust of the Mariana Trough which exceeds two watts per meter, *Eos* **64**:315.

Hobart, M. A., Langseth, M. G., and Anderson, R. N., 1985, A geothermal and geophysical survey on the south flank of the Costa Rica Rift: Sites 504 and 505, *Initial Reports of the Deep Sea Drilling Project* **83**:379–404.

Honda, S., 1985, Thermal structure beneath Tohoku, northeast Japan—A case study for understanding the detailed thermal structure of the subduction zone, *Tectonophysics* **112**:69–102.

Horai, K., and von Herzen, R. P., 1985, Measurements of heat flow on Leg 86 of the Deep Sea Drilling Project, *Initial Reports of the Deep Sea Drilling Project* **86**:759–777.

Hussong, D. M., and Uyeda, S., 1981, Tectonic processes and the history of the Mariana arc: A synthesis of the results of Deep Sea Drilling Project Leg 60, *Initial Reports of the Deep Sea Drilling Project* **60**:909–929.

Hutchison, I., von Herzen, R. P., Louden, K. E., Sclater, J. G., and Jemsek, J., 1985, Heat flow in the Balearic and Tyrrhenian Basins, western Mediterranean, *J. Geophys. Res.* **90**:685–701.

Hyndman, R. D., 1983, Geothermal heat flux, Juan de Fuca Plate Map: JFP-10, Pacific Geoscience Centre, Sidney, British Columbia.

Hyndman, R. D., Davis, E. E., and Wright, J. A., 1979, The measurement of marine geothermal heat heat flow by a multipenetration probe with digital acoustic telemetry and *in situ* thermal conductivity, *Mar. Geophys. Res.* **4**:181–205.

Jessop, A. M., Hobart, M., and Sclater, J. G., 1976, World heat flow data compilation—1975, Geothermal Series No. 5, Department of Energy, Mines and Resources, Ottawa, Ontario.

Karig, D. E., 1971, Structural history of the Mariana island arc system, *Geol. Soc. Am. Bull.* **82**:323–344.

Kimura, M., Kaneoka, I., Kato, Y., Yamamoto, S., Kushiro, I., Tokuyama, H., Kinoshita, H., Isezaki, N., Masaki, H., Oshida, A., Uyeda, S., and Hilde, T. W. C., 1986, Report on DELP 1984 cruises in the middle Okinawa Trough. Part V. Topography and geology of the central grabens and their vicinity, *Bull. Earthq. Res. Inst.* **61**:269–310.

Kimura, M., Uyeda, S., Kato, Y, Tanaka, T., Yamano, M., Gamo, T., Sakai, H., Kato, S., Izawa, E., and Oomori, T., 1988, Active hydrothermal mounds in the Okinawa Trough backarc basin, Japan, *Tectonophysics* **145**:319–324.

Klein, G. deV., and Kobayashi, K., 1980, Geological summary of the north Philippine Sea, based on Deep Sea Drilling Project Leg 58 results, *Initial Reports of the Deep Sea Drilling Project* **58**:951–961.

Langseth, M. G., and Herman, B. M., 1981, Heat transfer in the oceanic crust of the Brazil Basin, *J. Geophys. Res.* **86**:10805–10819.

Langseth, M. G., and von Herzen, R. P., 1970, Heat flow through the floor of the world oceans, in: *The Sea*, Vol. 4, Part I (A. E. Maxwell, ed.), Wiley–Interscience, New York, pp. 299–352.

Langseth, M. G., Le Pichon, X., and Ewing, M., 1966, Crustal structure of the mid-ocean ridges. 5. Heat flow through the Atlantic Ocean floor and convection currents, *J. Geophys. Res.* **71**:5321–5355.

Langseth, M. G., Hobart, M. A., and Horai, K., 1980, Heat flow in the Bering Sea, *J. Geophys. Res.* **85**:3740–3750.

Lee, C.-S., and McCabe, R., 1986, The Banda–Celebes–Sulu Basin: A trapped piece of Cretaceous–Eocene oceanic crust? *Nature* **322**:51–54.

Lee, C.-S., Shor, G. G., Jr., Bibee, L. D., Lu, R. S., and Hilde, T. W. C., 1980, Okinawa Trough: Origin of a back-arc basin, *Mar. Geol.* **35**:219–241.

Lee, W. H. K., and Uyeda, S., 1965, Review of heat flow data, *Geophys. Monogr. Am. Geophys. Union* **8**:87–190.

Letouzey, J., and Kimura, M., 1985, Okinawa Trough genesis: Structure and evolution of a backarc basin developed in a continent, *Mar. Pet. Geol.* **2**:111–130.

Lister, C. R. B., 1972, On the thermal balance of a mid-ocean ridge, *Geophys. J. R. Astron. Soc.* **26**:515–535.

Lister, C. R. B., 1977, Estimates for heat flow and deep rock properties based on boundary layer model, *Tectonophysics* **41**:157–171.

Lonsdale, P., and Becker, K., 1985, Hydrothermal plumes, hot springs, and conductive heat flow in the Southern Trough of Guaymas Basin, *Earth Planet. Sci. Lett.* **73**:211–225.

Lu, R. S., Pan, J. J., and Lee, T. C., 1981, Heat flow in the southwestern Okinawa Trough, *Earth Planet. Sci. Lett.* **55**:299–310.

Macdonald, K. C., Becker, K., Spiess, F. N., and Ballard, R. D., 1980, Hydrothermal heat flux of the "black smoker" vents on the East Pacific Rise, *Earth Planet. Sci. Lett.* **48**:1–7.

Matsubara, Y., Kinoshita, H., Uyeda, S., and Thienprasert, A., 1982, Development of a new system for shallow sea heat flow measurement and its test application in the Gulf of Thailand, *Tectonophysics* **83**:13–31.

McConachy, T. F., Ballard, R. D., Mottl, M. J., and von Herzen, R. P., 1986, Geologic form and setting of a hydrothermal vent field at lat 10°56′N, East Pacific Rise: A detailed study using Angus and Alvin, *Geology* **14**:295–298.

McKenzie, D. P., 1967, Some remarks on heat flow and gravity anomalies, *J. Geophys. Res.* **72**:6261–6273.

Menard, H. W., and McNutt, M., 1982, Evidence for and consequences of thermal rejuvenation, *J. Geophys. Res.* **87**:8570–8580.

Mrozowski, C. L., and Hayes, D. E., 1979, The evolution of the Parece Vela Basin, eastern Philippine Sea, *Earth Planet. Sci. Lett.* **46**:49–67.

Mrozowski, C. L., Lewis, S. D., and Hayes, D. E., 1982, Complexities in the tectonic evolution of the West Philippine Basin, *Tectonophysics* **82**:1–24.

Nagao, T., 1987, Heat flow measurements in the Tohoku–Hokkaido regions by some new techniques and their geotectonic interpretation, D.Sc. thesis, University of Tokyo, Tokyo.

Nagumo, S., Kinoshita, H., Kasahara, J., Ouchi, T., Tokuyama, H., Asanuma, T., Koresawa, S., and Akiyoshi, H., 1986, Report on DELP 1984 cruises in the middle Okinawa Trough. Part II. Seismic structural studies, *Bull. Earthq. Res. Inst.* **61**:167–202.

ODP Leg 110 Scientific Party, 1987, Expulsion of fluid from depth along a subduction-zone decollement horizon, *Nature* **326**:785–788.

Otofuji, Y., and Matsuda, T., 1984, Timing of rotational motion of Southwest Japan inferred from paleomagnetism, *Earth Planet. Sci. Lett.* **70**:373–382.

Parker, R. L., and Oldenburg, D. W., 1973, Thermal model of ocean ridges, *Nature Phys. Sci.* **242**:137–139.

Parsons, B., and McKenzie, D., 1978, Mantle convection and the thermal structure of the plates, *J. Geophys. Res.* **83**:4485–4496.

Parsons, B., and Sclater, J. G., 1977, An analysis of the variation of ocean floor bathymetry and heat flow with age, *J. Geophys. Res.* **82**:803–827.

Revelle, R., and Maxwell, A. E., 1952, Heat flow through the floor of the eastern north Pacific Ocean, *Nature* **170**:199–200.

RISE Project Group, 1980, East Pacific Rise: Hot springs and geophysical experiments, *Science* **207**:1421–1433.

Sclater, J. G., and Francheteau, J., 1970, The implications of terrestrial heat flow observations on current tectonic and geochemical models of the crust and upper mantle of the earth, *Geophys. J. R. Astron. Soc.* **20**:509–542.

Sclater, J. G., Ritter, U. G., and Dixon, F. S., 1972, Heat flow in the southwestern Pacific, *J. Geophys. Res.* **77**:5697–5704.

Sclater, J. G., Crowe, J., and Anderson, R. N., 1976, On the reliability of oceanic heat flow averages, *J. Geophys. Res.* **81**:2997–3006.

Sclater, J. G., Jaupart, C., and Galson, D., 1980, The heat flow through oceanic and continental crust and the heat loss of the earth, *Rev. Geophys. Space Phys.* **18**:269–311.

Shen, H.-C., 1982, Heat flow and sea-bottom sedimentary environment analysis of southwest Pacific, *Sci. Sin.* **25**:1354–1364.

Shih, T. C., 1980, Magnetic lineations in the Shikoku Basin, *Initial Reports of the Deep Sea Drilling Project* **58**:783–788.

Simmons, G., and Horai, K., 1968, Heat flow data, 2, *J. Geophys. Res.* **73**:6608–6629.

Tamaki, K., 1986, Age estimation of the Japan Sea on the basis of stratigraphy, basement depth, and heat flow data, *J. Geomagn. Geoelectr.* **38**:427–446.

Taylor, B., and Hayes, D. E., 1983, Origin and history of the South China Sea Basin, *Geophys. Monogr. Am. Geophys. Union* **27**:23–56.

U.S. Geological Survey Juan de Fuca Study Group, 1986, Submarine fissure eruptions and hydrothermal vents on the southern Juan de Fuca Ridge: Preliminary observations from the submersible Alvin, *Geology* **14**:823–827.

Uyeda, S., 1982, Subduction zones: An introduction to comparative subductology, *Tectonophysics* **81**:133–159.

Uyeda, S., and Ben-Avraham, Z., 1972, Origin and development of the Philippine Sea, *Nature* **253**:177–179.

Uyeda, S., and Horai, K., 1981, Heat flow measurements on Deep Sea Drilling Project Leg 60, *Initial Reports of the Deep Sea Drilling Project* **60**:789–800.

Uyeda, S., and Kanamori, H., 1979, Back-arc opening and the mode of subduction, *J. Geophys. Res.* **84**:1049–1061.

Uyeda, S., and Vacquier, V., 1968, Geothermal and geomagnetic data in and around the island arc of Japan, *Geophys. Monogr. Am. Geophys. Union* **12**:349–366.

Vander Zouwen, D. E., 1984, Structure and evolution of Okinawa Trough, M.Sc. thesis, Texas A&M University, College Station.

Von Herzen, R. P., 1959, Heat-flow values from the southeastern Pacific, *Nature* **183**:882–883.

Von Herzen, R. P., and Uyeda, S., 1963, Heat flow through the eastern Pacific Ocean floor, *J. Geophys. Res.* **68**:4219–4250.

Von Herzen, R. P., Detrick, R. S., Crough, S. T., Epp, D., and Fehn, U., 1982, Thermal origin of the Hawaiian Swell: Heat flow evidence and thermal models, *J. Geophys. Res.* **87**:6711–6723.

Wang, C.-Y., and Shi, Y.-L., 1984, On the thermal structure of subduction complexes: A preliminary study, *J. Geophys. Res.* **89**:7709–7718.

Watanabe, T., Epp, D., Uyeda, S., Langseth, M., and Yasui, M., 1970, Heat flow in the Philippine Sea, *Tectonophysics* **10**:205–224.

Watanabe, T., Langseth, M. G., and Anderson, R. N., 1977, Heat flow in back-arc basins of the western Pacific, in: *Island Arcs, Deep Sea Trenches, and Back-arc Basins* (M. Talwani and W. C. Pitman, III, eds.), Maurice Ewing Ser. (Vol. 1), American Geophysical Union, Washington, D.C., pp. 137–167.

Weissel, J. K., and Anderson, R. N., 1978, Is there a Caroline plate? *Earth Planet. Sci. Lett.* **41**:143–158.

Weissel, J. K., and Watts, A. B., 1979, Tectonic evolution of the Coral Sea basin, *J. Geophys. Res.* **84**:4572–4582.

Williams, C. F., Narasimhan, T. N., Anderson, R. N., Zoback, M. D., and Becker, K., 1986, Convection in the oceanic crust: Simulation of observations from Deep Sea Drilling Project Hole 504B, Costa Rica Rift, *J. Geophys. Res.* **91**:4877–4889.

Williams, D. L., von Herzen, R. P., Sclater, J. G., and Anderson, R. N., 1974, The Galapagos Spreading Center: Lithospheric cooling and hydrothermal circulation, *Geophys. J. R. Astron. Soc.* **38**:587–608.

Yamano, M., 1986, Heat flow studies in the circum-Pacific subduction zones, D.Sc. thesis, University of Tokyo, Tokyo.

Yamano, M., Uyeda, S., Aoki, Y., and Shipley, T. H., 1982, Estimates of heat flow derived from gas hydrates, *Geology* **10**:339–343.

Yamano, M., Honda, S., and Uyeda, S., 1984, Nankai Trough: A hot trench? *Mar. Geophys. Res.* **6:**187–203.

Yamano, M., Uyeda, S., Kinoshita, H., and Hilde, T. W. C., 1986a, Report on DELP 1984 cruises in the middle Okinawa Trough. Part IV. Heat flow measurements, *Bull. Earthq. Res. Inst.* **61:**251–267.

Yamano, M., Uyeda, S., Furukawa, Y., and Dehghani, G. A., 1986b, Heat flow measurements in the northern and middle Ryukyu arc area on R/V Sonne in 1984, *Bull. Earthq. Res. Inst.* **61:**311–327.

Yamano, M., Uyeda, S., Foucher, J.-P., and Sibuet, J.-C., 1988, Heat flow anomaly in the middle Okinawa Trough, (submitted for publication).

Yokota, T., Kinoshita, H., and Uyeda, S., 1980, New DSDP (Deep Sea Drilling Project) downhole temperature probe utilizing IC RAM (memory) elements, *Bull. Earthq. Res. Inst.* **55:**75–88.

Yoshii, T., 1973, Upper mantle structure beneath the north Pacific and the marginal seas, *J. Phys. Earth* **21:**313–328.

Yoshii, T., 1975, Proposal of the "aseismic front," *Zisin 2* **28:**365–367 (in Japanese).

Yoshii, T., 1979, Compilation of geophysical data around the Japanese islands (I), *Bull. Earthq. Res. Inst.* **54:**75–117 (in Japanese with English abstract).

Yuen, D. A., Fleitout, L., Schubert, G., and Froidevaux, C., 1978, Shear deformation zones along major transform faults and subducting slabs, *Geophys. J. R. Astron. Soc.* **54:**93–119.

Chapter 11C

CONTROLLED SOURCE SEISMOLOGY

K. Suyehiro

Department of Earth Sciences
Faculty of Science
Chiba University
Chiba 260, Japan

I. INTRODUCTION

In the Pacific region, all three types of plate boundaries—oceanic ridges, transform faults, and ocean trenches—exist. There are ocean basins, marginal seas, and bathymetric highs such as chains of seamounts and oceanic plateaus. Many seismic experiments have been made to delineate the seismic structure here. Compilations of published data can be found in Raitt (1963, 1964), Kosminskaya and Riznichenko (1964), McConnel *et al.* (1966), Lee and Taylor (1966), Shor *et al.* (1970), Hotta (1970a, 1972), Woolard (1975), and Christensen (1982), among others.

Figure 1 shows post-1970 temporary stations operated in the Pacific Ocean compiled from published data. Most of these data have not been treated in previous compilations. Still much of the area remains to be covered by investigations employing modern techniques, yet it can be seen that conspicuous features have been surveyed to a great extent. Figures 2–4 show enlarged maps of selected areas indicated in Fig. 1 for more details of various profiles.

It is impossible to objectively summarize and correctly assess the reliability of all the previous results spanning more than 30 years of observations. Differences in experiment design and procedure, actual operational conditions, data analysis and interpretation all affect the results. Therefore, to deduce something new from a

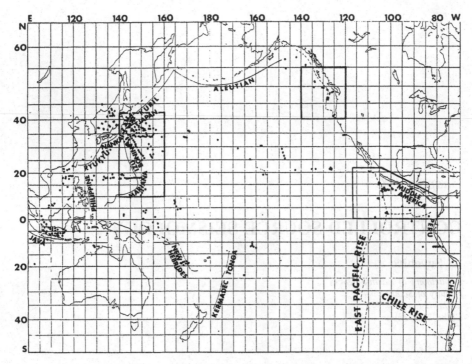

Fig. 1. Map of Pacific Ocean area. Major trenches and oceanic rises comprising plate boundaries are indicated by solid and broken lines, respectively. Dots indicate temporary stations mainly for seismic refraction studies operated after 1970. Many investigations carried out prior to 1970 collected data from a broader area with less spatial density. Areas enclosed by thick lines are shown in greater detail in Figs. 2–4.

statistical treatment of the data from different experiments becomes extremely difficult. A classic and one of the few successful examples may be that of Hess (1964).

Such an attempt will not be made here, and papers that focus on statistical analysis will be given little attention. Instead, more recent results from various tectonic regimes are presented in an attempt to point out implications either not known before the 1970s or which require revision of the conventional view. Also, this chapter focuses on seismic structures at depths resolvable mainly by methods termed "refraction technique." Therefore, details of very shallow horizons or deep upper mantle structures are little treated.

In the following section, possible methodological effects which should be considered when referring to derived models are briefly discussed. Then, the results from different tectonic regions are given with geophysical implications. Less weight is put on results from single profiles.

While many discoveries impossible to make from land-based observations have led to clearer understanding of the geodynamics, modern experiments also added detailed, and in some cases somewhat puzzling data which require further revision or refinement to advance the modern plate tectonics theory.

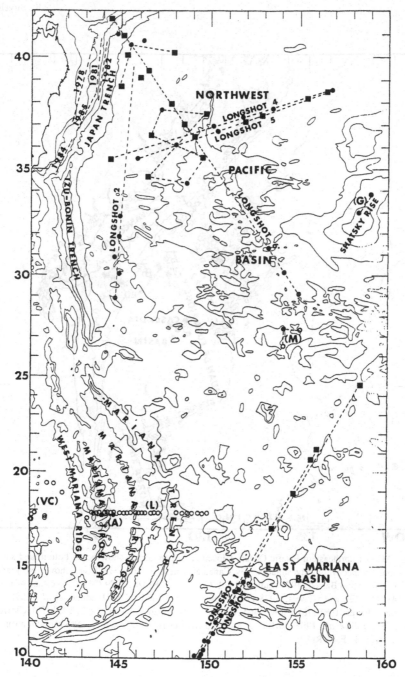

Fig. 2.   Map of western Pacific Ocean. Large explosive shots are indicated by closed squares. Closed circles are ocean bottom stations. Open circles are surface measurements. Profiles denoted LONGSHOT 1 through 6 are long-range explosion experiments carried out from 1974 to 1980 (Asada and Shimamura, 1976; Asada et al., 1983; Shimamura et al., 1983). Gettrust et al. (1980) studied the structure beneath Shatsky Rise (G). Multi-OBS–explosion/air gun profiles shot by Japanese scientists in the Japan Trench area (see Fig. 10) are denoted by years of operation. Explosion and airgun profiles across Mariana arc are by LaTraille and Hussong (1980) (L) and Ambos and Hussong (1982) (A). Profiles denoted M are by Malahoff et al. (1977). Others in the eastern Philippine Sea (VC) are by Mrozowski and Hayes (1979) and Langseth and Mrozowski (1981).

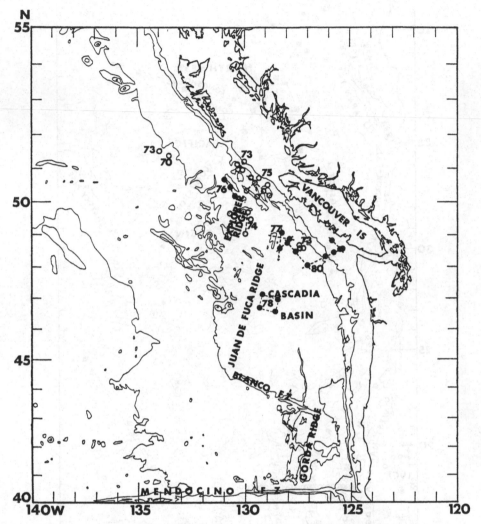

Fig. 3. Map of northeastern Pacific Ocean. Closed and open circles show ocean bottom and surface observations, respectively. Investigations by Canadian scientists since 1970 are shown by years of operations [Keen and Barrett, 1971 (indicated 70); Clowes and Knize, 1979 (73); Clowes and Malecek, 1976 (74); Malecek and Clowes, 1978 (74); Clowes et al., 1981 (75); Hyndman et al., 1978 (76); Cheung and Clowes, 1981 (76); Clowes and Au, 1982 (77); Au and Clowes, 1982, 1983 (77); Clowes et al., 1983 (80); Ellis et al., 1983 (80)]. K. J. McClain and Lewis (1982) made an OBS refraction study in 1978 over Juan de Fuca Ridge.

II. METHOD

A. Instrumentation

Advancement of instrumentation is the key in collecting new data which provide not only more detail and accuracy but more importantly a new view of the

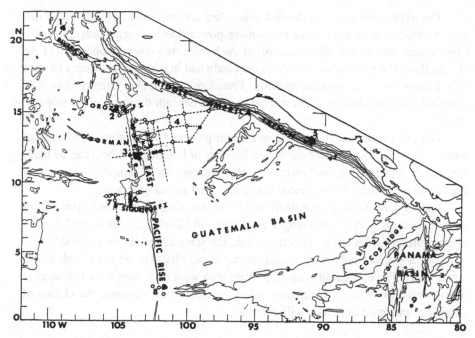

Fig. 4. Map of Cocos Plate and northern part of East Pacific Rise showing locations of stations and profiles discussed in the text. Water depth contours are given every 1 km. Open circles are surface stations. Closed circles are ocean bottom stations. Broken lines indicate seismic profiles. References are as follows: 1, Reid *et al.* (1977); 2, ROSE Phase II (see text); 3, ROSE Phase I (see text); 4, Snydsman *et al.* (1975), Lewis and Snydsman (1977, 1979); 5, Ibrahim *et al.* (1979); 6, Rosendahl *et al.* (1976); 7, Herron *et al.* (1978), Stoffa *et al.* (1980), Herron (1982); 8, Neprochnov *et al.* (1980); 9, Stephen (1983).

earth's structure. Probably the most basic is location accuracy which has improved to within a few hundred meters or even less than tens of meters with the most recent GPS positioning system.

Conventional methods in refraction seismology such as two-ship or sonobuoy techniques since the 1950s are now being replaced by more advantageous ocean bottom observations. Vertical and horizontal geophones and/or hydrophones housed in pressure vessels have been designed and put into use by a number of institutes.

These ocean bottom seismographs (OBS) or hydrophones (OBH) or their combinations (OBSH) suffer no drift due to ocean current at environment with less noise than at surface. Geophones record shear converted phases also, which add important information to constrain the structure. The number of OBSs used in each experiment is increasing for better resolution. These instruments, however, differ in design, reflecting differences in "philosophy," and improvements are still necessary for a better coupling to the seafloor, less susceptibility to sea bottom current noise, and a large storage of wide dynamic range and frequency window data.

The explosives as the controlled source are still the most powerful. The dispersed detonation of the charge gives more power than a single charge with the same weight and further allows control of the bubble frequency independent of the charge size. The explosives, however, are costly and in many cases cannot be used near a coast for environmental reasons. Therefore, a high shot density cannot be expected, nor can surveys of the crustal structure beneath the continental slope be made.

The pneumatic sound sources, such as air guns and water guns, used in large volume in array can penetrate down to a few tens of kilometers. Shots can be made with much higher density and precision. Also, the shot signal reproducbility is very good, which is suitable for various kinds of digital signal processing.

Although not treated extensively in this chapter, multichannel reflection survey using two ships also is becoming a powerful tool (Stoffa and Buhl, 1979). This method utilizes wide-angle reflections also, the same as refraction methods.

The above are only very simple examples of what can be done with newer instruments. The data quality can easily vary with what may seem trivial changes in instrumentation. Without continuous effort to improve instruments, the chance of obtaining important new data will decrease.

B. Data Analysis and Interpretation

Sophisticated analyses can reveal features not immediately discernible from raw data but basically this requires good-quality and high-density data. Conversely, efforts must be made to extract as much information as possible from a good data set.

The simple homogeneous plane layer model has been most frequently adopted after seafloor topography correction. This is somewhat unrealistic but permits important inferences such as structural differences among different regions relevant in tectonic studies. Efforts have been made to make reversed profiles over relatively flat bathymetry to increase confidence in wave velocities where a linear dip is present. Time term analyses can correct for complex shape of the layer interface.

Identification of wavespeed gradient is now frequently made from nonlinearity and amplitude consideration of a certain phase in record section. The transformation of a normal record section into so-called tau-p domain, namely intercept time and ray parameter, provides better phase identification and direct linear inversion (e.g., Bessonova et al., 1974).

Given a high-density coverage of both the sources and receivers, two-dimensional structure may be retrieved. Often in this field, a forward modeling employing ray tracing is made in an iterative manner starting from a conventionally obtained one-dimensional model. Now, synthetic seismograms can be computed for laterally heterogeneous media and this provides useful information on waveform behavior. It is important, however, to keep in mind the limits of validity in the methods of generating synthetics, no one of which is exact (e.g., Spudich and Orcutt, 1980).

The same data set should produce one wavespeed model with appropriate variance. Arguments within this variance, however, can result in different interpretations with different important geophysical implications. Actually, variance is in most cases not well defined as it is impossible to account for all the errors in data acquisition, reduction, and analyses.

III. SEISMIC WAVESPEED STRUCTURE

A. Ocean Basins

1. *Structure of the Oceanic Crust*

This section summarizes the results on normal oceanic crust in the Pacific Ocean Basin. The word "normal" implies that the crust being considered lies beneath normal ocean basins reasonably distant from major tectonic features. Often, the oceanic crust is divided into a number of homogeneous layers. A division into three is the classic one, where layer 1 is the sedimentary layer, layer 2 is the basement layer, and layer 3 is the oceanic layer. Table I gives a compilation of data gathered mainly by the Scripps Institution of Oceanography between 1950 and 1969 (Shor *et al.*, 1970).

The structure in Table I may be considered as a reference model in the following discussion. The sediment wavespeeds mostly represent those from abyssal plains with thin pelagic sediments. The layer 2 wave velocity has a large scatter due to difficulty in its determination. The scatter in Pn wave velocity arises from anisotropy (Raitt *et al.*, 1969, 1971).

After accumulation of sonobuoy measurements which give high sampling rate in space (Ewing and Houtz, 1969; Houtz *et al.*, 1970; Sutton *et al.*, 1971; Houtz, 1976; Houtz and Ewing, 1976), claims have been made for a further division of the crustal layers. Sutton *et al.* (1971) found a high-wavespeed basal crustal layer (7.1–

TABLE I
Average Structure of the Oceanic Basin

	Velocity (km/sec)	Thickness (km)	Depth (km) below seafloor
Layer 1	2.20 (0.31)	0.66 (0.90)	0.66
Layer 2	5.19 (0.64)	1.49 (0.98)	2.15
Layer 3	6.81 (0.16)	4.62 (1.30)	6.77
Mantle	8.15 (0.31)		

7.7 km/sec) about 3 km thick. This led to the division of layer 3 into layers 3A and 3B (Fig. 5b). Also, layer 2 which had a large scatter in both wavespeed and thickness was divided into two, 2A and 2B, or even three, adding 2C.

However, high-resolution and better-quality data should be able to resolve between plane layer models and gradient models. Indeed, as Spudich and Orcutt (1980) point out, when later arrivals and amplitudes are also taken into account, and allowing smoothly varying wavespeed, the gradient model better explains the data.

The lateral variability of the structure is not well known in the Pacific. As Purdy (1983) points out, the profiles must be designed so that they are within the same plate age to obtain a data set reliable for comparison with those from other areas.

As can be seen from Fig. 1, recent surveys are more often conducted in areas which may be considered not to possess normal oceanic crust. Therefore, few recent results from ocean basins can be presented. Young oceanic crust is treated in Section III.C.

First, we will look at recent data showing little deviation from the reference model (Table I). One of the earliest air gun–OBS measurements was carried out in the northwestern Pacific Basin. The profile consists of a part of LONGSHOT-4 in Fig. 2. An important parameter, the Poisson ratio, was determined to be 0.48 in the sediments. The observations of clear PmP arrivals at two OBSs constrained the crustal thickness to be 6 km. The layer 2 wave velocities vary between 5.1 and 6.2

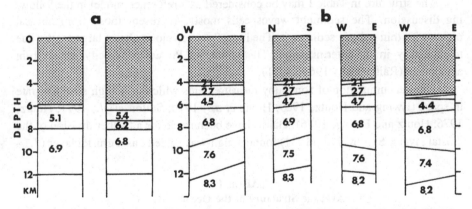

Fig. 5. Crustal structure models with and without high-speed basal layer. P-wave velocities are indicated in km/sec. Open triangles indicate positions of stations. Horizontal length of each model indicates relative profile length. Vertical exaggeration is 10 in most cases. (a) Topmost layer speed was assumed to be 2.0 km/sec. Vp/Vs ratios inside that layer were estimated to be 5.9 and 4.8 at two OBSs at 37°21.5′N, 151°49.3′E (left panel) and 37°49.0′N, 153°28.4′E (right), respectively (Takeda, 1978). (b) High-velocity basal layers are evident in these models derived from sonobuoy data (from Sutton et al., 1971). Rightmost panel is from the area north of Murray Fracture Zone and the rest from California Continental Rise.

km/sec and that of layer 3 is about 6.9 km/sec with 4-km thickness (Takeda, 1978) (Fig. 5a).

The oceanic crust in the central eastern Pacific between Molokai and Clarion Fracture Zone of Oligocene age was investigated by sonobuoy–air gun refraction technique shooting 12 unreversed profiles (Ludwig and Rabinowitz, 1980). Low-velocity sediments are less than 100 m thick overlying layers of 2B and 2C wave-speeds. Layer 3 with 6.7 km/sec and 4.8-km thickness overlies 8.2-km/sec mantle.

It has now become a matter of controversy whether layer 3B, the basal high-speed layer, is widespread, or an artifact due to a misidentification of the phase, which is actually the mantle reflection. The following studies identified the high-velocity layer from first arrivals from a sufficiently large energy source. It would be difficult to totally deny its existence, although its tectonic meaning is not clear.

Nagata (1984) used the OBS–explosion data recovered from the Mariana Basin (part of LONGSHOT-1 in Fig. 2) (Asada *et al.*, 1983) to suggest a rather unusual 7-km-thick 7.4-km/sec layer overlying 8.2-km/sec mantle. Malahoff *et al.* (1977) showed the presence of the high-speed basal layer (7.5 km/sec) with about 5-km thickness in the western Pacific Basin (1 in Fig. 2).

Amplitude consideration has led to more detail in the models, e.g., the inclusion of low-velocity zones. In the Nauru Basin, an OBS refraction study indicated that the Moho lies at a depth of about 14 km where the water is about 5.2 km deep. Sonobuoy wide-angle reflections were used to constrain the shallower structure, namely a 0.5-km-thick 1.7-km/sec layer overlaying a 0.4- to 0.5-km-thick 3.5-km/sec layer. Below it, the layer 2 wavespeed increases to 5–5.5 km/sec to con-stitute a rather thick layer 2. The wavespeed increases from 6.3 to 7.5 km/sec in layer 3. The mantle velocities differ little between the NS (8.2 km/sec) and EW (8.0 km/sec) profiles. It is to be noted that a low-velocity zone is required in the upper crust to explain the amplitudes (Wipperman *et al.*, 1981). The area between the Molokai and Murray Fracture Zones with crust ages of 60 Ma in the northeastern Pacific was studied by Kempner and Gettrust (1982a). In the lower crust a high-velocity basal layer or a nearly constant-velocity layer is not supported. Rather, a 2-km-thick 6.4-km/sec layer over a 2-km-thick 7.1-km/sec layer and 1-km-thick low-velocity zone fits the data and is consistent with the Bay of Island Ophiolite data.

Shearer and Orcutt (1985) describe the 1983 Ngendei Experiment in a region in the South Pacific where the crustal age is 140 Ma. A borehole seismometer and two OBSs were laid out to observe four split refraction lines and one 10-km-radius circular line. The anisotropy study was made in detail utilizing azimuthal depen-dence of travel time and also P-wave polarization anomaly. They found for the first time anisotropy at the base of layer 2, with velocity differences of 0.2–0.4 km/sec. The fast direction was N120°E, orthogonal to the upper mantle anisotropy which varied between 8.0 and 8.5 km/sec.

Whether wavespeed discontinuities exist in layer 2 or not is still unclear at the

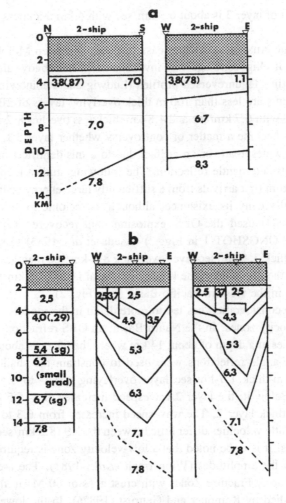

Fig. 6.   Crustal structure models for off the west coast of Canada (Fig. 3). Numbers in parentheses are velocity gradients, in which case the velocity immediately below the interface is given. Other notations follow that of Fig. 5. (a) Pacific Ocean basin southwest of Queen Charlotte Islands from Keen and Barrett (1971). (b) Winona Basin structure (northeast of Explorer Ridge) from Clowes *et al.* (1981). (c) Explorer Ridge area from Malecek and Clowes (1978). (d) Structure west of Explorer Ridge from OBS data (Cheung and Clowes, 1981). (e) Structure of Nootka fault zone (northeast of Juan de Fuca Ridge) from Au and Clowes (1982). (f) Northern part of Juan de Fuca Ridge. Evidence of magma chamber could not be found (from K. J. McClain and Lewis, 1982).

present time. Higher-frequency data with high spatial sampling rate are necessary to resolve this thin layer with a large wavespeed change. The ongoing Ocean Drilling Program allows penetration into this layer which will permit vertical seismic profiling together with acquisition of core samples.

Off the coast of British Columbia (Fig. 3), a large velocity change of 3.8 to 5.5

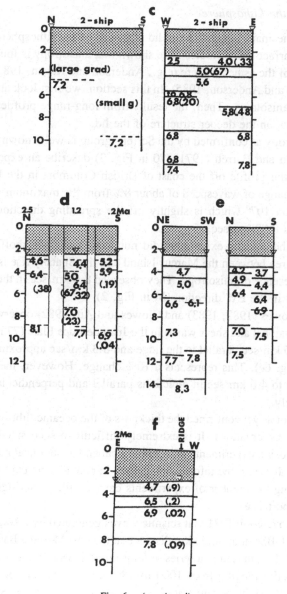

Fig. 6.   (continued)

km/sec in layer 2 occurs above the 4- to 5-km-thick layer 3 which has velocities of 6.7–7.0 km/sec (70 in Fig. 3; Fig. 6a) (Keen and Barrett, 1971). This may alternatively be interpreted as layer 2A (4.0 km/sec) and 2B (5.5 km/sec) (73 in Fig. 3) (Clowes and Knize, 1979). They also found the sedimentary layer to be 0.6 km thick with a velocity of 2.4 km/sec.

2. *Structure of the Lithosphere*

Probably the main interest in the lid structure of the lithosphere is its anisotropy. Recent surface wave studies have shown that anisotropy is found well down
into the depths of the asthenosphere (e.g., Anderson and Regan, 1983; Montagner,
1985; Tanimoto and Anderson, 1985). In this section, we first look at recent results
on Pn velocity anisotropy. Then, the results from long-range profiles are given to
permit inferences on the deeper structure of the lid.

The anisotropy as confirmed by the Scripps group is well known (e.g., Raitt *et
al.*, 1971). Keen and Barrett (1971) (70 in Fig. 3) describe an experiment which
included a circular profile off the coast of British Columbia in the Pacific Basin.
They found a change of wavespeed of about 8% from the maximum velocity along
azimuth direction 107° which is slightly off the spreading direction of 90°. The
mean value was 8.07 km/sec.

Although the circular experiment did not work well, Malahoff *et al.* (1977)
report that the area between the Marcus Island and the Shatsky rise is not inconsis-
tent with the presence of anisotropy. They observed 8.0 km/sec in the NS direction
and 8.4 km/sec in the EW direction (M in Fig. 2).

Au and Clowes (1982, 1983) and Clowes and Au (1982) observed anisotropy
of both compressional and shear waves in the Juan de Fuca Plate (77 in Fig. 3). The
P velocity is 7.5 km/sec parallel to the ridge and 8.3 km/sec approximately perpendicular to it (Fig. 6e). This represents a 10% change. However, the S speed only
varies from 4.5 to 4.6 km/sec in directions parallel and perpendicular to the ridge
axis, respectively.

We have not as yet confirmed the thickness of the oceanic lithosphere from on-
site body wave observations. It is extremely difficult to successfully carry out a
long-range explosion experiment for technological and economical reasons. As will
be seen in the following, the information gained in recent years may not seem to be
much considering the number of experiments carried out, yet its significance is of
first degree importance.

It was discovered in 1971 that seismic waves generated by a 1-kg charge could
be observed by OBSs at up to 90-km distance in the east Mariana Basin (Asada and
Shimamura, 1971). This led to a series of long-range OBS refraction experiments in
the western Pacific ranging from 1000 to 1800 km by the Japanese (Asada and
Shimamura, 1976, 1979; Nagumo *et al.*, 1981; Asada *et al.*, 1983; Shimamura and
Asada, 1983; Shimamura *et al.*, 1983; Shimamura, 1984).

In the course of several experiments at different regions (LONGSHOTs in Fig.
3), a great variability in lithospheric structure was revealed. First, the attenuation is
larger in the Marianas compared to the northwestern Pacific. Both LONGSHOT-1
and LONGSHOT-3 records are clearly different from those of LONGSHOT-2 and
LONGSHOT-6. Little seismic energy is transported across the east Mariana Basin
which may indicate that the low-velocity zone is more developed in the Mariana
Basin. While Asada *et al.* (1983) emphasize that further discussion is not possible

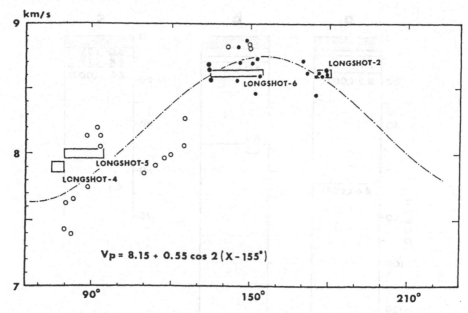

Fig. 7.   Large-scale azimuthal anisotropy inferred from long-range explosion profiles and two different apparent speed observations of natural earthquakes (closed circles in 1974 and open circles in 1977) in the northwestern Pacific (Shimamura and Asada, 1983) (Fig. 2).

from low S/N ratio records, Nagumo *et al.* (1981) claim a thin lithosphere, about 50 km in thickness, and suggest an alternation of a high-velocity and low-velocity zone underlying it.

Second, azimuthal dependence of velocity exists in the northwestern Pacific, which strongly suggests anisotropy over the whole thickness of the lithosphere cap extending over 100 km in depth (Fig. 7). The velocity change is 4–7% where the azimuth of the maximum velocity is 150–160° clockwise from north, perpendicular to the magnetic lineation. Furthermore, the low-velocity zone inferred from traveltime offset and spectral behavior from LONGSHOT-2 was not evident in other profiles (Fig. 8a,b).

Orcutt and Dorman (1977) report a 600-km-long explosion experiment over 70-Ma crust near the Clarion Fracture Zone carried out in 1976. The profile was shot along an isochron. Their OBS data exclude velocities in excess of 8.4 km/sec down to a depth of 60 km (Fig. 8c).

Bibee (unpublished data) describe a 500-km-long OBS/H explosion profile in the West Philippine Basin north of the Central Basin Fault also along an isochron. Some problems in the field were poor location and a small number of successful data recovered. They found anomalously large velocity gradient of 0.018 1/sec on the average for the first 32 km below the Moho where the velocity reaches 8.5 km/sec.

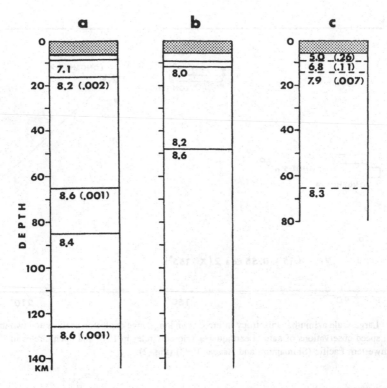

Fig. 8. Lithosphere structure models from long-range explosion profiles. (a) LONGSHOT-2 model from Asada and Shimamura (1976). Low-velocity zone is found at a depth of about 90 km. (b) LONGSHOT-6 model from Shimamura *et al.* (1983). No low-velocity zone is evident. (c) Lithosphere model near Clarion Fracture Zone from Orcutt and Dorman (1977).

As we have seen, instead of a clearer understanding of the tectonics, we now have more questions to address in light of the results which indicate diversity in lithosphere structure. Also there seems to be some discrepancy between thick (> 100 km) lithosphere in the northwestern Pacific and 50-km-thick lithosphere proposed by Anderson and Regan (1983). More large-scale, high-spatial-density experiments are necessary to clarify the degree and extent of anisotropy and the structure inside the lithospheric cap.

B. Plate Subduction Zones

All the plates in the Pacific area are subject to subduction. The seismic structure at each plate subduction zone is given in this section. A subduction zone is interpreted, here, to include the backarc region.

Numerous past investigations have revealed that crustal structures of oceanic type persist beneath the trenches. This is not surprising given today's knowledge.

We now want to know the extent of differences in seismic structure between the various subduction zones.

Some trench areas such as the Yap and Palau trenches remain to be investigated and many more require further modern investigations to reveal two-dimensional features. Such examples are the Aleutian trench (Shor, 1962, 1964) and the Kermadec trench (Shor *et al.*, 1971). These are only examples and it seems that the present data sets are insufficient to synthesize for discussion on a global scale.

1. *Pacific Plate*

(a) Kuril Trench. Recent air gun–multi-OBS seismic surveys near the southwestern end of the trench revealed that low-velocity (2.5–4.7 km/sec) material is more widely spread over the continental slope as compared with the Japan Trench (Fig. 9) (Nishizawa, 1985; Nishizawa and Suyehiro, 1986). This must reflect the difference in the transport and subduction of sediments from land. Beneath the continental slope at a water depth of 2000 m, velocities of 6.0–6.3 km/sec indicative of continental crust are found, suggesting that the continental material extends to about 40 to 80 km landward of the trench axis.

The subducting oceanic crust is normal. A large velocity gradient in layer 2 and relatively homogeneous layer 3 are found. An interesting feature is that the layer 3 velocity seems to be lower beneath the trench compared to that beneath the ocean basin despite the increase in depth (Fig. 9).

(b) Japan Trench. Since the first refraction survey in 1964 (Ludwig *et al.*, 1966), which revealed that the continental material existed about 30 km landward (west) of the trench and a normal oceanic crust seaward of the trench, many surveys have been conducted (Asano *et al.*, 1980; Murauchi and Ludwig, 1980; Suyehiro *et al.*, 1984, 1985, 1986), including multichannel profiles for Deep Sea Drilling Project site surveys which clearly indicated the subduction of the oceanic crust (Nasu *et al.*, 1980). The area is the best surveyed of the subduction zones.

Air gun-explosives and multi-OBS profiles give more detailed structure down to the subducting uppermost mantle as compared with the results from the 1960s (Fig. 10). The dip of the plate is less than 10° in the vicinity of the trench.

Thick low-velocity sediment is found just beneath the inner lower trench slope. This vanishes about 10 km or so westward and a continental-type crust appears. This is in contrast with the results from the Kuril Trench but agrees well with the gravity data (Nishizawa and Suyehiro, 1986).

It seems worthwhile noting that this factor does not necessarily discriminate the two trenches. At the southern end of the Japan Trench, where a seamount (Daiiti–Kashima) is subducting, low-velocity material does not seem to be present at all (Nosaka *et al.*, 1984).

In the last 5 years nearly all the latitude range spanned by the trench has been investigated using OBSs and it is expected that the degree of lateral variation along the trench axis will soon be resolved in more detail.

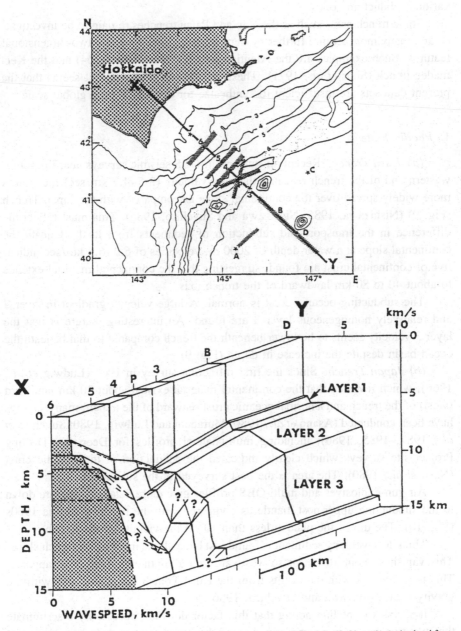

Fig. 9.   Detailed crustal structure model across southwestern Kuril Trench (X–Y section) obtained from multi-OBS and air gun experiment (from Nishizawa and Suyehiro, 1986). Top figure shows the studied region. Air gun profiles are indicated by thick lines. Asterisks are OBS locations.

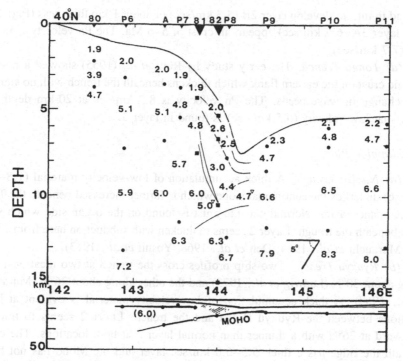

Fig. 10.   Crustal structure model across northern Japan Trench compiled from a number of studies in the area (from Suyehiro et al., 1986). Accumulation of low-velocity material beneath the lower inner slope differs from that beneath southwestern Kuril Trench (Fig. 9). Gradual change in dipping angle of subducting Moho is observed.

A number of large explosions in the northwestern Pacific Basin just east of the Japan Trench were observed by many land stations in the form of fan shooting profiles. It was found that the Pn speed changes rather abruptly from about 8.0 km/sec to 7.5 km/sec representative of normal oceanic mantle and that beneath Honshu Island, respectively (Research Group for Explosion Seismology, 1977; Yoshii et al., 1981). It is suggested that this sudden transition occurs at the Aseismic Front proposed by Yoshii (1975).

(c) Izu–Bonin and Mariana Trenches. The depression of the oceanic crust beneath the trench has been confirmed along 32°N (Hotta, 1970b), along 23.5°N (Murauchi et al., 1968), and along 18°N (Latraille and Hussong, 1980).

LaTraille and Hussong (1980) found an unusually thick layer 2A within 250 km east of the trench in their best profile (L in Fig. 2). They suggested some effect of volcanism by the Magellan Seamount.

Ambos and Hussong (1982) discuss crustal evolution in the Mariana Trough (A in Fig. 2). The layer 2 structure differs between two areas of crust of age 1 and 5–6 Ma. Layer 2A (3.3 km/sec) of about 2-km thickness seems to thin with age to 1 km and in turn to develop layer 2B (5.3 km/sec) of about 1-km thickness (Fig. 11). Also layer 3A (6.7 km/sec) appears in crust of 5–6 Ma. The Pn velocity is rather low (7.7 km/sec).

(d) Tonga Trench. The early study by Raitt *et al.* (1955) showed a normal oceanic crust on the eastern flank which thickens beneath the trench with no significant change in wavespeeds. The Pn velocity is 8.2 km/sec at 20-km depth. A somewhat low velocity (6.5 km/sec) is found in layer 3.

2. *Philippine Plate*

(a) Nankai Trough. A thick accumulation of low-velocity material (1.8–4.0 km/sec) underlies the continental slope which is further increased beneath the Tosa and Kumano basins. Normal oceanic crust is found on the ocean side which subsides beneath the trough. Layer 2 seems to thicken with subduction here from 1 to 3 km (Murauchi *et al.*, 1964; Den *et al.*, 1968; Yoshii *et al.*, 1973).

(b) Ryukyu Trench. Two-ship profiles cross the trench at two locations, one along about 30°N (Ludwig *et al.*, 1973) and the other along about 26°N (Murauchi *et al.*, 1968). A thick accumulation of low-velocity material is evident at both locations between the Ryukyu Ridge and the trench. Layer 2 seems to thicken westward at 26°N with a thinner than normal layer 3 at both locations. The crust beneath the ridge has a thick 5.5–6.0 km/sec layer and the Moho has not been detected.

In 1976, a US–China cooperative study was conducted in the Okinawa Trough

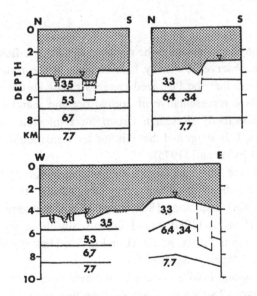

Fig. 11. Crustal structure model in the Mariana Trough indicating rather low Pn velocity (from Ambos and Hussong, 1982). See also explanations of Figs. 5 and 6.

(Lee *et al.*, 1980). From their five two-ship refraction profiles, 14-km-thick crust was recorded. Layer 2 has velocities of 5.5–6.1 km/sec and is 3–4 km thick, and overlies layer 3 which has velocities of 6.3–6.7 km/sec and is 7 km thick (Fig. 12).

In 1984, a cooperative OBS refraction study between West Germany and Japan was carried out, which achieved very high spatial density of both OBSs and shots in the northern part of the trench.

3. Australia–India Plate

(a) Sunda and Timor–Aru–Seram Trenches. In the Indonesian region, a number of expeditions have been carried out to delineate the complex structure there. Curray *et al.* (1977) presented results from the Monsoon, Eurydice, and Luciad expeditions of the Scripps Institution of Oceanography. Layer 3 thickens beneath the Sunda Trench. Kieckhefer *et al.* (1980) showed a wedge of 4.7 to 4.9 km/sec material thickening toward Nias Island beneath which the crust is 40 km thick although the Moho is found to be less than 20 km deep 30 km seaward.

(b) New Hebrides Trench. The New Hebrides arc–trench system was investigated by Ibrahim *et al.* (1980) using the OBS refraction method (Fig. 13). Unusually thick oceanic crust exists seaward of the trench. A wedge consisting of material with a wavespeed of 4.1 km/sec is found beneath the landward slope.

4. Cocos Plate

Middle America Trench. The pioneering refraction seismic study here was made by Shor and Fisher (1961). Their results showing the Moho depression beneath the trench and the continuation of layer 3 beneath the outer edge of the continental shelf are in agreement with plate subduction.

A more recent investigation off Guatemala (Ibrahim *et al.*, 1979) using OBS

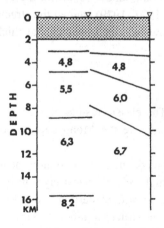

Fig. 12. Crustal structure model near Ryukyu trench (from Lee *et al.*, 1980).

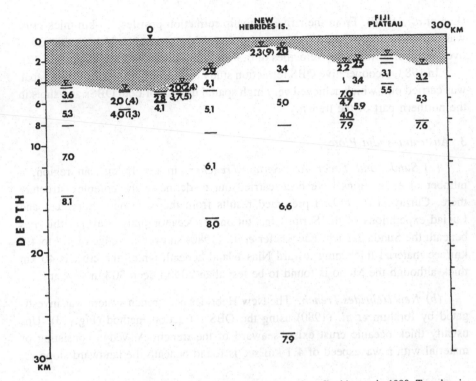

Fig. 13.   Crustal structure model across New Hebrides Trench (from Ibrahim *et al.*, 1980. Trench axis is indicated by closed triangle. See explanations of Figs. 5 and 6 also.

refraction and multichannel reflection techniques confirmed this in more detail. The top of layer 3 and the Moho can be traced about 30 and 15 km from the trench axis dipping landward, respectively. Furthermore, they inferred multiple slabs of oceanic crustal structure dipping landward within the upper midslope.

Lewis and Snydsman (1977, 1979) studied the structure of the Cocos Plate off Mexico and suggested a low-velocity lower crustal thickening toward the trench axis (Fig. 14a).

5. *Nazca Plate*

Peru–Chile Trench. The classic results of Fisher and Raitt (1962) from the DOWNWIND expedition indicate the Moho depression and the continuation of layer 3 landward.

Hussong *et al.* (1975) made air gun–sonobuoy measurements along 12°S and showed the existence of a high-velocity basal layer in the crust (7.1–7.4 km/sec). They suggested a low-angle thrust fault starting to dip 300 km seaward of the trench from the apparent offsets in the structural models. Subduction of the oceanic Moho, however, was not detected.

An experiment termed Oblique Seismic Experiment was carried out on Deep

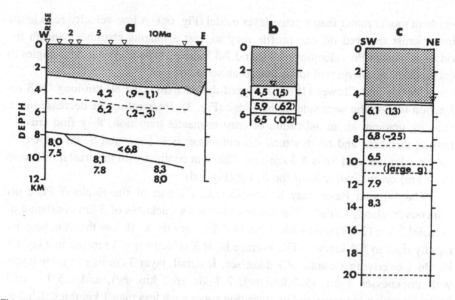

Fig. 14.   Velocity models of the Cocos Plate. (a) Bottom layer in the crust thickens toward Middle
American Trench (from Lewis and Snydsman, 1979). Two Pn velocities indicate the effect of azimuthal
anisotropy, fast in spreading direction. (b) Model from Oblique Seismic Experiment from Stephen
(1983). (c) Low-velocity lower crust and large-gradient uppermost mantle just before subduction (from
Meeder et al., 1977). See explanations of Figs. 5 and 6 also.

Sea Drilling Project Leg 70 using a borehole geophone clamped at 52 and 542 m
delow sea bottom. The site was of 6 Ma age. As shown in Fig. 14b, detailed
structure in layer 2 was revealed. Azimuthal dependence of velocity was not observed (Stephen, 1983).

Meeder et al. (1977) made an OBS refraction study at 18°S. They found a low-
velocity basal layer in the crust and a large positive velocity gradient in the uppermost mantle (7.9 Pn velocity to 8.5 km/sec at 20-km depth) prior to subduction
(Fig. 14c).

C. Ocean Rises as Plate Boundaries

In the Pacific region, a number of ocean rises divide the Pacific Plate from the
North America, Cocos, Nazca, and Antarctic Plates. The Nazca Plate is separated
from the Cocos and Antarctic Plates by the Nazca–Cocos Ridge and the Chile
Ridge, respectively. Among these, the Pacific–Antarctic Ridge, Nazca–Cocos
Ridge, and Chile Ridge remain to be investigated.

1. *Pacific–North America (Juan de Fuca)*

(a) Explorer Ridge. A two-ship refraction survey was carried out in 1974
(Malecek and Clowes, 1978) (Fig. 3). Consideration of amplitudes required a

gradient model rather than a plane layer model (Fig. 6c). A low-velocity zone in the lower crust required on one profile may suggest a magma chamber beneath the ridge. Reversed Pn velocities of 7.9 and 7.3 km/sec in directions at right angles to each other are interpreted to be due to anisotropy.

Cheung and Clowes (1981) analyzed the data from the 80-km-long OBS refraction line on the west side of the Ridge (Fig. 3). Interpreting the decrease in the velocity gradient as an indication of crust-to-mantle transition, they find crust of normal thickness and no structural discontinuities both for P and S. Pn velocity is only 7.4 km/sec and Sn is 4.3 km/sec. This is in contrast with the crustal thickness (> 9 km) on the east side of the Ridge (Fig. 6d).

The Winona Basin may be considered as a part of the Explorer Plate off Vancouver Island, Canada. The sediments reach a thickness of 5 km consisting of material 2 km (2.5 km/sec) and 3 km (4.3 km/sec) thick. Below this, the velocity rapidly rises to 5.4 km/sec. The average layer 3 velocity is 6.3 km/sec in a layer 8 km thick overlying a mantle of 7.8 km/sec. In detail, layer 3 consists of three layers with thicknesses 1 km (5.4 km/sec), 2.7 km (6.2 km/sec), and 2.5 km (6.7 km/sec), which are separated by transition zones each less than 1 km thick (Clowes et al., 1981) (Fig. 6).

(b) *Juan de Fuca Ridge.* The sediment thins and the Moho rises more rapidly than the seafloor toward the ridge, where the Pn wavespeed is low (Shor *et al.,* 1968). However, more recent study using deep tow reflection and OBS refraction data indicate little change across the ridge (Fig. 3) (K. J. McClain and Lewis, 1982). Clear PmP arrivals across the ridge showed no significant offset. A 5-km-thick crust explains the data (Fig. 6f).

Three profiles were shot in the Nootka Fault Zone area (Au and Clowes, 1982) (Fig. 6e). The zone is considered to separate the Explorer Plate and Juan de Fuca Plate. The seismic energy suffers attenuation across the zone.

The best velocity–depth models indicate a gradual change in the crust. Layer 2A velocity ranges from 3.7 to 4.7 km/sec, which is differentiated from layer 2B with an average velocity of 6.0 km/sec by a change in gradient. A crustal thickening due primarily to the change in layer 3 thickness is found. The crust thickens from 6.4 km at 1 Ma to 11.2 km at 2 Ma, suggesting rapid growth (Au and Clowes, 1982).

From the same profiles, an S wave structure was obtained by Au and Clowes (1983). In layer 2, the S wave velocity increases from 2.2 to 3.6 km/sec over 3.5-km thickness. Layer 3A has a rather constant velocity of 3.6 km/sec. The wave velocity then increases to 4.5–4.6 km/sec typical of mantle velocities without a clear discontinuity.

(c) *Gorda Ridge.* Shallow mantle (6.5-km depth) and low Pn wave velocities (< 7.9 km/sec) seem to characterize the ridge (Shor *et al.,* 1968).

2. *Pacific–Cocos, Nazca, Antarctica*

East Pacific Rise. The Hess Basin was first investigated by the air gun–OBS method, revealing a 5-km-thick crust beneath a water depth of 3.1–3.4 km. It consists of a layer 0.7–0.9 km thick with a velocity of 3.5 km/sec followed by a 1.5- to 2.0-km layer, velocity 5.5 km/sec, over a 2.5- to 3.5-km layer with a velocity of 6.8 km/sec over the mantle of velocity of 8.0–8.2 km/sec (Neprochnov et al., 1980) (8 in Fig. 4).

A tripartite OBS array was laid out at 21°N to detect the presence of a low-velocity zone about 2.5 km below the seafloor in the proximity of the rise axis. The zone seems to vanish within 10 km from the axis but the high attenuation of shear wave energy suggests partial melting occurs (Reid et al., 1977).

Beneath the rise crest at 9°N, a pronounced low-velocity zone has been found about 2 km below the seafloor by OBS refraction (Fig. 4). Sonobuoy refraction data indicated an average crustal model which includes a low-velocity zone thinning away from the rise axis as shown in Fig. 15b (Rosendahl et al., 1976). A higher wave velocity, from 5 to 6.7 km/sec, occurs above the low wave velocity zone. The Pn wave velocity was found to be 7.7 km/sec. Comparison of the structure with two other profiles over 2.9- and 5-Ma crust does not show a low-velocity zone, suggesting rapid changes occur in the upper crust with age (Fig. 15c) (Orcutt et al., 1976).

A multichannel reflection survey over the same area, which better resolves lateral variation, showed the Moho continuous across the rise crest but with lateral variability and no distinct reflector between layer 2 and layer 3 (Stoffa et al., 1980). Herron et al. (1978) and Herron (1982) further identified two reflectors R2 and R4 as the bottom of layer 2A (2.8 km/sec, 0.4 km thick) and the top of the low-velocity magma chamber, respectively. Hale et al. (1982) further analyzed the data to suggest an asymmetric roof to the magma chamber with respect to the ridge axis.

A multiinstitutional seismic experiment abbreviated ROSE, carried out in 1979 between the Orozco and Clipperton Frature Zones (Fig. 4) (Ewing and Meyer, 1982), confirmed a large wave velocity gradient in layer 2 (Purdy, 1982b; Lewis and Garmany, 1982; Gettrust et al., 1982; Kempner and Gettrust, 1982b). Lateral variability of layer 2 was tested by Purdy (1982b) using OBH data from a dense network of air gun shots over 0.5- and 4-Ma crust. After appropriate water path correction (Purdy, 1982a), no lateral heterogeneity on the scale of a few kilometers was detected. A small change to higher wave velocity for a given depth between layer 2 of 0.5- and 4-Ma crust may be due to undetected sediment lying over the basement of the older crust.

A low-velocity zone associated with the magma chamber, if it exists, must lie in the width range of 0.5 to 4 km at 3 to 6 km/sec, respectively (Lewis and Garmany, 1982). Pn amplitudes suggest anisotropy in the mantle velocity gradient, which is positive in the spreading direction and negative along the axis (Fig. 15d).

Gettrust et al. (1982) found along-isochron lateral variability to be larger than the change with age, somewhat in contrast with Purdy's (1982b) result. A normal crust–mantle transition is observed over the crust as young as 4.5 Ma (Fig. 15a).

A recent refraction OBS experiment described by McClaim et al. (1985) was carried out near 13°N. The uppermost crust wave velocity is highest beneath the rise crest decreasing with age to 0.1 Ma. A magma chamber of 4-km width starting 1.1 km from the seafloor down to the Moho is the preferred interpretation (Fig. 15d).

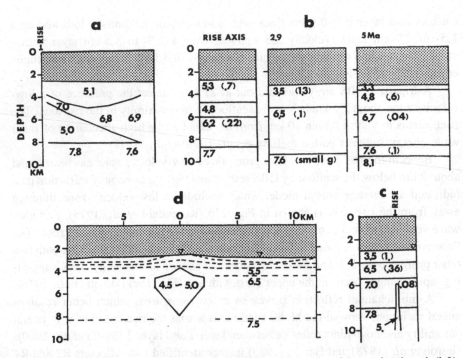

Fig. 15.   Crustal structure models of the East Pacific Rise. (a) Average model normal to rise axis (left). Low-velocity zone thins away from the axis (from Rosendahl *et al.*, 1976). (b) Models of crust of 0 (left panel), 2.9 Ma (middle), and 5 Ma (right) (from Orcutt *et al.*, 1976). (c) Magma chamber seems to be narrowly confined (from Lewis and Garmany, 1982). (d) Magma chamber model from McClain *et al.* (1985) from OBS data. See explanations of Figs. 5 and 6 also.

D. Marginal Seas

1. *Bering Sea*

Normal oceanic structure with a 4-km-thick sediment layer underlies the Bering Sea Basin. A 6-km/sec layer found beneath the Bowers Ridge implies a crust more than 20 km thick. Layer 3B seems to characterize the Bowers Basin (Ludwig *et al.*, 1971). In the Aleutian Basin (Shor, 1964; Helmberger, 1968, 1977), the sediment is also thick with oceanic features. Waveform analysis by Helmberger (1968) led him to conclude that the layer 3 wave velocity increased from 6.5 to 7.0 km/sec over a 7-km thickness, and that the crust-to-mantle transition occurs over a vertical thickness of 2 km. Kamchatka (Komandorskiye) Basin has a thinner sedimentary layer but is still thicker than average. The features found here are a thin (2 km) layer 3 and somewhat low Pn velocity of 8.0 km/sec (Shor and Fornari, 1976) (Fig. 16a).

2. *Okhotsk Sea*

The crust in the south Okhotsk basin is found to be oceanic (Gnibidenko and Svarichevsky, 1984). A 3- to 4-km-thick sedimentary layer of 2–4 km/sec followed

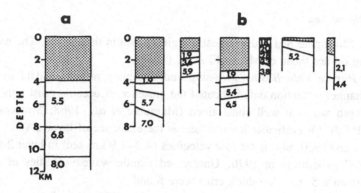

Fig. 16. Velocity models in (a) Kamchatka Basin (from Shor and Fornari, 1976) and (b) Japan Sea (from Ludwig *et al.*, 1975).

by a layer 2 thickness of 0.5–1.5 km with velocities of 4.5–5 km/sec constitute the upper crust. Layer 3 is about 5 km thick with a velocity of 6.5–7 km/sec.

Shibuya and Miyashita (unpublished), using the air gun–OBS data from the Derygin Basin at a water depth of 1600 m collected during the joint USSR–Japan experiment in 1974, determined that the crust has a 0.4-km-thick sedimentary layer (2.1 km/sec), 3.4-km-thick layer 2 (4.6 km/sec), and 6-km-thick layer 3 (6.4 km/sec) overlying 8.1-km/sec mantle.

3. *Japan Sea*

The Japan Sea mainly consists of the Japan Basin with water depths of 3000 to 3700 m, the Yamato Basin with water depths of 2000 to 2500 m, and the Yamato Ridge, which divides the two basins. Since the study by Andreyeva and Udintsev (1958), a number of refraction experiments have been made (Kovylin and Neprochnov, 1965; Murauchi, 1966; Murauchi *et al.*, 1968; Ludwig *et al.*, 1975) (Fig. 16b). The Japan Basin has an oceanic crust about 8–9 km thick with typical wave velocities for a three-layer solution. The Yamato Basin seems to have a thicker layer 3 and a low Pn mantle velocity. Layer 2A seems to typify the upper crust of the Yamato Basin. The crust beneath the Yamato Ridge is more than 18 km thick and the upper part is probably continental with velocities of 6.1–6.3 km/sec. The existence of anisotropy was proposed by Okada *et al.* (1978) who used land observation data of shots made at sea.

The most recent investigation was carried out by Japanese scientists using modern equipment—20 OBSs, a 20-liter air gun, and multichannel seismics over two profiles of 240- and 120-km extent in the Yamato Basin—to obtain detailed structure down to the uppermost mantle. Preliminary results indicate a very thick layer 3 with about 7 km/sec down to about a depth of 20 km. The Pn wavespeed is about 8.0 km/sec on the two profiles which were perpendicular to each other (DELP Shipboard Scientists, 1985).

4. *Philippine Sea*

The Philippine Sea is the largest marginal basin in the Pacific. The basic data of the region are those of Murauchi *et al.* (1968).

(a) Parece Vela Basin. Two-ship and sonobuoy refraction data as well as multi–channel reflection data indicated the presence of oceanic crust although the Moho depth was not well constrained (Murauchi *et al.*, 1968; Mrozowski and Hayes, 1979). Of particular interest here is the existence of layer 2A.

Louden (1980) also notes low velocities (4.2–4.9 km/sec) in layer 2 from the INDOPAC expedition in 1976. Unreversed mantle wave velocities of 8.0–9.3 km/sec from a 5- to 7-km-thick crust were found.

(b) West Philippine Basin. Henry *et al.* (1975) describe two reversed sonobuoy profiles. They found layer 2 to be less than 1 km thick with wave velocities higher than 5.7 km/sec in the northeastern part. The crust as a whole is thin—less than 5 km thick. In the area between the Central Basin Fault and the Benham Rise, a rather low layer 3 wave velocity of 6.3 km/sec was determined (Fig. 17a). Also, a remarkably thin upper crust and Pn wavespeeds of 7.7–8.7 km/sec were reported (Louden, 1980). In the northeast a total crustal thickness of only 3–4 km is suggested.

Among the bathymetric highs in the basin, the Amami Plateau was studied using an OBS tripartite array by Nishizawa *et al.*, (1983). Figure 17b shows the upper crust model with a thick layer 2.

(c) Shikoku Basin. Using OBS refraction, Poisson's ratio in the sediment was found to be 0.47 and 0.33 in layer 2A, which is about 0.5 km thick with a velocity of 3.7 km/sec overlying a layer of 4.8 km/sec (Nagumo *et al.*, 1980).

5. *East China and South China Seas*

The upper crust in the East China Sea was studied by Leyden *et al.* (1973). The sediment is thick but shows large variations.

In the South China Sea, the crust is 7–8 km thick and typically oceanic with layer 2 (5.0–6.2 km/sec) about 1 km thicker than usual but with half the usual thickness layer 3 (6.5–7.4 km/sec) (Ludwig *et al.*, 1979).

6. *Java, Flores, Banda, Molucca, and Arafura Seas*

The Java Sea, where the water depth is very shallow, was investigated in 1971 using expendable sonobuoys. However, penetration was not sufficient to reveal the whole crustal structure (Ben-Avraham and Emery, 1973).

In the west of Flores Sea, the crust shows a character intermediate between oceanic and continental. The Moho rises eastward through the Flores Sea to the Banda Sea to form a normal oceanic structure (Curray *et al.*, 1977). Oceanic crust persists beneath Weber Deep (Bowin *et al.*, 1980).

A further refraction profile shot in 1976 in the central Banda Sea using OBH

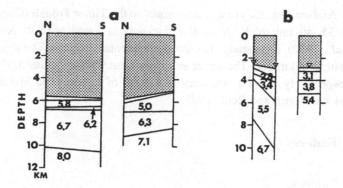

Fig. 17.   (a) Velocity models in Philippine Sea from Henry *et al.* (1975). (b) Crust beneath Amami Plateau from Nishizawa *et al.* (1983). See explanations of Figs. 5 and 6 also.

and sonobuoy also shows oceanic structure. Layers 2, 3A, and 3B have thicknesses of 1.5–2.0, 2.0–3.5, and 2.5–4.6 km, respectively. Layer 3B characterizes the crust here making the crust 9–10 km thick (Purdy and Detrick, 1978) (Fig. 18a).

The Molucca Sea is bordered by two island arcs, Halmahera and Sangihe. Eleven seismic sonobuoy refraction profiles were shot in 1976–1977. Abnormally thick (6–15 km) low-wave-velocity material overlies the 6- to 7-km/sec basement layer. The depth to the basement layer is several kilometers shallower to the west of Talaud Mazu Ridge. The attenuation is high and a 20-kg charge reached only a 60-km range and the Moho was not determined (McCaffrey *et al.*, 1980) (Fig. 18b).

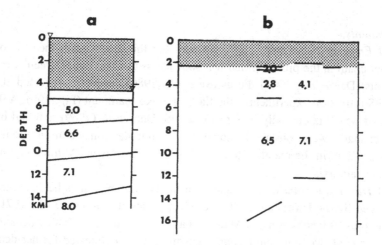

Fig. 18.   Velocity models in (a) Banda Sea (from Purdy and Detrick, 1978) and (b) Molucca Sea (from McCaffrey *et al.*, 1980). See explanations of Figs. 5 and 6 also.

In the Arafura Sea, the crust is continental to the Timor Trough (Curray et al., 1977) and 35–40 km thick. A similar thickness is found beneath Aru Trough (Bowin et al., 1980). OBS study also shows continental structure. The sediment is 2 km thick with 2–4 km/sec. The upper and lower crust with velocities 5.97 and 6.52 km/sec, respectively, have a boundary at a depth of 11 km. The Moho lies at a depth of 34 km (Rynn and Reid, 1983).

E. Other Features

1. Ocean Plateaus

(a) Shatsky Rise. Den et al. (1969) describe results from two-ship refraction profiles across the rise along 32°N. A broad feature is the deep Moho of at least 22-km depth near the crestal zone. The structure just beneath the crest was not well determined. They suggested the thickening of the crust to be due to a layer of 7.3–7.8 km/sec beneath layer 3.

An OBS reversed refraction profile of 130-km length was shot in 1977 (Gettrust et al., 1980) (G in Fig. 2). The authors could not obtain Pn arrivals but the Moho, if it exists, must lie at a depth greater than 26 km. They did not observe a layer 7.3–7.8 km thick but a velocity of 7.14 km/sec was recorded at a depth of 14 km.

(b) Ontong Java and Manihiki. Two-ship refraction surveys over the Ontong Java Plateau revealed the crust to be thick (35 to 42 km) with thick sediment (Murauchi et al., 1973; Furumoto et al., 1976) (Fig. 19a). Air gun–sonobuoy study shows wave velocities in the crust similar to oceanic but 5.0 and 3.1 times thicker than ocean basin crust in Ontong Java and Manihiki, respectively (Hussong et al., 1979).

2. Seamounts

(a) Emperor Seamount. The crust does not thicken beneath the axis of the seamount chain in the profile which was shot across the axis but between two large seamounts (Den et al., 1969). Furukawa et al. (1980) describe a reversed 80-km-long OBS refraction experiment at the flank of Koko seamount (Fig. 19b). A 9-km-thick crust which is basically in agreement with Den et al. (1969) is found but no lateral changes were observed, however. Amplitude consideration requires a positive gradient in the mantle. Houtz (1976) identified layer 2A thinning outward from the chain axis.

(b) Hawaiian Islands. The vicinity of the Hawaiian Islands has been mainly investigated by the Hawaiian Institute of Geophysics (Furumoto et al., 1971).

The crust thickens beneath Mauna Loa (Zucca and Hill, 1980; Zucca et al., 1982). A recent two-ship multichannel seismic survey delineated further detail of the structure of the thickened crust (Watts et al., 1985).

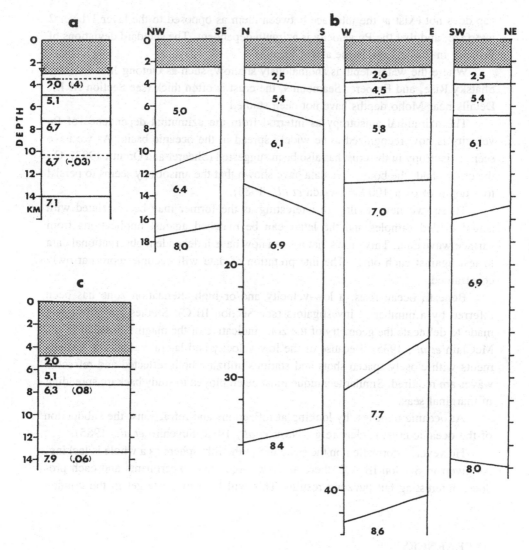

Fig. 19.   Velocity models of bathymetric highs. (a) Shatsky Rise by Gettrust *et al.* (1980). (b) Manihiki and Ontong Java Plateaus (from Furumoto *et al.*, 1976) and (c) crust near Koko Seamount (from Furukawa *et al.*, 1980). See explanations of Figs. 5 and 6 also.

IV. CONCLUDING REMARKS

The velocity structure model of the crust in the Pacific as given in Table I does not require major revisions in the light of recent results. However, some refinements may be made, namely, that the velocity gradient is larger in layer 2 (about 0.5 to 1.0 1/sec) than in layer 3 (up to a few tenths of a 1/sec), that a significant velocity gap does not exist at the interface between them as opposed to the layer 1/layer 2 interface, and that the Pn velocity is azimuth dependent. The standard deviations of velocity in Table I imply the above features.

Where the water depth is anomalously shallow, such as Ontong Java Plateau, Shatsky Rise, and Emperor Seamounts, the crust is often thick (see Section III.E). Details near Moho depths have not been studied.

The azimuthal anisotropy as inferred from the azimuthal dependence of Pn velocity is now recognized to be widely spread in the oceanic basin. As we have seen, anisotropy in the crust has also been suggested (Shearer and Orcutt, 1985). On the other hand, the body wave data have shown that the anisotropy seems to persist to a depth of over 100 km (Asada et al., 1983).

These two new results are interesting as the former may be correlated with actual drilled samples and the latter can be matched against implications from surface wave data. This shows that we can now have independent observational data to test against each other. The interpretation of data will become more narrowly constrained.

Beneath ocean rises, a low-velocity and/or high-attenuation zone has been inferred by a number of investigators (see Section III.C). Studies are now being made to delineate the geometry of the zone indicative of the magma chamber (e.g., McClain et al., 1985). Because of the low velocity and lateral variability, experiments with closely spaced shots and stations utilizing both reflected and refracted waves are required. Similar technique must be employed to study backarc spreading in marginal seas.

At oceanic trenches, by looking at reflections and refractions, the subduction of the oceanic crust is clear (e.g., Nasu et al., 1980; Suyehiro et al., 1985).

However, information on the structure of the lithosphere as a whole is lacking. As given in Section III.A.2, there are few large-scale experiments and each produces interesting but puzzling results. This will be a major target in the coming years.

REFERENCES

Ambos, E. L., and Hussong, D. M., 1982, Crustal structure of the Mariana Trough, J. Geophys. Res. 87:4003–4018.

Anderson, D. L., and Regan, J., 1983, Upper mantle anisotropy and the oceanic lithosphere. Geophys. Res. Lett. 10:841–844.

Andreyeva, I. B., and Udintsev, G. B., 1958, Bottom structure of the Sea of Japan according to data obtained by the research ship Vityaz, Izv. Akad. Nauk SSSR Ser. Geol. 10:3–20 (in Russian).

Asada, T., and Shimamura, H., 1971, Sur l'observation seismique au fond oceanique, Mer (Tokyo) (Bull. Soc. Fr. Jpn. Oceanogr.) 9:35–45 (in Japanese).

Asada, T., and Shimamura, H., 1976, Observations of earthquakes and explosions at the bottom of the western Pacific: The structure of the oceanic lithosphere revealed by Longshot experiments, in: The Geophysics of the Pacific Basin and Its Margin (G. H. Sutton, M. H. Manghnani, R. Moberly, and E. U. McAfee, eds.), Am. Geophys. Union Monogr. 19:135–154.

Asada, T., and Shimamura, H., 1979, Long-range refraction experiments in deep ocean, *Tectonophysics* **56**:67–82.

Asada, T., Shimamura, H., Asano, S., Kobayashi, K., and Tomoda, Y., 1983, Explosion seismological experiments on long-range profiles in the northwestern Pacific and the Marianas Sea, in: *Geodynamics of the Western Pacific–Indonesian Region* (T. W. C. Hilde and S. Uyeda, eds.), Geodynamics Series **11**:105–120.

Asano, S., Yamada, T., Suyehiro, K., Yoshii, T., Misawa, Y., and Iizuka, S., 1980, Crustal structure in a profile off the Pacific coast of northeastern Japan by the refraction method with ocean bottom seismometers, *J. Phys. Earth* **29**:267–281.

Au, D., and Clowes, R. M., 1982, Crustal structure from an OBS survey of the Nootka fault zone off western Canada, *Geophys. J. R. Astron. Soc.* **168**:27–47.

Au, D., and Clowes, R. M., 1983, Shear wave velocity structure of the oceanic lithosphere from ocean bottom seismometer studies, *Geophys. J. R. Astron. Soc.* **77**:105–123.

Ben-Avraham, Z., and Emery, K. O., 1973, Structural framework of Sunda shelf, *Bull. Am. Assoc. Pet. Geol.* **57**:2323–2366.

Bessonova, E. N., Fishman, V. M., Ryaboyi, V. Z., and Sitnikova, G. A., 1974, The tau method for inversion of travel times. I. Deep seismic sounding data, *Geophys. J. R. Astron. Soc.* **36**:377–398.

Bowin, C., Purdy, G. M., Johnston, C., Shor, G., Lawver, L., Hartono, H. M. S., and Jezek, P., 1980, Arc–continent collision in Banda Sea region, *Bull. Am. Assoc. Pet. Geol.* **64**:868–915.

Cheung, H. P. Y., and Clowes, R. M., 1981, Crustal structure from P- and S-wave analyses: Ocean bottom seismometer results in the north-east Pacific, *Geophys. J. R. Astron. Soc.* **65**:47–73.

Christensen, N. I., 1982, Seismic velocities, in: *Handbook of Physical Properties of Rocks* (R. S. Carmichael, ed.), Vol. 2, CRC Press, pp. 2–227.

Clowes, R. M. and Malecek, S. J., 1976, Preliminary interpretation of a marine deep seismic sounding survey in the region of Explorer ridge, *Can J. Earth Sci.*, **13**:1545–1555.

Clowes, R. M., and Au, D., 1982, In-situ evidence for a low degree of S-wave anisotropy in the oceanic upper mantle, *Geophys. Res. Lett.* **9**:13–16.

Clowes, R. M., and Knize, S., 1979, Crustal structure from a marine seismic survey off the west coast of Canada, *Can. J. Earth Sci.* **16**:1265–1280.

Clowes, R. M., Thorleifson, A. J., and Lynch, S., 1981. Winona Basin, west coast Canada; crustal structure from marine seismic studies, *J. Geophys. Res.* **86**:225–242.

Clowes, R. M., Ellis, R. M., Hajnal, Z., and Jones, I. F., 1983, Seismic reflections from subducting lithosphere?, *Nature* **303**:668–670.

Curray, J. R., Shor, G. G., Jr., Raitt, R. W., and Henry, M., 1977, Seismic refraction and reflection studies of crustal structure of the eastern Sunda and western Banda arcs, *J. Geophys. Res.* **82**:2479–2489.

DELP Shipboard Scientists, 1985, Detailed seismic structure in the Yamato Basin, Japan Sea, C2, Abstr. Fall Meet. Seismol. Soc. Jpn.

Den, N., Murauchi, S., Hotta, H., Asanuma, T., and Hagiwara, K., 1968, A seismic refraction exploration of Tosa deep-sea terrace off Sikoku, *J. Phys. Earth* **16**:7–10.

Den, N., Ludwig, W. J., Murauchi, S., Ewing, J. I., Hotta, H., Edgar, N. T., Yoshii, T., Asanuma, T., Hagiwara, K., Sato, T., and Ando, S., 1969, Seismic–refraction measurements in the northwest Pacific Basin, *J. Geophys. Res.* **74**:1421–1433.

Ellis, R. M., Spence, G. D., Clowes, R. M., and Waldron, D. A., Jones, I. F., Green, A. G., Forsyth, D. A., Mad, J. A., Berry, M. J., Mereu, R. F., Kanasewich, E. R., Cumming, G. L., Hajnal, Z., Hyndman, R. D., McMechan, G. A., Loncarevic, B. D., 1983, The Vancouver island seismic project: a CO-CRUST onshore-offshore study of a convergent margin, *Can. J. Earth Sci.* **20**:719–741.

Ewing, J., and Houtz, R., 1969, Mantle reflections in airgun–sonobuoy profiles, *J. Geophys. Res.* **74**:6706–6709.

Ewing, J. I., and Meyer, R. P., 1982, Rivera ocean seismic experiment (ROSE) overview. *J. Geophys. Res.* **87**:8345–8358.

Fisher, R. L., and Raitt, R. W., 1962, Topography and structure of the Peru–Chile Trench, *Deep Sea Res.* **9**:423–443.

Furukawa, K., Gettrust, J. F., Kroenke, L. W., and Campbell, J. F., 1980, Crust and upper mantle structure along the flank of Koko seamount, *Bull. Seismol. Soc. Am.* **70**:1161–1169.

Furumoto, A. S., Campbell, J. F., and Hussong, D. M., 1971, Seismic refraction surveys along the Hawaiian Ridge, Kauai to Midway Island, *Bull. Seismol. Soc. Am.* **61**:147–166.

Furumoto, A. S., Webb, J. P., Odegard, M. E., and Hussong, D. M., 1976, Seismic studies on the Ontong Java Plateau, 1970, *Tectonophysics* **34**:71–90.

Gettrust, J. F., Furukawa, K., and Kroenke, L. W., 1980, Crustal structure of the Shatsky Rise from seismic refraction measurements. *J. Geophys. Res.* **85**:5411–5415.

Gettrust, J. F., Furukawa, K., and Kempner, W. B., 1982, Variation in young oceanic crust and upper mantle structure, *J. Geophys. Res.* **87**:8435–8446.

Gnibidenko, H. S., and Svarichevsky, A. S., 1984, Tectonics of the south Okhotsk deep-sea basin, *Tectonophysics* **102**:225–244.

Hale, L. D., Morton, C. J., and Sleep, N. H., 1982, Reinterpretation of seismic reflection data over the East Pacific Rise, *J. Geophys. Res.* **87**:7707–7717.

Helmberger, D. V., 1968, The crust–mantle transition in the Bering Sea, *Bull. Seismol. Soc. Am.* **58**:179–214.

Helmberger, D. V., 1977, Fine structure of an Aleutian crustal section, *Geophys. J. R. Astron. Soc.* **48**:81–90.

Henry, M., Karig, D. E., and Shor, G. G., Jr., 1975, Two seismic refraction profiles in the west Philippine Sea, *Initial Reports of the Deep Sea Drilling Project* **31**:611–614.

Herron, T. J., 1982, Lava flow layer–East Pacific Rise, *Geophys. Res. Lett.* **9**:17–20.

Herron, T. J., Ludwig, W. J., Stoffa, P. L., Kan, T. K., and Buhl, P., 1978, Structure of the East Pacific Rise crest from multichannel seismic reflection data, *J. Geophys. Res.* **83**:798–804.

Hess, H. H., 1964, Seismic anisotropy of the uppermost mantle under oceans, *Nature* **203**:629–631.

Hotta, H., 1970a, Stability of the crust–mantle structures and tectonics of the island arc and trench system, *J. Phys. Earth* **18**:79–113.

Hotta, H., 1970b, A crustal section across the Izu–Ogasawara arc and trench, *J. Phys. Earth* **18**:125–141.

Hotta, H., 1972, Crustal structure in the Pacific from seismic exploration, in: *KAITEI BUTSURI*, pp. 31–66 (in Japanese).

Houtz, R. E., 1976, Seismic properties of layer 2A in the Pacific, *J. Geophys. Res.* **81**:6321–6331.

Houtz, R., and Ewing, J., 1976, Upper crustal structure as a function of plate age, *J. Geophys. Res.* **81**:2490–2498.

Houtz, R., Ewing, J., and Buhl, P., 1970, Seismic data from sonobuoy stations in the northern and equatorial Pacific, *J. Geophys. Res.* **75**:5093–5111.

Hussong, D. M., Odegard, M. E., and Wipperman, L. K., 1975, Compressional faulting of the oceanic crust prior to subduction in the Peru–Chile Trench, *Geology* **3**:601–604.

Hussong, D. M., Wipperman, L. K., and Kroenke, L. W., 1979, The crustal structure of the Ontong Java and Manihiki oceanic plateaus, *J. Geophys. Res.* **84**:6003–6010.

Hyndman, R. D., Rogers, G. C., Bone, M. N., Lister, C. R. B., Wade, U. S., Barrett, D. L., Davis, E. E., Lewis, T., Lynch, S., and Seemann, D., 1978, Geophysical measurements in the region of the Explorer ridge off western Canada, *Can. J. Earth Sci.*, **15**:1508–1525.

Ibrahim, A. K., Latham, G., and Ladd, J., 1979, Seismic refraction and reflection measurements in the Middle America Trench offshore Guatemala, *J. Geophys. Res.* **84**:5643–5649.

Ibrahim, A. K., Pontoise, B., Latham, G., Larue, M., Chen, T., Isacks, B., Recy, J., and Louat, R., 1980, Structure of the New Hebrides arc–trench system, *J. Geophys. Res.* **85**:253–266.

Keen, C. E., and Barrett, D. L., 1971, A measurement of seismic anisotropy in the northeast Pacific, *Can. J. Earth Sci.* **8**:1056–1064.

Kempner, W. B., and Gettrust, J. F., 1982a, Ophiolites, synthetic seismograms, and ocean crustal structure. 1. Comparison of ocean bottom seismometer data and synthetic seismograms for the Bay of Islands ophiolite, *J. Geophys. Res.* **87**:8447–8462.

Kempner, W. B., and Gettrust, J. F., 1982b, Ophiolites, synthetic seismograms, and ocean crustal structure. 2. A comparison of synthetic seismograms of the Samail Ophiolite, Oman, and the ROSE refraction data from the East Pacific Rise, *J. Geophys. Res.* **87**:8463–8476.

Kieckhefer, R. M., Shor, G. G., Jr., and Curray, J. R., 1980, Seismic refraction studies of the Sunda Trench and forearc basin, *J. Geophys. Res.* **85**:863–889.

Kosminskaya, I. P., and Riznichenko, Y. V., 1964, Seismic studies of the earth's crust in Eurasia, in: *Research in Geophysics,* Vol. 2, MIT Press, Cambridge, Mass., pp. 81–122.

Kovylin, V. M., and Neprochnov, Y. P., 1965, Structure of the earth's crust in the deep central part of the Japan Sea, according to seismic data, *Izv. Akad. Nauk SSSR Ser. Geol.* **4**:10–26 (in Russian).

Langseth, M. G. and Mrowzowski, C. L., 1981, Geophysical surveys for Leg 59 sites, Deep Sea Drilling Project, Init. Repts. DSDP, Washington, D.C., 487–502.

LaTraille, S. L., and Hussong, D. M., 1980, Crustal structure across the Mariana island arc, in: *The Tectonic and Geologic Evolution of Southeast Asian Seas and Islands* (D. E. Hayes, ed.), Am. Geophys. Union Monogr. **23**:209–221.

Lee, C.-H., Shor, G. G., Jr., Bibee, L. D., Lu, R. S., and Hilde, T. W. C., 1980, Okinawa Trough: Origin of a backarc basin, *Mar. Geol.* **35**:219–241.

Lee, W. H. K., and Taylor, P. T., 1966, Global analysis of seismic refraction measurements, *Geophys. J. R. Astron. Soc.* **11**:389–413.

Lewis, B. T. R., and Garmany, J. D., 1982, Constraints on the structure of the East Pacific Rise from seismic refraction data, *J. Geophys. Res.* **87**:8417–8425.

Lewis, B. T. R., and Snydsman, W. E., 1977, Evidence for a low velocity layer at the base of the oceanic crust, *Nature* **266**:340–344.

Lewis, B. T. R., and Snydsman, W. E., 1979, Fine structure of the lower oceanic crust on the Cocos, *Tectonophysics* **55**:87–105.

Leyden, R., Ewing, M., and Murauchi, S., 1973, Sonobuoy refraction measurements in East China Sea, *Bull. Am. Assoc. Pet. Geol.* **57**:2396–2403.

Louden, K. E., 1980, The crustal and lithospheric thicknesses of the Philippine Sea as compared to the Pacific, *Earth Planet. Sci. Lett.* **50**:275–288.

Ludwig, W. J., and Rabinowitz, P. D., 1980, Geophysical characteristics of ocean crust: IPOD candidate site PAC 5, central Pacific Ocean basin, *Mar. Geol.* **35**:111–127.

Ludwig, W. J., Ewing, J. I., Ewing, M., Murauchi, S., Den, N., Asano, S., Hotta, H., Hayakawa, M., Asanuma, T., Ichikawa, K., and Noguchi, I., 1966, Sediments and structure of the Japan Trench, *J. Geophys. Res.* **71**:2121–2137.

Ludwig, W. J., Murauchi, S., Den, N., Ewing, M., Hotta, H., Houtz, R. E., Yoshii, T., Asanuma, T., Hagiwara, K., Sato, T., and Ando, S., 1971, Structure of Bowers Ridge, Bering Sea, *J. Geophys. Res.* **76**:6350–6366.

Ludwig, W. J., Murauchi, S., Den, N., Buhl, P., Hotta, H., Ewing, M., Asanuma, T., Yoshii, T., and Sakajiri, N., 1973, Structure of east China Sea–west Philippine Sea margin off southern Kyushu, Japan, *J. Geophys. Res.* **78**:2526–2536.

Ludwig, W. J., Murauchi, S., and Houtz, R. E., 1975, Sediments and structure of the Japan Sea, *Bull. Geol. Soc. Am.* **86**:651–664.

Ludwig, W. J., Kumar, N., and Houtz, R. E., 1979, Profiler–sonobuoy measurements in the South China Basin, *J. Geophys. Res.* **84**:3505–3518.

McCaffrey, R., Silver, E. A., and Raitt, R. W., 1980. Crustal structure of the Molucca Sea collision zone. Indonesia, in: *The Tectonic and Geologic Evolution of Southeast Asian Seas and Islands* (D. E. Hayes, ed.), Am. Geophys. Union Monogr. **23**:161–177.

McClain, J. S., and Lewis, B. T. R., 1982, A seismic experiment at the axis of the East Pacific Rise, *Mar. Geol.* **35**:147–169.

McClain, J. S., Orcutt, J. A., and Burnett, M., 1985, The East Pacific Rise in cross section: A seismic model, *J. Geophys. Res.* **90**:8627–8639.

McClain, K. J., and Lewis, B. T. R., 1982, Geophysical evidence for the absence of a crustal magma chamber under the northern Juan de Fuca Ridge: A contrast with ROSE results, *J. Geophys. Res.* **87**:8477–8490.

McConnel, R. K., Jr., Gupta, R. N., and Wilson, J. T., 1966, Compilation of deep crustal seismic refraction profiles, *Rev. Geophys.* **4**:41–100.

Malahoff, A., Hussong, D., Odegard, M., Udintsev, G. B., Neprochnov, Y. P., Moskalenko, V. N., and Shishkina, N. A., 1977, The crustal structure in the Marcus Island area of the Pacific Ocean, *Oceanology* **17**:680–688.

Malecek, S. J., and Clowes, R. M., 1978, Crustal structure near Explorer Ridge from a marine deep seismic sounding survey, *J. Geophys. Res.* **83**:5899–5912.

Meeder, C. A., Lewis, B. T. R., and McClain, J., 1977, The structure of the oceanic crust off southern Peru determined from an ocean bottom seismometer, *Earth Planet. Sci. Lett.* **37**:13–28.

Montagner, J. P., 1985, Seismic anisotropy of the Pacific Ocean inferred from long-period surface waves dispersion, *Phys. Earth Planet. Inter.* **38**:28–50.

Mrozowski, C. L., and Hayes, D. E., 1979, The evolution of the Parece Vela basin, eastern Philippine Sea, *Earth Planet. Sci. Lett.* **46**:49–67.

Murauchi, S., 1966, Explosion seismology, in: Second Progress Report on the Upper Mantle Project of Japan (1965–1966), National Committee for UMP, Science Council of Japan, pp. 11–13.

Murauchi, S., and Ludwig, W. J., 1980, Crustal structure of the Japan Trench, *Initial Reports of the Deep Sea Drilling Project* **56, 57**:463–469.

Murauchi, S., Den, N., Asano, S., Hotta, H., Chujo, J., Asanuma, T., Ichikawa, K., and Noguchi, I., 1964, A seismic refraction exploration of Kumano Nada (Kumano Sea), Japan, *Proc. Jpn. Acad.* **40**:111–115.

Murauchi, S., Den, N., Asano, S., Hotta, H., Yoshii, T., Asanuma, T., Hagiwara, K., Ichikawa, K., Sato, T., Ludwig, W. J., Ewing, J. I., Edgar, N. T., and Houtz, R. E., 1968, Crustal structure of the Philippine Sea, *J. Geophys. Res.* **73**:3143–3171.

Murauchi, S., Ludwig, W. J., Den, N., Hotta, H., Asanuma, T., Yoshii, T., Kubotera, A., and Hagiwara, K., 1973, Seismic refraction measurements on the Ontong Java Plateau northeast of New Ireland, *J. Geophys. Res.* **78**:8653–8663.

Nagata, K., 1984, Crustal and upper mantle structure of Mariana Basin determined by ocean bottom seismographic observation, Master's thesis submitted to Hokkaido University.

Nagumo, S., Ouchi, T., and Koresawa, S., 1980, Seismic velocity structure near the extinct spreading center in the Shikoku Basin, north Philippine Sea, *Mar. Geol.* **35**:135–146.

Nagumo, S., Ouchi, T., Kasahara, J., Koresawa, S., Tomoda, Y., Kobayashi, K., Furumoto, A. S., Odegard, M. E., and Sutton, G. H., 1981, Sub-Moho seismic profile in the Mariana Basin-ocean bottom seismograph long-range explosion experiment, *Earth Planet. Sci. Lett.* **53**:93–102.

Nasu, N., Von Huene, R., Ishiwada, Y., Langseth, M., Bruns, T., and Honza, E., 1980, Interpretation of multichannel seismic reflection data, Legs 56 and 57, Japan Trench transect, Deep Sea Drilling Project, *Initial Reports of the Deep Sea Drilling Project* **56, 57**:489–503.

Neprochnov, Y. P., Sedov, V. V., Semenov, G. A., Yel'nikov, I. N., and Filaktov, V. D., 1980, The crustal structure and seismicity of the Hess Basin area in the Pacific Ocean, *Oceanology* **20**:317–322.

Nishizawa, A., 1985, Fine crustal structure at the southern part of the Kurile Trench by airgun–OBS system, Dr. Sci. dissertation to Tohoku University.

Nishizawa, A., and Suyehiro, K., 1986, Crustal structure across the Kurile Trench off southeastern Hokkaido by airgun–OBS profiling, *Geophys. J. R. Astron. Soc.* **86**:371–397.

Nishizawa, A., Suyehiro, K., and Shimizu, H., 1983, Seismic refraction experiment at the Amami Plateau, *J. Phys. Earth* **31**:159–171.

Nosaka, M., Nishizawa, A., Suyehiro, K., Tokuyama, H., and Kagami, H., 1984, Crustal structure off Kashima from OBS–airgun profiling, Abstr. Spring Meet. Seismol. Soc. Jpn. B83.

Okada, H., Moriya, T., Masuda, T., Hasegawa, T., Asano, S., Kasahara, K., Ikami, A., Aoki, H., Sasaki, Y., Hurukawa, N., and Matsumura, K., 1978, Velocity anisotropy in the Sea of Japan as revealed by big explosions, *J. Phys. Earth* **26**(Suppl.):491–502.

Orcutt, J. A., and Dorman, L. M., 1977, An oceanic long range explosion experiment, *J. Geophys.* **43**:257–263.

Orcutt, J. A., Kennett, L. N., and Dorman, L. M., 1976, Structure of the East Pacific Rise from an ocean bottom seismometer survey, *Geophys. J. R. Astron. Soc.* **45**:305–320.

Purdy, G. M., 1982a, The correction for the travel time effects of seafloor topography in the interpretation of marine seismic data, *J. Geophys. Res.* **87**:8389–8396.

Purdy, G. M., 1982b, The variability in seismic structure of layer 2 near the Pacific Rise at 12 N, *J. Geophys. Res.* **87**:8403–8416.

Purdy, G. M., 1983, The seismic structure of 140 My old crust in the western central Atlantic Ocean, *Geophys. J. R. Astron. Soc.* **72**:115–138.

Purdy, G. M., and Detrick, R. S., 1978, A seismic refraction experiment in the Central Banda Sea, *J. Geophys. Res.* **83**:2247–2257.

Raitt, R. W., 1963, The crustal rocks, in: *The Sea* (M. N. Hill, ed.), Vol. 3, Wiley–Interscience, New York, pp. 794–815.

Raitt, R. W., 1964, Geophysics of the south Pacific, in: *Research in Geophysics*, Vol. 2, MIT Press, Cambridge, Mass., pp. 223–241.

Raitt, R. W., Fisher, R. L., and Mason, R. G., 1955, Tonga Trench, *Geol. Soc. Am. Spec. Pap.* **62**:237–254.

Raitt, R. W., Shor, G. G., Jr., Francis, T. J. G., and Morris, G. B., 1969, Anisotropy of the Pacific upper mantle, *J. Geophys. Res.* **74**:3095–3109.

Raitt, R. W., Shor, G. G., Jr., Morris, G. B., and Kirk, H. K., 1971, Mantle anisotropy in the Pacific Ocean, *Tectonophysics* **12**:173–186.

Reid, I., Orcutt, J. A., and Prothero, W. A., 1977, Seismic evidence for a narrow zone of partial melting underlying the East Pacific Rise at 21 N, *Bull. Geol. Soc. Am.* **88**:678–682.

Research Group for Explosion Seismology, 1977, Regionality of the upper mantle around northeastern Japan as derived from explosion seismic observations and its seismological implications, *Tectonophysics* **37**:117–130.

Rosendahl, B. R., Raitt, R. W., Dorman, L. M., Bibee, L. D., Hussong, D. M., and Sutton, G. H., 1976, Evolution of ocean crust. 1. A physical model of the East Pacific Rise crest derived from seismic refraction data. *J. Geophys. Res.* **81**:5294–5304.

Rynn, J. M. W., and Reid, I. D., 1983, Crustal structure of the western Arafura Sea from ocean bottom seismograph data, *J. Geol. Soc. Aust.* **30**:59–74.

Shearer, P., and Orcutt, J., 1985, Anisotropy in the oceanic lithosphere—Theory and observations from the Ngendei seismic refraction experiment in the south-west Pacific, *Geophys. J. R. Astron. Soc.* **80**:493–526.

Shimamura, H., 1984, Anisotropy in the oceanic lithosphere of the northwestern Pacific Basin, *Geophys. J. R. Astron. Soc.* **76**:253–260.

Shimamura, H., and Asada, T., 1983, Velocity anisotropy extending over the entire depth of the oceanic lithosphere, in: *Geodynamics of the Western Pacific–Indonesian Region* (T. W. C. Hilde and S. Uyeda, eds.), Geodynamics Series **11**:121–126.

Shimamura, H., Asada, T., Suyenhiro, K., Yamada, T., and Inatani, H., 1983, Longshot experiments to study velocity anisotropy in the oceanic lithosphere of the northwestern Pacific, *Phys. Earth Planet. Inter.* **31**:348–362.

Shor, G. G., Jr., 1962, Seismic refraction studies off the coast of Alaska, *Bull. Seismol. Soc. Am.* **52**:37–57.

Shor, G. G., Jr., 1964, Structure of the Bering Sea and the Aleutian Ridge, *Mar. Geol.* **1**:213–219.

Shor, G. G., Jr., and Fisher, R. L., 1961, Middle America Trench: Seismic–refraction studies, *Bull. Geol. Soc. Am.* **72**:721–730.

Shor, G. G., Jr., and Fornari, D. J., 1976, Seismic refraction measurements in the Kamchatka Basin, western Bering Sea, *J. Geophys. Res.* **81**:5260–5266.

Shor, G. G., Jr., Dehlinger, P., Kirk, H. K., and French, W. S., 1968, Seismic refraction studies off Oregon and northern California, *J. Geophys. Res.* **73**:2175–2194.

Shor, G. G., Jr., Menard, H. W., and Raitt, R. W., 1970. Structure of the Pacific Basin, in: *The Sea* (A. E. Maxwell, ed.), Vol. 4, Wiley–Interscience, New York, pp. 3–27.

Shor, G. G., Jr., Kirk, H. K., and Menard, H. W., 1971, Crustal structure of the Melanesian area, *J. Geophys. Res.* **76**:2562–2586.

Snydsman, W. E., Lewis, B. T. R. and McClain, J., 1975, Upper mantle velocities on the Northern Cocos Plate. *Earth Planet Sci. Lett.*, **28**:46–50.

Spudich, P., and Orcutt, J., 1980, A new look at the seismic velocity structure of the oceanic crust, *Rev. Geophys. Space Phys.* **18**:627–645.

Stephen, R. A., 1983, The oblique seismic experiment on Deep Sea Drilling Project Leg 70, *Initial Reports of the Deep Sea Drilling Project* **69**:301–308.

Stoffa, P. L., and Buhl, P., 1979, Two-ship multichannel seismic experiments for deep crustal studies: Expanded spread and constant offset profiles, *J. Geophys. Res.* **84**:7645–7663.

Stoffa, P. L., Buhl, P., Herron, T. J., Kan, T. K., and Ludwig, W. J., 1980, Mantle reflections beneath the crestal zone of the East Pacific Rise from multichannel seismic data, *Mar. Geol.* **35**:83–97.

Sutton, G. H., Maynard, G. L., and Hussong, D. M., 1971, Widespread occurrence of a high-velocity basalt layer in the Pacific crust found with repetitive sources and sonobuoys, in: *The Structure and Physical Properties of the Earth's Crust* (J. G. Heacock, ed.), Am. Geophys. Union Monogr. **14**:193–209.

Suyehiro, K., Inatani, H., Kono, T., and Yamamoto, K., 1984, Upper crustal structure beneath the continental slope off the Joban coast, Honshu, Japan, *J. Phys. Earth* **32**:83–96.

Suyehiro, K., Kanazawa, T., Nishizawa, A., and Shimamura, H., 1985, Crustal structure beneath the inner trench slope of the Japan Trench, *Tectonophysics* **112**:155–191.

Suyehiro, K., Kanazawa, T., and Shimamura, H., 1986, Air gun–ocean bottom seismograph structure across the Japan Trench area, *Initial Reports of the Deep Sea Drilling Project* **87**:751–755.

Takeda, K., 1978, Seismic exploration studies using airguns and ocean-bottom-seismometers, M.Sc. Thesis, Hokkaido University.

Tanimoto, T., and Anderson, D. L., 1985, Lateral heterogeneity and azimuthal anisotropy of the upper mantle: Love and Rayleigh waves 100–250s, *J. Geophys. Res.* **90**:1842–1858.

Watts, A. B., ten Brink, U. S., Buhl, P., and Brocher, T. M., 1985, A multichannel seismic study of lithospheric flexure across the Hawaiian–Emperor seamount chain, *Nature* **315**:105–111.

Wipperman, L. K., Larson, R. L., and Hussong, D. M., 1981, The geological and geophysical setting near site 462, *Initial Reports of the Deep Sea Drilling Project* **61**:763–770.

Woolard, G. P., 1975, The interrelationships of crustal and upper mantle parameter values in the Pacific, *Rev. Geophys. Space Phys.* **13**:87–137.

Yoshii, T., 1975, Proposal of the "aseismic front," *Zisin 2* **28**:365–367 (in Japanese).

Yoshii, T., Ludwig, W. J., Den, N., Murauchi, S., Ewing, M., Hotta, H., Buhl, P., Asanuma, T., and Sakajiri, N., 1973, Structure of southwest Japan margin off Shikoku, *J. Geophys. Res.* **78**:2517–2525.

Yoshii, T., Okada, H., Asano, S., Ito, K., Hasegawa, T., Ikami, A., Moriya, T., Suzuki, S., and Hamada, K., 1981, Regionality of the upper mantle around northeastern Japan as revealed by big explosions at sea. II. Seiha-2 and Seiha-3 experiment, *J. Phys. Earth* **29**:201–220.

Zucca, J. J., and Hill, D. P., 1980, Crustal structure of the southeast flank of Kilauea volcano, Hawaii, from seismic refraction measurements, *Bull. Seismol. Soc. Am.* **70**:1149–1159.

Zucca, J. J., Hill, D. P., and Kovach, R. L., 1982, Crustal structure of Mauna Loa volcano, Hawaii, from seismic refraction and gravity data, *Bull. Seismol. Soc. Am.* **72**:1535–1550.

Chapter 11D

MAGNETIC ANOMALIES

Nobuhiro Isezaki

Department of Earth Sciences
Kobe University
Nada, Kobe 657, Japan

I. INTRODUCTION

To date, enormous volumes of magnetic data measured at sea level by ships, above the sea surface by airplanes, and by satellites have been accumulated. Satellite data, which have an almost uniform geographic coverage, have detected long-wavelength anomalies which are thought to be associated with magnetic structures lying some tens of kilometers deep. In contrast, surface data are not equally spaced, but are very dense in one area and sparse in another. The anomalies observed by ships are associated with very shallow magnetic structures, mainly above the oceanic layer. Magnetic anomaly lineations, which cannot be measured by satellite because of the altitude, have played a very important role in the earth sciences for more than two decades.

In the Pacific Ocean, magnetic anomaly lineations were first measured in detail in areas off the west coast of the United States by scientists of the Scripps Institution of Oceanography (SIO), University of California (Mason, 1958; Mason and Raff, 1961; Raff and Mason, 1961). Vacquier *et al.* (1961) showed great offsets (1420 km) of magnetic anomaly lineations across the Mendocino and Pioneer Fracture Zones.

Since Vine and Matthews (1963) and Vine (1966) proved that magnetic anomaly lineations were originated by the combined effect of seafloor spreading (Dietz,

1961; Hess, 1962) and reversal of geomagnetic polarity (e.g., Cox *et al.*, 1964), intensive magnetic surveys at sea have been carried out. Scientists from the Lamont–Doherty Geological Observatory (LDGO), Columbia University, established the fact that there are magnetic lineations in all oceans of the earth (e.g., Pitman *et al.*, 1968; Dickson *et al.*, 1968; LePichon and Heirtzler, 1968). The positions of Eulerian poles and the angular velocities with which the plates rotate around their Eulerian poles were first determined using magnetic anomaly lineations as well as transform faults or fracture zones (Morgan, 1968; LePichon, 1968). At the same time, seamount magnetic surveys in the Pacific were conducted and the results were interpreted in terms of movement of the seafloor (e.g., Uyeda and Richards, 1966; Vacquier and Uyeda, 1967).

In 1971, Schouten presented a mathematical method of analyzing magnetic anomaly lineations using Fourier transforms; since then, they have been interpreted by taking their skewness into account (e.g., Larson and Chase, 1972; Schouten and McCamy, 1972). This method has been used to compare anomaly profiles in different areas and to determine the ages of ocean basins and the reversal time scale of geomagnetic polarity extending to Mesozoic time (e.g., Larson and Pitman, 1972; Pitman *et al.*, 1974).

Since the 1970s, magnetic anomaly lineations have been found in the marginal basins in the western Pacific (e.g., Lee and Hilde, 1971; Isezaki *et al.*, 1971) and have played an important role in the study of the evolution of these basins. The distribution of magnetic anomalies has been defined for most of the Pacific (e.g., Pitman *et al.*, 1974).

Although magnetic surveys have been conducted on and above the sea surface and also near the seafloor, in this chapter I discuss only data collected at the surface in the Pacific.

II. AVAILABLE DATA

Figure 1 shows track lines in the Pacific for which magnetic data were available from the National Geophysical Data Center (NGDC) of the U.S. National Oceanic and Atmospheric Administration (NOAA) as of 1985. The total length of these track lines is 3.01 million nautical miles; 29% of these lines are from SIO, 26% from LDGO, 16% from NOAA, and 9% from the Hawaii Institute of Geophysics and various institutions in Japan. Because NGDC does not have all the data from all institutions, the amount of existing marine magnetic data should be much greater than that utilized here.

A. Magnetic Anomaly Lineations in the Pacific Basins

Figure 2 shows the distribution of magnetic anomaly lineations in the Pacific compiled by Karasik and Cochebanoba (1981). The main oceanic ridges presently

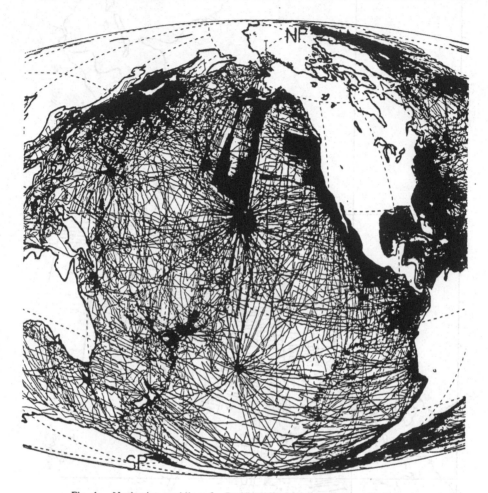

Fig. 1.   Navigation tracklines for Pacific NGDC geophysical data holdings.

active as spreading centers in the Pacific are the Juan de Fuca Ridge, the Gorda Ridge, the East Pacific Rise, the Pacific–Antarctic Ridge, the Galapagos Rift, and the Chile Rise. As mentioned above, the area off the west coast of North America is the area of the Pacific where magnetic anomaly lineations were first established. Because of their poor topography, two features in this area—the Juan de Fuca and Gorda Ridges (J and G in Fig. 2)—were not immediately recognized as ridges from bathymetric data, but were identified from the magnetic anomaly lineations (Vine and Wilson, 1965; Wilson, 1965b). The Juan de Fuca Ridge is connected to the Queen Charlotte Island Fault to the north and the Gorda Ridge and the San Andreas Fault to the south.

The San Andreas Fault, which is a typical transform fault (Wilson, 1965a), is connected to the East Pacific Rise via the Gulf of California, which started opening at anomaly 3 time (4 Ma) (Larson *et al.*, 1968; Moore and Buffington, 1968). The

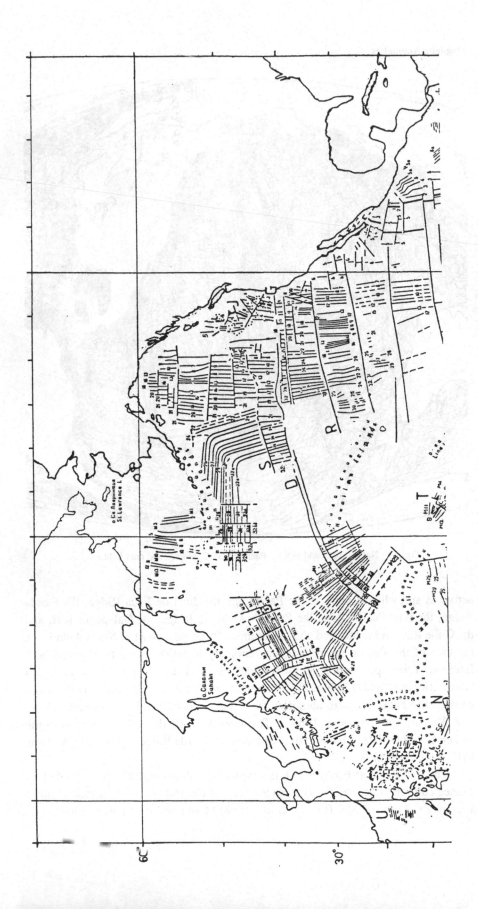

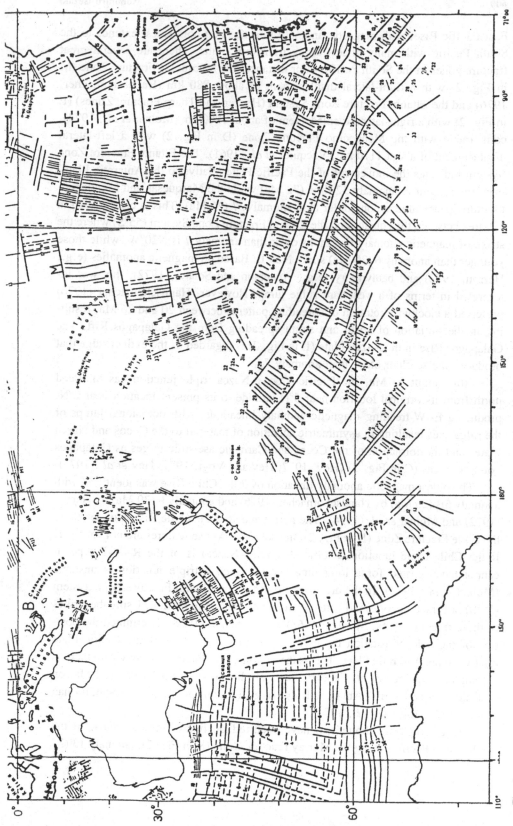

Fig. 2.   Magnetic anomaly lineations in the Pacific Ocean compiled by Karasik and Cochebanoba (1981). B = Bismarck Sea; C = Cocos and Carnegie ridges; D = Mendocino Fracture Zone; E = Eltanin Fracture Zone; G = Gorda Ridge; J = Juan de Fuca Ridge; L = Galapagos Rise; M = Marquesas Fracture Zone; N = Caroline Plate; R = Murray Fracture Zone; S = Surveyer Fracture Zone; T = Magellan Rise; U = South China Sea basin; V = Coral Sea Basin; W = Lau Basin.

East Pacific Rise, with magnetic anomaly lineations on both sides, extends to the South Pacific with many offsets along transform faults. There are several great fracture zones in the South Pacific—for instance, the Marquesas Fracture Zone (M in Fig. 2) with a right-lateral displacement of about 880 km (Handschumacher, 1976) and the Eltanin Fracture Zone System (Heezen and Tharp Fracture Zones) (E in Fig. 2) with a right-lateral displacement of about 1000 km (Molnar *et al.*, 1974), comparable with the Mendocino Fracture Zone (D in Fig. 2) with a left-lateral displacement of about 1170 km (Vacquier *et al.*, 1961). The Eltanin Fracture Zone System indicates the movement of the Pacific Plate relative to the Antarctic Plate, from the present to anomaly 32 time (72 Ma) while the Marquesas and Mendocino Fracture Zones indicate one older than anomaly, 7 (26 Ma). This is clearly seen in the area between the equator and the Eltanin Fracture Zone System (55°S) where the strike of magnetic anomaly lineations older than anomaly 7 is N20°W, while those younger than anomaly 6 (20 Ma) strike N20°E. Based on magnetic anomalies (e.g., Herron, 1972) and bathymetry (e.g., Anderson and Sclater, 1972), this was interpreted in terms of a westward ridge jump. However, Handschumacher (1976) preferred a clockwise rotation of spreading patterns between 20 and 26 Ma, resulting in the initiation of north and south spreading from the Galapagos Rift. The Galapagos Rise in the Nazca Plate (L in Fig. 2) is regarded as the extinct ridge that produced the seafloor older than anomaly 7.

From about 23 Ma, the Pacific–Cocos–Nazca triple junction has migrated north from its original location near 5°N latitude to its present location near 2°N, producing E–W-trending magnetic anomaly lineations, with occasional jumps of the ridge axes resulting in asymmetric accretion of material to the Cocos and Nazca Plates and the formation of the Cocos and Carnegie aseismic ridges as Galapagos hot-spot traces (C in Fig. 2) (Hey, 1977; Hey and Vogt, 1977; Hey *et al.*, 1977).

The oldest magnetic anomaly lineation over the Chile Rise was identified with anomaly 6B (23 Ma) by Handschumacher (1976) and anomaly 9 (29 Ma) by Herron (1972) and Molnar *et al.* (1974). The northern end of the Chile Rise is offset along the Chile Fracture Zone (H in Fig. 2), the details of whose features are not yet clear. If the Chile triple junction (Pacific–Antarctic–Nazca) is of the R–R–F type, it cannot have existed for a long time because such a triple junction is unstable (Molnar *et al.*, 1974). In the area to the south of the Eltanin Fracture Zone System (55°S), all fracture zones indicate the almost present plate motion. Using the strikes of these fracture zones, Morgan (1968) and LePichon (1968) determined the position of the pole of rotation representing the present motion of the Pacific Plate relative to the Antarctic Plate. This is the region of the Pacific Ocean where there is a complete sequence of all Cenozoic anomaly lineations from 1 to 32 (72 Ma) on both sides of the Pacific–Antarctic Ridge (e.g., Pitman and Heirtzler, 1966; Pitman *et al.*, 1968; Molnar *et al.*, 1974).

Ridge jumps are thought to have occurred on many ridge systems. In the northeastern Pacific, along the Murray Fracture Zone (R in Fig. 2), anomaly 13 (37

Ma) is offset by 150 km while anomaly 21 (48 Ma) is offset by 680 km. Although Malahoff and Handschumacher (1971) interpreted anomalies 8–13 as those formed by a secondary spreading center, Harrison and Sclater (1972) explained them in terms of an eastward ridge jump of 530 km at 40 Ma, resulting in a different identification of anomaly lineations in the area between 130°W and 140°W as shown in Fig. 3. The interpretation of Malahoff and Handschumacher (1971) in this area is adopted in Fig. 2. Shih and Molnar (1975) interpreted magnetic anomaly lineations to the east of the Surveyer Fracture Zone (S in Fig. 2) by a ridge jump which resulted in the elimination of the offset represented by the Surveyer Fracture Zone in the area (Fig. 4).

Spreading is not always symmetrical about the ridge axis, as with the Galapagos Rift mentioned above. Using detailed magnetic anomaly lineations, Elvers *et al.* (1973) demonstrated asymmetry about the Juan de Fuca and Gorda Ridges. However, Hey and Wilson (1982) explained the tectonic evolution in the Juan de Fuca Ridge area in terms of a continuously propagating rift (spreading center) model (Hey *et al.*, 1980) on the assumption of symmetric spreading. They claimed that ridge jumps and major seafloor lineations oblique to both relative and absolute plate motions in this area (e.g., offsets of magnetic anomaly lineations on the Juan de Fuca Ridge) resulted from rift propagation. This model would provide a simple explanation for a large class of previously unexplained seafloor features although it so far has proved successful only in such small areas as the Juan de Fuca Ridge area and the Easter microplate areas (Hey *et al.*, 1985).

In the Gulf of Alaska, N–S-trending magnetic anomaly lineations (e.g., Pitman and Hayes, 1968; Naugler and Wageman, 1973; Taylor and O'Neill, 1974) have been explained by the movement of the ridge triple junction into the Gulf of Alaska (Hayes and Pitman, 1970); however, anomalies older than anomaly 23 (52 Ma) trend N20°W to the north of the Mendocino Fracture Zone, and they trend E–W to the south of the Aleutian Trench (Grim and Erickson, 1969; Peter *et al.*, 1970) (Fig. 5). This great change in the trend of the lineations is called a magnetic bight— one of the most important features of magnetic anomaly lineations in the Pacific.

Hayes and Pitman (1970) explained the origin of the magnetic bight in the northeastern Pacific called the Great Magnetic Bight in terms of the triple junction of ridges that moved northeast relative to the Pacific Plate and subsided under the Aleutian Trench at early Paleocene time. Grow and Atwater (1970) named one limb of the triple ridges the Kula Ridge; it produced the lineations presently trending E–W to the south of the Aleutian Trench. Atwater (1970) described the Cenozoic plate motions between the North American Plate and the three oceanic plates—the Pacific, Kula, and Farallon Plates—providing an explanation for the tectonic evolution of western North America.

As indicated earlier by Raff (1966), the region of magnetic anomalies off the west coast of the United States is bounded on the west by a zone of very low-amplitude anomalies, the magnetic quiet or smooth zone. The most plausible origin

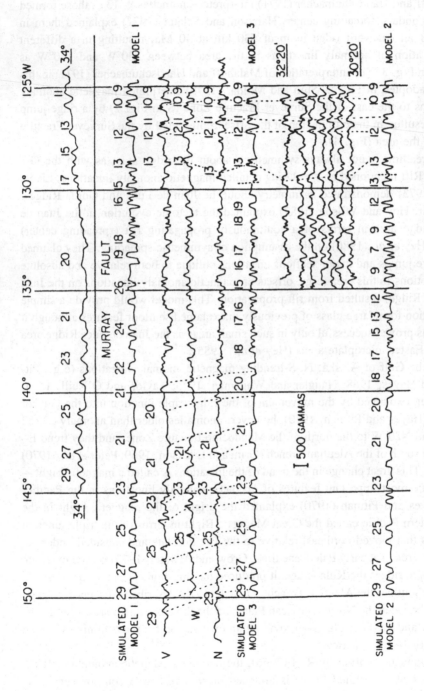

Fig. 3. Profiles of models 2 and 3 are simulated anomaly profiles from Malahoff and Handschumacher (1971) and Harrison and Sclater (1972), respectively. Model 1 is a simulated anomaly according to the model of Menard and Atwater (1969) (Harrison and Sclater, 1972). See the different identifications of anomalies from 13 to 21 among three models.

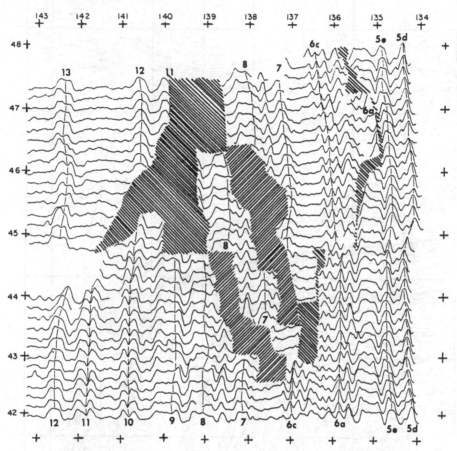

Fig. 4. Upper left to lower right diagonal rulings indicate locations where westward jumps of spreading centers have occurred. Upper right to lower left rulings indicate approximate locations where eastward jumps of spreading centers have occurred (Shih and Molnar, 1975).

of a magnetic quiet zone is a long period of dominantly normal geomagnetic polarity (e.g., Larson and Pitman, 1972). Since Heirtzler *et al.* (1968) numbered the magnetic lineations, new lineations have been found and numbered and, in the northeastern Pacific, there are well-defined lineations that have been the basis for the reversal time scale of geomagnetic polarity for Late Cretaceous and Cenozoic time (e.g., LaBrecque *et al.*, 1977). [The age of anomaly lineations is taken from Harland *et al.* (1982) throughout this chapter.] The number of the oldest definitely identified anomaly is 32 (72 Ma) in the Pacific (anomalies 33 and 34 are not always identified). The Cretaceous magnetic quiet zone, i.e., the area between anomaly 34 (83 Ma) and M0 (118 Ma), where the polarity of the seafloor magnetization is normal, covers a wide area in the central Pacific.

Mesozoic magnetic anomaly lineations have been identified in the north-

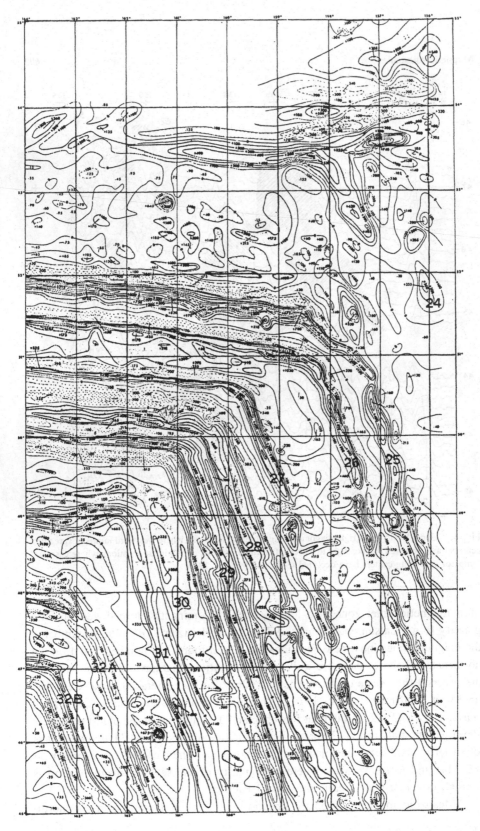

Fig. 5. Magnetic anomaly contours of the magnetic bight in the northeast Pacific. Contour interval is 100 nT. Large numerical orders from 24 to 32B are anomaly lineation numbers (Peter *et al.*, 1970).

western Pacific (e.g., Larson and Chase, 1972; Hilde *et al.*, 1976). Southeast of the Shatzky Rise, in particular, there is a complete sequence of well-defined anomaly lineations from M0 (118 Ma) to M25 (161 Ma), which Larson and Hilde (1975) used to revise the Mesozoic reversal time scale of geomagnetic polarity proposed earlier by Larson and Pitman (1972). In the same area, the magnetic anomaly lineations are offset along two long faults which seem to continue to the Mendocino Fracture Zone (Hilde *et al.*, 1976) (see Fig. 2). Southwest of the Shatzky Rise there exists a magnetic bight similar to that found in the northeastern Pacific (Hilde *et al.*, 1976). Mesozoic anomaly lineations with a N70°E trend occur along the Marshall and Kiribath Island chain in the western equatorial Pacific region (Larson *et al.*, 1972; Larson, 1976). In the Central Pacific Basin (Magellan Rise) area (T in Fig. 2), Tamaki *et al.* (1979) found Mesozoic lineations from M9 (139 Ma) to M11 (126 Ma), with a fan-shaped trend of N45°W. The area between the western equatorial Pacific lineations and the northwestern Pacific lineations is also occupied by a Jurassic magnetic quiet zone. Cande *et al.* (1978) showed that the Jurassic anomaly lineations from M26 (162 Ma) to M29 (164 Ma) have amplitudes smaller than 100 nT along the boundary of the Jurassic magnetic quiet zone.

In the westernmost equatorial Pacific, the Caroline Plate (N in Fig. 2) was postulated by Weissel and Anderson (1978) where spreading started at the time of anomaly 13 (36 Ma) and stopped at the time of about anomaly 8 (28 Ma). From 28 Ma to 1 Ma, no relative motion occurred between the Caroline and Pacific Plates; however, since 1 Ma, the Caroline Plate is thought to have behaved as a discrete plate (Hegarty *et al.*, 1983).

The following is a summary of the plate tectonics interpretation of the overall distribution of magnetic anomaly lineations in the Pacific.

At about 190 Ma, the Kula, Farallon, and Phoenix Plates met at a single point, a ridge triple junction, which developed into three R–R–R triple junctions within the Pacific Plate (Hilde *et al.*, 1977) (Fig. 6A). As the Pacific Plate grew, the Kula–Pacific Ridge moved northward relative to the Pacific Plate and, by Late Cretaceous time, subsided under the Japanese islands (Fig. 6B). The western segment of the Kula–Pacific Ridge, now called the Central Basin Ridge—the spreading center of the West Philippine Basin—had been offset by a transform fault which became a subduction zone due to the change in Pacific Plate motion in the Late Eocene (Uyeda and Ben-Avraham, 1972). The Kula–Pacific Ridge was subducted under the Aleutian Trench by late Tertiary time (Grow and Atwater, 1970), and one of the three triple junctions migrated into the Gulf of Alaska, forming the Great Magnetic Bight (Pitman and Hayes, 1968).

On the other hand, Woods and Davies (1982) presented another plate tectonics model wherein the Pacific–Kula ridge was born and began spreading and the Kula plate detached at about 85 Ma from the Pacific Plate because extrapolation of the Great Magnetic Bight backwards in time resulted in an implausible ridge configuration. Therefore, they proposed the Izanagi plate instead of the Kula plate in the

Mesozoic which consequently led to the new interpretation that the Pacific–Farallon–Izanagi triple junction formed the Mesozoic Japanese lineations and the Mesozoic magnetic bight in the Shatzky Rise area.

The Pacific–Farallon Ridge started subducting beneath the west coast of North America in the middle Tertiary (Fig. 7A), and almost the entire Farallon Plate was subducted by about 10 Ma (e.g., Atwater, 1970) (Fig. 7C). The Juan de Fuca Ridge, the Gorda Ridge, and the East Pacific Rise are the remnants of the Pacific–Farallon Ridge.

At the time when the subduction of the Pacific–Farallon Ridge beneath the North American continent started (about 26 Ma) (Fig. 7A), the southern segment of the ridge, a part of which may be the extinct Galapagos ridge, rotated and/or jumped to the present East Pacific Rise (e.g., Herron, 1972; Handschumacher, 1976) (Fig. 7B), and the remnant Farallon Plate was divided into the Cocos and Nazca Plates by the Galapagos Rift (e.g., Hey, 1977).

The Pacific–Phoenix Ridge migrated rapidly to the southeast by 65 Ma (Hilde et al., 1977) (Fig. 6B) and the Phoenix Plate vanished before the formation of the Tasman Sea (Fig. 8). After the formation of the Tasman Sea, between 37 Ma (anomaly 13) and 56 Ma (anomaly 25)—during which time Australia began to drift away from Antarctica—the direction of plate motion in the South Pacific changed. This is seen in the Eltanin Fracture Zone System, which changes in direction, offset, and spacing at about anomaly 25 time (e.g., Molnar et al., 1974).

B. Magnetic Anomaly Lineations in the Pacific Marginal Basins

A number of backarc basins have been developed along the western margin of the Pacific. There are several models for the origin of a backarc basin: a forceful diapir model (e.g., Karig, 1971); a secondary convection model (e.g., Hsui and Toksoz, 1981); a passive upwelling model (e.g., Uyeda and Kanamori, 1979); and a trapped basin model (e.g., Ben-Avraham and Uyeda, 1983). The nonuniqueness of these models seems to correspond to a variety of features of magnetic anomaly lineations in the backarc basins: one-sided, two-sided, trapped lineations, and so forth. The following is a brief review of the magnetic anomaly lineations in Pacific backarc basins.

In the West Philippine Sea, magnetic anomaly lineations were first reported by Lee and Hilde (1971); since then, they have been investigated by many others (e.g., Ben-Avraham et al., 1972; Uyeda and Ben-Avraham, 1972; Louden, 1976; Watts et al., 1977; Hilde and Lee, 1984). Identification of the ages of lineations, however, differed: for instance, they were regarded as Mesozoic by Ben-Avraham et al. (1972) and as Late Cretaceous to Tertiary by Louden (1976). Recently, Shih (1983) proposed that the opening of the West Philippine Basin started at the time of anomaly 25 (56 Ma) and ceased at the time of anomaly 7A (25 Ma). Shih (1983)

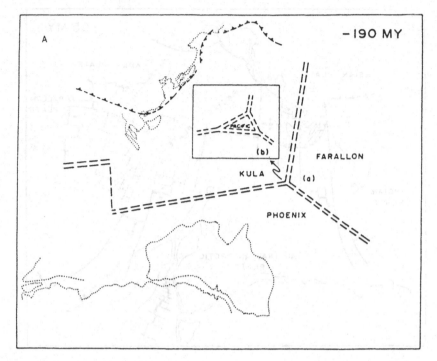

Fig. 6. (A)(a): Plate, ridge, transform faults, and subduction zone locations in the western Pacific at 190 Ma, (b) Pacific plate origin at about 185 Ma, and (B), those at 65 Ma (Hilde *et al.*, 1977).

and Hilde and Lee (1984) claimed, as first proposed by Uyeda and Ben-Avraham (1972), that the West Philippine Basin is a trapped basin that resulted from the change in Pacific Plate motion at about 40 Ma (e.g., Clague and Jarrard, 1973), while several authors (e.g., Karig, 1975) claimed that the West Philippine Basin originated by backarc spreading from the Central Basin Ridge (Fault).

In the Shikoku Basin adjacent to the West Philippine Basin, there are magnetic anomaly lineations with a N–S trend created by two-sided backarc spreading (e.g., Tomoda *et al.*, 1975; Watts and Weissel, 1975; Kobayashi and Nakata, 1978; Shih, 1980), whose age was estimated from about 24 Ma (anomaly 6C) to about 16 Ma (anomaly 5C) (Shih, 1980). Although the Shikoku and Parece Vela Basins seem to form one backarc basin between the Kyushu–Palau Ridge and the Izu–Bonin and West Mariana Arcs, Mrozowski and Hayes (1979) presented ages for the lineations in the Parece Vela Basin that are slightly different from those in the Shikoku Basin, i.e., ages ranging from about 30 Ma (anomaly 10) to about 17 Ma (anomaly 5D).

In the Sea of Japan, although Isezaki *et al.* (1971) found magnetic anomaly lineations in the Japan Basin, definite correlation with the reversal time scale has been difficult because there are only a few anomaly peaks in the basin (Isezaki and Uyeda, 1973). Recently, Isezaki (1986) compiled almost all available data of the

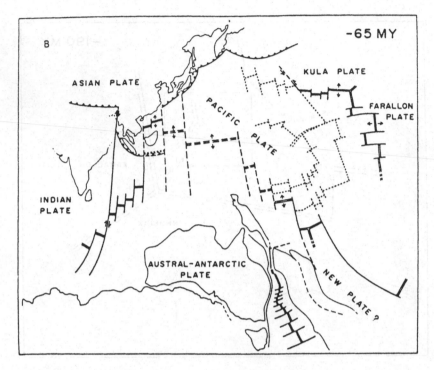

Fig. 6. (*continued*)

Sea of Japan, and two-sided lineated anomalies in the Japan Basin were correlated with the Cenozoic anomalies from 11 (32 Ma) to 5B (15 Ma).

Ben-Avraham and Uyeda (1973) recognized magnetic anomaly lineations trending almost east and west in the South China Sea basin (U in Fig. 2), and Taylor and Hayes (1980) correlated these with lineations from anomaly 11 (32 Ma) to 5D (18 Ma), and concluded that its genesis may have not been a simple backarc spreading but similar to the genesis of the Tasman and Coral Sea Basins.

In the Tasman Sea there are two-sided magnetic anomaly lineations from anomaly 33 (79 Ma) to 24 (52 Ma), with a N30°W trend (e.g., Hayes and Ringis, 1973), and these lineations were interpreted in terms of simply two separating continental blocks (plates), i.e., the Australian continent and the Lord Howe rise (see Fig. 8) (e.g., Weissel and Hayes, 1977).

Other areas with two-sided magnetic anomaly lineations include: the Bismarck Sea (B in Fig. 2), from anomaly 2A (3 Ma) to 1 trending N70°E (e.g., Taylor, 1979); the Coral Sea Basin (V in Fig. 2), from anomaly 26 (58 Ma) to 24 (52 Ma) trending N70°W (e.g., Weissel and Watts, 1979); the South Fiji Basin, fan-shaped anomalies from anomaly 12 (34 Ma) to 7A (26 Ma) (e.g., Weissel and Watts, 1975); and the Lau Basin (W in Fig. 2), from anomaly 2A (3 Ma) to 1 (e.g.,

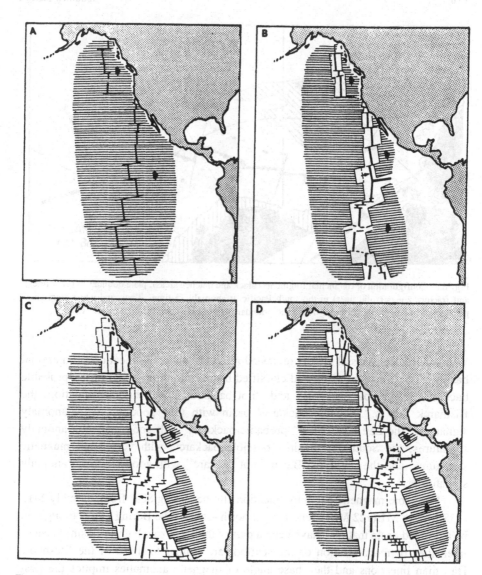

Fig. 7.   Reconstructions of spreading patterns and related tectonic events during the Miocene. Although these diagrams are schematic and contain scale distortions where necessary for clarity, they are consistent with the observed magnetic and bathymetric data. In each, the active spreading ridge is indicated by the heavy line; magnetic and bathymetric trends generated by preceding stage or stages, by light solid or dashed lines. The Pacific plate is held fixed; the relative motion of the other plate is shown by arrows. (A) ~ 26 Ma; Pacific–Farallon ridge intact but colliding with the Americas plate (Farallon–Americas trench system) off the west coast of North America. (B) ~ 20 Ma; a clockwise rotation of the southern portion of the Pacific–Farallon ridge, development of the Galapagos rift zone, and development of dual spreading from the Pacific–Antarctic Ridge and Chile Rise. (C) ~ 10 Ma; breakup of the ridge south of Baja California, northward extension of dual spreading of the Pacific–Antarctic Ridge and Chile Rise, and indicated ridge jumps (small arrows). (D) ~ 5 Ma; development of the Gorda–Juan de Fuca–San Andreas–Gulf of California spreading system, eastward jump of the East Pacific Rise from its former position at the crest of the Mathematicians ridge, and westward jumps of the Galapagos triple junction and of the East Pacific Rise from its former position on the Galapagos Rise (Handschumacher, 1976).

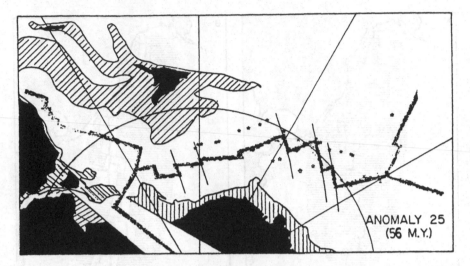

Fig. 8.   Configuration of continental fragments (black and hatched areas), plate boundaries (gray lines), and fracture zones (solid lines) in the south Pacific at anomaly 25 time. Black and white stars show position of anomaly 31. Australia–Antarctic boundary is not yet active (Molnar *et al.*, 1975).

Weissel, 1977). Weissel (1981) reviewed magnetic anomaly lineations in marginal basins of the western Pacific and classified basins into three categories: "probable backarc," "possible backarc," and "not backarc." After his classification, the Bismarck Sea is "probable backarc" basin with very good magnetic anomaly signatures, the Lau Basin also "probable backarc" with good magnetic anomaly signatures, the South Fiji Basin "possible backarc" with very good magnetic signatures, and the Coral Sea Basin "not backarc" with good magnetic anomaly signatures.

Mesozoic magnetic anomaly lineations in the Bering Sea, from anomaly M13 (138 Ma) to M1 (122 Ma), trend almost north–south. The Bering Sea is thought to be a trapped basin that may have been a part of the Mesozoic Pacific Basin, because intensities of magnetization of the seafloor are similar to those of the Mesozoic Hawaiian lineations and the phase angle of magnetic anomalies implies the possibility of northward migration of the seafloor of the basin (Cooper et al., 1976).

C. Magnetic Anomalies Due to Seamounts in the Pacific

Magnetic anomalies due to a seamount carry almost the same information about the motion of the seafloor or plate as magnetic lineations do; the difference between them is only in the dimensions of the magnetic source, i.e., two dimensions for a lineation and three dimensions for a seamount. From this point of view, Uyeda and Richards (1966) and Vacquier and Uyeda (1967) investigated magnetic anomalies of seamounts in the northwestern Pacific near the Japanese islands and obtained a northward movement of the Pacific seafloor by about 30° latitudinally since Cretaceous time.

Francheteau *et al.* (1970) compiled the results of more than 50 seamount surveys in the Pacific and obtained a geomagnetic polar wander curve for the northeastern Pacific (Fig. 9) and a paleomagnetic pole for the Cretaceous at 60°N, 16°E. Harrison *et al.* (1975) examined the Cretaceous seamounts and obtained a Late Cretaceous paleomagnetic pole at 58°N, 350°E from 26 Pacific seamounts, excluding the southern Japanese seamounts.

There are several seamount chains of possible hot spot origin in the Pacific (see Morgan, 1971). The hot spot hypothesis can be verified by seamount magnetizations—namely, all seamounts of one hot spot chain should, from their magnetization, give the same paleolatitude, which should be the present latitude of the hot

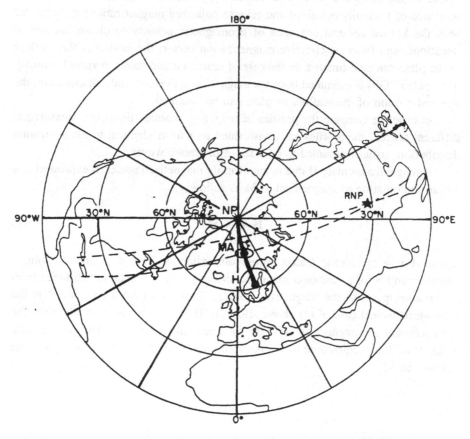

Fig. 9. Preliminary polar curve relative to the northeastern Pacific from the Cretaceous to the present (heavy line). NP, North Pole; MA, Midway atoll; H, Hawaiian seamounts. The ovals surrounding the poles define the 95% confidence limits in the position of these poles. The dashed lines are great circles defining the pivot point RNP represented by a star at 33°N, 97°E. Lambert equal-area projection (Franchetaeu *et al.*, 1970).

spot if the position of a hot spot is fixed to the rotation axis of the earth. Moreover, all polar wander curves obtained from all hot spot chains on the Pacific Plate should coincide if the relative positions of hot spots have not changed. Kodama *et al.* (1978) obtained the paleolatitude 17°N ± 5° of the Suiko Seamount of the Emperor Seamount Chain, and Sager (1984) obtained 17.5°N ± 4.4° for the Abott Seamount of the Hawaiian Ridge. These are not significantly different from the present latitude of the Hawaiian hot spot, which is approximately 19.5°N.

III. METHOD OF INTERPRETATION OF MAGNETIC ANOMALIES

Magnetic anomalies observed at sea are analyzed to obtain the magnetization vector of the magnetic source. In the case of magnetic anomaly lineations, the sequence of normally polarized and reverse-polarized magnetizations is compared with the known reversal sequence of geomagnetic polarity to obtain the age of lineations, and from the effective magnetization vector, the motion of the seafloor or the plate can be estimated. In the case of seamount anomalies, a virtual geomagnetic pole (VGP) is estimated from the magnetization vector, and consequently, the age and motion of the seafloor or plate can be obtained.

In order to compare the profiles of magnetic anomaly lineations measured at different places, the profiles are transformed to a form identical to magnetization distribution, usually assumed to be a series of square waves.

A magnetic anomaly **H** due to a uniformly magnetized source is expressed as a linear combination of magnetization **M** as follows:

$$\mathbf{H} = A \cdot \mathbf{M} \tag{1}$$

where A is a symmetric matrix whose dimension is 2 for a magnetic anomaly lineation and 3 for a seamount magnetic anomaly, and a function of distance from observation point to the magnetized source. Talwani's (1965) equation 2 is the three-dimensional case of (1) above. Because **H** is usually much smaller than the total regional geomagnetic field **T**, the total intensity anomaly of the geomagnetic field, F is almost equivalent to **H** projected onto **T**. F, therefore, also has a linear relation to **M**, i.e.,

$$F = \mathbf{B} \cdot \mathbf{M} \tag{2}$$

where **B** is A**T**. Using the relation **M** = M**K** where M and **K** are the intensity and the unit vector of magnetization, respectively, (2) is expressed as

$$F = M \cdot \mathbf{B} \cdot \mathbf{K} \tag{3}$$

If the magnetic source consists of blocks, a magnetic field at point i due to blocks is

$$F_i = \left(\sum_j M_j \cdot \mathbf{B}_{ij} \right) \cdot \mathbf{K} \tag{4}$$

where $\mathbf{K}$ is assumed to be constant in all blocks, but the intensity M is a function of a place.

In the first stage of analysis of magnetic lineations, (4) was used to obtain M_j using rectangular blocks as the magnetic source extending infinitely in the direction of lineation. If the depths of upper and lower surfaces and the width of each block are given, $\mathbf{B}_{ij}$ in (4) can be calculated. With a given constant $\mathbf{K}$ and observed F_i, M_j for each block can be determined by the linear inversion method (e.g., Bott, 1967; Emilia and Bodvarsson, 1969). Matthews and Bath (1967) and Harrison (1968) used (4) to simulate magnetic anomalies produced by dike injection at the spreading center (Fig. 10). Although the pattern of M_j obtained by the linear inversion method using (4) is essentially similar to that of F_i, an important feature is that positions of peaks and valleys of M_j do not agree with those of F_i.

Schouten (1971) presented the parameter θ, indicating the quantity of phase shifting between F_i and M_j mentioned above (see Fig. 11). If the magnetic source is divided into a unit block, (4) becomes

$$F(x) = \int_{\infty}^{\infty} M(\xi)\mathbf{B}(x - \xi) \cdot \mathbf{K} \cdot d\xi \tag{5}$$

where x and ξ axes are defined perpendicular to the strike of lineations. Schouten (1971) showed the Fourier transform of (5) as

$$f(s) = m(s) \cdot g(s) \tag{6}$$

where f, m, and g are the Fourier transforms of F, M, and $\mathbf{B} \cdot \mathbf{K}$, respectively, and s is a wavenumber. The parameter θ was defined as the phase spectrum of g. If $\theta = 0$, the phase spectrum of f is the same as that of m; therefore, the profile of F accords with that of M (Fig. 11). If $\theta \neq 0$, then both sides of (6) are multiplied by $e^{-i\theta}$. Because $f \cdot e^{-i\theta}$ has the same phase spectrum as m, the profile of the inverse Fourier transform of $f \cdot e^{-i\theta}$ accords with that of M. This operation is called deskewing. Thus, when we make a deskewing operation, we must know the skewness parameter θ. However, θ is usually unknown because it includes effective inclination and declination of magnetization of the magnetic source. Blakely and Cox (1972a) proposed one method of deskewing in which $\mathbf{K}$ is given and magnetic anomaly lineations are transformed to the pole where θ is always 0 because both effective inclinations of magnetization and of geomagnetic field are always 90°. Isezaki

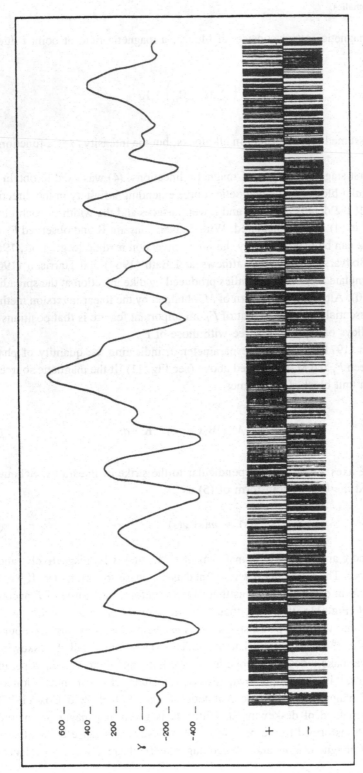

Fig. 10. The lower section shows the distribution of dikes each 0.1 km wide at the end of the injection process. A stroke above the line indicates normal magnetization. The upper section shows the resultant magnetic anomaly. The total length of section shown is 200 km (Matthews and Bath, 1967).

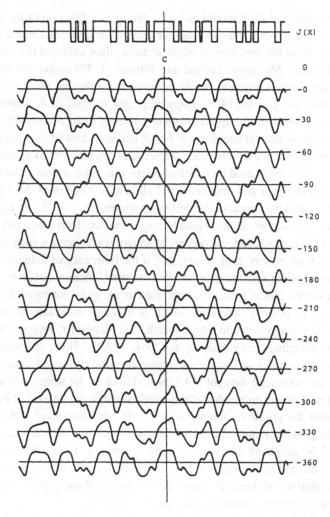

Fig. 11.   A model profile phase-shifted in 30° increments. $J(x)$ is a magnetization profile represented by square waves (Yoshida, 1985).

(1975) presented almost the same deskewing method such that, with a given **K**, magnetic anomaly lineations are not transformed to the pole but aligned parallel to the magnetic meridian in order that θ becomes 0. Schouten and McCamy (1972) tried to obtain M directly through the inverse Fourier transform of f/g $(= m)$ with a given **K**.

On the other hand, if m is given, then θ, the phase spectrum of g, can be obtained from (6) because f is observed. Larson and Chase (1972) compared Mesozoic magnetic anomaly lineations in the Pacific (the Japanese, Phoenix, and Hawaii lineations), where they assumed M and m was multiplied by $|g| \cdot e^{i\theta}$, varying θ until

$m|g| \cdot e^{i\theta}$ fits observed F. It should be noted that $|g|$ is a function of the geometry of the source alone. From θ thus obtained for each set of lineations which includes the information about the direction of magnetization, they obtained the paleomagnetic pole for the Late Mesozoic. Larson and Pitman (1972) correlated the Mesozoic anomaly lineations at different places in the world using the same method.

Schouten and Cande (1976) proposed a new method of deskewing from the standpoint that M should be approximated by a series of square waves; θ is selected so that the reverse Fourier transform of $f \cdot e^{-i\theta}$ becomes best fit for a series of square waves. This method is useful because neither M nor K needs to be assumed. Louden (1976) and Shih (1983) used this method on the magnetic anomaly lineations in the West Philippine Basin and obtained a 60° clockwise rotation of the basin.

Strictly speaking, because the deskewed anomalies, $f \cdot e^{-i\theta}$, are equal to $m \cdot |g|$, the deskewed anomalies are not equal to magnetization distribution but modified in amplitude by $|g|$. Moreover, the corners of square waves are rounded depending on the dimension of the spreading center; i.e., if the frequency of dike injection in unit time and area at the spreading center follows the Gaussian distribution about the distance from the central axis, M is expressed as the convolution of the distribution of geomagnetic polarity reversals, which is a series of square waves, with the Gaussian distribution function whose standard deviation (σ) represents the dimension of the spreading center (Blakely and Cox, 1972b). Then, although m is modified in amplitude by $|g|$ and $\exp(\sigma)$, resulting in round corners of square waves, the deskewed anomalies are thought to be most similar to the square waves as seen in Fig. 11. The problem in the deskewing method mentioned above is how to determine whether the profile of anomalies has really been deskewed—in other words, whether the profile of anomalies is transformed most similarly to a series of square waves. All investigators determine this by inspection. Recently, Yoshida (1985) proposed a method to determine θ more objectively, i.e., not by visual inspection but by calculating the mean deviation of amplitudes of anomalies, which seems to be the most effective method of those proposed so far.

As for the interpretation of seamount anomalies, the magnetization is usually assumed to be constant over the whole magnetic source of the seamount. Then (4) is expressed as

$$F_i = \left(\sum_j \mathbf{B}_{ij} \right) \mathbf{M} \tag{7}$$

Vacquier (1962) and Richards et al. (1967) presented the method to calculate $\mathbf{B}_{ij}$ of (7) in which a seamount is approximated by an assemblage of a prism small enough to express the irregular shape of the source and the magnetic fields due to the magnetic moments induced on the surfaces of prisms are added up. Uyeda and Richards (1966) and Vacquier and Uyeda (1967) applied this method on the seamounts in the northwest Pacific. Talwani (1965) presented the method in which a

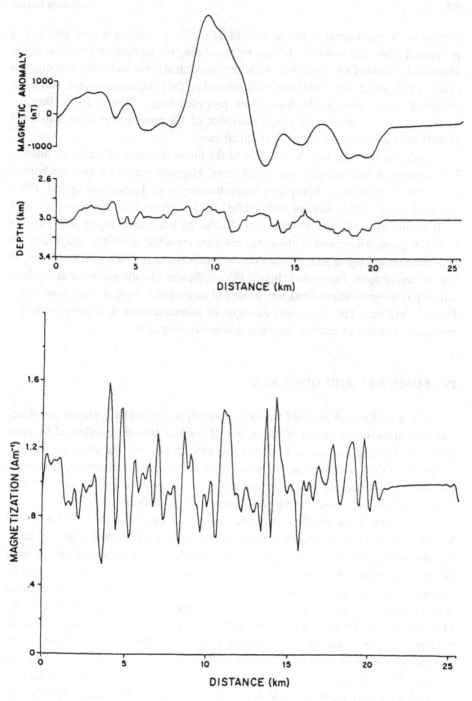

Fig. 12. (Top) Magnetic anomalies, reduced to a horizontal line 2475 m below the sea surface, and topographic relief of the seafloor. The vertical exaggeration is 6 : 1 in the topography. The profile bearing is 110°, the ambient field inclination −61.0°, and the declination 29.3°. (Bottom) The magnetic annihilator corresponding to the data in the top panel. This magnetization in a layer of constant thickness following the topography produces zero magnetic field above the layer (Parker and Huestis, 1974).

seamount is approximated by an assemblage of a polygonal plane. Because a polygonal plane can be easily defined by simulating the topographic contour of the seamount, Talwani's method is more widely used than other methods. Francheteau *et al.* (1970) and Sager (1984) used this method on the Pacific seamounts. Although analytical expressions of $\mathbf{B}_{ij}$ have been proposed (e.g., Plouff, 1976; Barnett, 1976), not much greater advantage than that of Vacquier's (1962) or Talwani's (1965) would be expected for the practical case.

It should be noted that the solution of the linear inversion of magnetic anomalies mentioned above is not unique in all cases. Magnetic surveys at near sea bottom have been carried out by a deep-tow magnetometer (e.g., Larson and Spiess, 1969; Klitgord *et al.*, 1972; Atwater and Mudie, 1973). Since observed anomalies have high amplitudes and short wavelengths affected by sea bottom topography, Parker (1972) presented a method to obtain the magnetization taking the topographic effect into account through a sum of the Fourier transformation of the profile of the sea bottom topography. Parker and Huestis (1974) showed that the magnetic annihilator which has magnetization, however, produced no magnetic field at some level using Parker's method. This is a good example of nonuniqueness of a solution to the inversion problem of marine magnetic anomalies (Fig. 12).

IV. SUMMARY AND DISCUSSION

It is a well-known fact that magnetic anomaly lineations have played an important role in the development of the theory of seafloor spreading followed by plate tectonics. Even at present, it is no exaggeration to say that the study of tectonic evolution of the seafloor cannot be accomplished without investigation of magnetic anomalies. It is because magnetic anomalies are easily measured at sea and bear precise information about the age and motion of the seafloor.

In the northwest Pacific, magnetic anomaly lineations were first found and have been rather easily correlated to the reversal time scale of geomagnetic polarity because the magnetic anomaly signatures are generally good and data distribution is dense. New ideas have been proposed to interpret the configuration of magnetic anomaly lineations in this area. Namely, the migration of an R–R–R triple junction was proposed as a natural consequence to interpret the Great Magnetic Bight (Pitman and Hayes, 1968), the ridge jump to interpret both the duplicated lineations between the Murray and Molokai Fracture Zones (e.g., Harrison and Sclater, 1972) and the disturbed lineations to the east of the Surveyer Fracture Zone (Shih and Molnar, 1975), the rift propagation to interpret the oblique lineaments in the Juan de Fuca Ridge area (e.g., Hey and Wilson, 1982), and so on. It is interesting that the latest idea of a propagating rift proved successful to explain the complex distribution of offsets of magnetic anomaly lineations oblique to plate motion in the Juan de Fuca Ridge area and also to explain very fast spreading of the Easter microplate ridges by the dual propagating rifts (Engebretson and Cox, 1984; Hey *et al.*, 1985). The idea is expected to explain the large-scale shift of the East Pacific Rise spreading system by the northward propagation of the East Pacific Rise replacing the old Pacific–Farallon plate (Hey *et al.*, 1985). The same shift was interpreted earlier by ridge rotation and/or ridge jump (e.g., Handschumacher, 1976; Herron, 1972) as mentioned in an earlier section.

In the northwest Pacific, the Mesozoic lineations are disturbed by seamounts, rises, troughs, and fracture zones so that the patterns of lineations are more complicated than in the northeast Pacific. The Mesozoic magnetic bight to the southwest of the Shatzky Rise does not have clear features compared to the Great Magnetic Bight. However, these two bights were interpreted as originating from the same R–R–R triple-junctioned ridges. In this case, the problem is how to interpret the present misalignment of the two bights. There will be two interpretations. One, proposed by Hilde *et al.* (1976), was that the ridge triple junction first jumped NE along the axis of the Shatzky Rise and thereafter SE along the axis of the Hess Rise accompanied with its progressive clockwise rotation which totally reached about 40° from anomaly M26 time to Cenozoic 32 time. All of these activities remained confined to the area north of the Mendocino Fracture Zone. The other interpretation, proposed by Farrar and Dixon (1981), was that between 67 and 40 Ma a NW–SE-trending fracture zone system along the Emperor–Line lineament over 8000 km long split the Pacific plate and accumulated at least 1700 km of dextral offset between the two magnetic bights.

In the latter interpretation, both bights are thought to have been formed through simple migration of the Pacific–Kula–Farallon triple junction before 67 Ma. However, Woods and Davies (1982) claimed that this proposal should not be the case as pointed out already by Hilde *et al.* (1976) because extrapolation of the Great Magnetic Bight backwards to Mesozoic magnetic quiet time results in an implausible ridge configuration. Then Woods and Davies (1982) proposed that the two magnetic bights originated by two independent R–R–R triple junction systems, namely, the Great Magnetic Bight began to be formed about 85 Ma by the Pacific–Kula–Farallon ridges and the Mesozoic magnetic bight had been formed already by that time by the Pacific–Farallon–Izanagi ridges. It seems to be impossible to conclude at the moment whether the two magnetic bights originated by a single R–R–R triple junction or by two independent ones.

REFERENCES

Anderson, R. N., and Sclater, J. G., 1972, Topography and evolution of the East Pacific rise between 5°S and 20°S, *Earth Planet. Sci. Lett.* **14:**433–441.

Atwater, T., 1970, Implication of plate tectonics for the Cenozoic tectonic evolution of western North America, *Geol. Soc. Am. Bull.* **81:**3513–3536.

Atwater, T., and Mudie, J. D., 1973, Detailed near-bottom geophysical study of the Gorda rise, *J. Geophys. Res.* **75:**8665–8686.

Barnett, C. T., 1976, Theoretical modelling of the magnetic and gravitational fields of an arbitrarily shaped three-dimensional body, *Geophysics* **41:**1353–1364.

Ben-Avraham, Z., and Uyeda, S., 1973, The evolution of the China Basin and Mesozoic paleogeography of Borneo, *Earth Planet. Sci. Lett.* **18:**365–376.

Ben-Avraham, Z., and Uyeda, S., 1983, Entrapment origin of marginal seas, in: *Geodynamics of the Western Pacific–Indonesian Region,* Geodynamics Series **11:**91–104, American Geophysical Union, Washington, D.C.

Ben-Avraham, Z., Segawa, J., and Bowin, C. O., 1972, An extinct spreading center in the Philippine Sea, *Nature* **240:**453–455.

Blakely, R. J., and Cox, A., 1972a, Identification of short polarity events by transforming marine magnetic profiles to the pole, *J. Geophys. Res.* **77:**4339–4349.

Blakely, R. J., and Cox, A., 1972b, Evidence for short polarity intervals in the Early Cenozoic, *J. Geophys. Res.* **77:**7065–7072.

Bott, M. H. P., 1967, Solution of the linear inverse problem in magnetic interpretation with application to oceanic magnetic anomalies, *Geophys. J. R. Astron. Soc.* **13:**313–323.

Cande, S. C., Larson, R. L., and LaBrecque, J. L., 1978, Magnetic lineations in the Pacific Jurassic quiet zone, *Earth Planet. Sci. Lett.* **41:**434–440.

Clague, D. A., and Jarrard, R. D., 1973, Tertiary Pacific plate motion deduced from the Hawaiian–Emperor chain, *Geol. Soc. Am. Bull.* **84:**1135–1154.

Cooper, A. K., Marlow, M. S., and Scholl, D. W., 1976, Mesozoic magnetic lineations in the Bering Sea marginal basin, *J. Geophys. Res.* **81:**1916–1934.

Cox, A., Doell, R. R., and Dalrymple, G. B., 1964, Reversal of the earth's magnetic field, *Science* **143:**1537–1543.

Dickson, G. O., Pitman, W., C., III, and Heirtzler, J. R., 1968, Magnetic anomalies in the south Atlantic and ocean floor spreading, *J. Geophys. Res.* **73:**2087–2100.

Dietz, R. S., 1961, Continent and ocean basin evolution by spreading of the seafloor, *Nature* **190:**854–857.

Elvers, D., Srivastava, S. P., Potter, K., Morley, J., and Spiedel, D., 1973, Asymmetric spreading across the Juan de Fuca and Gorda rise as obtained from a detailed magnetic survey, *Earth Planet. Sci. Lett.* **20:**211–219.

Emilia, A. A., and Bodvarsson, G., 1969, Numerical method in the direct interpretation of marine magnetic anomalies, *Earth Planet. Sci. Lett.* **7:**194–200.

Engebretson, D. C., and Cox, A., 1984, Relative motions between ocean plates of the Pacific basin, *J. Geophys. Res.* **89:**10291–10310.

Farrar, E., and Dixon, J. M., 1981, Early Tertiary rupture of the Pacific plate: 1700 km of dextral offset along the Emperor trough–Line Islands lineament, *Earth Planet. Sci. Lett.* **53:**307–322.

Francheteau, J., Harrison, C. G. A., Sclater, J. G., and Richards, M. L., 1970, Magnetization of Pacific seamounts: A preliminary polar curve for the northeastern Pacific, *J. Geophys. Res.* **75:**2035–2061.

Grim, P. J., and Erickson, B. H., 1969, Fracture zones and magnetic anomalies south of the Aleutian trench, *J. Geophys. Res.* **74:**1488–1499.

Grow, J. A., and Atwater, T., 1970, Mid-Tertiary tectonic transition in the Aleutian arc, *Geol. Soc. Am. Bull.* **81:**3715–3722.

Handschumacher, D. W., 1976, Post-Eocene plate tectonics of the Eastern Pacific, *Geophys. Monogr. Am. Geophys. Union* **19:**177–202.

Harland, W. B., Cox, A. V., Llewellyn, P. G., Pickton, C. A. G., Smith, A. G., and Walters, R., 1982, *A Geologic Time Scale,* Cambridge University Press, Cambridge.

Harrison, C. G. A., 1968, Formation of magnetic anomaly patterns by dyke injection, *J. Geophys. Res.* **73:**2137–2142.

Harrison, C. G. A., and Sclater, J. G., 1972, Origin of the disturbed magnetic zone between the Murray and Molokai Fracture Zones, *Earth Planet. Sci. Lett.* **14:**419–427.

Harrison, C. G. A., Jarrard, R. D., Vacquier, V., and Larson, R. L., 1975, Paleomagnetism of Cretaceous Pacific seamounts, *Geophys. J. R. Astron. Soc.* **42:**859–882.

Hayes, D. E., and Pitman, W. C., III, 1970, Magnetic lineations in the North Pacific, *Geol. Soc. Am. Mem.* **126**:291–314.

Hayes, D. E., and Ringis, J., 1973, Sea-floor spreading in the Tasman Sea, *Nature* **243**:454–458.

Hegarty, K. A., Weissel, J. K., and Hayes, D. E., 1983, Convergence at the Caroline–Pacific plate boundary: Collision and subduction, *Geophys. Monogr. Geophys. Union* **23**:349–359.

Heirtzler, J. R., Dickson, G. O., Herron, E. M., Pitman, W. C., III, and LePichon, X., 1968, Marine magnetic anomalies, geomagnetic field reversals and motions of ocean floor and continents, *J. Geophys. Res.* **73**,2119–2136.

Herron, E. M., 1972, Sea-floor spreading and the Cenozoic history of the East Central Pacific, *Geol. Soc. Am. Bull.* **83**:1671–1692.

Hess, H. H., 1962, History of ocean basins. in: *Petrologic Studies: A. F. Buddington Volume* (A. E. Engel, H. L. James, and B. F. Leonard, eds.), Geol. Soc. Am. Mem., pp. 559–620.

Hey, R. N., 1977, Tectonic evolution of the Cocos–Nazca spreading center, *Geol. Soc. Am. Bull.* **88**:1404–1420.

Hey, R. N., and Vogt, P., 1977, Spreading center jumps and sub-axial asthenosphere flow near the Galapagos hotspot, *Tectonophysics* **37**:41–52.

Hey, R. N., and Wilson, D. S., 1982, Propagating rift explanation for the tectonic evolution of the northeast Pacific—the pseudmovie, *Earth Planet. Sci. Lett.* **58**:167–188.

Hey, R. N., Johnson, G. L., and Lowrie, A., 1977, Recent plate motions in the Galapagos area, *Geol. Soc. Am. Bull.* **88**:1385–1403.

Hey, R. N., Duennebier, F. K., and Morgan, W. J., 1980, Propagating rifts on mid-ocean ridges, *J. Geophys. Res.* **85**:3647–3658.

Hey, R. N., Naar, D. E., Kleinrock, M. C., Phipps Morgan, W. J., Morales, E., and Schilling, J.-G., 1985, Microplate tectonics along a super seafloor spreading system near Easter Island, *Nature* **317**:320–325.

Hilde, T. W. C., and Lee, C. H., 1984, Original evolution of west Phillippine Basin: A new interpretation, *Tectonophysics* **102**:85–104.

Hilde, T. W. C., Isezaki, N., and Wageman, J. M., 1976, Mesozoic seafloor spreading in the North Pacific, *Geophys. Monogr. Amer. Geophys. Union* **19**:205–226.

Hilde, T. W. C., Uyeda, S., and Kroenke, L., 1977, Evolution of the Western Pacific and its margins, *Tectonophysics* **38**:145–165.

Hsui, A. T., and Toksöz, M. N., 1981, Back-arc spreading: Trench migration, continental pull, or induced convection? *Tectonophysics* **74**:89–98.

Isezaki, N., 1975, Possible spreading centers in the Japan Sea, *J. Mar. Geophys. Res.* **2**:265–277.

Isezaki, N., 1986, A geomagnetic anomaly map of the Japan Sea, *J. Geomagn. Geoelectr.* **38**:403–410.

Isezaki, N., and Uyeda, S., 1973, Geomagnetic anomaly pattern of the Japan Sea, *J. Mar. Geophys. Res.* **2**:51–59.

Isezaki, N., Hata, K., and Uyeda, S., 1971, Magnetic survey of the Japan Sea (part 1), *Bull. Earthq. Res. Inst.* **49**:77–81.

Karasik, A. M., and Cochebanoba, H. A., 1981, Map of oceanic paleomagnetic anomalies in the Pacific Ocean, Academy of Science, USSR.

Karig, D. E., 1971, Structural history of the Mariana island arc system, *Geol. Soc. Am. Bull.* **82**:323–344.

Karig, D. E., 1975, Basin genesis of the Philippine Sea, *Initial Reports of the Deep Sea Drilling Project* **31**:857–879.

Klitgord, K. D., Mudie, J. D., and Normark, W. R., 1972, Magnetic lineations observed near the ocean floor and possible implications on the geomagnetic chronology of the Gilbert epoch, *Geophys. J.* **28**:35–48.

Kobayashi, K., and Nakata, M., 1978, Magnetic anomalies and tectonic evolution of Shikoku inter-arc basin, *J. Phys. Earth* Suppl. **26**:S391–S402.

Kodama, K., Uyeda, S., and Isezaki, N., 1978, Paleomagnetism of Suiko seamount, Emperor seamount chain, *Geophys. Res. Lett.* **5**:165–168.

LaBrecque, J. L., Kent, D. V., and Cande, S. C., 1977, Revised magnetic polarity time scale for Late Cretaceous and Cenozoic time, *Geology* **5**:330–335.

Larson, R. L., 1976, Late Jurassic and Early Cretaceous evolution of the western central Pacific Ocean. *J. Geomagn. Geoelectr.* **28**:219–236.

Larson, R. L., and Chase, C. G., 1972, Late Mesozoic evolution of the western Pacific Ocean, *Geol. Soc. Am. Bull.* **83**:3627–3644.

Larson, R. L., and Hilde, T. W. C., 1975, A revised time scale of magnetic reversals for the Early Cretaceous and Late Jurassic, *J. Geophys. Res.* **80**:2586–2594.

Larson, R. L., and Pitman, W. C., III, 1972, World-wide correlation of Mesozoic magnetic anomalies and its implications, *Geol. Soc. Am. Bull.* **83**:3645–3662.

Larson, R. L., and Spiess, F. N., 1969. East Pacific rise crest: A near-bottom geophysical profile, *Science* **163**:68–71.

Larson, R. L., Menard, H. M., and Smith, S. M., 1968, Gulf of California: A result of ocean-floor spreading, *Science* **161**:781–784.

Larson, R. L., Smith, S. M., and Chase, C. G., 1972, Magnetic lineations of Early Cretaceous age in the western equatorial Pacific Ocean, *Earth Planet, Sci. Lett.* **15**:315–319.

Lee, C. H., and Hilde, T. W. C., 1971, Magnetic lineations in the western Philippine Sea, *Acta Oceanogr. Taiwan.* **1**:69–76.

LePichon, X., 1968, Sea-floor spreading and continental drift, *J. Geophys. Res.* **73**:3661–3697.

Louden, K. E., 1976, Magnetic anomalies in the West Philippine basin, *Geophys. Monogr. Am. Geophys. Union* **19**:253–276.

Malahoff, A., and Handschumacher, D. W., 1971, Magnetic anomalies south of the Murray Fracture Zone: New evidence for secondary sea-floor spreading center and strike-slip movement, *J. Geophys. Res.* **76**:6265–6275.

Mason, R. G., 1958, A magnetic survey off the west coast of the United States between latitudes 32°N and 26°N, longitudes 121°W and 128°W, *Geophys. J.* **1**:320–329.

Mason, R. G., and Raff, A. D., 1961, A magnetic survey off the west coast of North America, 32°N latitude to 42°N latitude, *Geol. Soc. Am. Bull.* **72**:1259–1270.

Matthews, D. H., and Bath, J., 1967, Formation of magnetic anomaly pattern of Mid-Atlantic Ridge, *Geophys. J. R. Astron. Soc.* **13**:349–357.

Menard, H. W., and Atwater, T., 1969, Origin of fracture zone topography, *Nature* **222**:1037–1040.

Molnar, P., Atwater, T., Mammerickx, J., and Smith, S. M., 1974, Magnetic anomalies, bathymetry and the tectonic evolution of the South Pacific since the Late Cretaceous, *Geophys. J. R. Astron. Soc.* **40**:383–420.

Moore, D. G., and Buffington, E. C., 1968, Transform faulting and growth of the Gulf of California since late Pliocene, *Science* **161**:1238–1241.

Morgan, W. J., 1968, Rises, trenches, great faults and crustal blocks, *J. Geophys. Res.* **73**:1959–1982.

Morgan, W. J., 1971, Convection plumes in the lower mantle, *Nature* **230**:42–43.

Mrozowski, C. L., and Hayes, D. E., 1969, The evolution of the Parece Vela Basin, eastern Philippine Sea, *Earth Planet. Sci. Lett.* **46**:49–67.

Naugler, F. P., and Wageman, J. M., 1973, Gulf of Alaska: Magnetic anomalies, fracture zones, and plate interaction, *Geol. Soc. Am. Bull.* **84**:1575–1584.

Parker, R. L., 1972, The rapid calculation of potential anomalies, *Geophys. J. R. Astron. Soc.* **31**:447–455.

Parker, R. L., and Huestis, S. P., 1974, The inversion of magnetic anomalies in the presence of topography, *J. Geophys. Res.* **79**:1587–1593.

Peter, G., Erickson, B. H., and Grim, P. J., 1970, Magnetic structure of the Aleutian Trench and northeast Pacific basin, in: *The Sea,* Vol. 4 (A. E. Maxwell, ed.), Wiley–Interscience, New York, pp. 191–222.

Pitman, W. C., III, and Hayes, D. E., 1968, Sea-floor spreading in the Gulf of Alaska, *J. Geophys. Res.* **73**:6571–6580.

Pitman, W. C., III, and Heirtzler, J. R., 1966, Magnetic anomalies in the Pacific–Antarctic Ridge, *Science* **154**:1164–1171.

Pitman, W. C., III, Herron, E. M., and Heirtzler, J. R., 1968, Magnetic anomalies in the Pacific and sea-floor spreading, *J. Geophys. Res.* **73**:2069–2085.

Pitman, W. C., III, Larson, R. L., and Herron, E. M., 1974, Age of the ocean basins determined from magnetic anomaly lineations, *Geol. Soc. Am. Charts.*

Plouff, D., 1976, Gravity and magnetic field of polygonal prisms and application to magnetic terrain corrections, *Geophysics* **41**:727–741.

Raff, A. D., 1966, Boundaries of an area of very long magnetic anomalies in the northeast Pacific, *J. Geophys. Res.* **71**:2631–2636.

Raff, A. D., and Mason, R. G., 1961, Magnetic survey off the west coast of North America 40°N to 52°N latitude, *Geol. Soc. Am. Bull.* **72**:1267–1270.

Richards, M. L., Vacquier, V., and Van Voorhis, G. D., 1967, Calculation of magnetization of uplifts from combining topographic and magnetic surveys, *Geophysics* **32**:678–701.

Sager, W. W., 1984, Paleomagnetism of Abbott seamount and implications for the latitudinal drift of the Hawaiian hot spot, *J. Geophys. Res.* **89**:6271–6284.

Schouten, H., and Cande, S. C., 1976, Paleomagnetic poles from marine magnetic anomalies, *Geophys. J. R. Astron. Soc.* **44**:567–575.

Schouten, H., and McCamy, K., 1972, Filtering marine magnetic anomalies, *J. Geophys. Res.* **77**:7089–7099.

Schouten, J. A., 1971, A fundamental analysis of magnetic anomalies over oceanic ridges, *Mar. Geophys. Res.* **1**:111–144.

Shih, J., and Molnar, P., 1975, Analysis and implications of the sequence of ridge jumps that eliminated the Surveyor transform fault, *J. Geophys. Res.* **80**:4815–4822.

Shih, T. C., 1980, Magnetic lineations in the Shikoku Basin, *Initial Reports of the Deep Sea Drilling Project* **58**:783–788.

Shih, T. C., 1983, Marine magnetic anomalies from the western Philippine Sea: Implications for the evolution of marginal basins, *Geophys. Monogr. Am. Geophys. Union* **23**:49–75.

Talwani, M., 1965, Computation with the help of a digital computer of magnetic anomalies caused by bodies of arbitrary shape, *Geophysics* **30**:797–817.

Tamaki, K., Joshima, M., and Larson, R. L., 1979, Remanent Early Cretaceous spreading center in the Central Pacific Basin, *J. Geophys. Res.* **84**:4501–4510.

Taylor, B., 1979, Bismarck Sea: Evolution of a back-arc basin, *Geology* **7**:171–174.

Taylor, B., and Hayes, D. E., 1980, The tectonic evolution of the South China Basin, *Geophys. Monogr. Am. Geophys. Union* **23**:89–104.

Tayler, P. T., and O'Neill, N. J., 1974, Results of aeromagnetic survey in the Gulf of Alaska, *J. Geophys. Res.* **79**:719–723.

Tomoda, Y., Kobayashi, K., Segawa, J., Nomura, M., Kimura, K., and Saki, T., 1975, Linear magnetic anomalies in the Shikoku basin, northern Philippine Sea, *J. Geomagn. Geoelectr.* **28**:47–56.

Uyeda, S., and Ben-Avraham, Z., 1972, Origin and development of the Philippine Sea, *Nature* **240**:176–178.

Uyeda, S., and Kanamori, H., 1979, Back arc opening and the mode of subduction, *J. Geophys. Res.* **84**:1049–1062.

Uyeda, S., and Richards, M., 1966, Magnetization of four Pacific seamounts near the Japanese Islands, *Bull. Earthq. Res. Inst.* **44**:179–213.

Vacquier, V., 1962, A machine method for computing the magnitude and the direction of magnetization of a uniformly magnetized body from its shape and a magnetic survey, Proc. Benedum Earth Magnetism Symp., pp. 123–137.

Vacquier, V., and Uyeda, S., 1967, Paleomagnetism of nine seamounts in the western Pacific and of three volcanoes in Japan, *Bull. Earthq. Res. Inst.* **45**:815–848.

Vacquier, V., Raff, A. D., and Waren, R. E., 1961, Horizontal displacements in the floor of the northeastern Pacific Ocean, *Geol. Soc. Am. Bull.* **72**:1251–1258.

Vine, F. J., 1966, Spreading of the ocean floor: New evidence, *Science* **154**:1405–1415.

Vine, F. J., and Matthews, D. H., 1963, Magnetic anomalies over oceanic ridges, *Nature* **199**:947–949.

Vine, F. J., and Wilson, J. T., 1965, Magnetic anomalies over a young oceanic ridge off Vancouver Island, *Science* **150**:485–489.

Watts, A. B., and Weissel, J. K., 1975, Tectonic history of the Shikoku marginal basin, *Earth Planet. Sci. Lett.* **25**:239–250.

Watts, A. B., Weissel, J. K., and Larson, R. L., 1977, Sea-floor spreading in marginal basins of the western Pacific, *Tectonophysics* **37**:167–182.

Weissel, J. K., 1977, Evolution of the Lau Basin by the growth of small plates, *Am. Geophys. Union Maurice Ewing Ser.* **1**:429–436.

Weissel, J. K., 1981, Magnetic lineations in marginal basins of the western Pacific, *Philos. Trans. R. Soc. London Ser. A* **300**:223–247.

Weissel, J. K., and Anderson, R. N., 1978, Is there a Calorine plate? *Earth Planet. Sci. Lett.* **41**:143–158.

Weissel, J. K., and Hayes, D. E., 1977, Evolution of the Tasman Sea reappraised, *Earth Planet. Sci. Lett.* **36**:77–84.

Weissel, J. K., and Watts, A. B., 1975, Tectonic complexities in the South Fiji marginal basin, *Earth Planet. Sci. Lett.* **28**:121–126.

Weissel, J. K., and Watts, A. B., 1979, Tectonic evolution of the Coral Sea basin, *J. Geophys. Res.* **84**:4572–4582.

Wilson, J. T., 1965a, A new class of faults and their bearing on continental drift, *Nature* **207**:343–347.

Wilson, J. T., 1965b, Transform faults, oceanic ridges, and magnetic anomalies southwest of Vancouver Islands, *Science* **150**:482–485.

Woods, M. T., and Davies, G. F., 1982, Late Cretaceous genesis of the Kula plate, *Earth Planet. Sci. Lett.* **58**:161–166.

Yoshida, K. K., 1985, Analysis of the parameter θ using mean deviation of anomaly pattern of marine magnetic anomalies, *J. Geomagn. Geoelectr.* **37**:443–454.

INDEX

Printed in the United States
by Baker & Taylor Publisher Services